Development and Implementation of RFID Technology

Edited by

Cristina TURCU

Development and Implementation of RFID Technology
Edited by Cristina Turcu

Republished by InTech
Janeza Trdine 9, 51000 Rijeka, Croatia

Technical Editor Teodora Smiljanic

Additional hard copies can be obtained from orders@intechopen.com

Development and Implementation of RFID Technology, Edited by Cristina Turcu
p. cm.
ISBN 978-3-902613-54-7

For EU product safety concerns:
IN TECH d.o.o., Prolaz Marije Krucifikse Kozulić 3, 51000 Rijeka, Croatia,
info@intechopen.com or visit our website at intechopen.com.

Meet the editor

Cristina Turcu is an Associate Professor of Software Engineering and Artificial Intelligence at Stefan cel Mare University of Suceava, Romania. She received her Diploma (M.Sc.) in Automatics and Computers and her Doctorate (Ph.D.) in Automatics, in 1991 and 2000, respectively, both from the Gheorghe Asachi Technical University of Iasi, Romania where she has been Head of Computer Department since 2004. Her research interests include software engineering, RFID applications for the end-consumer, and intelligent systems. Dr. Turcu is an Editor of four books and has served on various program committees of conferences in computing and RFID systems. She also has served as a reviewer for numerous referred journals and conferences. She is the Editor in Chief of the International Journal of Radio Frequency Identification & Wireless Sensor Networks. Dr. Turcu has published over 80 publications in books or book chapters, refereed journals, technical reports, and refereed conference/workshop/seminar proceedings.

Preface

The publication of Harry Stockman's 1948 paper entitled "Communication by Means of Reflected Power" is often related to a significant technological breakthrough: item identification. After 60 years this technology has started to re-assert its viability by boosting the development of most business activities. Thus, the new revolutionary technology has already allowed numerous organizations to obtain valuable security and safety benefits. Moreover, it has become increasingly prevalent in almost any field or area of activity such as transportation and logistics (e.g. transportation payments), animal identification, supply chain management, social retailing, monitoring systems in public institutions such as schools, universities, or museums, accounting applications (e.g. tying events to transactions and evaluating cause-and-effect relationships, item-level identification), the identification of patients and medical staff, human tagging and tracking in hospitals and emergency rooms, inventory systems, product tracking and tracing (e.g. the tracking and tracing of pharmaceuticals), lap scoring, and race timing.

As more and more manufacturers of RFID tags, readers, and software solutions enter the market, the price of implementing the technology is likely to continue its falling. At the moment, this technology enables a lot more manufacturers to improve their services to such an extent as to provide their customers and clients with the whole information necessary to manage complex inventory systems in large-scale building projects, for example, to simplify many business processes and reduce many inventory inaccuracies.

If more capabilities are added (e.g. the combination of RFID with sensor network technologies and real-time locating systems or product and service safety or authenticity), the RFID technology is certain to enhance its benefits by eliminating most manufacturing dysfunctions, simplifying inventory processes, eliminating inaccuracies, and even reducing the costs triggered by product theft, spoilage or obsolescence.

The decision to adopt the RFID technology lies with every organization which is to initiate thorough cost-benefit analyses. As a matter of fact, most decisions regarding when and how to adopt RFID are generally determined by the tradeoffs of several factors including cost, transmission rate, transmission range, and potential environmental interference. Consequently, any such decision stands in correlation with the tendencies and advances recorded in vital areas such as Research, Development and Technology Transfer, Business Process Innovations, Standardization and Legislation, Consumer Application Innovations and Environment Protection.

Following this conviction, the InTech team has decided to select and present some of the most recent research results of worldwide scientists, engineers, manufacturers and end-users interested in sharing and exchanging valuable information and ideas on the present development issues of and future trends in RFID technology.

The book generously covers a wide range of aspects and issues related to RFID systems, namely the design of RFID antennas, RFID readers and the variety of tags (e.g. UHF tags for sensing applications, surface acoustic wave RFID tags, smart RFID tags), complex RFID systems, security and privacy issues in RFID applications, as well as the selection of encryption algorithms.

A large section has been allotted to the presentation of numerous RFID applications that address a variety of aspects such as supply chain solutions, object recognition using a 3D RFID system, shared tag RFID system for multiple application objects, detection in RFID trails, and inter-tag communications, etc.

The book offers new insights, solutions and ideas for the design of efficient RFID architectures and applications. While not pretending to be comprehensive, its wide coverage may be appropriate not only for RFID novices but also for experienced technical professionals and RFID aficionados.

A special word of thanks goes out to all scientists who were kind enough to contribute to this book and to the team at InTech that has fulfilled its mission with the highest degree of dedication again.

Editor

Cristina TURCU
Stefan cel Mare University of Suceava
Romania

Contents

1

Development and Implementation of RFID Technology

Huiyun Li
Shenzhen Institute of Advanced Technology, Chinese Academy of Sciences, China

1. Introduction

Radio Frequency Identification (RFID) is an automated identification technology that uses tags to transmit data upon RFID reader queries. Compared to barcodes identification technology, RFID tags provide a unique identifier, which raises concerns over user privacy, such as clandestine tracking and inventorying [1]. In its original version, a RFID tag responds to a reader query with its fixed unique serial number. This fixed unique serial number enables tracking of tags and the bearers, possibly without the bearers' knowledge or consent. In addition to the unique serial number, some tags carry information about the objects they are attached to. Thus, a retail store or a person owning such tags might be under threat of clandestine inventorying.

Enormous research effort has been paid in attempt to solve the problem of consumer privacy and industrial espionage in the RFID world. However, most methods demand heavy or frequent cryptographic operations on RFID tags, which contradict the low cost demand of RFID tags ($0.05-0.10). Typically, a low-cost tag should only store hundreds of bits and have 5K-10K logic gates, only a fraction of the gates can be devoted to security tasks. The trade-off between cryptographic operations and low-cost has become a significant challenge in designing RFID tags, and this challenge has impeded RFID being the replacement of barcode technology for cost sensitive item-level applications, such as in supply chains, libraries and rental shops.

To solve this problem, a new RFID structure is proposed. Except the fixed unique serial number, tags carry only the IDs in disguise to avoid eavesdropping and clandestine tracking. The database, on the other hand, is responsible for protecting the information security, integrity and non-repudiation. This chapter discusses and presents the implementation of this passive ultra high frequency (UHF) RFID system, based on EPC Class 1 Generation 2 UHF RFID (abbreviate as Gen 2) protocols [1-2].

2. Communication between reader and tag

For a passive RFID system, the communication between the reader and the tag is fully controlled by the reader, i.e. the tag can't send data unless triggered by the reader [3]. The communication from the reader to the tag is referred to as the forward link, while the communication from the tag to the reader is referred to as the reverse link.

2.1 Forward link (from reader to tag)

A continuous RF wave is transmitted from the reader to the tag through the forward link. The data is sent from the reader to the tag as short gaps in this continuous wave in amplitude shift key (ASK) modulation with Pulse Interval Encoding (PIE). Figure 1 shows the Gen 2 protocol PIE encoding, where the duration of data '0' is T0, the duration of data '1' is between 1.5 T0 and 2 T0, the value of Pulse Width (PW) is from 0.265T0 to 0.525T0. The value of T0 is between 6.25μs and 25μs. So the data rate of forward link is between 26.7Kbps and 128Kbps. Figure 2 demonstrates the transmission example of amplitude shift key (ASK) modulation with Pulse Interval Encoding (PIE).

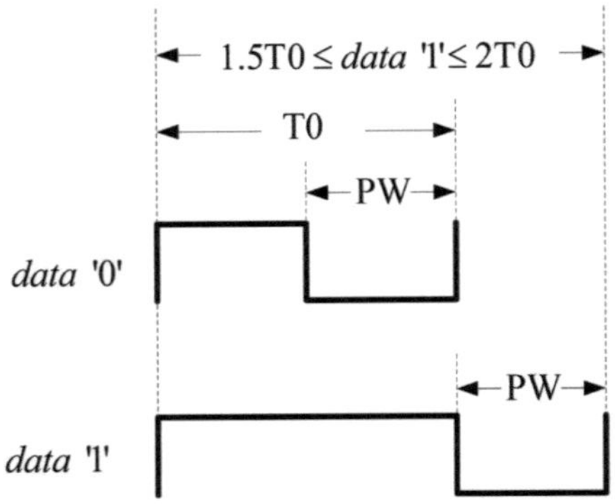

Fig. 1. Gen 2 Forward link PIE encoding

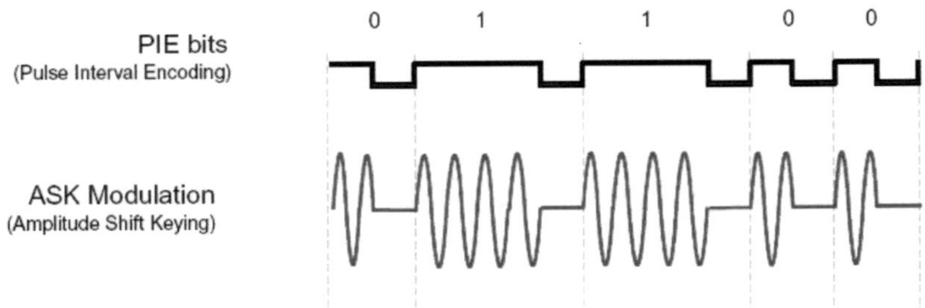

Fig. 2. Forward link transmission of ASK modulation with PIE encoding [3]

2.2 Reverse link (from tag to reader)

The reverse link in a RFID system is done using backscattering scheme. The modulation of the chip impedance can be done using either ASK or PSK. In ASK modulation, the chip impedance is varied between perfect match and complete mismatch. In PSK modulation, the real part of the chip impedance is kept in match with the antenna, while the imaginary part is varied between two capacitive and inductive values.

In Gen 2 protocol, Tags encode the backscattered data as either FM0 or Miller modulation. FM0 inverts the phase at every symbol boundary; a data '0' has an additional mid-symbol phase inversion, as shown in Figure 3.

Miller encoding inverts its phase between two data '0's, a data '1' has an additional mid-symbol phase inversion. The Miller subcarrier waveform is the baseband waveform multiplied by a square-wave at M times the symbol rate, and the value of M can be 2, 4 or 8 (selected by the reader). Figure 4 demonstrates Miller encoding when M is 2, 4 and 8 respectively [3]. Figure 5 illustrate the reverse link transmission of ASK and PSK modulation in Miller encoding.

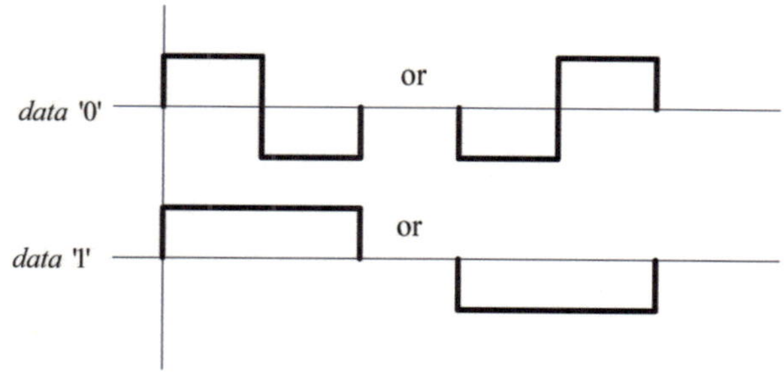

Fig. 3. Gen 2 reverse link FM0 encoding

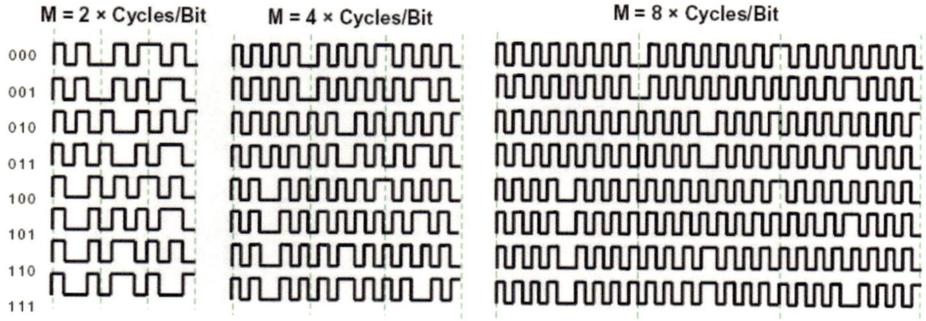

Fig. 4. Gen 2 Miller encoding [3]

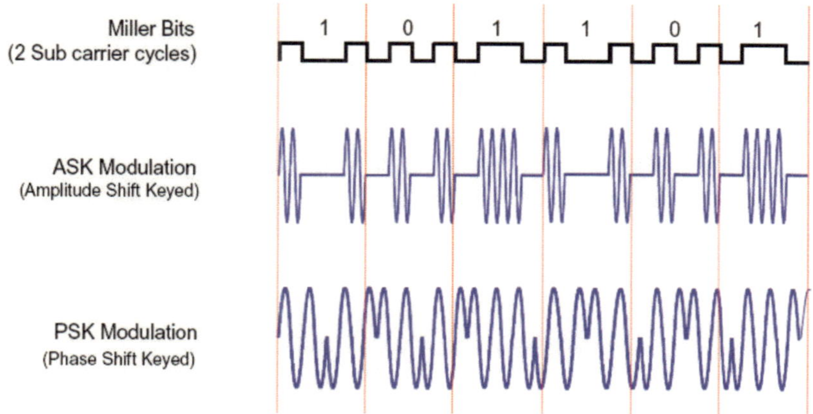

Fig. 5. Reverse link transmission in either ASK or PSK modulation with Miller encoding [3]

3. Tag implementation

In the following section, the implementation of a passive UHF RFID tag is discusses. Figure 6 shows a block diagram of RFID tag using backscatter modulation. The tag consists of tag antenna and tag chip. The tag chip contains a RF-analog front end, a digital control block, and a non-volatile memory.

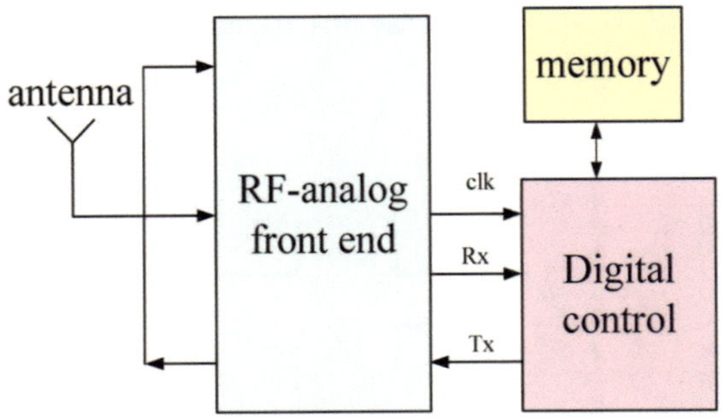

Fig. 6. Passive UHF RFID tag block diagram

3.1 RF-analog front end

The RF-analog front end includes a voltage rectifier, a demodulator, a clock generator, and a modulator [4], as shown in Figure 7.

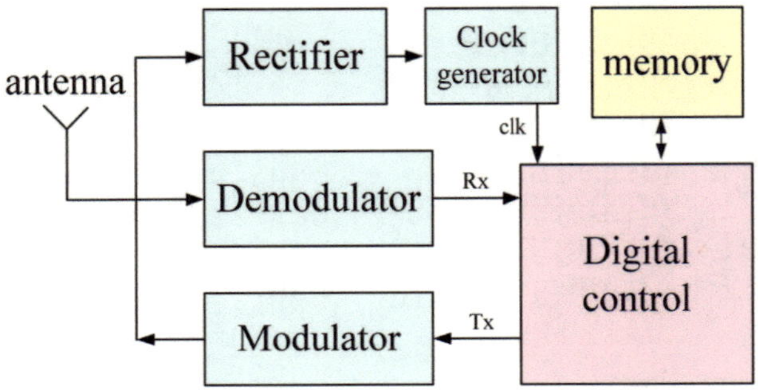

Fig. 7. RFID tag block diagram containing the RF-analog front end [4]

A. Rectifier

The rectifier has to supply the needed DC voltage with maximum efficiency possible. Figure 8 shows a multiple-stage rectifier that consists of diodes and capacitors [5]. All transistors and capacitors are set to be equal. The two input terminals V_{in} and *Gnd* are connected directly or via an impedance matching network to the UHF antenna (not shown), and is arbitrarily assigned as the ground node to the rectifier. The load capacitor C_L is large to store enough charge to complete signal processing tasks and to reduce the output ripple voltage. The output V_{out} is the input of clock generator.

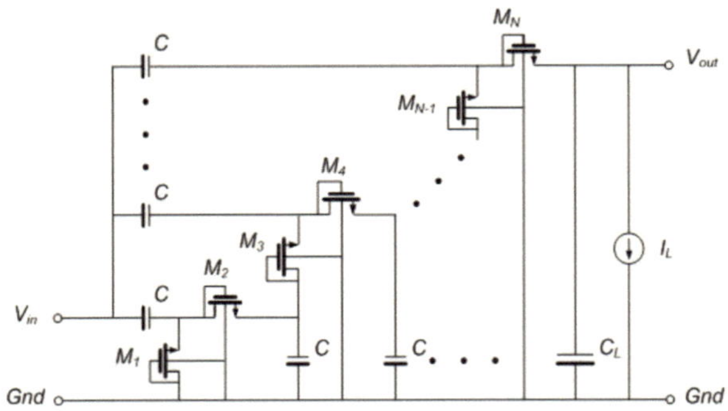

Fig. 8. Multi-stage rectifier [5]

B. Demodulator

Figure 9 shows the circuit schematic of ASK demodulator for forward link communication [6]. The ASK demodulator uses envelope detection and comparison with the average of the input voltage to recover baseband data. The envelope is transferred through a low pass filter to get its average value, and two values are then compared using a comparator. To deal with the voltage ripple from the envelope detector, the comparator needs hysteresis. The envelope detector uses 2-stage voltage multiplier to detect the envelope of the input RF signal. M3, which acts as a resistor, and the capacitor C make the low pass filter. The width and length of M3 determine the resistance. The input of the demodulator RF_{in} is the same as the difference of V_{in} and Gnd ($RF_{in} = V_{in} - Gnd$) shown in Figure 8, connected directly or via an impedance matching network to the UHF antenna. The output of the demodulator V_{out} is the input of later digital control block, and is the same as R_x shown in Figure 6.

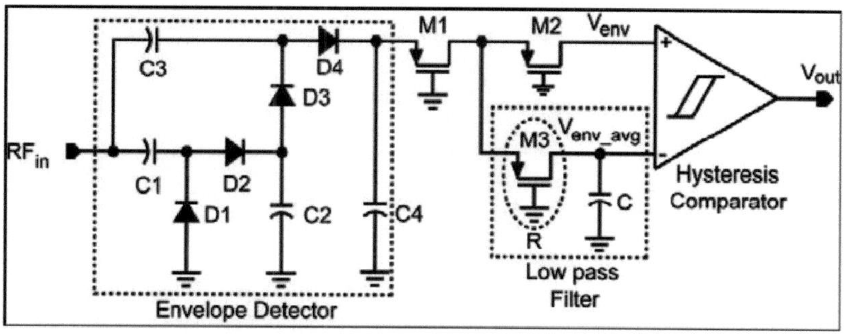

Fig. 9. Circuit schematic of the ASK demodulator [6].

C. Modulator

The PSK backscatter modulation is for reverse link communication. The circuit diagram is shown in Figure 10 [7]. Transistor M_1 is a MOS varactor that operates in inversion or in cutoff, depending on the input signal, causing the variation of the capacitance seen at the output of the modulator. Transistor M_2, instead, does not affect the output capacitance, since it has a small width with respect to M_1, but determines the resistance at the output of the modulator, which is, essentially, its drain-source resistance. As a consequence, the channel

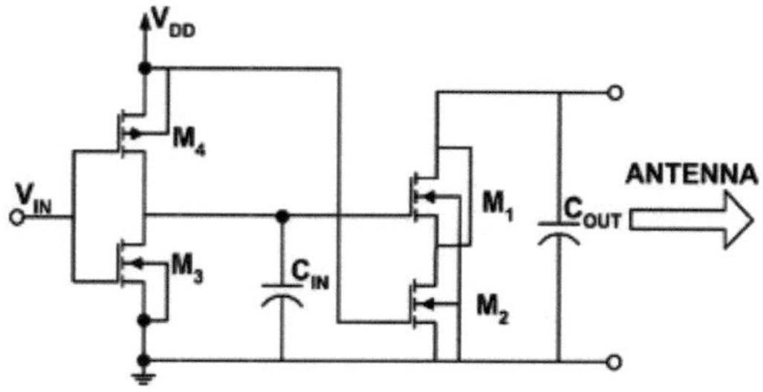

Fig. 10. PSK backscatter modulator.

length of M_2 has to be large enough so that the output resistance of the modulator is much larger than the antenna resistance. Such choice ensures that only a negligible fraction of the power at the antenna goes to the modulator, as required for the correct operation of the transponder. C_{OUT} is the capacitance seen at the output of the modulator and is due to the interconnections, the antenna and the input capacitance of the other stages which the modulator is connected to. The capacitance C_{IN} has to be larger than the gate-source and gate-drain capacitance of M_1, in order not to degrade the variation of the output capacitance of the modulator. Once the value of C_{IN} is chosen, the two transistors M_3 and M_4 of the inverter have to be dimensioned to fix the switching time of the varactor so that the channel bandwidth occupation of the backscattered signal complies with the requirements. The input V_{in} is from digital control block, the same as T_x as shown in Figure 6.

D. Clock generator

The clock generator circuit is based on RC relaxation oscillator [14]. The principle of operation of the circuit is shown in Figure 11. The capacitor C_{osc} charges when the output is low and discharges when the output is high.

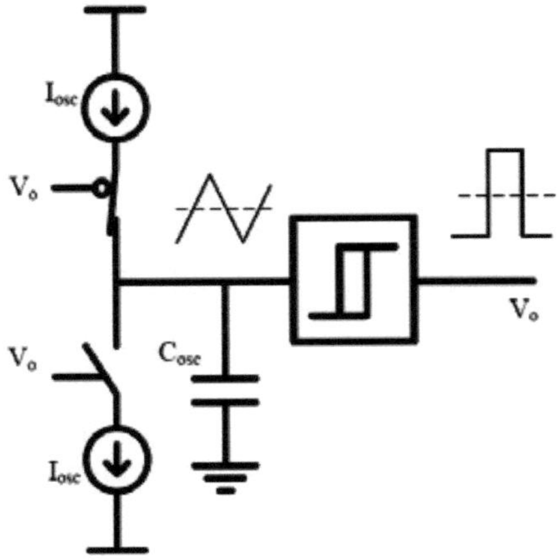

Fig. 11. Clock generation circuit [7]

3.2 Digital control block

The security of proposed secure low-cost RFID system depends largely on the digital control block of the RFID tags, which acts as an identification tag, carrying only the unique serial number and an ID number. The tag requires no secret key or PIN shared between tags and readers for authentication, thus eliminate the need of onerous work on key distribution and management.

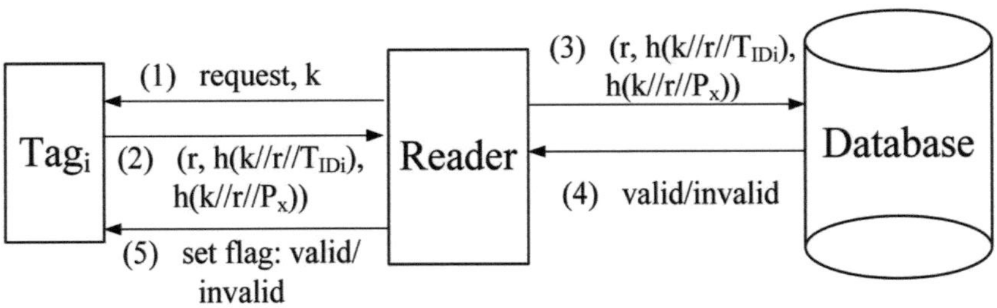

Fig. 12. Proposed Hash-block structure to enhance database searching efficiency and to prevent man-in-the-middle attacks.

The protocol is shown in Figure 12, the reader sends a random number "k" to tag_i at quering. Then tag_i generates a random number r and hashes it with random number k and the tag ID number "T_{IDi}"and P_x -- (r, $h(k//r//T_{IDi})$, $h(k//r//P_x)$), where // stands for concatenation. The reader receives it and passes to the back-end database with the random number k -- (k, r, h(k//r//T_{IDi}), h(k//r//P_x)). The database then computes the hash using k, r, T_{IDi} and P_x. The calculated hash is compared to the received hash. When there is a match, the correct identification is confirmed. Then the database retrieves the information of the confirmed tag T_{IDi}.

This protocol effectively prevents man-in-the-middle attacks by having reader sends a random number first, which a rouge reader can not mimic without notice by a legitimate reader in the later communication.

The proposed tag has the following features. 1) The tag is passive. 2) The RFID system provides semi-duplex communication mode between the reader and the tag. 3) The RFID system adopts "ALOHA" anti-collision mechanism [8]. 4) The tag sends and receives signals in serial.

The order of signal receiving, handling and sending in tag digital control block is: decode received packet → checkout cyclic redundancy check (CRC) → handle command (generate random number, hash information, run anti-collision mechanism, read/write non-volatile memory) → add CRC → encode and send packet.

The hardware implementation of the proposed RFID tag is straightforward, consisting of a RF/analog front-end, a digital control block and a non-volatile memory. The architecture of the tag is demonstrated in Figure 13. The digital control block is the major and most important part of the chip, since it needs to implement anti-collision algorithm and authorization scheme, including a PIE (pulse interval encoding) down-link decoder, a up-link FM0 encoder, a slot-counter for ALOHA-based anti-collision, a random-number generator (RNG) for slot-counter and hashing, a command handler as the central controller, and an Hash block [9-10].

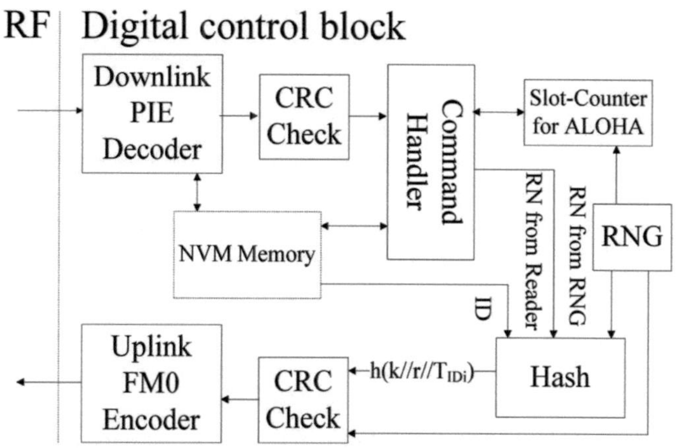

Fig. 13. Architecture of the tag digital control block

4. Reader implementation

An UHF RFID reader's structure is shown as Figure 14 [12]. From the function modules, UHF RFID reader consists of RF module and base-band module. It also consists of control part, transmission part and inception part. The transmission part and inception part can be known as RF module. The reader also includes I/O interface module and application program.

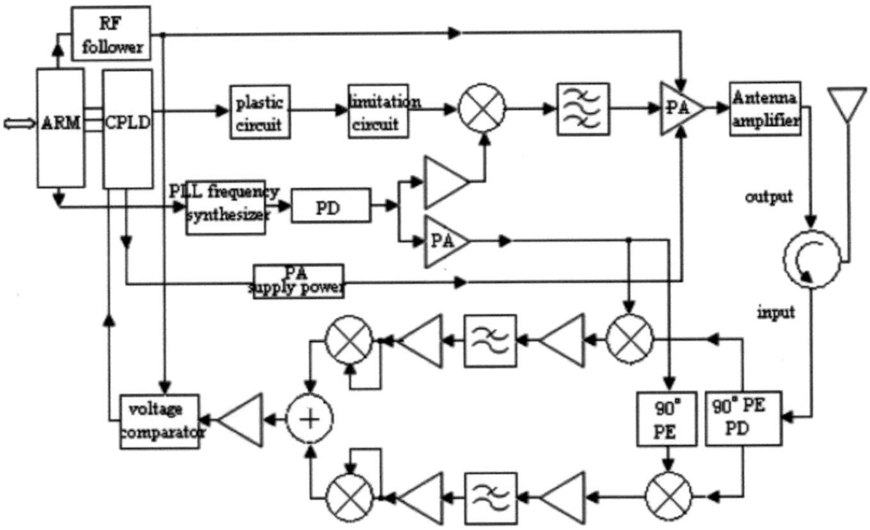

Fig. 14. UHF RFID Reader's structure [12]

4.1 Transmission part

Working flow of transmission part is shown as following:

1. The ARM Micro-controller receives operation commands from the computer, starts up application program and sends corresponding operation commands to the CPLD circuit.
2. The CPLD circuit forms base band signals to send to the plastic circuit and the limitation circuit to deal according to operation commands from the ARM Micro-controller. Then it sends the dealt signal to the mixer.
3. The mixer mixes the base band signal from the CPLD circuit and the LO signal and does ASK modulation.
4. The modulated signal is filtered by the filter, is amplified by the power amplifier, is amplified by the antenna amplifier and forms the final transmission signal.
5. The circulator sends the power signal from the antenna amplifier to the tag.

In that the frequency control of the LO signal created by the frequency synthesizer, setting the modulation depth, gain control of the power amplifier are made by ARM microcontroller according to communication protocols and system's working conditions.

4.2 Inception part

Working flow of inception part is shown as following:

1. After receiving the signal from the reader, the tag gets some energy and is activated. The tag starts up to perform the reader's command and sends the returned response information to the antenna of the reader in the way of the backscatter modulation.
2. After the antenna receives the signal, the circulator sends the signal returned from the tag to 90°phase-excursion power divider to split the signal into two orthogonal ways. These two signals are sent to two ways same demodulation circuits to deal. Two ways signals mixes with two orthogonal LO signals respectively. Mixed signals are amplified by the amplifiers, filtered by the filters and amplified again and sent to the multipliers to deal. The multiplier makes the sent signal squared to make pulse signals from the negative polarity into the positive polarity.

3. The signals dealt by the two demodulation circuits are amplified again after they add together. Then they are sent to the voltage comparator after passing the capacitor coupling.
4. The voltage comparator compares the voltage of the complete amplified modulated signal with the set norm voltage, forms base band signal returned from the tag, coordinates the signal and sends the signal to the CPLD circuit.
5. The CPLD circuit decodes the received base-band signal, does CRC check, forms the information about the tag ID and sends the information the ARM micro-controller.
6. The ARM micro-controller deals the received information about the tag ID. In these circuits, in order to ensure the accuracy of the demodulation circuit, use the amplifier to create the exactly 2.5 V virtual ground voltages as the middle voltage of the circuits such as the amplifier and the multiplier to use to ensure the stability of the inception circuit.

4.3 Main-control part

The main control of the system governs system parts to work in phase and realizes anti-collision control function. Its structure is shown as Figure 15. The control part of UHF RFID reader mainly realizes following functions: (1) communication with application software of the computer and execute the commands from the application system software; (2) complete quick real-time communication with tags; (3) the coding and the decoding of signals;(4) in some complex system, control part also executes anti-collision arithmetic; (5) encrypt and decrypt the transmitted data between tags and the reader, do ID validation between tags and the reader. Data exchange between control part and application software is done mainly by the communication interface of the reader. The interface can adopt RS-232 or RS-485. It also can adopt RJ45 or WLAN interface.

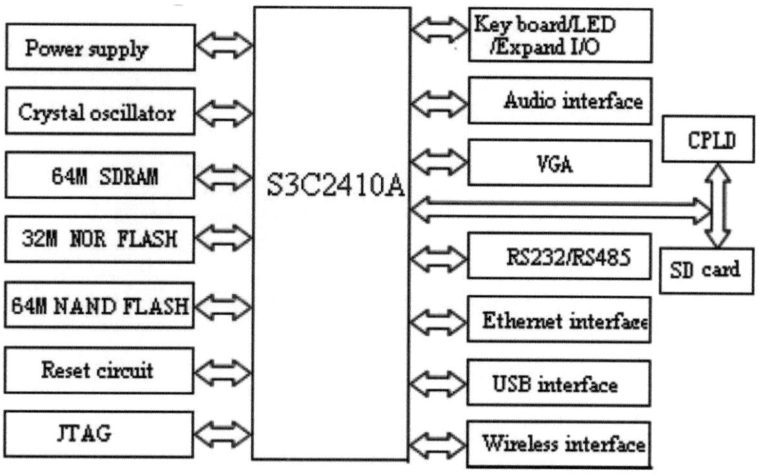

Fig. 15. Principle diagram of the main control part [12]

5. Single-chip reader implementation

To meet the demands of longer battery life and lower cost of mobile communication devices, the low-cost, low-power, single-chip reader draws great attention and thus the CMOS technology is believed to be the most promising candidate toward the system-on-a-chip (SoC), which integrates all the functions of a RF transceiver, data converters, a digital baseband modem, an MPU, memory, and host interfaces [13].

Figure 16 illustrates the block diagram of the single-chip RFID reader. Baseband modulator and demodulator are implemented in hardware logic. Most of the other digital functionalities are implemented in software to support multi-protocols and flexibility.

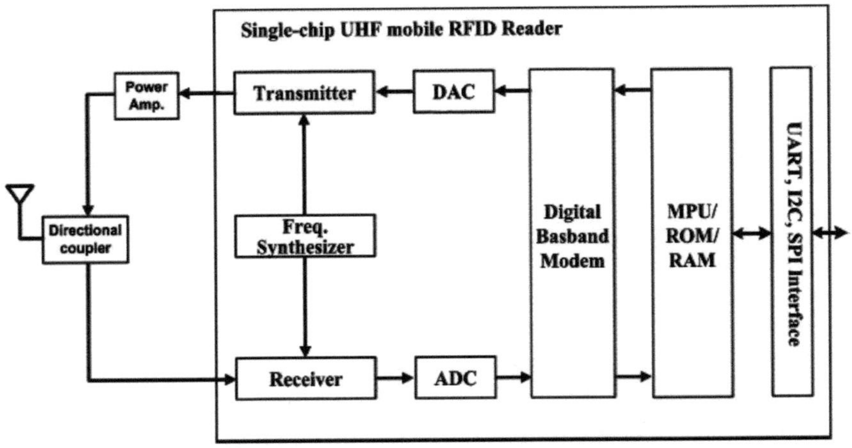

Fig. 16. Block diagram of the single-chip mobile RFID reader.

Figure 17 illustrates the receiver architecture of the RF transceiver. Among the various receiver architectures, the direct conversion receiver is adopted due to the backscattering communication solution. In the direct conversion receiver architecture, the transmitter carrier leakage to the receiver input is directly down-converted to DC. It can be removed by the DC offset cancellation (DCOC) feedback loop. However, the large transmitter carrier leakage leads to the saturation of the receiver RF front-end block. Hence, the low-noise amplifier (LNA) is bypassed in the backscatter detection mode for high P1dB characteristics to cope with very large transmitter carrier leakage.

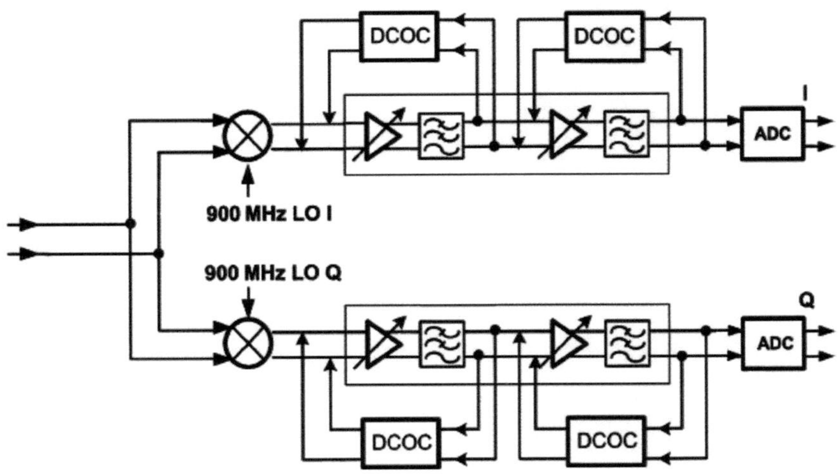

Fig. 17. Receiver architecture

The transmitter is implemented in the direct up-conversion architecture as illustrated in Figure 18. Baseband signals are transmitted to digital-to-analog converters (DACs) followed by the low-pass filters. Two identical mixers up-convert the baseband quadrature signals directly to the 900 MHz band, which is combined by current summing at the output. The transmitter supports both the SSB and the DSB modulation for the reader-to-tag communications and sends an unmodulated carrier for the tag-to-reader communications. In the DSB-ASK transmission, baseband signal is tied zero. In the SSB-ASK transmission, baseband signal is generated by the Hilbert transformer from the baseband signal. For generating 900 MHz LO signals with 200 and 500 kHz channel spacing, a frequency synthesizer based on a fractional-*N* phase-locked loop (PLL) derived from a 19.2 MHz crystal is implemented. A 1.8 GHz LO signal is generated by an integrated voltage-controlled oscillator (VCO) in the PLL and then the 900 MHz differential LO signals are obtained by a divide-by-two circuit.

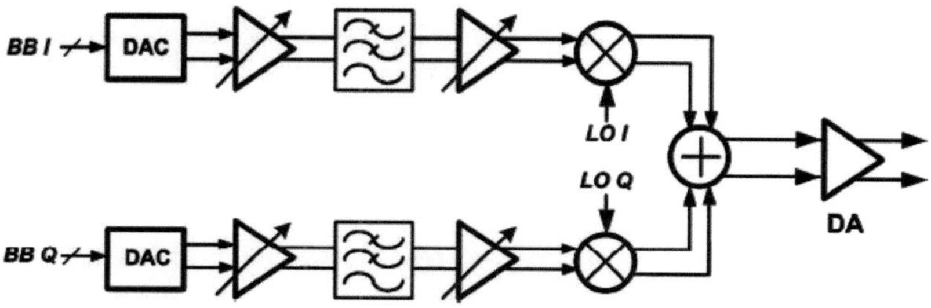

Fig. 18. Transmitter architecture

This single-chip UHF RFID reader for mobile phone applications has been implemented in a 0.18 m CMOS technology. It integrates an RF transceiver, data converters, a digital baseband modem, an MPU, memory, and host interfaces. Its die area is 4.5 mm x 5.3 mm including ESD I/O pads. The reader consumes a total current of 89 mA except external power amplifier with the 1.8 V supply voltage. The direct conversion RF transceiver architecture with the highly linear RF front-end circuit and DCOC circuit is used. It is suitable for the mobile phone reader with single-antenna architecture and low-power reader solution.

6. Reference

[1] Jin Li, Cheng Tao, "Analysis and Simulation of UHF RFID System", in proceedings of the 8th International Conference on Signal Processing, 2006.

[2] EPC™ Radio-Frequency Identity Protocols Class-1 Generation-2 UHF RFID Protocol for Communications at 860 MHz~960 MHz Version 1.0.9.

[3] Texas Instruments Proprietary Information, "UHF Gen 2. System Overview", 2005. http://rfidusa.com/superstore/pdf/UHF_System_Overview.pdf

[4] A. Ashry, K. Sharaf, "Ultra Low Power UHF RFID Tag in 0.13μm CMOS", in proceedings of International Conference on Microelectronics (ICM 2007), 2007.

[5] Department of Electronic & Computer Engineering, Chinese University of Hong Kong, "Single-Chip Passive UHF RFID Tags and Readers", 2005. http://www.ece.ust.hk /~rfid/phase1/power.htm

[6] N. Tran, B. Lee, JW. Lee, "Development of Long-Range UHF-band RFID Tag chip Using Schottky Diodes in Standard CMOS Technology", in proceedings of Radio Frequency Integrated Circuits (RFIC) Symposium, 2007.

[7] A. Facen, A. Facen, "A CMOS Analog Frontend for a Passive UHF RFID Tag", in proceedings of the 2006 international symposium on Low power electronics and design, 2006.

[8] EPC radio-frequency identi_cation protocols class-1 generation-2 RFID protocol for communications at 860 MHz-960 MHz Version 1.0.8.. EPCglobal, Dec. 2004.

[9] J. Wang, H. Li, F. Yu, "Design of Secure and Low-Cost RFID Tag Baseband", in proceedings of International Conference on Wireless Communications, Networking and Mobile Computing (WiCom), 2007.

[10] H. Li, F. Yu, Y. Hu, "A Solution to Privacy Issues in RFID Item-level application", in proceedings of IEEE International Conference on Integration Technology (ICIT), 2007.

[11] G. De Vita, G. Iannaccone, "Design criteria for the RF section of UHF and microwave passive RFID transponders", IEEE Transactions on Microwave Theory and Techniques, Vol. 53, pp. 2978 - 2990, 2005.

[12] W. Xiaohua, Z. Xiaoguang, S. Baisheng, "Design for UHF RFID Reader and Selection for Key Parts", in proceedings of 2007 2007 IEEE International Conference onAutomation and Logistics.

[13] I. Kwon, Y. Eo, H. Bang et al, "A Single-Chip CMOS Transceiver for UHF Mobile RFID Reader", IEEE Journal of Solid-State Circuits, Vol. 43, pp. 729-738, 2007.

[14] United States Government Accountability Office, "INFORMATON SECURITY: Radio Frequency Identification Technology in the Federal Government", 2005.

2

Design of Antennas for RFID Application

Ming-Tao Zhang[1], Yong-Chang Jiao[2], Fu-Shun Zhang[2] and Wu-Tu Wang[1]
[1]*China Academy of Space Technology (Xi'an), Xi'an, Shaanxi,*
[2]*National Laboratory of Antennas and Microwave Technology,*
Xidian University, Xi'an, Shaanxi,
P. R. China

1. Introduction

As a vital and integrated part of the radio-frequency identification (RFID) system, RFID antennas have been received much attention over years, and their design is very urgent and significant. In fact, the development of RFID antenna is of theoretical significance and practical value for the RFID system. In this chapter, the RFID technology is briefly introduced, and the operating principle of the RFID system is described. The antenna in RFID system is discussed, and the designing principle of the antennas for RFID applications is presented. Some commonly used antennas in the RFID system are also displayed.

2. RFID technology and antennas

As an automatic identification technique without touching, RFID technology uses radio waves carrying information stored about the identified object or commands to identify object via space coupling, such as inductive coupling or electromagnetic wave propagation. For the details about the RFID technology, refer to some web sites such as www.rfidchina.org, www.rfidinfo.com.cn, www.rfidofchina.com, www.cnrfid.net, www.superrfid.net/china/, www.rfidworld.com.cn, and www.kingant.com.

As a vital device for transmitting the RF power from the radio transceiver to the open space in the form of electromagnetic wave, or receiving it from space and transferring it to the next circuit, antenna is always the key part of the RF system, and its performance greatly affects the performance of the whole system. Thus design of antennas for the RFID system is very important. In the RFID system, according to their functions in the system, the antennas can be divided into two parts: tag antenna and reader antenna. The present RFID systems are applied at LF, HF (13.56MHz), UHF and microwave bands, and the antenna design is focused on these frequency bands. In fact, the system working at LF and HF bands is based on the magnetic field coupling between the tag coil and reader coil, whose operating principle is identical with that of the transformer. There is no radiation and wave transmission, and the antenna in the system is just a coil. The antenna discussed here is limited to the system that operates at UHF band, or microwave bands. Based on the different operating principles at different bands, design of the antennas in the system will be discussed at following sections.

2.1 Antennas in the RFID system

According to the different functions in the RFID system, the RFID antennas can be divided into two classes: the tag antenna and the reader antenna. The tag antenna not only transmits the wave carrying the information stored in the tag, but also needs to catch the wave from the reader to supply energy for the tag operation. Since the tag should be attached to the identified object, the size of the tag must be small enough, and the antenna should be small in size. In most cases, the tag antenna should have omnidirectional radiation or hemispherical coverage. Generally the impedance of the tag chip is not 50 ohm, and the antenna should realize the conjugate match with the tag chip directly, in order to supply the maximum power to the tag chip. In common applications, the tag antenna should be low-cost and easy to fabricate for mass production.

The reader antenna transmits the electromagnetic energy to activate or awaken the tag, realizes the data transfer and sends the instructions to the tag. Meanwhile, the reader antenna receives information from the tag. Generally the position or the orientation of the identified object is random, and the manner for attaching the tag to the identified object is unfixed. Thus the reader antenna should be a circularly polarized antenna, in order to avoid the polarization loss when the orientation of the identified object is changed. Meanwhile, the reader antenna should have low profile and realize miniaturization, some of which should operate at more than one band. In some special cases, multiple antenna technology or smart antenna arrays for beam scanning will be employed.

In passive RFID system, the energy for maintaining the tag operation comes from the electromagnetic wave transmitted by the reader antenna. Here the passive system is mainly discussed to show the impact of the antenna parameters on the system performance (Keskilammi, Sydanheimo & Kivikoski, 2003).

To double the reading range, the transmitted power, the antenna gain, or the sensitivity of the receiver should increase at least 12dB. First, the impact of the antenna gain on thte system performance is described. When the transmitted power is fixed, the maximum reading range of the RFID system is mainly limited by the antenna gain and the operating frequency. By the RF link analysis, the electromagnetic wave transmitted by the reader antenna radiates to the tag through the space loss, and then reversely propagates back to the reader, carrying the information stored in the tag. Suppose that the RF energy caught by the tag can be re-radiated into the space totally. Let the power transmitted by the reader be $P_{transmitted}^{reader}$, and the gain of the reader antenna be G_{reader}. The power density at distance R where the tag is placed can be expressed as

$$S_1 = \frac{G_{reader} P_{transmitted}^{reader}}{4\pi R^2} \tag{1}$$

The power received by the tag is calculated by

$$P_{received}^{tag} = S_1 A_{tag} \tag{2}$$

where

$$A_{tag} = \frac{G_{tag} \lambda^2}{4\pi} \tag{3}$$

Then, we have

$$P^{tag}_{received} = (\frac{\lambda}{4\pi R})^2 G_{reader} G_{tag} P^{reader}_{transmitted} . \quad (4)$$

The power density of the return wave from the tag at the position of the reader is

$$S_2 = \frac{G_{tag} P^{tag}_{received}}{4\pi R^2} \quad (5)$$

Thus the power received by the reader is

$$P^{reader}_{back} = S_2 A_{reader} = S_2 G_{reader} \frac{\lambda^2}{4\pi} \quad (6)$$

That is

$$P^{reader}_{back} = (\frac{\lambda}{4\pi R})^4 G^2_{reader} G^2_{tag} P^{reader}_{transmitted} \quad (7)$$

where G_{reader} stands for the gain of the reader antenna, A_{reader} the equivalent aperture of the reader antenna, G_{tag} the gain of the tag antenna, and A_{tag} the equivalent aperture of the tag antenna.

Define the equivalent transmitted power as

$$P_{(EIRP)} = G_{reader} P_{transmitted} \quad (8)$$

Then

$$P^{reader}_{back} = (\frac{\lambda}{4\pi R})^4 G^2_{tag} G_{reader} (P_{(EIRP)}) . \quad (9)$$

Denote by $P^{reader}_{sensitivity}$ the threshold power of the sensitivity. Then the maximum reading range is expressed as

$$R = \frac{\lambda}{4\pi} \sqrt[4]{\frac{P^{reader}_{transmitted} G^2_{reader} G^2_{tag}}{P^{reader}_{sensitivity}}} \quad (10)$$

Now we analyze the RFID system by using the radar principle. Suppose that the back-scattering section of the tag, including the antenna and the chip, is σ^{tag} , then the back-scattering power of the tag is

$$P_{BS} = S_1 \sigma^{tag} = \frac{G_{reader} P^{reader}_{transmitted} \sigma^{tag}}{4\pi R^2} \quad (11)$$

The power density of the back scattering wave at the position of the reader is

$$S_2 = \frac{P_{BS}}{4\pi R^2} = \frac{G_{reader} P_{transmitted}^{reader} \sigma^{tag}}{(4\pi)^2 R^4} \tag{12}$$

So we have

$$P_{back}^{reader} = S_2 A_{reader} = S_2 G_{reader} \frac{\lambda^2}{4\pi} = \frac{P_{transmitted}^{reader} G_{reader}^2 \sigma^{tag} \lambda^2}{(4\pi)^3 R^4}. \tag{13}$$

By adjusting the tag chip impedance according to the stored data in tag, σ^{tag} will be changed, and then the return wave coming from the tag and received by the reader will be changed such that the amplitude modulation and demodulation can be realized. In this manner, the tag information can be read, and the object detected by the tag can be identified. Generally, the operating frequencies of the normal RFID system based on the back-scattering include: 915MHz, 2.45GHz, and 5.8GHz, the corresponding wavelengths are 0.328m, 0.122m, and 0.051m. Obviously, the maximum reading range is directly proportional to the wavelength. In fact, for the same distance the space loss at higher frequency is greater than that at lower frequency. The space loss *SL* is defined as

$$SL = \left(\frac{4\pi R}{\lambda}\right)^2. \tag{14}$$

Commonly, the size of the antenna is relevant to its working frequency. For lower frequency, the antenna will be larger, and the size of the tag will increase. When the antenna size is fixed, the higher gain will be achieved for higher frequency. In most cases, the antenna size is a bottleneck for tag miniaturization. In order to appropriately choose the operating frequency for the RFID system, we should consider simultaneously many factors such as the space loss, the antenna gain, and the size of the tag.

There also exists another loss, called the polarization loss, which is caused by the polarization mismatch between the incoming wave and the antenna, or between the transmitting antenna and the receiving antenna. The polarization mismatch will make the antenna lose the ability to receive all the power of the wave.

Suppose $\vec{E}_i = \hat{\rho}_w E_i$ is the incoming wave, $\vec{E}_a = \hat{\rho}_a$ is the polarization orientation of the receiving antenna, and $\hat{\rho}_o$ is the vector that is orthogonal to the polarization vector of the receiving antenna. The polarization factor *PLF* is defined as

$$PLF = |\hat{\rho}_w \cdot \hat{\rho}_a|^2 = |\cos\varphi_p|^2, \text{ or } PLF(dB) = 10\lg PLF \tag{15}$$

Then, the power received by the antenna is denoted by

$$P_r = P_{\max} \cdot PLF, \text{ or } P_r(dB) = P_{\max}(dB) + PLF(dB) \tag{16}$$

where $P_{\max}$ stands for the power of the incoming wave, or the maximum power received by the antenna when the polarizations are matched, $\hat{\rho}_a$ the unit polarization vector of the receiving antenna, and $\hat{\rho}_w$ the unit vector of the incoming wave.
Assume that the incoming wave is circularly polarized. Then the unit vector $\hat{\rho}_w$ can be expressed as

$$\hat{\rho}_w = \frac{\sqrt{2}}{2}(\hat{\rho}_a \pm j\hat{\rho}_o) \tag{17}$$

$PLF = 1/2$, and $PLF(dB) = -3dB$.

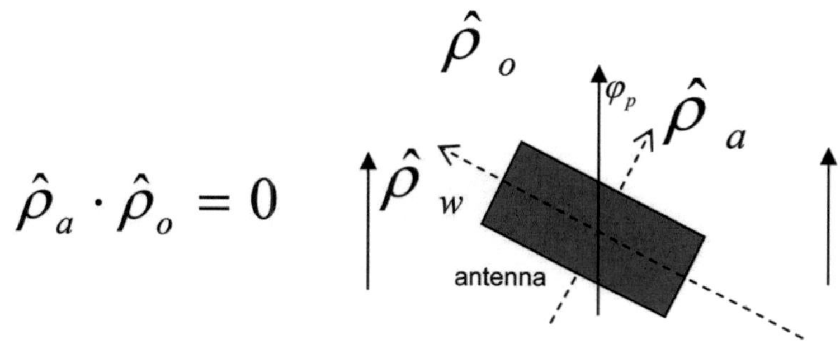

Fig. 1. Polarizations of the antenna and the wave

As shown in Fig. 1, the polarization mismatch between the antenna and the wave reduces the received power, and deteriorates the system performance. Thus choosing a suitable polarization is also an important step for designing the antenna.

2.2 Development of antennas in the RFID system

Potential applications of the RFID technology inspired the development of various antennas for the RFID systems. Lots of antennas with high performance for various requirements have been fabricated. As an identification system with huge market and potentials, RFID system requires the RFID antenna to meet some particular specifications. Design of the RFID antennas faces many challenges, such as the antenna structure, the antenna size, the operating mode, the bandwidth, the radiation pattern, the polarization, mutual coupling between multiple antennas, and the antenna scattering. In the present RFID system, the reader antenna is designed to be a circularly polarized antenna. Patch and spiral antennas are typical reader antennas. In some special cases, linearly polarized antennas can also be used. In the tag, the eroded or printed antennas are commonly used, and the dipole is the typical tag antenna structure. Some circularly polarized antennas for the tag may be required in some special applications.
In recent years, theory for matching the antenna with the tag chip is discussed, which guides the design of the tag antenna and the analysis of the tag configuration. Several tag antennas in common use are designed with simple impedance transformation for matching the chip with special impedance, especially for UHF band application. In the microwave band, some tag antennas are also designed to integrate with the already existing specific circuits with 50 ohm impedance.

Schemes for designing the circularly polarized reader antenna are also presented in some literature. Based on two ports for the dual circular polarization, the aperture-coupled patch antenna integrated with the microstrip branch line coupler is preferred. Some modifications are performed to achieve the wide band, or meet the practical requirements. The system, in which multiple reader antennas are used, is also discussed.

In the design of antenna for the RFID system, some other problems, such as the environmental effects on RFID tag antennas, especially surrounded by metallic objects, should be considered. Designing the RFID tag antenna, which is mounted on the metallic objects, also faces a challenge. The inverted-F antenna and its modifications are usually used in the tag for identifying the metallic objects, and other antenna structures can also be referred in designing antenna mounted on metallic surfaces. The electromagnetic scattering of the tag antenna is also introduced and discussed, and relative calculations have been performed.

2.3 Antenna design software for RFID application

Efficient numerical methods promote the antenna design. Modern antenna design becomes a manipulation of accurate computing based on relative theory and a design under the theory instruction or according to the calculated results. The antenna design method based on numerical methods has been applied to design antennas for various systems. Familiar numerical methods include Method of Moment (MoM), Finite Element Method (FEM), and Finite Difference Time Domain (FDTD). There already exist several design tools based on these methods, which are of different characteristic and are widely used. Fig. 2 shows some familiar methods and the design tools. These design tools can be chosen for different problems in designing antennas. The MoM can be used to calculate the antenna performance quickly and accurately, especially for some large antenna structures. Some optimization methods, such as the optimization tool used in Zeland IE3D, can be embedded into the analysis method to make the antenna achieve the excellent performance. The FEM and FDTD methods can be used directly to analyze the antenna performance. However, the FEM method gets more accurate results than the FDTD method. The FDTD method can be used to analyze some larger antenna structures, solve the wide band problems in time domain, and give a dynamic demo about the electromagnetic field distribution and radiation. Some tools such as HFSS, which are widely used to deign antenna for the RFID system, add the ability of automatically meshing to facilitate the user and improve the precision.

These design tools should be chosen properly for designing antenna, since they have different characteristics. Some tools can be used to analyze some types of antenna suitably but lose the ability for solving other antennas or affording the large memory requirement. In designing antenna, the antenna concept based on the electromagnetic theory should be mixed with the manipulating software skilfully, and the antenna prototype of the design scheme chosen for the system requirement is more important than the skill in applying the software. After the antenna scheme is decided, being familiar with the software and the relative numerical methods will help the designer to design antenna properly, and adjust the structure parameters to optimize its performance. To succeed in designing antenna, it is of great importance to apply software under the guidance of antenna principle and electromagnetic theory. Although the function of the software for designing antenna is more powerful, the basic theory and concept is also absolutely necessary. Both the antenna theory and the design software promote the design of antennas in the RFID system.

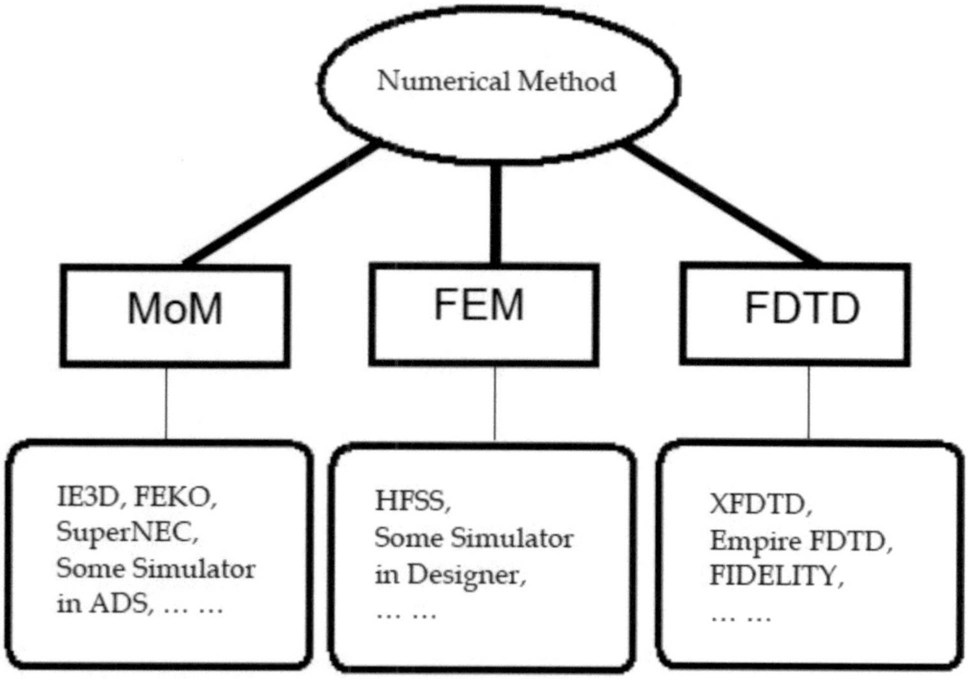

Fig. 2. Numerical methods and software

3. Power transmission between tag chip & antenna

Generally, the RFID system mainly consists of reader and tag. The tag design is the most important loop in the RFID application, and also the most difficult part in the function realization. Performance of the tag usually decides the performance of the whole system. The tag is composed of the tag antenna and the chip, between which good connection and power transmission directly impact on the system configuration, the relative function realization and also the system performance. Thus, it is necessary to analyze the connection of the tag antenna to the RFID tag chip, and to discuss the impedance match problem.

3.1 Theory of impedance match

The most important factor in the tag is the reading range, which is the maximum distance between the reader and the tag such that the reader can detect the backscattering signal from the tag. Compared with the tag, the reader is always of high sensitivity, and the reading range is mainly limited by the performance of the tag. Especially for the passive tag, both the energy for maintaining or arousing the tag and the power of signal retransmitted by the tag are from the RF energy, which is transmitted by the reader and caught by the tag. The impedance match between the antenna and the chip has a direct influence on whether the tag circuit can operate well and the chip is able to retransmit enough energy to implement the backscattering communication, and limits the reading range.

To maximize the power transfer between the antenna and the chip, the impedance of the chip connected to the antenna should be conjugate to the antenna impedance. When the working frequency comes into the microwave band, the impedance match problem becomes more serious. Ordinarily, the impedance of the antenna prototype designed for the tag is 50 ohm or 75 ohm, while the chip impedance may be a random value, or vary with frequency, and have a difference when the driving power is changed. It is extremely crucial to achieve suitable impedance match between the antenna and the chip. New integrated circuit chip design and development need large investment and long research period, however, designing antenna to match the existing chip is more convenient and practical. Due to the requirements such as easy manufacture, low cost and small size, adding the matching network is infeasible. To solve this problem, the antenna should be able to match the chip directly by adjusting its structure. How to design an antenna to match a chip of arbitrary impedance is an inevitable mission in designing antenna for the RFID system (Nikitin et al., 2005; Rao, Nikitin & Lam, 2005a).

By analyzing the tag, its equivalent circuit is shown in Fig. 3. Denote by Z_a the antenna impedance, and $Z_a = R_a + jX_a$, by Z_c the chip impedance, and $Z_c = R_c + jX_c$.

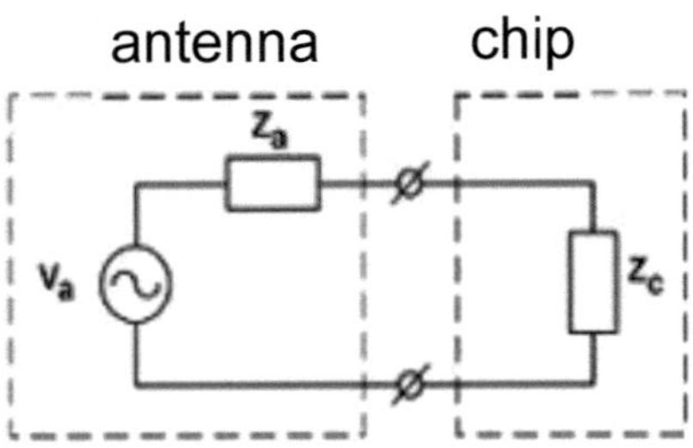

Fig. 3. Equivalent circuit of the tag

Define the complex power reflection coefficient s as

$$s = \frac{Z_a - Z_c^*}{Z_a + Z_c} \tag{18}$$

Then the power reflection coefficient is calculated by

$$|s|^2 = \left|\frac{Z_a - Z_c^*}{Z_a + Z_c}\right|^2 = \left|\frac{(R_a - R_c) + j(X_a + X_c)}{(R_a + R_c) + j(X_a + X_c)}\right|^2$$

$$= \left|\frac{[R_a + j(X_a + X_c)] - R_c}{[R_a + j(X_a + X_c)] + R_c}\right|^2 = \left|\frac{\left(\frac{R_a}{R_c} + j\frac{X_a + X_c}{R_c}\right) - 1}{\left(\frac{R_a}{R_c} + j\frac{X_a + X_c}{R_c}\right) + 1}\right|^2 \tag{19}$$

Let

$$\frac{R_a}{R_c} + j\frac{X_a + X_c}{R_c} = r + jy = \overline{Z}_a \tag{20}$$

be the antenna impedance normalized to the real part of the chip impedance, then

$$|s|^2 = \left|\frac{\overline{Z}_a - 1}{\overline{Z}_a + 1}\right|^2 \text{, or } |s| = \left|\frac{\overline{Z}_a - 1}{\overline{Z}_a + 1}\right|. \tag{21}$$

On the basis of the transformation, the traditional Smith Chart can be used to describe the impedance match between the antenna and the chip. $\overline{Z}_a$ can be marked according to its real part and imaginary part on Smith Chart like the traditional normalized impedance. The distance between the point of each $\overline{Z}_a$ and the centre point of Smith Chart expresses the magnitude of the complex power reflection coefficient s, while the trace of impedance points, which have a constant distance to the centre point, forms the concentric circle, which is called as the equivalent power reflection circle. The centre point of Smith Chart is the perfect impedance match point, while the most outer circle denotes the complete mismatch case, i.e. $|s| = 1$.

The power transmission coefficient (Rao, Nikitin & Lam, 2005b) can also be defined as τ, and $P_c = P_a\tau$, where P_a stands for the power from reader caught by tag antenna, P_c the power transmitted from the tag antenna to the tag chip. It follows from Fig. 3 that

$$\tau = \frac{4R_cR_a}{|Z_a + Z_c|^2}, 0 \le \tau \le 1 \tag{22}$$

$$\tau + |s|^2 = 1 \tag{23}$$

Let $x_a = \frac{X_a}{R_c}$, $r_a = \frac{R_a}{R_c}$, $Q_c = \frac{X_c}{R_c}$, then equation of the circle with constant power transmission coefficient is expressed as follows.

$$[r_a - (\frac{2}{\tau} - 1)]^2 + [x_a + Q_c]^2 = \frac{4}{\tau^2}(1 - \tau) \tag{24}$$

From equation (24), the impedance chart with the constant power transmission coefficient is draw, as shown in Fig. 4.

In Fig. 4, the x axis expresses the normalized real part $r_a = R_a / R_c$, and y axis the normalized imaginary part $x_a = X_a / R_c$. The circles with constant power transmission coefficients τ =1, 0.75, 0.5, 0.25 are draw in Fig. 4. The x axis is called as the resonant line with $X_a = -X_c$, while the y axis is called as the complete mismatch line. When τ's decrease, the radius of the circles with constant power transmission coefficient increase. While $\tau \to 0$, the circle with constant power transmission coefficient approaches to its tangent, that is the y axis, on which the impedance point cannot achieve the power transmission.

When the chip and the antenna are resonant, $X_a = -X_c$, and $x_a = -Q_c$, then equation (24) becomes

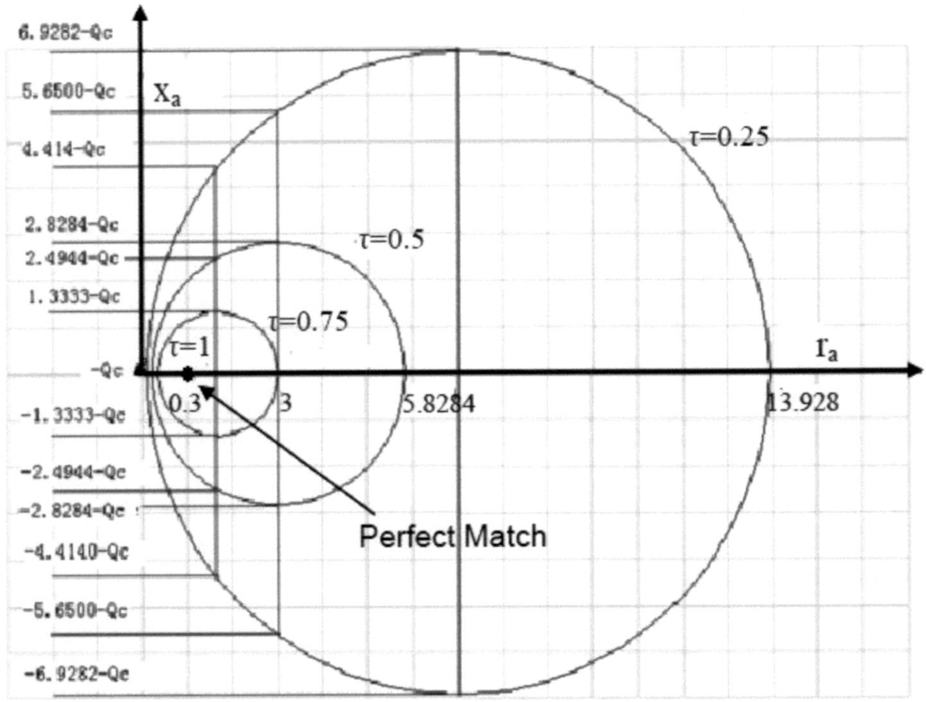

Fig. 4. The impedance chart with the constant power transmission coefficient

$$[r_a - (\frac{2}{\tau} - 1)]^2 = \frac{4}{\tau^2}(1-\tau) \tag{25}$$

$$[\tau r_a - (2-\tau)]^2 = 4(1-\tau) \tag{26}$$

Making the derivative for the both sides of equation (26), we have

$$2[\tau r_a - (2-\tau)](\tau + r_a \frac{d\tau}{dr_a} + \frac{d\tau}{dr_a}) = -4\frac{d\tau}{dr_a} \tag{27}$$

$$\frac{d\tau}{dr_a} = \frac{[(r_a+1)\tau - 2]\tau}{2r_a} \tag{28}$$

Obviously $\tau = 1$ means perfect match, and $\frac{d\tau}{dr_a} = 0$. $\tau = 0$ means complete mismatch, and $\frac{d\tau}{dr_a} = 0$. Thus either the perfect match or the complete mismatch is a steady point of τ with r_a, i.e. $\frac{d\tau}{dr_a} = 0$.

For the fixed $\frac{R_a}{R_c}$ and $\frac{X_a}{X_c}$,

$$\tau = \frac{4\frac{R_a}{R_c}}{|1+\frac{R_a}{R_c}+jQ_c(1+\frac{X_a}{X_c})|} = \frac{4\frac{R_a}{R_c}}{(1+\frac{R_a}{R_c})^2+Q_c^{\,2}(1+\frac{X_a}{X_c})^2} \tag{29}$$

$$\frac{d\tau}{dQ_c} = -8Q_c(1+\frac{X_a}{X_c})^2\frac{R_a}{R_c}[(1+\frac{R_a}{R_c})^2+Q_c^{\,2}(1+\frac{X_a}{X_c})^2]^{-2} \tag{30}$$

When the chip impedance is capacitive, i.e. $Q_c < 0$, it follows from (13) that $\frac{d\tau}{dQ_c} > 0$. While the chip impedance is inductive, i.e. $Q_c > 0$, $\frac{d\tau}{dQ_c} < 0$. When $Q_c = 0$, i.e. $X_c = 0$ and meanwhile $X_a = 0$, we have

$$\tau = \frac{4R_cR_a}{(R_c+R_a)^2} \tag{31}$$

The curve of τ versus Qc is shown in Fig.5. From this figure, we can see that for the fixed $\frac{R_a}{R_c}$ and $\frac{X_a}{X_c}$, Qc should be as small as possible from the power transmission point of view, when the tag antenna is connected to the tag chip.
For the tag antenna, the impedance chart can be used to guide the design or to describe the tag antenna. The chart is theoretically important and very useful for other applications.

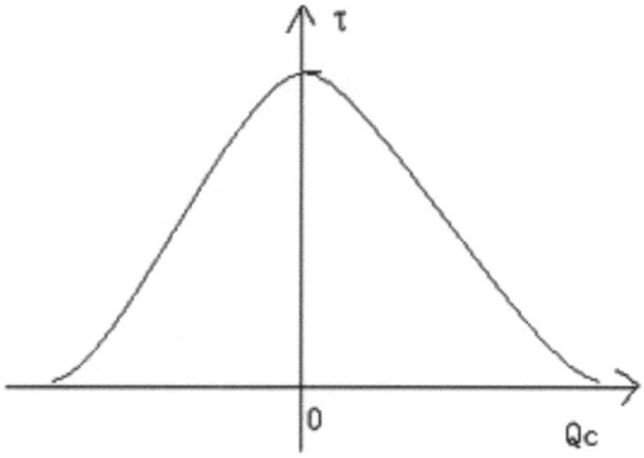

Fig. 5. Curve of τ versus Qc

3.2 Impedance design for the tag antenna

Aforementioned results indicate that the maximum power transmission can be realized only if the antenna impedance is equal to the conjugate value of the chip impedance. While the chip impedance is not normal 50 ohm or 75ohm, the structure of the tag antenna should be carefully chosen. In this section, a symmetrical inverted-F metallic strip with simple structure shown in Fig. 6 is proposed.

The antenna has the ability to realize several impedances. For UHF band application, the impedance of the antenna in four cases with different structure parameters is analyzed at 912MHz, whose real part is approximately 22ohm, 50ohm, 75ohm, 100ohm respectively. The simulated results for these four cases are shown in Fig. 7.

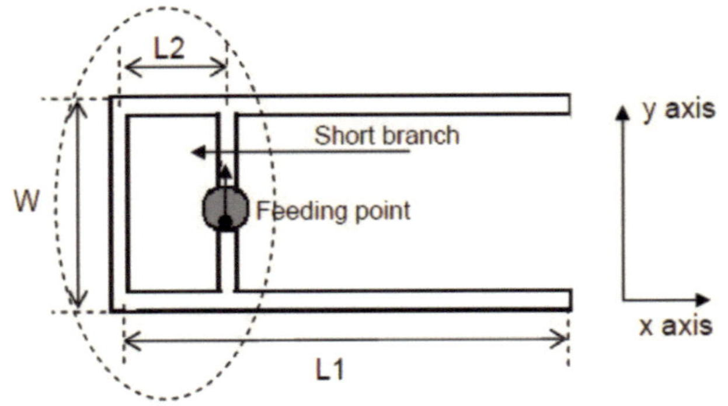

Fig. 6. The symmetrical inverted-F Antenna

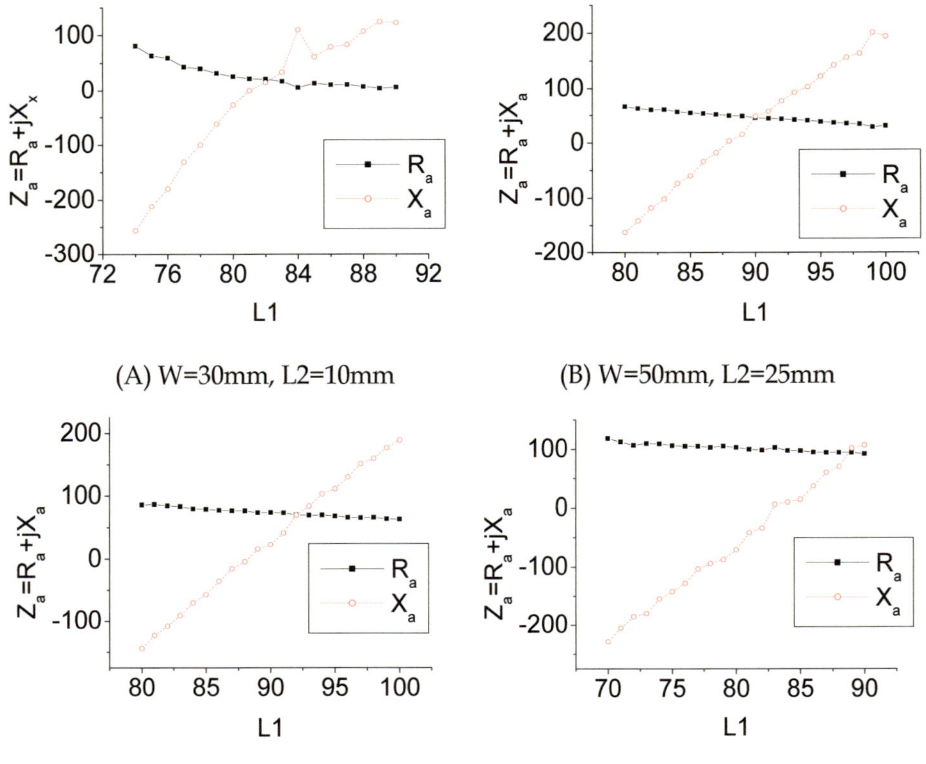

(A) W=30mm, L2=10mm (B) W=50mm, L2=25mm

(C) W=64mm, L2=32mm (D) W=73mm, L2=32mm

Fig. 7. Impedance results of the antenna in different cases

Fig. 7 shows that the symmetrical inverted-F metallic strip can realize several impedance values by adjusting its short branch. A lot of familiar types of tag antennas are the modifications or transformations of this structure (Dobkin & Weigand, 2005).
Fig. 8 shows the evolvement of several tag antennas. Antenna B has less influence on its performance than antenna A, when the antenna is curved (Tikhov & Won, 2004). Antennas C and D are fed by an inductively coupled loop (Son & Pyo, 2005).

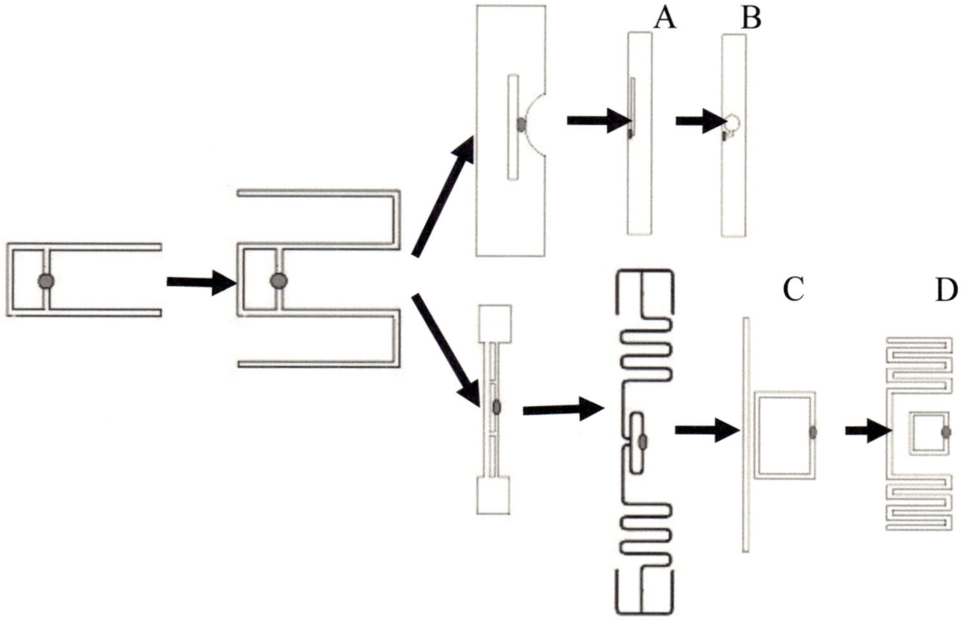

Fig. 8. Evolvement of the tag antennas

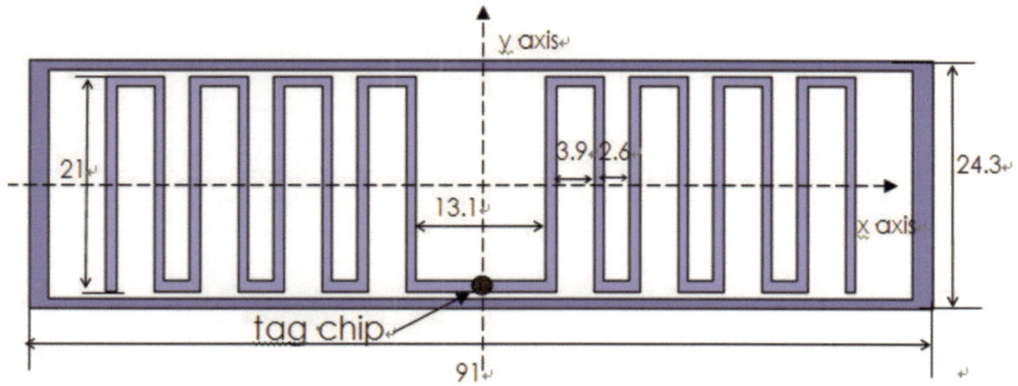

Fig. 9. Geometry of a meandered dipole antenna surrounded by the rectangular loop (dimensions in mm)

In our application, an UHF band tag chip with 43-j800 ohm impedance is used, and a tag antenna connected to this chip should match the tag chip. Meanwhile the tag antenna should be small in size and easily fabricated. In Fig. 9, a meandered dipole antenna is designed, and a pair of symmetrical meandered metallic strips surrounded by a rectangular loop is fed. The higher real part of the impedance can be realized by the meandered dipole, while its high imaginary part can be supplied by the coupling between rectangular loop and symmetrical meandered dipole. In this way, a tag antenna with higher absolute value impedance and higher Q value is designed and connected to the chip, to ensure the good power transmission. The gap of the feeding point is 0.1mm, the width of the metallic meandered strip and the horizontal part of the rectangular loop is 1mm, and the width of its vertical part is 2mm. The tag antenna has a thickness of 0.018mm.

The tag antenna is analyzed by the HFSS software, the performance of the antenna, including its impedance and radiation patterns, is calculated. The simulated results are shown in Table 1 and Fig. 10. These results show that the antenna with small size can be used as a tag antenna for the UHF band RFID chip application.

Freq(MHz)	Antenna impedance (ohm)	Power reflection coefficient $\|s\|^2$	Power transmission coefficient τ
900	36.6+j695.2	0.6365	0.3635
901	37.1+j701.6	0.6036	0.3964
902	37.7+j708.0	0.5670	0.4330
903	38.3+j714.5	0.5268	0.4732
904	38.9+j721.0	0.4833	0.5167
905	39.5+j727.7	0.4354	0.5646
906	40.1+j734.5	0.3840	0.6160
907	40.7+j741.4	0.3294	0.6706
908	41.3+j748.4	0.2728	0.7272
909	42.0+j755.5	0.2152	0.7848
910	42.7+j762.7	0.1593	0.8407
911	43.4+j770.0	0.1076	0.8924
912	44.1+j777.4	0.0632	0.9368
913	44.8+j785.0	0.0288	0.9712
914	45.5+j792.7	0.0076	0.9924
915	46.3+j800.5	0.0014	0.9986
916	47.1+j808.4	0.0107	0.9893
917	47.9+j816.4	0.0343	0.9657
918	48.7+j824.6	0.0707	0.9293
919	49.6+j832.9	0.1166	0.8834
920	50.4+j841.4	0.1695	0.8305
921	51.3+j850.0	0.2255	0.7745
922	52.2+j858.7	0.2822	0.7178
923	53.2+j867.6	0.3381	0.6619
924	54.1+j876.7	0.3923	0.6077
925	55.1+j885.9	0.4426	0.5574
926	56.1+j895.2	0.4890	0.5110
927	57.2+j904.8	0.5320	0.4680
928	58.3+j914.5	0.5710	0.4290
929	59.4+j924.4	0.6065	0.3935
930	60.5+j934.5	0.6387	0.3613

Table 1. The impedance and power reflection coefficient, power transmission coefficient for Tag antenna （chip impedance: 43-j800ohm）

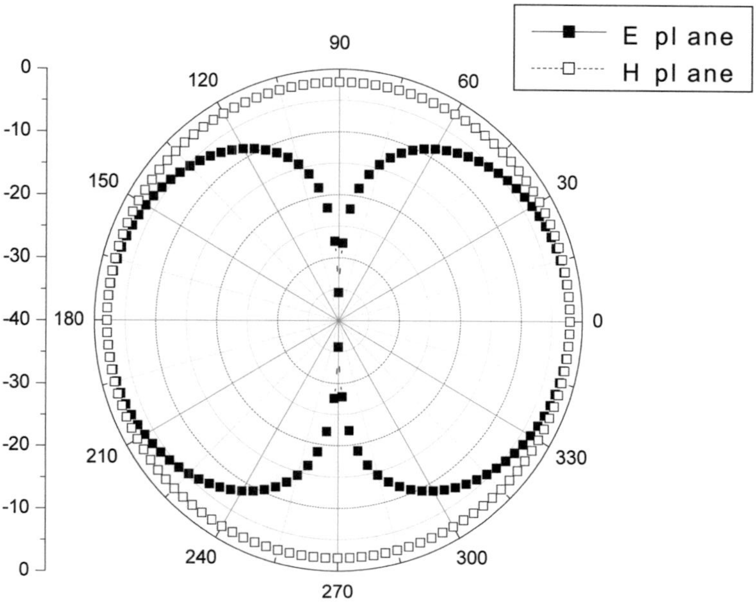

Fig. 10. Radiation pattern of the meandered dipole antenna

3.3 Tag antenna mountable on metallic objects

Since the RFID technology is applied in wide fields, RFID systems frequently appear in the metallic environment, and the effect of the metallic objects should be considered in designing the antenna (Penttilä et al, 2006). RFID antennas in microwave band have a defect of standing wave nulls under the impact of metallic environment. To solve the problem brought by the metallic objects, some special tag antennas should be designed. These antennas usually have a metallic ground. Some metallic objects, which make the performance of the RFID antenna worse, are modified to be as an extended part of the antenna to improve its performance. Some existing problems should be discussed.

When the traditional dipole antenna is attached to an extremely large metallic plane, its radiation will be damaged. In general, the tag antenna with a hemispherical coverage is required. In practical application, a tag antenna with low profile is frequently used, and its vertical current is limited. In Fig. 11, when a normal dipole antenna approaches closely the metallic surface, an inductive current in opposite direction is excited, and the radiation induced by the current will eliminate the radiation of the dipole, resulting in that the tag cannot be detected or read. As a class of antennas, the microstrip antenna may be a good choice for being mounted on the metallic surfaces and identifying the metallic objects. For ordinary tag chip, a balun or other circuit is needed to feed the antenna. Here, based on the dipole antenna, two design schemes for the metallic surfaces are proposed. One is a modification to the Yagi antenna, and the other is a dipole Antenna backed by an EBG structure. A substrate with high dielectric coefficient is sandwiched between the dipole and the metallic surface, its thickness will reverse the orientation of the inductive current, and the radiation is strengthened. An EBG structure can depress the primary inductive current, the radiation of the dipole will be available, and the metallic surface of the identified object is also the ground of the EBG structure.

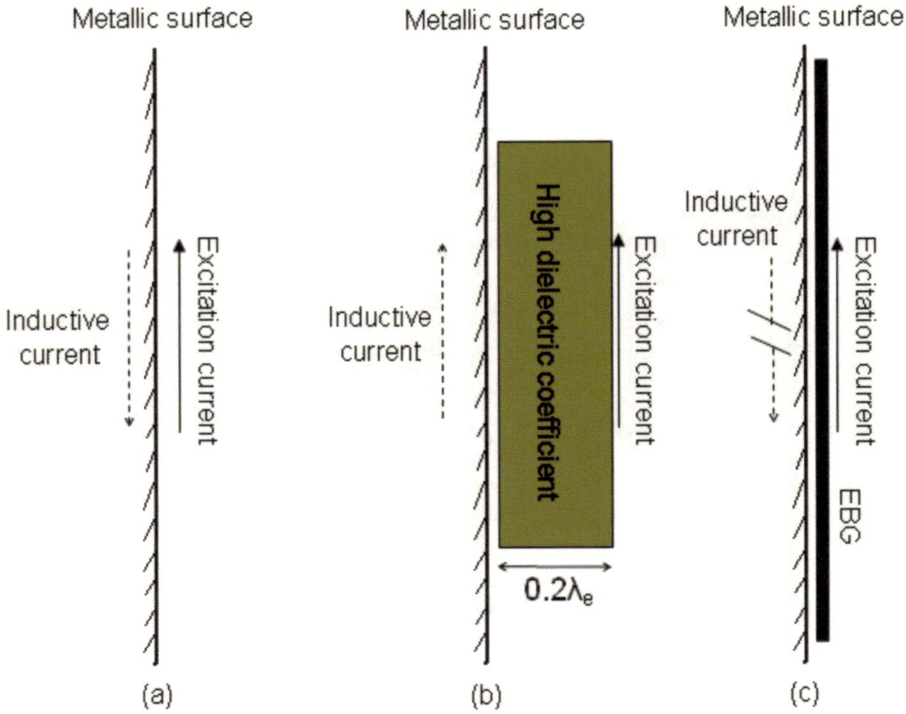

Fig. 11. Design scheme for the tag antenna on metallic surfaces
(a) Excitation current nearby the metallic surface; (b) Scheme based on the Yagi antenna
(c) Scheme based on the EBG structure

According to the introduced schemes, three tag antennas are designed for three tag chips with impedances 15-j20 ohm (chip 1), 6.7-j197ohm (chip 2), and 43-j800 ohm (chip 3), respectively. The tag antenna based on the Yagi antenna is shown in Fig. 12, and the geometry of the active dipole (Qing & Yang, 2004a) is also given in Fig. 13. In Fig.12, the active dipole is attached on the substrate with the relative dielectric coefficient εr=10.2. The width of the metallic strip is 0.8mm.

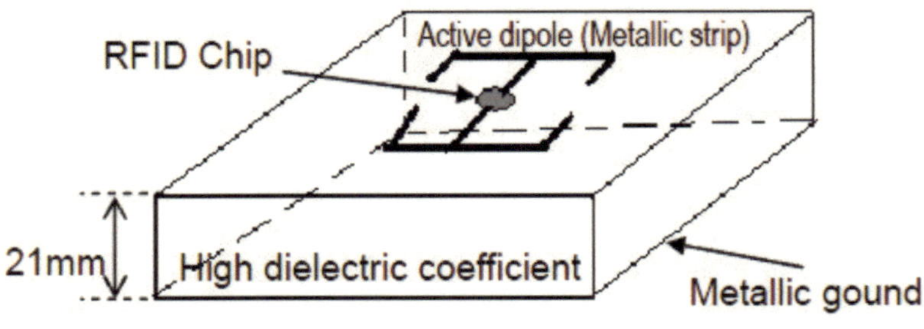

Fig. 12. The tag antenna for chip 1 based on the Yagi antenna

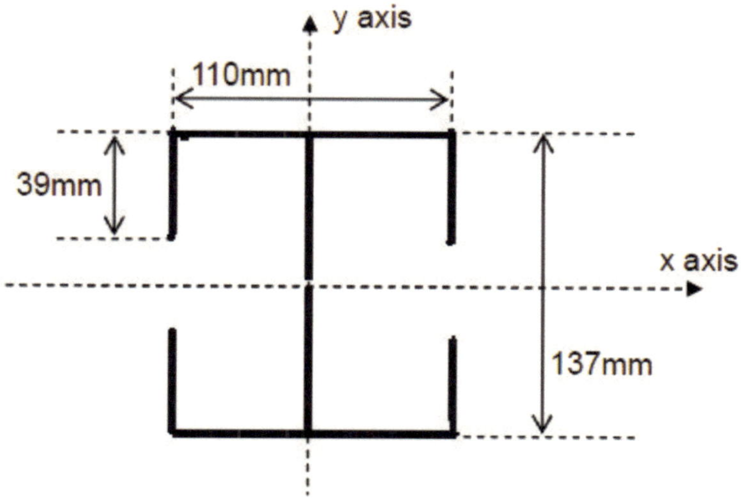

Fig. 13. Geometry of the active dipole (dimensions in mm)

The antenna shown in Fig. 12 is analyzed by the HFSS software. The calculated antenna impedance matches the chip impedance 15-j20 ohm in UHF band. Radiation patterns of the tag antenna are also calculated and shown in Fig. 14.

To design the antenna for chip 2 with 6.7-j197 ohm impedance, the structure parameters are adjusted. The designed dipole is shown in Fig. 15, and its simulated radiation patterns are presented in Fig. 16.

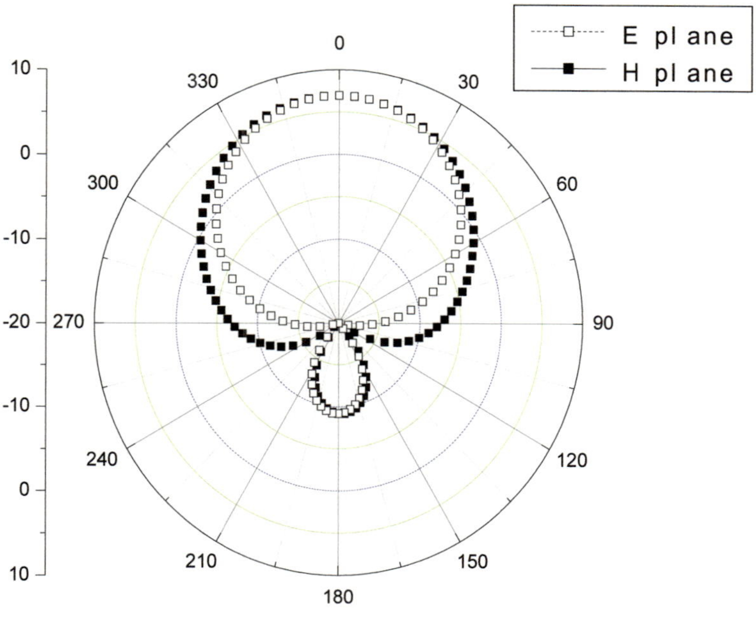

Fig. 14. Radiation patterns of the tag antenna for chip 1

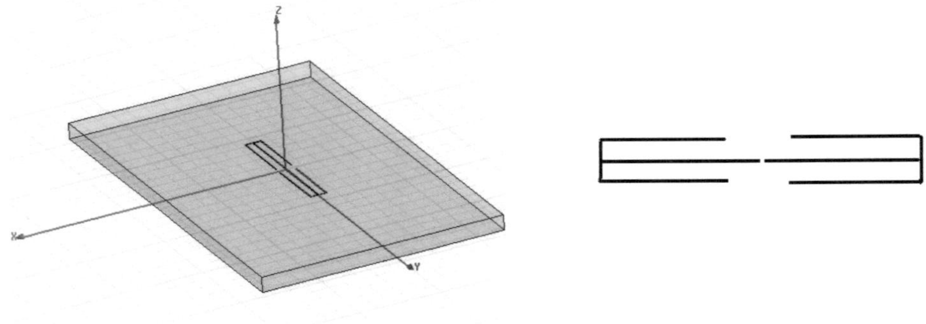

(1) The tag antenna and the substrate (2) The active dipole

Fig. 15. Geometry of the tag antenna for chip 2

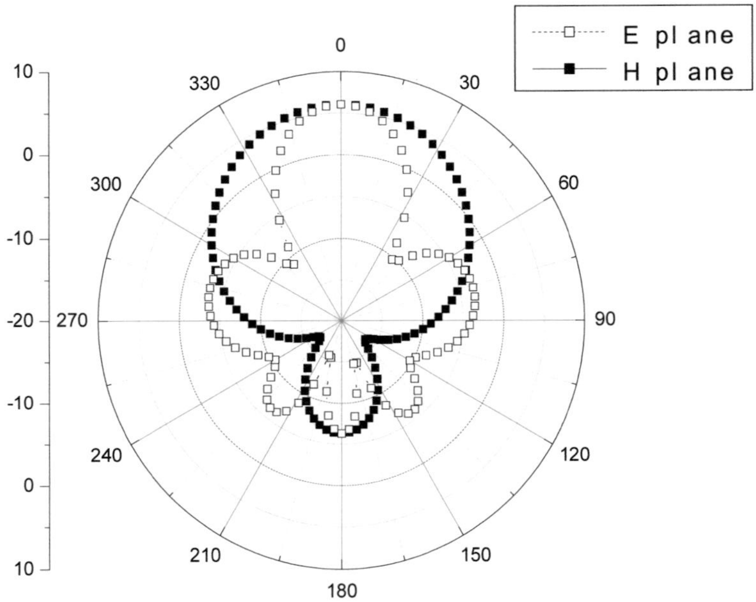

Fig. 16. Radiation patterns of the tag antenna for chip 2

Similar tag antenna can also be designed based on the EBG structure (Abedin & Ali, 2005a, 2005b, 2006; Yang & Rahmat-Samii, 2003) like the tag antenna shown in Fig. 12. The EBG structure is attached to the surface of the metallic object, and the tag dipole antenna like the active dipole in Fig. 13 is placed on the EBG structure formed by 5×7 elements, as shown in Fig. 17. This structure is analyzed at frequency 915MHz in the UHF band, and its radiation patterns are calculated, which are shown in Fig. 18. The simulated impedance values show that the tag antenna matches the chip 3 with impedance 43-j800 ohm. The relative dielectric coefficient of the substrate of the EBG structure is 2.65, its thickness is 2mm, and the total thickness of the tag antenna is 15mm. The low cost tag antenna with low profile will be fabricated.

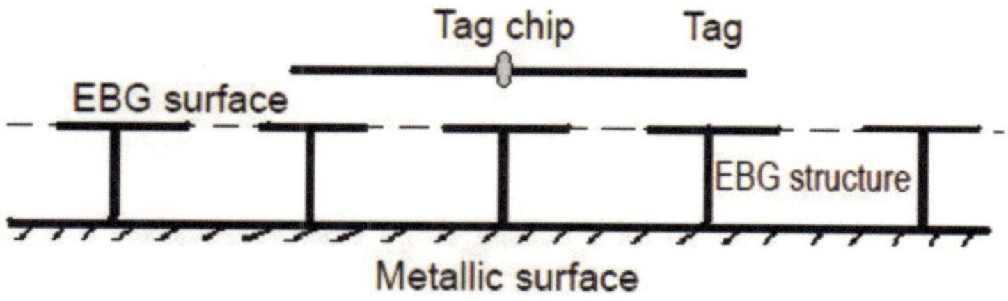

Fig. 17. The tag antenna backed by the EBG structure for chip 3

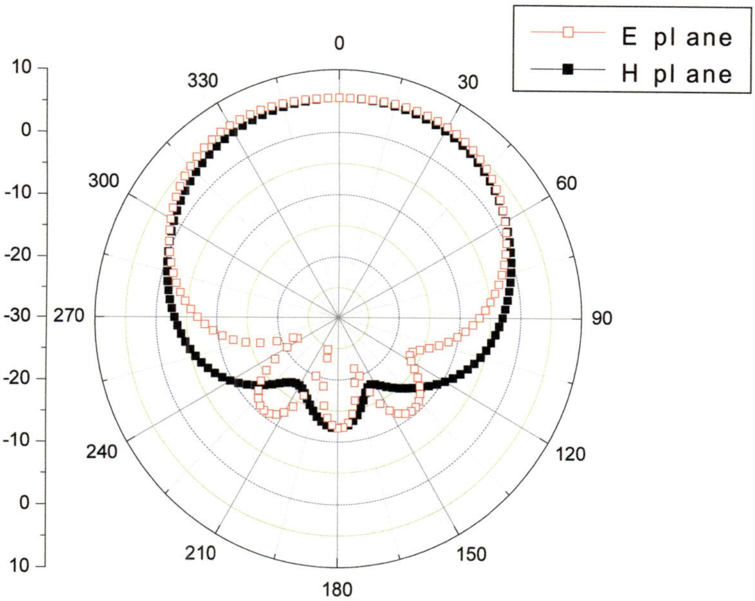

Fig. 18. Radiation patterns of the dipole backed by the EBG structure for chip 3

In this section, design of the tag antenna for the metallic surface is presented, and several cases are described and discussed. Other types of tag antenna mounted on the metallic objects, such as the inverted-F antenna and its modifications are also popular. For the details about these antennas, refer to Kim et al., 2005; Son et al., 2006; Ukkonen, Sydänheimo et al., 2004; Hirvonen et al., 2004; and Ukkonen, Engels et al., 2004.

4. Circular polarization modulation and design of the circularly polarized antennas

4.1 Circularly polarized reader antenna and circular polarization modulation

Generally the object to be identified or the tag does not point to a certain direction, so the circularly polarized reader antennas are usually used (Raumonen et al., 2004) to receive signals from all directions and do not miss the mismatched polarized signals of the moving object. The linearly polarized reader receives more than 3dB power, when the polarizations of the tag and the reader are matched. In some wireless communication systems, the circular polarization modulation (Fries et al., 2000; Kossel, Kung, et al., 1999), which is well adapted to the low rate RFID systems, is another choice that can reduce the requirement of the frequency band, and simplifies the data communication, as shown in Fig. 19. Therefore, the antennas, used for the reader and the tag, should be dual circular polarization antennas with two ports in the RFID system.

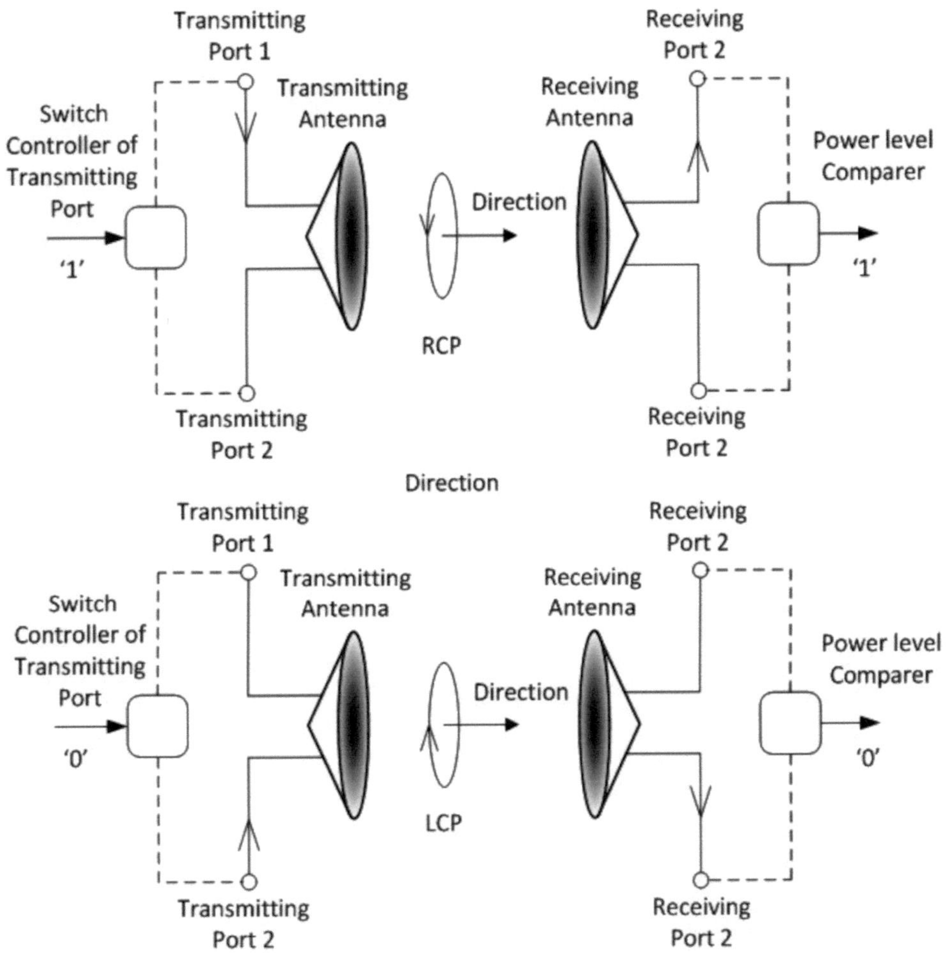

Fig. 19. Principle chart of the circular polarization modulation

Helix antennas and microstrip antennas are widely used as the circularly polarized reader antenna for one-port applications. The helix antenna has some advantages, such as low cost and simple design, except its larger physical size. The low profile helix antenna with the EBG structure instead of the metal ground plane can be used for the RFID reader (Raumonen et al., 2004).

The circular polarization modulation is always used in the RFID system, and its basic principle is that a logical zero is transmitted as the left-hand circularly polarized (LHCP) wave, and a logical one is represented by a right-hand circularly polarized (RHCP) wave. Both reader and tag can use circularly polarized antennas with switchable polarizations. Cross polarization isolation has the significant effect on the performance of the whole system. The maximum transmission distance can be expanded more than 20%, if the cross polarization level (XPL) reaches up to 20dB from 5dB. In the backscattering modulation system, the incident LHCP wave illuminated to the tag is modulated and backscattered into the RHCP wave, and then retransmitted to the reader. Relative to the system where the linearly polarized tag antennas are used, the signal received by the reader in the circular polarization modulation system will raise 6dB. In spite of what kind of the modulation is used, the system should have higher polarization isolation. At the same time, the tag antenna should have higher port isolation, which can reduce the interference between the transmission channel and the receive channel.

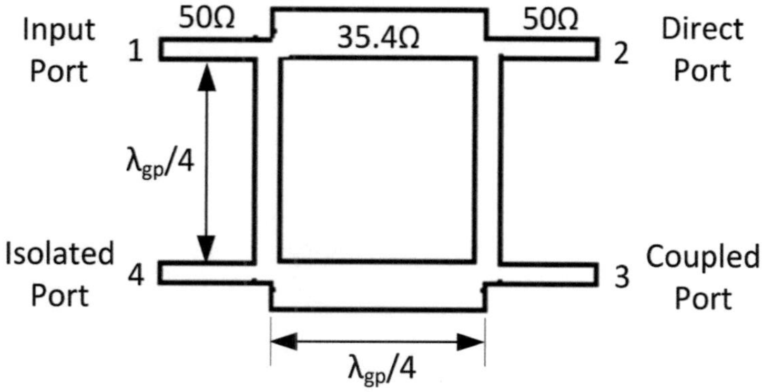

Fig. 20. The 3dB branch line directional coupler structure

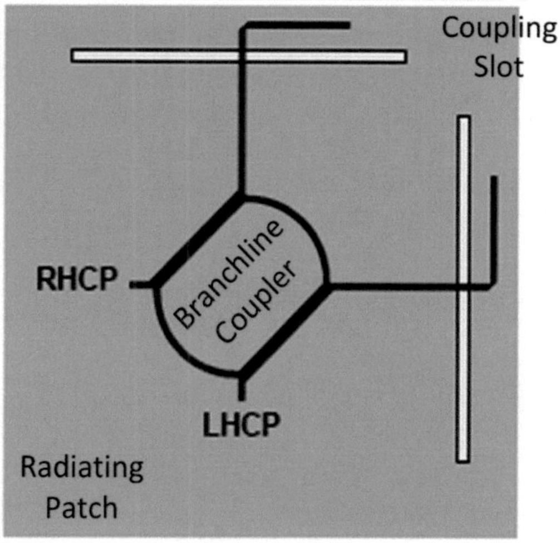

Fig. 21. Microstrip antenna with coupling slot based on the branch line coupler

The traditional design of the dual-port dual-polarization antenna (Kossel, Benedickte et al, 1999; Qing & Yang, 2004b; Sharma et al., 2004) is based on the branch line directional coupler, in which the electrical fields in two output branches have identical voltages and a 90° phase shift, and has high isolation between two output ports, as shown in Fig. 20. When the impedances of the four ports are matched very well and the signal inputs from Port 1, Port 4, called the isolation port, has no output signal, and there is a 90° phase shift between Port 2 and Port 3. The dual circularly polarized antenna, as shown in Fig. 21, is a microstrip patch antenna, which uses a branch line coupler to feed the orthogonal slot apertures and to realize the required 90° phase shift. Four different circularly polarized antennas are shown in Fig. 22. The multilayered antennas employ two substrates, the patch layer and the feed layer, and a ground plane with slot apertures between two substrates, as shown in Fig. 23. The patch antennas can realize the dual circular polarization by using the branch line coupler or the microwave branches to feed the slot apertures with the required phase shift.

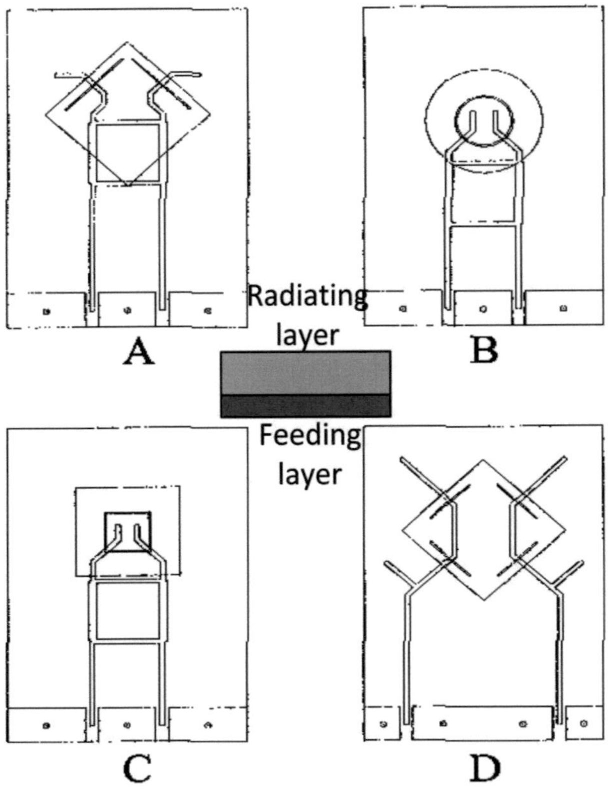

Fig. 22. Four dual-port dual circularly polarized antennas

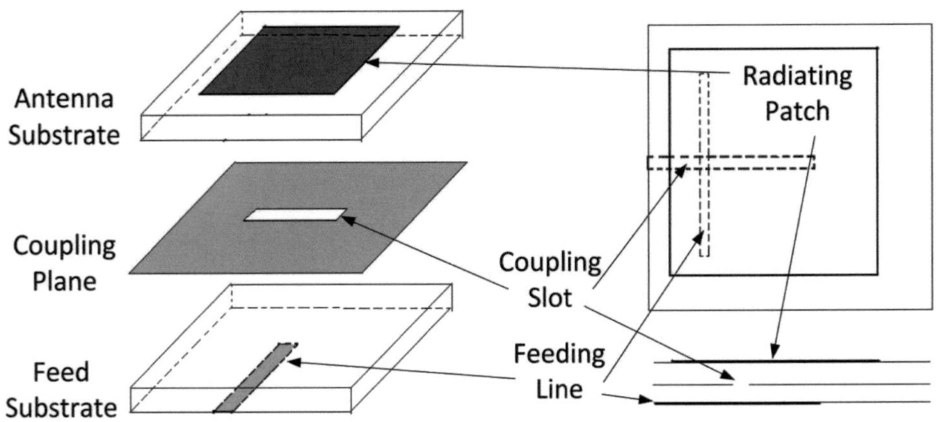

Fig. 23. Multilayered microstrip antenna structure

4.2 A compact dual circularly polarized antenna

In order to realize low profile and part the feed line from the patch, the slot aperture microstrip antenna is commonly used. For this kind of antennas, the designer could select different substrates for the feed and patch layers, according to the application requirements of the microwave integrate circuits. As shown in last section, dual circularly polarized antennas for the RFID system in microwave band are fed by two orthogonal and isolated slot apertures, based on the branch line directional coupler or other complex microwave networks. However, the configuration of the antenna presents a structural bottleneck, i.e. the isolated slots and feeding network limit the miniaturization of the antennas, and the microwave network with complex circuits occupies the larger space. It is well known that RFID antennas can achieve long distance propagation of electromagnetic waves, but sometimes have the problem such as standing wave nulls. Therefore, the antennas should be integrated with the loop, which could transmit power to the low frequency system through the inductance coupling, and reduce the size of the feed network. In order to get rid of the bottleneck on the miniaturization of the antennas, we should design the compact slot aperture microstrip antenna with simple feed network to accomplish the dual circular polarization. In this section, we present a compact dual circularly polarized antenna for RFID systems.

In the RFID system, the rate of the data communication is not so high, sometimes just a few bites. Therefore the circular polarization modulation can be used in the narrow bandwidth communication to simplify the data communication. It is necessary to design dual circularly polarized antenna with two well-isolated ports for the circular polarization modulation.

In order to miniaturize the dimensions of the antenna, as shown in Fig. 24, a dual circularly polarized microstrip antenna fed by crossed slots without the branch line coupler is proposed (Zhang, Chen., Jiao & Zhang, 2006), which is an optimal choice for the RFID system with larger bandwidth and the smaller size. The coupling aperture for the circularly polarized antenna comprises two crossed slots (Aloni E. & Kastener, 1994) in the ground plane, with four arms of the aperture fed serially by a single microstrip line located underneath the ground plane. The microstrip line feeds the four arms with 90° progressive

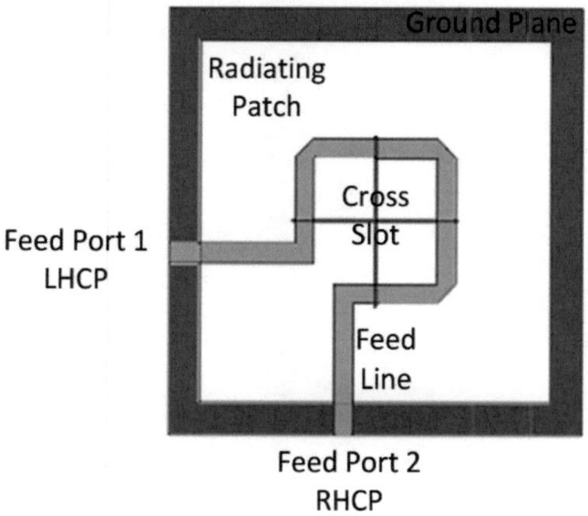

(a) Top view

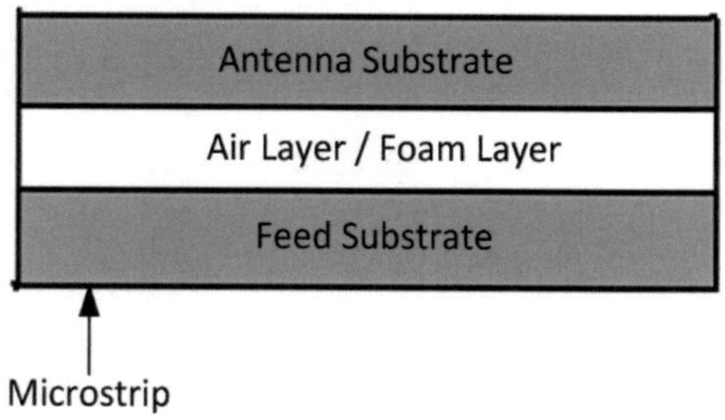

(b) Side view

Fig. 24. Structure of the dual circular polarized antenna

phase difference. The symmetric arrangement makes the antenna achieve easily the dual circular polarization. The design method has been widely used for the antenna at 2.45GHz in the RFID system.

In order to reduce the cost, the air layer used to replace the foam material, as shown in Fig. 24, is sandwiched between two substrate layers with the same dielectric constant εr=2.65. HFSS simulation results show that the performance of the antenna cannot satisfy the requirement for the RFID system. Thus, the structure of the antenna should be modified to improve its performance. As a result, a corner-truncated square patch (Wang, 1989) is used to replace the normal square patch, which will improve the circular polarization performance of the antenna and its port characteristics. At the same time, we cut a square aperture in the centre of the patch to restrict the current and to improve the port isolation. Steps of the patch evolution from the square to the corner-truncated square with a square aperture are shown in Fig. 25, and the final antenna structure is shown in Fig. 26. Simulated performance indices of these three patch antennas are given in Table 2, which indicate the effectiveness of the patch modifications.

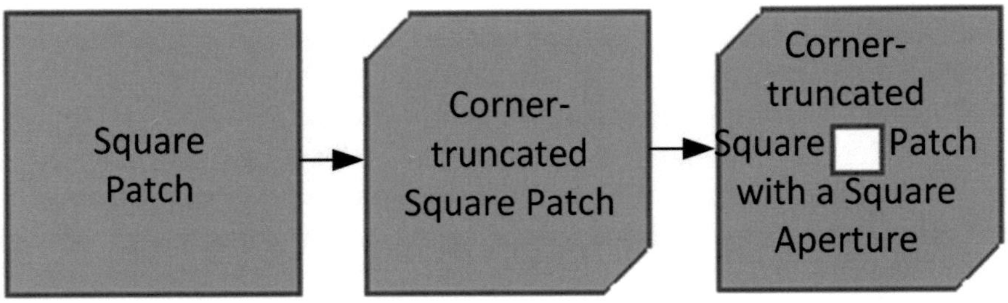

Fig. 25. Steps for the patch modifications

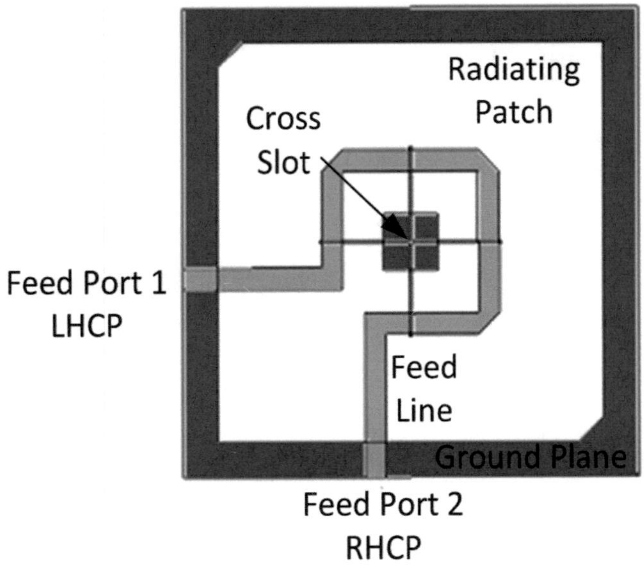

Fig. 26. Geometry of the antenna after modifications

We now determine the dimensions of the corner-truncated square patch antenna with a square slot. First we adjust the dimension of the patch to get the maximum gain, keeping the other parameters of the antenna fixed. We then adjust the length and width of the slots to improve the port characteristics. Because it affects the coupling between the microstrip and the patch more seriously, the length of the slots should be adjusted in advance. Finally, by adjusting the square aperture and the corner of the patch, the antenna with better performance is obtained. The total size of the antenna is 60mm×60mm× 3mm. The width of the corner-truncated square patch with a square aperture is 51mm, and each layer is 1mm thick. The microstrip line has a width that makes the transmission line have 50Ω characteristic impedance. The length and width of the aperture are 24mm and 0.316mm, respectively.

Patch modifications	*Axial Ratio* (dB)	*S11* (dB)	*S21* (dB)
Primary square patch	3.64	-11	-17
Corner-truncated square patch	0.73	-20.2	-19.2
Final patch structure	1.07	-20.4	-30

Table 2. Antenna parameters during the modification

We have tested a prototype of the compact dual circularly polarized antenna shown in Fig. 29. The measured results are shown in Figs. 27 and 28. Comparison between the simulated results and the measured data shows that the measured S parameters at two ports agree well with the simulated results. The measured S21 is better than the simulated one, however the measured S11 is worse than the simulated one. There is a tradeoff between the return loss and the port isolation. From the measured radiation patterns, we can see that the cross polarization levels better than -15 dB are achieved. Although the measured cross polarization levels cannot reach the simulated circular polarization performance, they meet the requirements of the RFID system. The antenna can be used to realize the circular polarization modulation for the RFID systems.

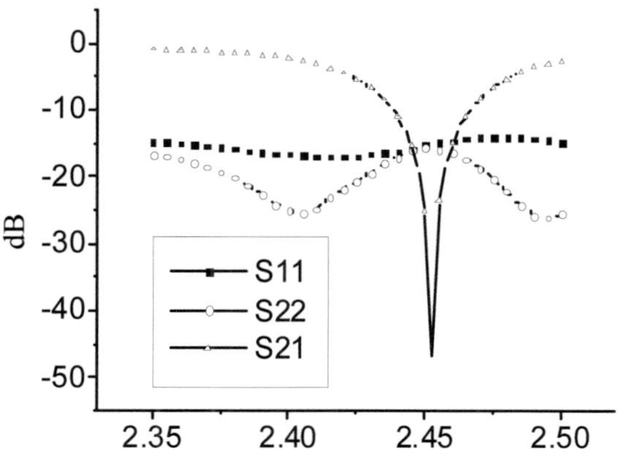

Fig. 27. Measured S parameters at two ports

Simulated and measured results for the compact dual circularly polarized aperture coupled patch antenna show that the compact structure meets the requirements for the RFID system. For the antenna with smaller size, a port decoupling better than 20 dB and a good circular polarization are achieved by the coupling and feeding technique, without using the microstrip branch line coupler or other complex feed networks. The design can save more space for the IC layout, and the miniaturization of the antenna is realized, which is very important for the integration of the RFID system at the microwave frequency and low frequency bands. The dual circularly polarized antenna with the compact structure is not only applicable to the normal RFID systems, especially in some identification card applications, but also suitable for using in some wireless communication systems.

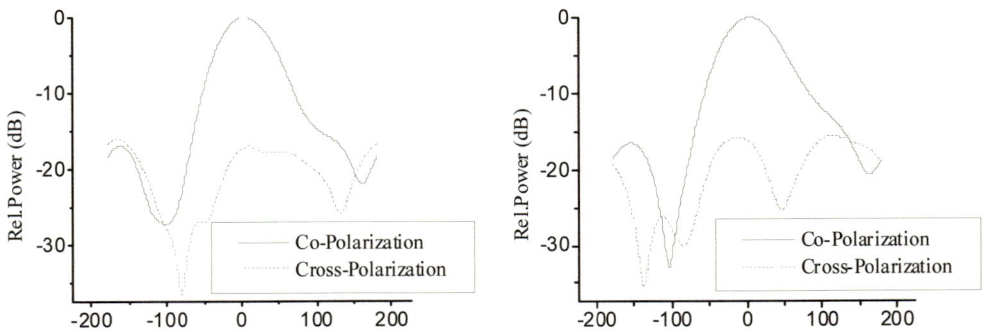

(a) (b)

Fig. 28. Measured radiation patterns at two ports
(a) Excitation in LHCP port; (b) Excitation in RHCP port

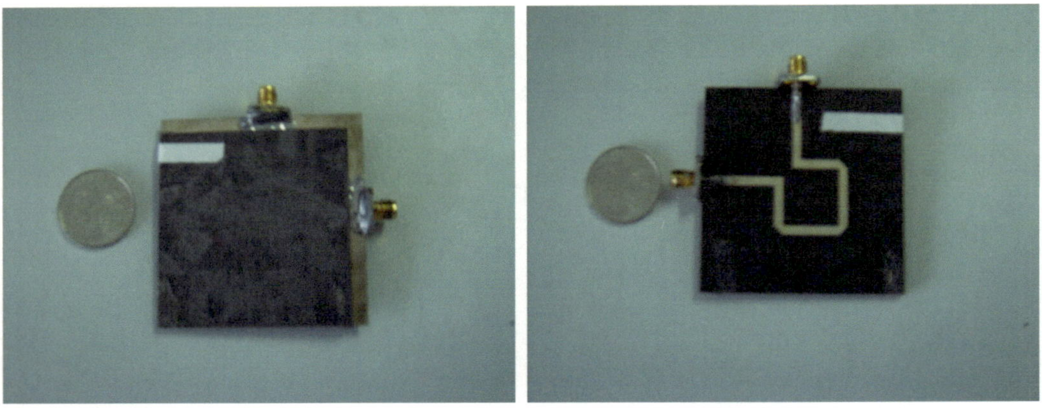

Fig. 29. Photograph of the compact dual circularly polarized antenna prototype

5. Design of antennas for the RFID tag in microwave bands

According to the design of the tag and the reader at the microwave band, special chips for the application of the RFID system are immature and seldom used in industry. Thus engineers usually use specific ASICs available in the market for some special RFID systems, and sometimes use the coaxial cable to connect the ASIC with the antenna, whose impedance is 50Ω.

A radio-frequency identification system consists of tags and readers, and readers communicate wirelessly with the tags to obtain or transfer the information. The data sent by the reader is modulated and backscattered from a number of tags. In some cases, the reader changes the data stored in the tag. Several frequency bands, such as 125 KHz, 13.56 MHz, 869 MHz, 902-928 MHz, 2.45GHz and 5.8GHz bands, have been assigned to the RFID applications. As the operating frequency for the RFID systems rises into the microwave bands, the antenna design becomes more acute and essential (Chen & Hsu, 2004; Liu & Hu, 2005).

The tag, which includes the antenna and a microchip transmitter, must be low in profile, low in cost and small in size for the valuable and easy use, when it is attached to an object to be identified. Therefore, a suitable antenna used in the tag becomes more and more important. As the RFID technology continues to be widely used in applications, especially meeting the reliability, the anti-interfere and the other special requirements, the systems that are able to work at two bands, such as 2.45GHz and 5.8GHz bands, are expected.

The reader with a single antenna is used for both data transmission and receiving, which needs severe operational requirements to the reader RF front end, since the communication takes place in both directions at the same time. The reader may use two antennas for the communication, one antenna for the data transmission and the other antenna for the data receiving (Penttilä et al, 2006). This choice can reduce the realization difficulty from the hardware point of view. In addition, there may have more than two antennas in the reader. In this case, the reader must follow a certain sequence to switch on an antenna at a time, while keeping other antennas switched off, to avoid interferences between these antenna signals.

The approach for using two antennas in the reader is based upon the following reasons:

1. The transmitting electromagnetic wave from the reader does not vanish, when the reflecting wave from the tag reaches the antenna of the reader in the single antenna systems.
2. The reader definitely has less sensitivity than the radar, and the transmitting wave of the reader has much more power than the receiving wave from the tag. Thus the circulator or the directional coupler should be designed to meet higher requirements.
3. The backscattered wave has the lower intensity than the transmitting wave, so the circulator or the directional coupler should meet higher isolation in order to separate the signals.
4. The reader must be inexpensive. Based on the aforementioned reasons, the reader is hard to realize.

Two antennas, one for transmitting and the other for receiving, can overcome these problems. However, a higher isolation between the antennas for the communication should be required, and the smaller tag makes the isolation hard to realize. Microwave frequency bands used in the RFID system include 2.45GHz and 5.8GHz bands, which have the similar transmission characteristic. Therefore the design method for the antenna operating in two frequency bands is also similar. If the transmitting antenna and the receiving antenna work at two frequency bands separately, it is easy to realize a narrow frequency band antenna for the reader, and then higher isolation can be achieved between the transmitting antenna and the receiving antenna. On the other hand, the RFID system only uses these two frequency bands, does not interferes signals at other close frequency bands assigned by global International Organization for Standardization (ISO), thus satisfies the EMC requirements. We present a two-antenna system which can operate simultaneously at 2.45 and 5.8GHz bands, as shown in Fig. 30. The tag antenna, worked at two frequency bands, not only can receive the transmitting signals from the reader at 5.8GHz, but also can transmit signals with the tag code at 2.45GHz, which are received and demodulated by the reader to obtain the tag information.

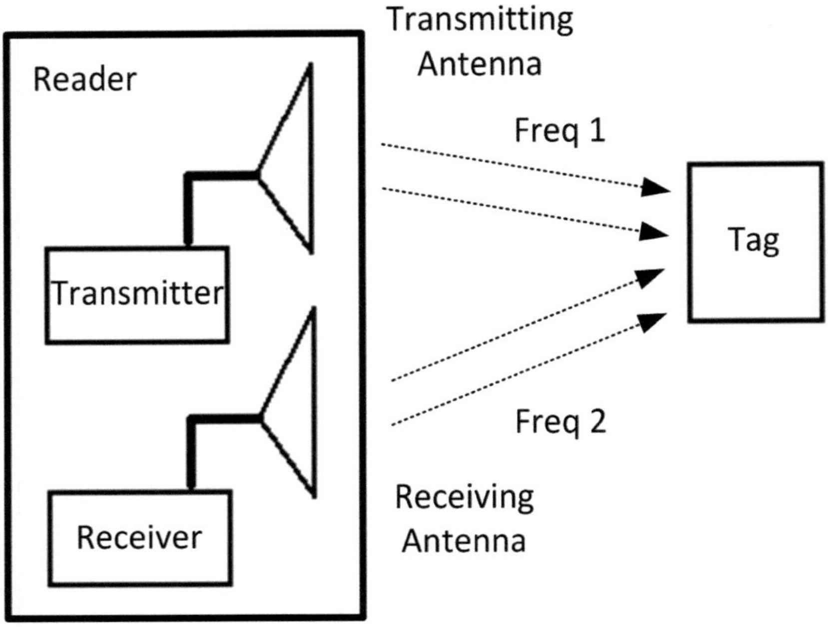

Fig. 30. The dual antenna system in the RFID system

We present a dual-band folded-slot antenna with RF performance suitable for the RFID tag use at 2.45/5.8 GHz (Zhang, Jiao & Zhang, 2006), which consists of a folded slot with a open end and a coplanar waveguide (CPW)-fed structure such that only a single-layer substrate is required for the antenna. By properly adjusting the folded slot on the rectangular patch, compact antenna size, good match at two frequency bands and the radiation characteristics suitable for the RFID application at 2.45 and 5.8 GHz could be achieved.

The geometry of the proposed CPW-fed folded-slot monopole antenna with the open end is shown in Fig. 31. The antenna has a simple structure with only one layer of FR4 dielectric substrate (thickness 1 mm and relative permittivity 4.4) and metallization. The antenna is symmetrical with respect to the longitudinal direction; a folded slot splits the rectangle patch into a double C-shaped ground, and a balance-shaped strip that is fed by the CPW and connects to an SMA forms a monopole structure. Clearly, as the radiating element of this antenna, the balance-shaped strip is thus separated from the ground plane by the folded slot with the open end. The strip can produce two resonant frequencies by adjusting the location of its double arms. The balance-shaped strip is chosen to be of height 31 mm, which is close to one-quarter wavelength in free space at 2.45 GHz, while the top part of the balance-shaped strip above the location of the double arms is chosen to be of height 12 mm, which is also close to one-quarter wavelength in free space at 5.8 GHz.

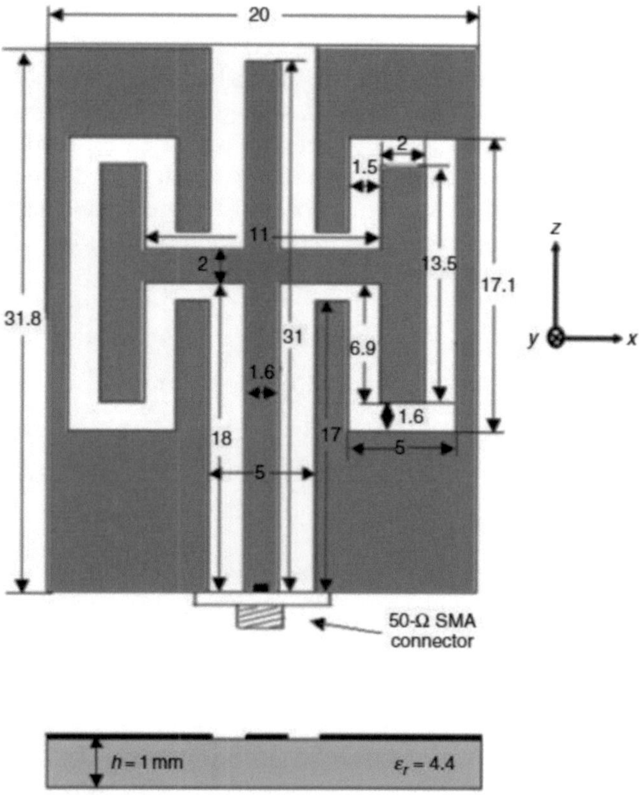

Fig. 31. Geometry of the proposed antenna with open end. The dimensions (in millimetres) shown in this figure are not to scale.

We first study the dimensions of the antenna by the simulation with the aid of HFSS electromagnetic software, analyze its performance by XFDTD simulation tool, and then adjust them by the experiment. Finally, the dimensions of the fabricated antenna are chosen with height 32 mm and width 20 mm, and details of the structure are shown in Fig. 31. For the balance-shaped strip, the top end of the vertical section with strip width 1.6 mm and length 31 mm is chosen to be open, and the other sections of the strip are adjusted to achieve good match at these two frequencies. The CPW feedline, with a signal strip of width 1.6 mm and length 18 mm, and a gap distance of 1.7 mm between the signal strip and the coplanar ground plane, is chosen to feed the dual-band monopole antenna centrally from its bottom edge.

The prototype of the proposed dual-band CPW-fed folded-slot antenna with optimal geometrical parameters, as shown in Fig. 32, is fabricated and tested. The performance of the antenna is also simulated with the aid of two electromagnetic simulators, HFSS and XFDTD. In Fig. 33, the measured and simulated frequency responses of the return loss at two bands for the proposed design are compared, and the measurement is made with a Wiltron 37269A network analyzer. As can be seen from the measured results, the antenna is excited at 2.45 GHz with a -10 dB impedance bandwidth of 320 MHz (2.36–2.68 GHz) and at 5.8 GHz with an impedance bandwidth of 260 MHz (5.73–5.99 GHz). However, the measured results show that the resonant modes are excited at 2.51 and 5.85 GHz simultaneously, which are almost the same as that from simulations. The measured radiation patterns at these two operating frequencies are presented in Figs. 34 and 35, respectively. The measured results show that the radiation patterns of the antenna are broadside and bidirectional in the E-plane and almost omnidirectional in the H-plane (x–y plane). The measured peak antenna gains of the antenna at 2.45 and 5.8 GHz are -1.8 and 2.3 dBi, respectively. Agreement between measurement and simulation is generally good, and the proposed design has sufficient bandwidth to cover the requirement of the RFID dual-band 2.45/5.8 GHz system.

Fig. 32. Photograph of the dual band tag antenna prototype

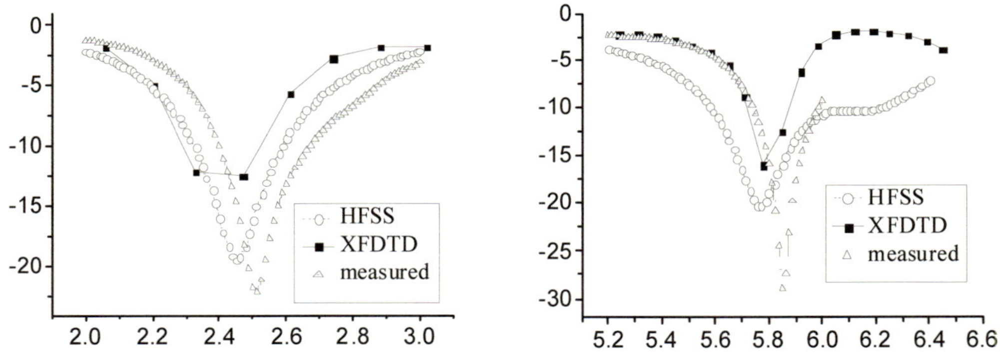

(a) 2.45GHz band (b) 5.80GHz band

Fig. 33. Measured and simulated frequency responses of the input return loss for the proposed antenna

A dual-band CPW-fed monopole antenna has been proposed and implemented. With the open end and the balance-shaped strip fed by the CPW connecting to an SMA, the proposed antenna can be designed to operate at the 2.45 and 5.8 GHz bands, and to have a corresponding bandwidth of 13.1% and 4.5%, respectively. A good radiation performance is also achieved. The low-cost antenna is only 32mm×20mm in size, mechanically robust, and easy to fabricate and integrate with the application-specific circuit. This design is not only suitable for the dual-band RFID systems, but also applicable to the dual-band communication systems for WLAN applications.

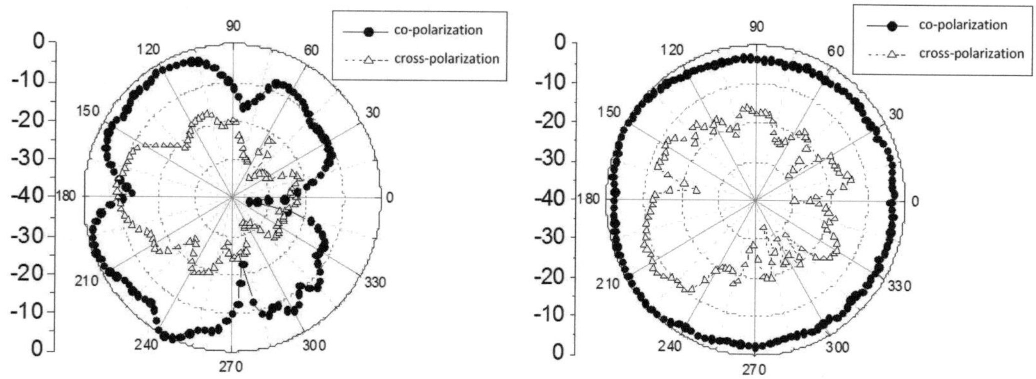

(a) E-plane pattern (b) H-plane pattern

Fig. 34. Measured far-field radiation patterns at 2.45 GHz for the proposed antenna

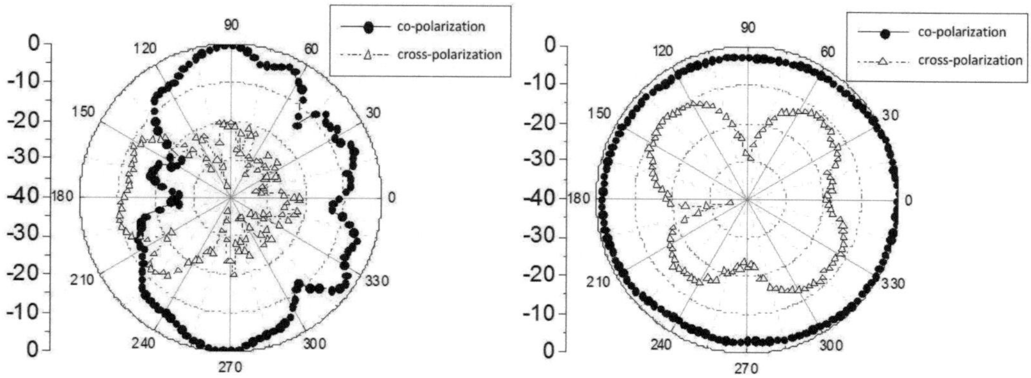

(a) E-plane pattern (b) H-plane pattern

Fig. 35. Measured far-field radiation patterns at 5.80 GHz for the proposed antenna

6. Summary and outlook

In this chapter, the antenna in RFID system is discussed, and the design of antenna is also described. The main contents include the status of the antenna in the RFID system, the design method for the antenna, the power transmission between the tag chip and the tag antenna, the tag antenna design, the scheme and design for the circular polarization, and the design of antenna for microwave band RFID tag. These researches almost cover all problems of the antenna encountering in the RFID application. The considerations and the design method are also significant for practical applications.

The development of the RFID technology for the practical applications impels the advancement of the antenna in the RFID system, and the progress of the antenna also promotes the spread of the RFID systems all over our life and society. In the future, the RFID system may require the multi-band antennas for the high reliability, or the integration of several antennas for multifunction. The unnoticed antenna is also preferred for some RFID applications. The miniaturization of the antenna is an eternal design target in designing the antenna, and reducing the cost of the antenna for large-scale applications is also a significant consideration in the antenna design, which will promote the development of the antenna technology. The RFID applications in special situations may initiate a new research field for designing the RFID antennas.

7. References

Abedin M. F. & Ali M. (2005a). Effects of a smaller unit cell planar EBG structure on the mutual coupling of a printed dipole array, *IEEE Antennas and Wireless Propagation Letters*, Vol. 4, pp. 274-276.

Abedin M. F. & Ali M.(2005b). Effects of EBG reflection phase profiles on the input impedance and bandwidth of ultra-thin directional dipoles, *IEEE Transactions on Antennas and Propagation*, Vol. 53, No. 11, pp. 3664-3672.

Abedin M.F. & Ali M. (2006). A low profile dipole antenna backed by a planar EBG structure, *2006 IEEE International Workshop on Antenna Technology Small Antennas and Novel Metamaterials*, pp. 13-16, March 6-8, 2006.

Aloni E. & Kastener R. (1994). Analysis of a dual circularly polarized microstrip antenna fed by crossed slots, *IEEE Transactions on Antennas and Propagation*, Vol. 42, No. 8, pp. 1053-1058.

Chen, S.Y. & Hsu, P. (2004). CPW-fed folded-slot antenna for 5.8 GHz RFID tags, *Electronics Letters*, Vol. 40, No. 24, pp. 1516–1517.

Dobkin D. M. & Weigand S. M. (2005). Environmental effects on RFID tag antennas, *2005 IEEE MTT-S International Microwave Symposium Digest*, pp. 135-138, June 12-17, 2005.

Fries M., Kossel M, Vahldieck R. & Bachtold W. (2000). Aperture coupled patch antennas for an RFID system using circular polarization modulation. *Proceedings of the Millennium Conference on Antennas and Propogofion*, p. 358, Davos, Switzerland, April 9 - 14, 2000.

Hirvonen M., Pursula P., Jaakkola K. & Laukkanen K.(2004). Planar inverted-F antenna for radio frequency identification, *Electronics Letters*, Vol. 40, No. 14, pp. 848-850.

Keskilammi M, Sydänheimo L. & Kivikoski M. (2003). Radio frequency technology for automated manufacturing and logistics control. Part 1: Passive RFID system and the effects of antenna parameters on operational distance, *The International Journal of Advanced Manufacturing Technology*, Vol.21, No. 10-11, pp. 769-774.

Kim S.-J., Yu B., Lee H.-J., Park M.-J., Harackiewicz F. J., & Lee B. (2005). RFID Tag Antenna Moutable on Metallic Plates, *2005 Asia-Pacific Microwave Conference (APMC 2005) Proceedings*, Vol. 4, pp. 2666-2668.

Kossel M., Benedickter H. & Baechtold W. (1999). Circular polarized aperture coupled patch antennas for an RFID system in the 2.4 GHz ISM band, *1999 IEEE Radio and Wireless Conference (RAWCON 99)*, pp. 235-238, August 1-4, 1999.

Kossel M.A., Kung R., Benedickter H. & Bachtold W. (1999). An active tagging system using Circular-polarization modulation, *IEEE Trans. Microwave Theory and Techniques*, Vol. 47, No. 12, pp. 2242-2248.

Liu, W.C. & Hu, Z.K. (2005). Broadband CPW-fed folded-slot monopole antenna for 5.8 GHz RFID application, *Electronics Letters*, Vol. 41, No. 17, pp. 5–6.

Nikitin P. V., Rao K. V. S., Lam S. F., Pillai V., Martinez R. & Heinrich H. (2005). Power reflection coefficient analysis for complex Impedances in RFID tag design, *IEEE Transactions on Microwave Theory and Techniques*, Vol.53, No.9, pp. 2721-2725.

Penttilä K., Keskilammi M., Sydänheimo L. & Kivikoski M. (2006). Radio frequency technology for automated manufacturing and logistics control. Part 2: RFID antenna utilization in industrial applications, *The International Journal of Advanced Manufacturing Technology*, Vol. 31, No. 1-2, pp. 116-124.

Qing X. & Yang N. (2004a). A folded dipole antenna for RFID, *IEEE Antennas and Propagation Society International Symposium*, Vol. 1, pp. 97-100, June 20-25, 2004.

Qing X. & Yang N. (2004b). 2.45GHz circularly polarized RFID reader antenna, *The Nine International Conference on Communication Systems (ICCS 2004)*, pp. 612-615, Sept. 6-8, 2004.

Rao K.V. S., Nikitin P. V. & Lam S. F. (2005a). Impedance matching concepts in RFID transponder design, *The Fourth IEEE Workshop on Automatic Identification Advanced Technologies*, pp. 39-42, Oct. 17-18, 2005.

Rao K.V. S., Nikitin P. V. & Lam S. F. (2005b). Antenna design for UHF RFID Tags: a review and a practical application, *IEEE Transactions on Antennas and Propagation*, Vol.53, No.12, pp. 3870-3876.

Raumonen P, Keskilammi M & Sydanheimo L. (2004). A very low profile CP EBG antenna for RFID reader, *2004 IEEE Antennas and Propagation Society International Symposium*, Vol. 4, pp. 3808-3811, June 20-25, 2004.

Sharma A.K., Singh R. & Mittal A. (2004). Wide band dual circularly polarized aperture coupled microstrip patch antenna with bow tie shaped apertures, *IEEE Antennas and Propagation Society International Symposium*, June 20-25, 2004, Vol. 4, pp. 3749-3752.

Son H.-W., Choi G.-Y. & Pyo C.-S.(2006). Design of wideband RFID tag antenna for metallic surfaces, *Electroics Letters*, Vol. 42, No. 5, pp. 263-265.

Son H.-W. & Pyo C.-S. (2005). Design of RFID tag antenna using an inductively coupled feed, *Electronics Letters*, Vol. 41, No. 18, pp. 994-996.

Tikhov Y. & Won J.H. (2004). Impedance-matching arrangement for microwave transponder operating over plurality of bent installations of antenna, *Electronics Letters*, Vol. 40, No. 10, pp. 574-575.

Ukkonen L., Engels D., Sydnheimo L. & Kivikoski M. (2004). Planar wire-type inverted-F RFID tag antenna mountable on metallic objects, *IEEE Antennas and Propagation Society International Symposium*, Vol. 1, pp. 101-104, June 20-25, 2004.

Ukkonen L., Sydänheimo L. & Kivikoski M. (2004). A novel tag design using inverted-F antenna for radio frequency identification of metallic objects, *2004 IEEE/Sarnoff Symposium on Advances in Wired and Wireless Communication*, pp. 91-94, April 26-27, 2004.

Wang B.F. (1989). Two-port circularly polarized microstrip antennas, *The Sixth International Conference on Antennas and Propagation (ICAP 89)*, Vol. 1, pp. 107-111, April 4-7, 1989.

Yang F. & Rahmat-Samii Y. (2003). Reflection phase characterizations of the EBG ground plane for low profile wire antenna applications, *IEEE Trans. Antennas Propagat.*, Vol. 51, No. 10, pp. 2691-2703.

Zhang M.-T., Chen Y.-B., Jiao Y.-C. & Zhang F.-S. (2006). Dual circularly polarized antenna of compact structure for RFID application, *Journal of Electromagnetic Waves and Applications*, Vol. 20, No. 14, pp. 1895–1902.

Zhang M.-T., Jiao Y.-C. & Zhang F.-S. (2006). Dual-band CPW-fed folded-slot monopole antenna for RFID application, *Electronics Letters*, Vol. 42, No. 21, pp. 1193-1194.

3

Design Fundamentals and Advanced Techniques of RFID Antennas

Sungtek Kahng
University of Incheon
South Korea

1. Introduction

The demand on the automated supply chain and logistics has been pervasive, aiming to replace the tedious bar-code labeling, and has driven an increasing number of research activities on the RFID to alternative and trustworthy solutions. The RFID takes the reader-and-tag paradigm where the interrogator(reader) uses its 'remote' correspondent(tag)s. To be sure about the reliable performance of an RFID system, though microelectronics for chip making and data acquisition are important, the antenna technologies for excellent wireless linkage have highly critical importance.

When it comes to the tag and reader antennas for the RFID, designers adopt the concept imitating the radar technology in which the reader transmits a signal to a tag and the tag sends back its recorded data to the reader. The considerations must be made with the frequency, the impedance of the chip and antenna, the constraints(overall size), the radiation pattern and gain, the reading range, and the tagged objects(geometry and materials). Especially, care must be taken of with regard to the realistic environment that affects the near-field region of reader- and tag antennas and the operational quality of the overall RFID system[1-10].

Prior to the design of the reader- and tag antennas, the basics of antennas are tapped to see the way the electromagnetic fields propagate from radiators for higher frequency regimes (860MHz-960MHz) along with the magnetic- and electric coupling mechanisms for lower frequency(125kHz-134kHz). And then as the first place in the UHF-band RFID antenna design, the impedance matching techniques are addressed with a variety of antenna structures apt to the size reduction and acceptable efficient radiation. In particular, a couple of design examples are practiced with the illustrations obtained by the electromagnetic field solver. As a matter of course, this is accompanied by the considerations of the tags' materials and relevant electromagnetic properties. And the advanced design schemes are introduced with the on-going topics such as multiple aspects in band and polarization as well as near-field UHF tags. It is followed by the remarks on the testing methodology of tag antennas' input impedance, gain, pattern and reading distance. Finally, conclusions are presented.

2. Principles of radio frequency identification & ABC's of RFID antennas

An RFID system comprises a reader and one tag or more. This is illustrated in Figure 1.

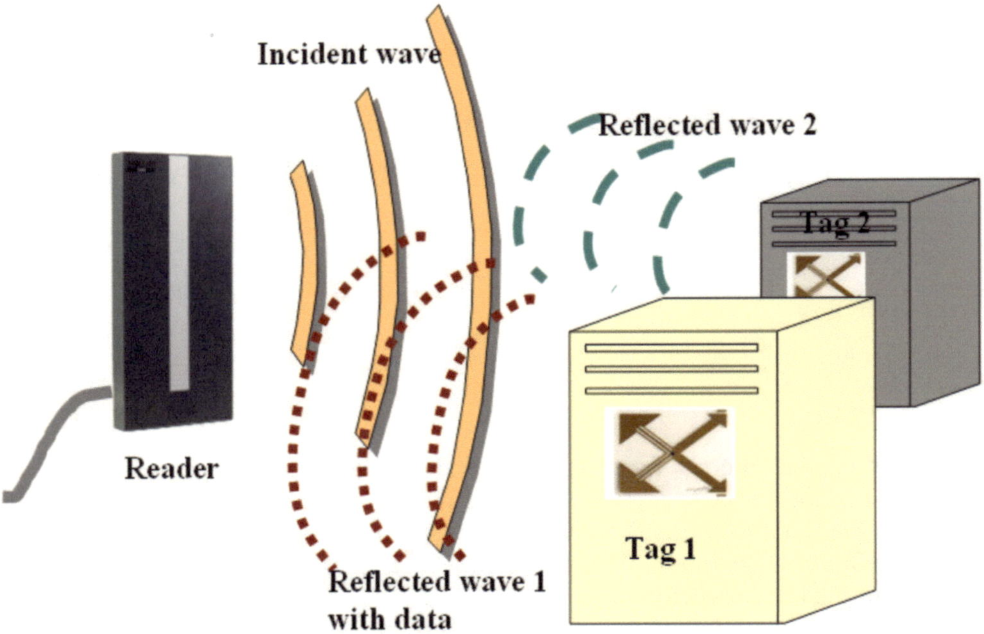

Fig. 1. Sketch of a typical RFID System

The reader sends the signal at a frequency of interest to its neighbor objects with the tags on. Each tag, which is composed of a chip and an antenna, should be responsive as efficiently as possible to the incoming RF signal. The received electromagnetic energy activates the chip through the antenna and the chip provides the stored information for the antenna sends the data conveyed in the RF energy back to the reader.

The interaction between the reader and the tag can be interpreted as what is made in the radar system. Actually, the things like the power, antenna gain and read-range of the antennas at step number 1 in the design are expressed by the so-called 'tag equation' looking pretty much the same as the formula of the radar cross-section(RCS). The only difference between the radar and RFID systems is that the RFID system concerns the impedance matching problem of the target, while the radar system doesn't. In other words, the target in the RFID application is an antenna which is not a simple scatterer.

Assuming the impedance and polarization matched between the reader and the tag, we derive the formulae on the power received by the chip in the tag and the power the reader will get as the re-radiation from the tag. We find them useful in determining the values of the antenna gains for the reader and the tag and input power at the beginning of the RFID system design.

$$P_L = P_{in} G_{reader} G_{tag} \lambda^2 / (4\pi R)^2 \tag{1}$$

$$P_{rec} = P_{in} G_{reader}^2 G_{tag}^2 \lambda^4 / (4\pi R)^4 \tag{2}$$

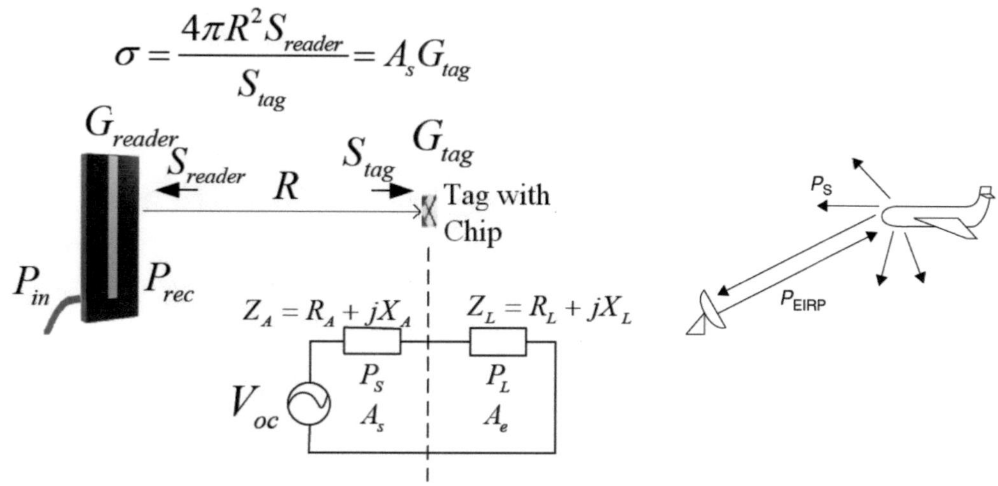

Fig. 2. Tag equation of an RFID system analogous to a radar cross-section problem where the symbols mean

A_s : effective scattering aperture

A_e : effective receiving aperture

P_L : power received by the chip in the tag

P_{rec} : power received by the reader(via the re-radiation)

S_i : power density at i

G_i : Antenna i's gain

σ : radar cross-section

If mismatch happens to the impedance and polarization, the equations are modified with the factors p (polarization mismatch factor) and Γ_{tag} (reflection coefficient from the tag).

$$P_L = (1-|\Gamma_{tag}|^2)\cdot p\cdot P_{in} G_{reader} G_{tag} \lambda^2 / (4\pi R)^2 \tag{3}$$

$$P_{rec} = p\cdot P_{in} G_{reader}^2 G_{tag}^2 \lambda^4 / (4\pi R)^4 \tag{4}$$

Using the power received by the tag and the reader, we can predict the read range as follows.

$$R_{\max} = \min[\{p\cdot P_{in} G_{reader}^2 G_{tag}^2 \lambda^4 / (4\pi)^4\}^{0.25}, \sqrt{(1-|\Gamma_{tag}|^2)\cdot p\cdot P_{in} G_{reader} G_{tag}}\,\lambda / (4\pi)] \tag{5}$$

Seeing this formula, it is clear to see what kind of parameter determines how far the RFID communication covers.

Now, let us look at the representative kinds of antennas that are frequently used and designed sticking to the specifications generated for the RFID system having $R_{\max}$ along with P_L and P_{rec} above. Firstly, there are tags and readers utilizing the magnetic coupling as follows.

The magnetic coupling is represented by $B_2(I_1)$ which means the magnetic flux density into loop 2 as the tag, stimulated by the current of the loop 1 'I_1' as the reader. The tags based on the magnetic coupling are used mostly in the LF(Low Frequecny), say, 125kHz or 134kHz for the near-field RFID like the animal tracking or access control. The coil tag can be sometimes used in a High Frequency(HF regime) like 13.56MHz.

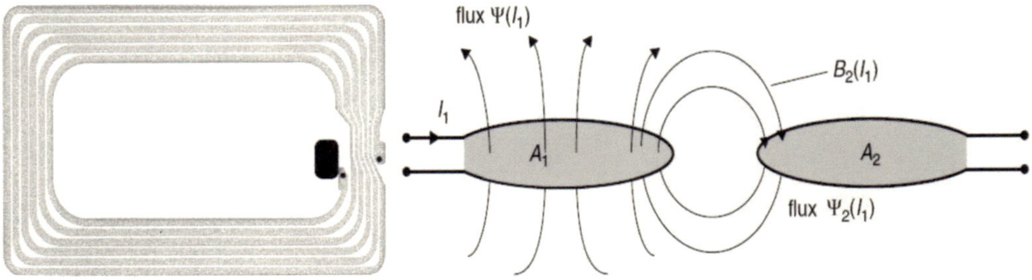

Fig. 3. Coil tag and magnetic coupling between two loops

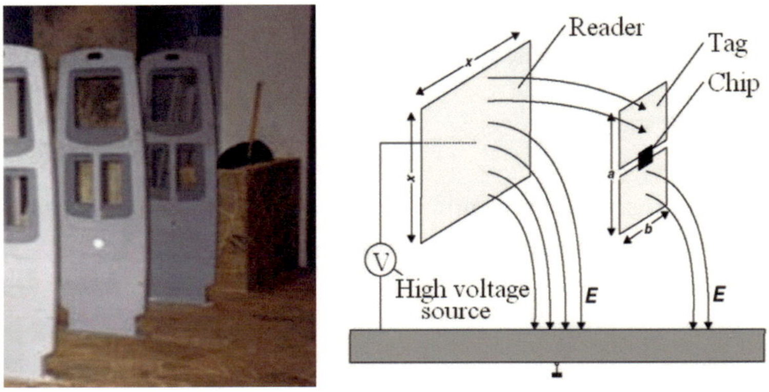

Fig. 4. Electric coupling based tags and electric field coupling in the system

As is seen from the left picture in Figure 4, the metal planes having different voltages form the electric field in between. In other words, the reader of the higher voltage is coupled with the tag of the lower voltage through the electric field as we see the right picture in Figure 4. This coupling is adopted for the LF RFID system such as theft prevention in a library.

For the RFID communication service above 30MHz, mostly the UHF band, instead of the magnetic or electric field coupling, the electromagnetic wave propagation is preferred to have an increased read-range and capacity of information exchange.

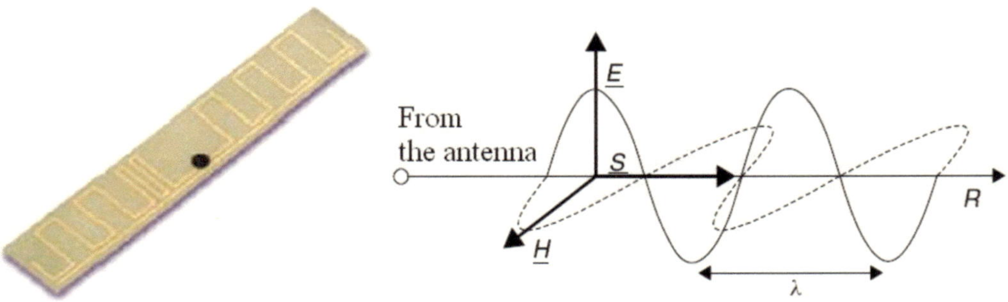

Fig. 5. A UHF RFID tag antenna and electromagnetic wave propagation

With the increasing frequency, the electric and magnetic fields are no more independent from each other, but become incorporated described by the Maxwell equations where the time-varying electric field leads to the space-varying magnetic field and vice versa. And they move in the air, carrying the power.

3. Conjugate impedance matching

The impedance matching condition has been addressed very essential regarding the quality of the RFID system with the power received by the tag and reader. In Figure 2, a simplified transmission line models the linkage of the reader antenna and the tag with the chip. The transmission line circuit is connected to the input impedance of the reader antenna and the tag, and it is ideal for the overall circuit to be matched to both the source and load. However, in practice, the impedance matching for the reader and the tag is split. And for the reader alone, the impedance matching is made to remove the reflection to the antenna from the feeding circuit. Simultaneously, as just for the tag, the impedance of the antenna should be matched with that of the chip. Though the matching problems of the reader and the tag are treated separately, the schemes are the same. Hence, throughout this section, we talk about the impedance matching techniques (Shunt stub matching, Inductive loop matching and Nested slot matching) for only the tag.

3.1 Shunt stub(T-) matching

The chip has the capacitive impedance which has the negative imaginary term. In order to have the best tag antenna efficiency, the input impedance of the antenna should have the inductive reactance, which cancels the capacitance reactance of the chip, when they are connected.

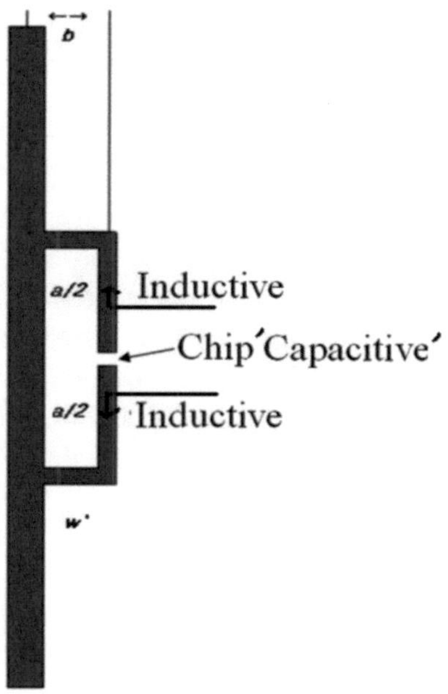

Fig. 6. Inductive stub matching of the antenna against the capacitive chip

As an example, the chip has the negative reactance -*j*100 in Figure 7. For the total input impedance of the tag to have only the real term 70Ω, the antenna with the stub should have the positive reactance +*j*100.

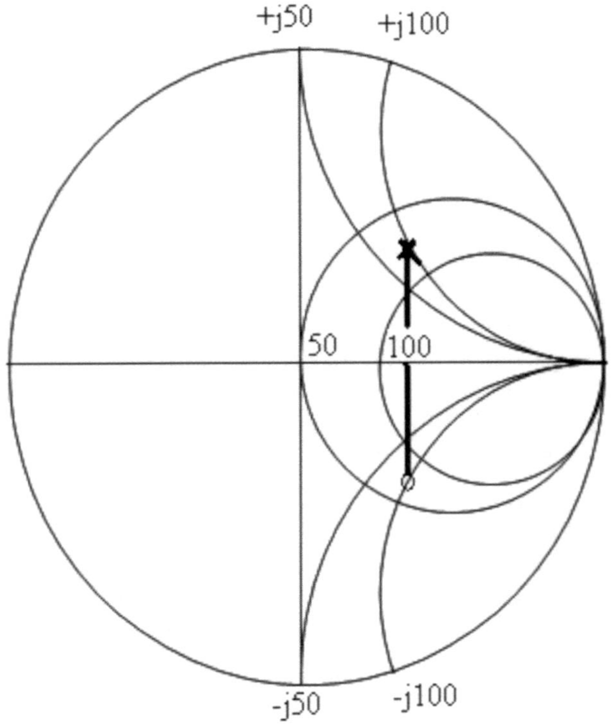

Fig. 7. Capacitive Chip(Z_L=70-*j*100,'**O**') meets the antenna of the Inductive reactance('**X**')

In reality, it is not true that all the capacitive reactance results from only the chip and the tag designers can use the antenna size as large as they want. Almost all the RFID tag antenna designs pursue the way they fit into the small spots specified on an object and this ends up with the decrease in the size from the original resonance length for the antenna radiation at the operating frequency. When the antenna size becomes smaller from the resonance(radiation) length, its input impedance becomes capacitive, namely, having the negative reactance or negative imaginary part. So the impedance matching must be carried out to consider the capacitive reactance due to not only the chip but also the shortened antenna.

3.2 Inductive loop matching

Related to the shunt stub matching, the inductive reactance of the antenna is needed to play the countermeasure of the capacitive reactance of a chip. This inductive reactance or the positive imaginary part of the antenna's input impedance can be alternatively introduced by the following configuration.

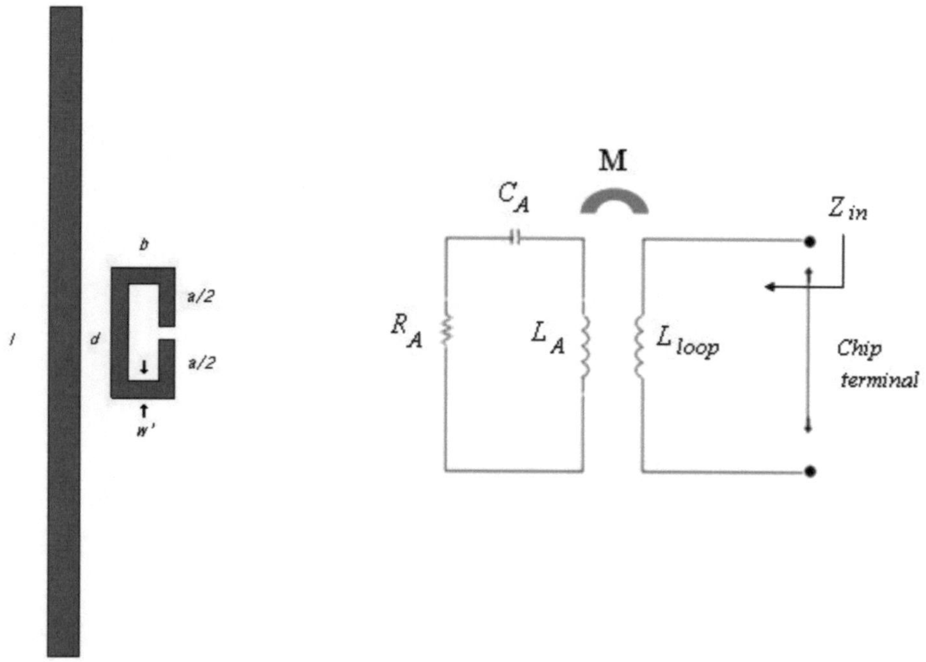

Fig. 8. An inductive loop near the main radiator and a circuit interpretation

Seeking a way to increase the inductive reactance, there are a variety of methods, and a metal loop and magnetic coupling can be used. If the spacing between the loop and the nearby radiator is not too small, the magnetic field from the loop forms the linkage to the radiator and it plays an important role in beefing up the imaginary part of the antenna's input impedance. The magnetic flux is represented by the transformer that renders the input impedance Z_{in}

$$Z_{in} = Z_{loop} + (2\pi fM)^2/Z_A \tag{6}$$

where Z_{loop} is $j2\pi fL_{loop}$.
It is a matter of course that M is the mutual inductance between the loop and the main radiator. And it should be noted that on the contrary to the shunt stub- or T- matching, the inductive loop matching is suitable for the case the input impedance's real term is far smaller than the imaginary term.

3.3 Nested slot matching

The former two matching approaches have added another metal geometry to the main one to use the concept of induction. However, a different scheme can be tried and practiced by forming slots inside the main radiator. For a short while, let us remind ourselves of the meaning of a slot[9]. A slot is an aperture or a window made by removing the area from the solid metal surface. For better understanding, the following figure has the slot area nested by the rest of the metal patch.

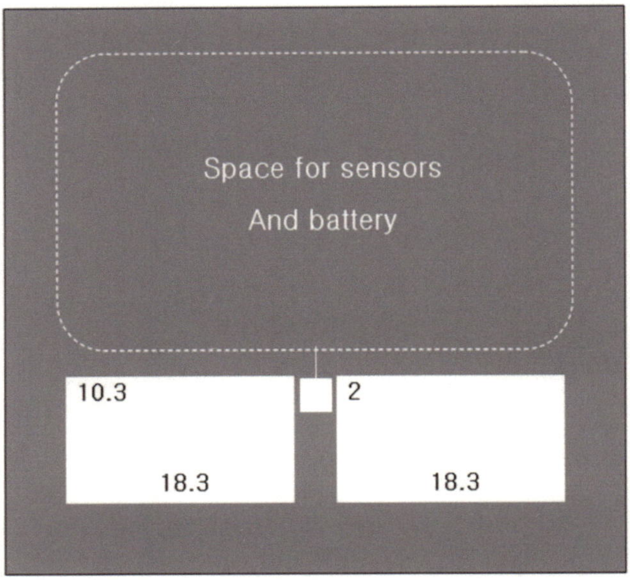

(a) Top-view

(b) Side-view showing different material layers(Air, Silicon slab, Metal patch, and Metal body)

Fig. 9. Top-and side-views of a metal patch with the nested slot

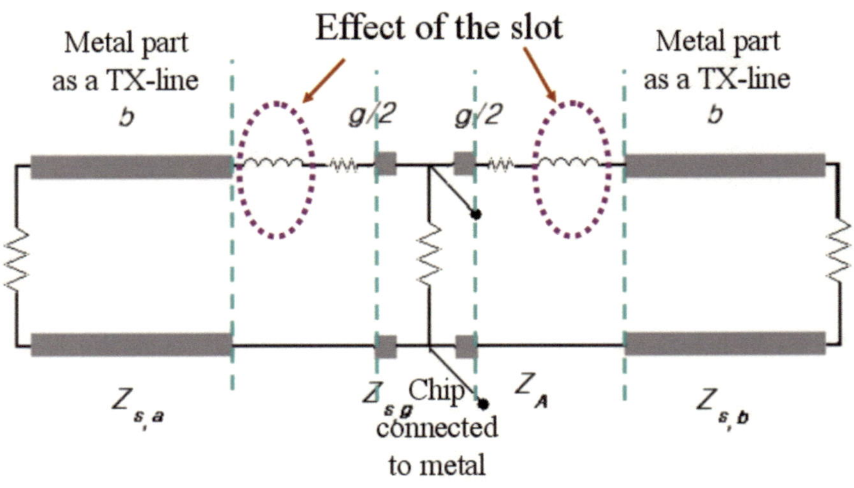

Fig. 10. Circuit model of the metal patch with the nested slot in Figure 9

The formation of the slot disturbs the surface electric current path of the original rectangular patch. The current starting from the driving point(chip connecting point) will flow along longer paths such as the edges of the slot. Especially, when the operating frequency is high enough, say, UHF or GHz band, each piece or segment(ΔL) of a conduction current path the current is equivalent to $j2\pi fL_{loop}\Delta L$ or the inductance.
As is shown in Figure 10, the inductance between the TX(transmission)-lines due to the slot can be used to cancel the capacitance of the chip. It is noteworthy that the real term of the impedance is as large as the imaginary term like the shunt stub matching case

4. Size reduction techniques

In the design of the RFID antennas, the size of the geometry is treated equally important as the impedance matching, since they tend to look for light, low-profiled and portable RFID equipment. Owing to the modern antenna technologies including materials, manufacturing, and electromagnetic field prediction, the size of an antenna has been successfully minimized, when it is needed to place in a portable device. In this section two ways of size reduction are presented.

4.1 Meandering

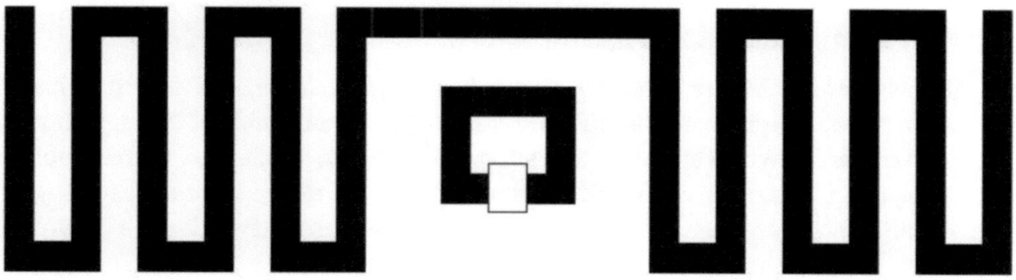

Fig. 11. Inductive loop with a meandered conducting wire

Figure 11 shows the inductive loop coupled with the meandered line[5-6]. Just remember the loop driven by the chip could bring the inductive reactance when it has the nearby main radiator which is as long as half-wavelength. Considering the straight main radiator stretches over the required area, the shape can be changed by meandering to fit the space or within the boundary, while its radiation and impedance performance meets the specifications. We will see later the design steps of a meandered tag antenna.

4.2 Inverted-F structure[7]

With the tendency that mobile phones are wanted to be less bulky, the wire antenna such as whipped(monopole) antenna is disappearing and the inverted-F type has been preferred from the outset of its invention. The word 'inverted-F' itself comes from its geometrical appearance that two conductors touch the metal ground and support a metal planar radiator. The two conductors are the feed and the shorting pin. The positions of the two conductors determine the impedance and bandwidth as well as the field pattern.

Fig. 12. Inverted-F antenna and its coplanar version

Seeing the structures in Figure 12, on your left side, the typical inverted-F antenna where the main planar radiator resides on the dielectric (or air) substrate, and on the right reversed letter 'F', which is the wire version of the inverted-F antenna, is attached to the ground on the same layer(coplanar). Basically, these antennas belong to the quarter-wavelength or monopole antenna. But the radiator is placed near the ground to keep a low-profile compared to the monopole which has its top end up in the air.

5. Other issues

As the wireless technologies are getting developed rapidly, the demands on the challenging requirements become more and more complicated such as the circular- or dual-polarization, the dual-band and so and so forth.

As we have seen before, the polarization mismatch lowers the power received by the tag and the reader and the resultant degradation of the RFID system's quality, with equations 3 and 4. This is why the antennas for the tag and the reader are deigned to have circular polarization which is less vulnerable to the mismatched polarization, or dual polarization which provides polarization diversity.

Regarding the frequency regimes adopted for the RFID applications, there exist a number of bands[8-10]. Different countries have different RFID industrial standards and different frequency bands. After a product is made in one country, it is shipped to another and it is assumed to have the RFID tag working at one frequency band. The tracking of the product will be failed as soon as it disembarks. In order to avoid this trouble, antennas are wanted to show dual-band operations. The following is an example of a dual-band RFID tag antenna.

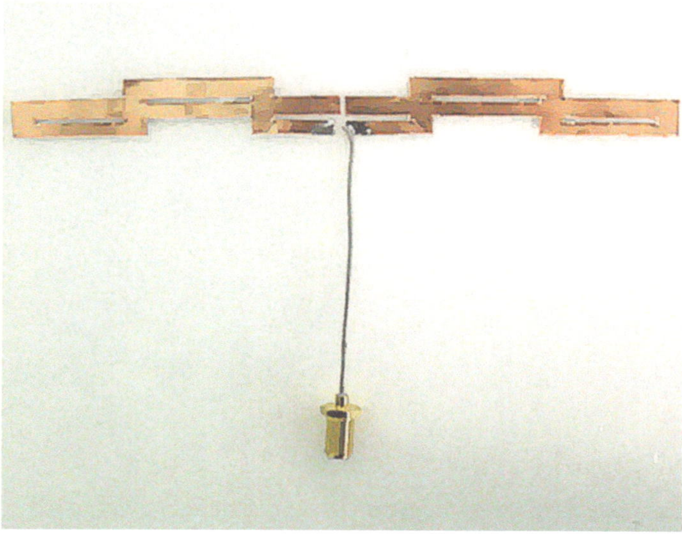

Fig. 13. A dual-band dipole antenna[11]

In Figure 13, the collinear-type antenna has up-and-down metal segments and slots together. Radiation occurs at two frequency bands 900MHz-UHF band and 2.4GHz-band. The lower band radiation results from the surface current resonant on the metal segments and the slots account for the higher band radiation.

6. Design example

We are going through the design procedure for a meander wire antenna with a parasitic element, using FEKO™ a full-wave EM simulator[11].
Step 1) Define Variables and Input Parameters
See Figure 14. Here, we set the operating frequency(= 916MHz), BW, constitutive parameters, etc

Fig. 14. Defining Variables and Input Parameters

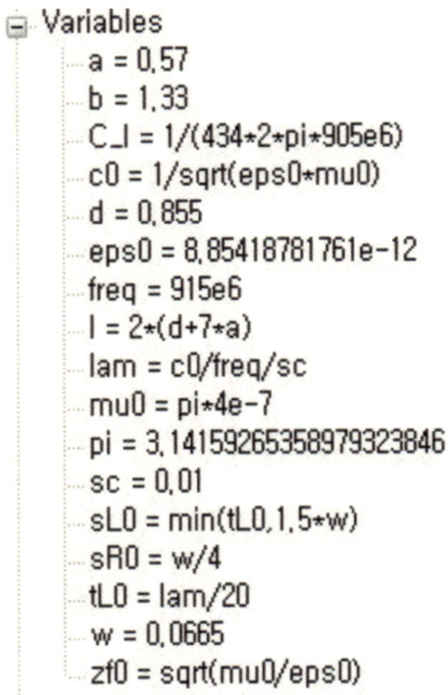

Fig. 15. Defined Variables and Input Parameters

Step 2) Draw the Geometry
Draw the shape of the antenna on your mind.
See Figure 16.

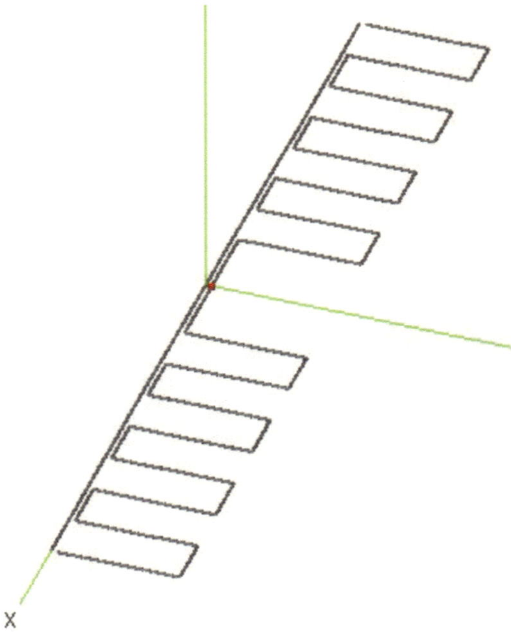

Fig. 16. Drawing the meander line

Step 3) Set the port or voltage source
Set the position of the driving point (port) and set the type such voltage gap source

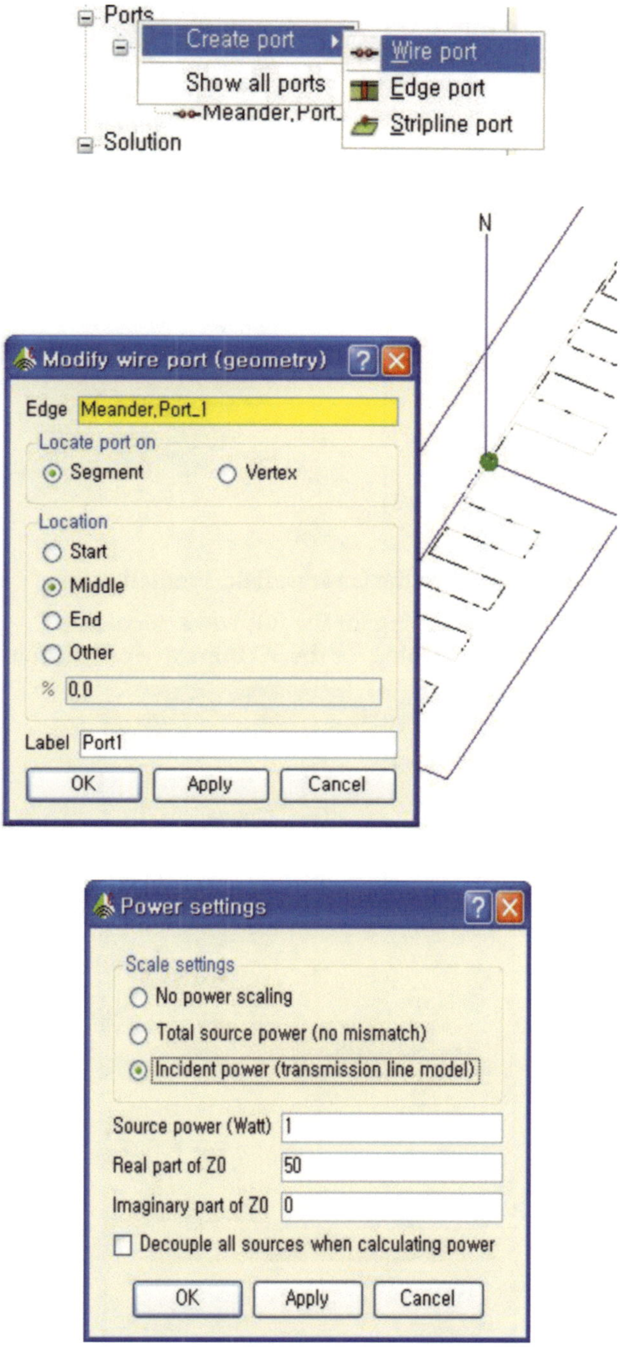

Fig. 17. Define the port and set the type

Step 4) Set the condition for the far-filed calculation
Determine the conditions of the far-field such as the pattern and the boundary of simulation

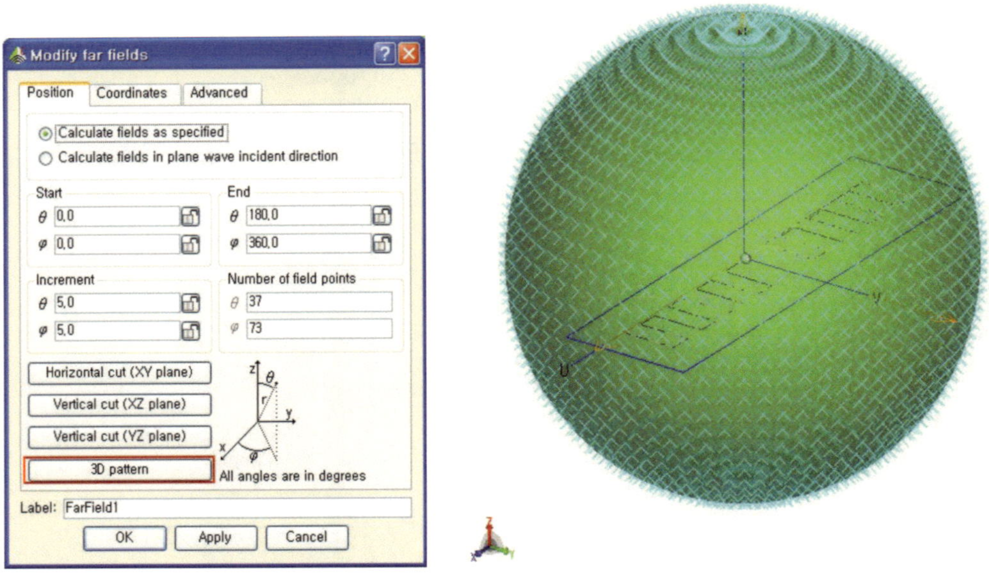

Fig. 18. Determine the conditions for the far-zone field simulation

Step 5) Set the condition for the meshing for the full-wave calculation
Determine the conditions of the meshing for the 3D full-wave calculation

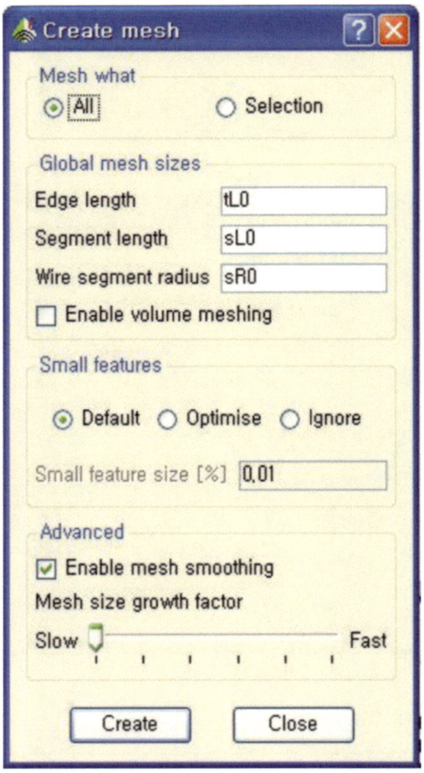

Fig. 19. Set the meshing condition for the meshing

Step 6) Running the program and getting the results

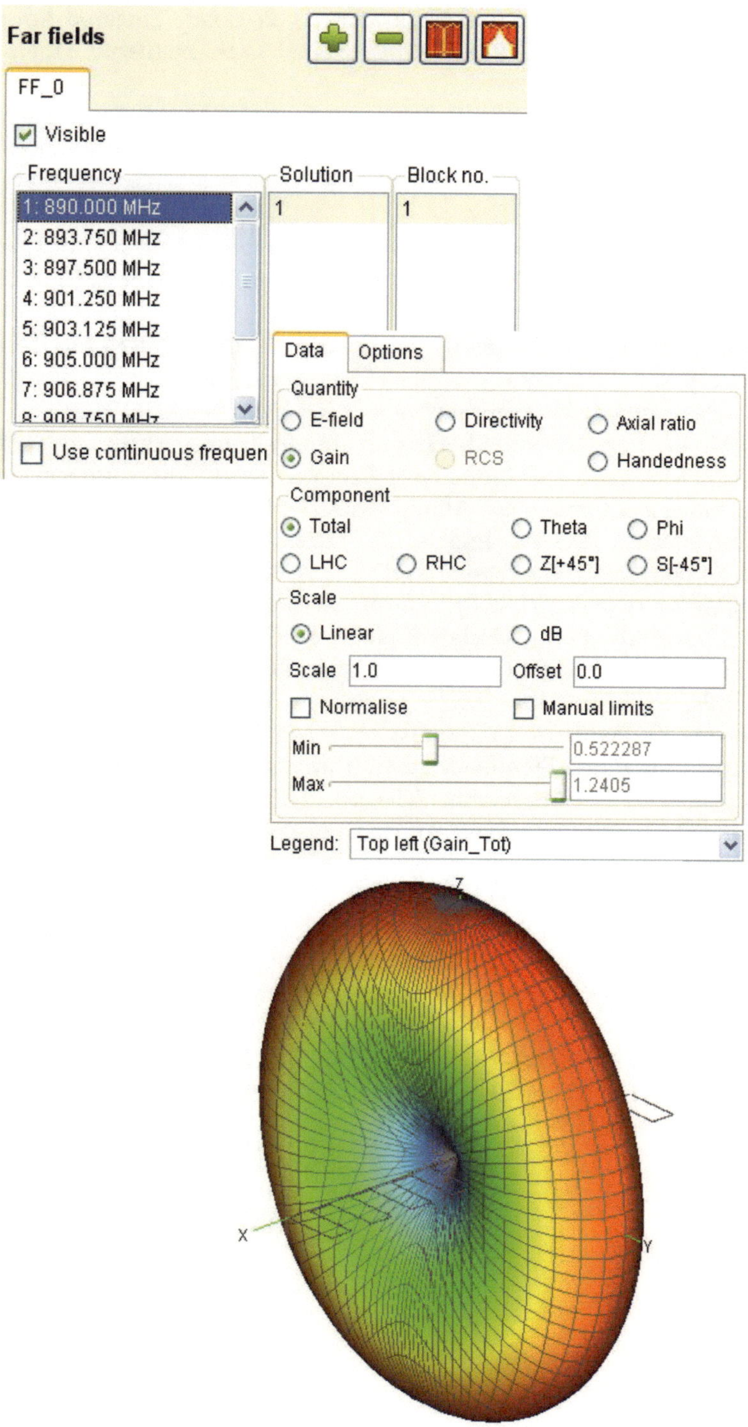

Fig. 20. Acquiring the data we need such as the 3D far-field pattern

7. References

[1] J. Curty, N. Joehl, C. Dehollain and M. J. Delercq, "Remotely Powered Addressable UHF RFID Integrated System," *IEEE Journal of Solid-State Circuits*, 40, 11, November 2005, pp. 2193-2202.

[2] S. Basat, S. Bhattachary, A. Rida, S. Johnston, L. Yang, M. M. Tentzeris and J. Laskar, "Fabrication and Assembly of a Novel High-Efficiency UHF RFID Tag on Flexible LCP Substrate," *Electronic Components and Technology Conference*, May-Jun 2006, pp. 1352-1355.

[3] Marrocco G. (2007), RFID antennas for the UHF remote monitoring of Human subjects, IEEE Transaction on. Antennas and Propagation, Vol.55, N. 6, June 2007, pp. 1862-1870

[4] Y. Lee, "Antenna Circuit Design for RFID Applications," Microchip Technology Inc., Application Note AP710, 2003, available at http://ww1.microchip.com/downloads/en/AppNotes/00710c.pdf.

[5] W. Choi, H. W. Son, C. Shin, J. H. Bae and G. Choi, "RFID Tag Antenna with a Meandered Dipole and Inductively Coupled Feed," *IEEE International Symposium on Antennas and Propagation*, Albuquerque, NM, July 2006, pp. 619-622.

[6] S. A. Delichatsios, D. W. Engels, L. Ukkonen and L. Sydanheimo, "Albano Multidimensional UHF Passive RFID Tag Antenna Designs," *Int. J. Radio Frequency Identification Technology and Applications*, , 1, January 2006, pp. 24-40.

[7] C. Cho, H. Choo and I. Park, "Design of Novel RFID Tag Antennas for Metallic Objects," *IEEE International Symposium on Antennas and Propagation Digest*, Albuquerque, NM, July 2006, pp. 3245-3248.

[8] M. Hirvonen, K. Jaakkola, P. Pursula and J. Saily, "Dual-Band Platform Tolerant Antennas for Radio-Frequency Identification," *IEEE Transactions on Antennas and Propagation*, AP-54, 9, September2006, pp. 2632-2636.

[9] S. Nambi and S. M. Wenthworth, "5.8 GHz Dual-Polarized Aperture-Coupled Microstrip Antenna," *IEEE International Symposium on Antennas and Propagation Digest*, Washington, DC, July 2005, pp. 235-238.

[10] S. Jeon, Y. Yu, S. Kahng, J. Park, N. Kim and J. Choi, "Dual-Band Dipole Antenna for ISO 18000-6/ISO 18000-4 Passive RFID Tag Applications," *IEEE International Symposium on Antennas and Propagation Digest*, Albuquerque, NM, July 2006, pp. 4285-4288.

[11] EM Software & Systems-S.A. (Pty) Ltd., FEKO User's Manual, Suite 5.1, Stellenbosch, South Africa. December 2005, available at http://www.feko.info

4

UHF RFID of People

Milan Polívka, Milan Švanda and Přemysl Hudec
Czech Technical University in Prague
Czech Republic

1. Introduction

Recently, the application of RFID systems is more and more extended on monitoring and identifying people, both in indoor and outdoor areas. This application concerns, for example, monitoring of employees in offices working with sensitive information, in large factories or power plants, personnel in an army storehouse, supervisors in prisons, or long distance runners at checking gates. In all of these cases it can be important to know who is where. Identification can be combined with fast access. The approved people can be allowed to enter specific rooms or an area, non-approved can be restricted from entering. All of this can be carried out over a moderate distances without the person having to worry or perform any action.

Fig. 1. Contactless and contact identification of people from the cartoonist's point of view

The identification distance of many RFID systems is relatively short, typically from several cm up to 1 m. This necessarily need not be a problem at a door access, but brings too many complications in wider corridors or open areas. For these purposes, it is important to identify the person at longer distances, convenient ranges can be from 2 to 10 m. For RFID purposes, a relatively wide range of frequencies is available. Operating frequencies used can be chosen in the low frequency band (LF, 125 kHz), high frequency band (HF, 13.56 MHz), ultra high frequency band (UHF, 860 - 930 MHz) or microwave bands (up to now dominantly 2.4 GHz). LF and HF RFID systems can be used for monitoring of people, the problem is that only over a relatively short identification distance. The main reason for that, is the coupling used between the reader and TAGs. At these frequencies, standard antennas are too large and have too low gain. That is why a simple inductive coupling is used and that is why LF and HF RFID systems can only provide a short identification distance, typically up to 1 m (Finkenzeller, 2003). The person to be identified must come into a narrow vicinity of the reader or even touch the reader. The UHF and microwave RFID systems can use standard matched antennas and electromagnetic wave propagation as a coupling mechanism. Antennas at these frequencies can have acceptable dimensions and gains and identification systems can provide the required 2 - 10 m maximum identification distance. That is why UHF and microwave RFID systems seem to be very good candidates for monitoring of people over moderate distances.

For these purposes, standard UHF RFID systems can be used, but without careful design, they can lead to relatively poor results. Efficient and reliable monitoring and identification of people in large buildings is influenced by a number of specific conditions and physical phenomena. Above all, this concerns the operating frequency, the behavior of propagating electromagnetic waves at this frequency and the specific parameters of the antennas used.

The principal conditions for correct identification can be described in a form of two power budgets. The area to be monitored is illuminated by a TX reader antenna. The person to be monitored wears a TAG including its own receiving/transmitting antenna. The RF power received by the TAG antenna must be higher than the TAG sensitivity. This power enables the TAG to transmit (reflect) back a wave bearing the identification code. The power of this wave at the input of the reader receiver must be higher than its own sensitivity. Both power budgets are influenced by the RF power transmitted by the reader, by the gains of all antennas, by the free-space loss between the reader antenna and the TAG antenna and by additional losses specific to the propagation of electromagnetic waves under the given conditions. This above all concerns multi-path fading and shadowing. These additional power losses must be described and evaluated and taken into account during system calculations. To a definite degree, they can be minimized by system optimization and setting.

As it has been already mentioned, in the UHF and microwave bands, standard antennas with impedance matching and positive dB gain values can be used. Since higher frequencies lead to smaller antennas, it may be seen to be beneficial to use a microwave band. It is true, but it also must take into account that higher frequencies lead to higher values of free-space loss. So UHF operational frequency can also be a very good choice.

Usually, the choice of reader antenna represents no substantial problem. Any antenna with an acceptable radiation pattern and gain can be used. However, in some cases it can be beneficial to use the antennas with higher directivity that enables one to focus the energy to the expected identification area and improve both the power budgets. More problems are usually associated with TAG antennas as they represent the biggest part of the TAG. For the identification of people, the TAG must be worn, and that is why they must be as small and lightweight as possible. The design of very small antennas is a problem by itself, due to the principal limitations. But an additional problem lies in the fact that the parameters of the majority of the standard RFID antennas are strongly affected by a nearby presence of a human body. The body can detune the antenna and absorb a substantial part of the received or transmitted RF power. Both phenomena can strongly affect power budgets and result in a wrong identification.

Efficient and reliable monitoring and identification of people in large buildings or outdoor areas belongs to a relatively difficult technical task. These difficulties are dominantly associated with operating frequency, with the behavior of the electromagnetic waves in the given frequency band and with the antennas used. All of these problems must be carefully solved and each RFID system must be "tuned" for the given application. The following sections include more detailed treatment of all of the above described problems.

2. TAG antennas for UHF RFID of people

The design of a UHF TAG antenna suitable for monitoring of people has several important requirements. Some of them are slightly different from demands made on standard communication antennas, this especially concerns:

- Impedance matching to complex impedance different from 50 Ω
- Immunity of antenna parameters from the nearby presence of a human body
- Small dimensions and weight, sometimes also flexibility

RFID chip impedances usually have a capacitive component (Finkenzeller, 2003) that is why, in order to ensure the impedance matching, the input impedance of the TAG antenna has to provide the corresponding inductive component. In the case of standard RFID dipole antennas, this inductive component is often realized by a small parallel loop (Sidén et al., 2006).

Furthermore, the design of the intended UHF TAG antennas has to take into account the close vicinity of a human body. In UHF band that represents a high-loss dielectric object with a relative permittivity $\varepsilon_r \sim 50 - 60$ and loss tangent $\tan \delta \sim 0.5 - 1.2$ (Gautherie, 1990). In case of the dipole type TAG antennas, the presence of such dielectric causes significant detuning and absorption of the radiated or received energy (Foster & Burberry, 1999), (Raumonen et al., 2003), (Dobkin & Weigand, 2005), (Griffin et al., 2006). This results in a low radiation efficiency and consequently in a short identification distance. Thus, antenna structures immune to an undesirable influence of a human body together with a small footprint, and a low profile are highly recommended (Ukkonen & Kivikoski, 2003).

The problem of frequency detuning and energy absorption can be, on principle, solved by insertion of a screening metallic plate. The plate can act as an additional shielding or can form an inherent part of the TAG antenna. The first following section concerns the problems of the operation of the standard dipole antennas above the metallic plane. The original solution using multiple-arm dipoles is presented. The other section describes the TAG antennas, newly designed especially for identification of people that employ metallic plates as an inherent part of their structures. All new TAG antennas were designed with respect to minimization of their profile, which is important for realization of wearable antennas. All were impedance matched at 869 MHz to the RFID chip described in Table 3.

2.1 Dipole type antennas

Dipole type antennas, for their simple design and manufacturing, are the natural and usual choice for TAG antennas in the UHF and microwave bands. To be able to efficiently radiate electromagnetic waves, their sizes have to be comparable with the half- or quarter-wavelength (approx. 160 - 170 and 80 - 85 mm in the UHF band, respectively). In case of any miniaturized version, such as the meander-shortened dipole, the radiation resistance

(Noguchi et al., 1997) and consequently the radiation efficiency and identification distance can decrease significantly. Unfortunately, electrical properties of this type of antenna are also very sensitive to any dielectric or metallic object situated in its near proximity. In the case of a closely spaced metallic plane, out-phase image currents induced in the plane decrease the antenna input resistance and consequently also the radiation efficiency. Thus, in order to operate efficiently, the dipoles usually require a relatively thick space pad (Ranasinghe et al., 2004), (Sidén et al., 2006).

In order to demonstrate impedance and radiation properties of a dipole closely spaced to a metallic plane, a planar dipole according to Fig. 2a is considered.

As it can be seen, the lower the relative distance h/λ_0 is, the lower the input impedance is, see Fig. 2b.

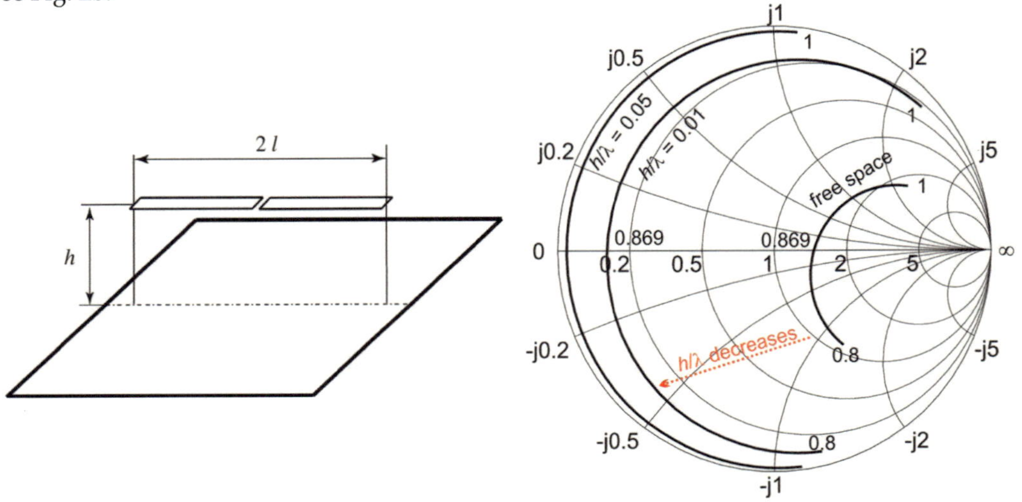

a) b)

Fig. 2. Geometry of a planar dipole, length $2l$ = 163.3 mm and width w = 2 mm placed above an infinite conductive plane a) and b) its input impedance curves in the Smith chart for the frequency range from 0.8 to 1.0 GHz as a function of the relative dipole distance over the conductive plane ($h/\lambda_0 = \infty$ (free space), 0.01, 0.005).

On the contrary to the decrease of the antenna efficiency, the decline of the radiation efficiency is not so progressive; see Fig. 3. As far as $h \geq 0.01\ \lambda_0$ (~ 3.5 mm at 869 MHz) the radiation efficiency is still slightly above 20 %, while the antenna efficiency due to the impedance mismatch (related to standard 50 Ω) drops to 1 - 2 %.

One of the possible ways how to increase the low input resistance of a dipole closely spaced above a metallic plane is to use the multiple-conductor (arm) folded dipole configuration (Best, 2004). The physical relation of the input impedance Z_{in} and the number N_{arm} of the dipole arms in the half wavelength, the multiple-arm folded dipole configuration may be expressed as follows (Balanis, 1997):

$$Z_{in} \approx N^2_{arms} Z_{11} \tag{1}$$

where Z_{11} is the self impedance of a single dipole. However, it is necessary to point out that the above-mentioned relation is uniquely valid provided that the dipoles are closely spaced and the magnitudes of all dipole currents are supposed to be the same. The effect of

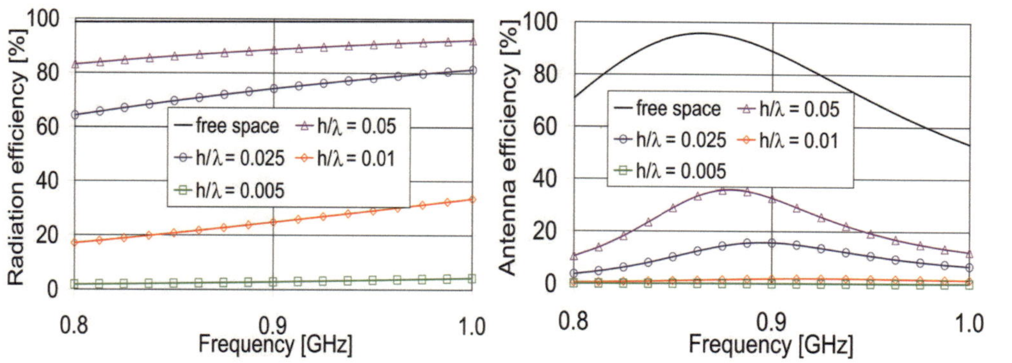

a) b)

Fig. 3. Simulated frequency dependencies of the a) radiation and b) antenna efficiencies of a planar dipole as a function of the relative distance h/λ_0 = {free space, 0.05, 0.025, 0.01, 0.005}

decreasing dipole Z_{in}, when closely spaced to a metallic plane, is thus compensated by the increase of Z_{in} caused by the multiple-arm folded dipole configuration. This results in an approximate linear dependence of $Z_{in} = f(N_{arms})$ as verified by the results of the EM field simulations (Polívka et al., 2008a). Using the operation in an inductive region above the half-wavelength resonance, a impedance matching of the imaginary part of the TAG antenna to the imaginary part of the RFID chip impedance can be easily achieved; see Fig. 4b. The same figure shows the increase in the input resistance of the multiple-arm folded dipole according to Fig. 4a as a function of the number N_{arms}, details can be found in (Polívka et al., 2008a). The simulated results corresponding to $N_{arms} = 33$ show approx. 50 % radiation and antenna efficiencies; see Fig. 5.

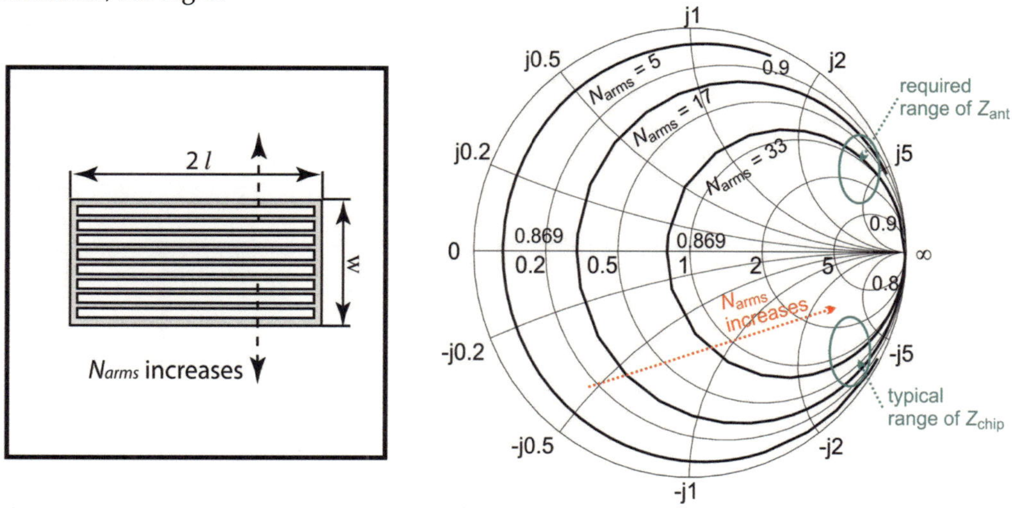

a) b)

Fig. 4. The geometry of the multiple-arm folded dipoles with dipole length $2l$ = 134 mm, width w = 98 mm, spacing of arms s = 1 mm, spaced h/λ_0 ~ 0.003 (d = 1.04 mm consisting of a 0.8 mm thin foam substrate and 0.24 mm thin GML 1100 woven-glass laminate) over a test metallic plane of size 242 × 120 mm a) and their input impedance in a Smith chart in the frequency range 0.8 - 1.0 GHz b) as a function of the number of arms (N_{arms} = 1, 5, 17, 33).

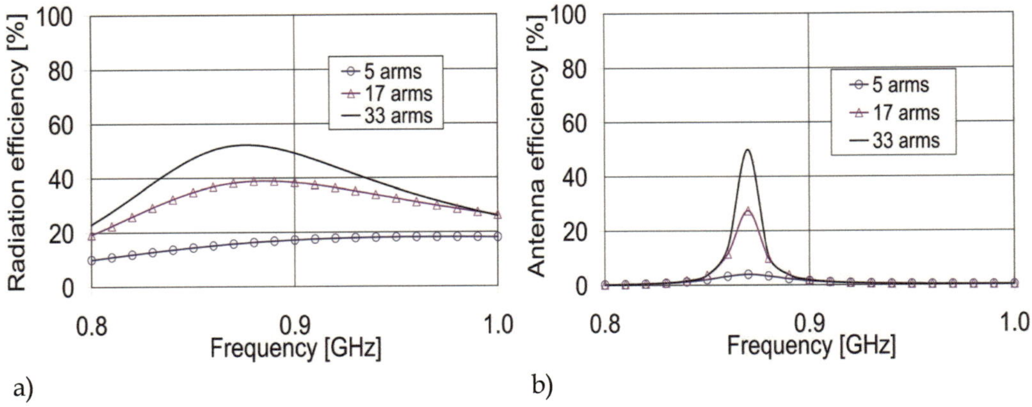

Fig. 5. The simulated a) radiation and b) antenna efficiency of the multiple-arm folded dipole from Fig. 4 (at f_0 = 869 MHz) as a function of the frequency with spacing h/λ_0 over the metallic plane as the parameter.

The main advantage of this solution is an extremely low achievable profile (h/λ_0 ~ 0.003). The drawbacks are a relatively large footprint size (~ 0.4 × 0.3 λ_0) and narrow operational bandwidth. But in the case of narrowband RFID applications, where antenna dimensions are not a limiting requirement, this multiple-arm folded dipole structure can provide the TAG antenna with acceptable parameters.

2.2 Patch and PIFA antennas

PIFA (Ukkonen et. al., 2004), (Hirvonen et. al., 2004) and patch (Ukkonen et al., 2005), (Švanda et al., 2007), (Švanda & Polívka, 2007) antennas represent radiating structures where the metallic ground plane is their inherent part. However, at relatively low operational frequencies (hundreds of MHz and lower), several potential difficulties must be taken into account. Firstly, when the substrate height is lower than ~ 0.01 λ_0, and relative permittivity ε_r is higher than that of the air or foam substrate, see Fig. 6, their radiation efficiency decreases significantly (Lee & Chen, 1997). Secondly, the basic patch resonant

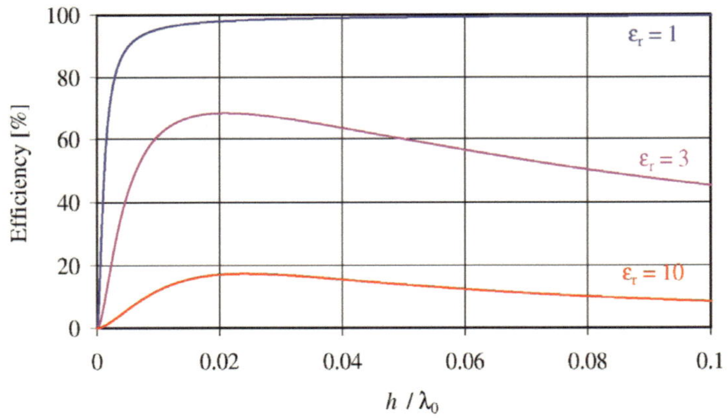

Fig. 6. Radiation efficiency as a function of the relative substrate height, according to (Lee & Chen, 1997)

frequency corresponds to $\lambda_0/2$ or $\lambda_0/4$ and, therefore, at UHF frequencies, the patch or PIFA antennas might not be sufficiently small for the intended application. However, in definite applications of person identification, larger patch antenna footprints need not be the most limiting parameter. For example, this concerns the identification of long distance racers in a more detailed way described in Section 3.4.1. where nearly $\lambda_0/2$ low profile and very lightweight antennas were integrated into their number labels.

This new RFID patch antenna (Švanda et al., 2007), see Fig. 7, was fabricated on a foam dielectric (G3 9568 foam $h = 4.8$ mm, $h/\lambda_0 \sim 0.014$) using a conductive fabric. The ground plane dimensions are 165 x 74 mm, the measured gain of an antenna placed on a human body is 5.0 dBi. The weight of the antenna is approx. equal to 20 g, it is flexible and, as it was mentioned before, it was easily integrated into the sportsmens' number labels.

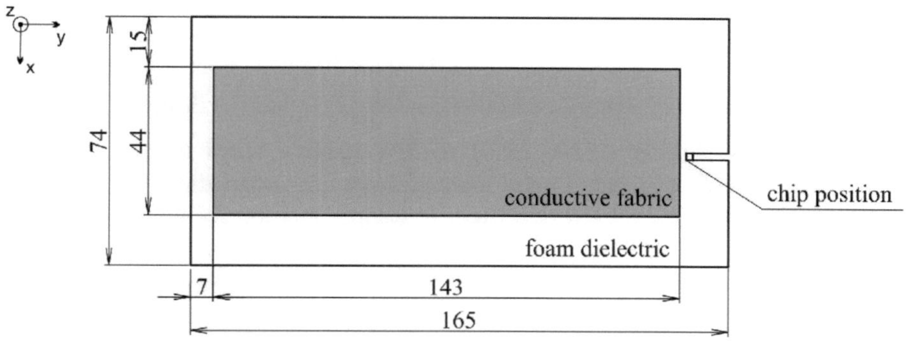

Fig. 7. Sketch of designed prototype of the patch TAG antenna

The comparison of the gains of the RFID patch antenna measured in a free space and on a human body phantom (a tank with 5 liters of salt water) is presented in Table 1. The parameter b represents the distance between the antenna and the phantom. Fig. 8 shows the measured radiation patterns in free space and fixed on the phantom and indicates the very low influence of a nearby human body on the antenna parameters.

	G [dBi]	D [dBi]	η [%]
free space	6.3	6.7	91
$b = 0$ mm	5.0	7.6	55

Table 1. Measured gain, directivity, and antenna efficiency of a patch TAG antenna

2.3 Small flat-loop antenna loaded by a sub-wavelength patch array backed by grounded dielectric slab

Generally, as electrical small loop antennas provide inductive input impedance (Hansen, 2006), they seem to be good candidates for RFID TAG applications. However, the properties of the loop antenna highly depend on the pad material, moreover the loop perimeter is comparable to the wavelength. If the circular loop is flattened to a low oval, its length is close to about a half-wavelength. Both of the above-mentioned problems can be eliminated by using a grounded substrate with relatively high permittivity ($\varepsilon_r \sim 10$). Due to the dielectric slab used, the loop size gets significantly smaller. There still remains a problem with the radiation efficiency, which drops in case of the thin substrate thickness (up to ~ 1.5 mm) to about 17 %; see Fig. 10. This phenomenon is caused by destructive interferences among the reflected and transmitted waves (Sievenpiper, 1999). Artificial

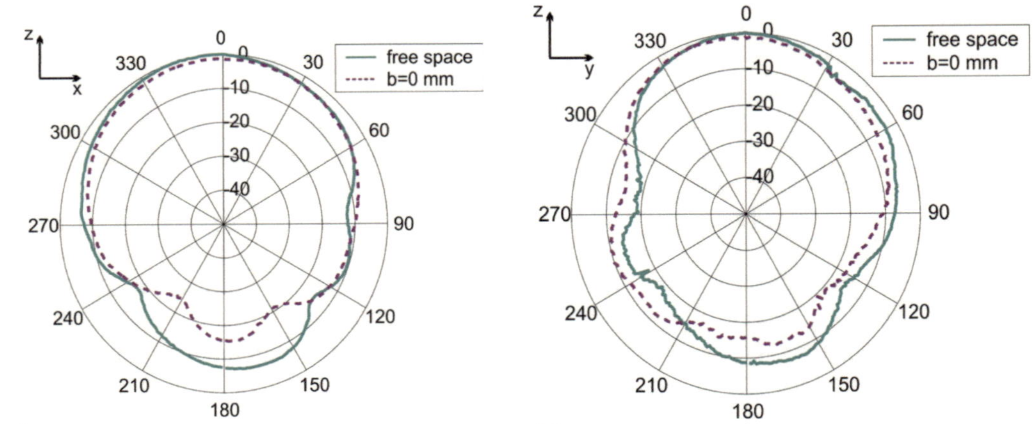

a) b)

Fig. 8. Measured radiation patterns of a patch antenna a) H-plane b) E-plane

magnetic conductors (Zhang et al., 2003) were a certain inspiration for designing a four-element sub-wavelength patch array backed by a grounded dielectric slab. This slab was used as a screening plane for a small flat dual-loop antenna (Švanda & Polívka, 2008); see Fig. 9. This solution minimizes the resulting antenna size, the dual-loop footprint is 70 × 11 mm. The total antenna size including the backing patch array etched on a 1.58 mm ($h/\lambda_0 \sim 0.005$) thick grounded dielectric substrate is 70 × 105 × 1.82 mm (relative size is 0.2 × 0.3 × 0.005 λ_0 at 869 MHz). The reason for using the dual-loop modification is to increase the real part of the antenna input impedance.

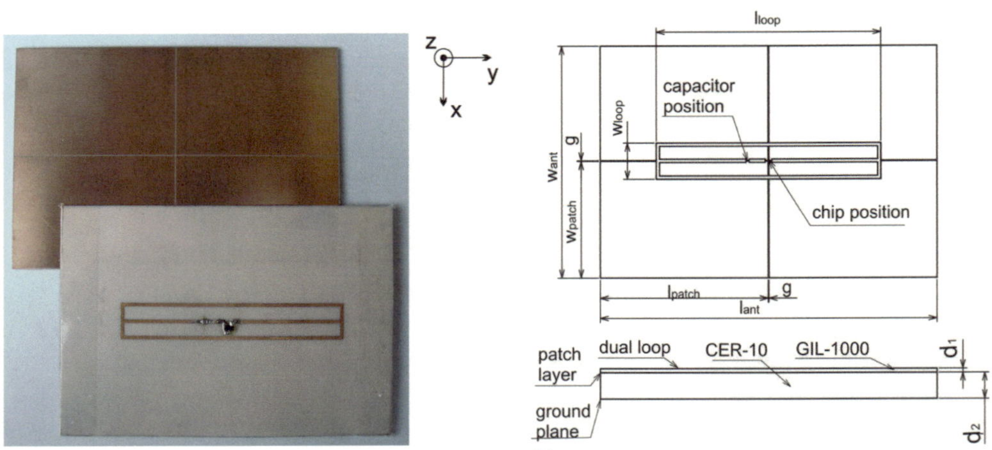

a) b)

Fig. 9. Photograph a) and drawing b) of the designed prototype of the flat dual-loop antenna closely spaced over a patch array surface, $d_1 = 0.24$ mm, $d_1/\lambda_0 \sim 0.0007$, $\varepsilon_{r1} \sim 3.05$, $d_2 = 1.58$ mm, $d_2/\lambda \sim 0.0046$, $\varepsilon_{r2} \sim 10$. The total relative dimensions are 0.2 × 0.3 × 0.005 λ_0 at 869 MHz.

The comparison of the radiation efficiency of the flat single loop antenna in the free space, above the metallic plane, and above the four-element patch array etched on the grounded dielectric slab can bee seen in Fig. 10.

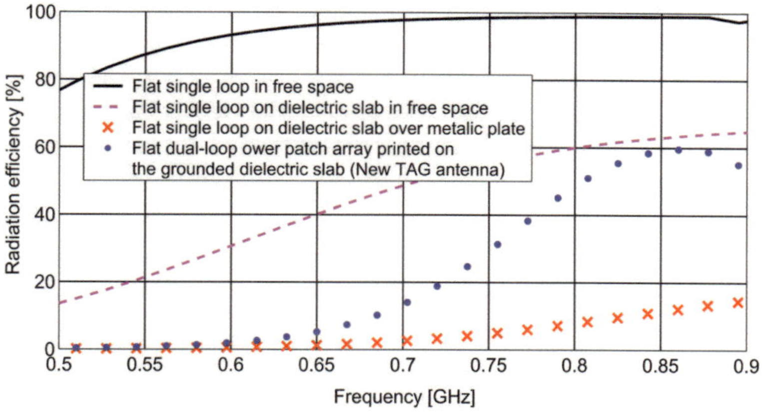

Fig. 10. Comparison of the simulated radiation efficiency for the different versions of flat loop antennas

In order to enable the measurement of the antenna input impedance, a half-loop (equivalent to the monopole) arrangement equipped an SMA connector was designed and realized; see Fig. 11a. The size of the mirror metallic plane is 145 × 145 mm. The half-loop antenna input impedance should have a half value of the loop antenna input impedance. The power transmission coefficient τ is calculated by means of the following equation:

$$\tau = 1 - |\Gamma|^2 = \frac{4R_{ant}R_{chip}}{(R_{ant}+R_{chip})^2 + (X_{ant}+X_{chip})^2}, \quad 0 \le \tau \le 1 \tag{2}$$

$$\Gamma = \frac{Z_{chip} - Z_{ant}}{Z_{chip} + Z_{ant}}, \quad 0 < |\Gamma| < 1 \tag{3}$$

where Γ is the reflection coefficient between the antenna and the chip impedances, R_{ant} and R_{chip} represent the antenna and the chip input resistances, X_{ant} and X_{chip} stand for the antenna and the chip input reactance, respectively.

a)

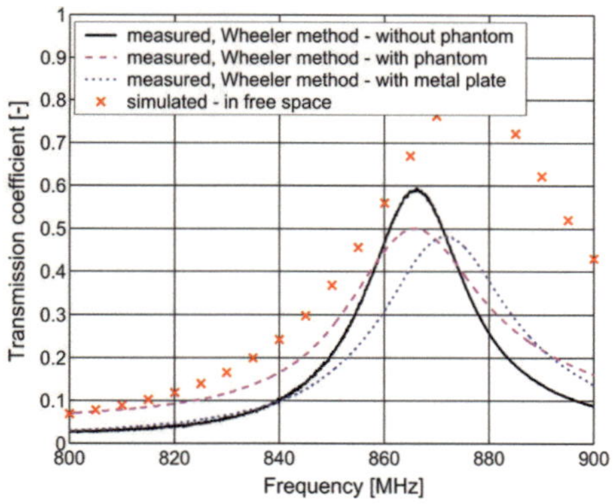

b)

Fig. 11. The manufactured prototype of the half-loop arrangement a) photograph b) transmission coefficient related to the chip impedance.

The transmission coefficient (see Eq. 2) of the half-loop arrangement has been simulated and measured in free space, and on a human body phantom (agar with $\varepsilon_r \sim 55$ and $\tan\delta \sim 0.5$). The above-mentioned arrangement enables a measurement of the antenna and radiation efficiencies by means of the Wheeler cap method (Wheeler, 1959). The cap dimensions were 123 × 123 × 123 mm. The measured and simulated values can bee seen in Fig. 12.

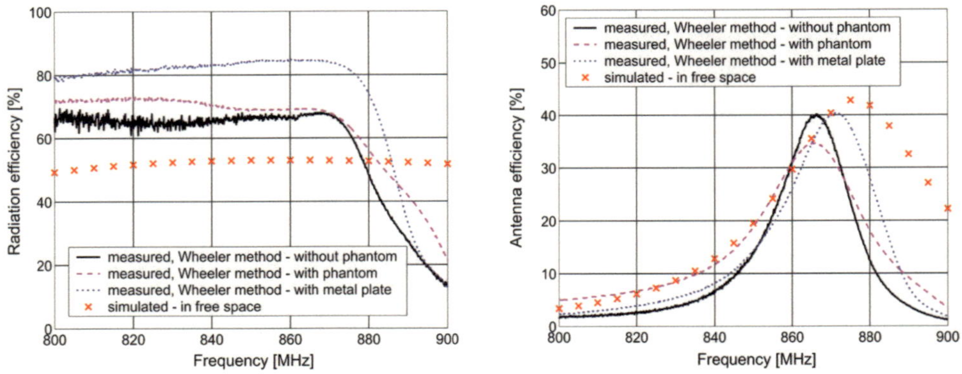

a) b)

Fig. 12. Comparison of the simulated and measured a) radiation efficiency and b) antenna efficiency of the half dual-loop antenna depicted in Fig. 11.

The antenna gain has been evaluated from the measured antenna efficiency (Wheeler cap method) and directivity (evaluated from the measured radiation patterns). The simulated and measured antenna gain and directivity are summarized in Table 2.

	Radiation efficiency [%]	Antenna efficiency [%]	Directivity [dBi]	Gain [dBi]
Simulated, free space	53	39	5.3	1.3
Measured, free space	68	38	5.0	0.8
Measured, with agar	70	33	5.4	0.6
Measured, with metal	83	39	5.3	1.3

Table 2. The TAG antenna efficiency and gain simulated and measured in free space and in the close vicinity of the human body phantom at f = 869 MHz

The comparison between the simulated and measured radiation patterns can be seen in Fig. 13. Radiation patterns were measured using the dual-loop antenna fed by coaxial cable with the non-perfect symmetrization. That is why the E-plane pattern shows approx. 15° tilt which is not expected to be present in real TAG antenna-chip arrangement.

Dual-loop backed antennas can provide advantageous dimensions 0.2 × 0.3 × 0.005 λ_0 while having acceptable antenna efficiency and gain. They are supposed to be used in the form of a standard badge also bearing the person's photograph and name fixed on the chest; see Fig. 25. A minor disadvantage can be relatively higher weight caused by the usage of higher permittivity dielectric substrate.

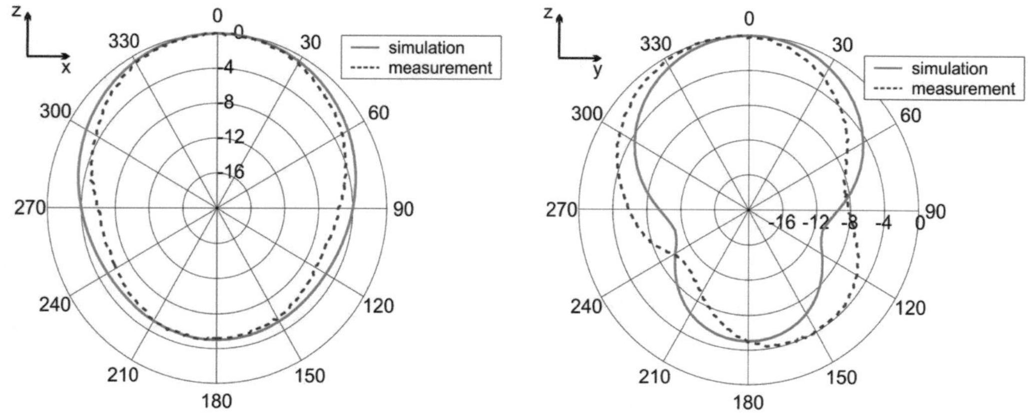

a) b)

Fig. 13. The simulated and measured radiation patterns of the flat dual-loop antenna backed by the patch array surface a) H-plane b) E-plane.

3. Power budgets

The behavior and functionality of any RFID system based on electromagnetic wave coupling mechanism, used in the UHF and microwave frequency bands, depend substantially on wave propagation effects and the corresponding reader-TAG and TAG-reader power budgets. The P_{rTAG} TAG input power must be higher than the TAG sensitivity $P_{rTAGmin}$, which provides energy for modulation of the reflected wave. In the same way, in order to ensure correct data processing on the side of the reader, the input power of the reader receiver $P_{rREADER}$ must be higher than the reader sensitivity $P_{rREADERmin}$. These two conditions must be fulfilled simultaneously and define the expected identification area, described in a more detailed way in 3.2.

3.1 Propagation phenomena in UHF band

Propagation of electromagnetic waves between the reader and TAG antennas is influenced by several effects, namely by interference of direct and reflected rays, by waveguide effects, by mutual shadowing among persons in the irradiated area and by the possible tilt of the identified person. The first two items describe the principal propagation phenomena, the second two represent random effects. All of them can significantly affect the total loss between the reader and the TAG antenna and, therefore, the functionality of the whole RFID system. This problem must be solved by as precise as possible evaluation of all these influences. In the case of the principal propagation phenomena, they can be included in propagation models, see Sections 3.2 and 3.3. The influences of random effects are usually measured and compensated for by creating sufficient backups, both in the reader-TAG and TAG-reader power budgets.

3.2 Modeling of propagation in open areas

In the case of open areas, evaluations of reader-TAG and TAG-reader power budgets can be performed relatively simply by means of the analytical two-ray path-loss model. This commonly known model works with one direct ray and one ray reflected from the ground.

In order to improve the relation of the model with practical RFID arrangements, suitable modification was designed and verified (Švanda et al. 2007); see Fig. 14. The modification takes approximated 3D radiation patterns and general tilt of both the reader and TAG antennas into account. According to practical experience, the modified model provides a sufficient agreement between the measured and simulated data.

The propagation of an electromagnetic wave from the reader to the TAG can be described by means of the following path-loss formula:

$$L = -20\log\left(\left(\frac{\lambda}{4\pi}\right)\left|\sqrt{G_{tV}(\alpha_d)G_{rV}(\beta_d)G_{tH}(\gamma)G_{rH}(\delta)}\cdot\frac{1}{r_1}e^{-j\cdot k\cdot r_1} + \right.\right. \quad (4)$$

$$\left.\left. + \sqrt{G_{tV}(\alpha_r)G_{rV}(\beta_r)G_{tH}(\gamma)G_{rH}(\delta)}\cdot\overline{R}(\vartheta)\cdot\frac{1}{r_2}\cdot e^{-j\cdot k\cdot r_2}\right|\right)$$

where r_1, r_2 are the lengths of the direct and reflected rays, $G_{rV}(\beta)$, $G_{rH}(\delta)$ stand for the angular dependences of the gain of the reader antenna in the vertical and horizontal planes, $G_{tV}(\alpha)$, $G_{tH}(\gamma)$ represent the angular dependencies of the gain of the TAG antenna in the vertical and horizontal planes, $\overline{R}(\vartheta)$ is the complex reflection coefficient of the ground.

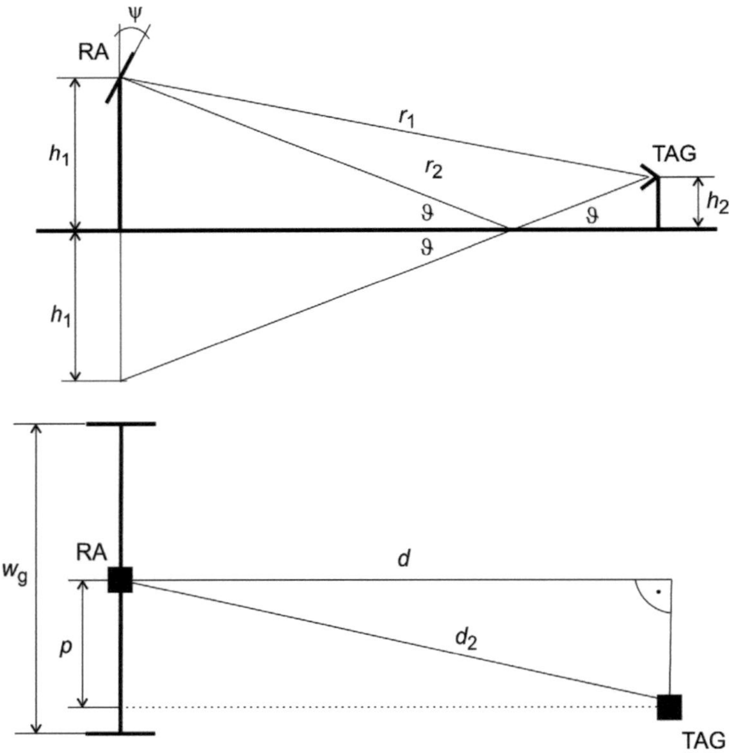

Fig. 14. Configuration of the two-ray model a) side-view b) top-view, with the following parameters: h_1 height of the reader antenna (denoted RA), h_2 height of the TAG antenna, r_1 direct ray length, r_2 reflected ray length, d_2 ground plane distance between the reader and TAG antennas, p reader and TAG antenna axis offset.

The P_{rTAG} power in dBm received by the TAG antenna can be expressed as follows:

$$P_{rTAG} = P_t - L - L_f \qquad (5)$$

where P_t is the power in dBm transmitted by the reader, L stands for the path-loss and L_f represents the attenuation of the feeder cable in dB. The peak power of the modulated signal reflected back from the TAG and received by the receiver within the reader $P_{rREADER}$ in dBm can be expressed as follows:

$$P_{rREADER} = P_{rTAG} - L - L_f - L_{conv} \qquad (6)$$

where L_{conv} is the conversion loss of the chip (typically 20 dB). The identification range can be defined as the area $\Delta d = d_{max} - d_{min}$ where d_{max} is the maximum identification distance and d_{min} is the minimum identification distance. Inside the identification range Δd, both received powers P_{rTAG} and $P_{rREADER}$ fulfill the following condition

$$P_{rTAG} \geq P_{rTAG\min} \wedge P_{rREADER} \geq P_{rREADER\min} \qquad (7)$$

as can be seen in Fig. 15.

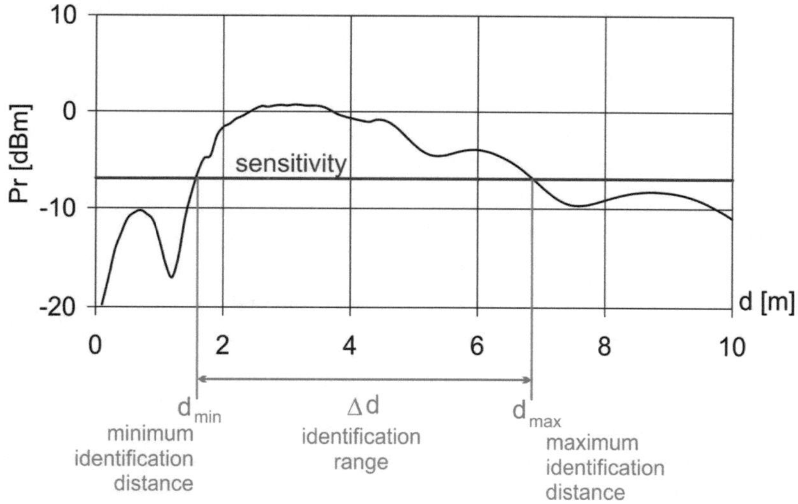

Fig. 15. Definition of the identification range

In order to cover the entire identification area, the identification range should be defined in more parallel lines with a p offset from the line axis. The reached values of the identification range Δd play an important role in the identification process. The reader can only perform a definite number (e.g. 70) of identifications per second. That is why the power budget conditions must be fulfilled for each TAG within a definite range Δd_{min} :

$$\Delta d_{\min} = v_{\max} / N_{rps} \qquad (8)$$

Where v_{max} is the maximum expected speed of a moving person, N_{rps} is the identification rate (readings per second). For example, $v_{max} = 10$ m/s corresponding to a running person and $N_{rps} = 70$ lead to $\Delta d_{min} = 0.142$ m, the maximum expected speed of a cyclist

v_{max} = 20 m/s leads to Δd_{min} = 0.285 m. From this point of view, the system optimization must also be focused on ensuring enough high values of Δd with respect to Δd_{min}. The longer the Δd is, the higher the probability of the correct identification is, even under more difficult conditions. The higher Δd range also leads to a higher probability of the correct identification in the case where random effects (shadowing, tilt) are present also.

3.3 Modeling of propagation in indoor areas

The path-loss in indoor areas (typically corridors) is also influenced by waveguide effects. Attenuation of a field strength $E(r)$ is proportional to the term $(1/r)^n$ where n is a slope or path-loss exponent affected by the geometry of the corridor and electrical properties of the ground and ceiling, and side walls (Zvánovec et al. 2003). In this case, modeling the electromagnetic wave propagation is more complicated due to multiple reflections, diffractions around vertical wedges, and shadowing of walls and usually some form of a semi-analytical or numerical approach must to be used. For modeling path-loss in a long test corridor, a ray tracing method implemented in WinProp® software (by AWE Communication) was used. The method takes into account all propagation paths which fulfill the following criteria: up to 6 reflections, up to 2 diffractions, up to a total number of 6 interactions with a combination of max. 6 reflections and max. 2 diffractions. The particular ray tracing model uses coherent superposition and includes the real radiation pattern of the reader antenna. The radiation pattern of the receiving antenna is omnidirectional. Results show that waveguide effects in the corridor can increase the received power at the expected identification distance by constructive summation of direct, reflected and diffracted rays. The increased received power results in the increased identification range in corridors compared to open areas. In closed rooms resonant effects, both constructive and destructive summation must be expected.

3.4 Optimization of RFID systems for identification of people

As it has already been mentioned, in order to identify people each RFID system must be optimized or "tuned". The tuning should be focused especially on:

a. The selection of a suitable operating frequency and RFID system (reader, TAG chips). Low $P_{rTAGmin}$ and $P_{rREADERmin}$ values are beneficial.
b. The choice or design of suitable reader antennas, optimization of their positioning and tilt with respect to the identification area.
c. The choice or design of suitable TAG antennas. The most important features are immunity against the influence of the human body, small dimensions and weight.
d. The selection of a suitable propagation model, power budget calculations.
e. Practical verification of the propagation model used, measurement of possible random influences (shadowing, tilt), ensuring of required power backups.
f. Practical testing of the tuned system.

The following paragraphs describe an example of the optimization of a standard RFID system for identifying sportsmen during mass races.

3.4.1 Basic system and measurement configurations

The chosen RFID system operates at 869 MHz, its detailed parameters are described in Table 3. The system was optimized for identifying sportsmen during mass races, see (Švanda et al., 2007) and Fig. 14. It is a typical open area task, the sportsmen are supposed to pass through the checking or finishing gates equipped with TX and RX reader antennas.

The TAGs used must be very light and flexible, on the other hand, since they can be included in relatively large label numbers, they need not be extremely small. That is why the RFID patch TAG antennas described in Section 2.2 were used. The RFID system is supposed to read the identification numbers of all sportsmen in the race for checking purposes only. The measurement of time and precise succession of racers is ensured otherwise (for example by an optical camera).

For system optimization, a series of computer simulations and practical measurements was performed, see Fig. 16. The gate height was $h_1 = 3$ m, its width $w_g = 6$ m, the measurement TAG antenna was fastened on the chest of the test person at a height $h_2 = 1.3$ m. All measurements were performed for $d = 0 - 10$ m with a step 0.2 m and $p = 0 - 2.5$ m step 0.5m, see Fig. 14. The gate was equipped with a TX reader antenna only. Since it was necessary to measure P_{rTAG} in wide dynamic ranges, a spectrum analyser with a standard 50 Ω input impedance was used. Consequently, it was necessary to face the fact, that the impedance of the TAG antenna is complex and different from 50 Ω. The TAG antenna equipped with an impedance transformer or a special test antenna of the same type as the TAG antenna but with a 50 Ω output, was used.

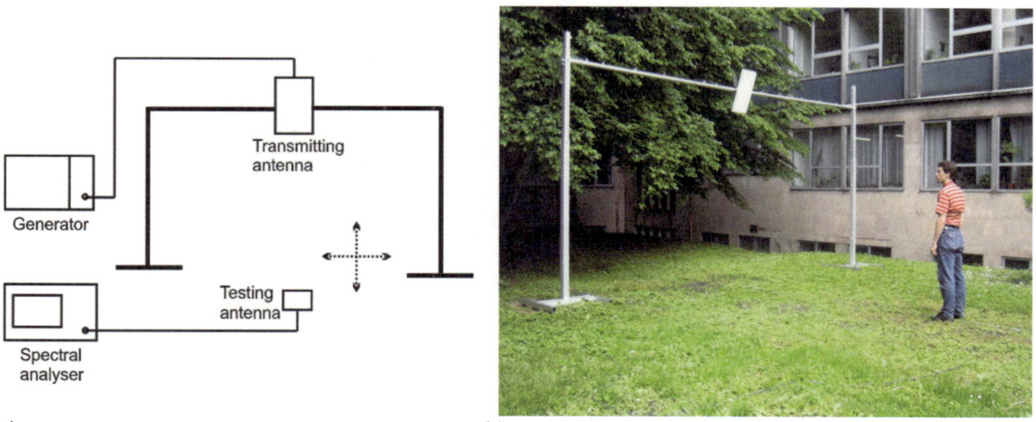

a) b)

Fig. 16. A practical measurement setup, a) scheme and b) a test person in front of a gate

3.4.2 New reader antenna

In order to enhance the effective radiated power of the transmitting antenna and to focus the energy to the expected identification area, the antenna with a microstrip patch collinear arrangement was used (Polívka et al., 2005). It provides a wide radiation pattern in the horizontal plane and narrower pattern in the vertical plane with a corresponding higher gain (11.7 dBi compared to 8.0 dBi of the original reader antenna). Fig. 17 shows its photograph and the measured radiation patterns. The same type was also used as the receiving antenna. The 3.7 dBi gain enhancement in both power budgets according to (5) and (6) was achieved.

3.4.3 Influence of tilt of the TX reader antenna

The first simulations were focused on finding the optimum tilt ψ of the reader antenna, see Fig. 14. Fig. 18 shows the simulation of the influence of ψ on the TAG input power P_{rTAG}.

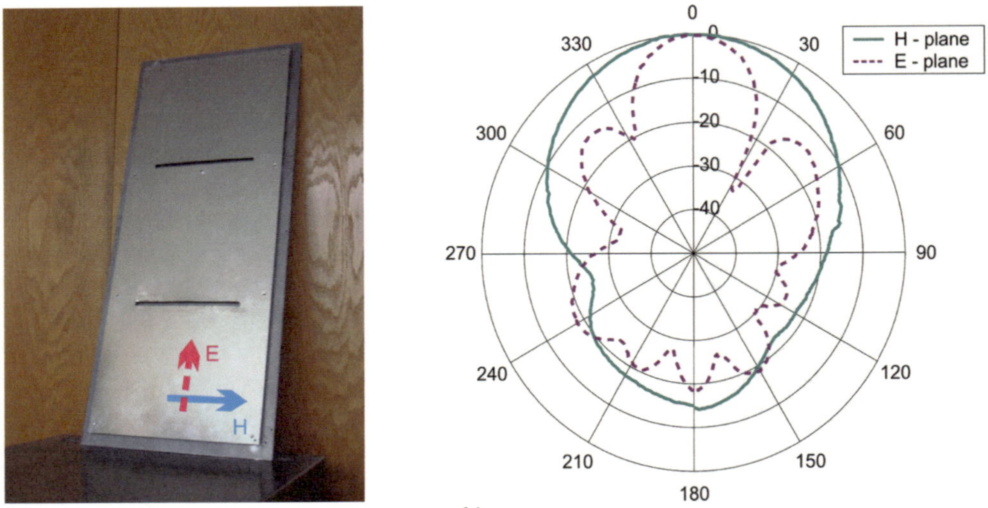

a) b)

Fig. 17. Photograph a) and measured radiation patterns b) of the new reader antenna

The plot indicates that the optimum tilt is $\psi = 30^{\circ}$. Higher ψ values result in a steep P_{rTAG} decline in the $3 \leq d \leq 4$ m range, whereas a lower ψ value provides a low TAG input power in an important region $d < 4$ m close to the gate, where a smaller influence of the shadowing of TAGs by neighboring sportsmen can be expected. The difference between the maximum receiver power on the axis ($p = 0$ m) and off the axis ($p = 2.5$ m) is about 4 dB. The $\psi = 30^{\circ}$ tilt was used for both other measurements and practical identification tests.

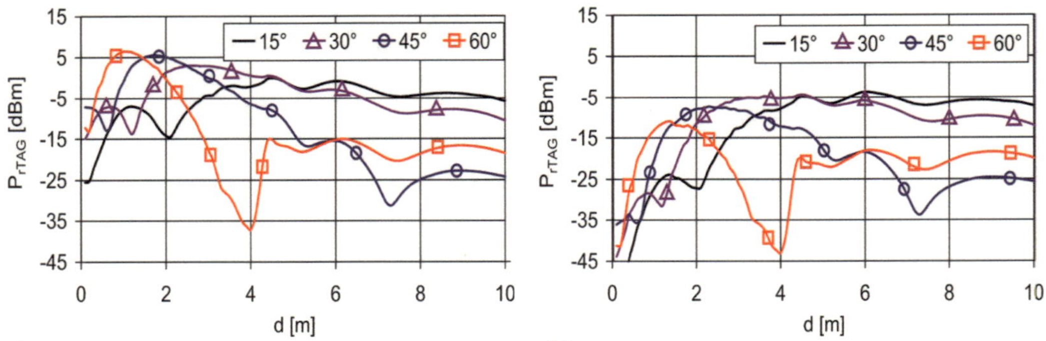

a) b)

Fig. 18. Simulation of the TAG input power P_{rTAG} versus distance d for different tilt ψ of the reader antenna ($P_t = 35.4$ dBm, $h_1 = 3$ m, $h_2 = 1.3$ m) on the axis ($p = 0$ m) and off the axis ($p = 2.5$ m)

3.4.4 Influence of tilt of the TAG antenna

One possible additional loss can be caused by more or less random tilt of the TAG antenna caused by the natural tilt of a human body. It can be especially significant in the case of people running. These additional link losses were investigated by means of practical measurements in the arrangement according to 3.4.1.

Fig. 19 shows the measured P_{rTAG} values as a function of the distance d from the gate (on the gate axis $p = 0$) and tilt φ of the person. The presented data indicates that the tilt $\varphi > 0$ can result in significant additional loss, especially in the ranges $d \leq 2$ m and $d > 7$ m. Nevertheless, in both of these regions, the simulated and measured P_{rTAG} values are below the $P_{rTAGmin}$ value even for $\varphi = 0$ (an erected person), and the identification is unlikely to be performed here. And on the contrary, the proper identification can be expected in the $2 \leq d \leq 7$ m range, where the influence of the person's tilt φ is relatively small and the additional loss caused by the tilt usually does not exceed 3 dB.

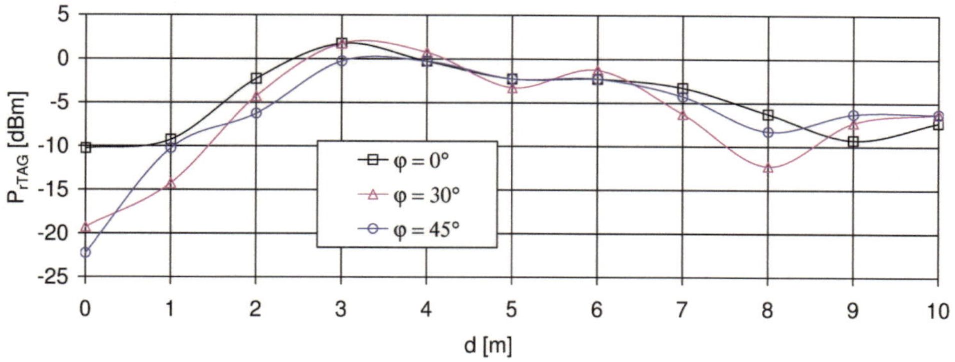

Fig. 19. Influence of tilt of a runner φ on the received TAG power P_{rTAG} (measurement, $P_t = 35.4$ dBm, $h_1 = 3$ m, $h_2 = 1.3$ m, $\psi = 30°$, $p = 0$ m)

3.4.5 Influence of mutual shadowing

If several people gather in a small area in an identification area, it may result in mutual shadowing and consequent additional path-loss. In practice, any configuration of people can appear. In order to get at least the basic information about this potential signal fading, a set of relatively simple measurements was performed. The P_{rTAG} values were measured at all positions described in 3.4.1 but one person was always standing erectly at a distance of 1 m in front of a person wearing the measurement antenna. The results of these measurements are presented in Fig. 20. As expected, the additional shadowing loss is strongly dependent on the distance d from the gate. Within the range $d \leq 3$ m, the influence of the shadowing is negligible. In the range between $3 \leq d \leq 5$ m, this shadowing loss is up to 3 dB, for $d > 5$ m the shadowing loss is around 6 dB.

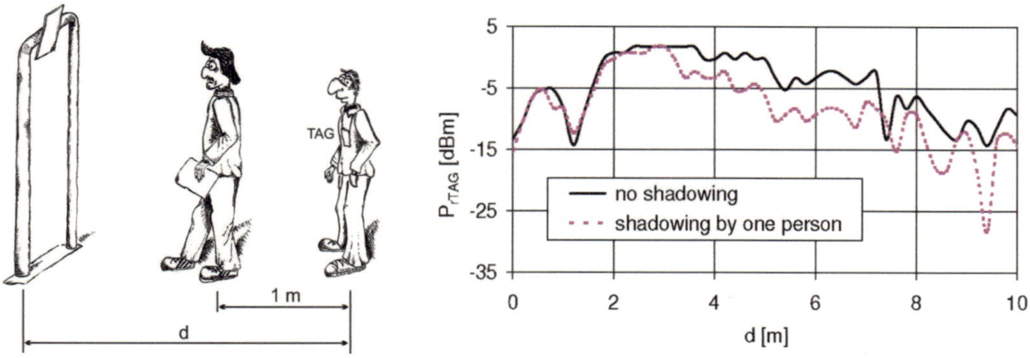

Fig. 20. Influence of shadowing on the TAG input power PrTAG (measurement, $P_t = 35.4$ dBm, $h_1 = 3$ m, $h_2 = 1.3$ m, $\psi = 30°$, $p = 0$ m)

3.4.6 Power budgets of the optimized system

The following figures show the P_{rTAG} and $P_{rREADER}$ plots as a function of distance d from the gate and p from the gate axis. To see the potential of the "tuning" procedure, the figures include the same dependencies measured with a standard meander dipole antenna (Polívka et al., 2006) fixed on a test person at a distance $b = 20$ mm without a screening metallic plate. Fig. 21 shows the plot of the simulated and measured TAG input power P_{rTAG} at the gate axis $p = 0$. Fig. 22 shows the simulated and measured P_{rTAG} values at the offset $p = 2.5$ m. Both figures include the corresponding $P_{rTAGmin}$ sensitivity value and compare the results obtained from the new RFID patch and the standard meander dipole. With the meander dipole antenna used, the P_{rTAG} values are above the $P_{rTAGmin}$ for $p = 0$ m, but with only 3 dB backup. For $p = 2.5$ m, P_{rTAG} is significantly below $P_{rTAGmin}$. That is why a non optimized system can only work close to the gate axis and with low identification reliability; see Table 4. The optimized system show min. 6 dB backup even at the $p = 2.5$ m offset. That is why its identification reliability is 100 % in the entire required identification area and even under worst expected conditions; see Table 4.

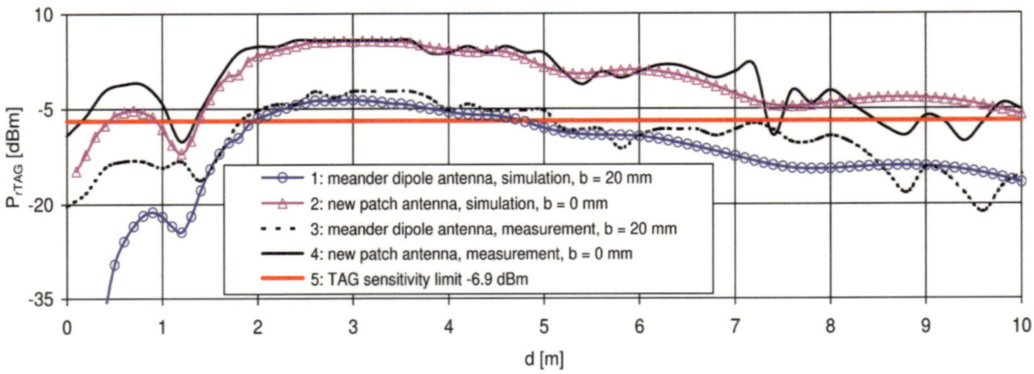

Fig. 21. Simulated and measured TAG input power P_{rTAG} on the reader-TAG trace ($P_t = 35.4$ dBm, $h_1 = 3$ m, $h_2 = 1.3$ m, $\psi = 30°$, $p = 0$ m)

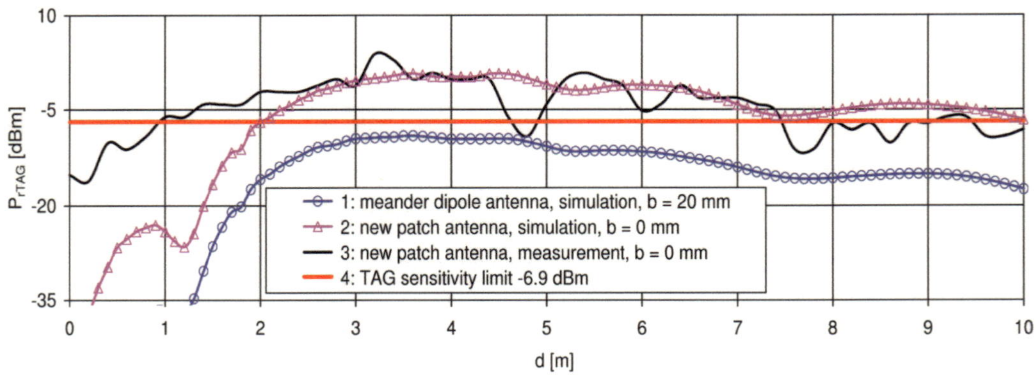

Fig. 22. Simulated and measured TAG input power P_{rTAG} on the reader-TAG trace ($P_t = 35.4$ dBm, $h_1 = 3$ m, $h_2 = 1.3$ m, $\psi = 30°$, $p = 2.5$ m)

Fig. 23 shows the simulated reader input power $P_{rREADER}$ values with respect to the reader sensitivity $P_{rREADERmin}$. Use of the standard meander dipole antenna leads to unacceptably low $P_{rREADER}$ values, especially off-axis. Employment of the RFID patch, optimized for operation on a human body, can guarantee high enough power even in this return TAG-reader link.

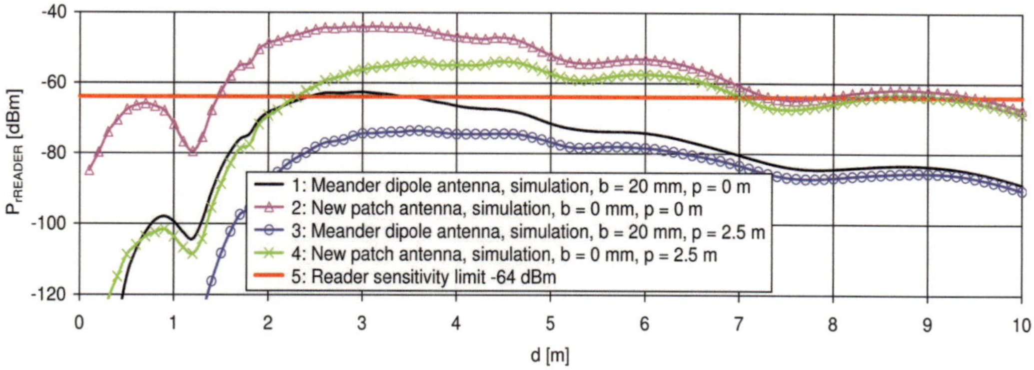

Fig. 23. Simulated reader input power $P_{rREADER}$ on the TAG-reader trace (P_t = 35.4 dBm, h_1 = 3 m, h_2 = 1.3 m, ψ = 30°)

4. Testing of RFID systems

According to 3.4 f) it is always necessary to perform practical testing of the optimized RFID system. As examples, testing of two RFID systems optimized for different applications are presented. The first system was optimized for identification of sportsmen; see Section 3.4. The second system was tuned for identification of employees in buildings and factory areas. Both arrangements employ the same reader and TAG chips.

4.1 Description of the RFID reader and chips

For the identification tests, a commercial RFID system (Trolleyscan Ltd, 2006) operating in the 869 MHz band was used. The main parameters of the system can be seen in Table 3.

System components	Parameter	Values
	Operating frequency (Europe)	869.5 - 869.7 MHz
Reader	Transmitted power	24.7 dBm - 36.0 dBm
	Receiver sensitivity	-64 dBm (200 pW)
	Identification rate	70 s^{-1}
Chip	Sensitivity	-6.9 dBm (200 μW)
	Impedance (measured value)	76 - j340 Ω
	Conversion loss	approx. 20 dB

Table 3. Standard UHF RFID system parameters, from (Trolleyscan Ltd, 2006)

4.2 Identification of sportsmen

In order to verify identification reliability of the optimized system, tests simulating real RFID system applications were performed. A group of racers moved on an asphalt surface in several different formations, see Fig. 24. In the first formation, 7 racers moved in a row, in the second formation 7 racers formed a kind of a matrix. Each formation moved with 3 different speeds simulating a walk (approx. 4 km/h), a fast walk (approx. 8 km/h) and a run (approx. 15 km/h). Each test was repeated 3 times. For comparison, tests were performed using both the RFID patch and standard planar dipole TAG antennas fixed on thick foam spacer (b = 20 mm). The results of performed tests are presented in Table 4.

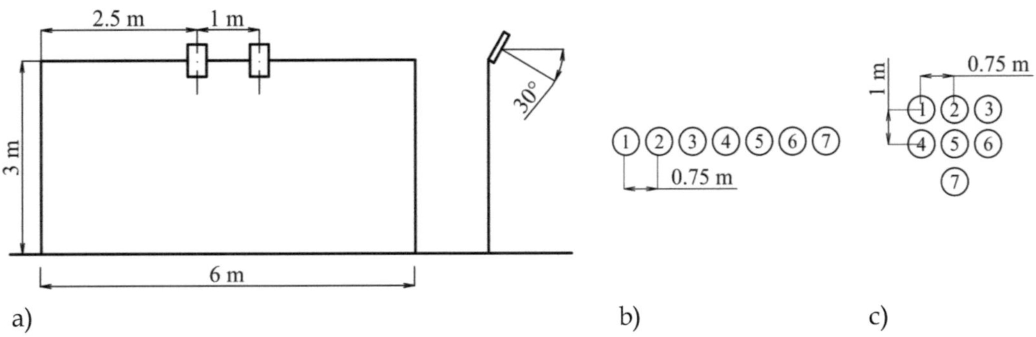

Fig. 24. Antenna gate dimensions a) and basic formations of racers used for testing of the optimized RFID system in a b) row, c) matrix

Configuration of racers	Speed of racers	Percentage of correct identification	
		Dipole antenna	RFID patch ant.
row	walk	66.7 %	100 %
	fast walk	52.4 %	100 %
	run	-	100 %
matrix p = 2.5 m	walk	61.7 %	100 %
	fast walk	52.4 %	100 %
	run	-	100 %
matrix p = 0 m	walk	85.7 %	100 %

Table 4. Reliability of identification of racers in open area obtained with dipole and the RFID patch TAG antennas

The presented results show that the optimization of the RFID system according to Section 3.4 can guarantee 100 % identification reliability, even under the worst expected conditions (shadowing tilt).

4.3 Identification of employees

The second test scenario was focused on identifying employees in buildings and factories; see (Polívka et al., 2008b). Identification tests were performed both in corridors and open areas in front of a building; see Fig. 25. Badge-type dual-loop TAG antennas, see Section 2.3, fixed on a person's chest and smaller 8 dBi reader antennas were used.

During these identification tests, values of the maximum identification distance d_{max} were measured. In order to enable comparison of the measured and expected d_{max} values, computer simulations of $P_{rTAG} = f(d)$ and $P_{rREADER} = f(d)$ were added; see Fig. 26 and Fig. 27. The d_{max} values are defined here as the most distant point where (7) is fulfilled.

a) b) c)

Fig. 25. Photograph of person with chest-fixed TAG in a) an open area and b) a corridor c) detail of TAG

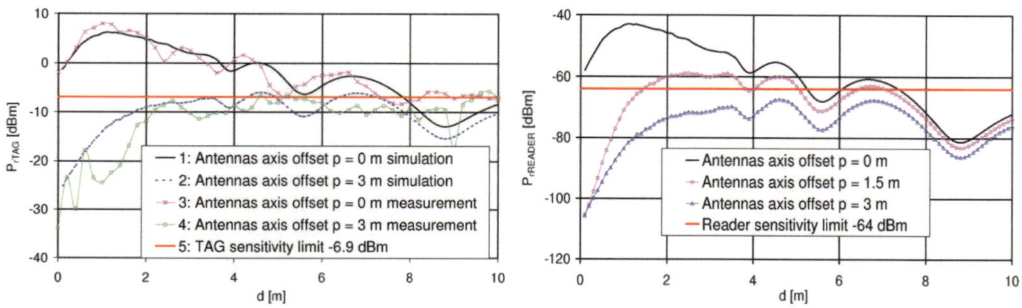

a) b)

Fig. 26. Simulation of the received power P_{rTAG} (a) and $P_{rREADER}$ (b) versus distance d ($P_t = 35.4$ dBm, $h_1 = 2.5$ m, $h_2 = 1.25$ m, $\psi = 30°$).

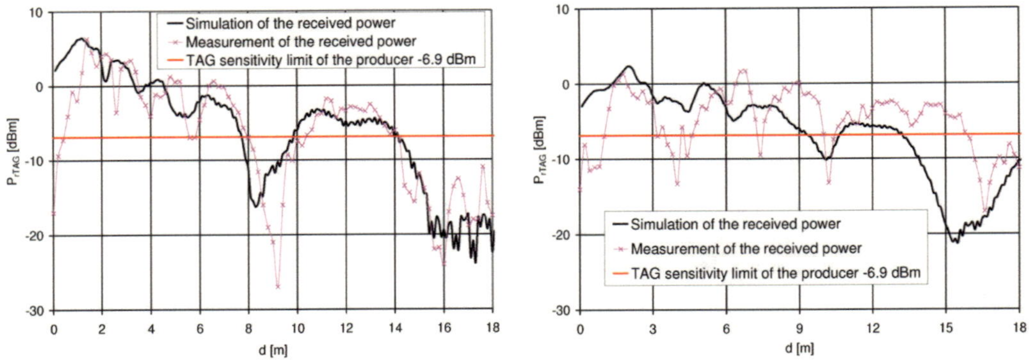

a) b)

Fig. 27. Simulated (3D ray tracing) and measured received power P_{rTAG} versus base antenna distance (P_t = 35.4 dBm, h_1 = 2.5 m, h_2 = 1.25 m, ψ = 30°) in case of persons identification in corridor. (a) In the axis p = 0 m, (b) in the antennas axis offset p = 1.8 m

Position of TAG antenna	Reader and TAG antenna axis offset - p [m]	Max. identification distance - d_{max} [m]
Person's chest, open area	0	9
	1	6.5
	2	4
	3	unreliable id.
Person's chest, 4 m wide corridor	0	9
	1	8
	1.8	8

Table 5. Maximum identification distance of test configurations

The measured d_{max} values are presented in Table 5. The agreement between the measured and simulated values is very good.

5. Conclusion

Reliable RF identification of people in moderate 2 – 10 m ranges must be based on electromagnetic wave coupling mechanism. Due to acceptable antenna dimensions, UHF or microwave operational frequencies must be used. At these frequencies, propagation of electromagnetic waves is influenced by several important physical phenomena, namely by interferences, shadowing or waveguide effects. Besides these, in this application, the functionality of TAG antennas can be negatively influenced by the presence of a nearby human body. The human body is able to detune definite antenna structures and to absorb a substantial part of the radiated or received signal power. All these effects can lead to the wrong or even no identification.

In order to guarantee reliable identification, the whole UHF RFID system must fulfil power budget conditions, both in the reader-TAG and TAG-reader radio paths. These conditions compare power levels received by the TAG and reader antennas with reader and TAG sensitivities. Primarily, power budgets are influenced by radiated power, gains of all antennas and by propagation loss. Beside these, several additional random attenuations of electromagnetic waves must be taken into account. Above all this concerns the tilt of the TAG antenna caused by the possible tilt of the person to be identified and by shadowing among persons in a group in an identification area. These additional random losses should be measured and added into power budget calculations as necessary reserves.

In order to reliably fulfil both power budget conditions, each RFID system intended for identification of people should be "tuned". Above all, this concerns both TAG and reader antennas. The TAG antennas must be as lightweight and small as possible, and must especially provide high immunity against the influence of a nearby lossy dielectric. Employment of antenna structures with metallic planes can be recommended. The metallic planes can form a mere screening or can form inherent parts of the antennas. In case of reader antennas, optimization of their beam shapes and tilt can also be very beneficial.

Two examples of RFID systems "tuned" for different purposes were presented. The first system was optimised for identification of runners in long-distance races. Initial tests using standard TAG and reader antennas provided unacceptably low identification probability (around 50 %). Designing new TAG and reader antennas and changing the system arrangements led to a 100 % identification probability in the whole required identification area under the expected rugged conditions. Similar power budget calculations and optimization steps were performed in the second system intended for identification of personnel in buildings or areas. The optimization was focused on designing a new TAG antenna with dimensions comparable with a standard photo and name bearing identification badge. Practical tests show, that this system is able to identify people walking in 4 m wide corridors or 5 m wide strips in open areas, in both cases the maximum available identification range is around 9 m. A suitable combination of open-area and corridor readers can cover the majority of standard buildings or nearby areas. The UHF RFID system can help with the organization of mass events or provide security information and services in many offices, warehouses or factories. All that without much inconvenience to personnel or time delays.

6. Acknowledgement

This work has been conducted in the Department of Electromagnetic Field at the Czech Technical University in Prague and supported and by two projects of the Grant Agency of the Czech Republic No. 102/08/1282 "Artificial electromagnetic structures for miniaturization of high-frequency and microwave radiation and circuit elements", and further by the Czech Ministry of Education, Youth and Sports in the frame of the Research

Project in the Area of the Prospective Information and Navigation Technologies MSM 6840770014, and by the research program "Research of Methods and Systems for Measurement of Physical Quantities and Measured Data Processing" MSMT6840770015, and by the COST project IC0603 "Antenna Systems & Sensors for Information Society Technologies".

7. References

Balanis, C. A. (1997). *Antenna Theory, Analysis and Design*, second edition, John Wiley & Sons, ISBN 0-471-59268-4, New York

Best, S. (2004). Improving performance properties of a dipole element closely spaced to a PEC ground plane. *IEEE Antennas and Wireless Propagation Letters*, Vol. 3, 2004, pp. 359-363, ISSN 1536-1225

Dobkin, D. M. & Weigand, S. M. (2005). Environmental effects on RFID TAG antennas, *Proceedings of IEEE Antennas and Propagation Society International Symposium 2005*, ISBN 0-7803-8845-3, Washington, USA, July 2005

Finkenzeller, K. (2003). *RFID Handbook: Fundamentals and Applications in Contactless Smart Cards and Identification*, 2nd edition, John Wiley & Sons, ISBN 0-470-844402-7, Chichester

Foster, P. R. & Burberry, R,. A. (1999). Antenna problems in RFID systems, *Proceedings of IEE Colloquium RFID Technology*, pp. 31-35, London, UK, October 1999

Gautherie, M (1990). *Biological Basis of Oncologic Thermotherapy*, Springer-Verlag, Conference, Berlin

Griffin, J. D.; Durgin, G. D.; Haldi,A. & Kippelen, B. (2006). RF TAG antenna performance on various materials using radio link budgets. *IEEE Antennas and Wireless Propagation Letters*, Vol. 5, 2006, pp. 247-250, ISSN 1536-1225

Hansen, R. C. (2006). *Electrically Small, Superdirective, and Superconductive Antennas*, John Wiley & Sons., ISBN 0-471-78255-6, New York

Lee, K. F. & Chen, W. (1997). *Advances in Microstrip and Printed Antennas*, John Wiley & Sons, ISBN: 978-0-471-04421-5, New York

Noguchi, K.; Syouji, H.; Mizusawa, M.; Yamaguti, T. & Okumura, Y. (1997). Impedance characteristics of small meander line antenna. *Technical report of IEICE*, pp 97-55, 1997

Polívka, M.; Holub, A. & Mazánek, M. (2005). Collinear Microstrip Patch Antenna. *Radioengineering*, Vol. 14, No. 4, 2005, p. 40-42. ISSN 1210-2512

Polívka, M.; Švanda, M. & Hudec, P. (2006). Analysis and Measurement of the RFID System Adapted for Identification of Moving Objects., *Proceedings of the 36th European Microwave Conference* [CD-ROM], pp. 729-732, ISBN 2-9600551-6-0, Manchester, October 2006, Piscataway: IEEE

Polívka, M.; Švanda, M. & Černý, P. (2008a). Multiple-Arm Folded Monopole Antenna Operating Extremely Close to a Conductive Plane, *Proceeding of COMITE 2008*, pp. 61-65, ISBN 978-1-4244-2137-4, Prague, May 2008, Czechoslovakia Section IEEE, Prague

Polívka, M.; Švanda, M. & Hudec, P. (2008b). UHF radiofrequency identification of persons in buildings and open areas. Submitted to *Transaction on Microwave Theory and Technique, Special Issue on Hardware and Integration Challenges of RFID's*, ISSN 0018-9480

Ranasinghe, D. C., Hall, D. M., Cole, P. H., Engels, D. W. (2004). An embedded UHF RFID label antenna for TAGging metallic objects, *Proceeding of Inteligent Sensors, sensor Networks and Information Processing Conference*, pp. 343 - 347, ISBN 0-7803-8894-1, December 2004

Raumonen, P., et al. (2003). Folded dipole antenna near metal plate, *Proceedings of IEEE Antennas and Propagation Society International Symposium 2003*, pp. 848-851, ISBN 0-7803-7846-6, June 2003

Sidén, J.; Nilsson, H.-E., Koptyug, A. &, Olsson, T., (2006). A Distanced RFID dipole for a metallic supply chain label, *Proceedings of IEEE Antennas and Propagation Society International Symposium 2006*, pp. 3229 - 3232, Albuquerque, New Mexico, July 2006

Sievenpiper, D. (1999). *High-impedance electromagnetic surfaces*, Ph.D. dissertation, Department of Electrical Engineering, University of California at Los Angeles, Los Angeles

Švanda, M.; Polívka, M. & Hudec, P. (2007). Application of the UHF RFID system for the identification of sportsmen in mass races, In: *Proceedings of the European Microwave Association*. Vol. 3, No. 4 (December, 2007), pp. 295-301, Edizioni Plus - Universita di Pisa, ISBN 88-8492-324-7, Pisa

Švanda, M. & Polívka, M. (2007). Dualband wearable UHF RFID antenna, *Proceeding of the 2nd European Conference on Antennas and Propagation* [CD-ROM], pp. 1-5, ISBN 978-0-86341-842-6, Edinburgh, October 2007, Stevenage, Herts: The Institution of Engineering and Technology (IET), Edinburgh

Švanda, M. & Polívka, M. (2008). Novel dual-loop antenna placed over patch array surface for UHF RFID of dielectric and metallic objects. Submitted to *Microwave and Optical Technology Letters*, ISSN 0895-2477

Trolleyscan, Ltd. (2006). http://www.trolleyscan.com/, 5. 6. 2006

Ukkonen, L. & Kivikoski, P. R., (2003). Challenges in the development of TAG antennas for passive RFID of metallic objects, *Proceedings of IV Finnish Wireless Communications Workshop*, October 2003

Ukkonen, L.; Engles, D. ; Sydanheimo, L. & Kivikoski, M. (2004). Planar wire-type inverted-F RFID TAG antenna mountable on metallic objects, *Proceeding of IEEE Antennas and Propagation Society International Symposium 2004*, pp. 101-104, Monterey, California, June 2004

Ukkonen, L.; Sydänheimo, L. & Kivikoski, P. R. (2005). Effects of metallic plate size on the performance of microstrip patch-type TAG antennas for passive RFID. *IEEE Antennas and Wireless Propagation Letters*, Vol. 4, 2005, pp. 410 - 413, ISSN 1536 - 1225

Wheeler, H. A. (1959). The radian sphere around a small antenna. *Proceedings of IRE*, (August 1959), pp. 1325 -1331

Zhang, Y.; Hagen, J. et al. (2003). Planar artificial magnetic conductors and patch antennas. *IEEE Transactions on Antennas and Propagation*, Vol. 51, No. 10, (October 2003), pp. 2704-2712, ISSN 0018-926X

Zvánovec, S., Pechač P. & Klepal M. (2003). Wireless LAN Networks Design: Site Survey or Propagation Modeling?. *Radioengineering*, Vol. 12, No. 4, (December 2003), pp. 42-49, ISSN 1210-2512

5

UHF Tags for Sensing Applications

Gaetano Marrocco
University of Roma Tor Vergata
Italy

1. Introduction

The recent advances in Wireless Sensor Networks (WSNs) (Kong & Kumar, 2003) for applications in mobile and time-varying environments are generating a growing attention to low-cost and low-power wireless nodes, equipped with radio/sensing ability, which are spatially distributed to ensure a cooperative monitoring of physical or application-specific conditions and parameters. Typical fields of applications for WSNs include environmental and habitat monitoring, disaster relief (Lorincz et al., 2004), health care, inventory tracking and industrial processing monitoring, security and military surveillance, smart spaces applications. A novel technological trend is the integration among wireless sensor networks and Radio Frequency IDentification (RFID) technologies. Such a convergence of sensing and identification features may enable a wide range of heterogeneous applications which demand a tight synergy between detection and tagging.

A new frontier is the wireless monitoring of people within Mobile Healthcare Services (Cheng-Ju et al., 2004) with the purpose to reduce the hospitalization of patients, to support disaster relief or to get an epidemic under control. An RFID system could provide real-time bio-monitoring and localization of patients inside hospitals or domestic environments, as well as in extreme conditions like a Space Capsule. In these cases the tag should be placed on the human body and equipped with bio-sensors (temperature, blood pressure, glucose content, motion) and, when activated by the reader, tag ID and bio-signals could be transferred to a remote units and then stored and processed.

Up to date, several approaches have been proposed to provide RFID devices with enhanced sensing and detection capabilities. The main solutions make use of active or passive RFID transponders and Surface Acoustic Wave (SAW) devices (Reindl et al., 2001). A significant example of enhanced passive RFID system is given by the Wireless Identification Sensing Platform (WISP) project (Sample et al., 2007) which introduced the concept of ID modulation making use of inertial switches and enhanced power harvesting units to mechanically toggle between two commercial RFID integrated circuits.

These devices could be *passive*, harvesting energy from the interrogating system, *semi-active* when a battery is included only to feed the sensors, or fully *active* where a local source directly feeds a microcontroller beside the transmitting radio. However, the large battery packs required for active techniques, in addition to the use of protruding antennas, could be suboptimal for some applications and additional issues have to be considered such as the compromise between a long battery-life and a miniaturized design.

In passive RFIDs, together with the microchip sensitivity, the tag antenna plays a key role in the overall system performance, such as the reading range and the compatibility with the tagged object. In case of RFID with sensing capability, the antenna should be additionally suited to electrical and physical integration with sensing electronics. Moreover, it is well experienced that the RFID system's performances are greatly dependent on the kind of object where the tag is attached on. When a same tag is placed onto different targets, the tag antenna's input impedance may in some case undergo a mismatch and may produce a change of the read distance. Conventional general-purpose tags are designed in free space, but when they are needed to be applied over objects having high values of the permittivity, as in the case of liquids and humans, the strong pattern distortion and the efficiency loss caused by energy dissipation and scattering, have to be taken into account in the first stage of the design.

Within this scenario, this chapter has a twofold purpose: the description of a new class of UHF tag layouts suited to host sensors and to be attached onto high dielectric and lossy targets, such as the human body, and to introduce the novel concept of *self-sensing* tags wherein the antenna itself acts both as a data transmitter and as a sensing device of some tagged body's features. These two arguments are both related to the modelling and handling of *dense* objects, and the *self-sensing* idea originates just from one of the main conventional drawbacks of UHF RFID, e.g. the dependence of reading performances on the dielectric value of the tagged object. The self-sensing tags are multi-chip antennas (multi-port system) exploiting the dependence of the tag's input impedance and radar cross-section on the physical and geometrical features of a real target.

The two subjects are here described both theoretically and corroborated by several preliminary prototypes and experimentations.

2. Basic definitions for RFID systems

At the beginning of the reader-to-tag communication protocol (Nikitin & Rao, 2006), the reader first *activates* the tag, placed over a target object, by sending a continuous wave which, on charging an internal capacitor, provides the required energy to perform actions. During this *listening mode*, the microchip (IC) exhibits an input impedance $Z_{chip} = R_{chip} + jX_{chip}$, with X_{chip} capacitive, and the antenna impedance $Z_A = R_A + jX_A$ should be matched to Z_{chip} ($Z_A = Z^*_{chip}$) for maximum power transfer. The maximum fraction $P_{R\to T}$ of the reader input power that is absorbed by the tag is

$$P_{R\to T} = \left(\frac{\lambda_0}{4\pi d}\right)^2 G_R \tau G_T P_{in} \tag{1}$$

$$\tau = \frac{4R_{chip}R_A}{\left|Z_{chip} + Z_A\right|^2} \tag{2}$$

where λ_0 is the free-space wavelength, d is the reader-tag distance, G_R is the gain of the reader antenna and G_T is the gain of the tag over the target, having assumed polarization-matched antennas aligned for maximum directional radiation. τ is the power transmission coefficient. The tag is activated when the absorbed power exceeds the tag's microchip sensitivity threshold: $P_{R\to T} > p_T$. Having fixed the effective power ($EIRP_R = G_R P_{in}$) transmitted by the reader, the tag antenna gain (G_{tag}) and the sensitivity (P_{chip}) of the tag microchip, e.g. the RF power required to the microchip electronics to turn on and complete its tasks, the maximum activation distance of the tag along the (θ,ϕ) direction is therefore given (Nikitin & Rao, 2006) by

$$d_{\max}(\theta,\phi) = \frac{c}{4\pi f}\sqrt{\frac{EIRP_R}{P_{chip}}\tau G_{tag}(\theta,\phi)} \tag{3}$$

During the next steps of the communication, the tag receives the command coming from the reader and it finally sends back the data through a back-scattered modulation of the continuous wave coming from the reader itself. The tag's IC acts as a programmable switching device which toggles between a low or high modulation impedances Z_{mod}. During the data transfer, the RFID system can be considered as a monostatic radar and therefore it can be characterized by the radar range equation which, for the case of typical RFID tags, can be expressed in the form

$$\frac{P_{R\leftarrow T}(d)}{P_{in}} = \left(\frac{\lambda_0}{4\pi d}\right)^4 G_R^2 G_T^2 \rho \tag{4}$$

$$\rho = \frac{4R_A^2}{\left|Z_{\mathrm{mod}} + Z_A\right|^2} \tag{5}$$

where $P_{R\leftarrow T}$ is the power received back by the reader and ρ is a modulation parameter related to the tag's radar cross section.

The maximum transmitted power allowed to the reader is constrained to local regulations. In Europe the relevant standards for UHF RFID applications are the ETSI EN330-220 and Draft TESI EN302 208-2. In particular within the 865.6-867.6MHz the maximum EIRP is 3.2W, which overcomes the previous limit 0.8W. In the U.S.A. the FCC allowed band is 902-928MHz with maximum transmitted EIRP=4W.

Microchip power activation threshold is continuously improving, reducing from 1mW in the year 2001 to some microwatts in today products or even less in the state of the art ASICS (Curty et al., 2005).

It is easy to show from equation (3) that antennas with averaged realized gain ($G_{tag}\tau$) not less than -10dB (when placed over the human body) could be in principle compatible with reading distances of the order of 5m if the microchip sensitivity is less than 10μW.

3. Energetic constraints for on-body RFID applications

The presence of the tagged object, and in particular of the human body, with its high permittivity and conductivity, will favour the antenna miniaturization but nevertheless will induce a strong power absorption in body tissues. The antenna gain, and hence the link distance, will be sensibly reduced with respect to the free space. Additionally, when the human body is exposed to electromagnetic field radiated by the reader, the power budget has to comply with the Safety standards, typically expressed in the UHF band in terms of maximum power absorbed by human tissues. The reference quantity is the Specific Absorption Rate (SAR), measured in [W/Kg], e.g. the power absorbed by the mass unit:

$$SAR(\underline{r}) = \frac{1}{2\rho_{mass}(\underline{r})}\sigma(\underline{r})\,|\,E(\underline{r})\,|^2 \tag{6}$$

where ρ_{mass} and σ are the mass-density and the electric conductivity, respectively, at point $\underline{r}$ of the body. Since the strength of electromagnetic field emitted by the reader decreases along with the distance, a *safety distance* may be introduced for the reader (having fixed the radiated power) or, from a different point of view, a *safety EIRP* when the distance is fixed. The useful link region is therefore theoretically bounded by the safety distance and the maximum range allowed by the transmitted EIRP.
To calculate the safety distance, the SAR has to be estimated by using computer simulations considering a realistic morphologic model of the human body and a simplified, but very general model of the emitted field from the reader. Although the data link could not be necessarily continuous, it is nevertheless useful to consider the extreme case of continuous interrogation, giving the maximum SAR over the period of the wave.

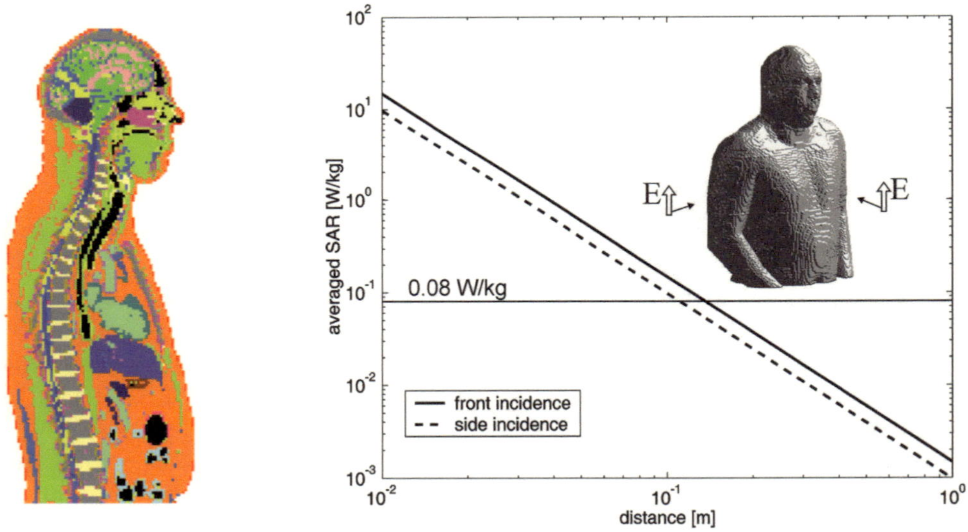

Fig. 1. Absorbed power density (SAR) in the realistic human body model exposed to a vertically polarized plane-wave coming from the reader and carrying a power 3.2W EIRP.

The voxel-based human body model used in the computer simulation has been derived by the Visible Human project (Ackerman, 1999) which comprises the thorax and the head with more than forty different tissues (Fig. 1). Physical parameters (relative permittivity ε_r, conductivity σ and mass density ρ) at 900MHz are retrieved by the tissue database in (Gabriel et al., 1996). Under the hypothesis that the human body is placed in the reader's far field, a plane wave exposure is considered. The plane wave impinging the side and the thorax carries a power density $S^{in} = EIRP_R/(4\pi d^2)$ and then the plane wave strength delivering power absorption into the human body is

$$|E^{in}| = \sqrt{60EIRP}/d \quad (7)$$

The numerical calculation, and the required dosimetric processing were accomplished by means of the tool in (Marrocco & Bardati, 1999), based on the Finite-Difference Time-Domain (FDTD) method (Taflove, 1998).

d [m]	Imperturbed r.m.s. |E| [V/m]	Maximum SAR in the head [mW/kg]	Maximum SAR in the trunk [mW/kg]	Body averaged SAR [mW/kg]
0.1	98	331	197	148
0.2	49	84	49	37
0.5	20	13	8	6
1.0	10	3	2	1.5
2.0	5	1	1	<1
3.0	3	<1	<1	<1
4.0	2	<1	<1	<1

Table 1. Human-body front exposure to Reader's field at 900 MHz, having assumed an emitted power of 3.2W EIRP

Fig. 1 shows the SAR averaged over the whole model with respect to the tag-reader distance, for a transmitted power 3.2W EIRP and two different wave incidences (toward the thorax and toward the torso's side), while Table 1 gives the detailed values of the impinging field strength as well as of the maximum and averaged SAR. According to the European regulations, which are more restrictive than the USA ones, the maximum allowed averaged SAR in the body is 80 mW/Kg (CENELEC, EN50364). It is hence possible to observe that the smallest safety distance is about d_{min}=10cm. Even assuming a more conservative value d_{min} =50cm, this safety distance is fully compatible with the concept of remote monitoring.

4. UHF Slot-Tags for on-body application

Three tag geometries are here described together with the related design methodologies. These layouts are slot-type antennas suited to be easily integrated with sensors and additional electronics and useful for placement on high dielectric targets such as human-body districts (Fig. 2). In particular the first two antennas are useful to integrate sensors in close contact with the human body (or any other dielectric), as for the detection of pressure, glucose or temperature while the third layout is engineered for non contacting sensors, for

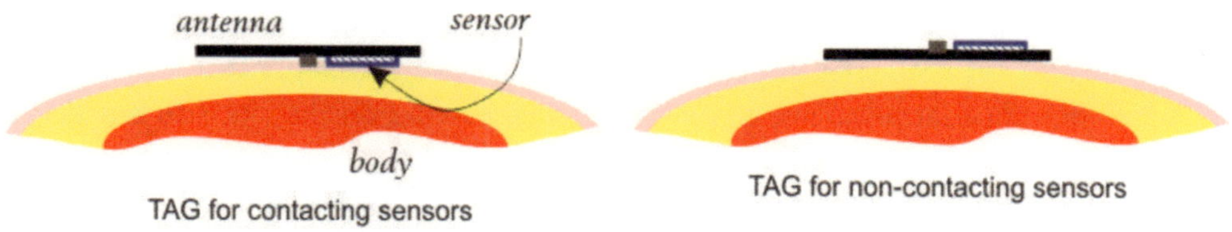

Fig. 2. Tag hosting sensor in close contact to the tagged body (temperature, pressure or chemical sensing) or shielded from it by the antenna itsself (motion-detection accelerometers).

instance accelerometers, and permits to achieve superior reading performance than the previous family.
All the presented tags are numerically modelled by the Finite-Difference Time-Domain solver as above, having considered the antenna placed onto realistic models of the tagged body.

4.1 Tags for contacting sensors

The first tag antenna family is a nested-slot suspended-patch (NSSP), (Marrocco, 2007). Small size slot antennas are naturally inductive and therefore appear more suited than dipoles to achieve conjugate impedance matching. The basic geometry is visible in Fig.3a and has been modelled as placed on a layered cylinder (Fig.3b) simulating the human torso (size and materials in Table 2). To prevent high losses on the skin, the antennas is attached onto the body through an insulating slab.

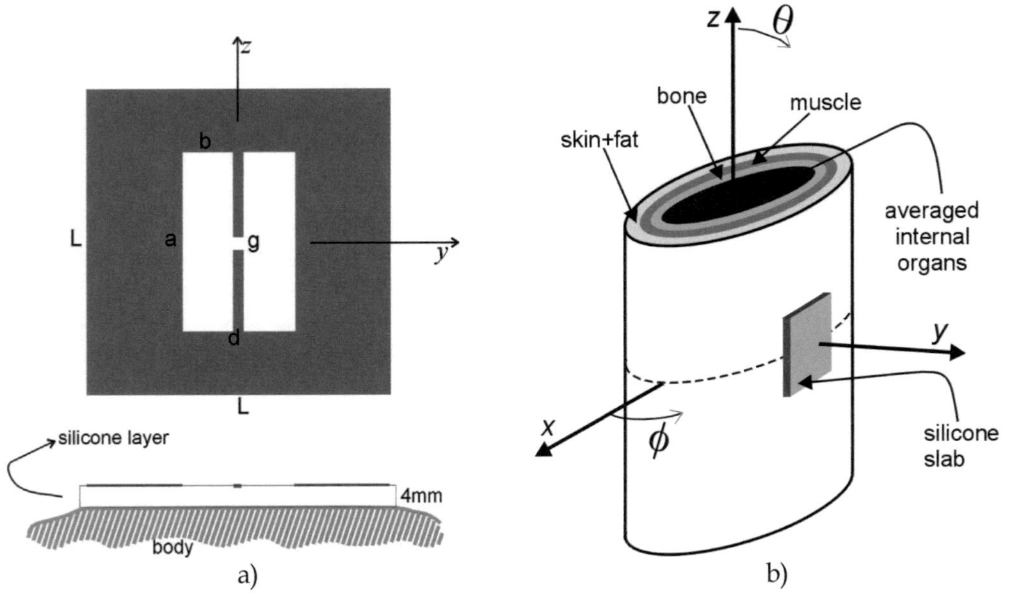

Fig. 3. a) NSSP tag: the microchip transmitter should be placed in the central gap of size *g* x *g*. b): Stratified elliptical cylinder model of the human thorax for the design of bio-compatible tags. The size of cylindrical cross-sections are reported in Table 2. Cylinder height: 40cm.

Since the slot sizes are comparable with the patch surface, the radiation features are related to both the objects. In particular, the maximum antenna gain is mainly fixed by the patch side *L*, while the impedance tuning can be changed by acting on the slot size *a* and *b*, as visible in the *Matching Chart* in Fig. 4a, which displays the iso-lines for antenna input resistance and reactance. Having chosen a particular microchip, this chart permits to design the H-slot shape factor such to achieve the conjugate impedance matching required for maximum reading distance.
Depending on the shape of the internal slot, the antenna mainly radiates either as a *dumbbell H*-slot or as a pair of rectangular loops sharing the sourced conductor (Fig. 4b).

Layer	ε_r	σ [S/m]	Ellipse axis *thin man* [cm]
Skin+fat	14.5	0.25	33.5 × 16.8
Muscle	55.1	0.93	31.0 × 14.2
Bone	20.8	0.33	28.4 × 10.5
Internal organs	52.1	0.91	27.2 × 8.4

Table 2. Physical and geometrical parameters of the layered anatomical model in Fig. 3 at 870MHz.

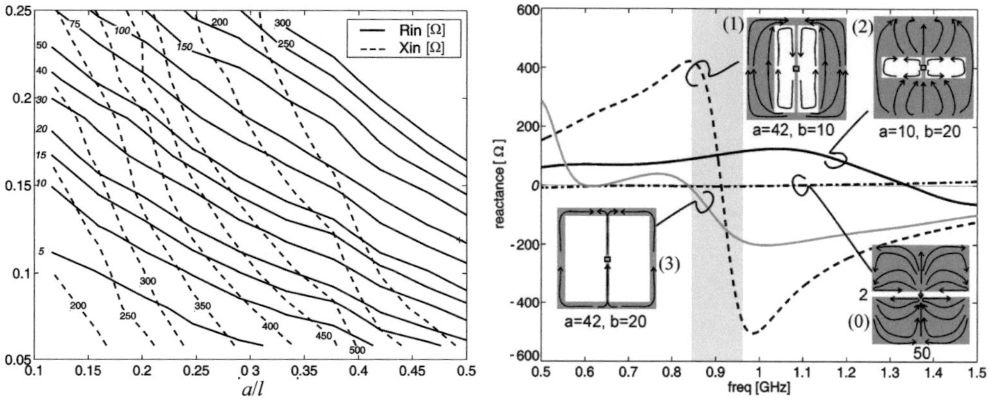

a) b)

Fig. 4. NSSP tag. a) Matching chart to design the H shape factor of the antenna such to match the particular microchip's impedance. b) Typical antenna input impedance for some choice of the H-slot parameters (in mm). In all the case the patch size is L=50mm.

Fig. 5 shows a fabricated prototype of the body-matched NSSP tag and the measurement set-up where the tag is attached onto a Perspex box (ε_r=2.7, σ=0) having 5 mm thickness and 20 cm width, filled with a muscle-type solution. The measured antenna input impedance and power matching, compared with simulations, are shown in Fig. 6. The experiment demonstrated a significant impedance tuning agility and a read region (Fig. 7) suited to small or even medium-size rooms.

The previously considered NSSP antennas are symmetric with respect to both the x and z axis. However, this geometry offers additional degrees of freedom in the position of the slot and in the connection to the microchip, provided that a larger number of slot discontinuities (Fig. 8a) are considered. This new layout is similar to a meandered slot and, when properly optimized, could permit to fulfill several electrical and geometrical constraints, such as the impedance matching to a particular microchip, dual-frequency operations, the embedding of a sensor of given size, and a stable response over a large variety of tagged dielectrics. The slot profile can be seen as a slot-line impedance transformer (Calabrese & Marrocco, 2008), where each discontinuity (tooth) provides energy storage and radiation. A Genetic Algorithm (Weile & Michielsen, 1997) optimization problem is hence formulated to shape the transformer layout, within input impedance and size requirements. As an example, Fig.8b shows the shape and the power transmission coefficient τ for some 870MHz slot-line antennas optimized to occupy only a fraction of the overall metallization. A preliminary experimental prototypes on FR4 has been fabricated and measured (Fig. 9). When compared with the NSSP tags, this layout permits to achieve a better gain and more space for the electronic payload.

Fig. 5. NSSP tag. Fabricated half--NSSP antenna in front of a Perspex cubic phantom filled with tissue-equivalent solution made of deionised water, saccharose and sodium chloride. The antenna and the box are placed over a 1m x 1m copper image plane simulating, by image effect,the other half structure.

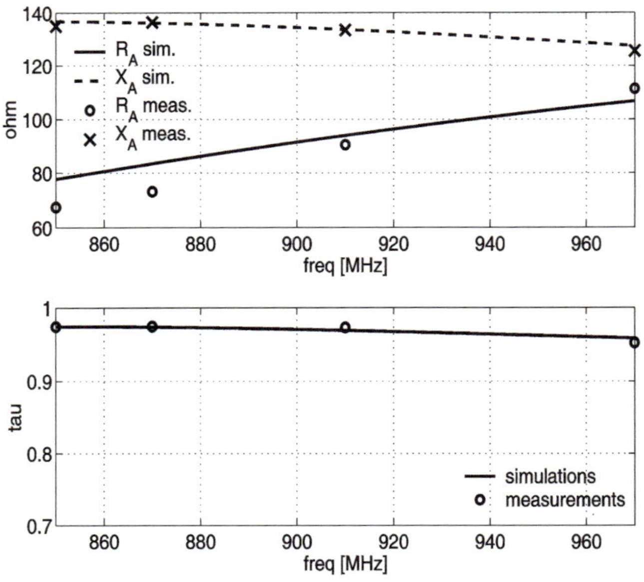

Fig. 6. NSSP tag. Measured and computer-estimated input impedance and power transmission factor τ.

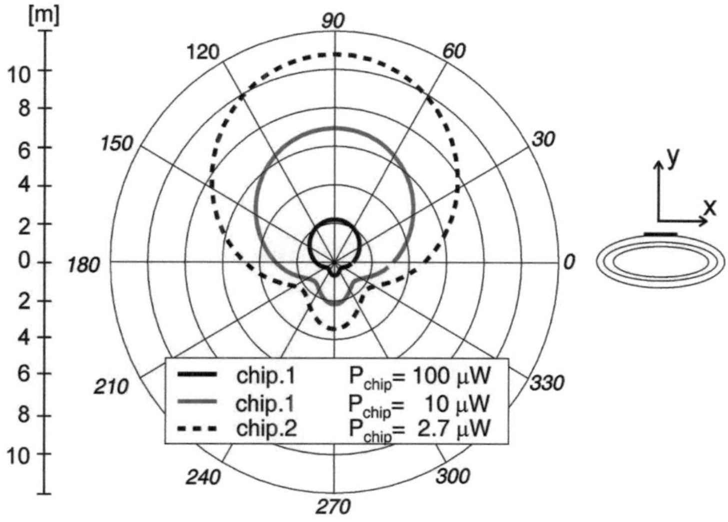

Fig. 7. NSSP tag. Estimated read distance for different kinds of microchip and 3.2 W EIRP emitted power.

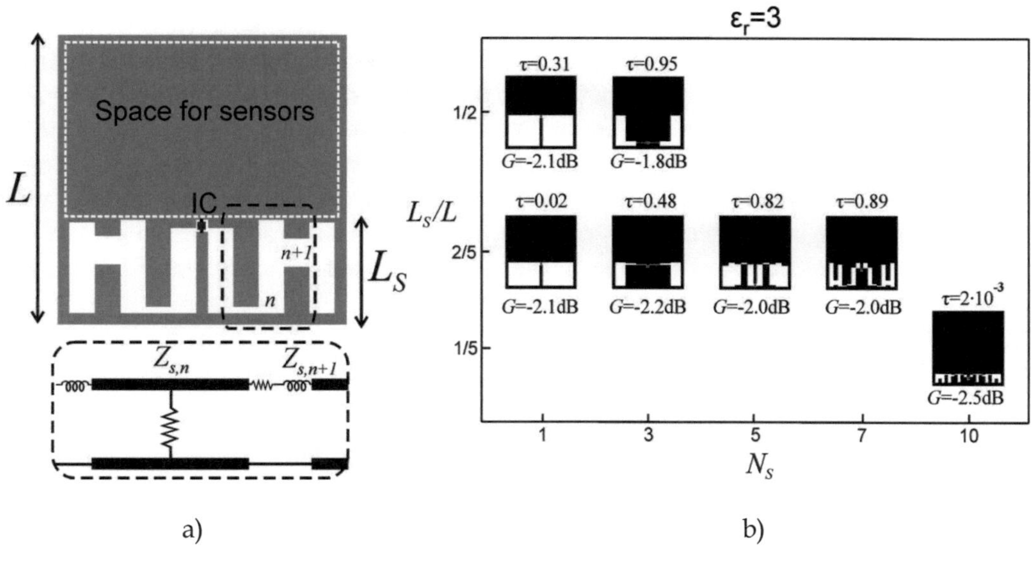

Fig. 8. Meandered Slot Antenna (MSA) tag. a) Layout and slot-line model. b) Examples with L=5cm, placed over a ε_r=3 dielectric half-space, which have been optimized for an IC with Z_{chip}=15-j450Ω, for different sizes L_s of the antenna region and for different number N_s of slot-line sections. A symmetric layout is assumed and therefore N_s represents half the overall slot transitions. G is the maximum gain in the air half-space.

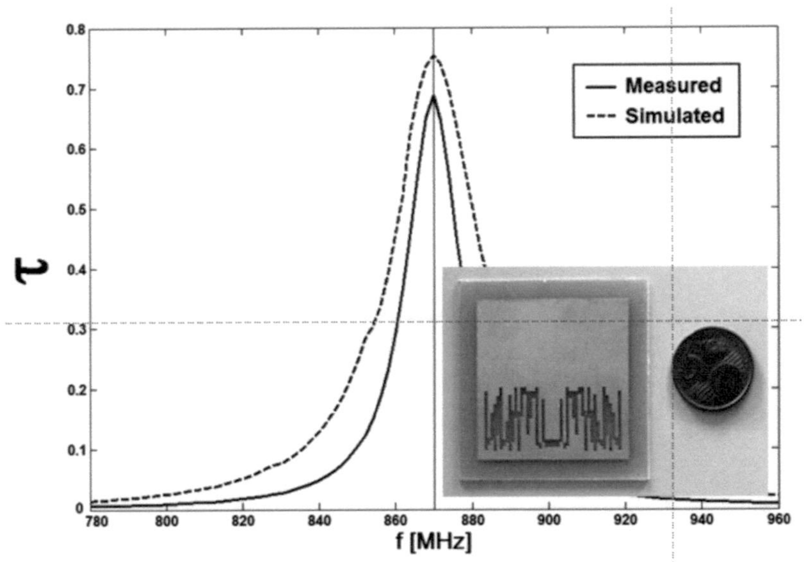

Fig. 9. MSA tag. Fabricated 5cm x 5cm prototype and in-air measurement of the power transmission coefficient.

4.2 Tag for non-contacting sensors

A further evolution of the slot-driven patch comprises an L-type patch folding (Fig. 10) with the purpose to increase the antenna radiation and in particular to reduce the power dissipation into the body district where the tag is placed. The folded region acts as a ground plane which partly isolates the antenna from the body. The radiation is now due to the H-slot itself, as in the previous layouts, but also to the current discontinuity in the folding and especially to the patch truncation. This configuration is referred to as *Slotted Clip Antenna* (SCA). When attached, for instance, onto a leg-like layered cylinder, this layout produces a larger gain than the NSSP and MSA tags, with maximum value of the order of 0dB with back radiation ranging within -10 ÷ -5dB. Fig. 11shows a fabricated prototype and the measured power transmission coefficient. The resulting read distance is sensibly improved in the front and at in the side regions of the antenna (Fig. 12).

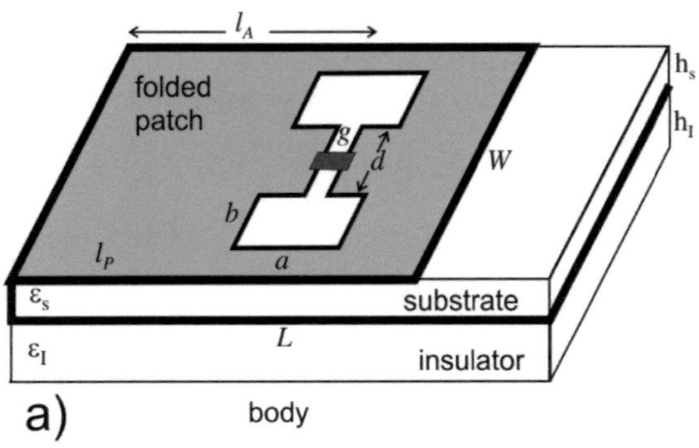

Fig. 10. SCA tag. Layout of inverted slot antenna.

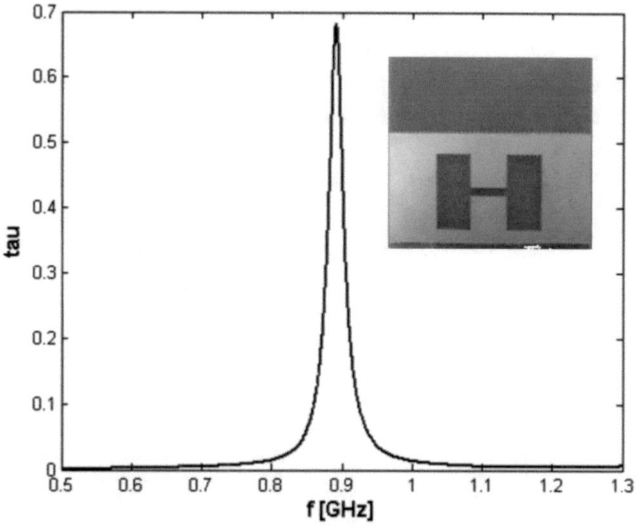

Fig. 11. SCA tag. Experimental prototype and measured power transmission coefficient for a microchip impedance Z_{chip}=10-j90 ohm.

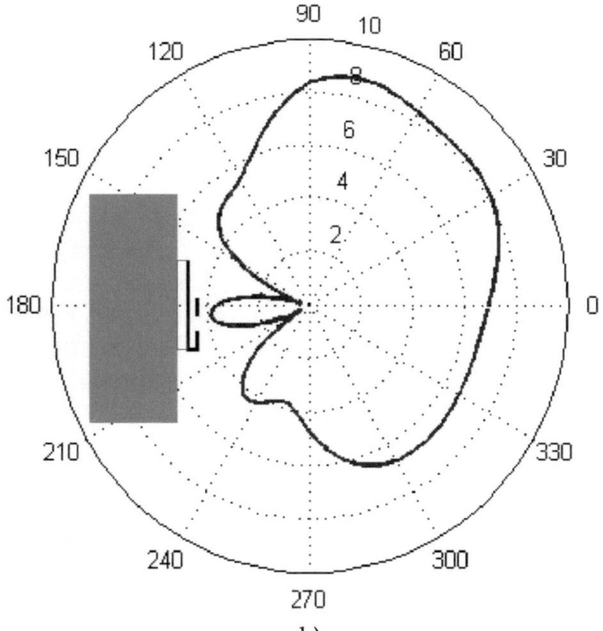

b)

Fig. 12. SILA Tag. Estimated activation distance for on-leg application when the reader emits 3.2 EIRP and the microchip has P_{chip}=10mW sensitivity.

4.3 An application example: monitoring the human motion

The SCA tag could be integrated with inertial switches for the monitoring of some Neural diseases such as the *Restless Legs Syndrome* (RLS) (Zucconi et al, 2006) and the *Periodic Limb Movements* (PLM). These sensor-motor disorders are clinically characterized by a compelling urge to move the limbs, accompanied by uncomfortable and unpleasant sensations in the extremities. Symptoms show a characteristic circadian evolution with a nocturnal worsening leading to insomnia and consequently daytime sleepiness and reduced quality of life. The diseases can be diagnosed by polysomnography (PSG) or by movement recording using actigraphs.

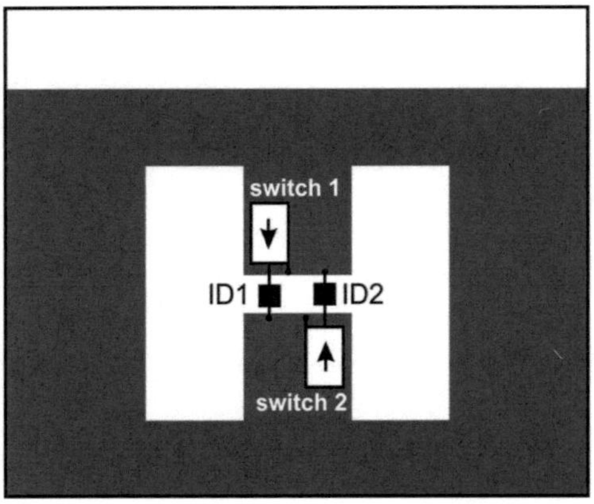

Fig. 13. Examples of integration between the SCA tag (front view), inertial switches (white boxes) and two microchips (black boxes). The arrows indicate the preferred direction of the switches.

The RFID technology offers interesting advantages over conventional diagnosis tools. The first one is the possibility to obtain very low-cost wireless devices, eventually single-usage, suited to be attached onto the body segments. Another interesting feature is that high-level aggregated data are easily achieved, just ready for the clinician.

In a possible set-up, one or more UHF RFID tags equipped with inertial switches will be attached onto the patient's limbs or onto other body regions. Depending on the verse of the applied acceleration, the switches react by changing their internal impedance from low (ideal short circuit) to high (ideally open circuit) impedance. A wireless one-axis sensor may be then achieved (Fig. 13) by means of a tag embedding two reverse-oriented inertial accelerometers, which turn between two RFID microchip transponder so that only one of the two possible IDs will be emitted (Fig. 13). Such an ID may be related to the acceleration's verse according to an ID modulation paradigm (Smith el al. 2005).

The tags are interrogated according to a proper repetition rate by an RFID reader placed at some place in the patient's room. The feasibility of this configuration is strictly correlated to the correct interrogation of the tags for any limbs position during the sleep within a typical hospital or domestic room, and with the compliance to the safety exposure regulations. In this perspective it is important to define the minimum features required to the body-antenna design, the number and the position of the tags, the sensibility of the inertial switches and the interrogation protocol.

A radio-mechanical model is introduced to simulate the biophysical signals collected by the reader during the legs motion, and to discuss the effect of other system parameters such as the inertial switch threshold and the interrogation rate. The model includes a simplified human phantom having 18 moving parts, individually controllable (Fig. 14.I). Some typical motion patterns in PLM episodes have been reconstructed by enforcing the trajectory of hip, knee and foot and calculated by a computer multi-body cinematic simulation solver. An example of ID-modulated data received by the reader is shown in Fig. 14.I for different choices of the interrogation frequency. Each tag is able to correctly capture the dynamics in term of number of predicted PML frequency and approximate duration.

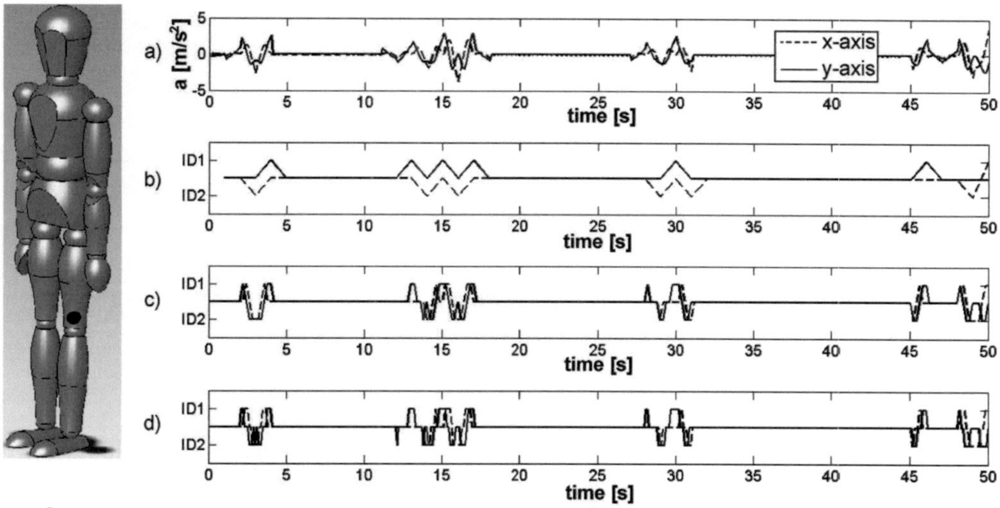

I. II.

Fig. 14. I) Simplified human phantom. II) Leg's acceleration at the leg position marked by a black circle and ID-modulated received data for switch's acceleration threshold 0.1G and some interrogation frequency: b) f_C=1Hz; c) f_C=5Hz; d) f_C=10Hz.

5. Self-sensing RFID tags

Like any antenna immersed or located close to a real object, the input and radiation characteristics of a passive RFID transponder placed on a target, as well as the strength of the back-scattered power, are closely related to the physical properties of the tagged object itself, e.g. on its constitutive material, shape, temperature, humidity or other. Let Ψ denote the set of the relevant target's features which could undergo changes along with the time, or could have to be monitored in someway. If the tag antenna has been designed for optimal performances when placed on a target with nominal set of features Ψ_T, e.g. such that the antenna impedance $Z_A(\Psi)$ equals in this condition Z^*_{chip}, a change of one or more target's parameters with respect to Ψ_T may produce a variation of the input impedance and hence the mismatch $Z_A \neq Z^*_{chip}$. Accordingly, also the back-scattered power collected at the reader port will be modified (Fig. 15). In the limiting case, the tag may be completely mismatched so that $P_{R\to T} < p_T$ and the tag is therefore inactive. For the sake of clarity, let us focus on the simplified case for which a single target's feature is subjected to change, and such a parameter be the relative dielectric permittivity (simply permittivity ε in the following). It is now useful to define the tag's *Activation Band* $A_\varepsilon(d)$ for a link length d, as the set of target's permittivity values for which the power harvested by the tag is enough to activate it: $A(d) = \{\varepsilon \mid P_{R\to T}(d,\varepsilon) \geq p_T\}$.

As suggested by equation (3), if the reader-tag distance were known, the change in the target permittivity could be theoretically detected by monitoring the power back-scattered by the transponder. Nevertheless, a single received data is not adequate to retrieve permittivity information in case of moving objects, or in applications in which the distance and the orientation of the tag with respect to the reader are not a-priori known (non-cooperative targets). To overcome these uncertainties, multiple independent back-scattered signals have to be collected by the reader. In the proposed platform, these signals are originated (Fig. 16) from either a *cluster* of N tags co-located onto a same target, or from a single tag provided with N input ports under the condition that each port or antenna has a different input impedance. In particular, we denote with $G_{T,n}$ the embedded radiation gain when only the nth port is fed and the others are connected to a reference load, and with $Z_{A,n}$ the input impedance at the nth port in the same conditions. Each port will be characterized by its Activation Band $A_{n,\varepsilon}(d)$.

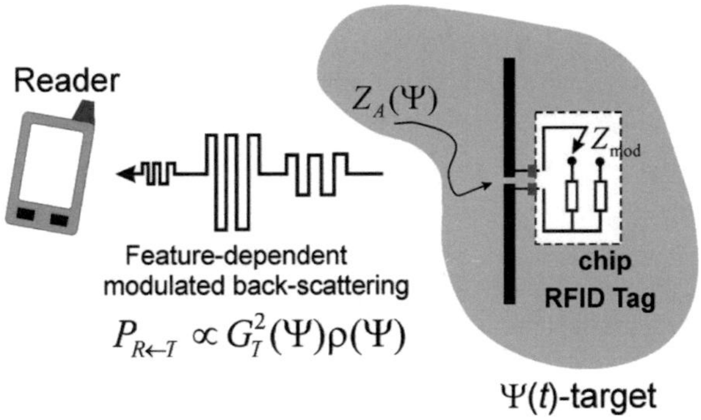

Fig. 15. Reader-tag scenario wherein the change of the target's features may produce a modulation of the backscattered power signal.

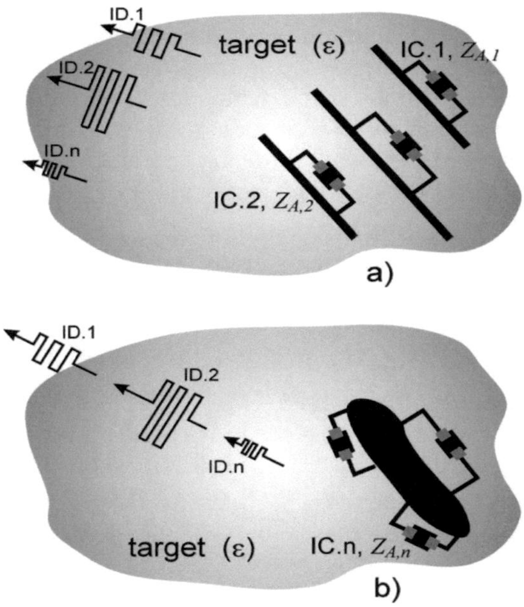

Fig. 16. Multi-port tag systems: a) a cluster of co-located single-port tags; b) a single multi-port tag provided with multiple chips.

The multi-port system has to be designed so that, having fixed a target geometry and having chosen N different *reference permittivities* $\{\varepsilon_1, \varepsilon_2, \; .. \; , \varepsilon_N\}$, the nth port impedance is matched to the microchip if the target's permittivity value is ε_n (e.g. $Z_{A,n}(\varepsilon_n) = Z^*_{chip}$). It means that, when the multi-chip system is placed on a real target, the ports will be differently mismatched ($Z_{A,n}(\varepsilon) \neq Z^*_{chip}$) and therefore they will originate independent back-scattered power signals, all of them carrying information about the target's permittivity. The resulting overall object is a *multi-port Sensor* RFID (S-RFID) *tag* that employs the same fabrication technology as the conventional RFID tags but, as shown later on, adds specific sensing capabilities to the typical identification features.

5.1 Sensing the target's permittivity

Depending on the link length d and on the particular design of the multi-port S-RFID tag, there will exist ranges of the target's permittivity for which either multiple ICs respond (overlapping of Activation Bands) and hence the reader is able to collect multiple backscattered signals, or only a port is at most activated and the reader may receive a single ID. Two different sensing modes can be correspondingly achieved: *analog sensing* (multiple responding ICs) and *discrete sensing* or *classification* (single responding IC). For both the cases, it is useful to introduce the *Sensing range* $\mathbf{S}(d)$ of the multi-port S-RFID tag, as *the set of all the possible values of the target's permittivity which could be detected, in some way, at a distance* d. The only analog sensing capability is here described, while a the complete theory could be find in (Marrocco et al., 2008).

If the tag has been designed for close reference permittivities { ε_n }, the port impedances will have similar (but not identical) power transmission coefficients τ_n so that multiple microchips will be turned on. In this case the multi-port system will have overlapped Activation Bands $\cap A_n \neq 0$. For any couplet of back-scattered signals received by the reader, each with a different modulation parameter $\rho_n(\varepsilon)$, it is possible to drop out the unknown reader-tag distance by calculating the *backscattered power ratio* $p_{i,j}$ between the received powers in equation (4),

$$p_{i,j}(\theta, \phi, \varepsilon) = \frac{P_{R \leftarrow T,i}(d, \varepsilon)}{P_{R \leftarrow T,j}(d, \varepsilon)} = \left[\frac{G_{T,i}(\theta, \phi, \varepsilon)}{G_{T,j}(\theta, \phi, \varepsilon)} \right]^2 \frac{\rho_i(\varepsilon)}{\rho_j(\varepsilon)} \tag{8}$$

However, $p_{i,j}$ is still affected by the uncertainty on the tag orientation (θ, ϕ) with respect to the reader. The multi-port tag design is therefore required to satisfy the condition of proportional gain patterns, e.g. such that $G_{T,i}(\theta, \phi, \varepsilon) / G_{T,j}(\theta, \phi, \varepsilon) = f(\varepsilon)$. This condition could be roughly satisfied considering a cluster of two antennas having a similar geometry.

The retrieval procedure is now described by means of an example involving a two-port system, e.g. able to backscatter two different IDs toward the reader. An overlapping configuration between the activation ranges is illustrated in Fig. 17. When both the ID_1 and the ID_2 are received by the reader, the unknown target dielectric permittivity ε_T will belong to the intersection of the two Activation Bands, e.g. $\varepsilon_T \in [A_1 \cap A_2]$, and therefore the p_{12} ratio can be calculated as in (8). The value of the target's permittivity is hence retrieved by using a *calibration curve* $\varepsilon(p_{12})$ which associates a target's permittivity to the actual backscattered power ratio, measured by the reader, (Fig. 18). Such a $p_{12} \rightarrow \varepsilon_T$ relationship is specific for the particular application, e.g. for a particular class (geometry) of targets and needs to be produced off-line through measurements or numerical simulations on simplified, or well representative, target models by operating a sweep of the parameter under observation and calculating the resulting backscattered power ratio. The application of such a technique therefore requires preliminary electromagnetic processing to produce calibration curves for the specific class of objects and the so obtained database, together with the retrieval procedure, have to be embedded in the reader's (post)processing unit. The S-RFID range **S**(d) is given by the merging of the Activation Bands shared by couplets of ports: $S(d) = \{d: \quad A_m(d) \cap A_n(d) \neq 0\}$.

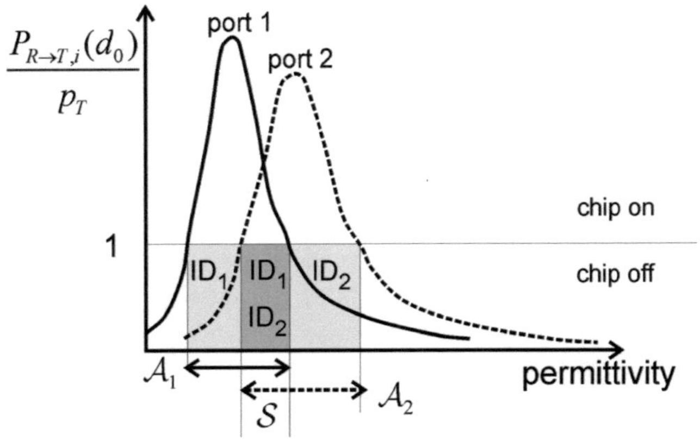

Fig. 17. Typical Activation Bands, and Sensing Range, of a two-ports RFID tag, designed to work in analog-sensing mode.

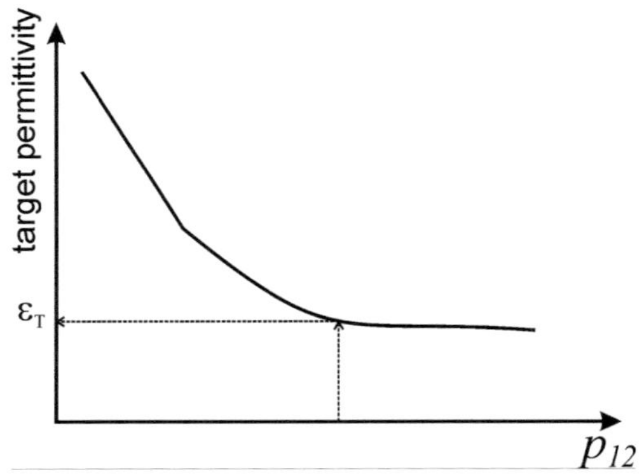

Fig. 18. Example of calibration curves $\varepsilon(p_{ij})$ relating the measured backscattered power ratio to a target's permittivity value.

5.2 An experiment: sensing the filling percentage of a container

A very preliminary laboratory experiment is here discussed. The purpose is to demonstrate the validity of the basic principle concerning the possibility to govern the variation of the two-port tag antenna features with respect to the change of a real tagged body. A two-MLA (Meander Line Antennas) tag (Marrocco, 2003) has been designed for the sensing of the filling level, *h*, of a box (Fig. 19). The variation of the shape of the target modifies the apparent permittivity sensed by the antennas and hence all their relevant parameters. With the aim to isolate and characterize the response of the antennas themselves, the experiment does not consider the RFID chip mounted on, and only the port impedances have been checked.

The target is again the perspex cubic box already used in Fig. 5, now filled up to a height *h* (changed during the experiment) by sugar powder (ε_r=3, σ=0). The box is placed over a large copper sheet (1m × 1m) acting as an image plane.

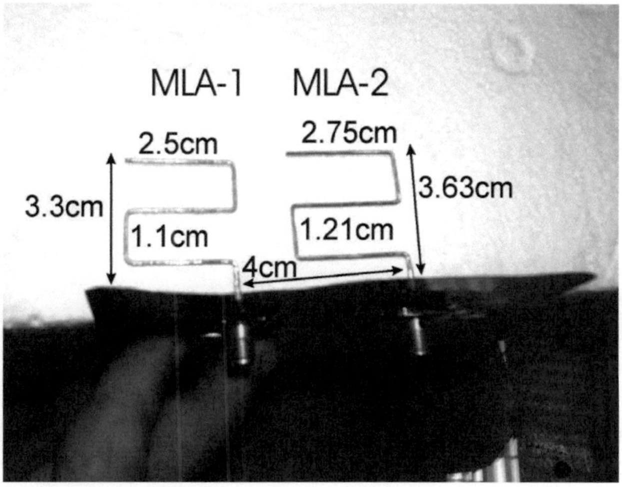

Fig. 19. Powder level sensing: meander-line-antennas prototype. Only half the structure is considered since the copper ground plane acts as an image plane.

The two MLAs are intended to be placed vertically on the external side of the box. Due to the presence of the ground plane, monopole configurations have been considered. Consequently, the impedance measurement results greatly simplified since no balun device is required. The antennas have regular meanders and they have been optimized for the best τ such that the MLA_1 and MLA_2 are matched, at 870MHz, to the microchip (Z_{chip}=50-j200Ω) when the sugar level is *h*=10cm and *h*=0cm (empty box), respectively. The two MLAs are scaled replicas. The overall antenna heights are 3.3cm and 3.63cm, respectively. The distance between the MLAs' gaps is 4cm. The tag prototype has been fabricated by 1mm-radius copper wire, and is shown in Fig. 19. The MLA monopoles are terminated on SMA connectors soldered on a 10cm copper sheet which is then placed in front of the perspex box as indicated in Fig. 20. At this purpose, the large ground plane was properly drilled to accommodate the SMA connectors for the connection to the HP 8753C Vector Network Analyzer by means of flexible coaxial cables.

The self-impedances Z_{11} and Z_{22} of the two-port tag are measured, having de-embedded the SMA connectors, when the filling level is increased in the range $0<h<10$cm with steps of 2cm. The measurements are repeated in the reverse order (by emptying the box) and the two resulting sets of data finally averaged.

The resulting matching diagram of the power transmission factor, estimated by FDTD and measured, is shown in Fig. 21. It is possible to appreciate, beside the nice agreement between simulations and measurements, that the τ-curves are monotonic with the change of *h* and that each port is rather mismatched in the condition for which the other one exhibits unitary τ.

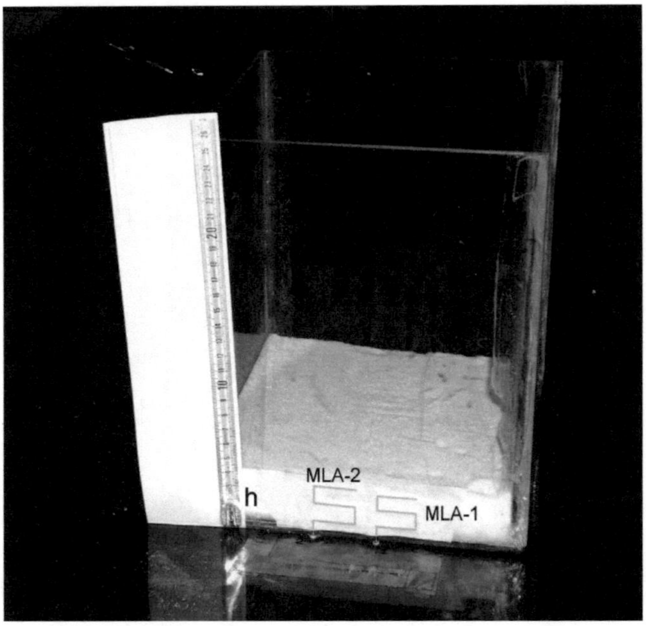

Fig. 20. Powder level sensing: experimental set-up comprising the two-MLA tags and a perspex cubic box of 20cm by 20cm cross-section partially filled with sugar up to a level *h*. The antennas are fixed to the box's vertical side by means of an adhesive ribbon.

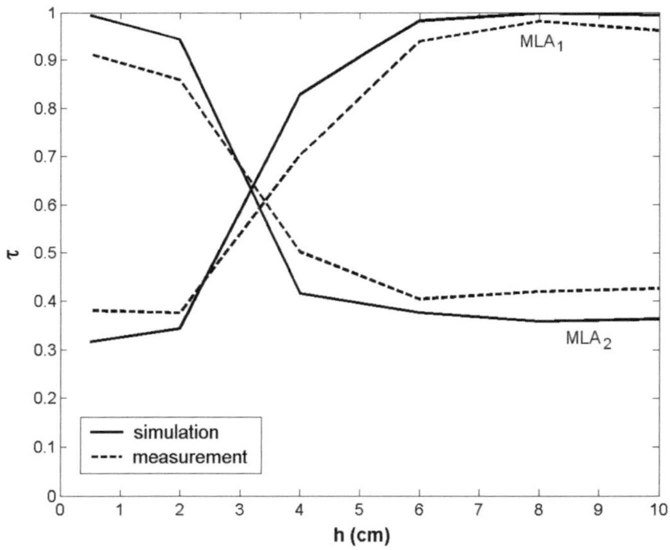

Fig. 21. Powder level: theoretical (simulated) and measured power transmission coefficients for the MLA_1 optimized for h=10cm, and MLA_2, optimized for an empty box (h=0).

The sensing effectiveness of the two-port tag depends on the calibration curve $p_{21} \leftrightarrow h$, and on the ratio in (8). The calibration curve $p_{21} \leftrightarrow \rho_1/\rho_2$ is monotonic, except for a very early short part with a good dynamic ($1 < p_{21} < 5$) when $2 < h < 8$. A saturation effect is clearly visible for levels higher than h=8cm, e.g. when the powder level greatly exceeds the vertical height of the antennas. In this condition, a further increase in h does not produce additional variation of the antenna responses and such a change of the target could not be sensed since the sugar powder acts as an infinite medium for the two antennas. The sensed dynamic of the powder level could be increased by designing longer tags or using a vertical arrays of properly tuned tags.

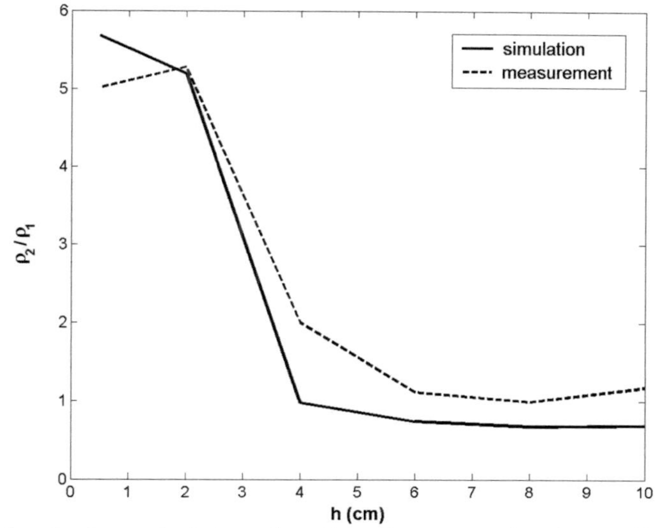

Fig. 22. Powder level: theoretical (simulated) and measured ρ_2/ρ_1 coefficients versus the sugar level h inside the box.

6. Conclusions

Designing low-cost antennas for sensing applications is still a great challenge, especially when the human body is involved. We are just at the beginning of the story and there is a significant margin of progress, both methodology, technological and experimental, to pursue in the next years. The reviewed tag configurations are only some of the many viable geometrical and electrical solutions, and are interesting for the relevant number of degrees of freedom and for the matching agility over different microchips and conditions.
Since the power consumption of the microchip transponder is continuously reducing, with a trend similar to the increase in the transistors' density in a chip (say the Moore Law), the concurrent research on antenna design, the use of smart materials embedding also sensorial capability, permits to figure out new classed of distributed and massive applications, mapping the physical phenomena into a virtual reality context, accessible from anywhere.

7. Acknowledgment

The author wishes to thank the many persons who gave significant support to this research in the last years, and in particular C. Stifano, C. Calabrese, D. Scarana, C. Occhiuzzi, L. Mattioni, G. Lovisolo, R. Pinto, S. Mancini, P. Tognolatti, L. Scucchia and S. Ricci.

8. References

Calabrese C., Marrocco G. (2008), Meandered-Slot Antennas for Sensor-RFID Tags, *IEEE Antennas and Wireless Propagation Letters*, Vol.7, N.1, Jan. 2008, pp. 5-8

Cheng-Ju L., Li L. Shi-Zong C., Chi Chen W. (2004), Chun-Huang W., Xin-Mei C., Mobile healthcare service system using RFID, *Proceedings of IEEE Int. Conf. Networking Sensing and Control 2004*, Vol.2, 2004, pp.1014-1019

Chong C.Y., Kumar S. (2003), Sensor Networks: Evolution, Opportunities and Challenges, *Proceedings of the IEEE*, Vol. 91, N.8, Aug. 2003, pp. 1247- 1256

Curty J., Joehl N., Dehollain C., Delercq M. J. (2005), Remotely powered addressable UHF RFID integrated system, *IEEE J. Solid-State Circuits*, Vol.40, N.11, Nov. 2005, pp. 2193-2202

Gabriel C., Gabriel S., Corthout E., The dielectric properties of biological tissues: I. Literature survey," *Phys. Med., Biol.*, Vol. 41, No. 11, Nov. 1996, pp. 2231-2249

Lorincz K., Malan D.J., Fulford-Jones T.R.F., Nawoj A., Clavel A., Shnayder V., Mainland G., Welsh M., Moulton S. (2004), Sensor Networks for Emergency Response: Challenges and Opportunities, *IEEE Pervasive Computing*, Vol. 3, No. 4, Oct 2004, pp. 16-23

M. J. Ackerman M. J. (1998), The visible human project, *Proceedings of the IEEE*. Vol.86, N.3, 1998, pp. 504-511

Marrocco G. Bardati F. (1999), BEST a finite-difference solver for time electromagnetics, *Simulation Practice Theory*, Vol.7, No.3, May 1999, pp. 279-293

Marrocco G. (2007), Rfid antennas for the UHF remote monitoring of Human subjects, *IEEE Transaction on. Antennas and Propagation*, Vol.55, N. 6, June 2007, pp. 1862-1870

Marrocco G. (2008), The art of UHF RFID antenna design: impedance matching and size-reduction techniques, *IEEE Antennas and Propagation Magaz.*, Vol.50, N.1, Feb. 2008, pp.66-79

Marrocco G., (2003), Gain-optimized self-resonant meander line antennas for RFID applications, *IEEE Antennas Wireless Propag. Lett.*, Vol. 2, 2003, pp. 302–305

Marrocco G., Mattioni L., Calabrese C. (2008), Multi-port sensor RFIDs for wireless passive sensing of objects - basic theory and early results", *IEEE Trans. Antennas Propagat.*, Vol.58, N.8 Part2, Aug. 2008, pp 2691-2702

Nikitin P.V., Rao K.V.S. (2006), Theory and Measurement of Backscattering from RFID Tags, *IEEE Antennas and Propagation Magazine*, Vol. 48, No. 6, Dec 2006. pp. 212-218

Reindl L.M., Pohl A., Scholl G., Weigel R. (2001), SAW-Based Radio Sensor Systems, *IEEE Sensors Journal*, Vol. 1, No. 1, Jun 2001, pp. 69-77,

Sample A.P., Yeager D.J., Powledge P.S., Smith J.R. (2007), Design of a Passively-Powered, Programmable Sensing Platform for UHF RFID Systems, *Proceedings of IEEE International Conference on RFID*, Mar, 2007, pp. 149-156

Smith J.R., Jiang B., Roy S., Philipose M., Sundara-Rajan K., Mamishev K. (2005), ID modulation: Embedding sensor data in an RFID Timeseries, *Lecture Notes in Computer Science*, Vol. 3727, Nov. 2005, pp. 234-246

Taflove A. (1998), *Advances in Computational Electromagnetics: The Finite Difference Method.* Norwood, MA, Artech House, 1998.

Weile D. S., Michielssen E. (1997), Genetic algorithm optimization applied to electromagnetics: A review, *IEEE Trans. Antennas Propag.*, Vol. 45, No. 3, Mar. 1997, pp. 343–353

Zucconi M., Ferri R., Allen R., Baier P. C., Bruni O. (2006), The official World Association of Sleep Medicine (WASM) standards for recording and scoring periodic leg movements in sleep (PLMS) and wakefulness (PLMW), *Sleep Medicine*, N.7, July 2006, pp. 175-183

6

Remotely UHF-Powered Ultra Wideband RFID for Ubiquitous Wireless Identification and Sensing

Majid Baghaei Nejad, Zhuo Zou, David S. Mendoza and Li-Rong Zheng
iPack Vinn Excellence Center, KTH-The Royal Institute of Technology, Sweden

1. Introduction

A new term of ubiquitous computing and communication is booming up which will transform our future corporate, community and personal life [1]. Early form of ubiquitous information and communication was happened in the use of mobile phones and nowadays it has become a vital part of everyday life for many million of people even more than internet. Recently, many research and developments are ongoing to bring this phenomenon more into everyday life by embedding smart systems into more objects which can interact to each other and people through a wireless link. It will provide connectivity for anything from anywhere, anyplace and for anyone. These connections create a network between items which lead to Internet of Things (IoT). Several kinds of information can be exchanged through the network such as environment status, and location which make a huge field of novel applications and market. To realize the IoT several technical innovation in different number of fields are essential. In order to have an embedded module in almost everything, first a simple and low cost system is vital. Second, embedding sensor technology into the items allows the system to detect changes in the physical status of things, which allows the system to changes or modifies some parameters of the system. And finally, system miniaturization allows smaller things have the ability of connection. A combination of all of these developments will create the IoT which connect the world's objects intelligently.

Different challenges need to be addressed. Energy issues such as low-power chipset design, energy harvesting, efficient and compact energy storage are some of the key issues. Many research need to be done in this area. Embedding sensor for data collection is another enabler for development of IoT. Low-power processing power and memory are important to process and store the sensor data. Different integration such antenna and passive integration need to be studied. Efficient communication protocol, modulation scheme and transmission speed is required to be studied. New methods for power management at different levels of the network are needed. Itegration of a smart device into the package or into the product is demanded. Different solutions such as system-on-chip and system-on-package should be investigated. Manufacturing challenges must be solved and the implementation cost must be lowered.

Radio frequency Identification (RFID) technology is a promising solution for IoT realization. It has been used mostly in supply chain management and logistic for several years [2].

However, recently RFID based ubiquitous identification and sensing systems are widely interested [3]. An RFID system identifies items using radio waves. A typical RFID system includes two parts: a transponder or tag which attached to the object to be identified, and an interrogator or a reader which identifies the tags. Active tags incorporate a battery which supplies the power for the operation. They offer long operation range and high performance but they are expensive and usually big size. On the contrary, Passive tags derive the required power from a reader using either inductive coupling or electromagnetic capture and communicate by utilizing load modulation or electromagnetic backscatter. They are more widely used than active tags because of their great advantages such as low cost, small size, and unlimited life time. Inductive coupling tag can offer high data-rate in proximity operation, while backscattering passive tags offer longer operation distance. Therefore they are more widely used. However, the returned signal power scales, unfortunately, as the inverse fourth power of the distance to the tag which makes the identification difficult, especially in a dense multipath and multi user environment. On the other hand, the data rate is limited to few hundreds of kb/s, and positioning accuracy is not better than 70 cm [4]. However, in new applications such as wireless sensing higher data-rate link with more accurate positioning capability is desirable. [2].

Ultra wideband impulse radio (UWB-IR) has been recognized as a promising solution for future wireless sensing and RFID because of its great advantages [5-9]. Information in I-UWB system is typically transmitted using a collection of short pulses with low duty cycle resulting in low power implementation. I-UWB technique has the possibility of achieving Gb/s data rate, several tens of meters operation range, low power consumption, subcentimeter accurate positioning, and low cost implementation [10]. It has been shown that UWB-IR can be a powerful candidate for the next generation of RFID with unique benefits such as longer operation range, fine localization and tracking, reliable tag reading in dense and metallic environment, small size and low cost.

This chapter describes potential applications of UWB-IR in future RFID, its advantages, and its design challenges. A brief introduction of UWB focused on low-power and low-data rate application will be given and possible types of UWB-RFID are presented. A special focus on an UWB/UHF hybrid passive RFID system with asymmetric wireless links is studied as an example. Unlike conventional RFID systems relying on backscattering and narrowband radio, UWB is introduced as the uplink for tag to reader communication. It enables a high network throughput (2000 tag/sec) under the low power and low cost constraint. The hardware implementation issues in silicon level are also considered. A 0.18μm single chip design for proof-of-concept is shown finally.

2. Introduction to ultra wideband radio

2.1 UWB-IR basics

This technology started 1893 when Heinrich Hertz used a spark discharge to produce electromagnetic waves. However, because of the significant superiority of continuous wave systems , UWB technique has not been used for several years. In 1960 short waves research started again in the impulse radar technology to achieve better imaging and localization capabilities. All UWB research performed prior 1994 were under classified US government programs until 2002 when the FCC (Federal Communications Commission) allowed the use of the 3.1 GHz -10.6 GHz band for unlicensed use with a maximum power emission of -41dBm/MHz. After some years, the FCC defined the power emission EIRP through out this band for different applications [11] such as indoor and outdoor communication, vehicle radar, ground penetrating radar, wall imaging systems, medical imaging systems, surveillance systems and law enforcements. Figure 1 shows the FCC emission mask for indoor and outdoor applications.

Ultra Wide Band has been defined by the FCC as a radio or wireless device where the occupied bandwith is greater than 20% of the center frequency or has a bandwidth higher than 500MHz. Two possible techniques for implementing UWB are Impulse Radio (IR) and multi-carrier UWB. Multi-carrier or multi-band UWB systems use orthogonal frequency division multiplexing (OFDM) techniques to transmit the information on each of the sub-bands. OFDM has several good properties, including high spectral efficiency, robustness to RF interference and to multi-path. It also has been proven in other commercial technologies such as IEEE 802.11a/g. However, it has several drawbacks. Up and down conversion is required and it is very sensitive to frequency, clock, and phase inaccuracy. On the other hand, nonlinear amplification destroys the orthogonality of OFDM. With these drawbacks MB-UWB is not suitable for low-power and low cost application.

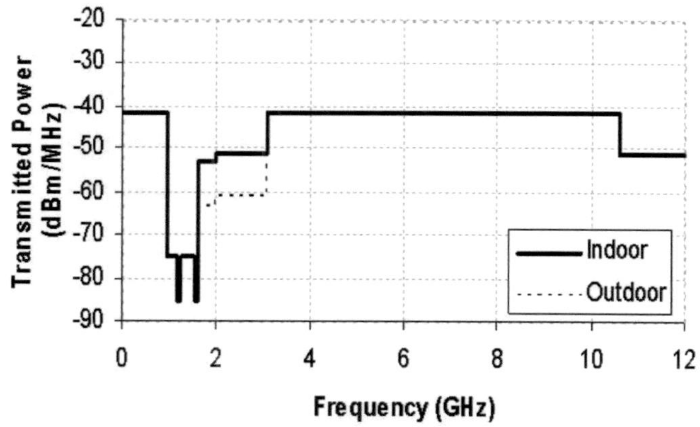

Fig. 1. FCC regulation mask for indoor and outdoor communications.

The main advantage of UWB-IR compared with narrowband systems can be described with Shannon's capacity equation (Eq. 1 where "B" is bandwidth, "S" is the signal power and "N" the noise power). The channel capacity is directly proportional to the bandwidth and has a logarithmic relation with the signal power. This means that increasing the Bandwidth higher data rates can be achieved keeping a small signal power.

$$C = B\log_2\left(1+\frac{S}{N}\right) \tag{1}$$

Information in impulse UWB techniques is send by modulating short pulses. In the literature is possible to find many waveforms that fulfill the spectral and power emission regulations stated in different parts of the world. Some of these signals are the Gaussian wave and its derivatives, Hermit pulses, Rayleigh and monocycle waveforms. Figure 2 shows the Gaussian pulses and its fifth derivative and Figure 3 shows the spectrum of them. In UWB-IR a non-carrier wave modulation is employed. The modulation is performed modifying some characteristics of the pulse such as amplitude, phase, and position. There are several modulation options which depend on application, design specifications and constraints, operation rage, transmission and reception power consumption, quality-of-service, regularity, hardware complexity, and capacity. Some of known modulation options in UWB-IR are ON-OFF Keying (OOK), Pulse Position Modulation (PPM), Pulse Amplitude Modulation (PAM) and Binary Phase Shift Keying (BPSK). In Figure 4, different modulation schemes have been illustrated.

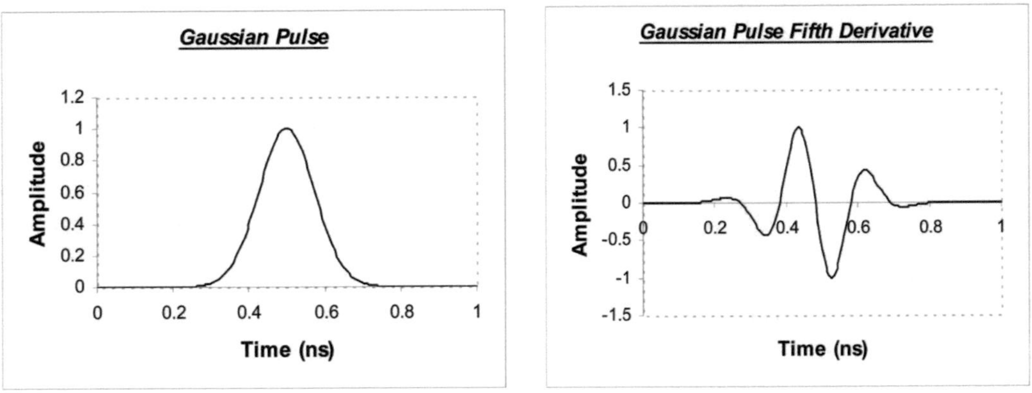

Fig. 2. Gaussian pulse and fifth derivative Gaussian pulse waveforms.

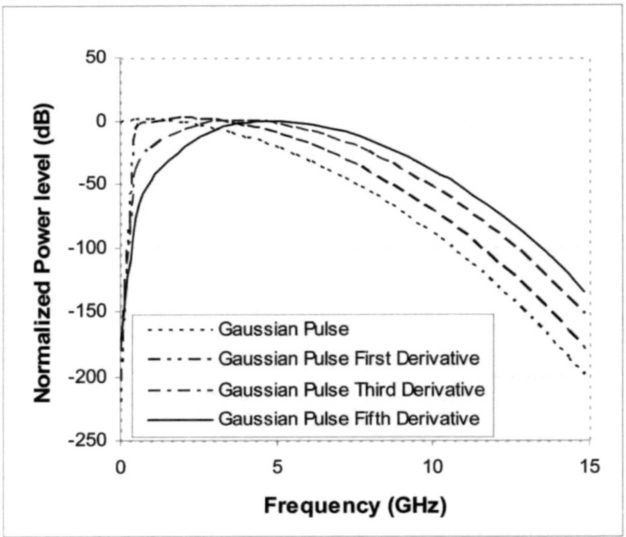

Fig. 3. Spectrums of different derivatives of Gaussian pulse with 500ps width

The transceiver complexity depends on the demodulation coherence. If the system uses OOK, PPM or M-ary PPM, a low complexity non-coherent demodulation scheme such as energy detection can be used. If the system use BPSK or M-ary PAM modulations, a coherent demodulation scheme is required increasing the hardware complexity and cost. Therefore, for low power and low-data-rate applications such as RFID and WSN lower-complexity modulation such as OOK or PPM is desired.

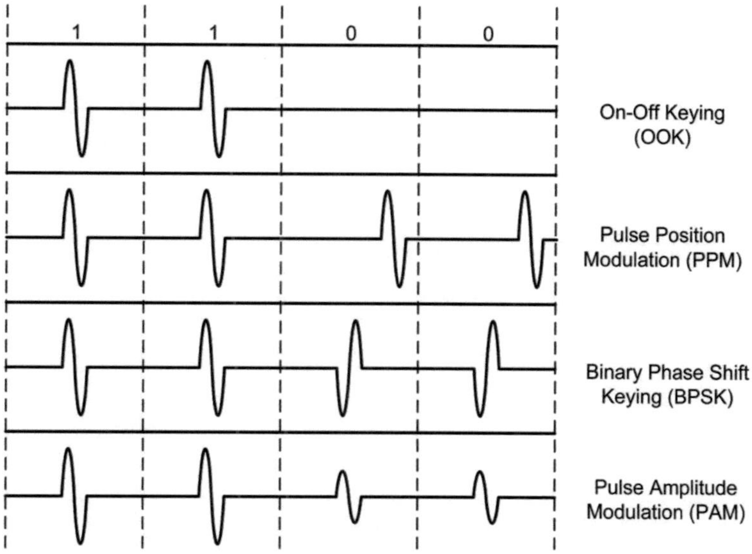

Fig. 4. Non-Carrier modulation schemes for I-UWB [12].

2.2 UWB-IR for RFID and WSN applications

Recently, the interest of the UWB to low-power low-data rate networks with ranging has been growing rapidly, along with the development of the IEEE 802.14.a. Applications such as RFID and wireless sensor network combine low data-rate (50kbps to 1Mbps), ranges 10 m to 100 m with accurate positioning capabilities.

UWB is attractive to RFID and WSN applications, which require low-power and low-cost implementation, due to the high node density of the network. Besides, some of applications need battery-free by energy scavenging. Therefore, an average power consumption on the order of 10 - 100 uW is expected, at the cost of conventional passive tags for identification and tracking. In contrast to conventional RF communication systems, UWB-IR uses very short pulses that are able to propagate without an additional RF mixing stage [13]. The baseband-like architecture with low duty cycle signal guarantees low complexity and low power implementation. Many studies on design of UWB transceiver show that UWB technology is a good candidate to achieve low power and low complexity implementation. Center for wireless communications in University of Oulu demonstrated a tag based UWB wireless sensor system for outdoor sport and lifestyle applications [14]. A VLSI implementation of low power, low data-rate UWB transceiver is designed for such applications. The transceiver based on non-coherent energy detection architecture is implemented in 0.35 μm SiGe BiCMOS technology with 134 mW power consumption at 5 Mbps data rate [8].

Security is a hot topic in RFID and WSN research and development. Noise-like UWB signals guarantee robustness against eavesdropping or jamming. Existing RFIDs using simple coding and modulation schemes are easily to be eavesdropped or jammed. On the other hand, higher level efforts for cryptography results to large area of digital blocks for ciphers, high power consumption and system latency. To address this problem, a research group from Virginia Tech introduced an RFID system replacing cryptography with UWB in high secure application [5]. TH-PPM UWB modulation is applied as their proposed solution. Because UWB signal is inherently with low duty cycle and low-power emission, it is very difficult to eavesdrop or jam and no extra cryptography block is required. It can simplify the hardware complexity, reduce the power consumption, and upgrade the system throughput. Excellent time resolution is another key benefit of UWB-IR signals for ranging and positioning application. Nanyang Technological University of Singapore developed an UWB-enabled RFID system which works with both active and passive tags to provide ranging and localization capabilities up to centimeter accuracy [15].

3. UWB/UHF hybrid system architecture

3.1 Design considerations of RFID and WSN systems

RFID and WSN applications hold some notable characteristics that are not shared with other communication systems:

- *System capacity:* A huge number of tags might appear in a reading zone simultaneously. Furthermore, multi-access (anti-collision) algorithm is essential for the system efficiency due to the massive tags environment.
- *Asymmetrical traffic loads and resources*: Unlike other RF communication systems, the traffic loads of RFID are highly asymmetrical between the uplink and the downlink. Data (e.g. synchronization, command) broadcasted from the reader is very few, but the traffic transmitted by a great number of tags in the field is rather heavy. In hardware perspective, tags have very limited resource such as memory, power supply, and computational ability, but a reader can be a powerful device.
- *Reading speed:* Reading speed in terms of processing delay is an important metric. High processing speed could be achieved by either a high data rate link for tag to reader communication, or an efficient anticollision algorithm.
- *Low power and low complexity hardware implementation:* Because RFID tags are resource-limited devices, the implementation upon the system specification must be simple and energy-efficient.

3.2 Asymmetric UWB-RFID architecture

On the basis of the considerations above, we propose an asymmetric UWB-RFID system architecture illustrated in Figure 5. Due to the nature of the impulse UWB radio, the UWB-IR transmitter integrated on the RFID tag provides a robust, high speed and high security uplink under a low power and low complexity implementation. Instead of the typical full-UWB system, the traditional RF transceiver is applied as the downlink. First, as the discussion in previous section, in full-UWB system, the wide-band RF receiver consumes too much power which makes it impossible for tag to be powered wirelessly in battery-less systems, whereas using battery in tag causes high maintenance cost and big size. Second, unlike other communication systems, RFID and WSN applications are dominated by uplink communication, where the low downlink traffic becomes insignificant for the system efficiency. As a result, the low data-rate narrowband radio is adequate [16, 17].

The reader broadcasts commands to tags using UHF (870MHz ~ 960MHz) signal. The modulation is ASK with pulse interval encoding (PIE). The data rate (clock frequency) is adaptive from 40Kbps (KHz) to 160 Kbps (KHz) controlled by the reader. A tag replies information by transmitting UWB signal with adaptive data-rate up to 10Mbps. The UWB pulse rate and data-rate are adapted by reader based on the available power and desired operation distance. In long range operation when the availavle power to the tag is low, lower pulse rate and data rate is chosen resulting lower power consumption. On the other hand, in short range applications, higher pulse rate with high data-rate can be transmitted since the available power is high enough.

In [12], BPSK achieves the best BER performance in the AWGN and Rayleigh fading channels simulations. However the circuit complexity is the highest and the receiver needs a coherent system to demodulate data. Thus, this modulation is not suitable for the RFID implementation. Either OOK or PPM modulation can be used in the tags to modulate the UWB pulses. Although, OOK modulation has less communication performance however, it results in simple and low power implementation. Therefore, in this design OOK modulation is utilized. As can be seen later, UWB pulses are transmitted synchronous with the incoming RF signal, which brings further simplification in synchronization and improves the detection performance in readers.

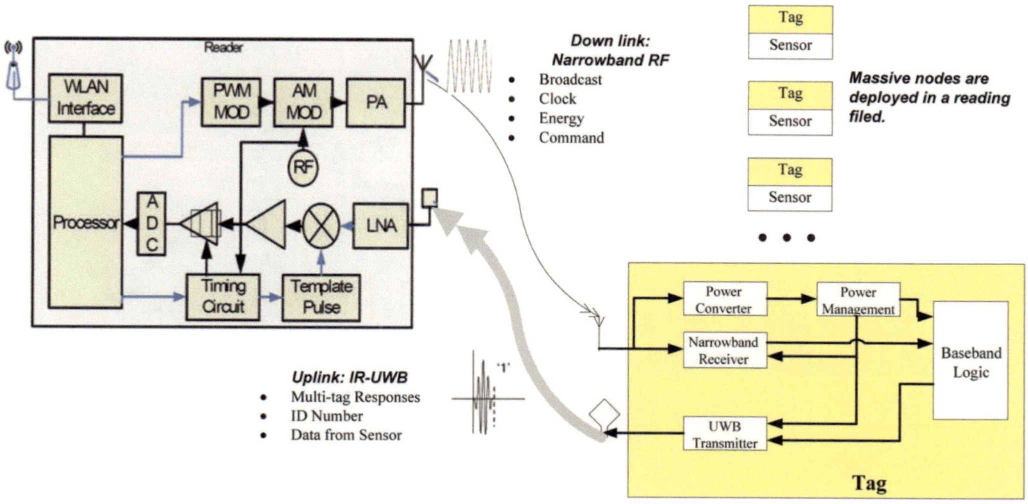

Fig. 5. Proposed system model of asymmetric UWB-RFID architecture

3.3 Data communication protocol

The specification in higher layers is a further issue that determines the energy efficiency as well as the system throughput. Hereby, we devise a specified data communication protocol for the proposed asymmetric UWB-RFID architecture. Multi-access is also considered in the proposed protocol.

3.3.1 Operation procedure

Because of the great asymmetric between reader and tags, the system works in a master-slaver communication mode. A reader initiates all the operations, followed by tags' responses. All the calculations are made by the reader and hence the tag implementation is very simple. Five operations are defined in the proposed protocol, namely Wakeup, Request, Write, Modify and Kill. The Wakeup and the Request are basic operations for identifying tags or gathering data. The Wakeup activates and identifies all tags in the reading field while the Request performs the similar function as the Wakeup, but does not affect the identified tags. The Write function is for initialization of the tag and the Modify is used to program a specific tag with access control. The tag can be deleted by using the Kill function. The frame format, also called round, which represents an operation initiated by readers, is composed by four phases: powering, start of frame (SOF), commands, and processing. In the powering phase, the reader radiates a continuous sinusoid wave to power passive tags. A SOF is used for frame synchronization. The sequence consists of ten continuous bit 0s and a bit 1. Afterwards, tags decode the received command and respond the reader.

An acknowledgement mechanism is employed to guarantee the successful receptions and to disable the identified tags. Unlike traditional RFID where data integrity (QoS) is controlled by both readers and tags such as CRC check, in the proposed system, only readers take charge of error handling. As a result, CRC checker can be removed from a tag which reduces the complexity. After each operation, the tag sends its current data and the reader checks the correctness of the operation. Because the uplink speed is high, this approach will not cause the processing delay even transmitting whole data.

Figure 6 shows the state transition diagram of the main state machine for tags.

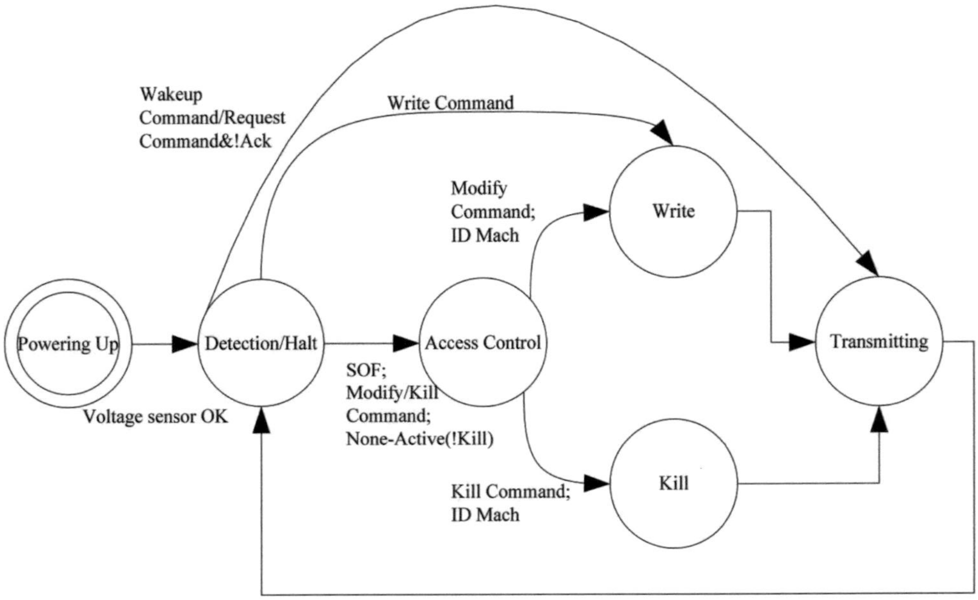

Fig. 6. The State Transition Diagram of UWB-RFID Tag

- ***Powering Up State:*** Passive tags capture power by the power scavenging units and store in a relatively big capacitor. This stored energy is used later for transmission.
- ***Halt/Detecting State:*** This is the initial state of each powered tag. In this state, tags are detecting incoming signals and capturing SOF and Command. After this state, tags enter a new frame to execute the corresponding operation.
- ***Transmitting State:*** A tag executes three procedures during transmitting state. First step is to load data into the cache and generate a PN code. Secondly, a slot counter in the tag counts down the PN code until it reaches 0. Finally, the tag sends the data and waits for ACK or NAK.
- ***Writing State:*** A tag programs its memory by receiving data from the reader.
- ***Access State:*** This state comes before an operation for a specific tag (Modify and Kill Commands). The tag compares its data with the incoming signals bit by bit. This state is interrupted by different bits. Only one tag with the same data completes the state.
- ***Kill State:*** It sets the Kill Flag to permanently disable the tag.

3.3.2 Anti-collisions

In contrast to conventional wireless system, massive nodes (tags) are deployed in a dynamic environment. Random access method is applied in our work, rather than current medium access control (MAC) protocols for UWB-IR including time division multiple access (TDMA), time hopping, or direct sequence UWB (DS-UWB) [18]. In [19] several versions of the ALOHA algorithm are presented in order to increase its feasibility and efficiency. Among them, the most widely used one in wireless sensor and identification systems is the framed slotted ALOHA algorithm. Time is divided into discrete time intervals, called slots. A frame is a time interval between requests of a reader and consists of a number of slots. A tag randomly selects a slot number in the frame and responds to the reader. A procedure called acknowledgment is required to resolve collisions or failed transmissions. Collided tags retransmit in the next frame [19].

The overall goal of the anti-collision algorithm is to reduce the identification period with simple hardware implementation and low power consumption. To improve network throughput, we propose a more efficient scheme to overcome the anti-collision problem. It is based on the framed slotted ALOHA algorithm by employing following improvements.

- ***The pipelined Communication Scheme:*** In conventional approaches, a time slot normally contains a tag's data packet and the acknowledgement from the reader. However, because there exist great asymmetry between the downlink and the uplink (UWB data rate is much higher than the narrowband radio data rate), the acknowledgement from the reader to tags becomes a bottleneck that decreases the network throughput. This problem can be solved by using a pipelined method that poses the data packet and its corresponding acknowledgement in two adjacent slots. As can be seen in Figure 7, a tag sends data in the K slot and receives the ACK in the K + 1 slot. Processing gain in slot is calculated in Eq.2.

$$Gain = T_{packet} + T_{ACK} - \max\{T_{packet}, T_{ACK}\} \tag{2}$$

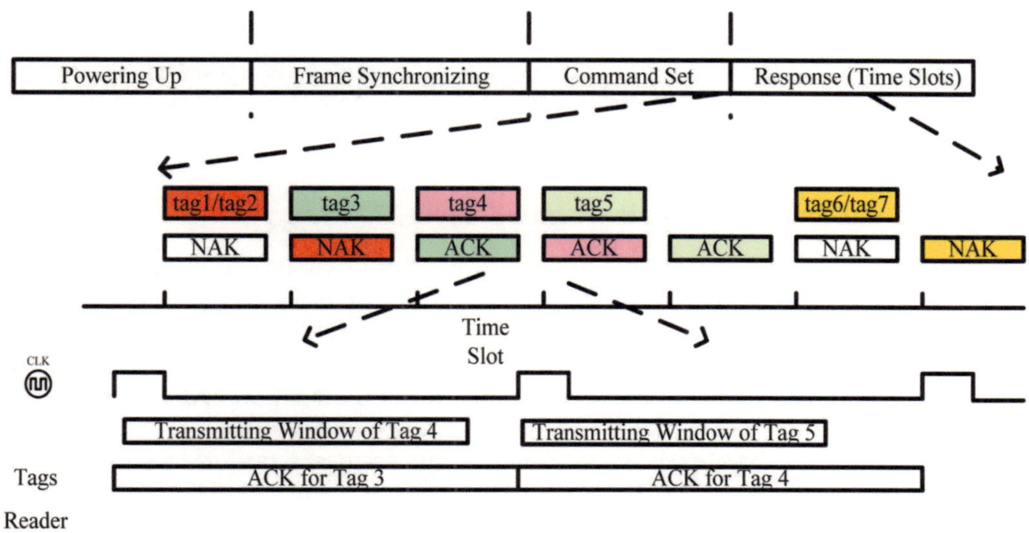

Fig. 7. The sketch of the pipelined communication protocol

- ***Skipping Idle Slots:*** Because the global clock is scalable controlled by the reader, it provides a possibility to skip idle slots. By detecting the incoming signals at the beginning of each slot, the reader can determine if there is any transmission in this time slot. If it is an idle slot (phase B in Figure 8), the reader skips this slot by adjusting the clock frequency and transits into the next cycle (slot) immediately.

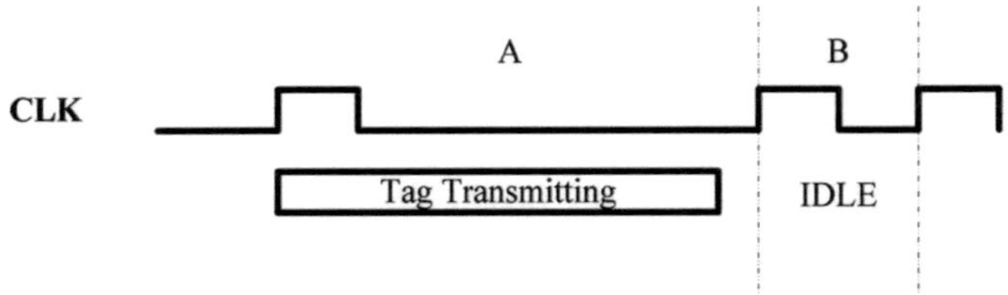

Fig. 8. Sketch of Idle Slot Skipping

- ***Adaptive Frame Size:*** The maximum system efficiency of the framed slotted ALOHA is achieved when the frame size (N) approximately equals to the tag number (n) [20]. Dynamic frame sizes allocation replaces the traditional fixed framed ALOHA. With the tag number estimation algorithm [21], the reader can estimate the number of tags, and optimized the frame size.

Hereby, the system efficiency is defined as the ratio of the successful transmission time to the frame size. Given N slots and n tags, the number of r of tags in one same slot is binomially distributed as Eq.3.

$$B_{n,\frac{1}{N}}(r)=\binom{n}{r}\left(\frac{1}{N}\right)\left(1-\frac{1}{N}\right)^{n-1} \tag{3}$$

If the frame size is small but the number of tags is large, too many collisions will occur and the fraction of identified tags will degrade. On the other hand, when the number of tags is much smaller than the number of slots, the wasted slots can occur. As described the dynamic frame size allocation can provide the optimal frame size to achieve the maximum throughput. Moreover, the idle skipping method can eliminate the delay caused by the empty slots. The simulation results of the system performance are shown in Figure 9. It demonstrates that more than 2000 tags/s can be processed. Table 2 presents the comparison result with some standardized RFID protocols.

4. Implementation of the remote-powered UWB-RFID tag

A Remote-Powered UWB-RFID tag is designed for proof of concept and implemented in UMC 0.18µm process (Figure 10). The module consists of five parts: an RF demodulator, an impulse UWB transmitter, a power management unit, a clock circuitry, and a digital baseband. The narrowband receiver receives RF signal and demodulates it into digital signal. The power management unit captures the incoming RF signal and rectifies to DC voltage and supplies the whole circuitry of the tag. A low frequency clock is recovered from the received data as the baseband control. Another high frequency clock for the UWB pulse generator is imported by dividing the carrier of the incoming signal. The digital baseband is responsible for control, i.e., decodes commands, programs memory, fetch data, and exports data to the transmitter.

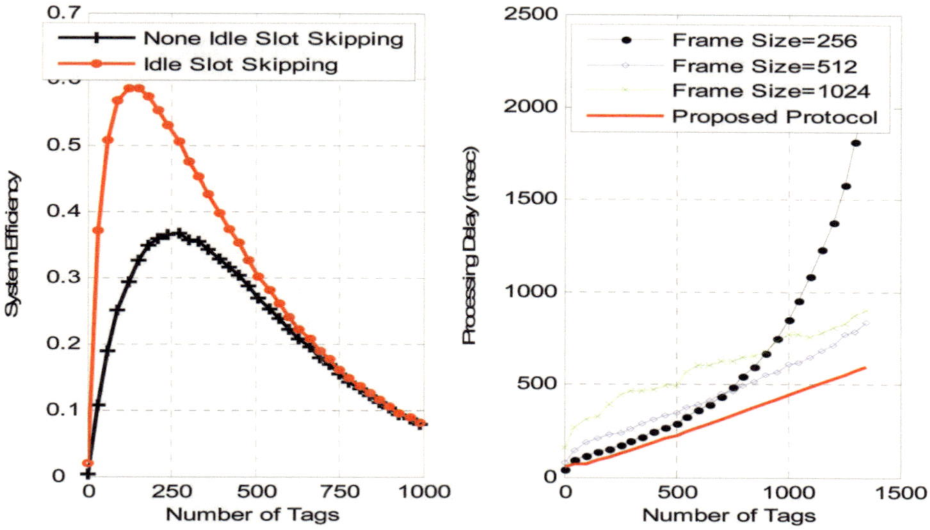

Fig. 9. Simulation Results of System Performance

Standards	Frequencies	Data Speed T→R	Processing Speed	Anti Collision
ISO 18000-3 Mode 2	HF	106 Kbps	1200 tags/sec	Not Specified
ISO 18000-6 A	UHF	40 Kbps	100 tag/sec	Framed Aloha
ISO 18000-6 B	UHF	40 Kbps	100 tags/sec	Binary Tree Search
EPC Class 0	UHF	Max 80 Kbps	200~800 tags/sec	binary tree search
EPC C1G2	UHF	Max 140 Kbps	1000 tags/sec	Framed Aloha
This Work	Semi-UWB	1Mbps	>2000 tags/sec	Enhanced Framed Aloha

Table 1. Comparison of different standardized protocols

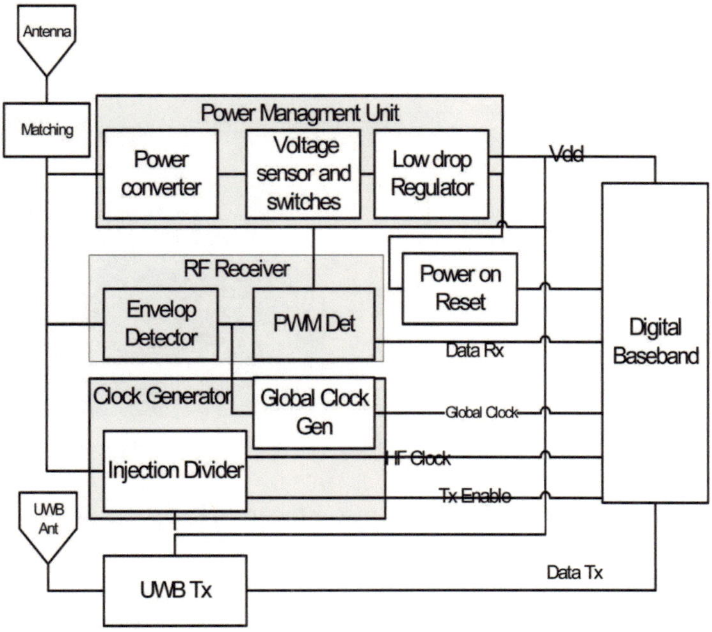

Fig. 10. Block diagram of the UWB-RFID Tag

4.1 Impulse UWB transmitter

Impulse UWB Transmitter generates 5th derivatives Gaussian pulses to modulate the baseband information into UWB signals. A tunable impulse UWB transmitter is shown in Figure 11. Duration and amplitude of the output pulse are controlled by two inputs that have capability to compensate the process and temperature variations, interconnection and packaging effects, and frequency response of the antenna. Furthermore, this ability allows the module to control output power and bandwidth in different pulse repetition rates. In short range applications, high repetition rate and low amplitude pulses are transmitted. On the contrary, to transmit data in longer distance, low repetition rate and high amplitude pulses are chosen. In both of two cases, amplitude and duration controls enable the module to transmit a signal comply the FCC regulation [22, 23]. The output impulse of the UWB transmitter and its power spectral density are shown in Figure 12. The power consumption of UWB-Tx at 10 MHz pulse repetition rate is 51 μA at 1.8V, and 252 μA at 50 MHz pulse repetition rate.

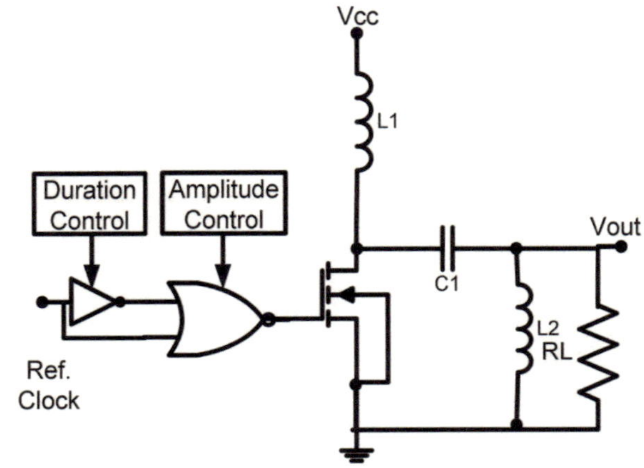

Fig. 11. Schematic of the I-UWB Transmitter

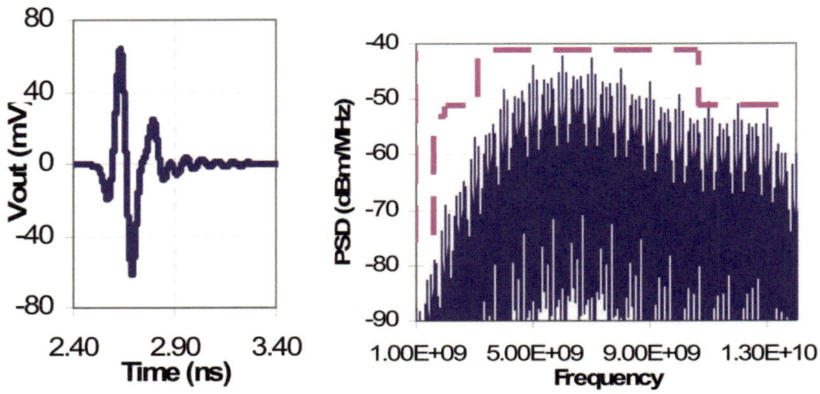

Fig. 12. Output pulse shape of the I-UWB Transmitter and ITS Spectrum

4.2 Power management unit

The power management unit provides power supply for the whole circuitry from incoming electromagnetic wave. Figure 13 shows the principle of operation. During the powering phase the Power-Switch is open, and thus the power consumption is very low (1uA). The power scavenging unit (PSU) converts the received electromagnetic wave to a dc voltage on an off-chip capacitor. When the voltage across the storage capacitor raises a certain value (e.g. 2.5V), a voltage sensor (Vsen) switches on the Power-Switch and the chip starts to operate. While the chip is working, voltage across the storage capacitor is degraded; therefore a low-dropoutput (LDO) voltage regulator is utilized to provide regulated voltage for the module. If the voltage becomes less than a threshold (e.g. 1.8V), the voltage sensor switches off the chip, and chip starts to gather energy for next run [24].

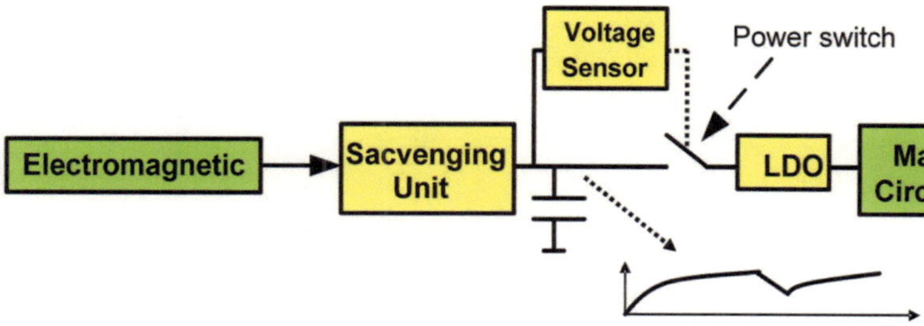

Fig. 13. Operation principle of power management unit

Figure 14 shows the schematics of different building blocks of the power management unit including of the power scavenging unit, the voltage sensor, and LDO voltage regulator. The minimum input power of 14.1 µW is achieved with this technique. It corresponds to 13.9 meters operation range which is great improvement compared with conventional RFID.

4.3 RF demodulator

Such as conventional RFID, a simple RF demodulator is utilized. It includes an envelope detector and a discriminator circuit which extract data and clock from the received signal. The envelop detector uses the same CMOS voltage multiplier topology than power scavenging unit, but with smaller capacitors and only 2 stages. The discriminator circuit decides whether a pulse is long or short and extracts data and clock. Extracted clock is used as the global clock for baseband control. Figure 15 depicts the schematic of the RF demodulator including of envelop detector, and clock and data recovery block diagram.

4.4 Clock generator

UWB transmitter requires high frequency clock with low skew and jitter. LC oscillators occupy large area and consume high power. On the other hand, ring oscillator show large variation across the process, temperature and voltage as well as huge phase noise [8]. Utilizing the PLLs which are used in communication systems are not applicable in RFID tag because of their high complexity and power consumption. In this work a low power harmonic injection locked (HIL) divide-by-3 is used to down convert the 900MHz carrier frequency [25]. Figure 16 shows the schematic of the divide-by-3 circuit and the output spectrum before and after locking. Simulation result of the harmonic injection locked divider shows total power consumption of 15.3µA. The minimum input voltage for locking is 100mv which is acceptable for this operation range. Phase noise of the output at 10Hz offset is -85dBc/Hz and jitter is 1.47ps.

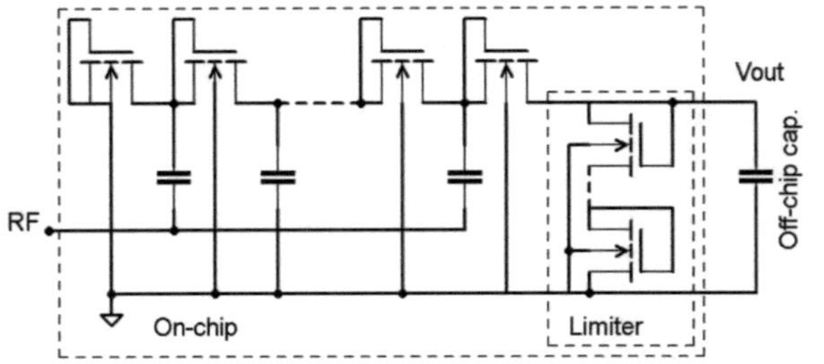

(a) Power scavenging unit

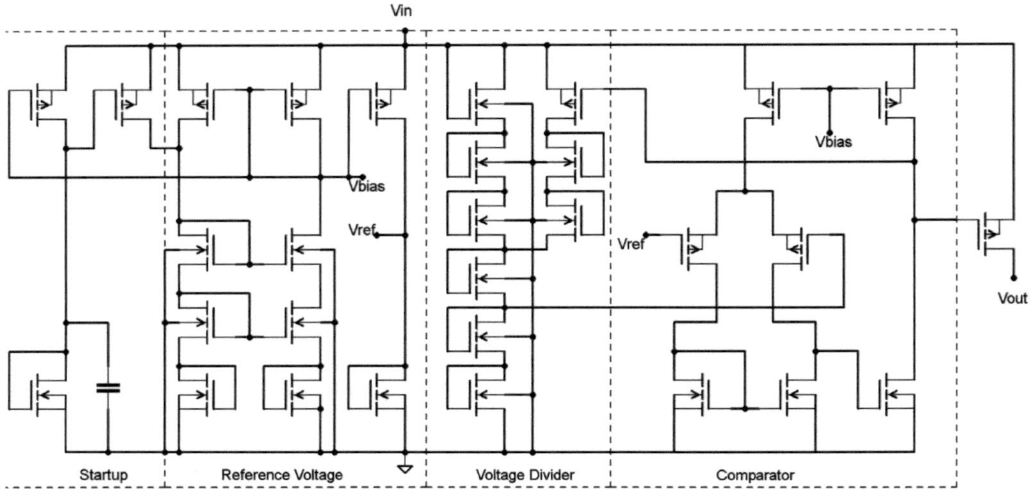

(b) Voltage sensor

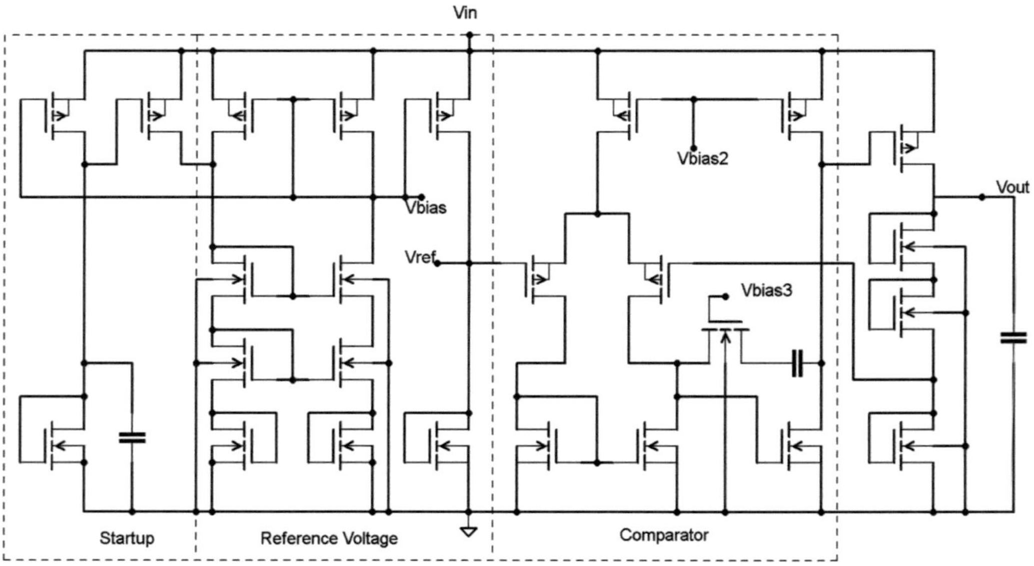

(c) Low-Drop-Out Voltage Regulator

Fig. 14. Schematics of power management unit

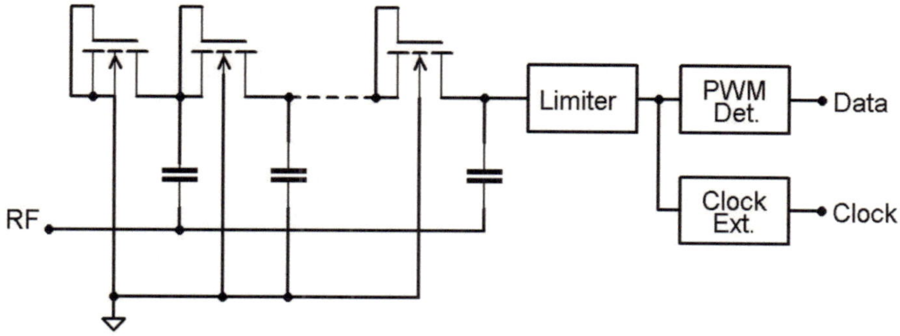

Fig. 15. Schematic of RF demodulator

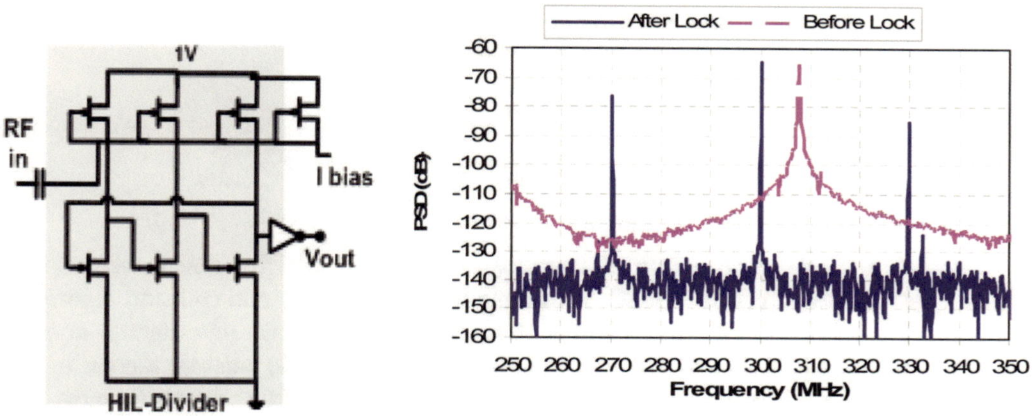

Fig. 16. Schematic of divide-by-3 and output spectrum before and after lock

4.5 Digital control logic

Digital control logic is used for baseband processing, medium access control, and power management. Figure 17 illustrates the architecture of the processor. The control unit is formed by several FSMs which generate control signals to each sub module whereas sub modules send status signals to the control unit. The pseudo number generator (PNG) and the slot counter are used to implement the transmission protocol and the anticollision algorithm. The circuit simulation is successful and the design is tested by FPGA prototype. We also map the design in UMC 0.18um process. The area is equivalent to 4000 NAND gates and the power consumption is around 800nW [26]

5. Conclusion

In this chapter, a novel system with asymmetric wireless links has been presented for ubiquitous wireless sensing and identification. Such as conventional passive RFID systems, nodes derive the power supply and receive data from the received RF signal transmitted by a reader. However, instead of backscattering, impulse UWB radio technique has been utilized in uplink from the nodes to the readers. It offers several advantages to the system such as high throughput, precise ranging and positioning, more security, long operation range, robustness to multipath, robustness to the narrowband interference and multi user

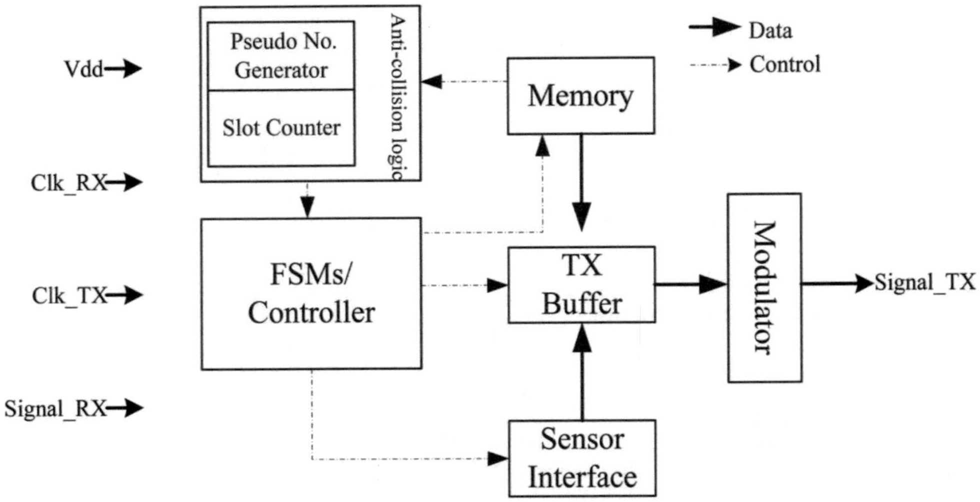

Fig. 17. Block diagram of baseband logic

interference. A new communication protocol is proposed for the novel system with asymmetric wireless links. It is based on Frame Slotted ALOHA anti-collision algorithm. Dynamic frame size allocation and idle slot skipping methods are investigated and the performance simulation results show a throughput more than 2000 tags per second for the system which great improvement compared to the conventional RFID systems (at most 1000 tags/s). To proof of the concept, a complete module for the tag has been implemented in 0.18 μm CMOS process. The measurement results shows the operation distance of 14 meters when 4W EIRP emission is allowed at 900 MHz frequency band. The impulse UWB transmitter consumes 51 μA at 10 MHz pulse rate which is low enough to be provided by the power management unit for 1.9 millisecond time. The results proof the validity of the proposed concept and show the great potential of impulse UWB radio for next generation of RFID for ubiquitous wireless sensing.

6. References

[1] "The Internet of Things," International Telecommunication Union (ITU) Internet Report 2005.

[2] K. Finkenzeller, RFID-Handbook: Fundamentals and Applications in Contactless Smart Cards and Identification, 2nd ed.: Wiley & Sons LTD 2003.

[3] R. Want, "Enabling ubiquitous sensing with RFID," *Computer,* vol. 37, pp. 84-86, 2004.

[4] J. Guang-yao, L. Xiao-yi, and P. Myong-Soon, "An Indoor Localization Mechanism Using Active RFID Tag," in *Sensor Networks, Ubiquitous, and Trustworthy Computing, 2006. IEEE International Conference on*, 2006, pp. 40-43.

[5] D. S. Ha and P. R. Schaumont, "Replacing Cryptography with Ultra Wideband (UWB) Modulation in Secure RFID," in *RFID, 2007. IEEE International Conference on*, 2007, pp. 23-29.

[6] X. Duo, T. Torikka, Z. Li-Rong, M. Ismail, H. Tenhunen, and E. Tjukanoff, "A DC-13GHz LNA for UWB RFID applications," in *Norchip Conference, 2004. Proceedings*, 2004, pp. 241-244.
[7] F. U. Dowla, "Long-Range Ultra-Wideband Radio-Frequency Identification," LLNL Engineering 2004.
[8] L. Stoica, A. Rabbachin, H. O. Repo, T. S. Tiuraniemi, and I. Oppermann, "An ultrawideband system architecture for tag based wireless sensor networks," *Vehicular Technology, IEEE Transactions on,* vol. 54, pp. 1632-1645, 2005.
[9] N. Patwari, J. N. Ash, S. Kyperountas, A. O. Hero, III, R. L. Moses, and N. S. Correal, "Locating the nodes: cooperative localization in wireless sensor networks," *Signal Processing Magazine, IEEE,* vol. 22, pp. 54-69, 2005.
[10] J. H. Reed, An Introduction to Ultra Wideband Communication Systems: Prentice Hall PTR, 2005.
[11] FCC, "First report and order," Available online http://hraunfoss.fcc.gov/edocs_public/attachmatch/FCC-02- 48A1.pdf, 2002.
[12] I. Guvenc and H. Arslan, "On the modulation options for UWB systems," in *Military Communications Conference, 2003. MILCOM 2003. IEEE,* 2003, pp. 892-897 Vol.2.
[13] K. Siwiak and D. McKeown, *Ultra Wideband Radio Technology*: John Wiley & Sons Ltd, 2004.
[14] I. Oppermann, L. Stoica, A. Rabbachin, Z. Shelby, and J. Haapola, "UWB wireless sensor networks: UWEN - a practical example," *Communications Magazine, IEEE,* vol. 42, pp. S27-S32, 2004.
[15] "http://www3.ntu.edu.sg/Centre/pwtc/research_projects_uwb.html."
[16] M. Baghaei Nejad, Z. Zou, H. Tenhunen, and L.-R. Zheng, "A Novel Passive Tag with Asymmetric Wireless Link for RFID and WSN Applications," in *Circuits and Systems, 2007. ISCAS 2007. IEEE International Symposium on*, 2007, pp. 1593-1596.
[17] M. Baghaei Nejad, Z. Zou, D. S. Mendoza, H. Tenhunen, and L.-R. Zheng, "Enabling UbiquitousWireless Sensing by a Novel RFID-Based UWB Module," in *First International EURASIP Workshop on RFID Technology*, Vienna, Ausria, 2007.
[18] R. Scholtz, "Multiple access with time-hopping impulse modulation," in Military Communications Conference, 1993. MILCOM '93. Conference record. 'Communications on the Move'., IEEE, 1993, pp. 447-450 vol.2.
[19] ETH, "RFID Multiple Access Methods," Zurich 2004.
[20] L. Su-Ryun, J. Sung-Don, and L. Chae-Woo, "An enhanced dynamic framed slotted ALOHA algorithm for RFID tag identification," in *Mobile and Ubiquitous Systems: Networking and Services, 2005. MobiQuitous 2005. The Second Annual International Conference on*, 2005, pp. 166-172.
[21] C. Jae-Ryong and K. Jae-Hyun, "Novel anti-collision algorithms for fast object identification in RFID system," in *Parallel and Distributed Systems, 2005. Proceedings. 11th International Conference on*, 2005, pp. 63-67 Vol. 2.
[22] S. M. David, M. Baghaei-Nejad, H. Tenhunen, and L.-R. Zheng, "Low Power Tunable CMOS I-UWB Transmitter Design," in *IEEE 2007 Norchip, 19-20 November 2007,* Aalborg, Denmark, 2007.
[23] M. Baghaei Nejad, H. Tenhunen, and L.-R. Zheng, "Chip-Package and Antenna Co-Design of a Tunable UWB Transmitter in System-on-Package with On-Chip versus

Off-Chip Passives," in *Electronics Systemintegration Technology Conference, 2006. 1st,* 2006, pp. 291-298.

[24] M. Baghaei Nejad, H. Tenhunen, and L.-R. Zheng, "Power Management and Clock Generator for a Novel Passive UWB Tag," in *System-on-Chip, 2007 International Symposium on,* Tampere, Finland, 2007, pp. 82-85.

[25] M. Motoyoshi and M. Fujishima, "43μW 6GHz CMOS Divide-by-3 Frequency Divider Based on Three-Phase Harmonic Injection Locking," in *Solid-State Circuits Conference, 2006. ASSCC 2006. IEEE Asian,* 2006, pp. 183- 186.

[26] Z. Zou, M. Baghaei Nejad, H. Tenhunen, and L.-R. Zheng, "Baseband Design for Passive Semi-UWB Wireless Sensor and Identification Systems," in *IEEE International SoC Conference SoCC 07,* 2007.

7

Development of Sensing and Computing Enhanced Passive RFID Tags Using the Wireless Identification and Sensing Platform

Alanson Sample[1,2], Daniel Yeager[1], Michael Buettner[1] and Joshua Smith[2]
[1]*University of Washington,*
[2]*Intel Research Seattle*
USA

1. Introduction

Passive RFID tags are becoming increasingly common in home and work environments. As RFID tags find new applications beyond shipment tracking, they are being embedded in objects throughout our environment. RFID tags are already being incorporated in credit cards for touch-free payments, in clothing for merchandise tracking, and in ID cards for building access control.

All these "non-shipping" RFID tags are powered wirelessly and are capable of wireless communication and rudimentary computation. Thus they can be viewed as micro-computing platforms with wireless power and communication capabilities. While the functionality of today's passive RFID tags is extremely limited, today's tags can already be thought of as a layer of invisible computing that is seamlessly embedded in objects throughout the environment. This primitive layer of embedded intelligence could grow in sophistication if additional sensing and computation capabilities could be added to RFID tags.

The authors' goal is to evolve this layer of passively powered embedded intelligence by creating RFID tags that support sensors and can execute general purpose computer programs. This chapter reviews several years' work on the development of our open, programmable passive RFID tag, the Wireless Identification and Sensing Platform (WISP). It also shows how to use the EPC Class 1 Generation 2 RFID protocol to implement advanced RFID sensing applications that go far beyond simple tag ID inventorying applications.

Our first venture into sensor-enhanced RFID was the α-WISP shown in Figure 1 (Philipose et al., 2005). With this device, one bit of sensor data was encoded by using anti-parallel tilt switches to multiplex one of two RFID tag ICs to a single antenna. Thus, a reader could infer three states about a tagged item (tag right side up, upside down, or not present). This simple example of overloading the EPC ID to encode sensor data allowed inference of very coarse orientation information. However, the use of commercial RFID tag ICs restricted our ability to control the RFID communication channel and in turn our ability to configure WISPs for new applications.

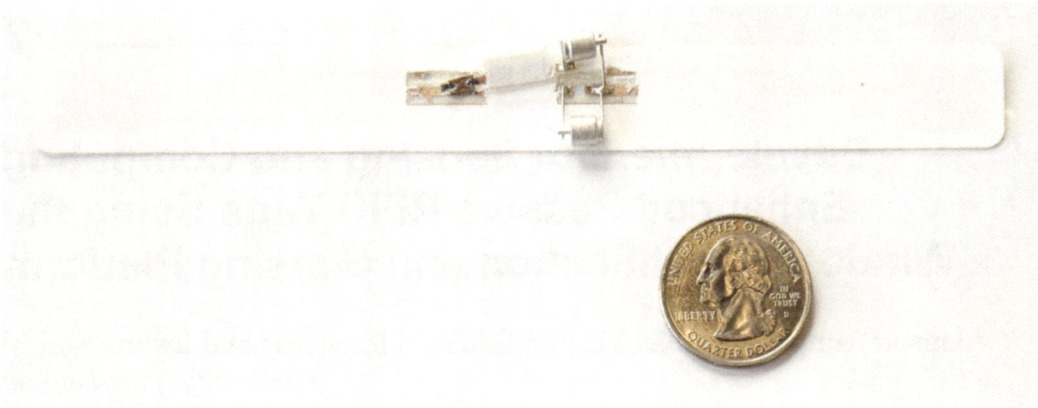

Fig. 1. The α-WISP uses two tilt switches orientated in opposite directions as a simple one bit RFID accelerometer.

In order to fully investigate passive RFID applications, we developed the general purpose Wireless Identification and Sensing Platform (simply called WISP) (Sample et al., 2008). Shown in Figure 2, the WISP is a battery-free, programmable RFID sensor device. Compliant with the Electronic Product Code (EPC) Class 1, Generation 2 protocol, the WISP can transmit multiple bytes of data per query and is fully configurable due to its ultra-low power 16-bit general-purpose microcontroller. Similar to conventional passive UHF RFID tags, the WISP has no batteries and is completely powered via the RF energy transmitted by an RFID reader.

The architecture of the WISP allows measurement of virtually any low power sensor which can also be wirelessly powered by the RFID reader. The WISP is implemented as a printed circuit board (PCB), which offers a flexible platform for exploring new sensor integration schemes and applications. To the authors' knowledge, the WISP is the first passive UHF RFID tag with an integrated microcontroller and has an operating range of several meters.

Fig. 2. Wireless Identification and Sensing Platform (WISP)

The first few section of this chapter presents an overview of the WISP platform including a detailed explanation of the architecture and power management algorithm. In particular, performance metrics describing operational range and real world performance are presented. Section 5 presents an over view of how the Electronic Product Code (EPC) Generation 2 Class 1 protocol is used to create a bi-directional communication channel for sending data to and from the a RFID reader. Section 6 explores the application space of the

WISP. This platform is intended to be a vehicle with which researchers can quickly investigate new and innovative applications in RFID. To highlight this concept several case studies of recent applications using the WISP are presented; for example, using the WISP to increase the security of RFID systems, and as a passive data logging device.

2. Prior work

To date there are several approaches to enhancing RFID tags with sensing capabilities. One method is to use standard commercial tag ICs and alter their functionality to transmit sensor data, as was done in the case of the α-WISP. The authors in (Johan et al., 2007) describe a humidity sensor for detecting moisture in walls of buildings and houses by placing a sponge in front of a tag. Moisture in the sponge detunes the tag's antenna, allowing the approximation of humidity levels from the read range of the tag. Another approach uses a custom tag with a built in fuse for sensing high temperatures in food products. The fuse melts above a particular threshold which enables or disables the tag (Watters et al., 2002). These passive tags based on physical properties are extremely limited in what they can report and are not reusable.

Other efforts have been made to retrieve richer, multi-bit sensor data from RFID tags for a wide variety of applications. Possible applications include infrastructure and object monitoring, automatic product tamper detection, identification of harmful agents, and biomedical devices for noninvasive monitoring (Want, 2004). To enable these applications two regimes have been explored: active battery-powered tags and passive battery-free tags.

Active tags, a subclass of RFID, are essentially wireless sensor nodes (Polastre & Szewczyk, 2005) (Savi Technology, 2006). They use batteries to power their communication circuitry, sensors, and microcontroller. Active tags benefit from a relatively long wireless range (approximately 30 m) and can achieve high data and sensing rates. An active tag with adaptive analog sensor thresholds for triggering sensor measurements was proposed in (Malinowski, et al., 2007). However, these devices require batteries which are a drawback when considering the cost, weight and volume of the device, and the need to replace the dead batteries.

In contrast, passive sensor tags receive their operating energy from the RFID reader which gives them a life time of years, if not decades. Examples of application-specific, non-programmable UHF passive tags with integrated temperature and light sensors, as well as an Analog to Digital Converter (ADC) can be found in (Namjun et al., 2005) and (Kocer & Flynn, 2006). One attractive feature of passive sensor tags is the prospect of permanently embedding them in objects for structural, medical, or product monitoring. Another advantage is their suitability for applications in which neither batteries nor wired connections are feasible, for weight, volume, cost, or other reasons. Of course, the limitation of purely passive sensor tags is the requirement of proximity to an RFID reader. However, methods such as solar, thermal, or kinetic energy harvesting could be used as a secondary power source if needed.

A further consideration is the configurability and computational power of RFID sensor tags. Existing devices are generally fixed-function with respect to sensory inputs and lack computational capabilities. A commercially available RFID tag with limited additional functionality is described in (Microchip Technology Inc, 2005); however, this device can only transmit one bit of sensor data in addition to its ID. Furthermore, it is limited by a short read range due to its 125 kHz operating frequency.

3. WISP architecture

The WISP is manufactured as a printed circuit board (PCB) which offers a number of advantages compared to traditional Integrated Circuit (IC) tag designs. Primarily low development cost, fast design cycles, and easy debugging and measurement of circuit parameters. The PCB implementation allows the flexibility to physically add and remove sensors and/or peripherals to create devices for new applications. In contrast, IC implementations offer the ability to customize components and decrease power consumption (yielding better range), as well as creating devices with a smaller form factor and at a lower cost when manufactured in high volume.

A block diagram of the WISP is shown in figure 2 and is similar in function to traditional IC RFID tags. The antenna is balanced by an impedance matching network and is fed into the RF power harvester. The Radio Frequency (RF) signal transmitted by the RFID readers is rectified into DC power to power the rest of the tag. The demodulator block converts the Phase-Reversed Amplitude Shift Keyed (PR-ASK) data that is superimposed on the RF carrier into a logic level steam of serial data. This extracted serial data is parsed by the MSP430 microcontroller (MCU) to receive downlink data from the reader. Uplink data is sent via the modulator circuit, which "back-scatters" the signal by changing the antenna impedance. Finally, the microcontroller's internal temperature sensor, as well as any external sensors, are powered and measured by the MCU.

As the power consumption of the microcontroller, sensors, and peripherals are much greater then that seen in traditional passive RFID technology, the WISP duty cycles between active and sleep mode. In sleep mode, the WISP shuts down and reduces its current consumption to a few micro-amps and energy is accumulated by the harvester block over multiple EPC queries. Once sufficient voltage is obtained, the WISP polls sensors and communicates with the RFID reader.

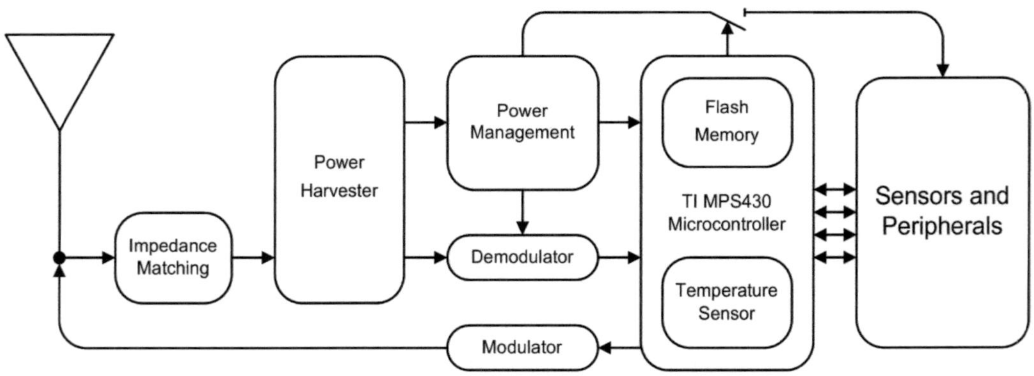

Fig. 3. Block Diagram of the WISP

Figure 3 depicts the WISP platform, made of a four layer FR4 PCB with components on both sides and an integrated dipole antenna. The WISP in its base configuration has several onboard sensors: a circuit for measuring the rectified supply voltage, a temperature sensor, and a 3D accelerometer. Small header pins expose all ports of the microcontroller for expansion to daughter boards, external sensors, and peripherals. Finally, a low current surface mount LED is included in the design.

3.1 RF power harvesting

The defining characteristic of far field RFID systems is that tags can be read at a significant distance, generally on the order of 2-10 meters. For passive RFID this requires that the RFID reader transmits sufficient energy to power the tag at large distances. However, due to regulatory limits on the amount of power that can be transmitted and the path loss associated with electromagnetic propagation, there is very little power that actually reaches the tags. Therefore, the power harvesting circuit must maximize the operating distance by converting the very limited incoming RF power to DC power with sufficient voltage to activate the tag.

The RF power received by the WISP's dipole antenna is fed to the analog front end depicted in figure 4. A discrete matching network is used to provide the maximum power transfer from the antenna to the rectifier. RF Schottky diodes specifically designed for 915MHz low power application were selected to make a five-stage voltage doubling circuit. This circuit converts the AC input signal to DC power which is fed into a storage capacitor.

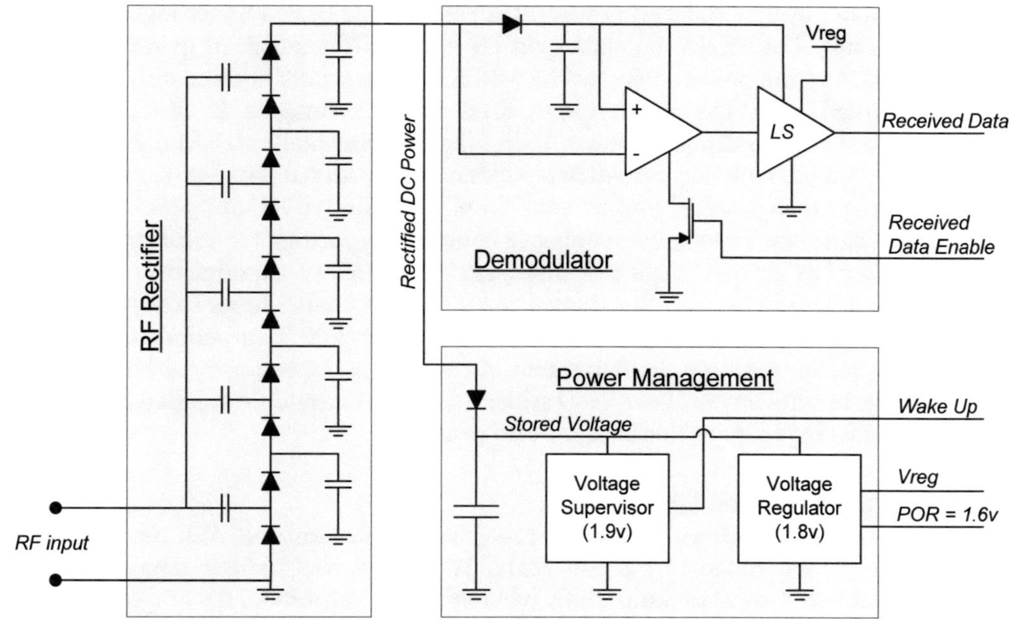

Fig. 4. Schematic of the Analog Front End

For RF rectifiers of this type, the input and output impedances are not well isolated. Further confounding the problem is that the output impedance of the rectifier is fairly high; an undesirable trait for any power source. This means that as the load on the rectifier changes the input impedance also changes, resulting in the analog front end becoming mismatched to the antenna. This leads to the problem of selecting values for the impedance matching network when it is not possible to guarantee constant input impedance.

To determine the correct values for the matching network the operating cycle of the WISP must be taken into account. First, the WISP is most effective at storing harvested energy when it is in sleep mode, as the current consumption is minimal. Second, the WISP will spend most of its time repeatedly charging up to 1.9v and then discharging to approximately 1.8v. Thus, to determine the correct values the WISP is put in sleep mode and

we find the impedance matching network that produces 1.9v for the lowest possible input power. Stated another way, the key parameter for maximizing the read distance of the WISP is minimizing the quiescent current consumption so that the minimum operating voltage of 1.9V (supervisor threshold) can be rectified with the lowest possible input power.
To characterize the system, a network analyzer was used to inject a continuous 915 MHz waveform into the antenna ports of the WISP. Using the minimum input power needed for activation, the expected operating distance for the WISP can be calculated with the logarithmic form of the Friis path loss equation (1), with a term for polarization mismatch included.

$$P_R = P_T - 20\log\left(\frac{4\pi d}{\lambda}\right) + G_T + G_R - L_P \tag{1}$$

The transmit power of the reader P_T = 1 W = 30 dBm. Its center frequency is 915 MHz, corresponding to a wavelength (λ) of = 0.33 m. The transmit antenna gain G_T = 6 dBi (this yields an effective isotropic radiated power (EIRP) of 4 W, the United States regulatory limit for this ISM band). The receive antenna gain G_R = 2 dBi (the standard gain figure for a dipole antenna), and the polarization loss L_P = 3dB. Loss L_P occurs because only half of the power transmitted from the circularly-polarized transmit antenna is received by the linearly-polarized receive dipole antenna. Using the experimentally determined operating thresholds of -9.5 dBm, equation 1 predicts a maximum operational range of 4.3m.
It should be noted that practical implementation of the WISP yields an operating range of approximately 3 meters. There are a number of contributing factors: the WISP antenna gain used in equation 1 is estimated not measured, and in the above experiment a continuous 915MHz signal was injected and the storage capacitor was charge to its steady state. Real RFID systems send out bursts of packets (power and data) with long periods of no signal between them. In this case the discharge rate of the storage capacitor must be taken into account. Finally, as with any far field RFID system, constructive and destructive interference due to multi-path plays a large role in real world results.

3.2 Demodulation and modulation

The EPC Gen 2 standard defines that reader-to-tag communication uses ASK modulation on a carrier wave in the range of 902-928 MHz. When not transmitting data the carrier waveform remains at a constant amplitude; when bits are transmitted, the amplitude of the carrier drops to at least ten percent of its normal value and the phase of the carrier may be reversed. The duration of the continuous waveform between these low amplitude pulses indicates logical "ones" or "zeros."
Figure 3 shows a schematic of the WISP's demodulator circuit. The output of the harvester is fed through the diode, which supplies power to the comparator and acts as a reference for the level shifter. A capacitor is used to filter out transients while allowing proper biasing at varying distance and receive power levels. When activated, the current consumption of the comparator functions as a constant-current source, pulling current through the diode. In this way, the voltage drop across the diode is used as a detector, where current supplied by the harvester (high amplitude RF modulation) results in positive voltage, and a lack of current (low amplitude RF modulation) yields negative voltage. The comparator is used to generate a rail-to-rail logic level waveform, and the level shifter converts the unregulated logic level to the regulated logic level. It is important to optimize current consumption and speed when

choosing a comparator. Further savings can be achieved by disabling the comparator when there is insufficient voltage to start up the MSP430.

Passive RFID tags do not actively transmit radio signals. Instead, they modulate the impedance of their antenna which causes a change in the amount of energy reflected back to the reader. This modulated reflection is referred to as back-scatter radiation. To change the impedance of the antenna a transistor is placed between the two branches of the dipole antenna. When the transistor conducts current, it short-circuits the two branches of the antenna which changes the antenna impedance; in the non-conducting state the transistor has no effect on the antenna, and thus the power harvesting and data downlink function as if it were not present.

3.3 Digital section and power conditioning

As the power available to RFID tags is extremely limited, careful component selection is critical to minimize current consumption. With advances in IC manufacturing that allow discrete components with current consumption in the range of 1 μA and operation at 1.8 V, it is now possible to construct functional, wirelessly powered RFID tags with discrete components.

The general purpose computation capability of the WISP is provided by an ultra low power microcontroller. This 16-bit flash microcontroller, the MSP430F2272, can operate at up to 4 MHz with a 1.8 V supply voltage and consumes approximately 600 μA when active at this frequency and voltage. Of particular interest for low power RFID applications, the MSP430 has a number of low power modes. Its minimum RAM-retention supply current is 0.1 μA at 1.5 V. The device provides over 8 kilobytes of flash memory, 256 bytes of RAM and a 10-bit, 200 kilo-samples-per-second Analog to Digital Converter (ADC). The low power consumption of this device is a critical factor in enabling a general purpose microcontroller in passive RFID systems.

Another critical design consideration is operation with uncertain power supply conditions. Because the available RF power varies greatly during device operation, supervisory circuitry is necessary to wake and sleep the device based on the supply voltage level. The WISP uses a 1.9 V supervisor and a 1.6 V power-on-reset to control device state and reset the microcontroller, respectively. The supervisor provides roughly 100 mV of headroom on the large storage capacitor above the 1.8 V regulator voltage. This serves to buffer the supply voltage from dropping below 1.8 V due to the large power consumption of the microcontroller in active mode.

4. Low level firmware and power management algorithm

The WISP is essentially a software defined RFID tag which uses the MSP430 to implement the EPC Gen 2 Class 1 protocol and performs sensing and computation tasks. There are significant challenges when developing applications on the WISP as compared to battery powered embedded systems. Primarily, there is no guarantee that a given task can be completed before running out of power. Although the voltage supervisor provides headroom above 1.8 V, the rate at which the energy stored in the supply capacitor is consumed is directly affected by the design choices of the programmer. Failure to properly manage sleep cycles, when the WISP harvests energy, or inefficient coding practices can result poor performance.

The WISP software can be described on three levels. At the lowest level is the power management algorithm which is responsible for managing the device state, including sleep vs. active modes. Built on that is the communication layer, which enables bi-directional communication by sampling downlink data bits, implementing a Gen 2 state machine, and generating uplink data bits. The third level is the application layer where users implement costume function and encoding data in the appropriate EPC packets.

4.1 Power management algorithm

Meeting the low power requirements of passive RFID tags requires that the MCU consumes, on average, as little power as possible. As mentioned previously, this is achieved by duty cycling between active and low power sleep states. The key is that the WISP receives a constant amount of power as defined by Friis path loss equation 1 for a set distance. When the WISP is in active mode the power consumption far exceeds the power harvested. However, when the WISP is in sleep mode the total current consumption of all the circuits is a few micro-amps and there is a net power gain which charges the storage capacitor. Therefore, duty cycling does not simply yield lower power consumption; it represents two different states, power harvesting and active operation.

The state diagram for the power management layer is shown in figure 5. State transitions are primarily driven by hardware interrupts from the voltage supervisor, which indicate if there is sufficient energy stored for operation. Initially the WISP is way from a RFID reader and is in a power down state. When the WISP is brought with in range of a reader it begins to harvest power and the voltage on the storage capacitors begins to rise. At approximately 1.6 V the MSP430 powers up in a reset state and begins executing code. Since this event is not driven by the supervisor it is important to enter sleep mode (LMP4) as quickly as possible in order to avoid browning out and thrashing on start up. Once in LMP4 the WISP waits for sufficient voltage (1.9 V) as indicted by the supervisor interrupt. Next, the state machine transitions to the application layer which performs user defined functions such as sensor measurements. Here an EPC packet is generated and the WISP sets up and waits from a commutation interpret which indicates the beginning of an EPC packet. In the communication layer the WISP processes the incoming data, executes the EPC Gen 2 protocol and transmits its response. While not shown in figure 5, the communication layer often reports the same date twice to increase communication reliability.

4.2 Communication and application layers

A considerable challenge when programming the MSP430 involves meeting the timing constraints of the EPC protocol while still maintaining a low clock frequency. RFID tags that have custom state machines are designed at the hardware level to receive and send using the EPC protocol. The general-purpose MSP430 must be carefully tuned to perform EPC communication, both for receiving and transmitting data. In particular, a mix of C and assembly language is used where the C code maintains ease of configurability for the firmware for different sensor applications and the assembly code allows fine-grained control of the timing of the MSP430 for EPC communication.

As previously described, the demodulator envelopes and thresholds the Phase-Reversed Amplitude Shift Keyed (PR-ASK) signal from the reader into a serial date stream representing the data bits 1 and 0 as long and short pulses, respectively. To interpret data from the reader, the MSP430 uses the periodic edge of the waveform as a hardware

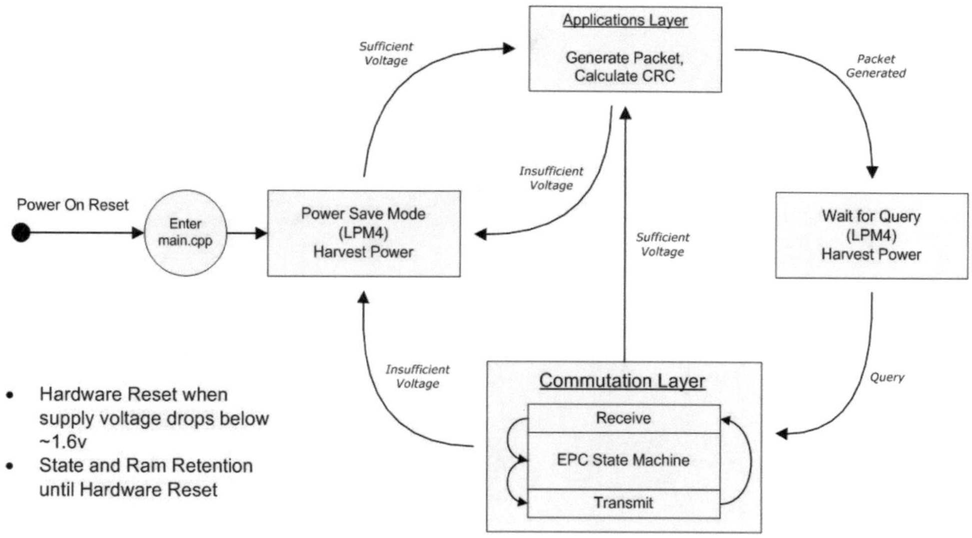

Fig. 5. Power Management Algorithm

interrupt, and then during the interrupt service routine re-samples the bit line to detect a 1 or 0 during the differentiated part of the waveform. This data is quickly shifted into memory before repeating this process. To detect the end of a transmission, a timer is refreshed during each bit. When bits are no longer received the timer expires, the packet is interpreted and, if appropriate, a response is sent to the reader. A detailed description of how the WISP uses and implements the EPC specification is described in section 5.

Figure 6 shows a set of EPC queries and responses along with the charge/discharge cycle of the WISP. Since the operating range of the WISP occurs between 1.9v-1.8v the rectified voltage appears to be nearly constant. In actually the WISP enters active mode at 1.9v, consumes the energy in the storage capacitor till ~1.8v, then enters a sleep state and harvests power until 1.9v is reached. This duty cycling can be seen in the packet transmitted plot. Here the WISP does not to responded to every packets sent by the reader, instead it spends most of its time in a sleep state.

Performing application level tasks such as sensor measurement is generally done in tight conjunction with the EPC protocol. In this scenario the completion of a receive/transmit cycle triggers the application layer to immediately take a sensor measurement, generator the desired EPC packet and setup for a Query. This protocol centric approach works well for sensor driven applications where data is requested from the RFID tag at regular intervals. However, applications which leverage the wirelessly powered computing capability of the WISP benefit from a loose coupling with the communication layer.

5. EPC class 1 generation 2 collecting sensor data

The Gen 2 MAC protocol used by the WISP provides primitives that can be used for gathering sensor data and transmitting queries. In this section, we give an overview of the Gen 2 protocol and discuss these primitives and their limitations.

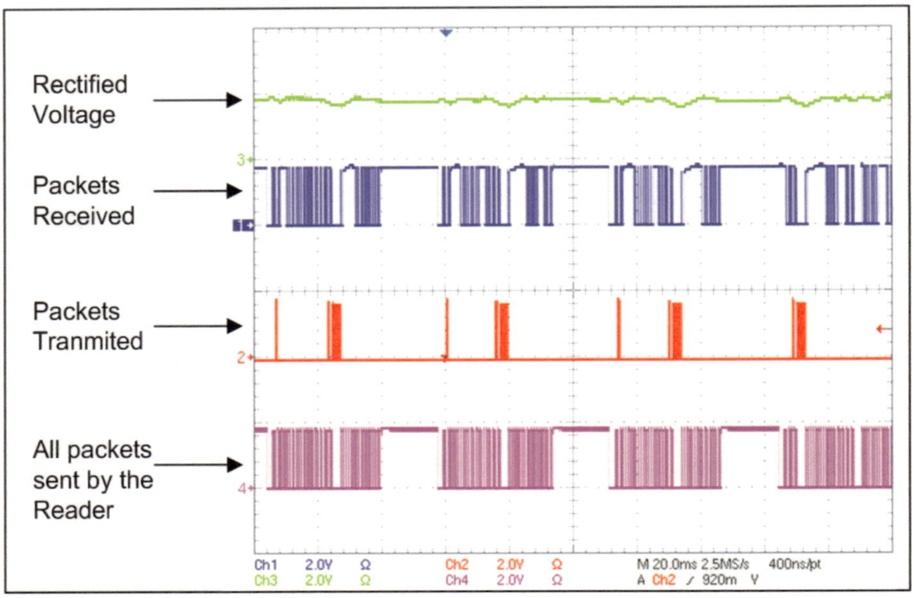

Fig. 6. Scope plot of the WISP responding to EPC queries along with the its rectified voltage.

5.1 Gen 2 background

5.1.1 MAC

The MAC protocol for Gen 2 systems is based on Framed Slotted Aloha (Roberts, 1975), where each frame has a number of slots and each tag responds in one randomly selected slot per frame. The number of slots in a frame is determined by the reader and can be varied on a per frame basis. Before starting a frame, a reader can optionally transmit a Select command which limits the number of active tags by providing a bit mask and a memory location, as only tags with IDs (or memory locations) that match this mask will respond in the subsequent frame.

To begin a frame the reader transmits a Query command which indicates the number of slots. Upon receiving a Query, each tag randomly chooses a slot in which to reply. If a tag chooses zero for its slot counter it responds immediately with a 16 bit random number (RN16). The reader echoes this RN16 in an ACK command and the tag responds with its ID. At this point, the tag is singulated. When a tag is singulated the reader can read and write tag memory as will be described in a later section.

After singulating a tag the reader transmits a QueryRepeat command which indicates the end of the slot. This signals to the tag that its ID has been read successfully and it should not respond in subsequent frames. Additionally, all other tags decrement their slot counter and transmit an RN16 if their counter reaches zero. Of course, tags may choose the same initial value for their slot counter. In this case, their transmissions will collide, the tags will not be singulated, and they will remain active in the next frame. A series of frames are conducted, each with a decreasing number of active tags, until all tag IDs have been read. This mechanism enables the rapid identification of tags and is supported by the WISP.

5.1.2 Tag memory

The Gen 2 standard specifies a memory architecture that includes banks for storing tag configuration information and the tag identifier, and also user memory with a size bounded only by the device hardware. In the case of the WISP, this user memory is used to store sensor data and the layout and semantics are defined by the WISP software. For instance, data for a particular sensor can be written to a given memory location or a time series can be written to a sequential range of memory locations.

5.2 Gathering sensor data

The Gen 2 protocol was designed to rapidly identify tags. However, when using the WISP for sensing applications the sensor data must be transmitted along with, or in lieu of, the simple identifier. The Gen 2 protocol provides two mechanisms that can be used for this: overloading the identifier and the Gen 2 Read command.

5.2.1 Overloading the identifier

The Gen 2 protocol efficiently reads tag identifiers, and by overloading the identifier to include sensor data a collection of WISPs can report data efficiently as well. In our initial applications the identifier was replaced with the sensor data of interest. However, when using more than one WISP, data from different devices cannot be differentiated. Additionally, this approach breaks the semantics of the protocol and limits the interoperability of WISPs and standard tags.

The Gen 2 specification allows for the transmission of up to 496 bits of identifier, while current tags generally have an identifier of only 96 bits. Hence, up to 400 bits of sensor data can be piggybacked along with the ID, enabling data from different devices to be differentiated and at least partially maintaining the original semantics. Unfortunately, by sending sensor data along with the identifier the read time per tag is increased and the time required to read data from a particular tag can be prohibitively high. For many sensing applications, particularly those that use a large number of devices, reading all sensor data from every tag will be undesirable and overloading the identifier may be insufficient to meet the application requirements.

5.2.2 Gen 2 read command

After singulating a tag, the reader can issue a series of Read commands to read the contents of tag memory, with each command eliciting up to 512 bytes of data. Before issuing a Read command the reader requests a temporary, random 16 bit handle from the tag. This handle is used in the Read command to address the tag, and an arbitrary number of Read operations can be issued in sequence. Using this mechanism, a reader can selectively read sensor data stored in the user memory of a single WISP.

Using the Read command to gather sensor data has drawbacks with respect to efficiency and flexibility. First, to read new data from a WISP the device must again be singulated and a new handle must be obtained. With a large number of tags, singulation is time intensive. Even in the best case where a single device is selected with the Select command, the singulation process must still be conducted, albeit with only a single tag responding, and a new handle must be obtained; only then can data be read from the WISP. This results in a large amount of the read time being protocol overhead. Additionally, the identifier of the device with the desired data must be known a priori, along with detailed knowledge of the memory layout with respect to sensor data location.

By overloading the identifier or using the Read command, basic sensing applications can be implemented using the WISP. However, when deciding which technique to use, the energy cost must also be considered. Specifically, using the Read command consumes more energy

than returning the data with the ID. This presents a trade-off between range and speed, with the proper balance being largely application specific.
However, for many deployments these mechanisms will likely be insufficient. This stems from the fact that such deployments are interested in sensor data and not simple object identities. Where object identification requires that all of the tags respond all of the time, the model for sensing protocols is very different. These protocols must express high level semantics explicitly in terms of sensor data, resulting in some of the tags relaying some of their data some of the time.

5.3 Transmitting data to the WISP

Along with gathering sensor data from the WISP, applications need to transmit data to the WISP, e.g. to actuate its behavior. While most of the Gen 2 commands implement specific functionality, the protocol does provide two commands which enable general purpose down-link communication: the Select command and the Write command.

5.3.1 Gen 2 select command

The Select command is intended to limit the number of tags that respond in a Query round. For example, a collection of retail items may have identifiers that indicate their model number, and the Select command can be used to inventory only items of a given model by providing a memory pointer and bitmask which matches only that model. However, this mechanism can be repurposed to function as a general purpose broadcast channel, with the pointer and mask being interpreted by the WISP software as opcodes and data. As an example, we have implemented software for the WISP which interprets Select commands as instructions to blink LEDs.

5.3.2 Gen 2 write command

Along with the general purpose broadcast facility of the Select command, the Gen 2 Write command can be used for unicast down-link communication. After a tag is singulated, the reader can write arbitrary memory locations on the tag in 2 byte words. Additionally, the BlockWrite command can be used to write up to 256 words at a time. This mechanism can be used to transfer data to the WISP, for example to store location information on the tag as it moves through a supply chain. Additionally, a WISP could be programmed to look to certain memory locations for parameters that affect its operation. For example, to modify the sampling rate of the WISP the Write command could be used to transmit the desired rate to a known memory location, and the WISP would refer to this value when setting its sampling rate.

5.4 Querying and tasking

To enable rich sensing applications, the reader must be able to query a collection of WISPs for particular types and ranges of sensor data. With this, a subset of the tags will reply and then only when their sensor data is relevant to the application. As a first approximation, the Select command can be used to match a bit mask on a particular user memory location. For integer valued sensors, a Select command could be used to select tags with specific sensor values over a given threshold and only tags that meet this criterion would respond in the following frame. Such an approach requires detailed knowledge of the memory layout and only works for queries that can be specified using a simple bit-mask.

As the WISP can perform general purpose computation, functionality at the device can be used to interpret abstract queries. For example, the WISP can be programmed to interpret SQL style queries, and the bit-mask of the Select command can be interpreted as commands instead of simple masks. Additionally, the Gen 2 protocol allows for the specification of Custom Commands which enable vendor specific functionality. Custom Commands are transmitted after singulation, and consequently can be used for sending unicast queries to a device.

More generally, by implementing high level functionality on the WISP, the Select command can be used as a broadcast channel to transmit opcodes and data to all devices, and Custom Commands can be used to send unicast messages to a single device. These two primitives enable the implementation of a wide range of protocols and a high degree of application flexibility and performance.

6. Applications

The longstanding goal throughout the development of WISP has been to facilitate RFID innovation. The WISP platform is designed as a research vehicle that allows people inside and outside the RFID community to explore new applications and usage models for RFID. Traditionally, RFID tag designers have been specialists in integrated circuit design (IC). They have generally focused on innovating CMOS circuit blocks such as RF rectification, power management, and low power state machines, with the goal of increasing tag read range. The process of manufacturing these custom IC tags presents a significant barrier to entry when considering the high cost of software, servers, chip fabrications, and specialized testing equipment; not to mention the long fabrication cycles.

Consequently, when researchers do add additional functionality, such as ADC and light sensors (Namjun et al., 2005) (Kocer & Flynn, 2006), the focus is on the device and is not driven by any particular applications. It is important to note that custom IC tag design will undoubtedly offer longer range, better performance vs. power consumption, and lower manufacturing cost in large volumes. However, it is difficult under this design paradigm to develop new applications that will take RFID beyond simple item tracking and identification.

In contrast, WISPs are flexible PCB based platforms that allow a relative novice to prototype both hardware and software RFID designs in a bench-top setting. The full-featured TI MSP430 allows for fast code development with debugging support. Sensors and peripherals can be easily added via the exposed headers or by using an optional daughter board. Testing equipment generally consists of an RFID reader and an oscilloscope. When compared to IC tags, probing and debugging circuit elements is easy and straightforward, as many of the signal lines are exposed by the PCB design.

The WISP fundamentally lowers the barrier to entry and allows people from a wide variety of fields to develop RFID technology. Whether it is students as part of a class project, security specialists, consumer electronics designers, or even artists, it is believed that a diverse group of people will be able to push RFID technology and find new and useful usage models. The hope is that this will lead to the discovery of compelling applications and that the IC tag designer will be able to draw upon the lessons learned from the WISP implementation. The following sections describe research being done with the WISP and focus on lessons learned.

6.1 Security

As RFID tags have become ubiquitous in the consumer marketplace for merchandise tracking and financial transactions, serious privacy and security concerns are being raised.

In many ways, the strengths of RFID are also its greatest weaknesses: RFID tags are mass produced, tiny, wireless transponders, which can be embedded in virtually any object and can be uniquely identified from a distance without the explicit consent of the owner. While this enables many valuable applications, it can also result in tagged items being used to identify and track individuals without their knowledge. In addition, RFID is increasingly being used to communicate sensitive data and not simple identifiers. Most notably, the banking industry has adopted RFID enhanced credit cards to enable fast, contactless payment. While the information transmitted via RFID is identical to that printed on the card, gathering this information no longer requires the conscious act of removing the card from a wallet and swiping the card through a magnetic reader. Consequently, thieves can steal the card information wirelessly, even while the card remains securely in the cardholder's wallet or purse; these attacks have already been seen in the wild.

To mitigate the privacy and security concerns inherent to RFID, there has been considerable interest in protecting the communication channel of RFID tags. One low-tech approach is to use a conductive sleeve to store RFID enabled devices, and the user must remove this sleeve for the device to be read. However, this relies on the diligence of the user and limits the usefulness of the technology. Consequently, more sophisticated approaches have been proposed that use cryptographic techniques to assure data authenticity and protect the data during transmission. Such techniques are generally beyond the capabilities of IC RFID tags, but are well matched to the WISP platform.

6.1.1 RC5 encryption

Conventional wisdom states that strong cryptographic algorithms are unrealistic for RFID considering the computational constraints and power issues of IC tags. As a result, various lightweight cryptographic protocols have been proposed and implemented. However, many of these protocols have serious vulnerabilities and were subsequently hacked or exploited (MBTA et al., 2008). However, the computational power and flexibility of the WISP enables the realization of stronger, more conventional cryptographic techniques designed to enhance both privacy and security.

In (Chat et al. 2006), the WISP was used to demonstrate RC5 based symmetric cryptography for use on UHF RFID tags. The particular RC5 variant implemented uses a 32-bit word, 12 rounds, and a 16-byte secret key which is stored in flash. While there were practical challenges in implementing RC5 on such a resource-constrained platform, the authors showed that with careful implementation strong cryptography is within the scope of UHF RFID. Additionally, their choice of RC5 was partly because RC5 can be efficiently implemented in both hardware and software, so their work can be used as a basis for IC implementations.

6.1.2 Context aware communication

Even when strong cryptography is used, RFID is still susceptible to "man in the middle" attacks. For instance, RFID is widely used for access cards where an RFID enabled employee badge uses a cryptographically strong challenge/response mechanism to open doors to a secured building. An attacker in this scenario does not need to break the encryption but only needs to generate the correct response to the RFID reader challenge. Attacks on such systems, referred to as "ghost and leech" attacks, involve one device near the reader (the "ghost") and one near the badge (the "leech"). The "ghost" receives the reader transmissions and forwards them over the internet to the "leech". The "leech" echoes these transmissions to the badge and when the badge responds, its response is forwarded back to the "ghost".

The "ghost" then transmits the badges response to the reader and the door is unlocked. In this scenario, the "ghost" may be at the building door while the "leech" is in a coffee shop down the street where the employee is having lunch.

The enabler of this attack is that the context of the badge is not factored into the access system protocol. This is largely because standard RFID offers no input vector to detect the context of the device. The sensing capabilities of the WISP can provide this context awareness, and in the case of "ghost and leech" attacks it can be used to detect user intent. In (Czeski et al., 2008), the 3D accelerometer of the WISP was used to implement a "secret hand shake" based authentication system to protect against "ghost and leech" attacks. When the user wants to authenticate a transaction or gain building access, they first perform a gesture with the card which unlocks the card and enables communication. The gesture could be a figure eight or any unique movement that the card would not experience in everyday activity. Only if this handshake is correct will the WISP unlock and transmits its ID to the reader. This approach leverages not only the computational power of the WISP, but also its sensing capabilities to provide a level of security that is not possible using standard IC tags.

6.2 Passive data logger

Several authors have demonstrated novel applications of passive RFID technology, which benefit from wireless, battery-free operation. However, these systems are inherently limited by the requirement of tag proximity to a reader for power and finite wireless range due to RF path loss over distance. One particularly interesting class of applications involves a tagged item that travels between two reader-equipped locations but does not have reader proximity during transit. For example, this situation occurs during cold chain transport for food and chemicals between warehouses. One may be interested in tracking the temperature or vibration of goods during transit where there is no reader coverage.

Fig. 7. WISP data logger with operational Super capacitor

To enable these applications, the authors have proposed a new tag device called a passive data logger (PDL). A PDL is a battery-free RFID tag with a large capacitor for energy storage. The PDL seamlessly recharges its capacitor when it is near a reader and uses the stored energy to measure attached sensors and log data to non-volatile memory (NVM) when it is away from a reader. Data is retrieved from the PDL using EPC Class 1 Generation 2 (Gen2) User-Memory "Read" commands. Additionally, the PDL can report an EPC ID like a conventional RFID tag.

There are several considerations in the design of a PDL. The most important design parameter is the reader-free runtime. Operating with a mean current I_{ave}, the expected runtime $\Delta t = C * \Delta V / I_{ave}$ where Δt is the runtime in seconds, C is the capacitance, ΔV is the difference between the maximum operating voltage V_{max} and the minimum operating voltage of the system, Vdd. Equation 2 defines the average current where T_{on} and T_{off} are the active and sleep mode durations, T_o is the period and I_{active} is the active mode current and I_{sleep} is the sleep mode current.

$$I_{ave} = (I_{active} * T_{on} + I_{sleep} * T_{off}) / T_o \quad (2)$$

Figure 8 illustrates several calculated operating times given a fixed capacitance and current consumption

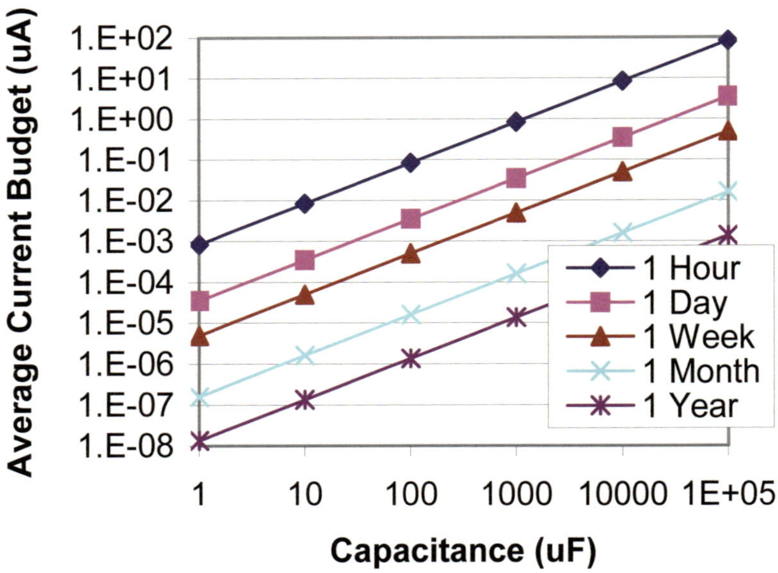

Fig. 8. Average current budget for various runtimes and storage capacitor

Another important parameter is the wireless charge time. The charge time is determined by capacitor size, Vmax and the available power at a distance d from the reader.

$$E_{stored} = \frac{1}{2} C \left(V_{charged}^{2} - V_{dd}^{2} \right) = P_{charge} * T_{charge} \quad (3)$$

The available power is modeled by Friis' transmission equation for path loss (equation 1), with terms for rectifier efficiency and polarization included.

The authors have implemented a PDL based on the WISP platform (WISP-PDL), which benefits from a programmable microcontroller for rapid, flexible prototyping. As a proxy for cold chain monitoring, a refrigerated milk container was instrumented with a WISP-PDL and monitored throughout its consumption (Yeager, et al., 2008). For this study, the WISP-PDL sampled and logged data in 10 second intervals and consumed 1.8 μA on average from a 1.8 V supply. Over the course of 24 hours, the temperature and fill level of the carton was measured and written to memory. At the end of the study, the data was read from the WISP-PDL using the Gen 2 Read command showing the complete history of the milk carton. As the refrigerator acted as a faraday cage, the WISP-PDL harvested energy only when removed from the refrigerator but continued to sense when not directly powered by a reader.

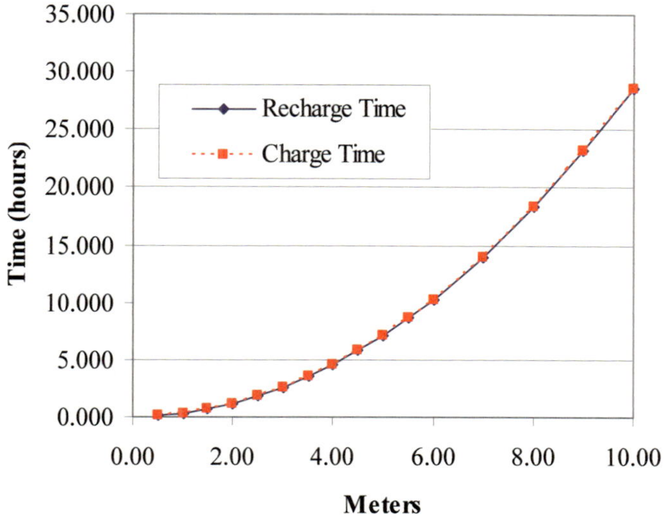

Fig. 9. Wireless charge rates for the data logger WISP

A second application of the PDL involved primate neural signal monitoring (Holleman, et al., 2008). To accurately measure neural activity, neural probes must be placed directly on the brain but the probes must also be powered. Consequently, transcutaneous wires are generally used which lead to a high risk of infection, or inductive power systems are used which require a bulky energy source that must be placed within a few centimeters of the implanted device.

The WISP platform provides key advantages for this application, as neural probes can be attached and the entire device can be implanted and powered wirelessly by an RFID reader. However, the RF transmissions from the reader overwhelm the sensitive neural probes, and thus the reader cannot be active while the neural probe is taking measurements. To overcome this limitation a WISP-PDL was used and a standard RFID reader charged the device for 3 seconds. The reader then powered down and the WISP-PDL began taking measurements for 5 seconds and storing the measurements to memory. After the sensing phase, the reader powered up and downloaded the measurement data from the WISP-PDL using the Gen 2 Read command. This process was repeated yielding high resolution neural pulse data.

7. Conclusions

This chapter presents the design of the Wireless Identification and Sensing Platform, a battery-free programmable RFID sensor device compliant with the Electronic Product Code Class 1 Generation 2 protocol. The WISP operates from wireless power at a distance of several meters and provides a robust bidirectional communication channel built on top of the RFID reader physical layer. The microcontroller allows the WISP to be reconfigured for new tasks and easily accommodating the integration of low power sensors. In a larger context the WISP demonstrates the feasibility of UHF RFID systems powering and reading complex tags which utilize compotes such as microcontroller and sensors.

The WISP has proven that is a flexible platform which allows research to quickly investigate new and innovative applications. Examples include the use of WISP for development of enhanced security measures on RFID tag, and for the development of a passive data logging device. It is believed that as more researchers use the WISP new and innovative RFID devices and applications will be discovers which can be later implemented in a integrated circuit design,

8. References

Holleman, J., Yeager, D., Prasad, R., Smith J., Otis, B., "Neural WISP: An Energy Harvesting Wireless Brain Interface with 1m Range", IEEE Biological Circuits and Systems (BioCAS) 2008, accepted, expected to publish November 2008.

Johan, S.; Xuezhi Zeng; Unander, T.; Koptyug, A.; Nilsson, H.-E., "Remote Moisture Sensing utilizing Ordinary RFID Tags," *Sensors*, 2007 IEEE , vol., no., pp.308-311, 28-31 Oct. 2007

Kocer, F. and M. P. Flynn. "A new transponder architecture with on-chip ADC for long-range telemetry applications," *IEEE Journal of Solid-State Circuits*, vol. 41, no. 5, May 2006, pp. 1142-1148

Malinowski, M., Moskwa, M., Feldmeier, M., Laibowitz, M., and Paradiso, J. A. 2007. "CargoNet: a low-cost micropower sensor node exploiting quasi-passive wakeup for adaptive asychronous monitoring of exceptional events." *In Proceedings of the 5th international Conference on Embedded Networked Sensor Systems* (Sydney, Australia, November 06 - 09, 2007). SenSys '07. ACM, New York, NY, 145-159.

MBTA vs. Anderson, et al (United States District Court - District of Massachusetts August, 2008)

Microchip Technology Inc., *125 kHz Passive RFID Device with Sensor Input*, August 25, 2005

Namjun, Cho, et al, "A 5.1-µW 0.3-mm2 UHF RFID Tag Chip Integrated With Sensors for Wireless Environmental Monitoring," *IEEE European Solid State Circuits Conference*, September, 2005, Grenoble, France. P. 279- 282.

Philipose, M.; Smith, J.R.; Jiang, B.; Mamishev, A.; Sumit Roy; Sundara-Rajan, K., "Battery-free wireless identification and sensing," *Pervasive Computing, IEEE* , vol.4, no.1, pp. 37-45, Jan.-March 2005

Polastre, J.; Szewczyk, R.; Culler, D., "Telos: enabling ultra-low power wireless research," Information Processing in Sensor Networks, 2005. IPSN 2005. Fourth International Symposium on , vol., no., pp. 364-369, 15 April 2005

Sample, A.P., Yeager, D.J., Powledge, P. S., Mamishev, A.V., and Smith, J. R., "Design of an RFID-Based Battery-Free Programmable Sensing Platform," *IEEE Transactions on Instrumentation and Measurement*, Accepted for future publication.

Savi Technology. Savi SensorTag ST-676. Datasheet, 11 June 2006. http://www.savi.com/products/SensorTag_676.pdf

Roberts, L. G., "Aloha packet system with and without slots and capture" *SIGCOMM Comput. Commun.* Rev., 5(2):28–42, 1975.

Want, R., "Enabling ubiquitous sensing with RFID," *Computer*, vol.37, no.4, pp. 84-86, April 2004

Watters, D. G., Jayaweera, P., Bahr, A. J., and Huestis, D. L., *Design and performance of wireless sensors for structural health monitoring*. In D. O. Thompson and D. E. Chimenti, editors, AIP Conf. Proc. 615: Quantitative Nondestructive Evaluation, pages 969–976, May 2002.

Yeager, D.J., Powledge, P.S., Prasad, R., Wetherall, D.; Smith, J.R., "Wirelessly-Charged UHF Tags for Sensor Data Collection," *RFID, 2008 IEEE International Conference on* , vol., no., pp.320-327, 16-17 April 2008

8

Surface Acoustic Wave RFID Tags

S. Härmä[1] and V. P. Plessky[2]
[1]*Helsinki University of Technology,*
[2]*GVR Trade SA,*
[1]*Finland*
[2]*Switzerland*

1. Introduction

The first radio identification systems appeared already during World War II for military applications, namely, for identification of planes. However, it is only now that the technical conditions are right for a wide use of radio frequency identification (RFID). The two key issues for the RFID technology are the number of different codes that can be 'written' on a tag and the possibility to store, transfer, and communicate information. Due to the ongoing miniaturization of semiconductor integrated devices, mass production of such devices at a low cost has become possible. More specifically, micro- and nanometer lithographic technology allows for the fabrication of very small tags (with a chip size on the order of 1 mm and smaller) operating at the GHz-range where sufficiently wide frequency bands are available. These industrial, scientific, and medical (ISM) frequency bands can be used without licensing with a limited radiated power. The wide frequency bands finally allow for a practically infinite number of different codes to be written and read at microsecond time intervals. As to the second key issue, the omnipresent internet, intranet, and similar communication networks enable the processing of databases and developing of smart systems that use the information automatically read from RFID tags.
It is interesting to point out that the dramatic development of mobile phones (that only combine a transmitter and a receiver, both used in radio communications for a century by now) was based exactly on the same two reasons: first, the development of technology allowing the use of high frequencies, wide frequency bands and, finally a large number of subscribers and, second, computer databases with high-speed data links enabling fast communication. The type of RFID tag introduced in this chapter, the surface acoustic wave (SAW) tag, is in many aspects similar to RF SAW filters, that are widely used in mobile phones. SAW tags and SAW filters use the same technology.

2. Active and passive RFID tags

RFID tags basically fall into two categories depending on whether they are passive or active. While active tags usually have an on-board battery, passive tags power their circuitry by using a part of the interrogation signal energy transmitted by an external reader. The incorporation of a battery makes a device expensive, limits its life-time, and furthermore, makes it questionable in environmental aspects. The application of a rectifier stage for extracting power from the interrogation signal, together with the limited licensed radiation power of the read-out signal, restricts the reading distance to a very limited range. SAW tags do not fit into either of the two categories. They do not require any power supply. They simply return (reflect) the interrogation signal in a coded form that carries the identification information. SAW tags employ SAW delay lines and feature low losses, large delay times, and small dimensions. In addition, they have a simple and robust structure.

As compared to the widely used barcode, both semiconductor-based and SAW-based RFID tags have the following obvious advantages:

- They can be read automatically, that is, without human presence. This allows for an unambiguous identification of objects, people, and animals.
- They do not need to be in line-of-sight to the reader nor is any particular tag orientation demanded.
- They can have a reading distance as large as 10 m and even larger, depending on the system used. For barcodes, reading distance is limited to about 30 cm.

3. SAW RFID tags

The principal characteristic of SAW tags is that their operation is based on microacoustics of piezoelectric crystals instead of semiconductor physics. The main advantage of these devices is their total passiveness: they do not require any DC power because they merely reflect the interrogation signal. Moreover, the interrogation signal can be about 100 times smaller (about 2 mV on the tag antenna) than for integrated circuit (IC) based tags. Another attractive feature is the simple structure. SAW tags are fabricated using single-metal-layer photolithographic technology. Admittedly, operation in the microwave region requires submicron lithography (about 0.3-μm-wide electrodes), which is a standard tool today in IC fabrication. This enables the fabrication of devices working at the 2.45-GHz frequency band reserved globally for ISM applications.

SAW tags utilize the unique nature of piezoelectric materials which allows for a transformation of electromagnetic waves into 100 000 times slower surface acoustic waves. SAW tags can hence function as delay lines and provide a sufficient delay (with a relatively small substrate length) for temporally separating the tag response signal from the read-out signal.

3.1 Principle of operation

The fundamental physical phenomenon lying behind SAW devices is piezoelectricity. This is, in general terms, a coupling between a material's electrical and mechanical properties: in certain dielectric crystals, the application of mechanical stress produces an electric polarization and, conversely, such a crystal undergoes a mechanical distortion when an electric field is applied. This property is used in SAW devices and in many other applications to produce a mechanical output from an electrical input or vice versa. In SAW devices, the transduction between an electrical signal and an acoustic wave is achieved by utilizing an interdigital transducer (IDT), consisting of two interlaced comb-like metal structures deposited on the surface of a piezoelectric substrate.

The principle of operation of a reflector-based SAW tag is shown schematically in Fig. 1. A reader emits an interrogation pulse, which is received by the tag antenna, directly connected to an IDT. The IDT transforms the electrical signal into a nano-scale surface acoustic wave, which is a mechanical wave of particle displacements. The generated SAW pulse then propagates along the surface of the substrate, which is usually made of a strong piezoelectric material such as lithium niobate ($LiNbO_3$). The SAW pulse is partially reflected and partially transmitted by each of the so-called code reflectors, placed at precisely determined positions on the chip. These reflectors usually consist of one or a few narrow aluminum strips. The reflected SAW returning to the IDT thus carries a code based on the positions of the reflectors. In other words, this encoding method is based on the time delays of reflected pulses. It is known as time position encoding or pulse position modulation (PPM) and is described in further detail in section 3.3. When the train of reflected SAWs finally returns to the IDT, the acoustic signal is reconverted into an electrical form and retransmitted by the tag antenna. The response signal is then detected and decoded by the reader. In SAW tags, a surface acoustic wave is hence used for 'reading' a sub-micron 'barcode' of properly arranged reflectors.

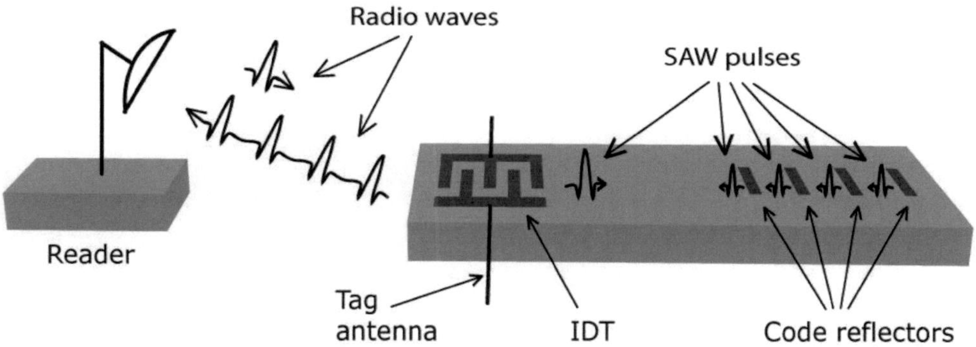

Fig. 1. Operating principle of a SAW tag system.

Figure 2 shows a typical time response of a SAW tag. In this case, FEM/BEM software has been used to simulate the performance of a tag having 14 code reflectors. As illustrated by the mask image in Fig. 3, ten of the reflectors are used for encoding itself; the first and the last reflector are used for calibration and are typically designed to have stronger responses than the others; while the two reflectors preceding the very last one are used for error control, for creating a checksum. The reflector array is designed to produce uniform amplitudes for code reflections, in order to help achieve a maximal read range. Amplitudes of response signals are adjusted by gradually increasing the reflectivity of code reflectors, by adding electrodes to reflectors and by increasing their width. This is done in order to compensate for the losses due to propagation on the substrate surface and to reflections from preceding code reflectors. As mentioned above, a SAW tag must provide certain time delay in order to separate the response signal from the read-out signal. The reflected signals must be received by the reader only after a delay sufficient for the environmental echoes (reflections from walls or other nearby objects) to die away. An adequate initial delay is typically about 1 μs and is facilitated by leaving about 2 mm of empty space on the substrate between the IDT and the code reflectors. The free-surface SAW velocity on $LiNbO_3$ is about 4000 m/s.

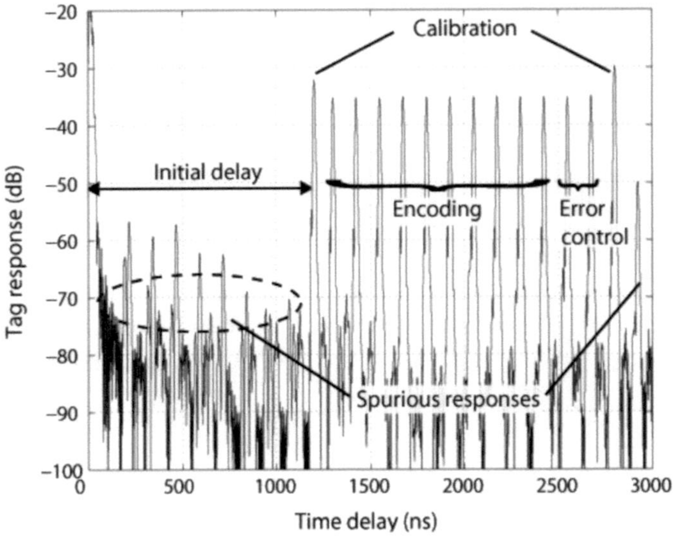

Fig. 2. Simulated SAW tag response.

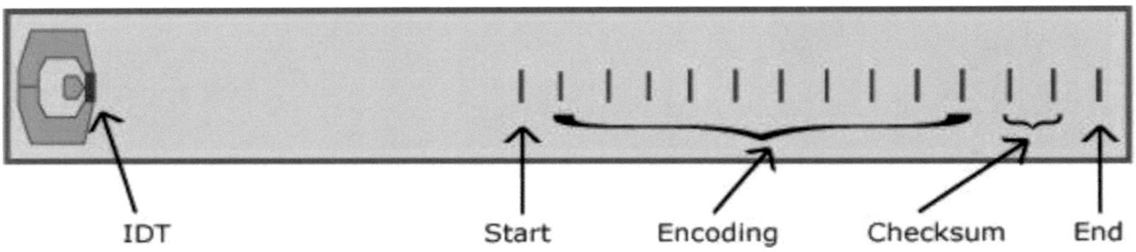

Fig. 3. Mask image of a reflector-based SAW RFID tag.

3.2 Data capacity of SAW tags

Although the idea of SAW tags was already proposed decades ago (Davies et al., 1975), its final commercial breakthrough has not yet realized. In order for the SAW tag to become a commercially attractive product for mass applications, its data capacity must be at least 20 to 32 bits, that is, a few million codes or, better, billions.

The number of different codes is determined by the *BT* product (*B* is the used frequency band and *T* is the coding time), as suggested by Shannon's formula (Shannon, 1948). As a SAW tag must be small and cheap, we cannot use more than 2 μs to 4 μs for coding. These delays correspond to propagation distances of 8 mm and 16 mm. If a data capacity of 32 bits (better 64 bits or 128 bits) is desired, a frequency band of 16 MHz (or 32 MHz, or 64 MHz) is needed. Such frequency bands are available only at relatively high frequencies. Effectively, the only suitable frequency range available globally is the ISM band from 2400 MHz to 2483.5 MHz. This band is now extensively used around the world for local communication systems: Bluetooth, WLAN, wireless keyboards, etc.

Achieving a sufficient number of codes in a SAW tag hence requires the use of the 2.45-GHz range. This calls for submicron photolithographic tools as the narrowest linewidths needed at this frequency range are on the order of 0.3 μm to 0.4 μm. It is to be noted that this

requirement is rather modest in comparison with the state-of-the-art IC technology operating with a resolution down to 0.05 μm. SAW tag technology can hence reuse equipment from older generations of IC, which decreases the fabrication cost.
An identification code can be written on the SAW tag in time positions, amplitude, phase, or other suitable signal characteristics of the reflected pulses. The reflected pulses represent the symbols of the tag response signal and can code for one or more bits each. The first commercial SAW tags, designed according to these principles, are currently used in demanding industrial environments, more specifically, for automation of car assembly lines. The number of unique codes commercially achievable at present is rather limited: on the order of 10 000. New ideas are currently being developed aiming at a radical increase in the data capacity of SAW tags to 64 or even 128 bits (Hartmann, 2002).

3.3 Time position encoding

SAW RFID tags can be encoded in several ways. Currently existing SAW tag products use the so called time position encoding (Plessky et al., 1995; Stierlin & Küng, 2002), which represents the most straightforward way of data encoding in SAW tags. This is the only method currently used in commercial SAW tags (Reindl & Shrena, 2004; Stelzer et al., 2004).
In this encoding scheme, the total time delay is divided into slots of certain duration. The slot width is roughly equal to the time width Δt of the pulses, that is, $\Delta t = 1/B$, where B is the frequency band of the overall system (actually determined by the band of signals radiated by the reader). At 2.45 GHz, a band of 40 MHz is typically used, and the corresponding slot width is thus 25 ns. The slots form groups of, for example, five slots. For a tag using such grouping, one of the first four slots of each group is occupied by a reflector while the fifth one, the guard slot, is always left empty (see Fig. 4). Each reflector thus has four possible positions (equal to 2 bits of data) and the total number of different realizable codes is 4^n for a tag having n reflectors. Ten code reflectors will thus yield about 10^6 distinct codes. When all the reflectors are placed in-line in one acoustic path, the chip space required by these ten reflectors is about 2.5 mm. The advantage of this encoding method is that one always has the same number of reflectors, which makes it easier to design a SAW tag with uniform amplitudes of response signals. Also, for the reader, the problem then simplifies to the search of a single response in a given group of time slots.

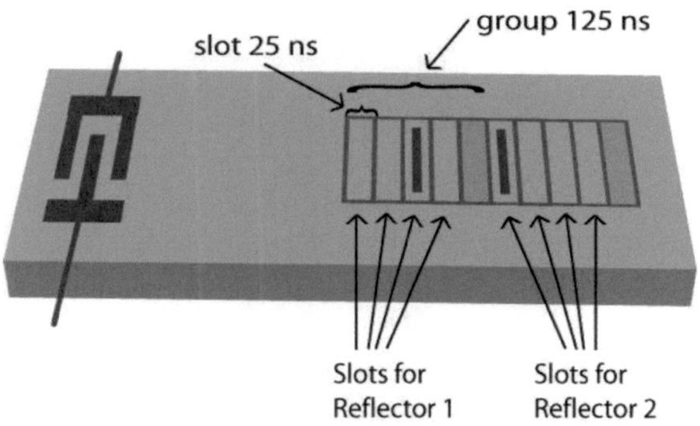

Fig. 4. Principle of time position encoding in SAW tags.

To maximize the number of codes (for a given total coding time) in the time position encoding scheme, about 3 to 4 slots per group must be used. However, in practical devices, as in that shown in Fig. 5, decimal groups are employed. In such a scheme, a reflector can occupy one of ten possible positions. Commercially available SAW tags have a data capacity of 10 000 different codes, which in the decimal time position system corresponds to four code reflectors.

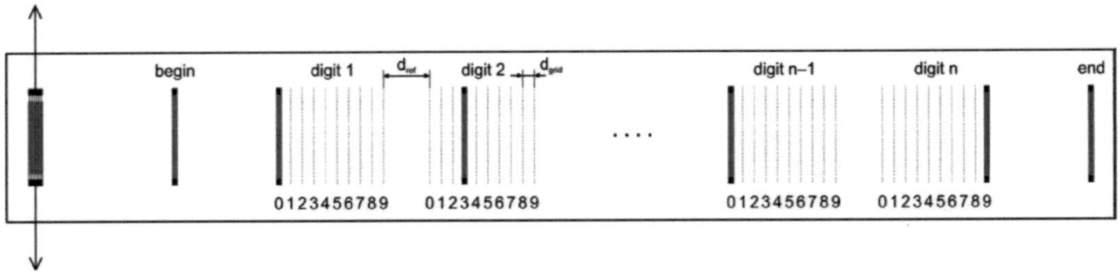

Fig. 5. Practical example of time position encoding.(Stelzer et al., 2004)

3.4 Phase encoding

For time position encoding, the exact coordinate of a particular reflector is not so important. It must be within a 25-ns time slot. The calibration reflectors help to account for inaccuracies in position due to temperature, technology variations, and other shifts. A code reflector often consists of only one or a few electrodes. A single reflector electrode has a width of about 0.4 μm to 0.6 μm, that is, it is significantly narrower than the slot it occupies. The slot width of 25 ns corresponds to about 50 μm.

If the phases of the reflected pulses could be measured accurately, the coding capacity would increase significantly. Phase encoding has been discussed for many years but not yet implemented in actual products. The idea of phase coding is simple: by displacing the reflectors slightly, phase shifts can be realized and phase coding implemented. Figure 6 illustrates the principle of introducing phase shifts of 90° by shifting reflector positions by multiples of $\lambda/8$ (Härmä et al., 2008a). In such a case, each reflector can have 4 phase positions, which adds 2 additional bits to time position encoding. The above described SAW tag with ten code reflectors will then have 2^{40} variants of codes, 40 bits, or about 10^{12} different codes. This is a large number: for every human being on Earth, there will be about 150 tags available with different codes never repeated.

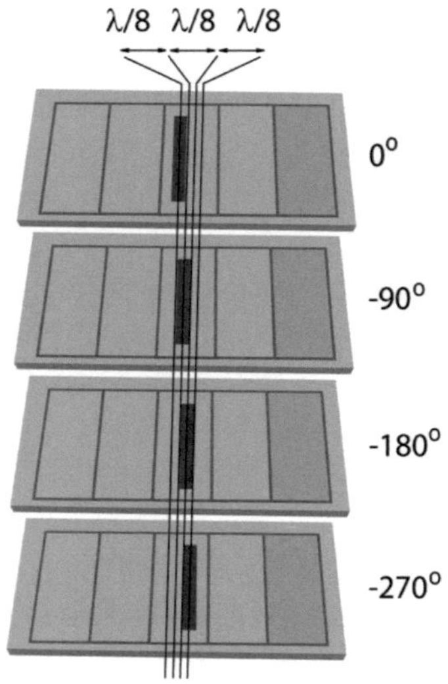

Fig. 6. Principle of phase encoding.

Phase coding can be combined with time position encoding in a more clever way (Hartmann 2002, 2004, 2005): instead of keeping time slots unchanged and introducing phase modulation of the reflectors, it is proposed to narrow the time slots and radically increase the number of slots in a group, keeping the duration of time for the whole group unchanged. Each slot is assigned a definite phase of the reflector if the reflector is placed there. In this modulation scheme, the phase is used to determine the time position of the reflected pulse. It is evident that the uncertainty of measurement of phase of reflected responses depends on the signal-to-noise ratio (Kuypers et al., 2008). The needed strength of signal increases with increasing accuracy of phase values used for encoding. Optimal methods for phase encoding and decoding are under intensive investigation.

3.5 Encoding technology

Whatever encoding scheme employed, each SAW tag produced has a unique physical appearance. Currently, for the manufacture of only 10 000 different codes, the images of all these tags are placed on a large - but still reasonable - number of photomasks. This technology will evidently be too expensive for 10^6 codes and totally unrealizable for 10^{13} different codes.

An idea of a double-stage photolithography process, where in the first stage all reflectors in all possible positions are produced (or at least exposed) with high accuracy, and subsequently all redundant reflectors are deleted, say, with some fast and programmable tool, was already proposed. However, it has not yet been implemented.

4. Developments in SAW tags

The main goals of SAW tag design include a reduction of device losses, a reduction of device size, and an enhancement of data capacity. A combination of time position encoding and phase encoding provides a means for increasing the information capacity, as described in section 3.4. This chapter presents ideas of further solutions and shows that a small device size can be achieved for SAW tags simultaneously with a sufficiently large data capacity.

4.1 Loss reduction in SAW tags

A standard IDT, as depicted in Figs 1 and 4, consists of electrodes with alternating polarities. As it transforms the electrical signal into an acoustic form, it generates surface acoustic wave propagation equally in both directions. When such a bidirectional IDT is used in SAW tags, half of the signal energy is already lost in transduction. This problem can be overcome by using a unidirectional IDT that only generates wave propagation in one direction. For a similar reason, SAW-tags with several parallel acoustic channels will have a higher loss level than a device wherein all reflectors are situated in the same channel.

However, typical unidirectional transducers (more specifically, single-phase unidirectional transducers, SPUDTs) include electrodes with a width of $\lambda/8$, where λ is the wavelength of SAW on the piezoelectric substrate. At 2.45 GHz, $\lambda/8$ is about 0.2 μm. This makes SPUDT-type transducers inaccessible for the photolithography currently used in SAW industry. Recently, however, a SPUDT especially designed for SAW tag applications was proposed (Hartmann & Plessky, 2007) exploiting the fact that, on 128°-$LiNbO_3$, the reflectivity of short-circuited electrodes can reach zero at some metal thickness and electrode width. The proposed transducer uses $\lambda/4$-wide (and wider) electrodes.

Currently used tags, that use a bidirectional IDT, have a loss level on the order of -55 dB for 10 000 codes. This will be reduced to about -40 dB for SAW tags with a unidirectional IDT and 10^6 codes (Härmä et al., 2008b). The unidirectional transducer may include 0.3-μm-wide electrodes. Also the reflectors must be rather narrow: for some cases, reflector electrodes must have a width of 0.3 μm to 0.4 μm. Therefore, a reliable photolithography capable of producing linewidths of 0.3 microns is needed. In addition to reduced losses, the use of a SPUDT in SAW tags has the advantage of a lower level of parasitic reflections, including reflections from the transducer itself.

4.2 Size reduction of SAW tags

Replacing the bidirectional IDT with a unidirectional IDT also serves to reduce the chip size. SAW tags using a bidirectional transducer are normally designed to have their reflectors on both sides of the transducer. In this case, space for the initial delay must also exist on both sides, which results in an inefficient use of the substrate area. When a unidirectional transducer is employed, all reflectors must be placed on the same side of the transducer and only one initial delay is needed.

A further reduction of chip size can be achieved by folding the channel used for SAW propagation. A Z-path SAW tag has been designed and fabricated (Härmä et al., 2008b) that uses two inclined, strongly reflecting mirrors (each consisting of an array of open-circuit metal strips), as shown in Fig. 7. Although such folding demands two additional reflectors (and four reflections of the signal), which inevitably results in additional losses on the order of -5 dB to -10 dB, the reading distance is reduced less than 2 times. This can be an acceptable price to pay for a significant reduction of size and cost of a SAW tag.

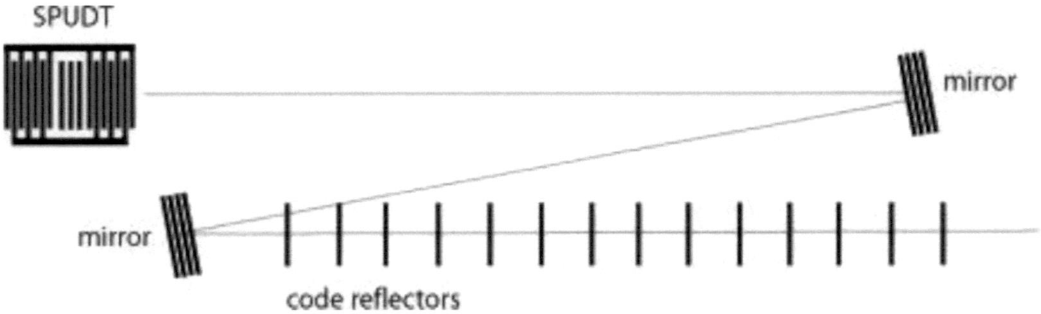

Fig. 7. Z-path SAW tag geometry with two inclined reflectors.

4.3 Ultra-wideband SAW tags

The currently emerging ultra-wideband (UWB) technology offers many attractive possibilities for the development of SAW RFID tags. According to the regulation of the United States Federal Communications Commission (FCC) (Breed, 2005), an UWB device is a device emitting signals with a fractional bandwidth greater than 20% or a bandwidth of at least 500 MHz. A SAW tag operating at 2.5 GHz with a band of 500 MHz would satisfy this criterion. The UWB band being much wider than the 2.45-GHz ISM band, a certain value of *BT* product, determining the data capacity of a tag, can now be achieved with a significantly shorter coding delay, which enables a considerable reduction of tag size. For example, with B = 500 MHz, a *BT* of 200 only requires a coding time of 400 ns instead of the 2 μs typical for

2.45-GHz SAW tags. The total chip size can then be smaller than 0.5 x 1.0 mm^2. A shorter coding time also implies lower losses. A propagation time of 400 ns only corresponds to about -3 dB propagation loss. Another interesting possibility is to have signal processing partly performed within a SAW tag using, for example, a chirp transducer (Ianelli & Koslar, 2004), as illustrated in Fig. 8. This will allow for a matched-to-signal processing of the tag response, which, after being modified within the tag, will be different from the environmental echoes of the interrogation signal, also received by the reader. This makes the system more resistant to environmental interference, as the reader is now able to distinguish between the signal reflected by the SAW tag and that reflected by objects outside the tag. As the principle of the ultra-wideband technology is to reuse an already occupied frequency spectrum but with very low power, an UWB SAW tag system will also have an additional advantage of very low transmitted power levels.

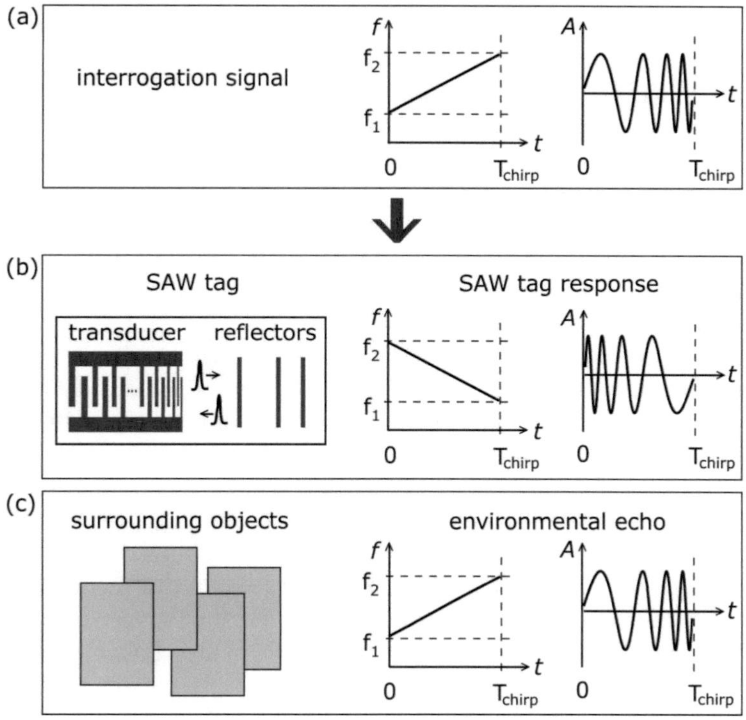

Fig. 8. Interrogation process for an ultra-wideband SAW tag. (a) An up-chirp linear frequency-modulated signal is used for interrogation. (b) The signal is compressed by the chirp transducer, reflected by code reflectors, and expanded by the transducer. The output signal has a dispersion opposite to the interrogation signal. (c) Reflections from surrounding objects have the same dispersion as the interrogation signal.

5. Discussion

The market demands small size, low cost, and environmentally sound RFID tags. That excludes devices consuming power from batteries. Both semiconductor-based tags and SAW tags can be read remotely, and both are small in size. They do not require maintenance and their life-time is limited only by the usual product time of the circuitry. However, SAW RFID tags and passive semiconductor RFID tags are based on fundamentally different physical principles. In this section, we compare in detail these two approaches.

5.1 Power issues in SAW tags and in IC tags

The main feature of SAW RFID tags (Reindl & Ruile, 1993; Schmidt et al., 1994; Plessky et al. 1995; Reindl et al., 1998; Hartmann, 2002) is that they do not use any autonomous power supply such as batteries. Moreover, they do not include any such circuitry that would need to be powered. SAW tags are truly passive devices that merely reflect the interrogation signal. This results in a linear operation at any signal level, even at a very low one. The signal energy of the SAW tag response must of course be sufficiently high compared to the noise level. However, with multiple readings, also tag signals with power below the noise level can be detected. The total power radiated by the reader typically is on the order of 10 mW. For high-speed long-read-range applications, only a fraction of a microwatt is needed at the tag position (RFSAW, 2004). This is the typical power level to which human beings will be exposed when in proximity of SAW tag systems. It is about a million times lower than the radiation exposure generated by mobile phones.

RFID systems based on semiconductor chips use an IC to receive and detect the signal sent by the reader, as well as to subsequently decode the signal and generate the response. The functional blocks of a typical IC tag include power accumulation, computation, and communication. The main feature of IC semiconductor tags is that they must include a proper DC power source for correct operation. The so-called 'passive' IC RFID tags, that do not carry a battery, are obliged to take this power from the RF interrogation signal. The main part of the signal sent by the reader serves for powering the IC and only a small modulation of this signal is for transmission of data. A rectifier circuitry is used to extract a sufficient power from the radio signal. The rectifier converts the signal into DC for storage in a capacitor and, ultimately, for powering the chip. The reading of the tag is performed using a predetermined protocol and is only possible if the necessary DC power level is maintained throughout the entire interrogation cycle. To this end, a minimum critical power of about 100 μW must be received continuously by the tag antenna during the whole time of decoding of the tag's signal (RFSAW, 2004). Below this signal threshold, rectification is not possible. This power restriction is imposed by the physics of semiconductors and thus is fundamental. For SAW tags, on the other hand, which are linear passive devices, no threshold exists. They generate a response at all power levels, usually orders of magnitude lower than what is required for IC tags.

5.2 SAW tags versus IC tags

SAW tags may be considered to have the following advantages over IC tags:

- SAW tag operation does not require continuous pumping of DC energy. SAW tags operate with low level RF pulses of about 10 mW. IC tags at the same distance require a continuous radiation by a reader on the order of a few watts.
- SAW tags operate with a sufficiently low reader power in the 2.45-GHz ISM band, which makes SAW tags compliant with RF emission regulations throughout the world. The use of semiconductor tags demands specific certification by authorities in each country.

- The low power required by SAW tags allows for longer reading distances and for the possibility to mount such tags on metal objects. SAW tag systems achieve greater penetration into pallets containing metal or liquid items (RFSAW, 2004). SAW tag reader systems are hence capable of reading interior items of a pallet, unlike systems based on semiconductor tags, which only allow for the read-out of tags on the corners and edges of pallets.
- SAW tag readers using low-power spread-spectrum signals have a substantially higher interference resistance and spectrum compatibility with other systems (RFSAW, 2004). Semiconductor tag readers radiating a few watts in the same frequency as Bluetooth, WLAN, etc. will inevitably cause strong interference to those systems.
- The reading process of SAW tags also may involve the determination of the individual phase shift of each symbol in the response signal. Comparison of the phases of the interrogation signal and the tag response signal permits a direct and accurate measurement of the tag temperature (Schmidt et al., 1994; Reindl et al., 1998; RFSAW, 2004). SAW tags thus have an inherent capability of functioning as sensors.
- SAW devices have a relatively simple structure. They only consist of a piezoelectric single crystal and a single layer of metal pattern. SAW tags thus are very robust and can be used in challenging environments (RFSAW, 2004). For example, they withstand high levels of alpha, beta, and gamma radiation as well as elevated temperatures. Semiconductor-based tags are more sensitive to such harsh conditions.

However, the semiconductor industry may also have arguments against SAW tags, such as:

1. Since IC tags include memory and a processor, any information in these tags can be re-written and the volume of information written in a tag is relatively large.
2. IC tags are small in size and relatively cheap.
3. IC tags can reach a reading distance of a few meters.

These three points may seem to comprise substantial advantages of IC-based tags. However, these advantages could be questioned as follows:

1. Is it really advantageous to keep valuable information in a small tag which costs only 10 cents and can easily be lost or thrown away? One would rather store on a tag just a sequence of numbers, a code, which indicates the access point to corresponding information reliably stored in a particular protected database. The possibility to re-write tags in every shop will inevitably make readers easily accessible and tag information easily read without authorization or even falsified. In contrast, SAW tags use a sub-micron technology to write in the code which is very difficult to falsify.
2. Since manufacturing SAW tags only requires one photolithographic step, they will, in mass production, become comparable in price or even cheaper than IC-based tags, which use an incomparably more complicated IC technology, many expensive masks, etc. Very importantly, however, the coding procedure for SAW tags must be developed to code millions and millions of different codes in a cost-effective way. This is one main challenge the SAW RFID technology is facing today. The second main challenge is to bring down the price of SAW tag readers to a level comparable to IC tag readers.
3. To achieve a reading distance comparable to that of SAW tag readers, IC tag readers have to radiate 100 to 1000 times higher RF power, up to a few watts. One can just imagine a supermarket wherein all customers are equipped with such readers! The

electromagnetic radiation level will be as high, or even higher, as if all customers continuously used two mobile phones each. Such a concentration of electromagnetic radiation may lead to health hazards. Furthermore, the readers would strongly interfere with each other and create strong interference with other communications systems using the same frequency range.

6. Conclusion

From the above comparison, it can be concluded that SAW tags have clear advantages on many accounts:

- SAW tags practically have an infinite number of codes sufficient for all reasonable applications.
- SAW tags have an incomparably larger reading distance with the same power radiated by the reader, when compared to IC-chip-based tags.
- SAW tags are small, robust, and can operate in harsh environments where IC-based tags fail.
- SAW tag readers using correlation techniques for signal processing can read several SAW tags simultaneously (Hartmann & Claiborne, 2003).

To sum up the above arguments, it is evident that the necessary technological tools as well as the necessary infrastructure and prerequisites are available for the development of smart SAW tags based systems. In a different direction, the development of internet offers conditions for efficient transfer of information to and from databases, which in turn is another pre-condition for the efficient use of RFID tags. In this respect, it is noteworthy that SAW technology has a clear analogy with the mobile phone technology. Exactly the same conditions are actually needed for SAW tags to become a mass product, namely, a large frequency band for having a sufficiently large number of subscribers and communications of large volumes of data between the base stations.

SAW tags offer an excellent technical solution. However, to convert this brilliant idea into a multi-billion business, a number of scientific and technological challenges must be solved, and the fabrication cost of reader devices and tags must be decreased drastically.

7. Acknowledgements

The authors thank E. Mayer (University of Freiburg, Germany) for multiple discussions on topics related to SAW tags.

8. References

Breed, G. (2005). A summary of FCC rules for ultra wideband communications. *High Frequency Electronics,* Vol. 4, pp. 42-44.

Davies, D. E. N.; Withers, M. J. & Claydon, R. P. (1975). Passive coded transponder using an acoustic-surface-wave delay line. *Electronics Letters,* Vol. 11, No. 8, (April, 1975) pp. 163-164.

Hartmann, C. S. (2002). A global SAW ID tag with large data capacity, *Proceedings of IEEE Ultrasonics Symposium*, pp. 65-69, Munich, Germany, October 2002.

Hartmann, C. S. & Claiborne, L. T. (2003). Anti-collision interrogation pulse focusing system for use with multiple surface acoustic wave identification tags and method of operation thereof, U. S. Patent no. 7084768, 2003.

Hartmann, C. S.; Brown, P. & Bellamy, J. (2004). Design of global SAW RFID tag, *Proceedings of 2nd Int. Symp. Acoustic Wave Devices for Future Mobile Commun. Syst.*, pp. 15-19, Chiba, Japan, March 2004.

Hartmann, C. S. (2005). Surface acoustic wave identification tag having enhanced data content and methods of operation and manufacture thereof, U. S. Patent no. 6966493, 2005.

Hartmann, C. S. & Plessky, V. P. (2007). Single phase unidirectional surface acoustic wave transducer and improved reflectors, U. S. Patent no. 7173360, 2007.

Härmä, S.; Arthur, W. G.; Hartmann, C. S.; Maev, R. G. & Plessky, V. P. (2008a). Inline SAW RFID tag using time position and phase encoding. *IEEE Transactions on Ultrasonics, Ferroelectrics, and Frequency Control,* Vol. 55, (August) pp. 1840-1846.

Härmä, S.; Plessky, V. P.; Hartmann, C. S. & Steichen, W. (2008b). Z-path SAW RFID tag. *IEEE Transactions on Ultrasonics, Ferroelectrics, and Frequency Control,* Vol. 55, (January) pp. 208-213.

Ianelli, Z. & Koslar, M. (2004). Surface-wave transducer device and identification system with such device, U. S. Patent no. 6788204, 2004.

Kuypers, J. H.; Reindl, L. M.; Tanaka, S. & Esashi, M. (2008). Maximum accuracy evaluation scheme for wireless SAW delay-line sensors. *IEEE Transactions on Ultrasonics, Ferroelectrics, and Frequency Control,* Vol. 55, (July) pp. 1640-1652.

Reindl, L. & Ruile, W. (1994). Programmable reflectors for SAW-ID-tags, *Proceedings of IEEE Ultrasonics Symposium*, pp. 125-130, Baltimore, MD, USA, November 1993.

Reindl, L.; Scholl, G.; Ostertag, T.; Scherr, H.; Wolff, U. & Schmidt, F. (1998). Theory and application of passive SAW radio transponders as sensors. *IEEE Transactions on Ultrasonics, Ferroelectrics, and Frequency Control,* Vol. 45, (September) pp. 1281-1292.

Reindl, L. M. & Shrena, I. M. (2004). Wireless measurement of temperature using surface acoustic waves sensors. *IEEE Transactions on Ultrasonics, Ferroelectrics, and Frequency Control,* Vol. 51, (November) pp. 1457-1463.

RFSAW, Inc. (2004). The Global SAW tag - a new technical approach to RFID. http://www.rfsaw.com.

Schmidt, F.; Sczesny, O.; Reindl, L. & Mágori, V. (1994). Remote sensing of physical parameters by means of passive surface acoustic wave devices ("ID-tag"), *Proceedings of IEEE Ultrasonics Symposium*, pp. 589-592, Cannes, France, November 1994.

Shannon, C. E. (1948). A mathematical theory of communication. *The Bell System Technical Journal,* Vol. 27, (July, October) pp. 397-423, 623-656.

Stelzer, A.; Pichler, M.; Scheiblhofer, S. & Schuster, S. (2004). Identification of SAW ID-tags using an FSCW interrogation unit and model-based evaluation. *IEEE Transactions on Ultrasonics, Ferroelectrics, and Frequency Control,* Vol. 51, (November) pp. 1412-1420.

Stierlin, R. & Küng R. (2002). Process for carrying out a non-contact remote inquiry, U. S. Patent no. 6407695, 2002.

Plessky, V. P.; Kondratiev, S. N.; Stierlin, R. & Nyffeler F. (1995). SAW tags: new ideas, *Proceedings of IEEE Ultrasonics Symposium*, pp. 117-120, Seattle, WA, USA, November 1995.

9

Smart RFID Tags

Nadine Pesonen, Kaarle Jaakkola, Jerome Lamy,
Kaj Nummila and Jouko Marjonen
*VTT Technical Research Centre of Finland,
Finland*

1. Introduction

RFID has been identified as one of the cornerstones of the upcoming Internet of Things (IoT) and the focus is moving from conventional RFID towards next generation pervasive networked and interconnected systems. In the future IoT billions of objects are envisioned to report their identity, location, environmental conditions and history over wireless connections.

An on-going effort that will support this gradual change is the development of "smart RFID tags", tags that are able to sense, monitor, and adapt to their changing environment. RFID systems alone provide item and product visibility within the supply chain. This visibility can further be translated into actionable data and predictive changes with additional information attained through sensing capabilities. Intelligent RFID tags can combine sensing, computation and communication into a single, small device. The need for new sensing solutions is highlighted further by the fact that legislation, regulatory and quality demands are setting requirements for certain branches (pharmaceuticals, explosives, transportation of dangerous goods, foods, etc). Cold chain compliance is a key requirement for pharmaceuticals, hospital transfusions, clinical trials, foods and perishable items. RFID can be used to fight counterfeiting and RFID can provide the electrical pedigree of a product.

Most efforts on RFID tag design so far have been concentrated on the ultra low price tag segment. This has led to compromised performance when label tags have been applied in "unsuitable" environments and has been evidenced as low reading accuracy in many RFID pilots. In order to reach a high reading accuracy and reliable long distance operation, RFID tags ought to be immune or adapt to their environment (e.g. presence of metals, liquids, gas...) to avoid detuning and other impairing effect caused by their surroundings. New solutions to implement platform insensitive/platform tolerant and platform adaptive RFID tags are emerging.

2. Platform tolerant and platform adaptive tags

2.1 Introduction

Although antenna theory as such is independent of the field of applications, the central role played by the antenna in the tag design makes the designing of RFID tag antennas different from any other antennas. The antenna design must comply with many competing tag design criteria such as: the overall cost including manufacturing issues, compact size and shape of the tag, frequency bandwidth, antenna gain, robustness and reliable operation over long distances. For example in a mobile phone there is always the handset structure into which the antenna is to be integrated. It means that one can take advantage of the already existing parts and use them as a part of the antenna. In many other electronic applications there is a printed circuit board that provides at least the ground plane for the antenna and consequently, the additional cost of the antenna is often negligible when compared to the rest of the device.

In an RFID tag, the antenna basically defines the tag itself, the only other parts being the tag substrate or matrix and the microchip that costs a few eurocents and is less than one cubic millimetre in size. The reading distance of a passive RFID system depends on the realized gain of the antenna and the antenna gain is a more critical parameter than for an active short range device such as a Bluetooth accessory. RFID tags are to be attached directly onto several different kinds of articles, which set some special requirements for the platform tolerance of the tag antennas. In the following, platform tolerant antenna solutions mainly for UHF will be examined. In terms of platform tolerance, the problems with HF (13.56 MHz) RFID are somewhat different. This is due to the fact that at HF frequencies the coupling between the reader and the tag is based on magnetic near field and the mutual inductance between the respective antenna coils. With some limitations, platform tolerance can be implemented for HF as well, but with different types of solutions. These will be shortly described in section 2.4.

2.2 Platform tolerance

The inexpensive labels utilizing an electric dipole type antenna are by far the most common type of UHF tags. Owing to their simple structure and suitability for cost-effective large-scale manufacturing, they have set the minimum price for passive RFID tags and rendered the use of RFID possible for many new applications. On the other hand, their application area is still quite limited. These tags cannot be put on metal surfaces or e.g. on surfaces of a liquid container. Also certain materials that do not necessarily prevent the operation of the tag, may limit the read range so severely that the reliability of the whole system is harshly reduced.

The inapplicability of label tags on certain surfaces is a fact that is purely based on physics – a two-dimensional antenna cannot work when placed on a conducting surface. A platform tolerant operation always requires some thickness. As many important applications require tags to be put on this kind of challenging surfaces, there has been a growing need for new types of tags. The fact that the current label tags are indiscriminately used in very different type of environments has lead to a situation where the tag typically is the weakest link of an RFID system and causes the whole system fail. For a reliable and robust RFID system, a certain read range of the tag should be guaranteed independently of the physical environment.

To overcome this, the ideal tag should be small in size, inexpensive, mechanically durable, should provide long operation range and should be possible to be attached into various objects, without any significant effects on its performance. [Foster et al., 1999] The so-called platform tolerant or metal surface tags can provide at least the last three of these properties of an ideal tag. This type of a tag is not a new concept. The need for such tags was already encountered in the very beginning of the development of UHF RFID systems. The simplest possible and commonly used solution to obtain a somewhat platform tolerant operation was to raise a conventional label dipole a few millimetres above the problematic surface, using e.g. a plastic spacer. This allows the tag to work, but the reading distance becomes typically only a fraction of what was originally intended for the same tag in free space.

There are some tag solutions in which this kind of raised dipole type tag antennas have been optimized to operate on a metal surface. This optimization can only be made in terms of antenna impedance in order to provide the conjugate matching between the microchip and the antenna. This so-called detuning effect can then be compensated, but the operation principle of such an antenna is still inappropriate to operate as a platform tolerant structure. It results in the so-called on-metal tag. On-metal tags and platform tolerant or platform insensitive tags are not necessarily the same thing. In many cases, an on-metal tag is only tuned for a metal surface, which means that for any other surfaces, their performance is compromised. Another approach has been to optimize tags for specific environments and mounting platforms. This approach results in many tag models i.e. one for each possible platform. Technically, this can already give good results. But there are two main problems. Firstly: the application environment needs to be very well known in order to select the right tag. For example, when tagging a cardboard box in a warehouse, one would have to know what the content is and how far from the surface of the box this content is. Secondly: a large variety of tag models is needed and thus it will be more difficult to reduce the tag price by production volumes.

The conclusion is that uncompromised operation can be best guaranteed by implementing a tag design that is really platform tolerant. Importantly, platform tolerance should be based on the physical operation principle of the antenna. Because of its general-purpose nature, such a tag could then be fabricated in large quantities, reducing the unit price. It is already possible to implement such tags and some models are already available on the market.

The main effects of the near environment on the antenna performance are the near field losses and the so-called detuning effect. [Rao et al., 1999] Losses in the near field reduce the radiation efficiency of the antenna. The detuning effect is due to the change of the antenna feed impedance visible to the microchip. The detuning of the antenna ruins the power matching between the antenna and the chip. These two effects affect the measured effective aperture of the antenna. The effective aperture also contains the directivity of the antenna. [Pursula et al., 2007] The realized gain, determined by the effective aperture of the antenna, in the direction of the reader is, together with a certain sensitivity of the microchip, what basically determines the read range of the tag.

By thinking how the world around us looks at the UHF frequencies, i.e. around 900 MHz, we see that it is magnetically quite neutral but very heterogeneous in terms of electric field. This means that the environmental disturbances of the antenna operation, reducing the realized gain, take place mainly via the electric near field of the antenna. Consequently, in a platform tolerant antenna, the magnetic near field of the antenna can be allowed to burst out of the outer dimensions of the tag antenna, whereas the electric field should be kept within the antenna structure. It also means that in all of the platform tolerant structures, the magnetic dipole is always of special significance as a radiator. By concentrating the near electric field inside the antenna itself and taking advantage of radiators based on magnetic dipole, both effects of the antenna near environment can be minimized, thus implementing a platform tolerant structure.

As HF RFID is based on utilizing magnetic coupling, the system is less sensitive to electrical disturbances and changing permittivities in the near environment of the tag. Metal platforms, however, are problematic also for HF. In fact, magnetically coupled near field tags are also available for UHF. These tags are typically simple one-round metal loops of 10 mm in diameter and they are used for item-level tagging. With those, some tens of centimetres of reading distance can be achieved. Metal platforms, however, are problematic also for these near field tags. One solution for HF platform tolerance is described in chapter 2.4.

As the feed impedance of a platform tolerant antenna is very stable from a mounting platform to another, the antenna is allowed to be narrowbanded, whereas with other types of antennas the sensitivity to external disturbances is typically reduced by broadbanded impedance matching. This difference is illustrated in Fig.1. The goal impedance, which is the complex conjugate of the chip impedance, is marked with a red spot. As broadbanded tuning is based on a compromised match (Fig.1 left), using platform tolerant antennas, we can implement a better match and gain in the read range.

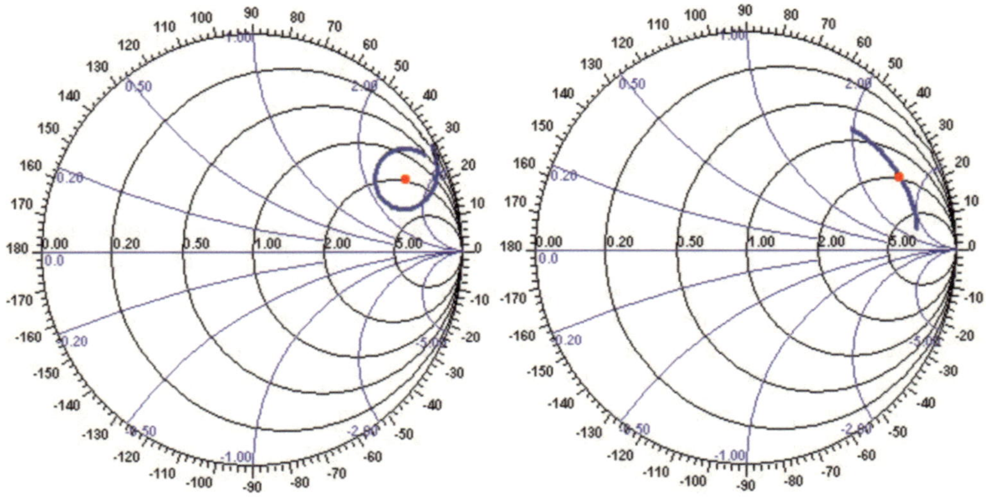

Fig. 1. Broadbanded (left) and narrowbanded impedance match between the antenna and the microchip.

2.3 PIFAs and PAFFAs

Most of the platform tolerant UHF tags are based on the antenna structure known as PIFA, which stands for 'Planar Inverted F Antenna'. [Hirvonen et al., 2004] PIFA is a very popular antenna type in general. In addition to the platform tolerant RFID tags, PIFA-type antennas are commonly used in mobile phones and e.g. for wireless networking in laptop computers. PIFA type RFID tags are shown in Fig. 2. There is a ground plane placed at the bottom and a radiating patch on top of the antenna. The length of the patch is about a quarter of the wavelength. The metal layers are short-circuited in one end and the chip is connected between them at the second point. By selecting the feed point, different feed impedances can be realized. This is important, because the RF front end of the microchip has to be optimized in terms of the efficiency of the RF rectifier; the resulting input impedance is then a secondary parameter. [Facen et al., 2006] Consequently, the RFID transponder chips and antennas are practically never 50 Ohm or even self-resonant structures. Instead, a very typical feed impedance of the transponder antenna is about (20 + j150) Ohms at 900 MHz.

Technically, PIFA is a very good solution: platform tolerant operation can easily be achieved and the main direction of radiation is exactly where we would like to have it - outward from the mounting surface. As the electric field of the PIFA is enclosed between the metal layers, the detuning effect of a well-designed PIFA is almost negligible. The directivity of the antenna, instead, increases as the tag is put onto a conducting surface. This is due to concentration of all the radiation in to the open hemisphere, whereas in free space the PIFA has a back lobe in its radiation pattern. The problematic issues about PIFA are the high price and the large size. The price is mostly due to a complicated structure and thus a complicated fabrication process. Especially the number of steps in the fabrication process is an important parameter here. PIFA tags have typically a separate plastic casing and a separate antenna inlay.
A typical fabrication process consists of injection moulding of the casing, etching of the antenna inlay and finally combining the two parts. And this last phase is typically the most challenging one, because of two things: firstly, the tuning of this antenna type is quite sensitive to the dimensional tolerances. Secondly, handling this flexible inlay requires special tools in the assembly line. In some cases, an additional ground plane, which is a metal plate, is also used beneath the antenna. The longest outer dimension of the antenna is a little over a quarter of a wavelength that is typically from 6 to 12 centimetres depending whether there is air or some plastic between the layers. The property of platform tolerance requires that the ground plane and thus the whole antenna to be somewhat bigger than the radiating top patch of the antenna.

Fig. 2. PIFA tags, 2005. Courtesy of VTT, Technical Centre of Finland.

Because of the constant pressure on the cost of an RFID tag, concepts such as PAFFA ('Planar Asymmetrically Fed Folded Antenna') have been developed. [6] The idea has been motivated by overcoming some of the structural complexity of PIFA - making a platform tolerant hard tag economically more competitive with label tags. The difference between PIFA and PAFFA structures is that in a PAFFA structure only one connection between the two metal layers of the antenna is needed. Because this required connection is in the end of the antenna, the antenna can be implemented e.g. by bending a flexible inlay around a piece of plastic. As vias and thus PCB processes are not needed, the metallization can also be

made using new techniques, e.g. by printing directly on a plastic substrate. [Allen et al., 2007] Technically, the idea is based on replacing the feed point connection between the metal layers by a quarter-wave open-ended microstrip line. The other possibility is to use a half-wave line that is short-circuited from one end. To reduce the size of the tag, these lines can also be meandered. As the microstrip lines also take part in the radiation of the antenna, their dimensioning can be optimized in terms of the desired radiation pattern. They also provide a degree of freedom when finding the right feed impedance for the microchip. A PAFFA type tag is shown in Fig. 3.
The PAFFA structure can be defined to consist of two quarter-wave microstrip lines between which the microchip has been connected. One of theses lines is short-circuited and the other is open-ended. The short circuited end of the first forms the magnetic dipole needed for platform tolerant operation. The resonant behaviour of the microstrip lines can also be utilized for some new features. By implementing multiple resonances in one tag antenna, a so-called global tag can be made. [Hirvonen et al., 2006] [Hirvonen et al., 2006] This global or triple frequency tag provides optimal operation at all of the frequency bands allocated for UHF RFID around the world: 867 MHz in Europe, 915 MHz in North America and 953 MHz in Japan. Compared to some global tag solutions that are based on utilizing one broadbanded resonance, the three separate resonances provide remarkably better performance, combined with platform tolerance.

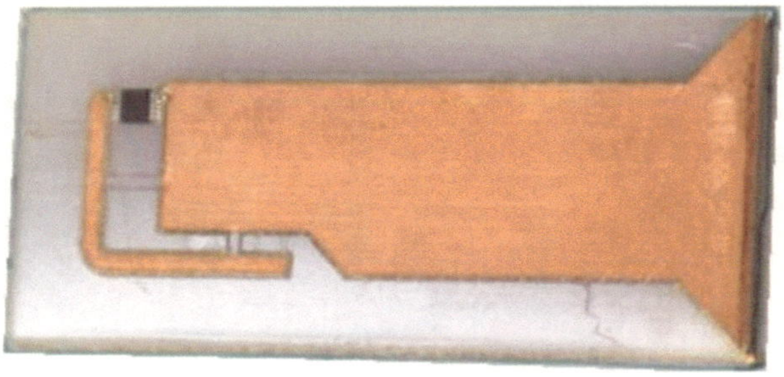

Fig. 3. PAFFA tag. Courtesy of VTT, Technical Centre of Finland, 2006.

2.4 Tag miniaturization

Physical size is one of the key parameters for RFID tags. In theory, microstrip antennas can be miniaturized by meandered lines. As long read range and thus good radiation efficiency is also required, meandered microstrip structures are not the best solution. Instead, some special materials can provide at least partial solutions. The basic idea with these is to reduce the effective wavelength λ inside the antenna structure. This is because the antenna should have a particular electrical length to operate efficiently. For PIFA and PAFFA type structures, the required electrical size is about a quarter of a wavelength. Compared to the physical size, the electrical size can be increased by raising either permittivity or permeability of the tag substrate material, according to the equation:

$$\lambda = \frac{c_0}{f\sqrt{\varepsilon_r \mu_r}}, \tag{1}$$

where c_0 is the speed of light in free space, f is the operation frequency, ε_r is the relative permittivity and μ_r is the relative permeability of the material. There are some materials with high permittivity available on the market. Typical plastics and materials for printed circuit boards have relative permittivity values of 2…5. Materials with relative permittivity over 5 can be often considered as "high permittivity" materials. But additionally to the material permittivity and permeability, the material losses are important parameters concerning their applicability for tag antennas. PIFA and PAFFA structures are especially sensitive to electrical losses and hence the material for antenna substrate should be carefully selected. If the inter-metal space in the antenna is to be filled with this material, the electric loss tangent ($tan\delta$) or loss factor of the material should in practice be below 0.005.

The main problems in utilizing high-permittivity materials are that the loss factor typically increases together with the relative permittivity and, even in the case of low loss factor; the high permittivity also increases the current densities and thus the conductive losses in the antenna metallization. A further problem with high permittivity materials is the resulting narrow frequency bandwidth. In terms of bandwidth, a better way would be to use high permeability instead of high permittivity. This can be understood in terms of physical equivalents: as increasing permittivity increases the effective area of the antenna, high permeability increases the effective height, making the magnetic dipole more efficient. Consequently, the high-permeability material could then be used for making the tag thinner. Comparing these two ways of miniaturizing the antenna, high permeability would be at least theoretically a more beneficial way. Utilizing a PIFA type basic structure, the ultimate solution would be to combine both: high permeability for the magnetic dipole part of the antenna and high permittivity for the end in which the electric field of the antenna is concentrated. Unfortunately, all the high permeability materials for UHF that are currently available are quite lossy i.e. their magnetic loss tangent is relatively high. [Martin et al., 2007] Therefore, practically all of the small tags utilizing special materials are so far based on the use of high permittivity substrates. [Hwang et al., 1997] However, the development for UHF applicable high-permeability materials is in progress. So far some applicable high frequency ferrites have been used in low-profile UHF near field tags.

For HF frequencies high permeability ferrites with a relative low loss factor are available and can be used between the tag coil and the surface. Here the idea is, instead of miniaturizing the tags, to direct the magnetic flux between the coil and the mounting platform, resulting in an on-metal HF tag. [Bovelli et al., 2006]

2.5 Platform adaptive tags

In addition to the platform tolerant tag antennas, another possible approach for implementing platform tolerance of the tag is the use of an adaptive RF front end of the tag IC. In this case, as the environment changes, the microchip automatically changes its input impedance for conjugate matching between the IC and the antenna. With this method, more simple tag antennas can be used. Technically, this solution can be based on a varactor diode or a pattern of switchable capacitors on the microchip. [Rostbakken et al., 1995] Theoretically, also more complex adjustable impedance matching networks can be used, but in most cases an adequate impedance match can be achieved by just tuning the parallel capacitance. Typically, some tens of picofarads can be implemented on a CMOS chip within an area that is still realizable.

Compared to platform tolerant antennas, the approach of adaptive RF front end has its limitations. The most critical thing is that the system cannot cope with the reduced radiation efficiency of the antenna in the proximity of lossy substances. One challenge about this type of a system is that the matching network requires some electrical power already in a phase when the impedance matching between the antenna and the chip may be poor. Therefore this type of a solution is more suitable for semi-passive than passive tags. The simple parallel-type tuning also suits best with simple reactive antennas such as near field UHF or HF loop antennas.

3. RFID sensor tags

3.1 Introduction

Along with the RFID technology, a new focus area placed on RFID sensor tags is emerging. RFID sensor tags are associated with a product, person or a location through a simple ID and are capable of measuring and acquiring data from the users' behaviour and the environment such as temperature, pressure, tampering, shock, humidity, etc. RFID sensor tags also allow the connection of data loggers and remote controls as well as the connection of displays, printed sensors, and biosensors via the traditional RFID tags.

Various user cases related to RFID-based smart sensing have emerged: in March 2006, the Japanese Ministry of Internal Affairs and Communications (MIC) started developing a system that allows for detailed information gathering about a disaster area by sprinkling RFID sensor tags from the sky. In June 2006, British Petroleum began a trial of an RFID-based sensor network to help it better manage chemical inventory, increase stock visibility and reinforce safe-handling business rules. In January 2007, the global shipping company DHL announced the deployment of RFID sensor tags to its pharmaceutical customers to track the temperature and shelf life of products being shipped from warehouse to store. And lately, in December 2007 Motorola and Intelleflex announced a strategic relationship in extended capability RFID; Motorola Ventures co-leads $15 million series C investment in Intelleflex. Specializing in battery-assisted RFID, the Silicon Valley-based company targets the aviation, yard management, logistics, hospitality sectors, and asset-tracking applications.

The HF range is currently the most widely used operation frequency for RFID and the unlicensed ISM band around 13.56 MHz is globally available for RFID devices. Its use is expected to grow further in the near future driven by the newly established NFC (Near Field Communication) technology that allows users to read small amounts of data from tags, as well as to communicate in a peer to peer fashion with other devices, by a simple touch with a handheld remote control such as a cellular phone. In January 2007 Nokia launched the world's first fully integrated NFC phone. With devices such as the Nokia 6131 NFC phone, users can make contactless payments and access mobile services with ease. In the future, users will be able to pick up information from their environment using NFC technology and sensor tags.

On the other hand, ultra high frequency (UHF 860-960 MHz) systems are gaining acceptance in many application areas. For example, the retailer Marks & Spencer has launched the largest UHF item level tagging in its department stores deploying more than 100 million UHF RFID tags with embedded EM Microelectronics passive 869 MHz RFID chips. The technology has helped Marks & Spencer increase both efficiency and customer service. Compared to lower-frequency systems the major difference and advantage of a UHF system is that it is able to operate in the radiating (propagating) far field of the reader. Lower-frequency systems utilize inductive coupling between reader and transponder, i.e. non-propagating magnetic field decreasing rapidly with distance from the reader device. In practice, operational ranges of several meters together with moderate reader sizes, such as handheld readers, can only be obtained by operating in the far field region. At HF the reading distance is comparable to the size of the reader. The establishment and acceptance of a world wide standard in the form of EPC Global has paved the way for faster implementations. Lately, the transponder price has undercut the critical 10 cent barrier, which makes them attractive for high volume applications.

3.2 RFID sensor tags types

Among the two dominating frequencies of RFID operation, i.e. HF and UHF, three categories of RFID sensor tags can be distinguished: active, semi-passive, and passive solutions. The available commercial passive and semi-passive solutions for wireless sensors and data acquisition systems mainly work at low operation frequencies and are based on short range inductive coupling. The active solutions based on UHF can provide much longer operation ranges and enhanced features at a higher cost.

Active RFID sensor tag transponders are constantly powered by a battery and contain an active radio transceiver. For example, Telepathx ITS [14] has developed an active system that integrates RFID tags, crash sensors, and wireless mesh networks to automatically detect crashes, assess the severity, and report the incident to public safety forces. The RFID sensor tags can transmit 100-125 meters in any direction. The sensor has a small battery installed for transmission purposes that has a life span of 10 years. Different RFID sensor tags can operate at the 433 MHz, 900 MHZ, or 2.4 GHz frequencies.

Semi-passive RFID sensor tags are hybrid implementations of active and passive transponders. A semi-passive tag contains a battery for powering the IC circuitry operations and backscatters the incoming reader field for tag to reader communication. Such semi-passive tags have three main advantages 1) Better sensitivity than passive tags (up to 30 meters) 2) Longer battery life than active tags 3) Can perform active functions under their own power, even when no reader is present. For example, Alvin Systems [15], located in Turkey, has launched a cold chain solution with credit card size smart labels that combine RFID operation compliant with the ISO 15693 standard, temperature sensors, and mobile Pocket PCs. The RFID sensor tags monitor the condition of temperature-sensitive objects during transportation or storage - for quality assurance and enhanced cold chain operations. Lately, Montalbano [16], an Italian-based company has started commercializing MT RFID sensor tags that have a credit card size and shape, are ISO 15693 compliant and can be applied to pallets and boxes of goods as well as on rounded surfaces. MTshock tags detect and store every single collision or shock that art-works, explosives, vintage wine, or any kind of fragile goods can suffer. MT humidity tags store relative humidity changes while MTSense tags monitor the thermal history of the monitored product.

Passive RFID sensor tags do not contain a power source and consequently, can only perform active measurements when powered by a nearby RFID reader. The existing passive RFID sensor tags are mainly based on proprietary solutions. For example, Bioett [17], a Swedish company has developed a system based on a biosensor for temperature monitoring. A chemical Time Temperature Biosensor is activated when the label is placed on the product to be traced and the Bioett System monitors the accumulated effect of temperature on products over time. Since this solution is based on bar code reading, the amount of information that can be coded is limited and reading distances are in the cm-range. Efforts to develop passive RFID based sensor transponders working at UHF frequencies in order to achieve longer reading ranges and smaller transponder sizes are growing but so far the results are at demonstration level.

Different solutions have been envisioned in the past. (i) Changing the polarization of the tag antenna as a function of the sensor response but this method requires two different antennas at the interrogator side and the sensor response can only contain 1-bit information. (ii) Using two tags with one tag being activated when the sensor response has reached a threshold (iii) Redesigning the RFID chip to include the sensor interface. For this last solution, papers have been published presenting working passive sensor tags [Pursula et al., 2007], [Cho et al., 2005] but these demonstrators only support their own proprietary protocol, which impedes their penetration to the market. A prototype of a passive UHF chip compatible with the ISO 18000-6c standard and capable of accessing four different external sensors was developed by VTT. The operational range of the first generation temperature measuring device was 80 cm.

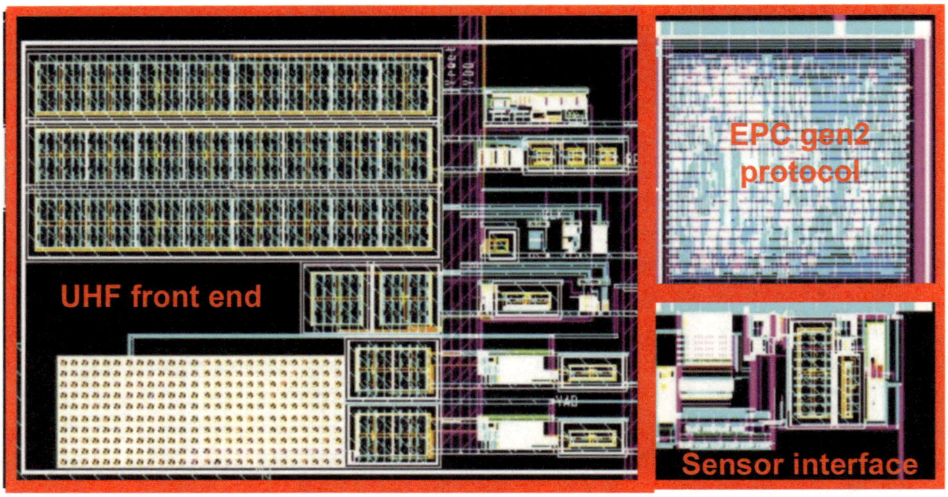

Fig. 4. Passive UHF RFID sensor able to monitor external capacitive and resistive sensors. Courtesy of VTT, Technical Research Centre of Finland, 2008.

3.3 RFID sensor tag architecture

3.3.1 Introduction

The current development of RFID and other wireless communication based sensor solutions have so far been mostly concentrating on active, rather high cost and high power solutions. Consequently, the sensors used in these solutions have not been optimised for low power consumption and the concept does not adopt well for low cost solutions. Semi-passive RFID sensor solutions based on an extension of the EPC Global protocol and equipped with proper power management, a low power sensor interface and tailored low power sensors offer a unique opportunity to develop commercially viable, low cost and low power solutions. Using cheap integrating platforms makes even disposable sensor tags viable. New solutions will be introduced into the market within a few years and the compliance with EPC Global can pave the way for a fast acceptance. Passive RFID sensor tags complement the solution and provide long lifetime, maintenance free solutions for appropriate applications, e.g. moisture monitoring in construction materials.

For such applications, efforts are needed on sensor development. Indeed, passive RFID sensor tags require low power sensors as well as low power interface electronics. Typically, about 100 μW (microwatts) of RF power is available for a passive UHF tag at a distance of 5 meters (assuming an antenna gain of about 0 dB and a 2 W equivalent radiated power). Product sheets indicate that typical tag integrated circuits currently require around 30 μW of RF power to operate. With a good RF rectifier efficiency of 20 % [Facen et. al, 2006], around 20 μW of the 100 μW is available for the DC electronics. This 70 μW marginal in RF power then equals 14 μW for the sensor electronics, if the reading distance of 5 m is to be maintained. Existing sensor interfaces, such as sigma-delta converters, pipelined or folded converters are usually high power architectures (up to 10 mA) as they are targeting high precision and fast implementations (greater than 10 bits and 30 MHz sampling rates), and thus are not suited for low power applications.

Consequently, new solutions are needed to develop mixed signal sensor interface architectures that would combine high resolution and low power operation with target DC power consumption of less than 20 μW in active mode, and less than 1 μW in power down mode. A tag IC could contain a generic mixed signal sensor interface that would allow multiple external sensors to be connected.

3.3.2 RFID sensor tag architecture

The RFID sensor tag is composed of:

a. An antenna directly matched to the tag's front end impedance to communicate with the reader
b. An analogue RF front end that typically contains rectifier circuitry to convert RF power into DC, a clock, a modulator and a demodulator
c. A logic part that is the translator between the front end and the sensor interface by coding, decoding, commanding, processing, and storing information. The logic implementation usually follows a defined standard and a certain associated protocol.
d. The signal interface that adapts the external signals (sensor reading, data logging, microcontrollers, display, keyboard...) to the standardized RFID tag. The signal interface can be of several nature:
   - A bus interface such as SPI, I2C to connect directly the logical part of the RFID tag to an additional block such as a data logger, microcontroller, display, etc. In this case, the RFID sensor tag can be either semi-passive or active.
   - A sensor interface that converts the change of value of a sensor into something that can be properly treated. The sensor interface is composed of a sensor readout circuitry (charge amplifier, resistive bridge...) and an analogue to digital converter (ADC). The sensor interface can either be passive (the readout electronics and the analogue to digital converter are fully powered by the reader's field) or semi-passive: an additional battery powers up the interface as well as the logic.

Different solutions exist for antennas, the RF front end and the logical part which are treated in the literature, whereas a low power signal interface is a relatively new concept in the RFID context. In the case of sensor reading, the mixed signal interface should implement low power architecture for passive, semi-passive, and even for active solutions for life span purposes. Solutions for low power architectures are detailed in the following paragraphs.

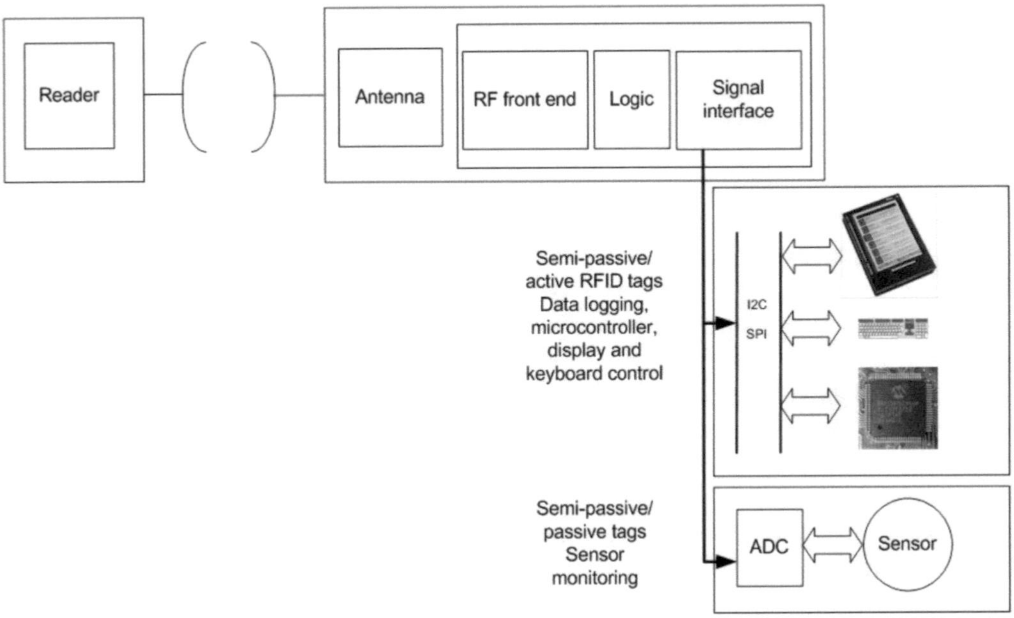

Fig. 5. A possible RFID sensor tag architecture

3.3 Analogue to digital converters architecture review

In this chapter different analog to digital converter architectures will be reviewed and their suitability for low power RFID based sensor solutions will be discussed.

Successive Approximations A-to-D Converter

The successive-approximation-register (SAR) analogue-to-digital converters [20], [Black], [Sauerbrey, 2003] are frequently the architecture of choice for medium-to-high-resolution applications, typically with sample rates below 5 Mega samples per second (MS/s). SAR ADCs most commonly range in resolution from 8 to 16 bits and provide low power consumption as well as a small form factor. This combination makes them ideal for a wide variety of applications, such as portable/battery-powered instruments, pen digitizers, industrial controls, and data/signal acquisition. The input signal does not need to be continuous, because the ADC takes a “snapshot” of the signal. The common blocks included in a SAR ADC are a sample and hold stage, a comparator, a successive approximation register, and a digital to analogue converter (DAC), see Fig. 6. At first, the SAR sets the most significant bit (MSB) of the DAC to be a logical one. If the output of the comparator indicates that the generated voltage is smaller than the actual analogue voltage, the MSB bit remains in the logical 'one' state. On the other hand, if the result of the comparison is the opposite, the MSB is set to zero. This "program" goes through all the bits until the least significant bit (LSB) is set.

Redundant Signed Digit (RSD) converter

RSD converters [Heubi, 1996] are rather similar to successive approximation converters but use an x2 amplifier instead of a DAC, hence doubling the voltage at each step and comparing to a same reference level. Such converters are less prone to the amplifiers' settling time and can be driven faster as the comparisons are made on scaled up voltages via the amplifier. Since the precision elements of the RSD architecture is only based on the x2 amplifier and the adder, less accurately matched components are required. As the same components are used at each step of the comparison, the design offers great linearity.

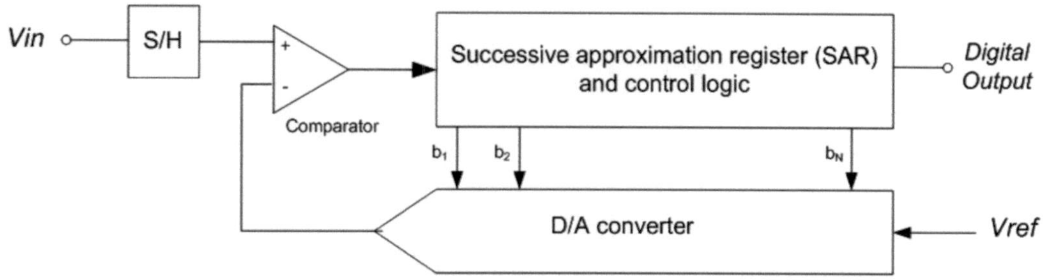

Fig. 6. Block level schematics of the SAR ADC

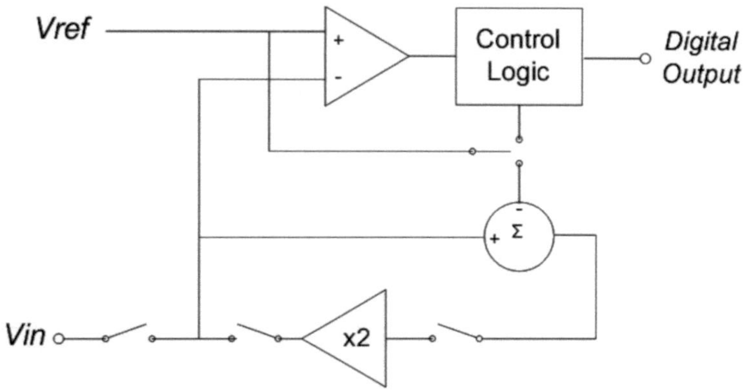

Fig. 7. Example of a Redundant Signed Digit ADC

Parallel or flash ADCs

Parallel or flash ADCs are very high-speed ADCs that are suitable for applications requiring very large bandwidths. They use parallel techniques to achieve short conversion times. Some only use one clock cycle as the conversion speed. Flash types ADC are fast but require a large number of comparators. Indeed, with an *N* bit resolution the flash ADC requires 2^N-1 comparators, which typically is area consuming. In addition, flash converters are power hungry and have a relatively low resolution.

Folding interpolating analogue to digital converter

A smart replacement of the flash architecture is the folding interpolating ADC [Kim et al., 2003]. The number of input amplifiers can be reduced compared to the flash architecture through the use of an interpolating architecture while the number of latch comparators can also be reduced by using a folding architecture. Together, the converter forms a folding interpolating converter that is similar to a two-step converter where a group of LSBs are determined separately from a group of MSBs. These converters belong to the class of medium-high fast converters and do not require a sample and hold circuit. They have smaller area and power dissipation than flash converters, but have a large input capacitance similar to flash converters. The main disadvantages of the folding interpolating converter are its high distortion at high frequencies due to the high frequency components generated internally, as well as a more complex architecture than that of flash converters.

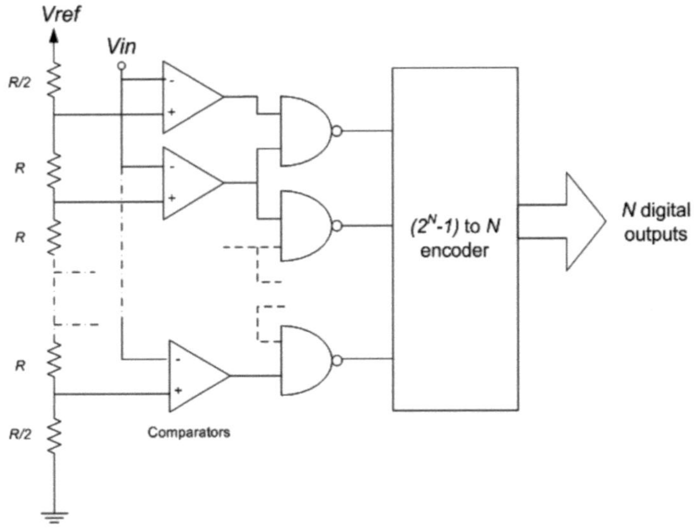

Fig. 8. Parallel or flash ADC architecture

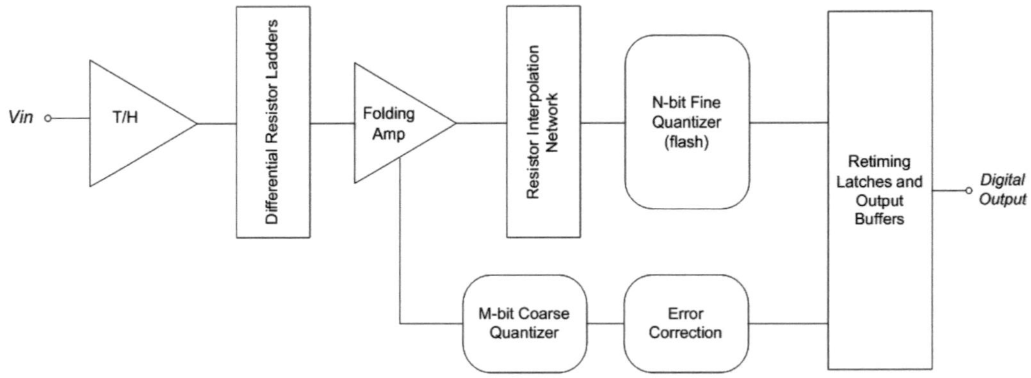

Fig. 9. A 6-bit conventional one-stage (direct) folding structure

Pipelined algorithmic analogue to digital converter

The pipelined algorithmic analogue-to-digital converter (ADC) [Vazl, 2003], [Miyazaki et al., 2001] has become the most popular ADC architecture for sampling rates from a few mega samples per second (MS/s) up to 100MS/s+, with resolutions ranging from 8 to 16 bits. The algorithmic ADC is based on *N* identical stages and *N* comparators that determine the sign of *N* outputs. Each ith stage multiplies its input by 2, and adds it to or subtracts it from the voltage reference depending on the sign of the $(i-1)^{th}$ output stage giving the ith bit at the output of the comparator.

Delta Sigma analogue to digital converter

Delta sigma converters are part of a different class of converters as they are based on oversampling and noise shaping rather than direct quantization of signal amplitudes. Delta Sigma analogue to digital converters are nowadays the most widely used oversampling ADCs found predominately in applications such as instrumentation, digital voice, and audio applications. Typically, their bandwidths are less than 1MHz with a range of 12 to 18 true bits [Dessouky et al.]. Because delta-sigma converters over sample their inputs, they can perform most anti-aliasing filtering in the digital domain. The sigma delta converter measures the input signal for a certain period of time and outputs a digital code corresponding to the signal's average over that time, thus the input signal of interest should be continuous.

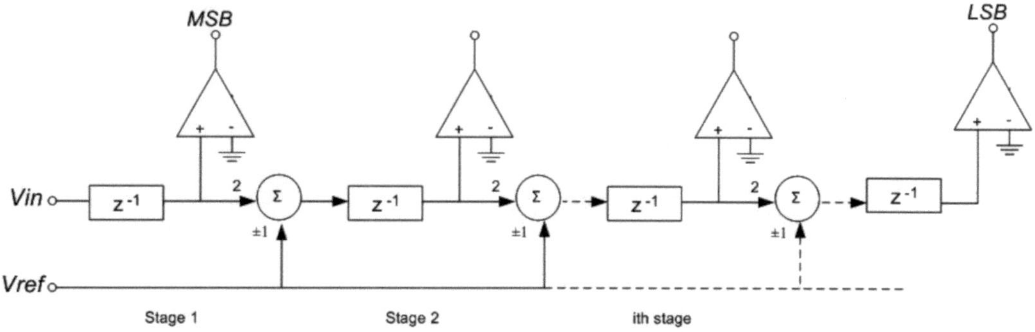

Fig. 10. Block diagram of a general pipelined analogue to digital converter

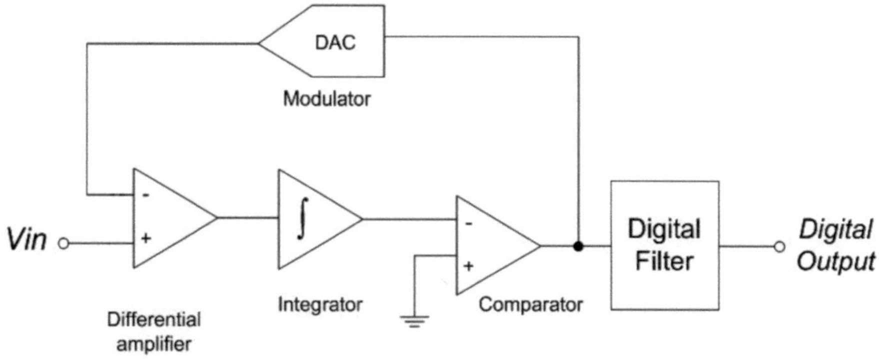

Fig. 11. General delta sigma analogue to digital converter

Capacitance to frequency converter

The capacitance to frequency converter [Chatzandroulis et al.], [Krummenacher, 1985] generates a modulated signal varying with capacitance changes. The capacitance to frequency converter is based on a relaxation oscillator that converts the capacitance into a period-modulated output signal. The converter includes an amplifier, a comparator, the capacitances C_1 and C_2, and a controlled current source I_{curr}. The bias source V_{ch} is used to discharge the capacitive sensor.

Conclusion

In contrast to wireless communication for mobile terminal (GSM/WCDMA) where high-speed ADC are required and where sigma delta converters prove to be robust and offer good analogue performance, ADC with medium resolution (8-12 bits) at samples rates up to a few mega samples have emerged to provide low power consumption, size reduction in applications such as RFID sensor tags.

The following table gives a comparison of all the above ADC structures applicable to RFID sensor applications. Both the delta sigma and the pipeline ADC have a longer latency than the SAR. It can be seen from the following table [Zheng, 1999], [33] that the pipelined analogue to digital converters, folding interpolating analogue to digital converters and flash converters are not suitable for the low power architectures required for RFID sensor tags.

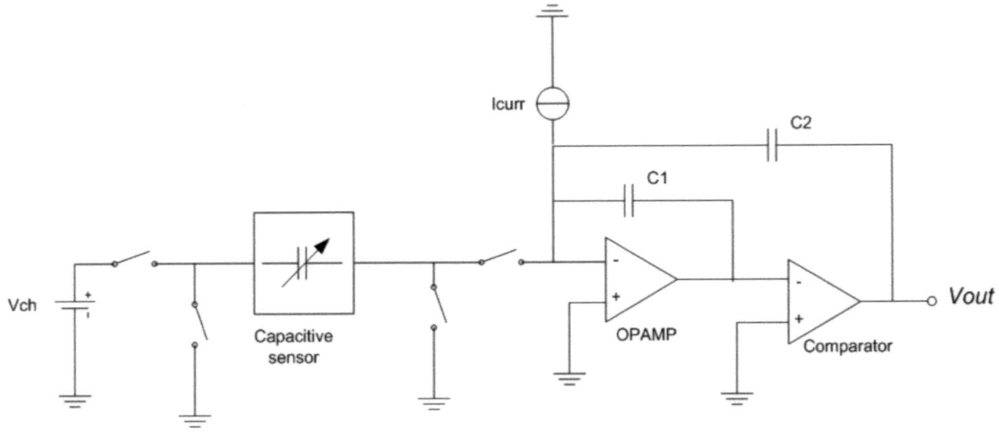

Fig. 12. Example of a capacitance to frequency converter

ADC Type	Resolution	Sampling rate	Silicon area	Power consumption
Successive approximation analogue to digital converter [20], [21], [22], [23]	9 bit; SNR=51.2dB	4.1kS/s (to 150kS/s)	0.18μm CMOS	0.85μW@4.1kS/s (to 30μW@150kS/s) @ 1V
	~8 bit (DC) , 7 bit (4.6kHz)	100kHz	$0.053mm^2$ in 0.25μm CMOS 2P5M	3.1μW (41pW stand by) @1V
Redundant Signed Digit (RSD) converter [24]	60 dB SNR for large input levels. 13.5 bits at 1 V and 14.5 bits at 2.5 V	16 kHz	$0.8mm^2$ in 2μm CMOS technology	125μW total power dissipation (@2.4V)
Folding analogue to digital converter [25]	10 bits	40 MHz	0.25μm CMOS 1P5M	62mW (@2.5V)
Pipelined analogue to digital converter [27], [28]	SNR=87dB; DR= 88dB	5MHz	$0.63mm^2$ in 0.35μm CMOS	950μW (@1V)
	10bit	30MHz	$3.12mm^2$ in 0.3μm CMOS 2P3M	16mW (@2V)
Delta Sigma analogue to digital converter [26], [29]	SINAD 64dB and Dynamic Range 82dB	1.28MHz with 8kHz bandwidth (OSR 64)	$0.23mm^2$ in 0.35um CMOS, (2 poly, 3 metal)	~ 60μW (from 0.8V to 1.5V)
Capacitance to Frequency converter [30], [31]	35Hz/mmHg (160kHz @ 0 mmHg)	49Hz data bandwidth?	$1.44mm^2$ in 0.8μm CMOS	4mW (@ 4V)
	10 bit	40kHz clk with 20Hz bandwidth	$1.1mm^2$ in 4μm CMOS 2P	80μW (@3 V)

Table 1. Comparison of different types of ADC structures.

The integration of sensors to tags requires the following:

- Proper sensor interface to offer good analogue performance while meeting the stringent power requirements
- Power down techniques to activate the sensor interface only when a measurement is to be taken which increases the reading range and battery life time of RFID sensor tags.
- Calibration method via the use of EEPROM to store calibration values to obtain a better measurement resolution.
- Need of a standard to address the integration of sensors/actuators to a tag

3.4 RFID sensor tags and standardization

3.4.1 Discussion on standardization

One challenge in implementing RFID sensor tags is the compliance with the RFID standards in use. Indeed, none of the main actual protocols (ISO/IEC 14443 and ISO/IEC 15693 at the 13.56 MHz HF frequency and ISO/IEC 18000-6c incorporating the EPCglobal Class-1 Generation-2 UHF standard) define the access to additional functions such as sensor interface of more generally microcontrollers and buses.

The IEEE Instrumentation and Measurement Society's Technical Committee on Sensor Technology TC-9 has developed the IEEE 1451 suite of standards for the design and implementation of "smart transducers". The standard defines a network capable application processor (NCAP) and a transducer interface module (TIM) that can have a maximum of 255 transducers (sensors/actuators) [34 - 38]. This suite of standards defines the interface between the TIM and the NCAP as either wireless (WiFi, BlueTooth, Zigbee, Future technologies) or wired (Point-to-point, SPI, serial interfaces, multi-drop interfaces, mixed-mode interfaces, CAN bus and CANopen interface, sensor to RFID communication) all under a specific separate standard (IEEE 1451.X). RFID "smart transducers" fall under the standards IEEE 1451.5 for wireless transducer interface and IEEE 1451.7 that is a proposed sensor-integrated RFID tag standard for battery-assisted tags.

An important issue is whether the IEEE 1451.5 concept is compatible with the ISO RFID standards. If not, a new series IEEE 1451.X could be developed to be compliant with the ISO/IEC standard. In the meanwhile, a working group on the IEEE 1451.7 for interfacing sensors to a battery-assisted RFID tag and using the existing ISO 18000 interfaces for data communication was formed in April 2007 to come up with a finalized version of the IEEE 1451.7 in 2008. Merges have already taken place with the ISO 18000-6(E) that now incorporates the IEEE 1451 as a transport-independent set of common sensor commands based on IEEE 1451.0. The ISO/IEC is also working on defining the sensor integration with corresponding sets of commands for sensors that would be part of an amendment for the ISO/IEC 18000-6c. Results are still is under discussion. [39].

A common effort between the standardization bodies, as well as leading manufacturers is therefore needed to encourage and facilitate the development of sensor-based RFID solutions and hence enable wireless networked solutions where RFID sensor tags would share and transfer information between each other.

3.4.2 Use of existing standards

An RFID sensor tag should operate and read the sensor without interfering with the standard protocol. Whereas the current ISO/IEC protocols do no yet specify explicitly the sensor interface, they however include the possibility to define proprietary commands.

These commands can include parameters, for example to select the sensor to be measured in a multi sensor smart tag. Different commands can be specified to control a sensor interface, an external interface, etc. To avoid breaking the protocol specification, in particular turn-around time, it is possible to split a command in to two successive commands. In the case of a sensor measurement, the reader can send a sensor measurement request, then wait a determined time and send a measurement value read command. In this way the sensor selection and acquisition are not time-constrained by the protocol timings. Table 2 shows the proprietary commands defined for a HF sensor tag developed at the Technical Research Centre of Finland.

description	*Byte 0*	*Byte 1*	*Byte 2*	*Byte 3*
Measurement request	0xF0	Sensor number	CRC	CRC
Read value	0xF1	CRC	CRC	

Table 2. Proprietary commands of an ISO 14443 smart tag

Using proprietary commands is an easy method to add functionalities to a tag, but requires designing new types of tags for every new implemented function, as well as using readers with custom software.

Another approach consists of using the standard *READ* and *WRITE* commands to control the tag interfaces. It is therefore possible to use standard tags and readers. To operate, glue logic is placed on the tag memory bus. This simple logic listens to all the requests from the tag core, and triggers some actions when some activity is detected on a special address. For a sensor interface, a simple protocol can be designed as:

- the reader issues a WRITE command at the specific address 1
- the glue logic intercepts the write action in the memory, and triggers the sensor measurement. The written word can carry information, such as the sensor number.
- when the measurement is done, the glue logic writes the result in the memory at specific address 2.
- the reader then issue a READ command at specific address 2 to gather the result

The same protocol can be used to control a serial interface master peripheral (I2C, SPI, 1-wire…). With such interface, the smart tag can be used with a variety of external sensors, displays, or integrated in a more complex system including a micro controller. Fig. 13 shows a schematic view of the UHF smart tag developed at the Technical Research Centre of Finland, including the glue logic and interfaces.

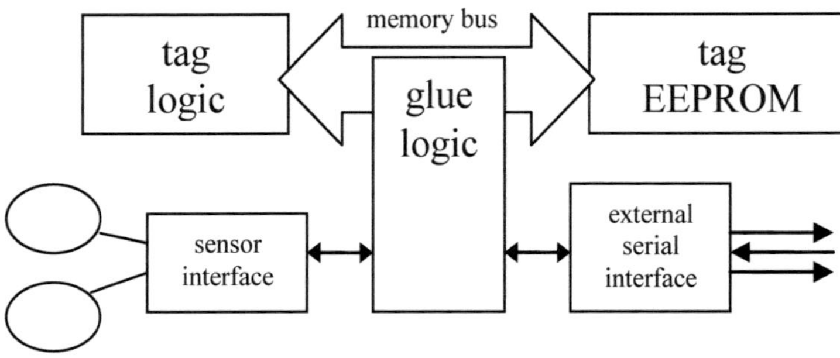

Fig. 13. UHF smart tag with external interfaces

4. References

P. R. Foster and R. A. Burberry, "Antenna problems in RFID systems," in Proc. Inst. Elect. Eng. Colloquium RFID Technology, Oct. 1999, pp. 3/1-3/5.

K.V.S. Rao, P.V. Nikitin, S.F. Lam, "Antenna Design for UHF RFID Tags: A Review and a Practical Application", *IEEE Transactions on Antennas and Propagation*, Vol. 53, pp. 3870 - 3876, Dec. 2005

P. Pursula et al. "Antenna effective aperture measurement with backscattering modulation", accepted for publication in IEEE Transaction on Antennas and Propagation, Oct. 2007.

M. Hirvonen, P. Pursula, K. Jaakkola, and K. Laukkanen, "Planar inverted-F antenna for radio frequency identication," Electron. Lett., vol. 40, pp. 848–850, Jul. 2004.

Facen, A Boni, "A CMOS Analog Frontend for a Passive UHF RFID Tag", International Symposium on Low Power Electronics and Design, 2006. ISLPED'06. Proceedings, 4-6 Oct. 2006 p: 280 - 285.

"RFID ANTENNA THAT CAN BE ATTACHED TO ANY SURFACE" http://www.vtt.fi/liitetiedostot/innovaatioita/rfid_antenna_for_all_surfaces.pdf

M. L. Allen, K. Jaakkola, K. Nummila and H. Seppä, "Applicability of metallic nanoparticle inks in RFID applications", *submitted to IEEE Transactions on Antennas and Propagation*, 2007.

M. Hirvonen, K. Jaakkola, P. Pursula, J. Säily, "Dualband Platform Tolerant Antennas for Radio Frequency Identification", *IEEE Transactions on Antennas and Propagation*, Vol. 54, pp. 2632-2637, Sept. 2006

M. Hirvonen, K. Jaakkola, J. Säily, "Triple-band pifa for radio frequency identification", *Electronics Letters*, Vol. 42, pp. 958-959, Aug. 2006.

L. Martin, D. Staiculescu, H. Li, S.L. Ooi, C. P. Wong and M. M. Tentzeris, "Investigation of the Impact of Magnetic Permeability and Loss of Magnetic Composite Materials on RFID and RF Passives Miniaturization", Workshop on Computational Electromagnetics in Time-Domain, 2007. CEM-TD 2007. 15-17 Oct. 2007 Page(s):1 - 4

Y. Hwang, Y.P. Zhang, Terry K.C. Lo, K.M. Luk and E. K.N. Yung, "Miniaturization on Planar Antennas with very High Permittivity Materials", Microwave Conference Proceedings, 1997. APMC '97., 1997 Asia-Pacific, Volume 1, 2-5 Dec. 1997 Page(s):217 - 220 vol.1

S. Bovelli, F. Neubauer, C. Heller, "A Novel Antenna Design for Passive RFID Transponders on Metal Surfaces", Microwave Conference, 2006. 36th European, Sept. 2006 Page(s):580 - 582

O. Rostbakken, G.S. Hilton, C.J. Railton, "Adaptive feedback frequency tuning for microstrip patch antennas", Antennas and Propagation, 1995., Ninth International Conference on (Conf. Publ. No. 407), 4-7 Apr 1995 Page(s): 166 - 170 vol.1

http://www.telepathx.com/

http://www.alvinsystems.com/

http://www.montalbanotechnology.com/

http://www.bioett.se/

P. Pursula, J. Marjonen, H. Ronkainen, K. Jaakkola, "Wirelessly powered sensor transponder for UHF RFID", submitted to TRANSDUCERS'07: The 14th International Conference on Solid-State Sensors, Actuators and Microsystems.

N. Cho, S. Song, S. Kim, S. Kim, H. Yoo, "A 5.1 μW UHF RFID tag chip integrated with sensors for wireless environmental monitoring", Proceedings of ESSCIRC, Grenoble, France, 2005, pp. 279-282.

Dallas Semiconductor, MAXIM; http://www.maxim-ic.com/an2094

B. Black; *"Analog-to-Digital Converter Architectures and Choices for System Design"*; Analog Devices; http://www.analog.com/library/analogDialogue/archives/33-08/adc/index.html

J. Sauerbrey, D. Schmitt-Landsiedel, and R. Thewes, "A 0.5-V 1-μW Successive Approximation ADC": (2003)

M. D. Scott, B. E. Boser, and K. S. J. Pister, "An Ultra-Low Power ADC for Distributed Sensor Networks":

A. Heubi, P. Balsiger and F. Pellandini, "Micro Power "Relative Precision" 13-bit Cyclic RSD A/D Converter", ISLPED'96, Aug. 12-14 1996, Monterey CA, USA, pp. 253-257:

T. Kim, J. Sung, S. Kim, W. Joo, S.-B. You and S. Kim, "A 10-bit, 40Msamples/s Cascading Folding & Interpolating A/D Converter with Wide Range Error Correction":

R. Révérend, I. Kale, G. Delight, D. Morling and S. Morris, "An Ultra-Low Power Double-Sampled A/D Mash ΣΔ Modulator":

B. Vaz1, N. Paulino, J. Goes, R. Costa, R. Tavares, and A. Steiger-Garção, "Design of Low-Voltage CMOS Pipelined ADC's using 1 pico-Joule of Energy per Conversion": (2003)

D. Miyazaki, S. Kawahito, and M. Furuta, "A 10-b 30-MS/s Low-Power Pipelined CMOS A/D Converter Using a Pseudodifferential Architecture": (2001):

M. Dessouky and A. Kaiser, "Very Low-Voltage Digital-Audio Δ Σ Modulator with 88-dB Dynamic Range Using Local Switch Bootstrapping":

S. Chatzandroulis, D. Tsoukalas, and P. A. Neukomm, "A Miniature Pressure System with a Capacitive Sensor and a Passive Telemetry Link for Use in Implantable Applications":

F. KRUMMENACHER: "A High-Resolution Capacitance-to-Frequency Converter":1985

Z. Zheng, "Low Power High Resolution Data Converter in Digital CMOS Technology", Oregon State University, 1999

Texas Instruments, "Amplifier and Data Converter Selection Guide", 2005, http://focus.ti.com/analog/docs/selectionguides.tsp?familyId=2

IEEE, *Standard 1451.1-1999, IEEE Standard for a Smart Transducer Interface for Sensors and Actuators – Transducer to Microprocessor Communication Protocols and Transducer Electronic Data Sheet (TEDS) Formats* (Institute of Electrical and Electronic Engineers, New York, 1998).

IEEE, *Standard 1451.3-2003, IEEE Standard for a Smart Transducer Interface for Sensors and Actuators – Digital Communication Protocols and Transducer Electronic Data Sheet (TEDS) Formats for Distributed Multidrop systems* (Institute of Electrical and Electronic Engineers, New York, 2004).

IEEE, *Standard 1451.4-2004, IEEE Standard for a Smart Transducer Interface for Sensors and Actuators – Mixed-mode Communication Protocols and Transducer Electronic Data Sheet (TEDS) Formats* (Institute of Electrical and Electronic Engineers, New York, 2004).

IEEE, *Standard 1451* (NIST; Washington DC, 2004)

IEEE, *Standard 1451.5-2004, IEEE Standard for a Smart Transducer Interface for Sensors and Actuators – Mixed-mode Communication Protocols and Transducer Electronic Data Sheet (TEDS) Formats* (Institute of Electrical and Electronic Engineers, New York, 2004).

International Organization for Standardization (ISO), *Information Technology – Radio Frequency Identification for Item Management – Part 6: Parameters for Air Interface Communication at 860 MHz to 960 MHz*, ISO/IEC 18000-6:2004 (ISO; Geneva, 2004).

10

Advances in RFID Components Design: Integrated Circuits

Arjuna Marzuki[1], Zaliman Sauli[2] and Ali Yeon Md. Shakaff[2]
[1] *Universiti Sains Malaysia*
[2]*Universiti Malaysia Perlis*
Malaysia

1. Introduction

Many RFID Components are either discrete components or integrated circuits components. With high demand and huge RFID market in supply and chain management, more RFID systems will employ integrated circuit components. This is due to cost competitiveness in using integrated circuit components. Understanding of these advanced components is therefore essential for the development of advanced RFID system.

This chapter emphasizes on the practical design of advanced RFID components namely the tag and the reader. This chapter discusses the design methodology of integrated circuit. The technology such as Silicon Bipolar and Silicon CMOS are studied in this chapter; the focus is on understanding the advantages and disadvantages of using these technologies for the design of advanced RFID components. The architecture or circuit topology of reader and tag are also thoroughly discussed; the integration of features which normally reduces the number of discrete components is the focus of this topic. The circuit technique in designing the tag and reader is studied and analyzed. The current research trend in RFID integrated circuit is more on reader than tag. SiGe BiCMOS (Chiu et al., 2007), 0.18 µm CMOS (Wang et al., 2007, Khannur et al., 2008) are amoung technologies used in the latest reader resarch and product. The measurement methodology of advanced RFID components is also discussed in the last topic. Overall this chapter provides basic knowledge to the readers for further research in RFID integrated circuits.

The understanding in advances of RFID components design, namely the technology, design techniques and test enables one to pursue in effective design of the state of the art RFID systems.

2. Design methodology of integrated circuit

RFID System consists of two main hardware components, RFID tags or transponder and RFID readers or receiver. Both components have analog and digital circuitry. The reader would have more complex digital circuitry than tag. Fig. 1 shows integrated circuit (IC) design flow which can be used in designing both components.

Specification: In all IC designs, the first step is to identify the specification or requirement. Price or cost is normally considered at this stage as well.

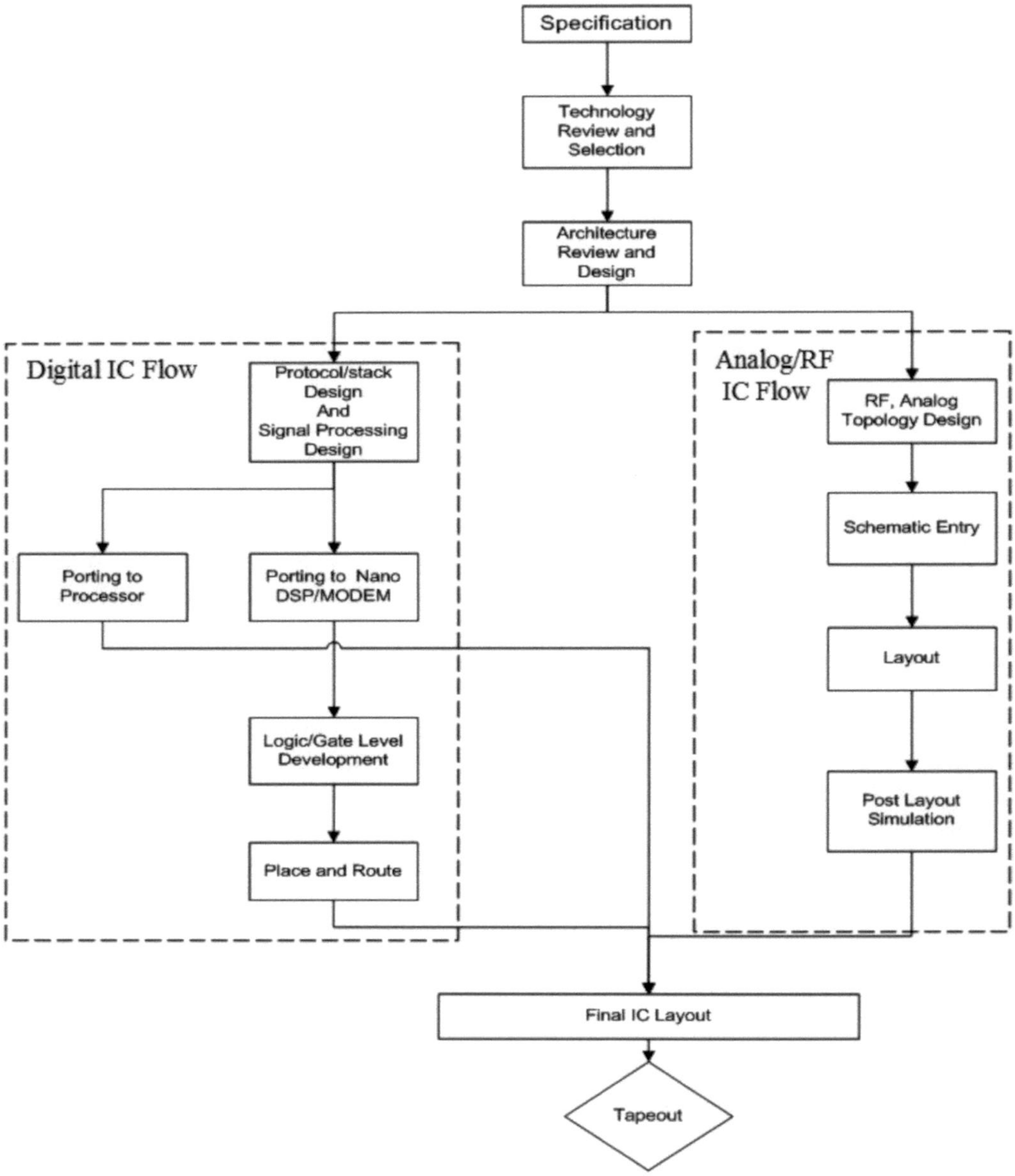

Fig. 1. IC design flow

Technology Review and Selection: Silicon, GaAs or Silicon on Insulator (SOI) is decided based on its capability to meet the requirement and specification.
Architecture Review and Design: To review the suitable design architecture which can meet both selected technology and specification.
Protocol/stack Design and Signal Processing Design: Flow of state diagram development, Signal processing architecture design, Matlab simulation.
Porting to Processor: Port the program to processor/microcontroller such as ARM (for Reader), to Finite State Machine or simple general purpose process, 6502 compatible (for Tag) (De Vita & Iannaccone, 2004).

Porting to Nano DSP/MODEM: Conversion of Numerical model or state diagram to hardware language. The hardware language such as Verilog or VHDL is normally used at this stage.
Logic/Gate Level Development: Conversion of the hardware language to logic level (standard cell) by using EDA tools such as design compiler. Behavioural language such as Verilog or VHDL is normally used at this stage.
Place and Route: Conversion of standard cell to layout. A synthesizer is used at this stage.
RF, Analog topology Design: Topologies of RF and analog circuits are decided and designed.
Schematic Entry: Applying the schematic design using EDA tools and simulation or verification is made to verify the performance against the specification.
Layout: To design the mask for Analog and RF IC.
Post Layout Simulation: Simulation is done together with parasitic components which are extracted from the layout.
Final IC layout: To design or combine masks for both Analog, RF and digital IC
Tapeout: The mask design is sent to mask shop and later wafer fab or foundry for IC fabrication.

3. Technology

There are at least two main figure of merit parameters which characterize technology. They are transconductance, g_m and transition frequency, f_T. Equation (1) is basic transconductance formula, (2) is f_T of NPN Bipolar Transistor.
g_m is proportional to the voltage gain, high voltage gain is required for amplifier block, which is an important block for RF and analog circuit. For CMOS device, current has to be increased in order to increase g_m . f_T is the frequency where current gain is unity, high frequency RFID IC requires high f_T device which is normally ten times of the operational frequency.

$$g_m = \frac{I_{OUT}}{V_{IN}} \tag{1}$$

Where I_{OUT} is output current of device, V_{IN} is input voltage to the device.

$$f_T = \frac{g_m}{2\pi(C_\pi + C_\mu)} \tag{2}$$

Where C_π is parasitic capacitance between base and emitter, C_μ is parasitic capacitance between collector and base.

3.1 Bipolar

From Fig. 2, NPN Bipolar transistor, is not a symmetrical device, this can been seen from the difference of doping level of emitter and collector. Buried layer is used to reduce the collector resistance. This will make RFID IC design more complex, where collector and emitter cannot be swapped. The scaling of Bipolar technology is not aggresive as CMOS technology (Semiconductor Industry Association, 1997). These factors have hindered the use of Bipolar technology for RFID tag where simple design is normally chosen.

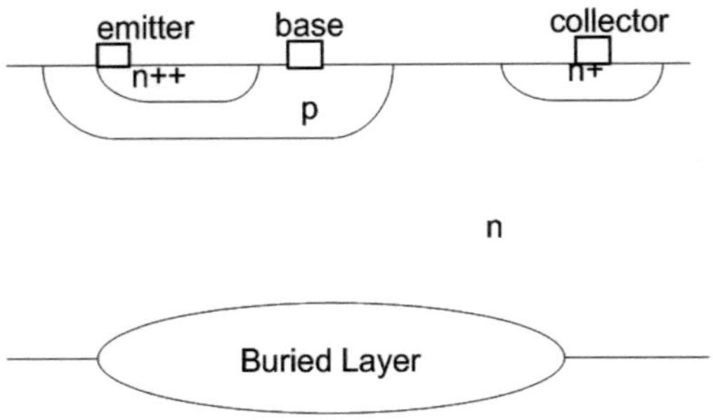

Fig. 2. Cross Section of NPN Bipolar Transistor

$$I_{OUT} = I_s \exp(\frac{V_{BE}}{V_T}) \tag{3}$$

Equation (3) shows current equation of NPN Bipolar transistor, where I_S is saturation current, V_T is thermal voltage and V_{BE} is voltage between base and emitter . By replacing (3) into (1), Transconductance of NPN Bipolar transistor is therefore,

$$g_m = \frac{I_{out}}{V_T} \tag{4}$$

Where $V_T = \frac{kT}{q}$, k is Boltzmann's constant and T is temperature in Kelvin and q is a unit electron charge. At room temperature, V_T = 25 mV. Replace (4) into (2) yields $f_T = \frac{I_{OUT}}{V_T 2\pi(C_\pi + C_\mu)}$.

3.2 CMOS

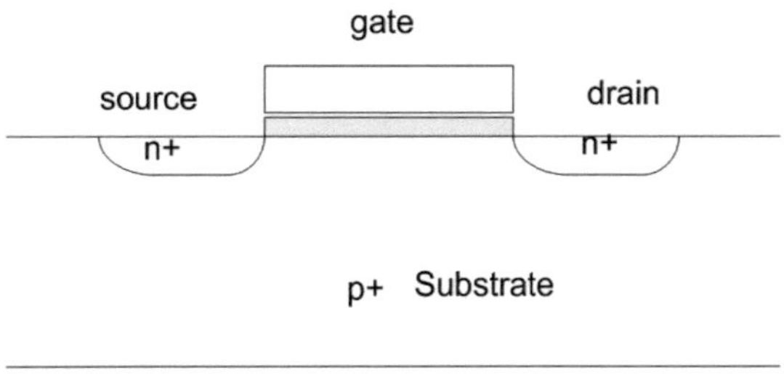

Fig. 3. NMOS Cross Section

From Fig. 3, NMOS device is a symmetrical device, this will ensure easier design and integration. The scaling of CMOS technology is aggressive and straight forward (Semiconductor Industry Association, 1997).

$$I_d = k'\frac{W}{2L}(V_{gs} - V_T)^2 \tag{5}$$

Equation (5) is drain current equation of NMOS device, where W is channel width, L is channel length, V_T is threshold voltage and k' is transconductance parameter. Transconductance for NMOS is

$$g_m = \frac{2I_d}{V_{gs} - V_T} \tag{6}$$

V_{gs}-V_T is normally in the range of hundreds millivolt.
From (6) and (4), bipolar devices tend to have higher g_m and f_T. Smaller dimension of CMOS technology would have almost similar f_T as Bipolar technology. Smaller dimension of CMOS technology is usually more expensive than Bipolar technology. Equation (7) is f_T of CMOS transistor.

$$f_T = \frac{g_m}{2\pi(C_{gs} + C_{gd})} \tag{7}$$

Where C_{gs} is parasitic capacitance between gate and source, C_{gd} is parasitic capacitance between gate and drain.

3.3 BiCMOS

BiCMOS is the technology where Bipolar and CMOS devices are in the same substrate. Designer can employ high g_m devices for RF and analog circuit, at the same time high integration of analog and digital circuit by using CMOS devices. This process/technology could provide a quick and cost effective solution to the design.

4. RFID Reader IC

Complete Reader IC for UHF application is still new, the motivation behind the development is because of high demand in portable RFID Reader for supply chain management application.
Fig. 4 shows block level of reader IC. Passive UHF RFID system is usually half duplex system. Interface block in RFID Reader IC is normally composed of serial to parallel (SPI), Inter-Integrated circuit (I2C) and Universal Asynchronous Receiver/Transmitter (UART) type. Digital baseband modem is normally some form of Digital Signal Processing (DSP). Standards or protocols are determined or configured by the DSP. Digital to Analog Converter (DAC) receives digital data from DSP and convert it to analog signal. The signal is then later converted to RF by the RF transmitter or modulator. Frequency sythesizer is used to provide accurate high frequency resolution, especially for very high accuracy application. Directional Coupler provides one way direction of signal. Demodulator receives signal which is downconverted by downconverter to lower frequency. This lower frequency signal is later converted to digital signal by Analog to Digital Converter (ADC) for further processing in Digital Baseband Modem. Memory unit such as Read Only Memory (ROM) and Randon Access Memory (RAM) are employed in the whole design. Micro-Processor Unit (MPU) is used to receive command from the Interface and control Digital Baseband Modem. Like other ICs, RFID Reader IC also requires biasing circuitry.

4.1 RFID reader analog and RF receiver section

Receiver section is normally a direct conversion receiver (DCR). This topology will reduce the number of components in the receiver design.

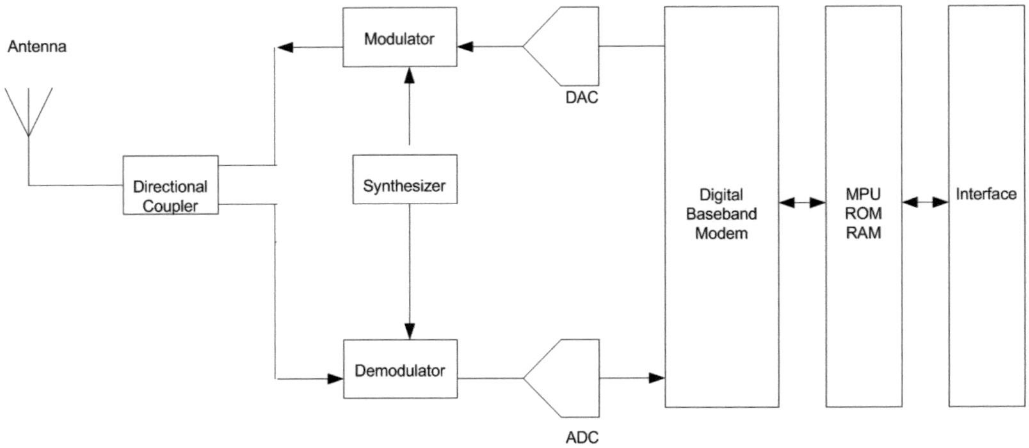

Fig. 4. RFID Reader IC block kevel

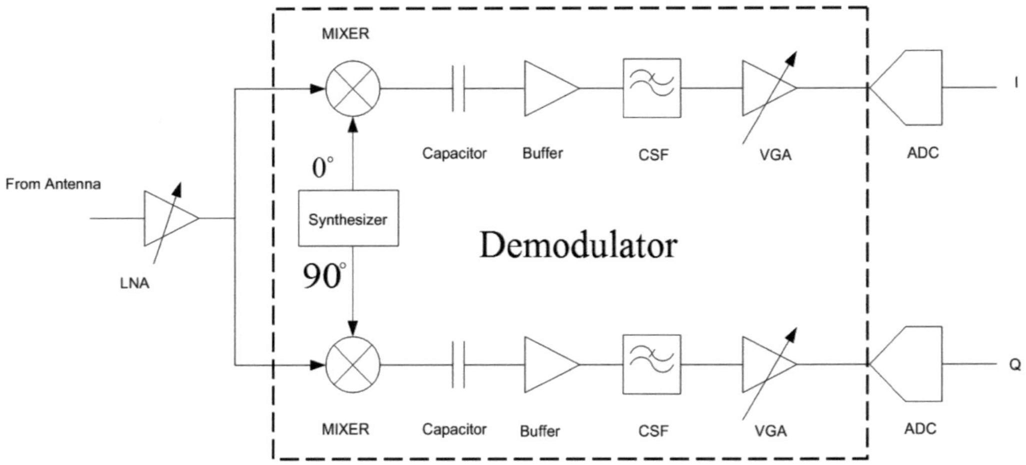

Fig. 5. Receiver section

Demodulator block discussed in Fig.4 is now clearly shown in Fig. 5. It consists of I/Q Mixer and Analog Baseband Circuit (Buffer, Channel Select Filter (CSF) and Variable Gain Ampifier (VGA)), Synthesizer is a separate block and not part of Demodulator. Self jammer is a problem to receiver, this is when a continuous wave (CW) signal is sent to a tag, a leakage of this transmit signal at receiver input leads to saturation of receiver block and degradation of sensitivity. For direct conversion architecture, the leakage to the receiver is directly down converted to DC. To deal with self jammer, two antennas sometimes are suggested to maximize the isolation between receiver and transmitter (Wang et al., 2007).

4.1.1 Low noise amplifier

The purpose of Low Noise Amplifier (LNA) as shown in Fig. 5 is to amplify weak received signal without adding too much noise to the signal. In the receiver chain, LNA determines the sensitivity. Fig. 6 shows 2-stage single ended input and differential output LNA. The fist stage is cascode LNA amplifier with capacitor, C_{ext} for Power-Constrained Simultaneous Noise and Input Matching (PCSNIM) (Nguyen et al., 2004). With the capacitance, the circuit can achieve low noise, high gain at an optimum low current consumption. Second stage is a differential amplifier with inductive degeneration, L_{S2}. L_g, C_{ext} and L_S are used to achieve PCSNIM. L_d and C_c is for output matching of the first stage. R_B is used to bias transistor M4. I_s set DC current for this stage. R_O is load for the second stage.This topology can achieve high linearity with low noise peformance. High linearity is required to tolerate self-jamming signal.

For some LNA's output matching, capacitive transformer is sometimes used. The capacitive transformer is implemented using MOS capacitors. These capacitors achieve high Q-factors and are compatible with standard CMOS technology (Marzuki & David, 2006).

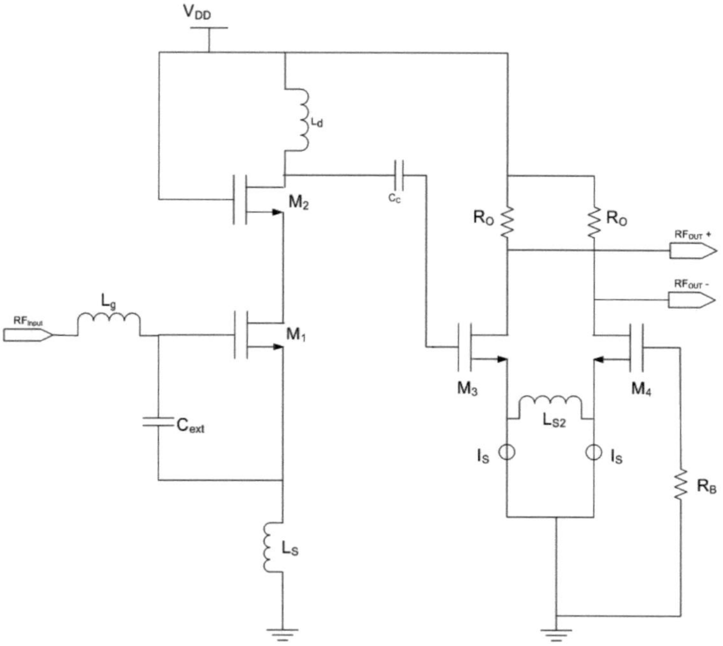

Fig. 6. Low Noise Amplifirer Circuit

4.1.2 I/Q mixer

I/Q Mixer is used to convert RF frequency signal to Intermediate Frequency (IF) frequency signal Passive mixer as in Fig. 7 has high linearity compared to conventional gilbert-cell mixer (Lee, 1998). The high linearity of the mixer due to linearity of NMOS switches (low impedance and no g_m non linearity). The passive switching mixer has very low $1/f$ noise because there is no DC current through the switching stage (Zhou & Chang, 2005). Local Oscillator (LO) signal is provided by synthesizer block. R_b is used to bias the drain of NMOS switches.

The output of the mixer are AC coupled to block DC offset. The DC blocking capacitor value is chosen depending on the data rate of tag.

4.1.3 Analog baseband circuit

The purpose of this block is to process the analog IF signal to digital singal. Analog Baseband Circuit composes Variable Gain Amplifier (VGA), filter and ADC. For DCR, transmitter carrier leakage at the receiver input is down converted to DC. It can also be further removed by a DC offset cancellation circuit in the baseband section.

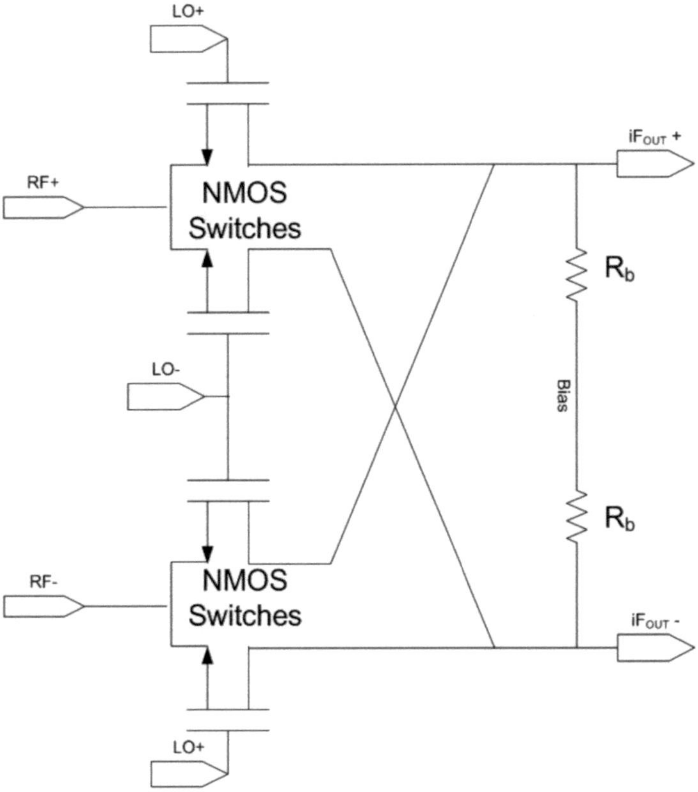

Fig. 7. Passive Mixer

For Class 1 and ISO 1800-6C, channel select filter is configured as low pass filter with zero IF. Example of the filter is 3 rd order, chebychev filter using Output Transconductance Amplifier (OTA) is shown Fig. 8.

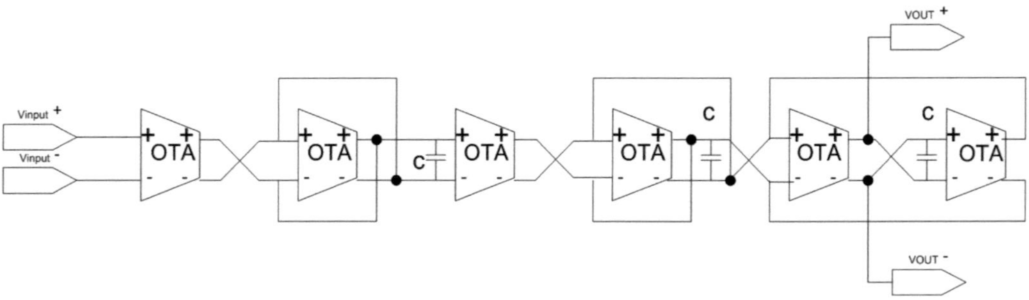

Fig. 8. 3rd Order Chebychev filter

A Simple OTA is shown in Fig. 9. This OTA uses 6 inverters.

Opamp-based programmable gain amplifiers are normally used as VGA for the excellent linearity performance. The resolution is 1 dB with large control range can be achieved. The amplifier is used to increase the signal voltage for easier analog to digital conversion. 10 bit pipe-lined ADC is used to convert analog signal to digital signal (Khannur et al., 2008).

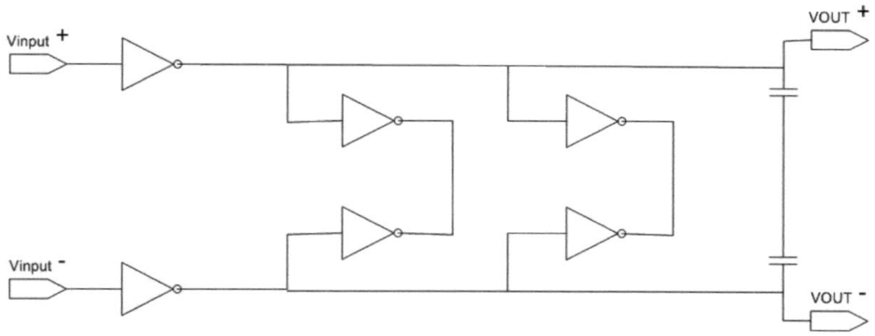

Fig. 9. Inverter-Based OTA

4.2 Receiver digital baseband demodulator

Fig. 10 shows receiver digital baseband demodualtor or modem. Digital Filter block provides a bandpass digital filtering to preserve the information contents of the in-band signal whilst altenuating out-of-band noise and interference.

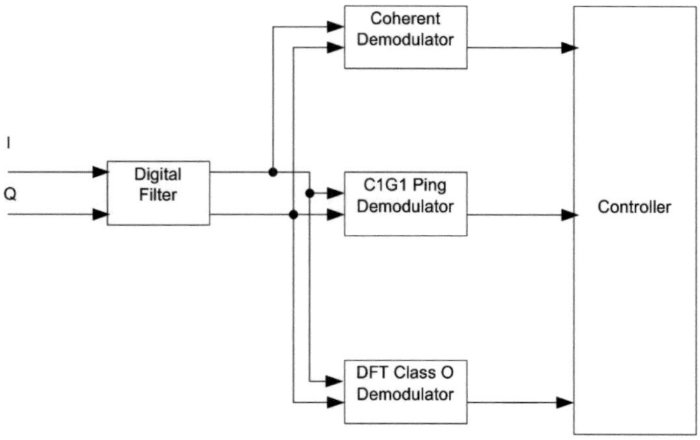

Fig. 10. Receiver Digital Baseband Demodulator

Class 0 Demodulator computes 16-bin DFT and compares the peak power in the data-1 bins. The soft decision is passed to bit slicing. Ping demodulator (Class 1 Gen 1) uses Zero crossing detector followed by matched filter for all 32 possible combination of a ping reply. This signal is with out preambles or synchronization

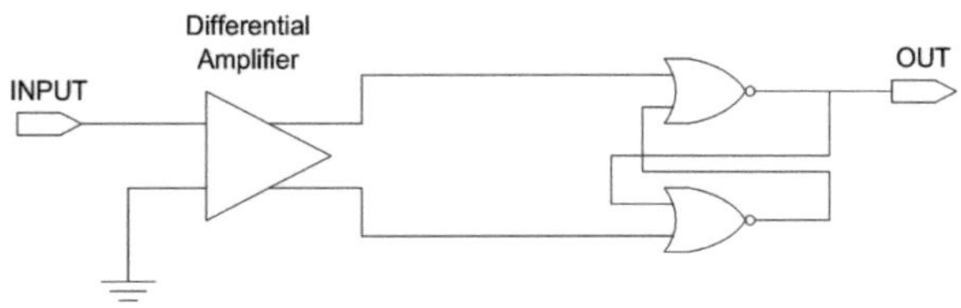

Fig. 11. Zero Crossing Detector

Fig. 11 shows a differential amplifier and a flip-flop as zero crossing detector. Coherent Demodulator uses a correlator and estimator to measure the incoming tag data rate utilizing the pilot tone as the preamble. Controller block will process digital data for application or host processor.

4.3 RFID reader analog and RF transmitter section

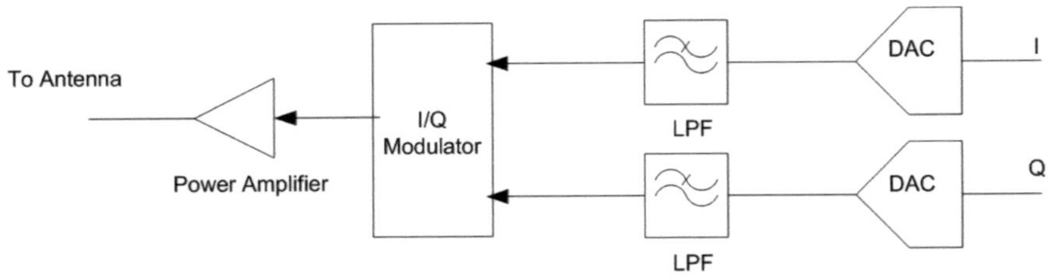

Fig. 12. RFID Reader Analog and RF Transmitter Section

Fig. 12 shows Analog and RF Transmitter of RFID Reader. The purpose of this block is to convert digital I and Q data to RF signal. Current steering 10 bit DAC (Khannur et al., 2008) are used to convert digital signal to analog signal. The low pass filter (LPF) is used to eliminate spectral spur and to ensure the transmit spectrum fits well into transmit mask. I/Q modulator is used to convert low frequency signal to 900 MHz band. Class A PA is used to produce to maximum 30 dBm output signal.

4.3.1 Power amplifier

Cascode topology (Khannur et al., 2008) is normally used for Power Amplifier (PA). Attenuator is used to control the gain or output power of PA. Matching circuit is needed to match for optimum power transfer between input and transistor gate. Biasing of the transistor is done through RF choke. L_d and capacitor tune the circuit to the UHF frequency. PA circuit is shown in Fig. 13

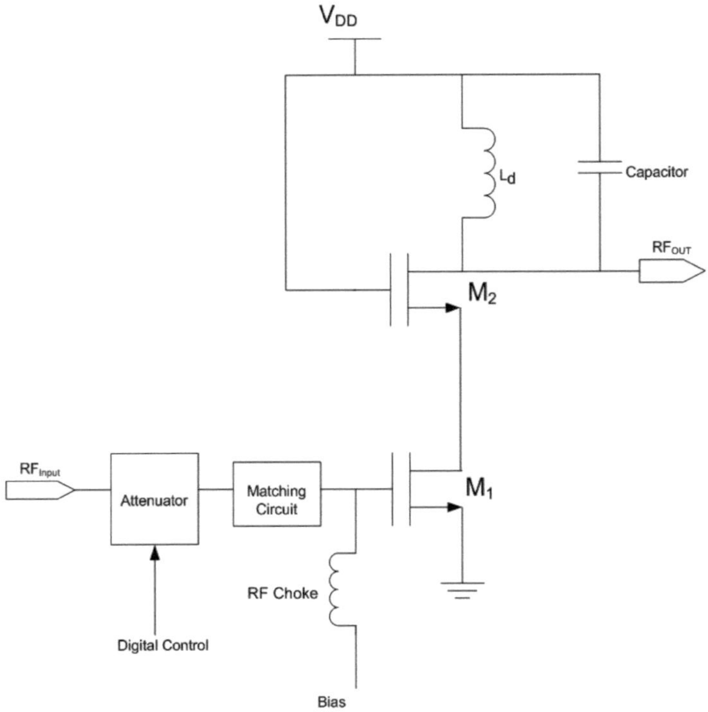

Fig. 13. Variable Power Amplifier

4.3.2 I/Q modulator

Gilber Cell mixers are used for I/Q mixer. Both I and Q sections are used for generating SSB -ASK Modulation. Input I/Q signal is from Digital to Analog Converters.
I/Q Modulator is shown in Fig. 14. Source degeneration, R_E is needed to increase the linearity to satisfy transmitter spectrum mask. Local oscillator (LO) signal is provided by the synthesizer block. I_s is used to set DC current for the modulator.

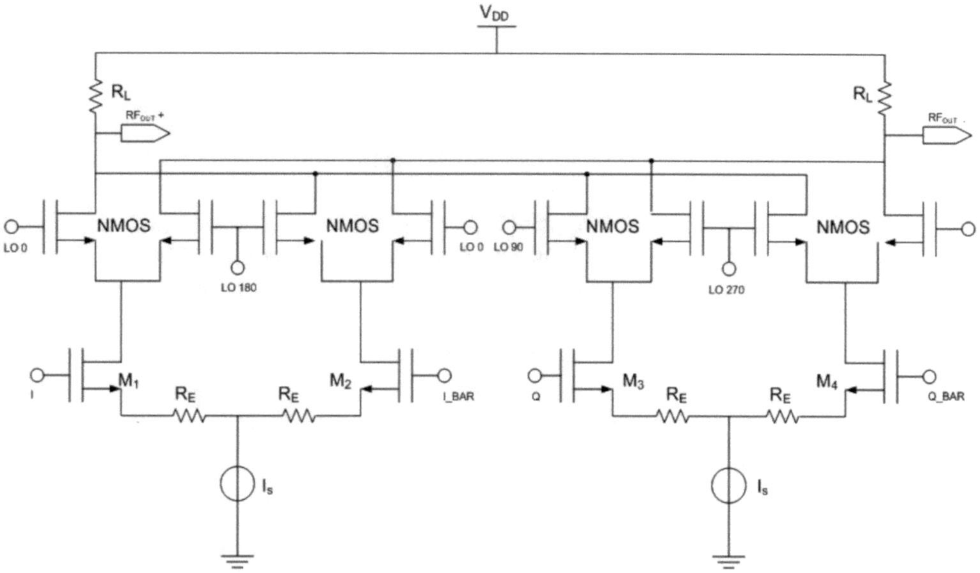

Fig. 14. I/Q Modulator

4.4 Transmitter digital baseband modulator

Fig. 15 shows Transmitter Digital Baseband Modulator. Hilbert Transform filter and cordic frequency shifter are used to provide real modulation or complex modulation. This is to support different modulation schemes such as ASK, SSB ASK or PR ASK. Predistorion block is needed to cope with the nonlinearities by PA when supply modulation is used. Controller is a programmable microcode sequencer. It receives non-real time commands from a protocol processor which is preambles and frame sync.

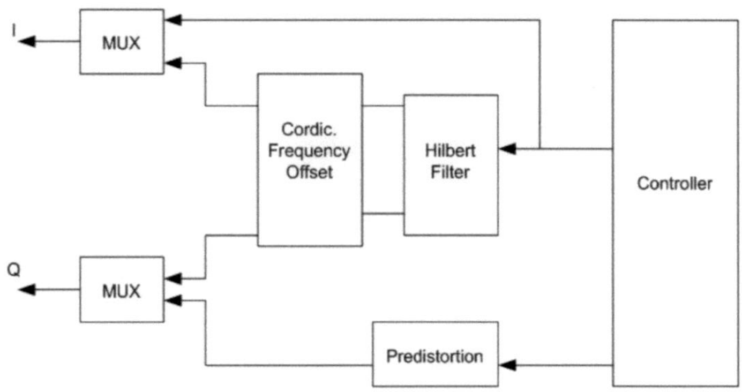

Fig. 15. Digital Transmit Modulator

Output I and Q are connected to DAC for analog and RF processing.

4.5 Synthesizer

Synthesizer provides LO signal for both receiver and transmitter. Dual loop PLL Synthesizer can achieve good phase noise and low spurious level with good frequency resolution (Razavi, 1997). Integer-N PLL with a divide-by- 2/3 with external loop filter can also meet the requirement (Chiu et al., 2007). This type of synthesizer is small, and this leaves space for more complex digital block to be implemented in reader IC.

4.6 Biasing circuitry

One important requirement in integrated circuit is constant performance across temperature. This is especially critical for analog integrated circuits such as radio frequency integrated circuits (RFIC). One way to control the performance is to control the current consumption.

A typical RFIC circuit employs a circuit which provides stable voltage or current source across temperature. A Popular circuit is bandgap circuit which provides stable voltage reference to RFIC circuit. The bandgap circuit employs diodes as internal reference, the diode array tends to increase the size of the layout and model accuracy is scarce.

A circuit called Voltage for current source (VCS) circuit provides voltage for the current source in all RF IC circuits. The circuit is shown in Fig. 16. VCS circuit employs three transistos. IPTAT is current proportional to absolute temperature and I_s is the required current. Equation (8) shows that I_s can be designed to be constant, proportional or complimentary across temperature variation.

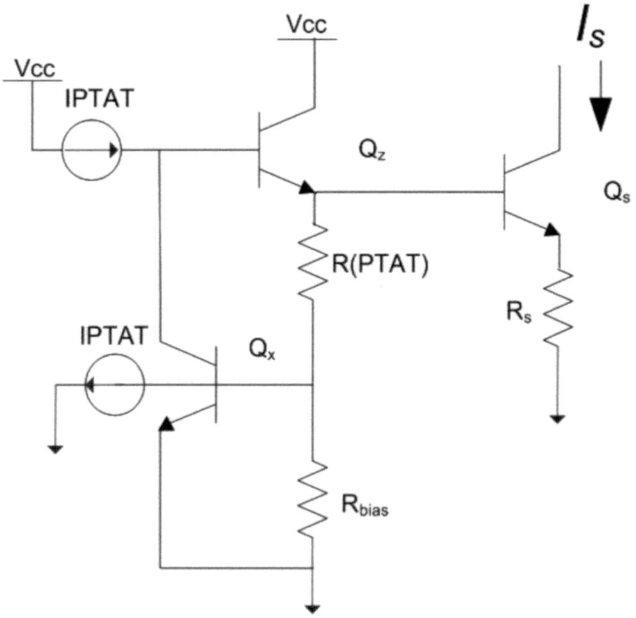

Fig. 16. Simplified VCS with Current Source (Marzuki et al., 2008)

$$I_S = \frac{(IPTAT \times R(PTAT) + V_{BE}(Q_x) - V_{BE}(Q_S)) + V_{BE}(Q_x) \times \dfrac{R(PTAT)}{R_{bias}}}{R_s} \tag{8}$$

5. RFID Tag IC

Unlike reader IC, tag IC development is already mature. The focus on the tag IC development is on reducing the cost. Fig. 17 shows RFID Tag block main diagram. A rectifier is employed to establish supply voltage to the rest of the circuits. The supply voltage is regulated using protection and voltage regulation circuit. Demodulator receives signal or information from the RFID reader, while modulator is used to convert digital data to higher frequency signal. Clock recovery and divider circuit accurately produce 50-50 duty cycle clock which is generated from RFID reader. Digital block is normally composed of MPU and memory block. Information identification is normally stored in memory block.

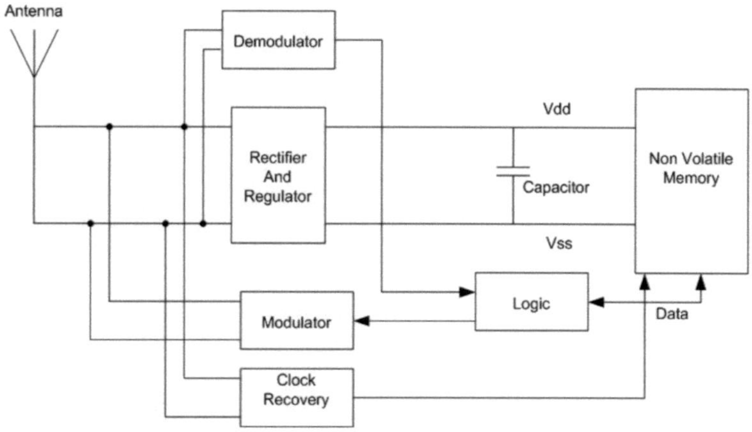

Fig. 17. RFID Tag

The design of UHF tag IC is normally targetted for long distance application. This proves to be a challenge. Design tradeoff must be made and some techniques must be done to accomodate the requirement. For UHF RFID tag, only (dipole) antenna is external component, while the rest of the blocks are in one IC.

5.1 Rectifier

The received signal, sent by an interrogator device, generates a sinusoidal voltage which is then rectified, filtered and multiplied if necesary to generate a DC voltage that can be used to supply and activate internal circuitry.

The concept of rectifier is to transfer the charge from C_{in} to C_{hold} during the cycles of incoming strong signal. The basic circuit of single stage rectifier is shown in Fig. 18

Low V_{th} PMOS (Fig. 18 (c)) is used to get low voltage drop of around 150 mV on each diode-connected transistor.

5.1.1 Efficiency

Efficiency is very important, high effeciency indicates that less power will be wasted, and longer detection range can be achieved. Basic efficiency (Kocer & Flynn, 2005) equation

$$Efficiency = \frac{DCV_{OUT}}{V_{INPEAK} \times n} \tag{9}$$

Where DCV_{out} is the generated output voltage. V_{INPEAK} is the peak incident RF amplitude, n is number of stages.

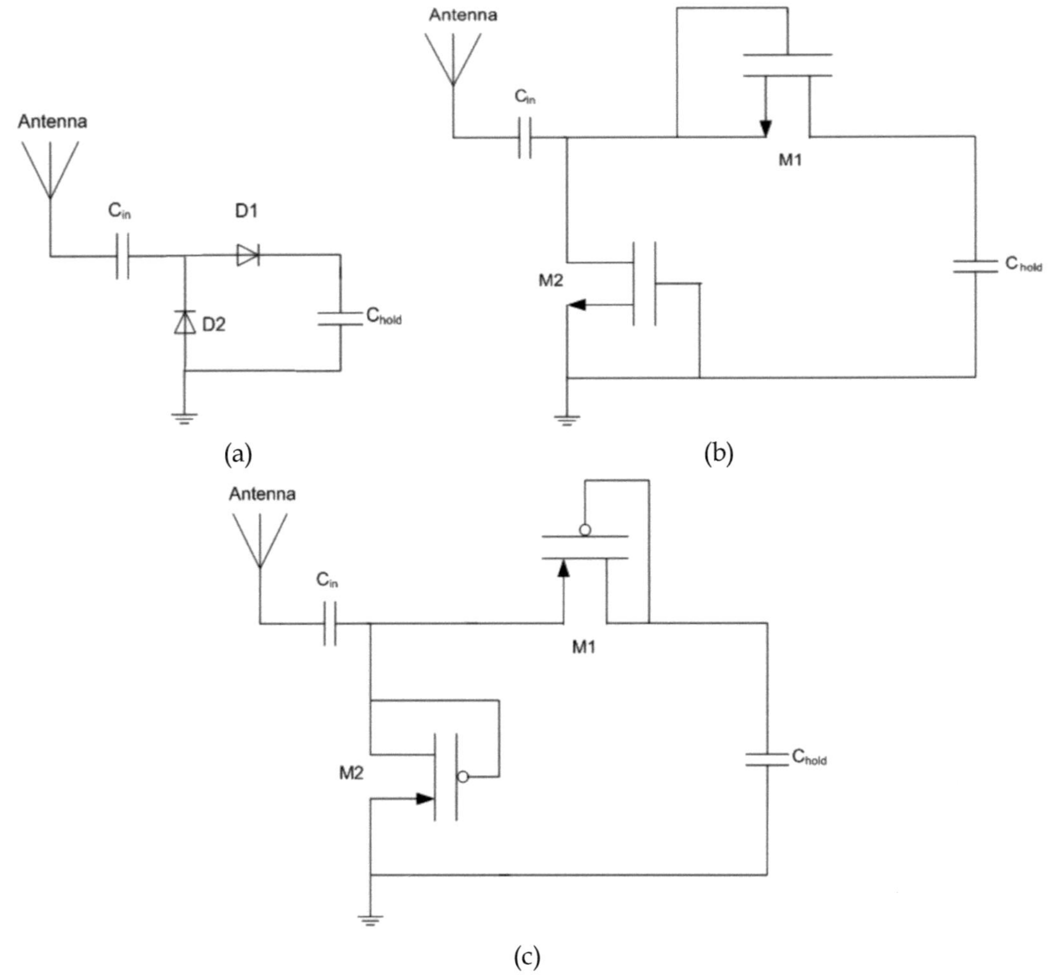

Fig. 18 (a). Rectifier using diode (b). Rectifier using NMOS (c) Rectifier using PMOS

One of the loss factors is the threshold voltage, which is associated with the diode-connected MOS. The specially designed Schottky diodes with low resistance allow for a high-efficiency conversion of RF energy to DC supply voltage, but not many IC technologies come with this diode. For standard CMOS technology, to reduce or eliminate the threshold voltage, an external voltage must be applied between gate and source. An example of the concept is shown in Fig. 19.
V_{bth} is external voltage source. The storage capacitor stores the V_{th} voltage for the MOS diode.

5.1.2 Full wave rectifier

Fig. 20 shows a single stage mirror-stacked stucture of two CMOS half-difference rectifier circuits, this can optimize power efficiency due to eliminating of parasitic capacitances a the input terminal IN -. This is due to IN – is AC ground. The circuit multiply the input voltage.

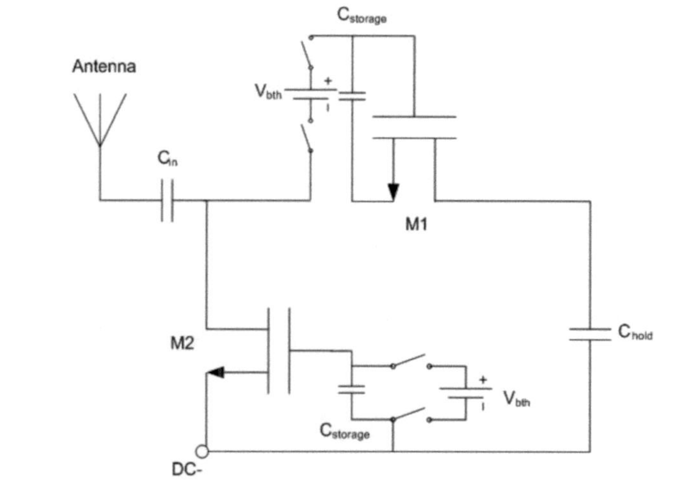

Fig. 19 Rectifier circuit with External Voltage Source (Nakamoto et al., 2007)

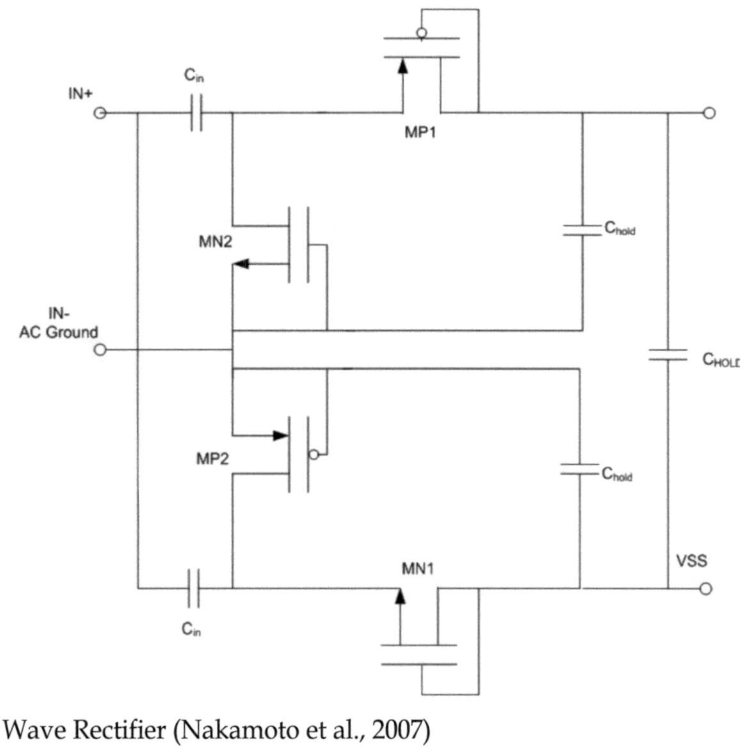

Fig. 20. Full Wave Rectifier (Nakamoto et al., 2007)

5.1.3 Power matching

At RF frequency, power matching is important for optimum power transfer between antenna and input of rectifier. If mismatch occur, smaller DC power is converted by the rectifier. The equivalent input impedance of the rectifier might be represented, as a zero-order approximation, by the parallel of a resistance and a capacitance. In the high frequency analysis of the rectifier, all the capacitances of the rectifier can be considered as short-circuited and so all diodes can be considered in parallel or anti-parallel with the input. As a consequence, at the input of the rectifier, the capacitances of all diodes are in parallel. The equivalent input resistance, R_{eq}, of the rectifier is the resistance calculated from power consumption. Using an LC power matching network, the $L= QR_A/\omega_o$, and $C = Q / R_{eq}\ \omega_o$, (Razavi, 1997), where, R_A is the resistance of the antenna and Q is the quality factor of the LC network $Q = \sqrt{R_{eq} / R_A - 1}$. An inductance L', in parallel with the input of rectifier, can be used to compensate the equivalent input capacitance of the rectifier. Fig. 21 shows the matching network.

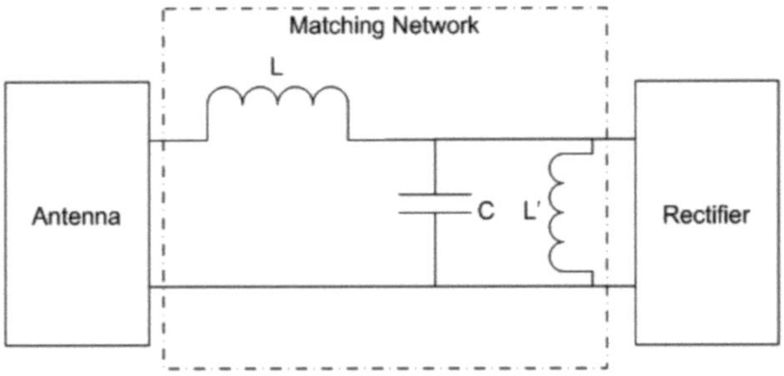

Fig. 21. Matching Network (De Vita & Iannaccone, 2004)

5.2 Modulator

Backscatter modulator is normally used for tag IC. The modulated backscattered signal happen when IC input impedance changing upon receiving CW carrier with small notches for data transfer from the reader. A design trade-off must be made in order to ensure a larger power still availabe for (rectifier) power supply. Fig. 22 shows PSK modulator circuit.

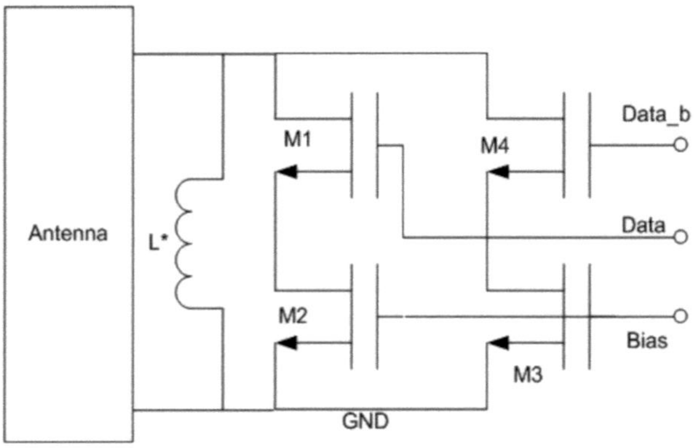

Fig. 22. PSK Modulator (De Vita & Iannaccone, 2004)

Depending on the data signal level, either MI or M4 conducts leading to a different output capacitance, since the two transistors have different sizes. The inductor **L*** makes the variation of the imaginary part of the impedance seen by the antenna symmetric with respect to zero. Transistors M2 and M3 allow us to obtain an output resistance of the modulator much larger than the antenna resistance so that only a negligible fraction of the power at the antenna goes to the modulator. The condition for ASK modulation is explained in (Finkenzeller, 2003), where input impedance must be in resistance.

5.3 Demodulator

The demodulator detects the ASK modulated signal from Reader. Fig. 23 shows basic circuit of ASK demodulator.

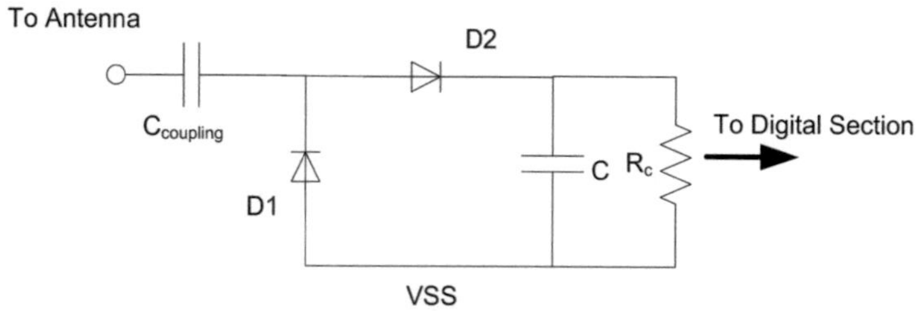

Fig. 23. ASK demodulator (De Vita & Iannaccone, 2004)

D1 and D2 can be replaced by diode connected MOS transistor.

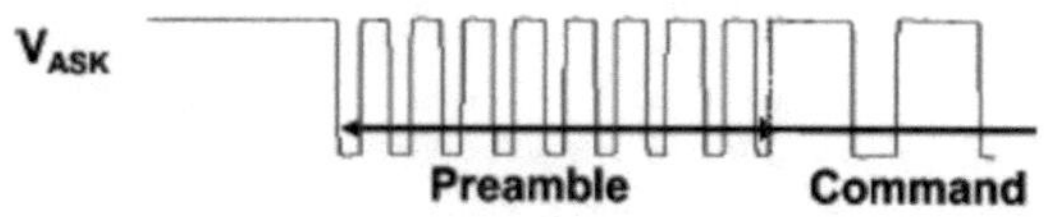

Fig. 24. Demodulated ASK (Nakamoto et al., 2007)

Fig. 24 shows an example of demodulated signal, output of demoulator. The first serial signal is preamble signal, the second serial signal is the command from reader.

6. Measurement methodology

For RFID Reader, the output of transmit after filter carrier phase noise is important. The receiver sensitivity to the noise due to leakage from the transmit signal is normally measured. While power efficiency measurement is crucial for tag IC.

6.1 RFID reader receiver sensitivity measurement

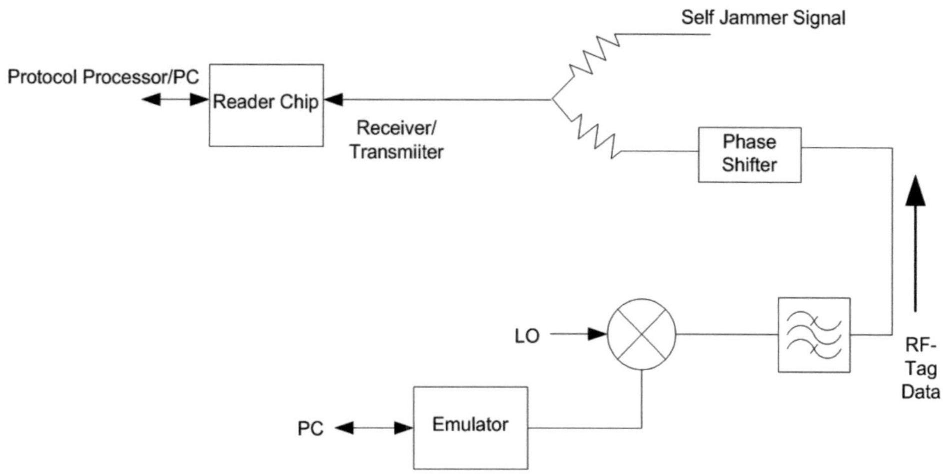

Fig. 25. Reader Receiver Sensitivity

Fig. 25 shows test bench measurement for reader IC. Emulator such as FPGA can be used to imitate RFID tag data. Phase Shifter is used to change phase of tag data.
Example measurement result is shown in Fig. 26. This is example of receiver sensitivity measurement with different incoming tag phase. The tag phase is orthogonal to the jammer noise is the best case phase.

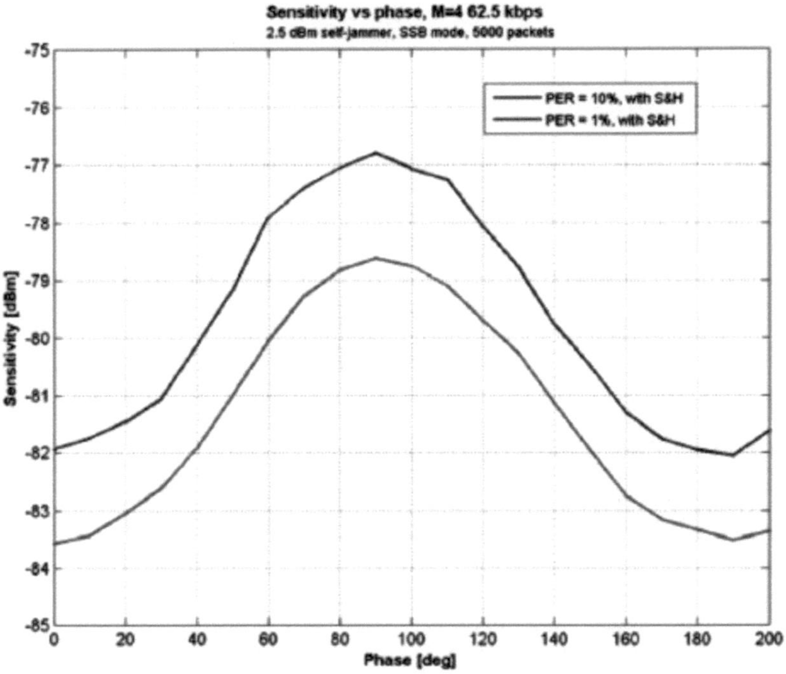

Fig. 26. Measured Packet Error Rate (PER) vs tag phase

6.2 RFID reader transmitter measurement

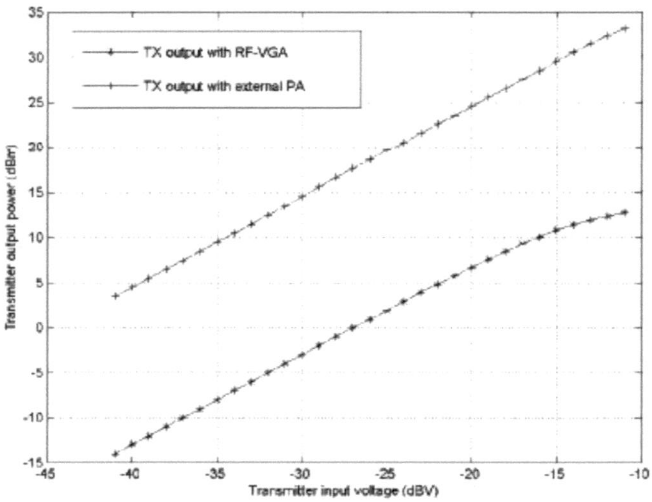

Fig. 27. P1dB Measurement.

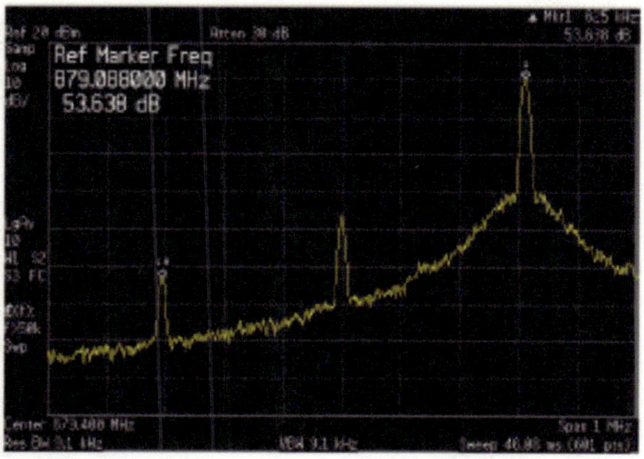

Fig. 28. Sideband Measurement

Fig. 27, 28 and 29 are transmitter measurement results. The transmitter output (Fig. 25) is normally connected to a spectrum analyzer. P1dB measurement shows measurement with and without PA. P1dB is when the power gain is decreased by 1 dB. Obviously from the measurement result, TX output with PA has higher linearity than TX output without PA. Sideband measurement shows sideband rejection ration of ~ 54 dB.

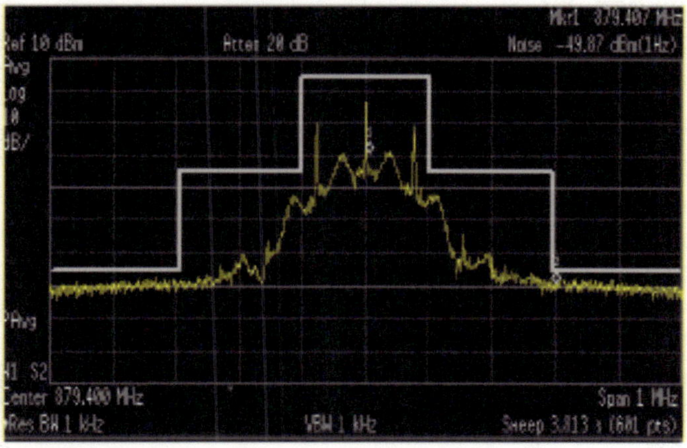

Fig. 29. Spectrum Mask

Fig. 29 shows the measured transmitter output spectrum for DSB-ASK and the requirement mask. The center frequency is 879.4 MHz.

6.3 RFID tag IC measurement

Unlike RFID Reader IC, RFID Tag measurement is tedious and difficult because of several factors namely the die area is small which prohibits test pads. This has required IC designers to endeavour to design and support custom test equipment and dramatically lengthening IC development.

On wafer measurement is a chosen method if anechoic chamber is not available, where antenna loss is not present and mismatch can be eliminated.

Example of power efficiency measurement is shown in Fig. 30.

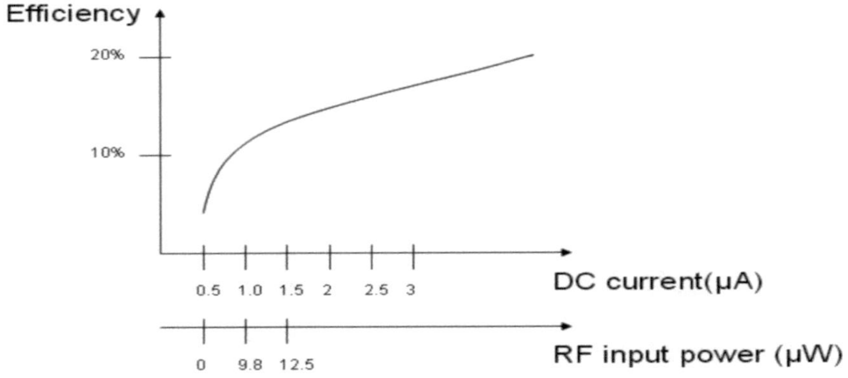

Fig. 30. Efficiency vs Tag DC current consumption

Say Fig. 30 shows an example of RFID tag measurement, say by design 1.5μA is total current consumption for tag IC in Read mode therefore the required RF input power is 12.5 μW. The power required to Write tag (EEPROM) is normally twice compared to Read mode (Karthaus & Fischer, 2003). Average RF input power (Karthaus & Fischer, 2003) to the tag,

$$P_{RF} = (1 - m^2).P_{available} \quad (10)$$

Where m is modulation index, $P_{available}$ is available power to the tag. Assuming m = 0.5, calculated $P_{available}$ is 16.7 μW. The difference between P_{RF} and $P_{available}$ is utilized as backscattered modulated power.

Reader with 500 mW Effective Radiated Power (ERP), more than 4.5m reading distance can be achieved with $P_{available}$ of 16.7 μW (Karthaus & Fischer, 2003).

6.3.1 More challenges with RFID tag IC testing for mass production (Murfett, 2004)

RFID Tag IC test measurement is a challenge due to its high volume and low cost requirement. Below are two main challenges.

6.3.1.1 Test visibility and digital logic

Due to the critical size constraints, dies may only contain the two bond pads for connection to the antenna. Addition of other pads for testing will cause a rapid expansion of the die size. Thus it is can often be difficult to find room for post-rectifier Vdd and Vss test points, let alone the pads required for conventional boundary scan chain type testing. Furthermore, the logic complexities of the IC's are often quite small. Function control is implemented via state machine logic, and never by something as complex as the most basic micro-controller.

6.3.1.2 Non-volatile memory issues

EEPROM memories must be properly initialised at wafer level. It must be possible to stress the memory to its performance limits to test endurance and data retention. It must be possible to adjust the operating state of the memory to set it to sensitive states. It also must be possible to perform write and erase all, or write and erase of odd or even pages, to perform memory initialisation cycling and cross-talk checkerboard tests. All of these commands add significant overhead to the command set of the IC. Furthermore, baking tests for data retention may last up to hours, which adds a significant time and hence cost to the overall testing of these dies.

7. Future research on RFID reader IC

In order to increase the function and application of RFID system, multiband tag IC has been developed. The tag can be operated for many application and places. An example of tag IC which can communicate from 13.56 MHz to 2.45 GHz (http://www.toppanforms.com /eng/news/mmchip.asp) is open for all applications. Once the demand is high for this multiband tag IC, new form of multiband RFID Reader IC need to be developed. Fig. 31 shows IC layout of dual band RFID Reader IC.

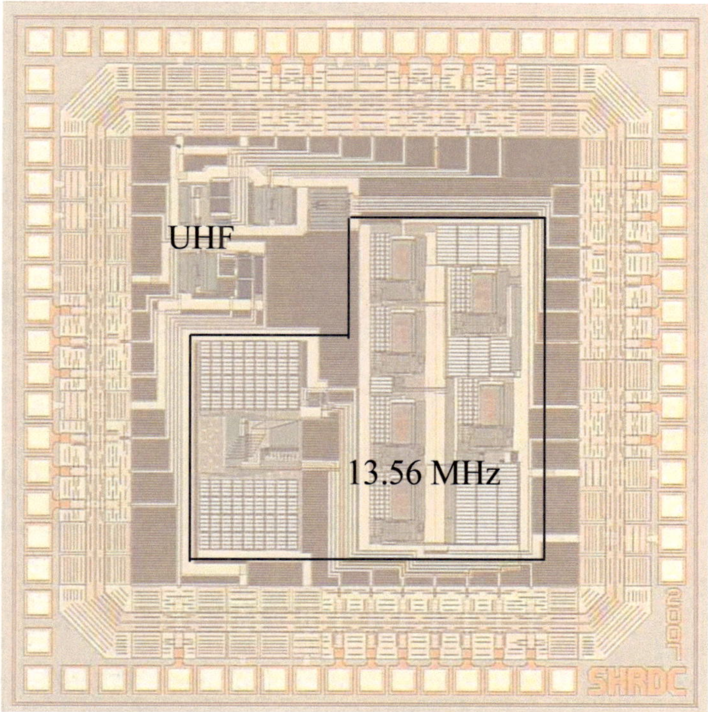

Fig. 31. Dual Band RFID Reader IC (Courtesy of C-RAD Technologies and SHRDC)

The Reader IC is fabricated in MIMOS using 0.35 µm CMOS technology.

8. Conclusion

This chapter has discussed the design methodology of integrated circuit. The technology such as Silicon Bipolar and Silicon CMOS are also studied in this chapter. The architecture or circuit topology of reader and tag are also thoroughly discussed. The circuit technique in designing the tag and reader is studied and analyzed. Digital circuit is not throughly discussed in tag due to its simplicity in nature. The main problem in tag IC development is only at Analog/RF front-end. The measurement methodology of advanced RFID components is also discussed. Challenges of high volume testing is explained for future improvement development. Finally, a new type of reader IC, i.e. dual band is presented.

9. References

Chiu, S. et al., (2007). A 900 MHz UHF RFID Reader Transceiver IC. *IEEE Journal of Solid State Circuits,* Vol. 42, No. 12, (December and 2007) (2822-2833).

De Vita, G. & Iannaccone, G. (2004). Design Criteria for the RF Section of Long Range Passive RFID Systems, *Norchip,* pp. 107-110, Norway, 2004, Oslo.

Finkenzeller, K. (2003). *RFID Handbook,* Wiley, Chichester

Karthaus, U. & Fischer, M. (2003). Fully Integrated Passive UHF RFID Transponder IC with 16.7 µW Minimum RF Input Power. *IEEE Journal of Solid State Circuits,* Vol. 38, No. 10, (October and 2003) (1602-1608).

Khannur, P. B. et al., (2008). A Universal UHF RFID Reader IC in 0.18 µm CMOS Technology. *IEEE Journal of Solid State Circuits,* Vol. 43, No. 5, (May and 2008) (1146-1154).

Kocer, F. & Flynn, P. M. (2005). A Long Range RFID IC with On-Chip ADC in 0.25 µm CMOS, *IEEE Radio Frequency Integrated Circuits Symposium,* pp. 361-364, USA, 2005, Long Beach California.

Lee, T. H. (1998). *The Design of CMOS Radio Frequency Integrated Circuits,* Cambridge.

Marzuki, A. et al., (2008). A voltage reference circuit for current source of RFIC blocks. *Microelectronics International,* Vol. 25, Issue. 3, (2008) (26-32).

Marzuki, A. & David, S. (2006). Design of 920 MHz, 0.8 µm CMOS Low Noise Amplifier for UHF RFID Reader, *Proceeding of RF and Microwave Conference,* pp. 149-151, Malaysia, 2006, Kuala Lumpur.

Murfett, D. (2004). The Challenge of Testing RFID Integrated Circuits, *Proceeding of the 2nd IEEE International Workshop on Electronic Design, Test And Application,* pp. 410-412, Australia, 2004, Perth Western.

Nakamoto, H. et al., (2007). A Passive UHF RF Identification CMOS Tag IC Using Ferroelectric RAM in 0.35 µm Technology. *IEEE Journal of Solid State Circuits,* Vol. 42, No. 1, (January and 2007) (101-109).

Razavi, B. (1997). *RF Microelectronics,* Prentice Hall.

Semiconductor Industry Association. (1997). *The National Technology Roadmap for Semiconductors: Technology needs.*

Wang, W. (2007). A Single Chip UHF RFID Reader in 0.18 µm CMOS, *IEEE Custom Integrated Circuits Conference,* pp. 111-114, USA, 2007, San Jose California.

Zhou, S. & Chang, M-C. F. (2005). A CMOS Passive Mixer with Low Flicker Noise for Low Power Direct-Conversion Receiver. *IEEE Journal of Solid State Circuits,* Vol. 40, No. 5, (May and 2005) (1084-1093).

11

A Low Cost Anticollision Reader

Dan Tudor Vuza[1], Reinhold Frosch[2], Helmut Koeberl[2] and Damien Boissat[2]
[1]Institute of Mathematics of the Romanian Academy,
[2]Frosch Electronics OEG, Graz,
[1]Romania,
[2]Austria

1. Introduction

The chapter presents some aspects related to the design of a low cost anticollision reader based on the HTRC110 reader chip from NXP Semiconductors (NXP Semiconductors, 2006 a) and the AT91SAM7S64 controller from Atmel.

The HTRC110 reader chip was mainly intended for integration into systems that need to identify one single RFID tag at a time. At the request of the host the chip sends the adequate command to the tag, retrieves the answer of the latter in analog form, digitizes the data and offers to the host the data retrieved from the tag in binary format. The chip does not provide support for collision detection and anticollision identification procedures in case that several tags in the antenna field respond to the same command at the same time.

Nevertheless, the chip does provide access to the analog demodulated signal representing the response sent by the tags via load modulation of the carrier. This access was meant by the producer for the purposes of antenna tuning and testing. In the present design, it is by exploiting this feature that one achieves the anticollision functions.

In section 2 of the chapter one describes the general configuration of the reader at block level, and particularly how the internal ADC of the AT91SAM7S64 controller is used for converting the test signal of the HTRC110 chip to numeric format and what signal-processing procedures are used for extracting the bit information from the signal and detecting collisions on bit positions. Included is also a brief description of the communication facilities of the reader with a host PC, such as RS232, USB and wireless.

In section 3 one addresses some aspects of the concern whether the analog signal provided by the HTRC110 chip is adequate enough for collision detection. The main point here is that, while the other members of the anticollision readers produced by Frosch Electronics use constant amplitude current drive for the antenna and they decode the antenna voltage variations caused by load modulation at the tag, in the present case the HTRC110 chip uses constant amplitude voltage drive and it decodes the voltage variations at the junction between the antenna coil and the tuning capacitor. Because of this, the load modulation employed by the tag causes transients in the reader antenna that are longer than in the current drive case and are directly influenced by the antenna Q. In turn these transients cause the demodulated signal from the tag to have slower raising/falling edges than in the current drive case, which may interfere with collision detection if these transients are too long. It is the purpose of the section to analyze the dependence on Q of this effect and to compare it with the current driven case.

The next two sections complement the discussion of transients exposed in section 3. In section 4 one shows how these transients can be simulated with the aid of Spice and observed in isolated form, that is, separated from the signal on which they are superimposed. Section 5 presents the tag simulator, which can be a useful tool for the designer assisting him in the process of antenna Q optimization.
Finally, section 6 describes how besides anticollision identification of several tags, the reader was also designed for identification of a single tag in public B mode, according to ISO standards 11784 and 11785 for animal identification. The problem is here that the tag sends its data in a continuous and cyclic manner, without needing any reader command, so the alignment of the bit boundaries with respect to the reader clock is not known in advance to the reader. A bit synchronization procedure is needed and one such procedure is described in the section.

2. Description of the reader

Figure 1 presents the block schematic of the new reader. The signal lines connecting the HTRC110 chip to our design are divided into three groups. The lines in the first group are connected to the antenna circuit as explained in the next section. The lines in the second group, Clock, Data In and Data Out, are digital lines and serve to the exchange of binary information with the Atmel AT91SAM7S64 micro-controller (uC). The lines in the third group represented as dashed on the figure form the "non-standard" group as they are related to our particular design and would not be present in the standard usage intended by the producer.

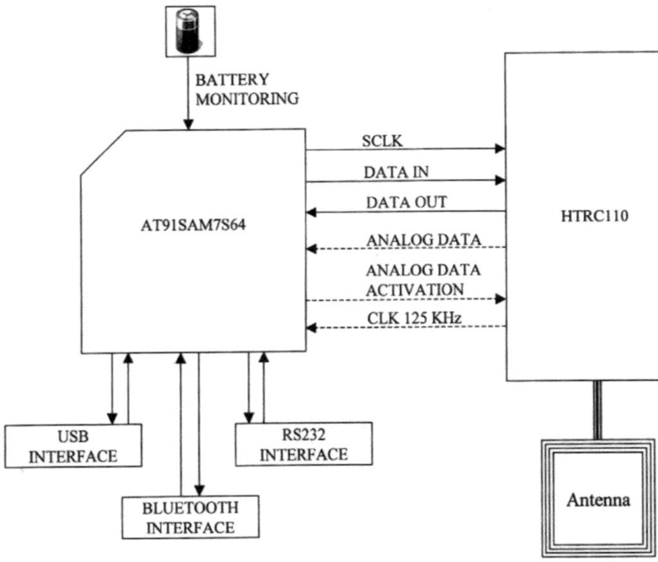

Fig. 1. Block schematic of the reader.

The HTRC110 chip may work with carrier frequencies of 125 KHz or 134 KHz. The antenna circuit provides both the power to the tags and the communication with the latter. Commands are sent to tags by 100% modulation of the carrier. The Data In line may be used by uC for modulating the carrier between on and off. A command to the tags is sent as a sequence of bits, each bit consisting of a space followed by a mark. The carrier is off during spaces and on during marks. A longer mark corresponds to a one, a shorter one to a zero. In our design, we use one of the internal uC timers for setting the duration of a mark. During marks, the timer is clocked by signal coming from one of the antenna drivers of the HTRC110 chip. During spaces, the timer is clocked from one of the frequency generators of uC since the carrier is off. The timer is programmed to generate an interrupt to uC after a certain number of clocks; the interrupt routine writes the Data In line of the HTRC110 chip with the appropriate value and programs number of clocks after which the next interrupt should be triggered.

So far this is standard usage of the HTRC110 chip. For us the significant part starts with the reception of the answer from the tags. The tags send their answer at a command from the reader by switching an internal load. The bits are encoded as shown in figure 2. When communicating with a single tag, Manchester (MC) encoding is used. Anticollision encoding (AC) is used when responses from several tags are expected. However, the HTRC110 chip was mainly intended for communication with a single tag and for this reason it has no provision for collision detection. In a standard usage, after demodulating and filtering the tag response, the chip digitizes it according to being lower or higher than a certain threshold and presents to the host this binary version of the signal on the Data Out line. The host has thus access to only a part of the information contained in the analog signal. In the absence of collisions, this information should suffice, in principle, for recovering the answer from the tag. The detection of collisions would be however quite questionable without having full access to the analog signal.

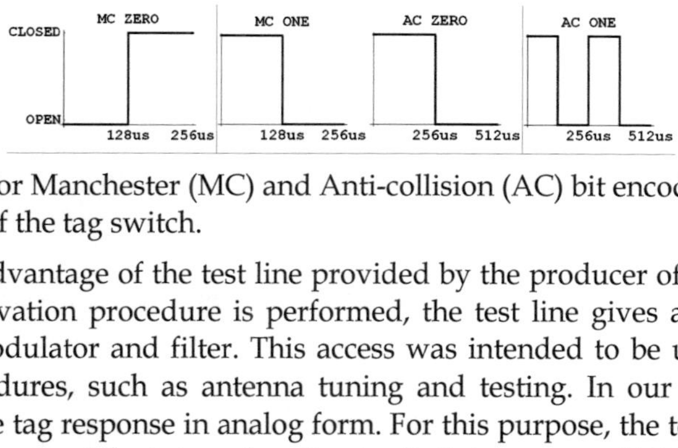

Fig. 2. Waveforms for Manchester (MC) and Anti-collision (AC) bit encoding. The Y-axis indicates the state of the tag switch.

Our design takes advantage of the test line provided by the producer of the HTRC110 chip. After a certain activation procedure is performed, the test line gives access to the analog output of the demodulator and filter. This access was intended to be useful especially for maintenance procedures, such as antenna tuning and testing. In our case, we use it for getting access to the tag response in analog form. For this purpose, the test line is connected to one of the 8 channels of the internal analog-to-digital converter (ADC) of uC. The signal coming from the antenna driver goes to another internal timer of uC where its frequency is divided by two. The ADC is programmed to use the output of the named timer as the start of conversion signal. Thus one sample of the analog signal is taken every two carrier cycles.

Following a command from the reader, the tag starts sending its answer after a certain number of carrier cycles since the end of the command, specified by the producer. Again by using the uC timer for counting carrier cycles, our reader waits for that specific number, after which the automatic recording of the tag answer starts by programming the interface between ADC and the internal RAM memory of uC. With its 16 Kbytes of RAM, uC provides plenty of storage space for the tag answer. After having recorded the latter, uC decodes it by computing the correlations between each bit and the encoding waveforms shown in figure 2. Let us consider here the AC encoded case. Denote by C_0 and C_1 the absolute values of the correlations between a certain bit and the AC waveforms for zero and one, and let C_{max} and C_{min} be the greater and the lower of C_0 and C_1 respectively. The uC firmware considers that a response has been indeed received if C_{max} exceeds some threshold, in which case the bit in question equals zero or one according to whether C_0 is greater or lower than C_1. A collision on that bit is considered to have occurred if the ratio C_{min}/C_{max} exceeds the collision threshold. In such situation, the reader sends more selective commands until a collision-free response from a single tag is received. With the aid of anticollision algorithms based on this principle, all tags in the antenna field are identified by the reader. All computations are done exclusively with integers and therefore are fast enough, so the timing requirements of the identification procedures are mainly determined by the tags without significant overhead added by uC.

Figure 3 left shows a sequence of bits in analog format together with their digitized form that would be provided by the HTRC110 chip in a standard usage. One can distinguish in the figure the AC coded bits 0, 1, 0, 0, 1, 1, 0, 1. Figure 3 right shows the same sequence but the 4-th, 5-th and 6-th bits are now affected by collisions. One sees that there is hardly any difference in the digitized output; so this signal cannot be used for collision detection. On the other hand, the collision ratios C_{min}/C_{max} as computed from the analog signal for the bits in question are 0.274, 0.629, 0.621 for the collision case and 0.002, 0.032, 0.035 for the collision-free case, the presence of collisions being thus identified without any doubt.

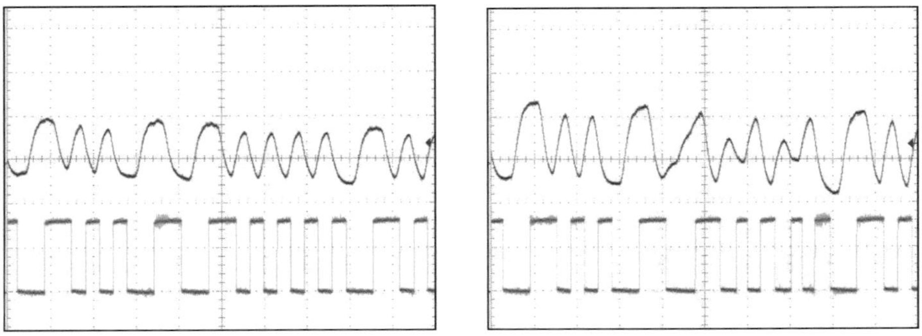

Fig. 3. Baseband signal demodulated by reader. Data from tags received without collisions (left) and with collisions (right).

Our reader provides communication with a host PC via three channels: RS232, USB and wireless. The RS232 channel facilitates the connection to a PC serial port; the serial protocol is supervised by the embedded serial interface of uC and needs an on-board component that translates between the logic levels employed by the uC and PC serial ports. The Bluetooth wireless module is connected to the RS232 channel and realizes the conversion between the serial and wireless protocols. No additional component is needed for the USB channel since all the protocol is managed by one of the embedded peripherals of uC.

The reader may be powered either by battery or from the PC via USB cable. A 3.3V regulator provides the power to the HTRC110 chip, the uC and to all other components on-board. When powered from battery, another channel of the ADC is used for monitoring the battery voltage. During the intervals when the ADC is not used for decoding the answer from tags, a sample of the battery voltage is periodically taken. The uC monitors these samples and triggers an alarm in the form of a blinking led when the voltage starts going low; it can also send these samples at the request of the host PC if it is wished to display the battery voltage on the screen.

In figure 4 one sees the reader being driven by a program run on a personal digital assistant. The communication between the latter and the reader is wireless. The computer screen shows the result of identifying three tags with the anticollision algorithm.

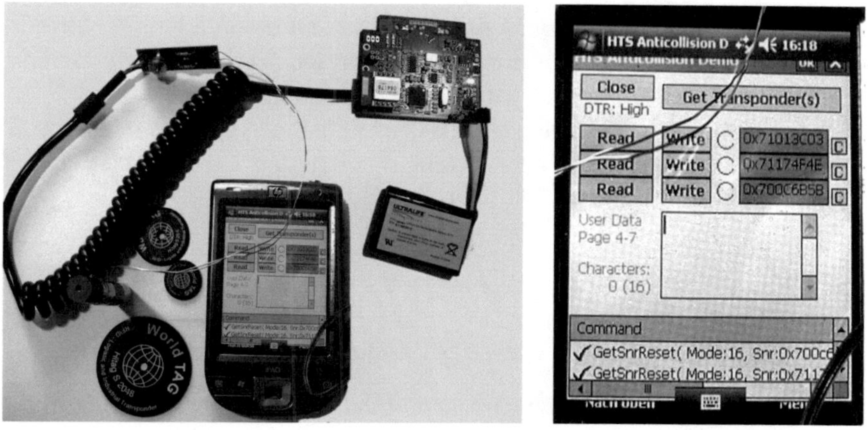

Fig. 4. The low cost reader driven by a PDA via wireless communication (left) and the screen of the PDA showing the result of identifying three tags (right).

3. Effects of transients on data reception

The reference book on RFID (Finkenzeller, 2003) discusses the influence of the reader antenna quality factor Q. A high Q leads to high current, hence larger reading distance. On the other hand, a too high Q would limit the bandwidth for transmission and reception. This means that transmission and reception act both in the direction of imposing an upper bound on Q. Since these processes are based on different effects, one cannot expect that the constraints they impose would be similar; in fact they can be quite dissimilar, as we shall see. In the mentioned book one gives an estimation formula for Q based on the constraints imposed by transmission, with just a few hints about how to proceed for reception. In the following we shall analyze the influence of Q on the process of reception by the method of transients. We shall see that, from the point of view adopted here, there is a fundamental difference between the so-called voltage driven readers and current driven readers as far as the constraints imposed by reception are concerned. The example reader design of (Finkenzeller, 2003) is a voltage driven reader, as it is the low cost reader discussed now; current driven readers are produced by Frosch Electronics and seems to be less known, therefore it would be worthwhile to be analyzed here.

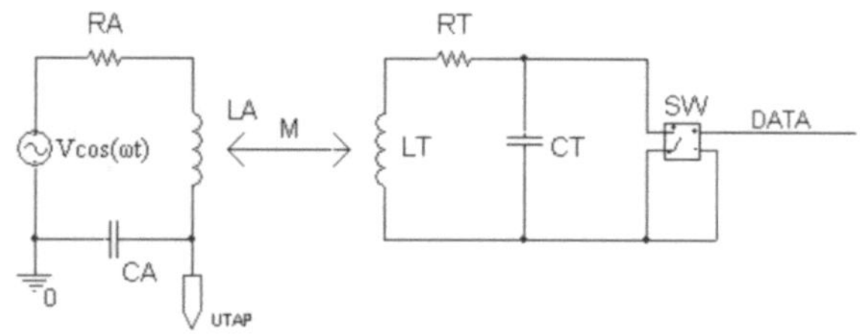

Fig. 5. Interaction between reader and tag.

Consider the principle schematic of the reception circuit of the low cost reader as shown in figure 5. During reception, the reader powers the antenna with a sinusoidal voltage of constant amplitude at the carrier frequency of f_C = 125 KHz. The tag transmits data by opening and closing the switch SW, which, due to the magnetic coupling M, modulates the current through the antenna. The reader senses the current modulation (or equivalently, the modulation of the voltage across an impedance traversed by the antenna current, such as the voltage U_{TAP}) and extracts the baseband signal that contains the data.
In the following we shall use the letter s as the argument in the Laplace transform. Denoting by U_A the driving voltage, by I_T the current flowing into the tag coil and by Z_T the impedance seen by the tag coil, we have the following relations:

$$U_A(s) = \left(R_A + L_A s + \frac{1}{C_A s} \right) I_A(s) + MsI_T(s),$$
$$-Z_T(s)I_T(s) = LsI_T(s) + MsI_A(s).$$

By eliminating I_T we find the antenna current and the voltage at the tap point in function of the antenna voltage and the tag impedance:

$$I_A(s) = \frac{U_A(s)}{R_A + L_A s + \dfrac{1}{C_A s} - \dfrac{M^2 s^2}{L_T s + Z_T(s)}}, \quad U_{TAP} = \frac{I_A(s)}{C_A s}. \tag{1}$$

The impedance Z_T is formed by the tag coil resistance R_T in series with the parallel combination of the resonance capacitor C_T and the resistance of the switch R_{SW}. In the following we shall consider only tags tuned to the carrier frequency. Using the equalities $L_A C_A = L_T C_T = (\omega_C)^{-2}$ where $\omega_C = 2\pi f_C$ and introducing the coupling constant $k^2 = M^2/L_A L_T$, the antenna quality factor $Q_A = L_A \omega_C / R_A$, the series and parallel quality factors of the tag $Q_s = L_T \omega_C / R_T$, $Q_p = R_{SW} / L_T \omega_C$ and the frequency normalized Laplace variable $x = s/\omega_C$, we arrive at

$$U_{TAP}(s) = \frac{P_T(x)}{P_A(x)P_T(x) - k^2 x^2 (x^2 + Q_p^{-1} x)} U_A(s) \tag{2}$$

where

$$P_A(x) = x^2 + Q_A^{-1} x + 1,$$
$$P_T(x) = x^2 + (Q_p^{-1} + Q_s^{-1})x + Q_p^{-1} Q_s^{-1} + 1.$$

We are interested to see what happens to U_{TAP} when the tag changes the state of the switch SW. This can be inferred from the following general principle. Consider a circuit described by the linear system

$$\frac{dX(t)}{dt} = SX(t) + Y\exp(j\omega t) \tag{3}$$

where X is the vector of the state variables and $Y\exp(j\omega t)$ is a harmonic signal that drives the circuit. The general solution of such a system is the sum between the harmonic solution $(j\omega - S)^{-1}Y\exp(j\omega t)$ and the general solution of the homogeneous system, for which $Y = 0$. Assuming the circuit stable, the solution of the homogeneous system will approach zero and will therefore represent the transient part of the complete solution. Now assume that some change in the matrix S happens at time t_0, such as opening or closing a switch that would cause a change in the component parameters and/or the configuration of the system. Suppose that $X(t)$ is a solution of (3) up to t_0. After t_0, the state vector of the circuit will evolve according to system (3′) obtained from (3) by replacing S with the new matrix S'. The new vector $X'(t)$ is uniquely determined by solving (3′) with the initial condition $X'(t_0) = X(t_0)$. As before, $X'(t)$ will be the sum between the new harmonic solution $(j\omega - S')^{-1} Y\exp(j\omega t)$ and a transient solution uniquely determined by the initial condition. As times goes past t_0, the evolution of the state vector will approach the harmonic solution. Thus, the change of configuration at moment t_0 results in changing the evolution of the system from one harmonic solution to another, but has also the side effect that transients will manifest themselves for some time after the change. The time constants of these transients are determined by the linear system in effect *after* the change, that is, system (3′). As well known from Laplace transform theory, if one is interested in the time constants of the transients that affect an output of the system, one has to look for the roots of the denominator of the transfer function from the driving input to that output and take the inverses of the real parts of those roots, provided that the degree of the denominator equals the order of the system (in order to be sure that the denominator includes all characteristic roots of the system).

In the case of our reader, we see that when the tag acts on the switch SW, U_{TAP} will be affected by a transient whose time constants are computed by finding the roots of the denominator $P(x)$ of the transfer function in (2). Specifically, for any such root x_0, $-1/(\omega_C \operatorname{Re} x_0)$ will be the time constant for a transient. In the following we shall consider the limiting case of small coupling constants k, which is the usual case in real situations. With this assumption, the roots of $P(x)$ will be in close vicinity of the roots of the equation $P_A(x)P_T(x) = 0$. There are now two cases to be distinguished. In the first case, the tag closes the switch. Two of the time constants are determined by the roots of P_A while the other two are determined by the roots of P_T after having substituted Q_p with its value corresponding to the closed state of the switch. Since R_{SW} is low in the closed state, the time constants determined by P_T will be short, leaving the transients determined by P_A to manifest themselves for a longer time. In other words, when the tag closes the switch, the transients affecting U_{TAP} will be mainly determined by the quality factor of the reader antenna. In the second case, the tag opens the switch. Since Q_p is high in this case, $P_T(x)$ essentially reduces to $x^2 + Q_s^{-1}x + 1$, an expression similar to $P_A(x)$. In other words, transients due to both the reader antenna and to the tag will be present; which of them will be longer depends on how the quality factor of the reader antenna compare to the series quality factor of the tag resonant circuit.

The above conclusions can be confirmed by experiment. Figures 6 and 7 show the voltages on U_{TAP} of our voltage-driven reader measured with a digital oscilloscope and displayed in Spice. Since a real tag would not allow for a clear display of the transients, these were obtained with the tag simulator to be described in section 5, which imitates a real tag by using the same electrical model as in figure 5. The switch pulse in figures 6 and 7 is the pulse applied at the data line of the switch in figure 5. For a better display of transients, the named figures show the envelopes of the voltages on U_{TAP} obtained by joining the maxima over positive half cycles; the voltage itself is shown in figure 6, which corresponds to the case of normal working conditions. In this case, the measured value of Q_A was found to be 9.45; this was done by measuring U_{TAP} when no tag was present and using the relation $Q_A = |U_{TAP}| / |U_A|$, which follows from (2) by substituting $x = j$ and $k = 0$ (one has also to take into account that, for an antenna circuit driven by a square pulse that toggles between U_{MAX} and $-U_{MAX}$, the amplitude U_A of the sinusoidal voltage source in our model is found to be 1.27 U_{MAX} by Fourier analysis). In figure 7, Q_A has been lowered to 1.41 by inserting an additional 162 Ohms resistance in the antenna circuit. One sees that the transient after closing the switch has been greatly shortened, confirming thus that this transient is mainly due to the reader antenna. On the other hand, the transient after opening the switch could not be reduced below 100 μs since it is mainly determined by Q_s of the tag when the transient due to the reader is short.

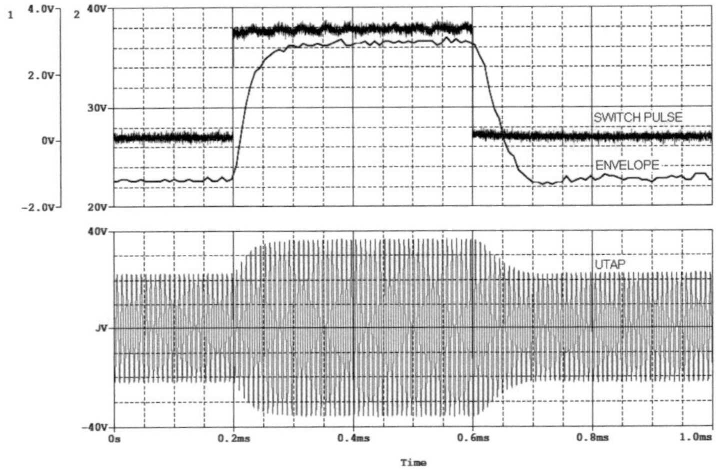

Fig. 6. Transients on U_{TAP} of the voltage-driven reader for normal Q_A.

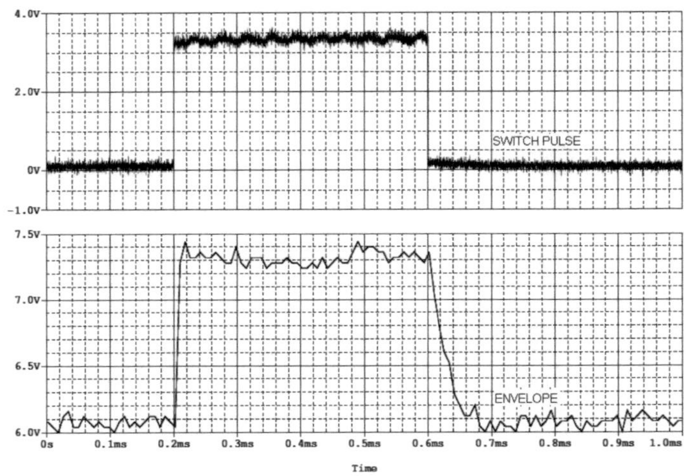

Fig. 7. Transients on U_{TAP} of the voltage-driven reader for lowered Q_A.

When k is small and the quality factors are not too small, the transients oscillate with a frequency close to f_C. Consequently they pass through the filters of the reader receiver and are demodulated, resulting in a baseband signal with slower transitions between the states corresponding to the open and closed positions of the tag switch. If the time constants of the transients are comparable to the duration of the bits transmitted by the tag, deleterious effects on bit decoding and especially on collision detection can result, since such transients can distort the waveforms that encode the bits. Therefore, an upper bound of the reader antenna quality factor Q_A should be imposed. By the above discussion, any decrease in Q_A will shorten the transient that occurs when the tag closes the switch. This will be true only to lesser extent for the transient after the opening of the switch: even if Q_A is made very low so that the transient due to the reader antenna vanishes quickly, there still remains the effects of the transient related to the serial quality factor Q_s of the tag.

From the above discussion, it is advisable that the 95% - 5% decrease time of the antenna transient, equal to 2.94 times its time constant, should be less than the smallest interval T_{SW} between changes of state of the tag switch. The time constant being equal to $2Q_A/\omega_C$, it results that the following upper bound should be imposed on Q_A:

$$Q_A \le \frac{\pi f_C T_{SW}}{2.94}.$$

For our reader, T_{SW} = 128 µs, corresponding to the case of Manchester encoding. It follows that an upper bound of 17 should be imposed on Q_A.

It is interesting to compare the above situation with that of other readers designed or produced by Frosch Electronics for which the antenna circuit is not voltage driven but current driven (Frosch Electronics, 2004; Gelinotte et al., 2006; Vuza et al., 2007). For those readers, it is the amplitude of the antenna current that is kept constant by the reader during reception, the action of the tag on the switch resulting in the modulation of antenna voltage that is sensed and decoded by the reader. In figure 5, one replaces the voltage source by a current source and instead of U_{TAP} one senses the voltage U_A across the current source. From (1) and (2) one obtains

$$U_A(s) = \frac{P_A(x)P_T(x) - k^2x^2(x^2 + Q_p^{-1}x)}{\omega_C C_A x P_T(x)} I_A(s). \tag{4}$$

One may think that the above method would not apply, since the degree of the denominator is less that the order four of the system; and moreover the root $x = 0$ would seem to suggest an infinite time constant. However, on closer look, we see that the current through L_A and the voltage across C_A do not count as state variables, as far as the transients caused by load modulation are concerned. Indeed, the former is fixed by the current source, while the latter is determined by the current source and the initial voltage at system start. The remaining state variables are attached to L_T and C_T so that the order of the system is two, not four. Rewriting (4) as

$$U_A(s) = \frac{P_A(x)}{\omega_C C_A x} I_A(s) - \frac{k^2x^2(x + Q_p^{-1})}{\omega_C C_A P_T(x)} I_A(s), \tag{5}$$

we see that only the second term in the right hand matters, the first term being independent of the part of the system where the state variable are located. The conclusion is that the time constants of transients are here completely determined by the tag and are neither dependent on the antenna circuit nor on the coupling. The antenna quality factor Q_A has no influence on the reception transients.

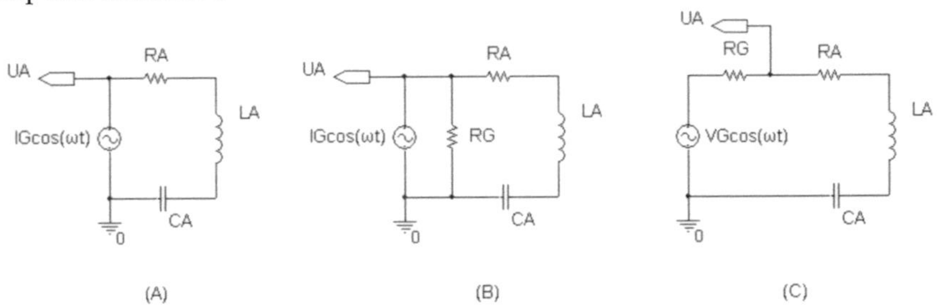

Fig. 8. Current driven reader: ideal current source (A), non-ideal current source (B), Thévenin equivalent of non-ideal current source (C).

The above argument is correct for an ideal current source. A non-ideal current source has a resistance R_G across it. We may replace the source, as in figure 8, by its Thévenin equivalent consisting of a voltage source producing the voltage $U_G = R_G I_G$ in series with resistance R_G. When this replacement is made, the current driven reader has been transformed into a voltage driven reader to which the above discussion apply; one has only to notice that R_A is now replaced by the composite resistance $R_A + R_G$. The current driven reader senses the voltage U_A across the antenna, which in our case equals $U_G - R_G I_A$. The expression for I_A in function of U_G is obtained by multiplying by $C_A s$ the voltage U_{TAP} from (2), in which we have replaced $P_A(x)$ by $R_G \omega_C C_A x + P_A(x)$ in order to take into account the composite resistance. After simplifications the result is

$$U_A(s) = \frac{P_A(x)P_T(x) - k^2 x^2 (x^2 + Q_p^{-1} x)}{R_G \omega_C C_A x P_T(x) + P_A(x)P_T(x) - k^2 x^2 (x^2 + Q_p^{-1} x)} R_G I_G(s).$$

As R_G grows larger, the denominator is essentially dominated by its first term and hence becomes less and less dependent on Q_A and k. In the limit of R_G growing to infinity, the above expression reduces to (4), as it should.

The conclusion is that, in the case of current driven readers, the process of reception imposes fewer restrictions on Q_A, amounting to no restriction at all for the limiting case of an ideal current source. Therefore, the upper bound of Q_A of such a reader is mainly imposed by the transmission.

4. Spice simulation of reception transients

As discussed above, the transients in the solution of the system (3) caused by switching from S to S' are obtained by subtracting the periodic solution corresponding to system S' from the actual solution. By using this principle one can display the reception transients in Spice. In figure 9, T, T0 and T1 are hierarchical blocks that model tags; each of them is coupled, with the same constant k, to a voltage driven reader. The model for the tag is the same as in figure 5. The inputs of those blocks correspond to the data lines that drive the switches. Through the simulation, blocks T0 and T1 have their switches set to the open and closed positions respectively; the switch of block T is changed from open to closed and back to open. After a short interval from the beginning of the simulation, the voltages $U_{TAP,0}$ and $U_{TAP,1}$ of the readers coupled to T0 and T1 will stabilize to the periodic solutions corresponding to the respective positions of their switches. The transient on voltage U_{TAP} of the reader coupled to T caused by closing the switch is then obtained by subtracting $U_{TAP,1}$, the periodic solution for the closed switch, from U_{TAP}, which is effected by the difference block in the figure. Similarly, the transient on U_{TAP} caused by opening the switch is obtained by subtracting $U_{TAP,0}$ from U_{TAP}.

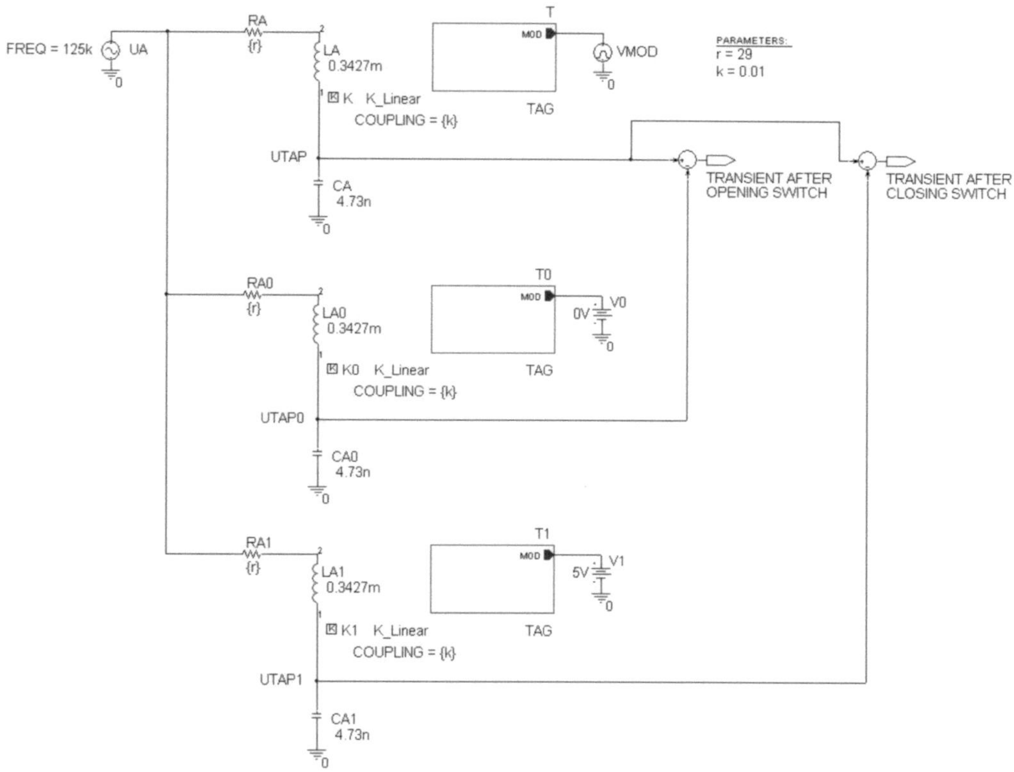

Fig. 9. Spice schematic for simulation of reception transients.

Figures 10 and 11 show the results of the simulations. In both figures, a value of 12.73 was used for Q_s; the switch was closed at time 2 ms and opened at time 3 ms. Figure 10 corresponds to a high value of 67.29 for Q_A. One sees that the transient after switch closing, which is due mainly to the reader antenna, resembles closely the transient after switch opening, with the exception of the first 100 μs when the transient due to the tag also manifests itself. This is confirmed by the simulation of figure 11 where a low value of 1.35 was used for Q_A. One sees that the transient after switch closing is indeed very short, while the transient after switch opening, which is now mainly determined by Q_s of the tag, lasts for about 100 μs. As a check, the 95% - 5% decrease time corresponding to Q_s is $2.94\frac{2Q_s}{\omega_C} = 95.3$ μs.

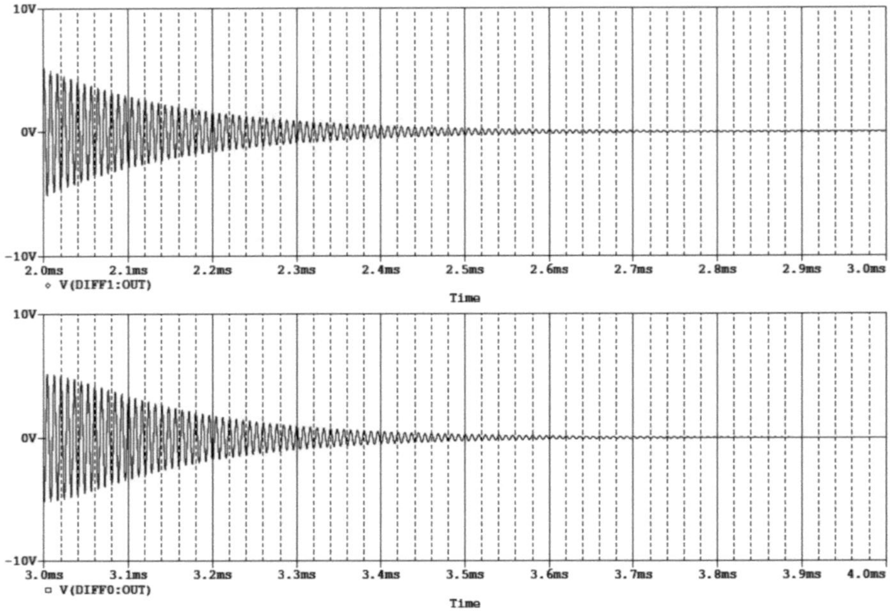

Fig. 10. Simulated transients for high Q_A.

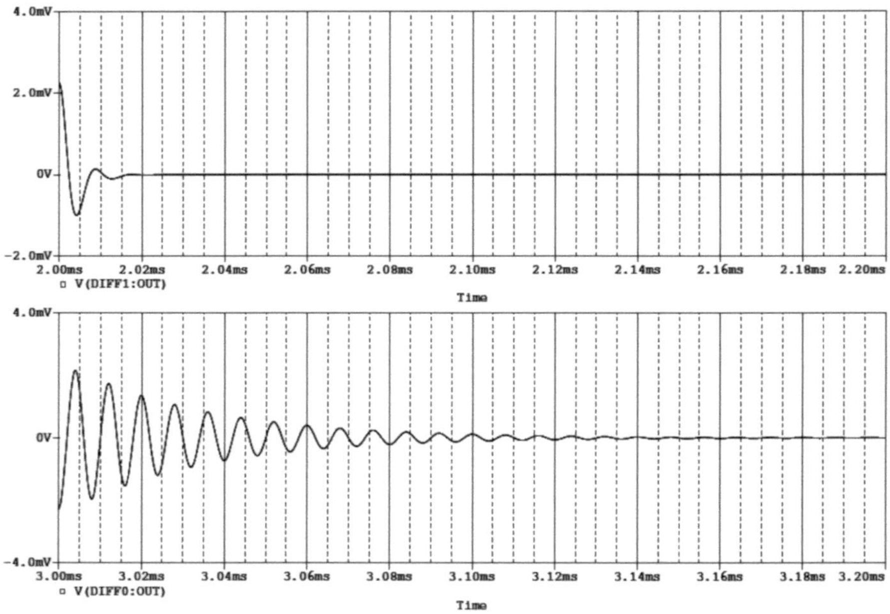

Fig. 11. Simulated transients for low Q_A.

5. A short description of the tag simulator

An ongoing cooperation between NXP Semiconductors and Frosch Electronics targets the development of a reader for a new member of the Hitag family of RFID transponders complying with the ISO/IEC 18000 standard. Since at the time of the reader development the tag itself was under development, a simulator has been designed in order to substitute the real tag and simulate the communication between reader and as many tags as needed. With its aid, several fast identification algorithms have been designed and tested on the reader.

Figure 1 presents the principle schematics of the tag simulator, which is built around the evaluation board for the Atmel AT91SAM7S64 micro-controller (uC). In the figure the simulator is interacting with a reader and its antenna.

For the purpose of receiving the commands intended for tags, two logic signals of the reader, namely the signal that drives the carrier modulator and the output of the carrier generator, are connected to two inputs of uC. Via the former signal uC is made aware of carrier interruption during spaces; the latter signal clocks one of the internal timers of uC with the aid of which uC counts the carrier cycles during marks, just as the tag would do.

The simulator transmits data to the reader via load modulation, by switching in and out resistor *RM*. This causes the variation of the current through the simulator coil *L* placed near the reader antenna, which results in modulating the voltage on the reader antenna via magnetic coupling *M*. Thus, from the electrical viewpoint, data transmitted by the simulator appears to the reader as if transmitted by a real tag, since the principle of load modulation employed by the latter is also used with the simulator.

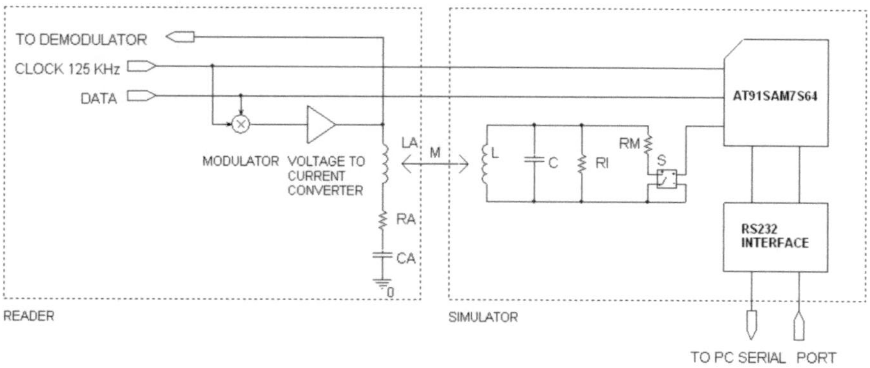

Fig. 12. The simulator and its interaction with a reader.

The uC keeps in its internal RAM a list of all tags to be simulated. Each tag is represented as a record containing the serial number and the status of the tag that can be powered off, active or halted. When responding to a Get ID command, the uC first builds the list of tags that in reality would respond to the command. Then, for every bit position of the response to be sent, uC looks whether the tags included in the list would transmit the same bit or not. In the former case, uC transmits that common bit according to the AC load modulation waveforms of figure 13. In the latter case, uC uses a modulation waveform called Collision in the named figure. At the reader, the reception of such a waveform will be interpreted as a collision, since it has a strong correlation with the encoding waveforms for both a zero and a one. In this way, the reader will be in the same situation as when receiving data simultaneously from several tags and collisions are detected on some bit positions.

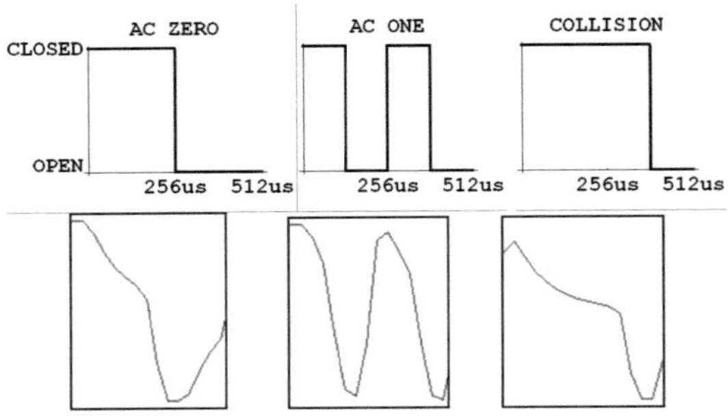

Fig. 13. The waveforms used by simulator for AC bit encoding via load modulation. The Y-axis indicates the state of the simulator switch. Below is shown the resulting baseband signal demodulated at the reader.

Via the RS232 interface of the Atmel board, a PC can program the tag simulator for performing various tasks. For instance, it can load the list of tag serial numbers into uC memory, set the number of tags participating to the simulation or choose the type of transponder to be simulated. In addition, the PC can program the simulator to perform tasks that would not be possible with a real tag, such as generating a sequence of load modulated pulses of a prescribed frequency or a single long pulse for determining the step response of the system antenna - reader.

The simulator can be a useful tool in the design of a reader, contributing to the visualization of transients and therefore to the choice of antenna Q according to what has been said in section 3.

6. Bit decoding for animal identification

In addition to the capability of tag identification with anticollision, our reader has been designed for data retrieval from single tags in Transponder Talk First (TTF) mode (NXP Semiconductors, 2006 b), which is used especially in animal identification. In contrast to the Reader Talk First (RTF) mode considered in section 2, where the tag sends data only following a reader request, in TTF mode the tag starts sending data as soon as it is powered by the antenna field. Data are sent continuously as long as power is maintained, the transmission of the last bit being followed by the retransmission of the first bit without pause between them. For the reader, this circumstance has the following implication. In RTF mode, the reader assumes that the start of the answer of the tag is separated from the end of the reader command by a known number of carrier cycles specified by the tag producer. This makes that bit boundaries are known to the reader. In TTF mode however, the reader has no indication about bit boundaries.

We shall describe in the following the procedure that is used in our reader for determining bit boundaries in TTF mode. The presentation of the principle will be clearer if we assume a continuous time axis instead of the discrete time imposed by sampling. We consider therefore a baseband signal $f(t)$ that Manchester encodes an infinite sequence of bits. The signal is formed by superposing translated copies of the basic waveform $w(t)$, which is defined as equal to 1 for $0 \leq t < T/2$ and to -1 for $T/2 \leq t < T$, where T is the bit duration. Specifically, $f(t) = \sum_{i=-\infty}^{\infty} c_i w(t - iT)$ where c_i equals 1 if the i-th bit is a one and -1 if the i-th bit is a zero. Normally we would decode the i-th bit by computing the correlation $\frac{1}{T}\int_{-\infty}^{\infty} f(t)w_i(t)dt$ between $f(t)$ and the test function $w_i(t) = w(t - iT)$. However this would work only if the test functions w_i are aligned with the bit boundaries in the signal $f(t)$; this would no longer be the case if instead of $f(t)$ we would have a translated version $f(t - d_0)$. To align the test functions with the bit boundaries in the latter case, we proceed as follows. We compute the "moving correlations sums"

$$C(d) = \sum_{i=0}^{N-1}\left|\frac{1}{T}\int_{-\infty}^{\infty} f(t-d_0)w_i(t-d)dt\right|.$$

It turns out that the function $C(d)$ attains its maximum at d if and only if *either* the test functions $w_i(t - d)$ are aligned with the bit boundaries in f (that is, $d = d_0 + mT$ for some integer m) *or* they are displaced by half a bit duration with respect to bit boundaries in f (that is, $d = d_0 + T/2 + mT$ for some integer m) and all the N+1 bit waveforms in f whose supporting intervals intersect the union of the supporting intervals of $w_0(t - d), \ldots, w_{N-1}(t - d)$ encode the same bit.

The proof of this result is very simple. Without narrowing generality we may assume $d_0 = 0$ and we may restrict our attention to the variation of $C(d)$ over $[0, T]$. In this situation, the supporting interval of $w_i(t - d)$ intersects only the bit intervals corresponding to bits i and i+1, so that $\left|\frac{1}{T}\int_{-\infty}^{\infty} f(t)w_i(t-d)dt\right|$ actually equals

$$\frac{1}{T}\left|c_i\int_{-\infty}^{\infty} w(t)w(t-d)dt + c_{i+1}\int_{-\infty}^{\infty} w(t-T)w(t-d)dt\right|.$$

As a function of d, the latter expression equals either $I(d)$ or $J(d)$ according to whether c_i coincides or not with c_{i+1}, the functions $I(d)$ and $J(d)$ being presented graphically in figure 14.

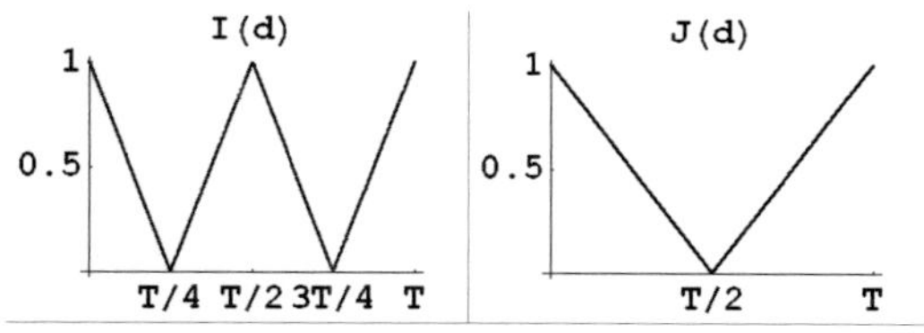

Fig. 14. The graphs of the functions $I(d)$ and $J(d)$.

It follows that $C(d)$ can be written as $kI(d) + (N - k)J(d)$, where k is the number of elements in the set $\{i \mid 0 \leq i < N, c_i = c_{i+1}\}$. In case that bits 0 up to N all coincide, we have $k = N$ and we see from the graph of $I(d)$ that indeed it attains its maximum at 0, $T/2$ and T. If this is not the case, we have $k < N$; from the symmetry of graphs of I and J we may restrict ourselves to the interval $[0, T/2]$. We have to prove that $C(d) < C(0)$ for every d in that interval different from 0. This is certainly true if $0 < d \leq T/4$ as both I and J are strictly decreasing on that interval. Now on the interval $[T/4, T/2]$ both I and J are linear and so is C, so that it suffices to prove that $C(T/2) < C(0)$; but this is immediate as $C(T/2) = k < N = C(0)$, which follows from the relations $I(0) = J(0) = 1$, $I(T/2) = 1$ and $J(T/2) = 0$.

We apply the described method in our reader as follows. Suppose we want to decode N bits transmitted by the tag and there are k samples per bit. We record in uC memory the samples for $N + 1$ bits, that is, $(N + 1)k$ samples. Then we compute the moving correlations sums as above, replacing the integrals by sums over discrete samples. Specifically, we start by computing the correlations for N bits as if the starting boundary of the first bit would be at the first recorded sample. Then we add together the absolute values of these correlations, which gives our first moving correlations sum. We memorize the decoding of N bits so obtained and the sum. Then we repeat the procedure, assuming now that the starting boundary of the first bit is situated at the second recorded sample, so that all correlations are computed starting from the second sample. We do this several times advancing by one sample, until we cover at least a bit interval; this is the reason of having recorded one bit in addition to the N required bits. Finally we estimate that the step that produced the maximal moving correlations sum among all computed sums gives the correct decoding of the N bits.

7. Conclusion

Low cost devices represent a significant trend of today's mass production, and the high competition places heavy demands on the ratio performance/price. We strived to face these demands by coming on the market with a design in which the non-standard usage of a chip allows one to obtain from it more than what was originally intended by the producer. We presented here tools and simulations that were useful to us in the design of the product.

8. References

Finkenzeller, K. (2003). *RFID Handbook*, John Wiley & Sons Ltd, ISBN 0-470-84402-7, West Sussex, England

Frosch Electronics (2004). *HITAG Long Range Reader Module HT RM801/901*. Available online at www.froschelectronics.com

Gelinotte, E., Frosch, R., Vuza, D.T. & Pascu, L. (2006). An RFID Reader Based on the Atmel AT91SAM7S64 Micro-Controller, *Proceedings of the 1st Electronics Systemintegration Technology Conference*, pp. 1158-1165, ISBN 1-4244-0552-1, Dresden, September 2006, Institute of Electrical and Electronics Engineers, Piscataway, NJ

NXP Semiconductors (2006 a). *HTRC110 Hitag Reader Chip*, Revision 3.0. Available online at www.nxp.com

NXP Semiconductors (2006 b). *Hitag S Transponder IC*, Revision 3.1. Available online at www.nxp.com

Vuza, D.T., Frosch, R. & Koeberl, H. (2007). A Long Range RFID Reader Based on the Atmel AT91SAM7S64 Micro-Controller, *30th ISSE 2007 Conference Proceedings*, pp. 445-450, ISBN 1-4244-1218-8, Cluj, May 2007, Institute of Electrical and Electronics Engineers, Piscataway, NJ

12

A Scientific Approach to UHF RFID Systems Characterization

Ulrich Muehlmann
NXP Semiconductors
Austria

1. Introduction

RADIO Frequency Identification (RFID) technology gained a lot of attention in the last few years. Especially, RFID at ultra high frequencies (UHF) is very attractive for various applications in supply chain management or logistics (see Glidden & Schroeter, 2005). The long-range power transportation feature and the ultra low power consumption property of current transponder chips enable a remotely powered communication along several meters of distance in terms of energy transport and data exchange.

In supply chain management or other applications that rely on high volume throughput, passive transponders are used because this type of transponder can be attached and manufactured in high volume at extremely low cost in the range of a few $ cents (see Glidden et al., 2004). Furthermore, an advanced protocol layer (see ISO Standards, 2007) enables the inventory of hundreds of items per second (see Glidden et al., 2004 and Vogt, 2002) having a sophisticated anti-collision algorithm.

From the system perspective, a passive transponder must obtain its energy for operation out of the RF field generated by an interrogator. Aside from the market relevant aspects such as low system costs and high transparency of stock inventories, UHF seams to be an appropriate frequency range for long-range passive RFID systems because of the moderate path-loss of RF field propagation and acceptable insertion-loss of silicon (see Finkenzeller, 1999; De Vita & Iannaccone, 2005; Karthaus & Fischer, 2003). This is a conclusion coming out of several technical analyses. Known issues coming from classical radio communication apply to UHF RFID as well. The finite propagation velocity of electromagnetic waves causes compulsorily a superimposition of the radiated power commonly known as multi-path propagation with the consequence of signal fading and superimposed interference during communication (see Saunders, 1999). Without appropriate measures, UHF RFID systems will suffer from read performance degradation and unwanted interrogation in the surrounding area.

Applications in the supply chain management, like the inventory process during the flow of pallets carrying hundreds of items through multiple gates or dock-doors, must ensure high operational reliability because missing reads cause mismatch between in-stock databases and actual item count and would make the entire identification system unacceptable for RFID deployment. In order to ensure the operational reliability one has to identify all performance degradation factors causing interrogation errors. J. Mitsugi et al. have published several studies dealing with readability degradation and field characterization related to UHF RFID applications (see Mitsugi & Hada, 2006; Mitsugi & Shibao, 2007; Mitsugi & Tokumasu, 2008). All these approaches are based on an extensive evaluation of the actual recorded field measurement data, illustrating the reading performance associated with a defined tag sensitivity threshold and with the maximum permitted interrogation power of the interrogator-to-tag forward link. Several theoretical and practical investigations (see Aroor & Deavours, 2007) have shown that the forward link limits the operational range and causes interrogation failures of actual UHF RFID technology in terms of tag sensitivity, standards and regulations (see ISO Standards 2007; ETSI, 2007a; ETSI, 2007b; ETSI, 2007c; IDA, 2008; FCC, 2007; SRRC, 2007). Based on these and on the following information, system integrators should have the ability to perform a proper RFID systems characterization and performance analysis.

2. Interrogation zone sensing

Interrogation zone sensing is a key process not only essential for the preparation of human visible data which give us a general picture of the actual performance level but is also important for computer aided analysis to determine improvement potentials and to define proper optimization strategies.

2.1 Sensing devices

In fact, it is not easy to determine the field strength in consideration of multi-path propagation within the relevant gate or portal area analytically. A better approach to this problem is not only the utilization of statistical models that are already successfully used to describe field propagation effects and indoor radio propagation channels (see Rappaport & McGillem, 1989), but also the use of ray tracing methods (see Bosselmann & Rembold 2006a; Bosselmann & Rembold 2006b) to simulate the field distribution in passive UHF RFID applications.

However, all these models require an extensive study of the actual field distribution obtained from the results of the measurement data analysis. For instance, the characterization of the field strength distribution inside a typical portal application in combination with different arrangements of tagged items on a pallet requires a complex measuring system. The measurement equipment should fulfil specific requirements. First, it should be neutral to the ambient conditions in order that the field distribution is not affected by the measurement setup itself. As second requirement, it should be capable to measuring the field characteristics along tag moving trajectories, on tag positions and optionally in the pallet for a better comparison to the actual available tag power. Repeatability should also be ensured in order to align all measurement data properly.

There are several measurement devices presented in the literature which are suitable for interrogation zone sensing. An interesting measurement device is proposed that allows the determination of the field strength on and inside individual pallets (see Redemske & Fletcher, 2005). This device operates as a battery powered active "tag emulator" and is part of the inventory and anti-collision process of the interrogator. Therefore, multiple sensor modules can be operated simultaneously. The individual field strength measurements are encoded in the EPC number (see ISO Standards, 2007), which is requested by an interrogator during the inventory process. This characteristic enables a wireless transmission of the measurement result. Another more sophisticated measurement devise (see CISC, 2006), also suitable for interrogation zone sensing, allows high-speed continuous data sampling of several sensor modules mounted along a specified moving direction, when triggered and moved through a gate for instance. Unfortunately, these devices suffer from additional drawbacks, on the one hand, the complicated cabling of sensor modules and main controller makes the handling difficult, on the other hand, the dipole measurement antenna is sensitive to material properties and may influences the measurement accuracy, especially in setups where the field strength inside pallets is of particular interest. As mentioned already in the introduction, J. Mitsugi et al., have already published a similar mobile field recorder for UHF RFID interrogation area measurement (see Mitsugi & Tokumasu, 2008).

For system integrators or non-professionals, it is very important to have a low-cost and easy to handle measurement device available. Furthermore, the measurement device should feature comparable size and shape of standard tags in order to be placed on real tag positions and to be operated on similar trajectories.

We have adopted these ideas and have designed and developed two contactless battery powered measurement devices. One of which is equipped with a dipole based field probe and is more suitable for free space field strength measurement and the other one equipped with a loop based field probe antenna which is more insensitive to different material properties for measurements inside pallets. See a photograph of these devices in Fig. 1 for detail.

Fig. 1. Photograph of NXP mobile field strength recorders. One of which equipped with dipole-antenna based field probe and the other with loop-antenna field probe

2.2 Test environments

UHF RFID gates and portals are deployed in deferent geographical regions operating under different local regulations. In open loop applications, goods are transported from one country to others passing different types of gates, but normally only with an individual tag attached during its production or packing process. Therefore, it is important that all gates, doors and portals have a standard shape in order to fulfil system reliability along the entire supply chain. In this context we propose a novel characterization method of UHF RFID portals based on RF field strength recording, RFID tag channel modeling, and qualification

according to the derived channel model. This method can also be extended to doors, portals or other high power UHF RFID applications with comparable characteristics.
Two basic principles of the measurement procedure for portal interrogation zone sensing are illustrated in figure 2 and figure 3.

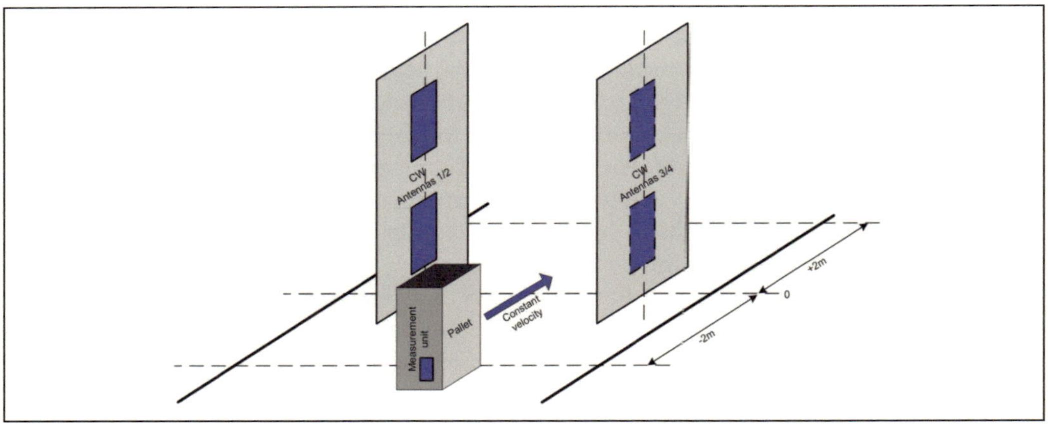

Fig. 2. Laboratory setup for interrogation zone sensing. The pallet under test is passed through the portal with constant velocity by using an automated transport system

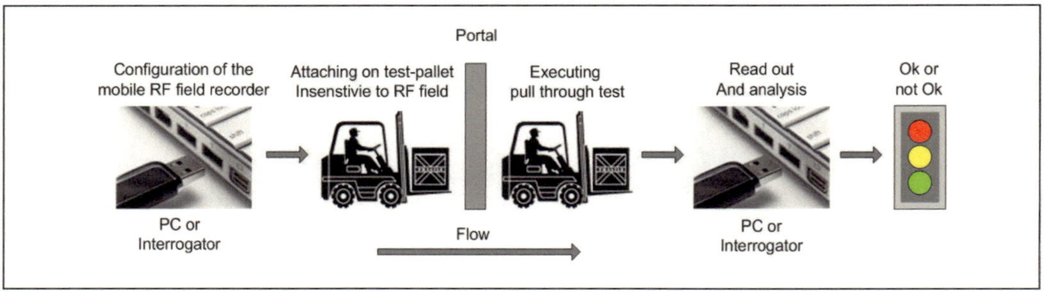

Fig. 3. Direct test in industrial environments. The FSRs are configured by a PC or interrogator, attached and passed through the portal. Later measurement data analysis gives information about portal quality .

In principle, there are three possibilities to perform interrogation zone measurements. One is a setup under test operated in a controlled environment. An experimental study that addresses the readability degradation in UHF RFID systems within an anechoic chamber is proposed in (Mitsugi & Hada, 2006). Such a setup is only useful to identify base performance and to use the measurement data as a reference to real-life environments since the influence of field propagation cannot be determined properly. Comparable real-life conditions can be achieved within a laboratory environment which refers as second possibility of evaluation. In section 3, the results of an application specific study of a passive UHF RFID system examined under laboratory conditions is presented. Furthermore, a direct test in industrial environments should be fast and reliable. One evaluation procedure for RFID portals can be as illustrated in figure 3. The presented FSR is specifically suitable for this kind of application tests. Once the FSR is configured by an interrogator or PC it can be attached or integrated in a pallet under test, triggered and moved through the gate. This operation is repeated several times depending on the number of FSR simultaneously used and to increase the accuracy of classification. In section 4, a model based classification method for RFID portals is presented.

3. Probability of interrogation degradation

Different application tests have shown that material properties of the tagged items, beam-width and quantity of the interrogator's transmit antennas, tag sensitivity and position of the tag have significant impact on the occurrence of missing reads. This study addresses also the influence of the pallet density and its absorption characterizes with respect to the electromagnetic wave propagation effects on field coverage and readability. All measurements were managed within a laboratory environment with the help of an automatic transportation device of pallets to ensure repeatable measurement conditions. The test portal was a replica of a typical portal used in a distribution centre (DC) within a real-life RF-environment.

3.1 Portal and pallet setups

The portal consists of two chambers that are 3m apart, each containing two circular-polarized interrogator antennas, the pallets are setup using three individual arrangements of items are chosen.

3.1.1 Pallet setup A

Setup A (see figure 4) is a type of skeleton pallet that is used to determine the field strength in a quasi free-space environment inside the portal area. A set of 45 sensor modules at different locations on the skeleton pallet are used to measure the field strength along the moving path through the portal. The portal area is restricted to ±2m from the antenna symmetry axis by definition (see figure 2). Previous measurements had shown that regions outside this area are not relevant for the interrogation zone analysis since the RF power levels outside are below the sensitivity threshold of typical tags.

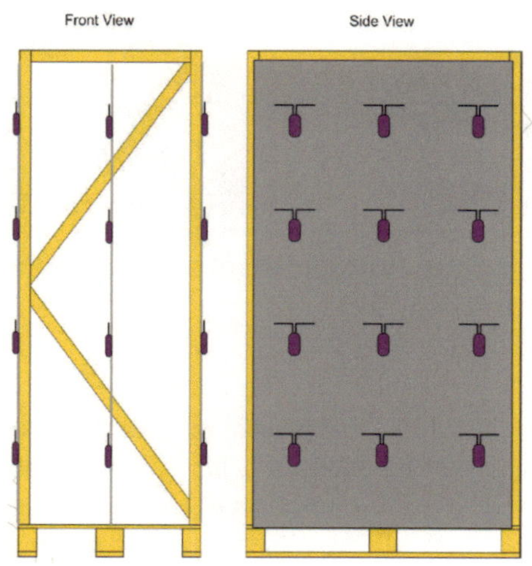

Fig. 4. Wooden skeleton pallet equipped with sensor modules for quasi free-space field strength analysis inside the portal area.

3.1.2 Pallet setups B and C

Setups B and C are sub-sets of the pallet arrangement illustrated in figure 5, indexed as pallet A and B, respectively. Pallet A consists of coffee packs, soap and liquids, whereas pallet B consists of toilet paper, chocolate bars and swaddling bands.

Fig. 5. Pallet with typical items often transported in a supply chain (mineral water, fish, toilet paper, toys and snacks) placed on the automated transport system

3.1.3 Pallet setup D

Setup D (see figure 6) consists of typical items that are often transported in a supply chain like mineral water, fish, toilet paper, toys and snacks. The alignment of the 45 sensor modules is similar to all previous setups.

Fig. 6. Pallet with typical items often transported in a supply chain (mineral water, fish, toilet paper, toys and snacks)

3.2 Measurement analysis

As already mentioned, the test portal consists of four interrogator antennas (Intermec IA39B) at distinct positions. Only one antenna is active during one test scenario of all listed pallets. This corresponds to the regular operation of common interrogators since there is just one RF transmitter integrated and all the attached antennas to it are multiplexed in time. The operating frequency was chosen in the European UHF RFID frequency band at channel 4 (867.5MHz) and the power level was set to the maximum permitted value of $2W_{ERP}$ for the entire measurement period. One measurement cycle consists of the following procedure.

First, the selected pallet was equipped with the sensor modules at well-defined locations on the items, where normally the tags are placed, and the pallet under test was placed on the automatic transportation device afterwards. In the second step, the selected pallet was automatically triggered and moved back and forth with constant velocity of v_0=0.544m/s to ensure measurement repeatability. The chosen velocity is a trade-off between maximum sampling rate of the field recorder and spatial resolution along the moving direction. In the final step, the average of five runs at every single sensor module position was calculated and used for further measurement analysis.

3.3 Measurement results

The field characteristics of two sensor modules of pallet setup A associated with the defined moving path through the portal and with antenna 1 as power source are illustrated in figure 7a. The characteristic of module 40 shows a significant peak in the centre of the portal. In contrary, module 44 shows a flat distribution overlaid by some degree of interference.

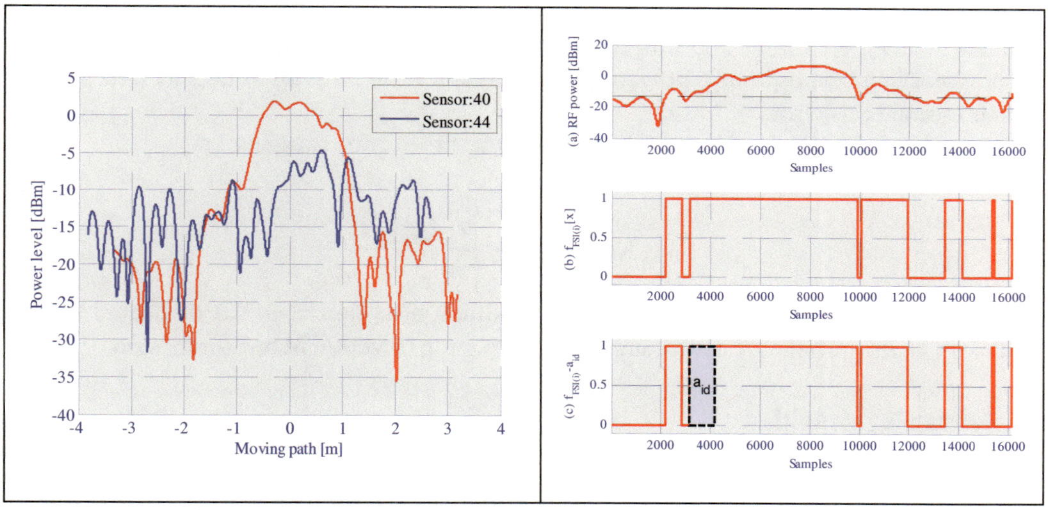

Fig. 7. (a) Field characteristics of two sensor modules of pallet setup A associated with the defined moving path through the portal and (b) graphical representation of the basic calculations used for field coverage and readability associated with a specified sensitivity threshold of -13dBm; the influence of a_{id} is illustrated at only one piecewise constant region for clarification

In general, missing reads will occur when the available power level is lower than the sensitivity threshold of the tag. For this reason, we propose to introduce two quality factors for gates and pallets defined as readability and field-coverage associated with a dedicated section or read zone of the maximum portal area. The readability represents a quality indicator for a dynamic environment and field-coverage for a static environment, respectively. We note that the expected readability and field coverage obtained in one section are identical for a static environment. These indicators are defined as follows:

3.4 Field-coverage definition

First, we define a field strength indicator function $f_{FSI}[x]$ as:

$$f_{FSI(i)}[x] = \begin{cases} 1, \text{for } p_i[x] \geq s_{th} \\ 0, \text{for } p_i[x] < s_{th} \end{cases}, s_0 \leq x \leq s_1. \tag{1}$$

Where $p_i[x]$ are the field strength measurement samples along the moving path of the i-th sensor module, s_{th} represents the sensitivity threshold of the tag, and s_0, s_1 define the section boundaries on the moving direction and it holds that $s_1>s_0$. The sensor boundaries are related to the valid interrogation zone. All the measurements were taken from the four interrogator antennas independently without varying the positions of the sensor modules. Therefore, we have to distinguish between four different sample sets recorded from every single module. The RF power levels measured depend on the effective distance between the sensor modules and the distinct interrogator antennas. In order to obtain the maximum field coverage, $p_i[x]$ is the set chosen, which produces the maximum on average power distribution along the moving direction generated by the individual interrogator antennas. This corresponds to the minimum effective distance from the individual sensor module to the operating interrogator antenna. In reference to the portal application, this is an issue of antenna switching strategy and this is not addressed in this context. However, this selection is comparable to the optimum powering conditions for the individual tags located at these sensor module positions.
Whereas, the field-coverage of a dedicated section or read zone is defined as:

$$I_{FC} = \frac{1}{N(s_1 - s_0)} \sum_{i=1}^{N} \sum_{x=s_0}^{s_1} f_{FSI(i)}[x] \tag{2}$$

Where N represents the number of sensor modules, s_1-s_0 describes the dedicated section boundary of the portal area associated with the moving direction in one dimension.

3.5 Readability definition

For the determination of the readability in a dynamic environment, additional parameters are required. One parameter is the inventory duration of a signal tag. For instance, if we rely on the ISO 18000-6C standard (see ISO Standards, 2007), this inventory duration requires on average 6ms in a dense interrogator environment. Furthermore, the moving speed of the pallet and the back-up lifetime of the tag after power loss are required, whereas the back-up duration can be neglected in this context, since this period is normally shorter than several µs and lies below the resolution of the field recorder used.
The readability of one single tag is defined as:

$$I_{RA(i)} = \sum_{j=1}^{M_i} \max\left[0, \sum_{x=n_{j0}}^{n_{j1}} f_{FSI(i)}[x] - a_{id}\right], \tag{3}$$

and the expected readability in a dedicated read zone is defined as the average readability calculated on the basis of the individual results:

$$I_{RA} = \frac{1}{N} \sum_{i=1}^{N} I_{RA(i)}. \tag{4}$$

Where n_{j0} and n_{j1} are the lower and upper bounds of the j-th piecewise constant region of $f_{FSI(i)}[x]=1$ within the section boundaries. M_i is the number of piecewise constant regions inside the section boundaries, N represents the number of sensor modules, a_{id} denotes the required field coverage for one tag inventory cycle during the moving of the pallet through the portal in a dynamic scenario which is defined as:

$$a_{id} = \left\lceil v_0 t_{id} n_{spm} \right\rceil . \tag{5}$$

Where v_0 is the moving velocity of the pallet through the portal, t_{id} represents the inventory duration and n_{spm} denotes the number of measurement samples per meter. See figure 7b for clarification.

3.6 Field coverage results

If we refer to figure 8a, we can see that the field coverage is nearly identical for the given sensitivity thresholds within the section boundaries ±0.5m in reference to the portal centre. If we consider the section boundaries ±1.5m, we can determine a field coverage reduction of about 8% when the sensitivity threshold of the tag is increased from -15dBm to -13dBm. These values correspond to sensitivity values of current commercial tags. That means, if the interrogator is triggered at these specified section boundaries, the probability of a successful inventory will decrease by 8% in a static (no motion) free space environment in consideration of multi-path propagation (pallet setup A).
The comparison of all specified pallet configurations at sensitivity thresholds -13dBm and -15dBm leads to the following results. It can be shown that the field coverage associated with the section boundaries measured in all specified pallet configurations has an equal tendency with different offsets mainly caused by the damping properties of the items on the pallet (see figure 8b). This damping characteristic is investigated in (Fletcher & Marti, 2005) by using fundamental electro-magnetic propagation principles. For instance, setup B leads to a field coverage of about 80% in the section boundaries ±0.5m whereas setup D achieves 86% coverage and setup C achieves 92% coverage.

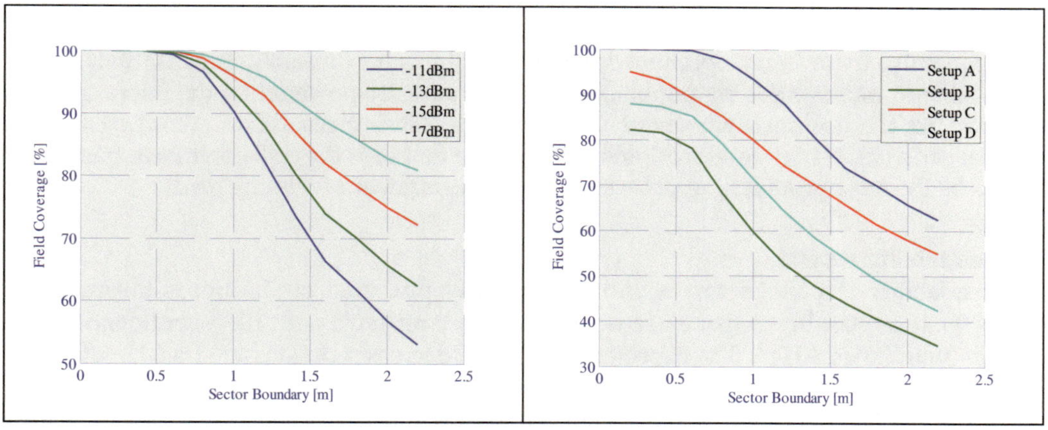

Fig. 8. (a) Field coverage associated with the dedicated section boundaries dependent of different tag sensitivity thresholds (-11, -13, -15, and -17dBm) based on the pallet setup A (free-space portal area); (b) Field coverage associated with the dedicated section boundaries of all specified pallet configurations based on a sensitivity threshold of -13dBm

The reflection characteristics of the electro-magnetic waves on and inside different pallets are independent of the material properties according to the measurement results. In contrary, the absorption characteristics depend significantly on the material properties. This means that the radiated RF power absorption by the items on the pallet and not the random multi-path interference causes reading performance degradation. A reduction of the sensitivity threshold from -13dBm to -15dBm (see figure 9a) is leading to a slight improvement of the field coverage. The probability of inventory increases by 2.5% in pallet setup B, by 3.5% in pallet setup C and by 3% in pallet setup D. If the read zone is expanded up to the gate boundaries, the relative field coverage will increase but will not exceed 10% in all configurations.

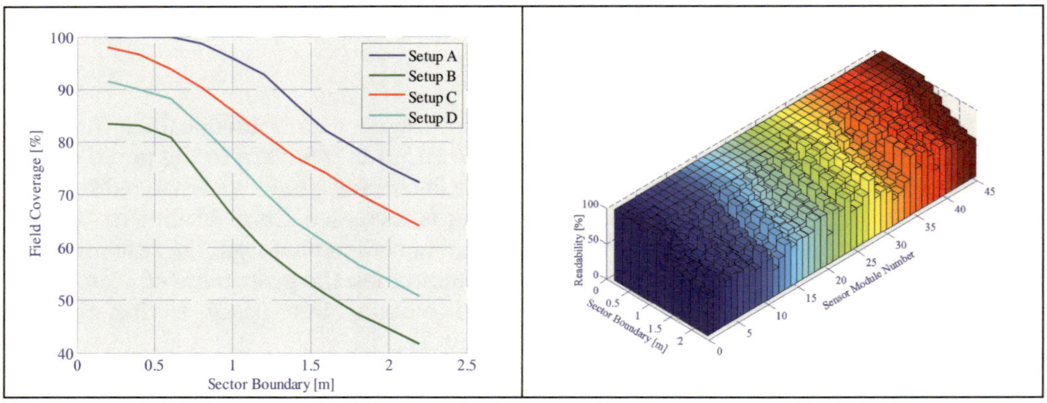

Fig. 9. (a) Field coverage associated with the dedicated section boundaries of all specified pallet configurations based on a sensitivity threshold of -15dBm; (b) Readability at the position of all sensor modules of pallet setup A (free space) with a sensitivity threshold of -13dBm associated with the dedicated section boundaries (read zones)

In conclusion, a reduction of the sensitivity threshold will increase the field coverage insignificantly in comparison to the absolute field coverage in the specified interrogation zone. The analysis has shown that the field coverage is less than 100% except in a free-space environment (setup A) and can lead to missing reads in all other configurations. Consequently, there is no optimum inventory optimization feasible, because it is nearly impossible to estimate the tag population, which is actually powered by the interrogator in order to set the optimum slot-count value for anti-collision resolution according to (ISO Standards, 2007). A trial-and-error polling technique that uses the collision rate as parameter would be the most promising approach to reduce the number of missing reads.

3.7 Readability results

The readability of pallet setup A with a sensitivity threshold of -13dBm is illustrated in figure 9b. It can be shown that all sensor modules are readable with 100% confidence in the section boundaries ±0.5m. The expected readability decreases on average by 60% when the section boundaries are set to ±2.2m. This characteristic is mainly caused by lower field coverage outside the portal centre and is a consequence of higher path-loss of the electromagnetic energy. This means, when the interrogator is triggered at these section boundaries, the probability of field coverage of on particular tag along its moving direction is lower. Therefore, the readability of tags decreases in combination with the random occurrence of its inventory instant inside the specified interrogation zone.

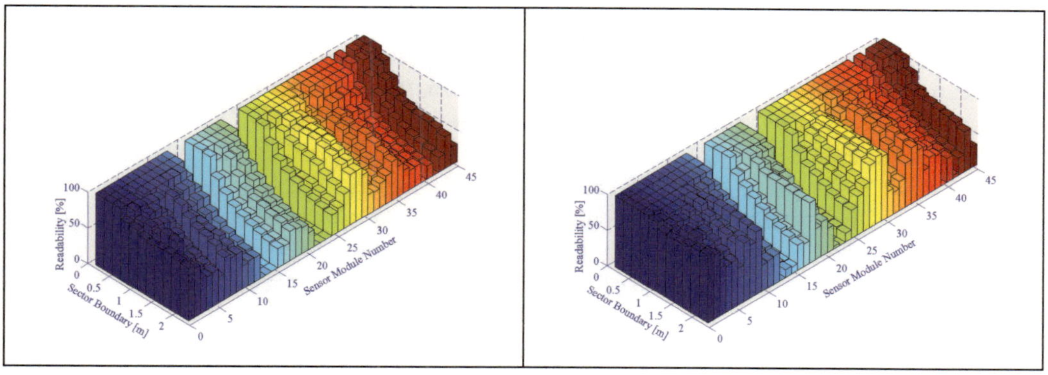

Fig. 10. (a) Readability at the position of all sensor modules of pallet setup B with a sensitivity threshold of (a) -13dBm and (b) -15dBm associated with the dedicated section boundaries (read zones)

The readability of tags is not ensured at several sensor module positions in pallet setup B (see figure 10a). The power levels at the sensor module positions 13, 14, 15, 22, 23, 24, 40, and 42 are permanently below the sensitivity threshold of –13dBm and tags attached to those particular items cannot be read in any case. This behaviour is mainly caused by the absorption characteristic of items, especially water or metal food casings, arranged in the line-of-sight direction to the operating interrogator antenna. For instance, when the sensitivity threshold is reduced from -13dBm to -15dBm the readability at sensor module positions 14, 15, 22, 24 and 42 can be achieved, but the readability is still not achievable at the positions of the remaining three modules 13, 23 and 40 (see figure 10b).

In conclusion, the readability of all other pallet configurations showed comparable results. The readability is critical at positions where metal food casings are in the line-of-sight direction to the powering antenna. A sensitivity threshold reduction helps to overcome the powering problem at certain positions, but under some extreme conditions, the readability is still not achievable.

4. Interrogation zone characterization

Basically, it is not possible to accomplish an interrogation zone characterization of all significant pallet configurations involved in terms of known performance degradations factors. Instead we propose a method that requires only free space measurement data recorded in the appropriate interrogation zone. This data is correlated with an RFID tag channel model valid for portal applications to determine a statistical value of the expected field coverage within the interrogation zone. An additional safety margin between supposed tag sensitivity and expected field coverage ensures the readability of pallets carrying items with high RF absorption characteristics. The safety margin will come out from experimental test in real life portal application and will be treated as an additional loss added to the free space path-loss model.

4.1 Statistical modeling of the interrogator-to-tag powering channel

The primary goal of the channel modeling is to identify the fading characteristics of the interrogator-to-tag energy transfer along a typical trajectory through the portal and to determine the mean path loss on the portal cross-section. Many of these channel modeling approaches are based on the evaluation of the probability or cumulative density function (CDF) of the transfer signal amplitude over time. Rappaport (see Rappaport & McGillem, 1989) can be mentioned for his pioneering work in the research of UHF signal fading characteristics in factory environments. During the investigation of UHF cellular radio a variety of different channel models were propose for distinct indoor or outdoor scenarios (see Nikookar & Hashemi, 1993; Hashemi, 1993). These channel models are not directly applicable to UHF RFID due to its heterogeneous mode of operation. Little research is carried out in this area so far. What is already published are channel models dealing with short range indoor applications (see Mayer et al., 2006) or statements that a Rician distribution fits UHF RFID applications (see Kim et al., 2003) because of its strong line-of-sight mode of operation. But this assumption is not always valid because tags are not only attached on the pallet surface. It is not promising to apply a statistical channel model to the entire read volume since tag trajectories and power density are not considered properly. Therefore, we propose a two step approach leading to a general channel model for portal applications. The basic idea is illustrated in figure 11a. The read volume is divided into two orthogonal planes defined as TAG-plane and LOS-plane. The TAG-plane is aligned to the tag moving path through the portal, perpendicular to the interrogation antennas, and is associated with a signal amplitude fading model and the LOS-plane is aligned to the mean power distribution of the cross-section in the centre of the gate associated with an appropriate path-loss model. Measurements have shown that the path-loss along the LOS-plane has a significant impact on the mean power distribution.

4.1.1 TAG plane fading model

Once the tag is moving through the portal along the TAG-plane, it will pass a RF power distribution comparable to a characteristic illustrated in figure 7a. A statistical probability density function of such characteristic can be modelled by means of a bimodal distribution function.

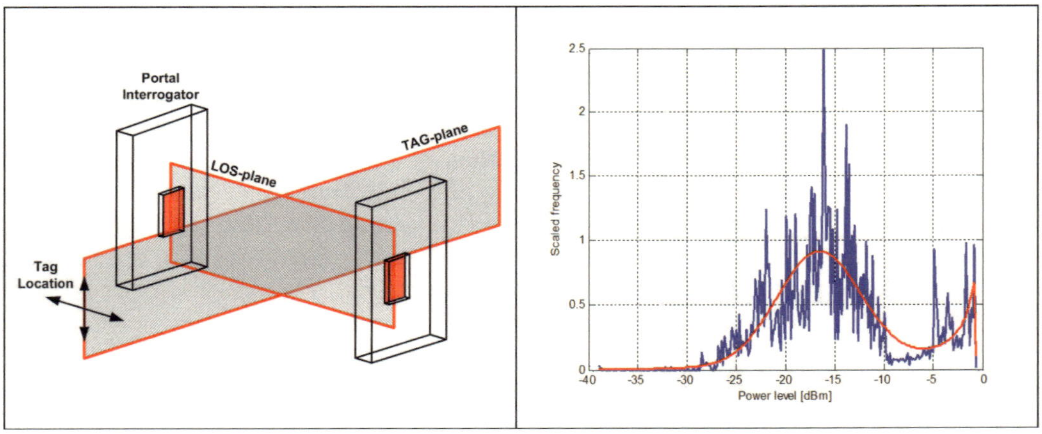

Fig. 11. (a) Two step approach to a portal channel model. TAG-plane is aligned in tag moving trajectory providing field coverage distribution and the LOS-plane is aligned to the portal cross-section for maximum mean power distribution modeling and (b) Bimodal RF power level probability distribution of one continuous measurement along a typical tag trajectory through the portal

The strong line-of-sight component occurring in the centre of the gate is modelled with a log-normal distribution and the fading components outside the portal centre are modelled with a normal distribution function. The general bimodal distribution function can be expressed with:

$$F_{bm}(x) = pF_1(x) + (1-p)F_2(x), \tag{6}$$

where p is the fraction of the deferent distribution functions $F_1(x)$ and $F_2(x)$. When we expand (6) with the defined distribution functions, we get the proposed statistical fading model for the TAG-plane as:

$$F_{TAG}(x) = \begin{cases} p\frac{1}{\sigma_1\sqrt{2\pi x}}\mathrm{e}^{-\frac{(\ln x-\mu_1)^2}{2\sigma_1^2}} + & \\ +(1-p)\frac{1}{\sigma_2\sqrt{2\pi}}\mathrm{e}^{-\frac{(x-\mu_2)^2}{2\sigma_2^2}} & x > 0 \\ 0 & x \le 0 \end{cases} \tag{7}$$

For example, fitting this distribution model to the amplitude fading characteristic illustrated in figure 11b yields values p=0.256, μ1=1.790, σ1=1.929, μ2=0.416, σ2=0.111, and scaling factor sc=0.327. The scaling factor is only important for the nonlinear least square fitting algorithm and has no influence on the cumulative distribution function that indicates probability of signal strength lower a specified threshold.

4.1.2 LOS-plane mean power model

The LOS-plane mean power model describes the expected mean power level of the cross-section in the centre of the portal. This model consists of a general path-loss model and a multi-ray interference model. General path-loss models are described in (Rappaport, 2002; Leong et al., 2006), but the Friis transmission equation (8) is used for the present problem formulation.

$$P_r = P_t G_t g_r \left(\frac{\lambda}{4\pi r}\right)^2, \tag{8}$$

where P_t is the transmitted power, G_t is the gain of the interrogation antenna, g_r is the gain of the tag antenna, λ represents wavelength of the carrier and r is the distance to the target point. On the other hand, the multi-ray interference model considers first order ground reflections and the total reflection property of the opposite portal chamber. The first order ground reflections are modelled by means of the well known Fresnel's reflection coefficients (9) valid for an interface between air and ground.

$$\rho_h = \frac{\cos\phi - \sqrt{n^2 - \sin^2\phi}}{\cos\phi + \sqrt{n^2 - \sin^2\phi}}, \quad \rho_v = \frac{n^2\cos\phi - \sqrt{n^2 - \sin^2\phi}}{n^2\cos\phi + \sqrt{n^2 - \sin^2\phi}}, \tag{9}$$

$$\rho_c = \frac{\rho_h + \rho_v}{2},$$

where, ρ_h, ρ_v, ρ_c represent perpendicular, parallel and circular incidence of the electric field, n is the complex refractive index of concrete and Φ is the angle of incidence. The thickness of the concrete ground is taken into account according to (Sato et al., 1996). The total reflection property of the opposite portal chamber has its impact only on the free-space measurement scenario. Once passing a pallet through the gate, the line-of-sight direction is blocked and most of the energy is absorbed or reflected by the pallet. We have set up a measurement procedure to identify the absorption characteristics by detection of the S_{21} scattering parameter using a network analyzer and opposite antenna pairs.
In many cases, the opposite chamber is made of metal plates without shielding in order to achieve robustness and cost efficiency. This property must be considered during the qualification procedure since the measured power levels in an empty portal are higher due to the virtual gain of the opposite chamber. We propose to model the reflection property according to the behaviour of a flat metal plate with dimensions of the chamber outline. This is just a rough estimation that accounts for the worst-case scenario and provides maximum safety margin with respect to the normal case of operation. The reflection property of a metal plate can be described with the well-known monostatic radar cross section (RCS) formula (10). This approach is also used for indoor radio channel modeling (see Kajiwara, 2000; Fenn & Lutz, 1993) and to determine a theoretical RCS value dependent on the incidence angle (see Ross, 1966).

$$G_r = \frac{\sigma_{RCS}}{4\pi d_p^2}, \quad \sigma_{RCS} = \frac{4\pi A^2}{\lambda^2}, \tag{10}$$

where, G_r is the equivalent gain, σ_{RCS} is the monostatic RCS, d_p is the portal width, A is the equivalent are of the chamber and λ is the wave length of the power carrier. Finally, we define a similar UHF RFID radio channel model as proposed in (Han et el., 2004).

$$P_r = P_t g_r \left(\frac{\lambda}{4\pi}\right)^2 \left| \sqrt{G_t(\vartheta_0)} \frac{1}{r_0} \mathrm{e}^{-jkr_0} + \sqrt{G_t(\vartheta_1)} \rho_c \frac{1}{r_1} \mathrm{e}^{-jkr_1} + \sqrt{G_r} \frac{1}{r_2} \mathrm{e}^{-jkr_2} \right|^2, \tag{11}$$

where P_r is the power level at the target point, P_t is the transmit power of the interrogator, λ is the wave length of the power carrier, $G_t(\vartheta_{0,1})$ is the appropriate interrogation antenna gain dependent on the E-plane angle, $r_{0,1,2}$ represents path length.

4.1.3 Joint TAG/LOS model

In order to obtain a 3D model that covers the desired interrogation zone, we propose the following approach. The TAG plane fading model describes the statistical distribution of the signal amplitude without contribution to the signal level. Referring to figures 7 and 12, the maximum value of the signal amplitude will be different at different tag locations and when the mean distribution is fitted over all tag locations, all the individual signal level distributions result in one CDF associated with the overall tag readability. In contrast, the LOS plane mean power model gives a value for the maximum signal value in the portal cross section and defines the abscissa maximum of the tag readability CDF. The abscissa minimum is defined by the resolution of the FSR, equal to -40dBm. Finally, the tag readability is obtained by application of a dedicated tag sensitivity threshold.

4.2 Results and pallet readability

We apply this approach to two different portal setups, one is operated according to EU (European) and the other according to US regulations. The EU setup consists of four individual Kathrein 25-180 circularly polarized directional antennas, 10.5 dBic, 70° H-plane 3dB-beamwidth and 30° E-plane beam width, arranged in [0.7, 1.4, 1.4, 0.7] meters from ground plus a Sirit Infinity 510 UHF RFID interrogator set to 27 dBm conductive power, 866 MHz, and continuous wave output. The US setup consists of 4 individual Symbol Andrew RFID-900-SC high performance area antennas, 6.0 dBi, 70° in both H and E-plane 3dB-beamwidth, arranged in [0.65, 1.75, 1.75, 0.65] meters and the Sirit Infinity 510 set to 27 dBm conductive power, 915 MHz, and continuous wave output.

4.2.1 TAG plane fading model results

According to the proposed method, the TAG-plane measurements are accomplished by means of the FSR devices. A set of 18 measurements where taken per antenna. Therefore, the FSR position is varied in dimensions of a typical pallet outline and moved through the portal under test utilizing an automated transportation device (see Muehlmann & Witschnig, 2007). The CDFs derived from the measurement data show similar characteristics and do not depend on the antenna position (see figure 12). Hence, it can be concluded that the portal interrogation zone does not depend on the portal surroundings. It can be noted, that the free space field coverage is a bit higher in the US setup, 50% achieved at around -10 dBm compared to -15dBm achieved in the EU setup. This is probably caused by the broader antenna beam width used in the US setup (the broader beam width combined with scattering from metal object surrounding thus the portal may generate a higher level of reflections and resulting in an increasing of the field).

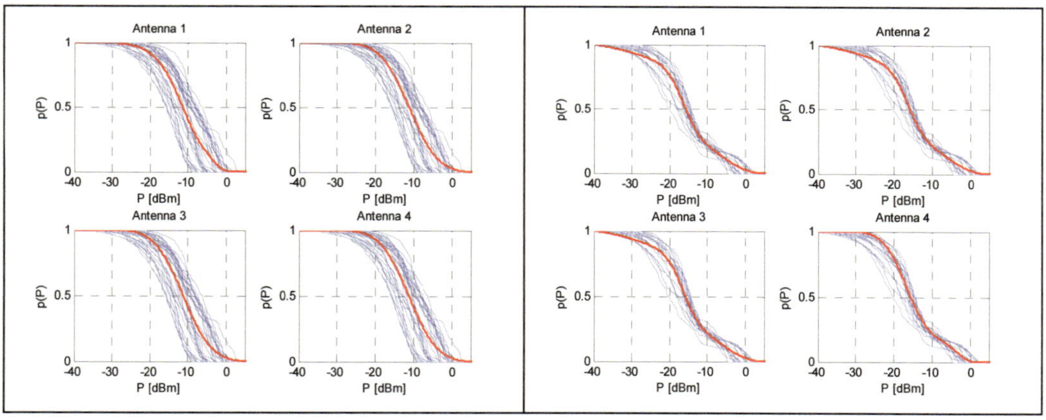

Fig. 12. (a) CDF of the used US portal derived from the FSM measurements and fitted model function showing p(P) as probability of power level greater a threshold and (b) CDF of the used EU portal derived from the FSM measurements and fitted model function showing p(P) as probability of power level greater a threshold.

4.2.2 LOS-plane mean power model results

According to the proposed method, the LOS-plane mean power model is used to describe the field strength distribution on the portal cross-section. Figure 13a shows the simulation result of the EU setup where antenna 3 is the interrogation antenna and which defines the x-axes origin. The simulation result matches well to the real life situation. The comparison is performed by taking the field strength values of the TAG-plane measurement data when the FSM is passing through the portal cross-section. The US setup is analyzed in an equal manner. The simulation result is illustrated in figure 13b and shows a slightly lower mean power distribution compared to the EU setup.

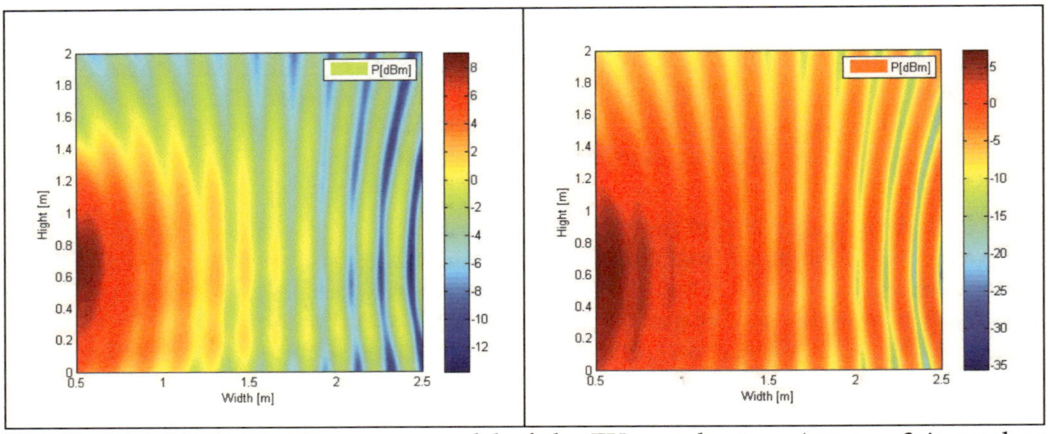

Fig. 13. (a) LOS-plane mean power model of the EU portal setup. Antenna 3 is used as interrogation antenna and defines the x-axes (vertical) origin and (b) LOS-plane mean power model of the US portal setup. Antenna 3 is used as interrogation antenna and defines the x-axes origin

4.2.3 Pallet readability

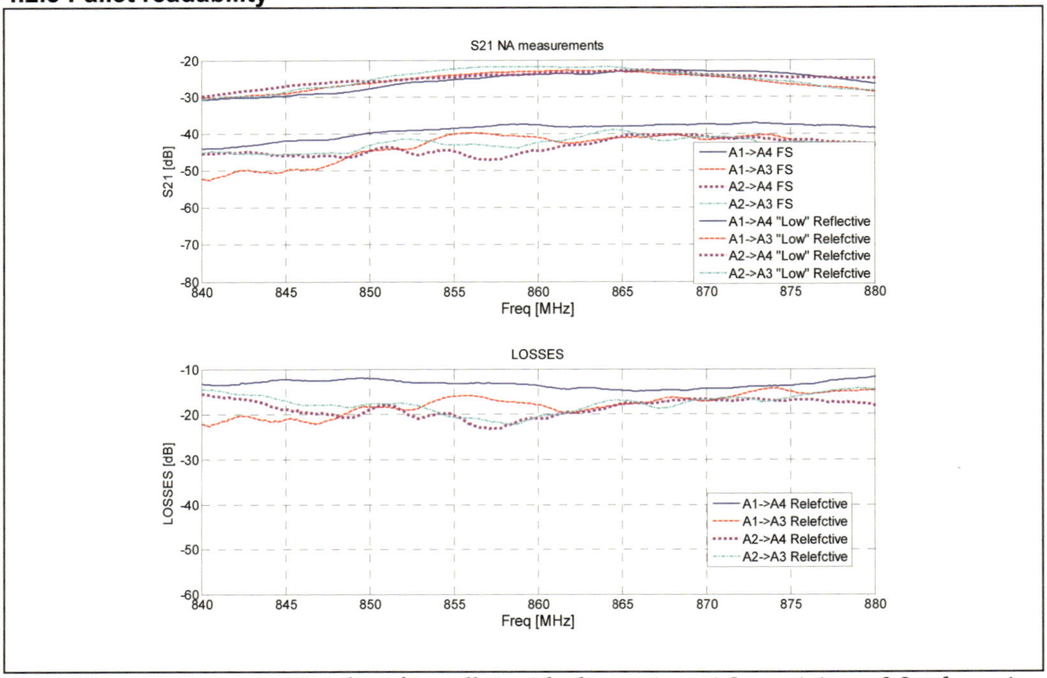

Fig. 14. S_{21} measurement results of a pallet with dimensions 1.2m x 1.4m x 2.2m by using opposite antenna pairs for radiation and reception. A network analyzer is used to determine S_{21} versus frequency.

The pallet readability depends not only on the prevalent field coverage of the interrogation zone and the pallet density but also on the operational sequence of the anti-collision protocol (see ISO Standards, 2007). There has been extensive research carried out in the optimization of such ALOHA anti-collision protocols (see Jin et al., 2007; Floerkemeier & Wille, 2006; Vogt, 2002; Wang & Liu, 2006) which impact on the reading performance is beyond the scope of this study. A practical test of these two portals with a pallet (1.2m,1.4m,2.2m) containing 200 tagged items (see figure 5) has shown that the EU portal setup reaches 86.8% read-rate whereas the US portal setup 80%. Referring to figure 13, the mean LOS power level is about 2dB higher in the EU compared to the US setup which explains the different read-rates. The pallet loss characteristic was measured and illustrated in figure 14b. Assuming a linear path-loss through the portal and that all four antennas are in the interrogation sequence involved, a path loss of -11dB can be expected from the pallet outline to its centre.

5. Conclusion

Two quality factors for gate and portal applications are proposed in this text, which are defined as field coverage and readability. Both indicators are in reference to the dedicated interrogation zone specified as sections with defined boundaries on the pallet moving path. The expected field coverage of different setups has similar tendency associated with the section boundaries and depends on the damping characteristics of the different pallet configurations and on the sensitivity threshold of the tag. It can be enhanced up to 10% by increasing the sensitivity from -13dBm to -15dBm. However, the sensitivity improvement is insufficient in reference to the absolute field coverage that is achieved in particular pallet arrangements. In contrary, the readability of tags at particular positions can be achieved by increasing their sensitivity.

The readability as well as the field coverage depends on the section boundaries. The closer the section boundaries to the centre of the gate the higher the expected field coverage and readability will be. This characteristic is mainly caused by the gain pattern of the interrogator antenna, which shows normally a dominant main lobe in the direction to the portal centre.

The probability of missing reads from the perspective of field coverage and readability can be reduced by defining the appropriate interrogator triggering position in combination with the main lobe of the interrogator antennas on the one hand. On the other hand, the improvement of the tag sensitivity will lead to higher readability and increases the probability of a successful inventory accordingly. However, this experimental study has shown that the readability is not guaranteed at certain positions on the pallet with state of the art technology, where extreme conditions prevent the activation of the affected tags.

The sensitivity enhancement up to a certain level must be investigated properly. Therefore, two conflicting factors that influence the overall system performance must be considered. These factors are the receiver sensitivity and dynamic range of the interrogator and the occurrence of unwanted reads in close proximity.

In conclusion, a novel interrogator-to-tag channel model has been presented that describes the field strength distribution in the portal interrogation zone. The model parameters are derived from the measurement data and a custom-made FSR is used to determine the actual field strength along typical tag trajectories.

Further investigations are needed on how to interpret the model parameters p, μ_1, σ_1, μ_2, and σ_2 with respect to an optimization of the portal setup, beam-width and selection of the antenna, etc. Furthermore, the reflection characteristic of the opposite chamber needs to be studied in different setups to derive general numbers. Based on the LOS-model it should be possible to predict this reflection characteristic out of the measurement data. In order to predict the read-rate out of the model parameters, it is essential to know absorption and reflection figures of possible pallet configurations as well as actual tag locations on the tagged items. These parameters are mainly customer related and no work to this subject is presented in this text accordingly. In addition, it is essential to incorporate the influence of the anti-collision algorithm in order to make a statement about the overall read-rate.

6. References

Aroor, S. R. & Deavours, D. D. (2007). Evaluation of the State of Passive UHF RFID: An Experimental Approach. *IEEE Systems Journal*, vol 1(2), December 2007, pages 168-176

Bosselmann, P. & Rembold, B. (2006a). Ray Tracing Simulations for UHF Passive RFID Applications, *15th IST Mobile and Wireless Communications Summit*, Mykonos, Greece, 4-8 June 2006

Bosselmann, P. & Rembold, B. (2006b). Ray Tracing Method for System Planning and Analysis of UHF-RFID Applications With Passive Transponders, *2nd ITG/VDE Workshop on RFID*, Erlangen, Germany, 4-5 July

CISC. (2006). RFID Field Recorder R 1.0, www.cisc.at.

De Vita, G. & Iannaccone, G. (2005). Design Criteria for the RF Section of UHF and Microwave Passive RFID Transponders, *IEEE Transactions on Microwave Theory and Techniques*, vol. 53, No. 9, September 2005, pages 2978-2990

ETSI. (2007a). European Telecommunications Standards Institute (ETSI), EN 300 220 (all parts): Electromagnetic compatibility 2007 EPCglobal Inc. Page 6 of 41, 11 June 2007, and Radio spectrum Matters (ERM); Short Range Devices (SRD); Radio equipment to be used in the 25 MHz to 1000 MHz frequency range with power levels ranging up to 500 mW

ETSI. (2007b). European Telecommunications Standards Institute (ETSI), EN 302 208: Electromagnetic compatibility and radio spectrum matters (ERM) - Radio-frequency identification equipment operating in the band 865 MHz to 868 MHz with power levels up to 2 W, Part 1 - Technical characteristics and test methods

ETSI. (2007c). European Telecommunications Standards Institute (ETSI), EN 302 208: Electromagnetic compatibility and radio spectrum matters (ERM) - Radio-frequency identification equipment operating in the band 865 MHz to 868 MHz with power levels up to 2 W, Part 2 - Harmonized EN under article 3.2 of the R&TTE directive

FCC. (2007). Federal communication commission, Radio Frequency Devices Intentional Radiators, Radiated emission limits, general requirements, Part 15 Subpart C, § 15.245, 15.246, 15.247

Fenn, A. J. & Lutz, J. E. (1993). Bistatic radar cross section for a perfectly conducting rhombus-shaped flat plate: simulations and measurements, *IEEE transactions on antennas and propagation*, vol. 41, pages 47-51

Finkenzeller, K. (1999). *RFID Handbook: Fundamentals and Applications in Contactless Smart Cards and Identification*, 2nd ed. New York: Wiley

Fletcher, R.; Marti, U.P. & Redemske, R. (2005). Study of UHF RFID Signal Propagation through Complex Media, *IEEE Antennas and Propagations Society International Symposium*, vol. 1B, July 2005, pages 747-750

Floerkemeier, C. & Wille, M. (2006). Comparison of transmission schemes for framed ALOHA based RFID protocols, *Applications and the Internet Workshops, 2006. SAINT Workshops 2006, International Symposium on*, Jan. 2006, pages 23-27

Glidden, R. & Schroeter, J. (2005). Bringing long-range UHF RFID tags into mainstream supply-chain applications, *RFDESIGN, RF and Microwave Technology for Design Engineers*, www.rfdesign.com

Glidden, R. et al. (2004). Design of ultra-low-cost UHF RFID tags for supply chain applications, *Communications Magazine, IEEE*, vol. 42, pages 140-151

Han, Y.; Li, Q. & Min, H. (2004). System modeling and simulation of RFID, *In Auto-ID Labs Research Workshop*, Zurich, Switzerland

Hashemi, H. (1993). The Indoor Radio Propagation Channel, *Proceedings of the IEEE*, vol. 81, no. 7

IDA. (2008). Infocom Development Authority of Singapore (IDA), IDA TS SRD Technical Specification for Short Range Devices, Issue 1 Rev 3, January 2008, Singapore

ISO Standards. (2007). ISO 18000-6C Standard - RFID UHF Air Interface, *Information technology – Radio frequency identification for item management – Part 6: Parameters for air interface communications at 860 MHz to 960 MHz*

Jin, C.; Cho, S. H. & Jeon, K. Y. (2007). Performance Evaluation of RFID EPC Gen2 Anti-collision Algorithm in AWGN Environment, *International Conference on Mechatronics and Automation*, 5-8 Aug 2007, pages 2066-2070

Kajiwara, A. (2000). Circular polarization diversity with passive reflectors in indoor radio channel, *IEEE Transactions on Vehicle Technology*, May 2000, vol. 49, no. 3, pages 778–782

Karthaus, U. & Fischer, M. (2003). Fully integrated passive UHF RFID transponder IC with 16.7 uW minimum RF input power, *IEEE Journal of Solid-State Circuits*, vol. 38, No. 10, October 2003, pages 1602-1608

Kim, D.; Ingram, M.A. & Smith, W.W., Jr. (2003). Measurements of small-scale fading and path loss for long-range RF Tags, *IEEE Transactions on Antennas and Propagation*, vol. 51, No. 8, August 2003, pages 1740-1749

Leong, K. S.; Ng, M. L. & Cole, P. H. (2006). Positioning Analysis of Multiple Antennas in a Dense RFID Reader Environment, *International Symposium on Applications and the Internet Workshop 2006*, 23-27 Jan 2006, pages 56-59

Mayer, L. W.; Wrulich, M. & Caban, S. (2006). Measurements and Channel Modeling for Short Range Indoor UHF Applications, *Proceedings of The European Conference on Antennas and Propagation*, EuCAP 2006, 6-10 Nov. 2006, Nice, France

Mitsugi, J. & Hada, H. (2006). Experimental Study on UHF passive RFID Readability Degradation, *SAINT Workshops 2006*, pages 52-55

Mitsugi, J. & Shibao, Y. (2007). Multipath Identification using Steepest Gradient Method for Dynamic Inventory in UHF RFID, *International Symposium on Applications and the Internet Workshops 2007 (SAINT Workshops 2007)*

Mitsugi, J. & Tokumasu, O. (2008). A Practical Method for UHF RFID Interrogation Area Measurement Using Battery Assisted Passive Tag, *IEICE Transactions on Communications*, vol. E91-B, No.4, pages 1047-1054

Muehlmann, U. & Witschnig, H. (2007). Hard to read tags: an application-specific experimental study in passive UHF RFID systems, *elektrotechnik und informationstechnik*, vol. 11, pp. 391-396, Vienna, Austria: Springer

Nikookar, H. & Hashemi, H. (1993). Statistical Modeling of Signal Amplitude Fading Of Indoor Radio Propagation Channels, *Proc. of Int. Conf. on Universal Personal Communications*, vol. 1, pages 84-88

Ramakrishnan, K. & Deavours, D. (2006). Performance benchmarks for passive UHF RFID tags, *Proceedings of the 13th GI/ITG Conference on Measurement, Modeling, and Evaluation of Computer and Communication Systems*, pages 137-154

Rappaport, T. S. (2002). *Wireless Communications – Principles and Practice*, Prentice Hall, Second Edition

Rappaport, T.S. & McGillem, C.D. (1989). UHF fading in factories, *IEEE Journal Selected Areas of Communications*, Vol. 7, No. 1, January 1989, pages 40-48

Redemske, R. & Fletcher, R. (2005). The Design of UHF Tag Emulators with Applications to RFID testing and Data Transport, *Proceedings of 4th IEEE Conference on Automatic Identification Technologies*, October 2005

Ross, R.A. (1966). Radar cross section of rectangular flat plates as a function of aspect angle, *IEEE Transactions on Antennas and Propagation*, July 1966, vol. 14, no. 3, pages 329–335

Sato, K.; Manabe, T., Polivka, J., Ihara, T., Kasashima, Y. & Yamaki, K. (1996). Measurement of the Complex Refractive Index of Concrete at 57.5 GHz, *IEEE Transactions on Antennas and Propagation*, vol. 44, no. 1, pages 35-40.

Saunders, S. R. (1999). *Antennas and Propagation for Wireless Communication Systems*, ISBN: 978-0-471-98609-6, 426 pages, 10/1999

SRRC. (2007). State Radio Regulation Committee (SRRC), Ministry of Informatics Industry (MII), P.R.China, 800/900 MHz Radio Frequency Identification (RFID)

Vogt, H. (2002). Multiple object identification with passive RFID tags, *IEEE International Conference on Systems, Man and Cybernetics*, vol: 3, 6-9 Oct. 2002

Wang, L. C. & Liu, H. C. (2006). A Novel Anti-Collision Algorithm for EPC Gen2 RFID Systems, Wireless *Communication Systems, 2006. ISWCS '06*, Sept. 2006, pages 761-765

13

Security and Privacy in RFID Applications

Paweł Rotter
Joint Research Centre of the European Commission,
Institute for Prospective Technological Studies
Seville, Spain
Currently at:
AGH-University of Science and Technology, Automatics Department
Kraków, Poland

1. Introduction

RFID technology raises a number of security and privacy concerns, which may substantially limit its deployment and reduce potential benefits. Public consultations led by the European Commission with citizens, RFID manufacturers, system integrators, academic institutions and public bodies confirm that privacy and security is a major concern (www.rfidconsultation.eu). Features which make RFID especially vulnerable among information systems are:

1. Wireless transmission between tag and reader:
   Most of the attacks on RFID systems described in the next part of this chapter exploit the air interface.
2. The limited resources of the tag:
   The low power supply and small memory of low-cost passive tags limit the extent to which security measures can be applied.
3. The small size of tags:
   RFID tags can be almost invisible,[1] which allows them to be attached to items carried by people without their consent or even their knowledge.

The most common threat is unauthorised access to the data stored on the tag or sent via the air interface. Attackers can achieve this either by reading the tag with an unauthorized reader *(rogue scanning)* or by *eavesdropping* on a legitimate communication. Access to the data on the tag is a threat in itself, but it can also be the first step to other types of attack. For example, in a *replay* attack, the attacker repeats the authentication sequence captured when it was emitted by an authorized tag, and in this way he may usurp the identity of another person. The attacker can also make a *duplicate of the tag,* with has the same functionality. Another threat is the malicious *modification of the memory content* of the RFID tag, with a view to changing attributes reported by the tag or using the tag as a carrier of malware. Denial of service can be avoided by *blocking* (putting the anti-collision protocol in a practically infinite loop) and frequency *jamming*. By *reverse engineering* and *side channel attack,* the attacker may discover algorithms and data on the tag (including the cryptographic key). Moreover,

[1] The smallest passive tags commercially available in 2006 are of size 0.15×0.15×0.0075 mm (Harrop et al. 2008).

protection measures for RFID-based cards are more difficult to apply than for contact cards. Finally, RFID systems may be the subject of attack to backend, like any other information system.

Depending on the application in which an RFID system is commercialized, security and privacy threats should be differently treated. Some applications demand high levels of security (like access control systems) and privacy (like e-documents), while for others, like livestock tracking or some manufacturing processes, these concerns are less important. Also, types of risk depend on the application. For presentation in this chapter, we have selected the set of application areas where the most relevant privacy and security issues arise. (However, where the same issues appear in different applications, we have not tried to discuss all of them.) We have looked especially at those applications which are large in economic terms and involve a large number of users. Detailed criteria are presented at the beginning of Section 3. The four selected application areas are: item-level tagging, electronic ID documents, contactless smart card and RFID implants.

Item-level tagging is foreseen to be the main RFID application in terms of market value and number of tags, and the most pervasive one. The main privacy concern here is unauthorized tag reading. When tagging at item level becomes common, if appropriate countermeasures are not applied, attackers will be able to find out what items a person has in a bag (e.g. what type of medicine), the price and brand of clothes, etc. A set of tags attached to items usually carried by a person may allow his identification and tracking. There are many countermeasures, which can reduce and even eliminate the risk, but just the possibility of massive invasions of privacy and a "big brother" scenario has an important impact on image of RFID and its social acceptance.

Electronic identity documents may use different technologies. Nevertheless, for electronic passports, RFID has been selected, as it is more appropriate for the booklet form of e-passports than, for example, contact smart cards. The combination of two privacy-sensitive technologies - i.e. RFID and biometrics - brings particular concerns about privacy. The main threats are: secret reading of personal data and biometrics, copying the passport, tracking the passport's owner, and theoretically even the construction of a bomb which could be triggered by a passport of a specific nation or individual. Though several security measures have been proposed in the ICAO specification (Basic Access Control, Active Authentication, and Extended Access Control) there is ongoing discussion as to whether the protection they offer is sufficient.

Contactless smart cards and single-use RFID-based tickets increase convenience and efficiency in public transport and allow additional services to be offered. They provide detailed information about traffic patterns which can be used in traffic management (schedule optimisation) and enable new payment plans, like fee per kilometre. Apart from security risks typical to each RFID application based on wearable tokens, privacy is a special issue for public transport applications, since travel patterns of individuals can be recorded and stored in a central database.

RFID implants for identification and authentication of people are probably the most controversial among RFID technologies. They provide a permanent and physical link between the person and the tag. The first implant was approved for commercial use by the FDA in 2004. Since then, about two thousand people were injected with tags, mostly in order to be included in a healthcare information system. This system provides online access to medical record of a patient based on ID number communicated by the implant. In the future RFID implants may have a wide range of applications. However, privacy and security issues, as well as possible health risks, may limit or even stop further deployment of this technology.

Our purpose was not to give a complete discussion of all applications where privacy and security is important, which would be rather repetitive. Instead, we provided four examples, which cover the most of issues. Threats and measures in, for example, access control systems or electronic payment will be similar to those which are discussed here.
In this chapter, we focus mostly on the technical aspects of security and privacy and the technical countermeasures, but there are also legal, social and economic challenges related to security issues. Moreover it is important to bear in mind that security and privacy protection need to be followed by the creation of user trust and awareness. Even a secure system will not be successful if the user's perception of security and privacy protection is low.
This chapter is structured as follows: in Section 2, we present in more detail the threats mentioned above and corresponding countermeasures. In Section 3, we discuss selected applications. We provide a summary and conclusions in Section 4.

2. Threats to RFID systems – state of the art

In this section, we present the threats to RFID and corresponding countermeasures - see Fig. 1. We focus on those risks which are not an issue in other information systems. We do not

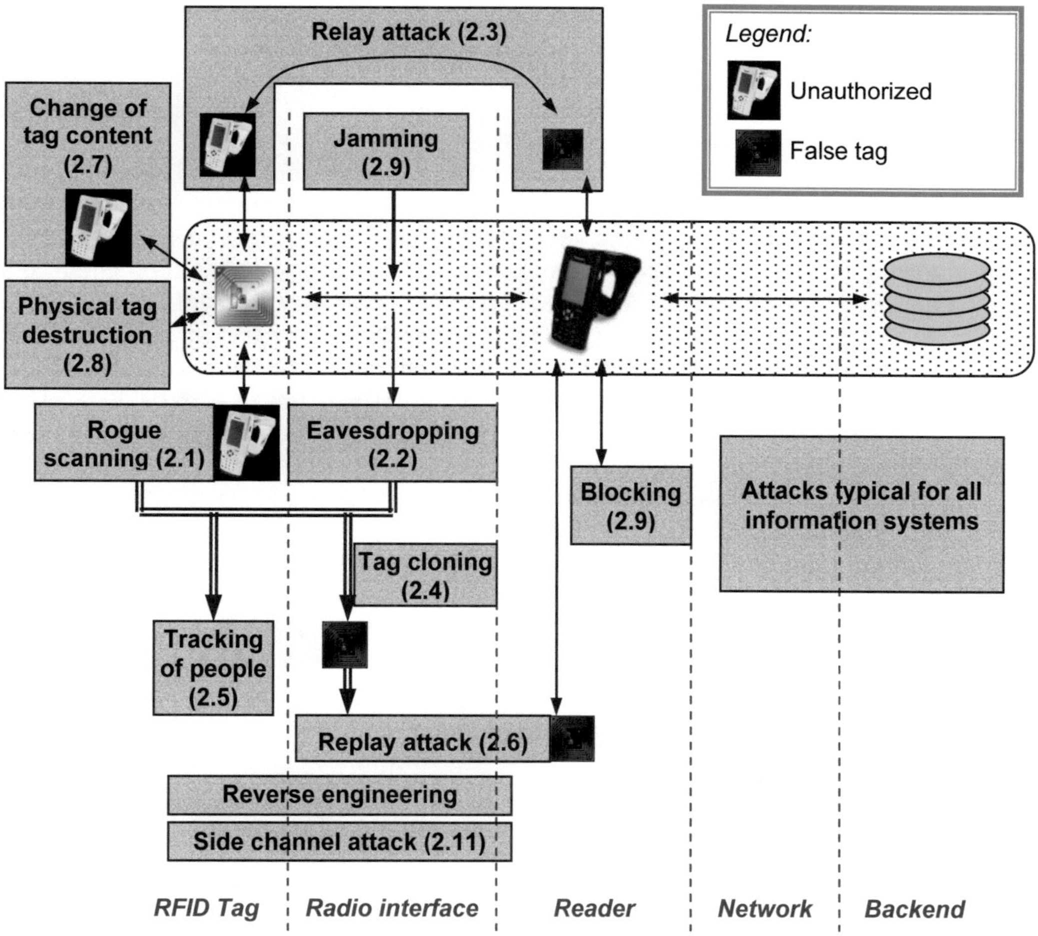

Fig. 1. Threats to RFID systems and number of subchapters where they are discussed

discuss *attacks on the backend* of the RFID system, which are similar to attacks on non-RFID information systems. Exhaustive information about risks and countermeasures in information systems can be found in, for example (Hansche et al., 2004).
It is interesting to observe that one type of attack may be a preparatory step for another one. For example, eavesdropping may enable cloning of the tag; this may then result in a replay attack and the final consequence may be unauthorized access to a restricted area. These kinds of relations imply that a single vulnerability of the system, even if it is not perceived as a problem in itself, may threaten security and privacy in areas which are not directly related to it.

2.1 Rogue scanning

A fake reader can be used for unauthorized reading of information from a tag. The range of a reader may be extended several times beyond the standard communication distance. For example for standard ISO 14443, used in proximity cards like MIFARE and in electronic passports, the standard communication range is 10 cm. Kirschenbaum & Wool (2006) built a "home-made" reader able to operate from 25 cm at a cost of $100. Further extension of the range up to about 35 cm is possible, probably at a similar cost. Fortunately, range increase is not only a matter of reader parameters. Simulations led by Kfir & Wool (2005) show that ISO 14443 cards can be read from maximum distance of 55 cm in the worst-case scenario, where there is only man-made noise and sophisticated signal processing by the attacker. For larger distances, it is not possible to separate the signal from the noise. However, even 25 cm is enough to read a card in someone's pocket.
Using *short-range tags* wherever possible makes rogue scanning more difficult. *Shielding* with an anti-skimming material (e.g. aluminium foil) when the tag is not in use, protects it from scanning. A specific and common countermeasure against unauthorized tag reading is the *authentication of the reader*. Risk can also be reduced by moving sensitive information to a *protected database* in the system's backend. In this case, in order to retrieve information based on an ID number read from the tag, the user must authenticate himself to access the backend part of the system, where authentication methods are not limited by the constraints of RFID technology. However, it should be noted that keeping personal data in a central database is generally perceived as more privacy invasive than when they are kept only on tokens owned by users. Moreover, although the back office can include stronger security than RFID tags, there is always some risk of compromising all the records in one attack. Other concerns related to central vs. local storage are discussed in Section 5.1 of the report (Snijder 2007).
Another countermeasure against rogue scanning is to let the tag send information only when it is *activated by the user* (e.g. by pressing a button), thus the possibility of unauthorized reading is limited to moments when a legitimate communication is demanded. This solution is appropriate for active tags, like car remotes, where the communication can be initiated by the tag. However, for most low-cost passive tags or smart cards, this solution is not practical. Also, in many applications, the full automation of the process is RFID's main asset.
Many privacy concerns can be avoided by *permanent deactivation* of tags which are not going to be used any more. This possibility has been foreseen in the EPC Global standard and will probably become common with the massive deployment of RFID in retail.

2.2 Eavesdropping

Eavesdropping on a legitimate communication is a secret monitoring of data sent via the air interface between an RFID tag and a reader. The attacker does not need to power the tag, which is already powered by a legitimate reader. Because of this, the maximum range for eavesdropping may be significantly larger (for the same type of tag) than for rogue scanning. Eavesdropping is a passive action - the attacker does not emit any signal - and is therefore very difficult to detect.

The most common countermeasure is encryption of data transmitted between tag and reader, so the signal can still be eavesdropped but not understood. There are, however, several challenges. As we mentioned in the introduction, RFID tags have limited resources. In low-cost passive tags, the total number of gates is about 500-5,000 (Weis, et al., 2004) and not more than half of them can be dedicated to security.[2] Realization of advanced cryptographic algorithms requires from several thousand to about 25 thousand gates. Small amount of power that can be harvested by a tag antenna is also a limitation for processing data. Another issue is related to protection and administration of keys. If symmetric cryptography is applied, all tags and readers share the same secret, and there is a risk that it can be retrieved from any tag. Tags are generally not tamper-resistant and even if a cryptographic algorithm is well defined and does not allow an attacker to obtain the key from a communication, there is a risk that the key will be revealed by spying into the manufacturer's documentation, reverse engineering (of tag or reader) or by a side-channel attack. Advanced asymmetric cryptography algorithms are often too heavy for RFID, and neither are they free from problems with key management. Another possible countermeasure is shielding the tag and reader during information exchange. However, this is rarely applied, as it is not very practical. It is also important to use the standard with the smallest communication range sufficient for a given application.

2.3 Relay attack

Relay attack is a type of man-in-the-middle attack (Kfir & Wool 2005), where the attacker creates a connection between a legitimate reader and the victim's legitimate tag, as shown in Fig. 2. From the point of view of the RFID system, the communication looks as if the legitimate tag and the reader are close to each other when, in fact, they are communicating through the communication channel, usually wireless, established by the attacker. In this way, the attacker may authenticate himself in an access control system or a payment system. The maximum distance between a legitimate tag and an attacker's reader (called sometimes a "leech") is the same as in the case of rogue scanning, but the distance between a legitimate reader and an attacker's device which simulates a legitimate tag ("ghost") is much longer - up to 50 m. A successful relay attack against an RFID system complying with the ISO 14443A standard has been proven to be feasible (Hancke 2005).

Since the attacker only re-transmits information, without the need to understand it, the authentication protocol (e.g. challenge-response) does not protect against this kind of attack. This threat can be countered by using short range tags and by shielding tags (e.g. by keeping them in bags containing aluminium foil, when not in use). There is also a specific countermeasure against relay attack - distance bounding protocol - which estimates the distance between the reader and the tag, based either on response time (Hancke & Kuhn, 2005; Reid et al., 2006) or signal-to-noise rate (Fishkin & Roy, 2003).

---

[2] The number of gates in tag increases from year to year but still memory and power harvested by the antenna are strong limitations to the security on the tag side. In most applications the manufacturers focus rather on reduction of tag costs than increasing memory size.

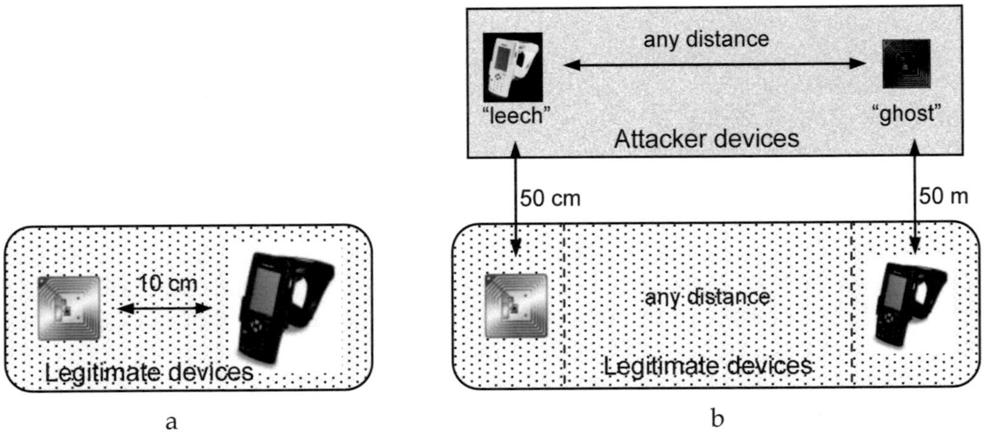

Fig. 2. A legitimate communication (a) and relay attack (b). Maximum ranges refer to ISO 14443 and are based on theoretical results received by Kfir & Wool (2005)

2.4 Cloning the tag

'Cloning' means making a duplicate of an RFID tag. A clone may be similar in form to the original or be a larger device with the same functionality. Duplicates can be used to access a restricted area, abuse private data or make an electronic transaction on behalf of a victim.

Cloning can be prevented by the use of cryptographic methods for authentication of the tag. If a challenge-response protocol is used, information which can be obtained by the attacker using the air interface (e.g. by eavesdropping) is not sufficient to duplicate the tag. Although reverse engineering, in theory, may allow duplication of any electronic circuit, these methods require special equipment and a very high level of knowledge. Moreover, there are countermeasures which can be applied at the circuit manufacturing stage.

Authentication of the tag should be based on well established cryptographic algorithms, which are constantly analysed by researchers. Although their security has not been mathematically proved, it can be assumed that their vulnerabilities are well known. The use of proprietary methods, where security is supposed to be based on secrecy of the algorithm, is generally not recommended. There are at least several examples where RFID authentication protocols, developed in laboratories of big companies, have been cracked. The best known cases are the cracking of Digital Signature Transponder (Texas Instruments) and of MiFare (Philips), described in Section 3.3. On the other side, looking at almost twenty years of contact smart card history, we cannot agree with popular opinion that security should be based only on the secrecy of the key. Especially when it comes to chip design, public chip schemes would make it much easier to retrieve the key directly from the circuit and therefore manufacturers make a considerable effort to hide the structure and mislead those who try to discover it (see section on reverse engineering).

Another frequent reason for security gaps (in the two cases mentioned and many others) is too short encryption keys. Short keys mean lower power consumption and lower cost, so manufacturers try to use the shortest keys which, at the moment, seem safe. However, the lifetime of a solution like this is often longer than foreseen and, due to progress in technology, the size of the key is no longer sufficient. Unfortunately, when the system is already deployed on a large scale (like DTA and MiFare), the cost of security updates is enormous.

2.5 Tracking of people

Tracking of people takes place when an attacker follows the movements of individuals through the RFID tags they carry with them. Tracking can be performed with rogue readers placed, for example, in doors, or by the deployment of eavesdropping devices in the proximity of legitimate readers.

Many countermeasures to reduce the risk of tracking have already been mentioned, like using short range tags, shielding them, authentication of readers and disabling tags when not used. However, we can foresee that, in the future, people will carry many RFID tags with them and therefore a personal device which controls access to them, possibly integrated in their mobile phones or PDAs, may be very useful - like the one proposed by Rieback et al. (2005). There are also countermeasures which can be implemented at tag-design stage, such as: pseudonyms (changing identifiers) or estimation of distance from the reader (Garfinkel et al. 2005).

2.6 Replay attack

In the case of replay attack, the attacker abuses another person's identity by repeating the same authentication sequence as the one provided by an authorized person. A replay attack may be led by a clone of the legitimate tag or by re-sending the eavesdropped signal from a PC equipped with an appropriate card and antenna.

In order to perform a replay attack, an attacker has to obtain some information which is sent by the tag during normal communication. The first line of defence is therefore to counter eavesdropping and unauthorized tag reading. A specific countermeasure against replay attack is authentication of the tag e.g. with a challenge-response protocol. If the protocol is well designed, the key necessary for calculation of response cannot be deduced from information exchanged through the air interface.

2.7 Malicious change of the tag content

As a result of malicious change of the tag content, the attributes of an item described by the tag may be distorted or an authorized person may be falsely rejected by the access control system. Furthermore, writable tags may become carriers of malware, e.g. data on RFID tag can be maliciously modified in such a way that they are interpreted by the system as a command. An example of a successful attack of this type is the SQL injection described by Rieback et al. (2006).

In some writable tags, memory content can be protected by temporarily or permanently disabling writing capability (*'lock'* and *'permalock'* functions in standard EPCglobal Class 2 Gen 2). Malware on RFID tags cannot affect the system if the implementation excludes the possibility of interpretation of the tag's data as a command. This is similar to switching off macros in MS Office which protects the system from running malicious code embedded in documents.

Using sophisticated equipment, like a focused ion beam, it is also possible to change the content of memory (EEPROM or ROM) in non-writable tags. This technique can be used to set a secret key to a known (zero) value, but it also requires that the location of the key in memory is known, expensive equipment, a high level of knowledge and considerable effort. In high security applications, measures like protective layers on chips and memory scrambling make this kind of attack impractical.

2.8 Physical tag destruction

Physical tag destruction, e.g. by heating in a microwave or hitting with a hammer, is the easiest and the cheapest way to disrupt RFID systems. This is a particular issue for applications where RFID tags are used not only for identification purposes, but also for the protection of items against theft, like in retail or in libraries. RFID tags in e-passports can be destroyed by owners who have concerns about possible abuse of their privacy - especially as an e-passport with a non-working RFID tag is still valid (Wortham 2007).

2.9 Blocking and jamming

Blocking is performed with a 'blocker' tag, which simulates the presence of an enormous number of tags and causes a denial of service (non-ending interrogation of physically non-existing tags by the reader). However, blocking may also be a useful mechanism and serve, as originally proposed, for the protection of consumer privacy, when a blocker tag protects from unwanted scanning (Juels et al. 2003). Another threat to the air interface is jamming, which paralyses the communication of an RFID system by generating a radio noise at the same frequency as that used by the system.

Blocker tags and jamming devices are easy to detect and localize immediately after starting operation and appropriate warning functionalities can be built into a system.

2.10 Reverse engineering

The term 'reverse engineering' is usually used for invasive methods of discovering circuit structure and even values of voltage at different points of the circuit during its operation. The goal is to retrieve the algorithm or the cryptographic key, often with the final purpose of copying the tag. This kind of attack requires a high level of knowledge and experience, as well as specialized and expensive equipment, like micromanipulators, focused ion beams, laser cutters, microscopes and chemical etching equipment.

The manufacturers of contact smart cards apply a wide variety of measures, which can also be used in contactless solutions, although with some limitations resulting from limited power supply. Typical measures are: dummy structures which do not have any function except to mislead attackers, scramble buses and memory cells, form protective shields on the top of chip (especially memory) and encrypt memory content. Active protection is also possible: sensors included in the circuit can detect symptoms of attack like change of voltage, clock frequency, temperature, etc. - for details, see Chapter 8.2.4 of a monograph (Rankl & Effing 2004). Due to resource limitations, RFID-based cards allow only limited protection and especially active methods are rather beyond this limit.

There are also methods of reverse engineering at the logical level, without any physical manipulation of the circuit. For example, details of the algorithm used in DST were discovered from a general outline which was published, together with observed challenge-response data for different values of the key, which could be arbitrarily set on blank tokens available from the manufacturer.

2.11 Side channel attacks

Channel side attacks are based on information gained from physical implementation of cryptosystem, like power consumption, time of computations or electromagnetic field (Bar-El 2003). *Power analysis attack* is based on the fact that different operations consume different power. Analysis of power changes can provide information which, combined with other

cryptanalysis methods, can help to recover the secret key. In *timing attack,* the attacker analyses time needed to perform operations. For example, in straightforward implementation, PIN comparison is done byte by byte and returns no-match result after the first difference. Based on time, it can be deduced which byte caused the rejection of a PIN number and a guess can be made, byte by byte. Analysis of the *electromagnetic field* around the chip during its operation is more difficult for RFID than it is for contact chips, because of the interference with a stronger field which comes from the communication with the reader. However, as shown in (Carluccio at al. 2005), after separation of the antenna from the chip, the electromagnetic field generated by operation of the chip can be analysed.
A basic countermeasure against side channel attacks is to design hardware and software to keep power consumption steady and ensure that the time taken by calculations does not depend on data or partial results of the operations. This can be achieved by avoiding conditional execution of any part of the code, even if the result of the calculation is not going to be used. In hardware design, manufacturers can add dummy registers and gates, which balance the consumption of energy but, again, resources for this kind of measure are very limited. An exhaustive list of references on side channel attacks can be found at http://www.crypto.ruhr-uni-bochum.de/en_sclounge.html.

3. Discussion of selected applications

In this section, we will discuss the application areas which we found especially important and sensitive to privacy and security threats. Our selection is based on several criteria:

- The importance of the application in terms of economics (market value, number of tags) and social impact (number of users, social implications).
- Security and privacy-related criteria, proposed in (Rotter 2008):
    - Range of deployment of the system
      In systems operating locally within a restricted area, information between readers and the backend of the system is exchanged through a local network. Applications of this type, like some manufacturing processes or access controls, are generally less sensitive to security risks, as the physical security of the place is the first barrier to attacks. At the other extreme are global systems, where breaking security gives access to the data on millions of tags worldwide, or to a central database.
    - Type of link between an RFID tag and identity-related data
      Privacy risks only exist in systems where it is possible to establish a link between the RFID tag and the identity of a person. Systems where it is not possible to link a tag to the identity of a person, for example most industrial and livestock tracking systems, do not raise any privacy concerns. In item-level tagging for example, or in anonymous tickets in public transport, a tag can be temporarily linked to identity. In some other applications, this link is fixed and defined in the system – like e-Passports, payment systems, (e.g. Speedpass) and personal tokens for access control. Future applications of this type include credit card systems, location-based services and mobile phones equipped with Near Field Communication. Finally, systems based on RFID implants are the most privacy-sensitive as the link between a person and an RFID tag is physical and not very easy to remove.
    - Demand for security
      Demand for security depends mostly on two factors: a) the size of potential damage, in terms of loss of money, loss of customers or, for example, disclosure of

privacy-sensitive information, and b) the level of motivation of attackers, related to the potential prize they could win if they are successful. These two factors are often correlated but not always: for example, in medical information systems, wrong treatment may cause serious damage. In general, however, attacker motivation is much lower than it is, say, in payment systems or e-passports.

In the case of security (not privacy)-demanding applications, we pay more attention to the public sector, as we believe that the business sector will more easily find a proper balance between expenses for security measures and losses caused by insufficient security.

- Coverage of the most relevant issues related to security and privacy in the set of selected applications.

We do not offer a complete overview of all the application areas where privacy and security is relevant - for example, we do not discuss e-payment and access control. However, the privacy and security issues in these areas are similar (at least qualitatively) to those related to transport or other presented applications.

3.1 Item-level tagging

RFID is becoming very popular in logistics and the supply chain (Bose & Pal, 2005), where it is employed as a kind of barcode with new, very desirable features. For example, unlike printed barcodes, RFID tags do not have to be in line-of-sight to be read, and they enable multiple scanning (e.g. the whole truck or basket at once) allowing for further automation in many industrial processes. In contrast to a barcode, which replicates an identification number only, tags may contain other information e.g. product details or, if combined with sensors, the history of storing conditions, mechanical shocks, etc.

Threats to the privacy and security of users

Item-level tagging brings privacy threats, which may limit its deployment. RFID tags attached to objects people have bought can be interrogated by someone to reveal what items they have in their shopping bags (including, for example, medicines) or the prices they paid. Moreover, although the set of things a person carries changes, it does not usually change completely. Such a set, called the "RFID shadow" or "RFID constellation" of a person (Garfinkel et al., 2005), if regularly updated, may serve to effectively track that person. RFID tags used for retail cannot be read from more than several meters, even if the standard reading distance is extended by a more powerful reader. However, if attackers placed readers at the entrances of shops, metros, airports, etc., they would be able to track individuals. This possibility has raised concerns for some privacy organizations and individuals, like those presented in (Albrecht, McIntyre 2005).

Moreover, there is a potential risk of physical attack on a specific individual, based on his/her automatic identification. In the case of electronic passports some attention has been paid to the possibility of constructing a bomb triggered by information received from the RFID chip in the e-passport of a specific person or citizen from a specific nation ("American-sniffing bomb"), see e.g. (Juels et al., 2005). An RFID constellation could be used in a similar way and some features of tags used for item-level tagging make them even easier to exploit for potential attackers. First, they have a longer standard range, typically 30-70 centimetres, compared with 10 cm for the standard 14443A tag used in e-passports. In both cases, the standard range can be extended: for e-passports to about 30-40 cm, but for tags used in retail considerably further. Second, the e-passport has security protection mechanisms, which make unauthorized identification of the owner more difficult, which are not included in tags used in retail. Another concern of some consumers and privacy organizations is 'function creep', i.e. using a large amount of data obtained by RFID systems for different purposes than original ones intended by the system. For example, the data collected by retailers could be used for unsolicited targeted advertising, customers could be discriminated against on the basis of their purchase history, and the police or intelligence agencies could request the data.

Fig. 3. The consumer privacy problem. Privacy concerns around RFID and the vision of society under surveillance may significantly influence future deployment of item-level tagging. Source: (Juels 2006)

Concerns about privacy and security are the main reason for low public acceptance of item-level tagging. Even the big retailers, which for economic reasons are definitely interested in quick deployment of RFID, must consider public opinion. Benetton's plans to attach RFID tags to items of clothing caused a boycott of the company's products, organized by CASPIAN[3] (http://www.boycottbenetton.com). Protest campaigns have been organized against some retailers - for example, WalMart. Undoubtedly, the concerns of consumers and their low acceptance of RFID in item-level tagging have slowed down its deployment. Another important implication of privacy and security issues for the RFID market is the need for the application of technical and legal measures, which make RFID (both single tags and whole systems) more complex, and therefore more expensive.

On the other hand, the demand for security can be seen as a market opportunity. Apart from the need for security to be built into RFID systems, we can foresee the demand for personal devices which help the user to keep control over the tags he owns. Such devices, for example the RFID guardian mentioned in the paragraph on countermeasures, can be integrated into mobile phones or PDAs.

Security threats – the retailers point of view

Item-level tagging is related to a number of privacy concerns, but there are only a few threats related to system security. An attacker who can change the memory content of an

[3] Consumers Against Supermarket Privacy Invasion and Numbering

RFID tag can modify information about the product. This action could falsify the price of the product and this could lead to small fraud or, if maliciously applied on a large scale to all products in a supermarket, could cause considerable losses. Writable tags, even those as simple as EPCGlobal tags, can be carriers of malware (e.g. SQL injection). Physically destroying the tag, or tearing it off the object, is the simplest and the cheapest way to disrupt RFID systems. This vulnerability may be exploited when an RFID system is used to protect items against theft. Blocking and jamming are threats to the air interface and may result in paralysing RFID system communication.

Generally, the demand for security in item-level systems is not very high and the risk is mostly related to material losses on the part of retailers, which are able to apply corresponding countermeasures and ensure an adequate level of security at reasonable cost.

Countermeasures

The basic security measure against unauthorized reading of RFID tags attached to items is deactivation of the tag at the supermarket check-out. A "Kill" command, foreseen in EPCGlobal standard (EPCglobal 2004), permanently and irreversibly disables the tag. Another method, which gives full control over deactivation to the user, is a design of tag which facilitates its easy mechanical destruction by the owner (Karjoth & Moskowitz 2005). Unfortunately, deactivation of the tag also disables post-sales services. For example, clothes tagged with RFID could automatically set the appropriate programme in a washing machine, a refrigerator could be "aware" of its content and report what kind of food should be bought (or even make an order on the Internet), and microwaves could prepare food according to instructions. If tags are deactivated when products are sold, none of this would be possible. A "killed" tag cannot be used if the item is returned to the shop or if the product is recalled, which can be essential for some products. For example, a tracking capability which facilitates recall in the case of safety defects is one of the main drivers for the introduction of RFID in tyres (Garfinkel et al., 2005). Disabling of tags after item purchase will also squander the chance to use RFID for automatic segregation of waste and recycling.

Researchers have therefore proposed several methods which give the user full control over the tags in his possession, so it is not necessary to deactivate them. RFID guardian, proposed by Rieback et al. (2005), is a device which the user carries with him, possibly embedded in mobile phone. It allows tag information to be read only if the user agrees and warns him about unauthorized reading attempts. However, this device has not been commercialised as yet.

In addition to technical aspects, legal privacy measures should also be applied. For example, retailers should be obliged to give customers at least the option to deactivate tags, and to mark places where RFID readers are operating with special signs.

3.2 Electronic identity documents

In order to make the identification of people more resistant to falsification, faster and more convenient, there is a need to store the data on identity documents in a form which allows automatic reading. Different technologies are used for this purpose, like cards with magnetic strips, contact smart cards or even optical memory, like in Italian ID cards. Although these technologies are not as convenient as RFID, privacy and security aspects and the low acceptance of RFID technology are sufficient arguments against its use. The situation is different in the case of electronic passports. The booklet form of the passport makes the use of contact solutions difficult. On the other hand, although the air interface of

RFID creates potential threats, this technology, due to data processing on chip, allows for much more sophisticated and robust security measures than, for example, magnetic or optical data storage. RFID-based e-passports have been recently introduced in many countries, including all the European Member States. Each e-passport contains personal data and a digital photo of the owner. The second generation (introduction in European Union is planned for 28 June 2009) will include also fingerprints. In the future, other biometrics, especially iris data, could be added.

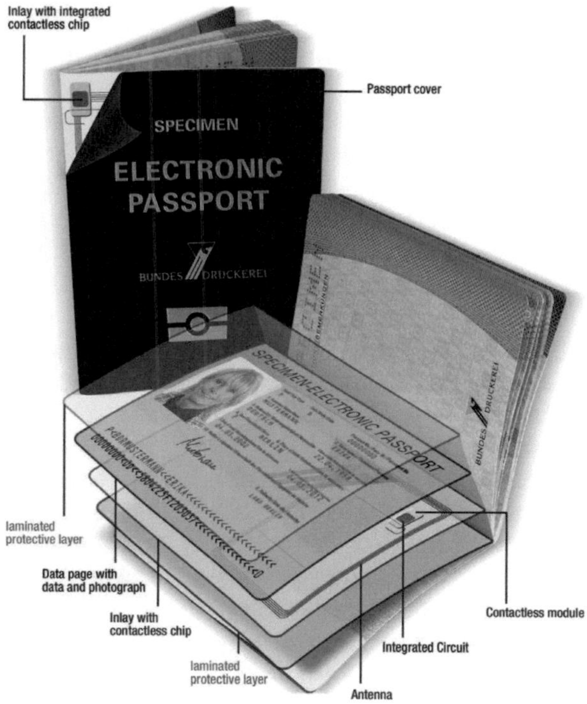

Fig. 4. The Physical form of an electronic passport is the same as a traditional one. Passports with RFID are marked with a sign " "on the cover. Source: Bundesdruckerei GmbH

Privacy and security of electronic passports

Personal and biometric data are particularly sensitive and the possibility of unauthorized access to these data by rogue scanning of passports in owner's pockets, or eavesdropping at border checkpoints, is a major privacy concern. The maximum range for rogue scanning is about 30 cm, while for eavesdropping it is at least several meters.[4] Another privacy threat is people tracking with extended-range readers built, for example, into door frames. Identification of the owner of a passport or of an issuing country might lead to the construction of a bomb triggered by the proximity of citizens with defined nationalities (Juels et al., 2005), see video at: http://pl.youtube.com/watch?v=-XXaqraF7pI.

---

[4] E-passports are based on standard ISO 14443, details on maximum range for scanning are discussed in section *Rogue scanning*. At the Black Hat 2005 Security Conference in Las Vegas, NV, a company called Felixis, demonstrated eavesdropping from over 20 m (Thornton et al. 2006).

Countermeasures

The standard security mechanisms offered by electronic passports is called Basic Access Control (BAC). The data printed on the last page of a passport (passport number, expiry date, name and date of birth of the owner) are scanned at the checkpoint and, on the basis of this data, the 128-bit key is calculated. The size of the key would be sufficient (the National Institute of Standards and Technology recommends 112 bit as safe till 2015) but the information which serves as a base for key calculation has limited entropy. Moreover, the data on the last page of a passport are interrelated, e.g. the passport number is related to issue date. As demonstrated in (Hoepman et al., 2006), the total entropy of the key can be decreased to merely 41 bits (an example which has been calculated for Dutch passports), which is definitely not sufficient.

Storing fingerprints in e-passports will require stronger security than BAC. As recommended by ICAO, fingerprints in European passports will be protected by Extended Access Control (EAC), which is based on asymmetric cryptography. EAC includes the authentication of both the passport and the reader and limits access to additional biometrics (other than face image) to countries which have an agreement with the issuing country, see (Gemalto, 2007) for details. Generally, EAC offers strong security but it has some weak points:

- Additional biometrics will be used only to authenticate citizens of "friendly" countries (authorized by the issuing country). Border controls, if any, between such countries are usually not very strong anyway. Identification of citizens of other countries will not be facilitated by additional biometrics.
- As an e-passport contains a passive chip, it does not have an internal clock and must rely on date information received from the reader. Therefore it cannot effectively verify if the reader's certificate is up-to-date. According to the standard, passports should keep the date sent by the reader in the most recent authentication, which mitigates the problem but does not solve it definitively.
- Revoking the authorization of a reader to read e-passports is technically impossible. This means that a stolen reader will keep its certificate until it expires. Even after this, it is possible to use the reader to read passports which had not updated the date after the expiry date of the certificate.[5]
- As noted by (Hoepman et al., 2006), shallow certificate hierarchy makes it difficult to use e-passports for many applications (problems with the management of certificates). On the other hand, reserving additional biometrics exclusively for border control facilitates user privacy.

As regards the tracking of people with rogue readers, the attacker would either need to break BAC security or use the tag identifiers which are part of the anti-collision protocol. The second possibility, discussed in (Hoepman et al., 2006), can be relatively easily eliminated by using a random number as an anti-collision identifier. Still, the information exchanged between the RFID tag and the reader *before* authentication allows, in many cases, the identification of the issuing country.

---

[5] Additional measures are possible: the certificate does not need to be kept physically on the reader but can be sent to the reader though a secure connection when it is needed. It must be also noted that a stolen reader would not be sufficient for secret scanning anyway, as BAC is additional barrier.

At the moment, Basic Access Control seems to be a weak point in e-passport security. The introduction of Extended Access Control will not solve this issue, as BAC will remain the main way of protecting access to personal data and digital photos. Simple ways of increasing the security of BAC, as proposed by Avoine et al. (2008), are: a) the introduction of progressive time delay[6] when several queries are received in a short period of time and b) increasing the entropy of BAC keys by random numbering of passports and by filling in the optional (usually not used) field on the last page of a passport with a random number. Apart from sophisticated cryptographic measures, shielding seems a simple, effective and inexpensive solution. It has already been introduced in the United States: one passport cover contains the chip and the other contains anti-skimming material, so the passport cannot be read when it is closed. Common introduction of shielding in e-passports would substantially increase the level of security. In general, as pointed out in (Snijder, 2007), there is a need for an integrated approach to privacy and security for e-passports, harmonized at international level.

Deployment of electronic passports is still in the early stages. They have demonstrated some vulnerabilities, which should be improved. On the other hand, it is also important to understand the security offered by electronic chips in the broader context (Kefauver, 2007). A single instance of the vulnerability of RFID in passports does not necessarily imply the vulnerability of the whole system. For example, though data from the chip can be copied relatively easily, they cannot be easily modified. The use of biometrics will therefore ensure that a clone will not be very useful for illegal border crossing. RFID and biometrics are additions to security measures used before and there is no doubt that the introduction of RFID substantially increases overall security.

3.3 Transport

The first widespread applications of RFID are related to cars. Remote control devices that open/close cars are nothing other than active RFID tags. Immobilizers, a fairly efficient way protecting against theft, are RFID passive tags embedded in a key, which communicates with the car reader to authenticate a key. Tags mounted in cars allow automatic collection of tolls. The Speedpass System facilitates fast payment at ExxonMobil petrol stations and McDonalds in the US (Garfinkel & Rosenberg 2005, chapter 10).

Contactless smart cards and single-use RFID-based tickets have been used for several years in mass transport, making it more efficient and effective. The throughput of passengers through metro gates has increased considerably in cities where RFID-based travel cards are used. Precise data about travel patterns help to optimize the schedule and number of vehicles to increase the system performance. Contactless cards make a big difference to convenience for passengers: it is much easier and faster to pass a metro gate or to cancel a ticket on a bus, if they do not even need to take the cards out of their wallets. At the end of 2007, in a trial programme, Oyster cards were built into mobile phones. Introduction of RFID creates opportunities for new services, like e-purse, rental of bicycles, and facilitates the use of special offers (e.g. holiday tickets). As such systems provide exact information about routes taken by each passenger, they enable new payment schemes, like for example payment per kilometre. Finally, RFID systems, if properly implemented, can provide high reliability and promise a more efficient fight against fraud.

---

[6] With upper bound, to prevent denial-of-service attack

Privacy concerns around RFID use in public transport

Privacy concerns about tracking of people through rogue scanning or eavesdropping in proximity of legitimate readers are similar to those which apply to electronic documents. Cards used in ticketing have similar range (10 cm with a standard reader) and work on the same frequency as e-passports, so we can expect that attackers would have the analogical maximum ranges of about 30 cm (theoretically up to 50 cm) for rogue scanning and several meters for eavesdropping.

There are some concerns about data which are legally collected by public transport companies. RFID systems provide precise data about each passenger's travel trajectories, which are kept in the system for some time, e.g. 8 weeks in the case of the London system. Although these data are considered confidential, the fact of their collection raises consumer worries about potential abuse. The Metropolitan Police regularly request journey information about Oyster card users. The information has been used as an investigative tool to track movements of criminals; however the rapid increase of the number of queries has attracted press attention (7 requests in the whole year 2004, 61 in January 2006 and 243 in March 2006). On the other hand, it seems that most users do not mind their travel data being collected since the convenience, lower prices and additional services compensate for this. In the Oyster system, users can choose between personalized and anonymous cards, which do not allow direct assignation of travel trajectories to a passenger name. In practice, many more people choose personalized cards, as these provide more services.

Security issues

As previously mentioned, the use of proprietary solutions may cause security gaps in the system. Nevertheless, due to the limited resources of RFID tags, many companies try to develop their own security algorithms, in order to provide security at lower computational or memory cost than well known and researched solutions. This was the case of the Digital Signature Transponder (DST), used in many immobilizers, for example in Ford and Toyota cars and in the Speedpass system. In 2004, researchers from the John Hopkins University and RSA Laboratories managed to break DST security. They used a general outline of the algorithm published on a website by a Texas Instruments researcher and found out the details by reverse engineering.[7] Having discovered the algorithm, they were able to break a 40-bit key in a brute force attack based on two input-output pairs (Juels 2005), see Bono et al. (2005) for details.[8] The story does not imply that systems which use DST with 40 bit keys are entirely unsafe. The challenge-response protocol of tag authentication is only one of several layers in car anti-theft protection and in Speedpass security. Moreover, cracking requires specialized equipment and knowledge, while most car thieves are opportunists. On the other hand, the level of protection is undoubtedly significantly lower than intended by the developers.

Another successful attack against proprietary encryption was reported at the beginning of 2008. Researchers were able to recover, in an algebraic attack, a 48-bit key used in the MiFare Crypto-1 algorithm. This algorithm has been implemented in about one billion RFID tags, mostly in public transport: London Oyster Card, Dutch public transport OV-

---

[7] They use so-called blank tags – tags where a secret key is programmable, and analysed authentication sequences with different key values. They did not use any invasive methods.

[8] Some photos and videos are available at: http://www.carthiefstoppers.com/About-RFIDs-and-the-Texas-Intruments-DST.html

Chipcard and Boston Charlie Card, and also some access control applications. According to preliminary results, published in (Courtois et al., 2008), the attack can take only several minutes and can be based on a single eavesdropped transaction. Although the researchers published only general information and the details needed for a repetition of the attack were not revealed, it is highly probable that they will be discovered and used in a malicious attack soon. Public transport systems, built at high cost with the promise of fraud reduction, may even increase it. Moreover, the fraud can be more troublesome, as free journeys with cloned cards would be charged to the accounts of particular passengers.

The security issues described in this section apply largely to other application domains, especially access control systems and electronic payment.

Countermeasures

Cases like DST and MiFare Classic show that security measures applied at the production stage may suddenly become insufficient. Unfortunately, if the system based on an insecure solution has already been developed, it may be extremely costly to upgrade it, especially if the security gaps exist at tag design level. Therefore, special attention should be paid at the manufacturing stage in order to avoid errors like:

- Security gaps in proprietary encryption algorithms.
- Insufficient key size - this can be long enough while the tag is being designed but, due to technological progress, become too short after several years.
- Insufficient key entropy - for example, a 32-bit nonces used in MiFare Classic has, in fact, only a 16-bit entropy, due to a weakness in the pseudo-random generator (Nohl & Plötz, 2007).

If security gaps are reported when the system has been already developed, there are still solutions which can help to make it more secure, and avoid the need to immediately replace the tags. De Koning Gans et al. (2008) propose the use of strong encryption in the backend and the storage of encrypted information only on the tags. In any case, systems should not rely only on the security of the tag and it is important to include fraud detection in the backend, as has been done in the DST-based Speedpass system.

In order to ensure privacy a number of privacy-enhancing technologies (PET) can be applied, like those proposed by Heydt-Benjamin et al. (2006). However, they make public transport systems even more complex and costly, and it seems that, in the near future, the main goal of developers will be to reduce costs and decrease organizational complexity by providing security at the minimum level necessary, rather than to deploy advanced PET methods.

3.4 RFID implants

RFID implants are passive tags implanted under the skin, to provide a means of personal identification. As they operate without a battery, they can be operational for many years once implanted. The use of RFID implants for the identification of people provides some advantages compared to established methods. The identification process is practically immediate and fully automatic - and therefore extremely convenient: the user is not required to take any action. Implants cannot be lost, stolen or forgotten. They are a reliable method of identification, especially when compared to biometrics, where due to the statistical nature of the matching process, there is always some error probability. Implants are more durable than tokens and many types of biometrics, which usually change during a person's life. RFID implants can be used by everyone without exception, including people with cognitive impairment. The user can always be identified, even if he is unconscious or not carrying any identity documents.

Present commercial applications

In 2004, the first and, until now, the only RFID implant - the VeriChip - obtained approval from the U.S. Food and Drug Administration. The VeriChip implant (www.verichipcorp.com), which stores only an identification number, can be read from a distance of about 10 cm with a handheld reader and 50 cm with a door reader. The ID number is long enough to identify uniquely everybody in the world. Other data related to the owner are not stored in the implant, but in a centralized database.

Fig. 5. VeriChip implant (original size 3×13 mm). Electronics is encapsulated in bioglass in order to make it biocompatible.

The first commercial application, called VeriMed, is designed to identify patients in healthcare. An authorized doctor can access a patient's medical files through a password-protected website, using the patient's ID number which he can get from an RFID reader (www.verimedinfo.com). Another commercial application based on the same type of implant is VeriGuard - a system for access control to physical spaces like, for example, offices. Implants are also used in entertainment: for example, members of the Baja Beach Club in Barcelona and a Rotterdam club, who have RFID chips implanted, benefit from a quicker VIP service. To date, about 2,000 people have been implanted with VeriChip tags. RFID implants can be potentially used in the future for identification and authentication in many application areas, either as the only ID technology or in combination with other methods - a detailed discussion can be found in (Rotter et al., 2008).

Security and privacy concerns

In spite of their high potential, RFID implants raise some serious concerns, largely related to security and privacy. The permanent and physical link between an RFID tag and a person makes RFID implants more susceptible to privacy risks than any other kind of contactless tokens. The user can be identified any time, without his consent or awareness. Here, the tracking of people, already a concern in item-level tagging and e-documents, is much easier, effective and more difficult to prevent. RFID implants used for authentication are particularly vulnerable to coercive attacks, where attackers force authorised users to provide their credentials. RFID implants carry the risk of physical harm, as attackers could cause injuries by extracting the implants from the victims' bodies. For this reason, the use of RFID implants for secure authentication is questionable, regardless of any technical security solutions. It is even argued that implants should not include high security in order to make their extraction by an attacker unnecessary (Halamka et al., 2006). However, lack of security reduces the reliability of the identification and therefore limits possible areas of application. In addition, RFID implants, especially in their current form, are susceptible to cloning and replay attacks - for a detailed description of VeriChip cloning, see: http://cq.cx/verichip.pl.

Security measures for present and future use of implants

Their lack of an internal power source and the small size of their antennae limit the processing power of RFID implants. It is therefore difficult to include advanced authentication methods in their design. Currently deployed RFID implants do not include even basic security. The tag, when interrogated, sends back an identification number without any type of encryption. It is, however, possible to include some security measures like encryption of the identification number and authentication of the reader. There are examples of individuals who have been implanted with RFID tags, which were originally manufactured for industry or supply chain purposes, and are equipped with cryptosecurity features (Graafstra 2006).

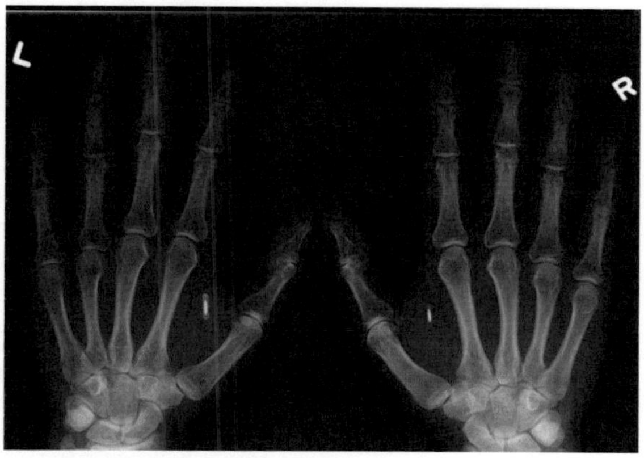

Fig. 6. An X-ray of Amal Graafstra's hand. The chip in the right hand is a Philips HITAG S 2048 and is equipped with crypto-security. Source: http://www.amal.net/rfid.html

Implants, even if not equipped with strong security features (authentication protocol), can be used as an element of access control systems safely, increasing the security and the efficiency of the overall system. Combined in multimodal systems, they protect against spying for passwords or against stealing tokens. In systems with authentication based on passwords and tokens, implants as an additional modality counteract unauthorized delegation of privileges to colleagues. In secure environments, implants could be used for continuous detection of presence in the sense that access (e.g. to control boards, or computers) is blocked immediately when authorized people leave and then can be re-established through other, more secure authentication methods. In any case, when strong security is required, implants should be used only as an *additional* means of authentication.

It is worth noting that security and privacy issues are not the only concerns related to implants. Social acceptance of implants is, at the moment, very low. Unclear health implications, especially the possible relation between implants and cancer (Lewan, 2007; Wustenberg, 2007; Rotter et al., 2008), understandably limit the number of people who would like to use them and may even stop their further deployment.

4. Conclusions

Concerns about privacy and security may limit the deployment of RFID technology and its benefits, therefore it is important they are identified and adequately addressed. System developers and other market actors are aware of the threats and are developing a number of countermeasures. RFID systems can never be absolutely secure but effort needs to be made to ensure a proper balance between the risks and the costs of countermeasures.

The approach taken to privacy and security should depend on the application area and the context of a specific application. In this chapter, we selected and discussed four application areas, but there are many others where privacy and security issues are relevant. In Table 1, we list the main threats and the application areas in which they arise.

Threats	Application areas	Main countermeasures
Rogue scanning of confidential data from personal documents	E-documents, e-payment, mass transport, access control cards, healthcare	Using short-range tags, shielding, authentication of the reader, moving sensitive information to a protected database, activation of tag by the user
Rogue scanning of data of items carried by a person	Item-level tagging (retail)	Permanent deactivation ("kill" command), RFID privacy management devices like RFID guardian
Eavesdropping of confidential data from personal documents	E-documents, e-payment, mass transport, access control cards, healthcare	Data encryption, using short-range tags, shielding tags with reader during information exchange
Rogue scanning or eavesdropping of other non-public data	Logistics, administrative process, industry	Data encryption, using short-range tags, shielding, authentication of the reader, moving sensitive information to a protected database, restricted physical access
Relay attack	E-payment, e-documents, access control	Using short-range tags, shielding, distance bounding protocols
Tag cloning and replay attack	E-documents, e-payment, mass transport, access control cards	Tag authentication with challenge-response protocol, tag design which counters reverse engineering
People tracking	Item-level tagging, e-documents, public transport, RFID implants, e-payment, access control cards	Reader authentication, 'kill' command (in some applications, mostly retail), random identifiers in anti-collision protocol, changing pseudonyms, using short-range tags is possible, shielding (in some applications)
Change of tag content (e.g. registers value)	E-payment, transport (ticketing), some of applications for access control, e-documents and administrative process	Limited use of re-writable memory in tags, disabling writing feature ("lock" and "permlock" commands)
Physical tag destruction	Item-level tagging (anti-theft protection), e-documents (possibility of destruction by citizens concerned about their privacy)	Adequate physical location of tags on objects in retail
Blocking and jamming	Applications where attacker can benefit from denial of service (e.g. security-related)	Facilities for detection and localization of jamming devices.
Reverse engineering and side channel attacks	High security applications: e-payment, access control, e-documents	Protective layers, dummy structures, memory and bus scrambling, encryption of memory content; design of the tag which ensures data-independent time and power consumption

Table 1. Threats to privacy and security in RFID systems, application areas where they exist and the main countermeasures

Security and privacy must be considered in the early stages of RFID system development; a large part of technical security measures should be taken into account at the stage of tag design. Developers should consider not only present but also future levels of risk resulting from foreseen improvements in the technology used by attackers. Updating security later is very costly, much more so than it is in traditional information systems. Here, when a new vulnerability is discovered, it is rarely possible to solve the problem with a software upgrade like a security patch. Special attention should be paid to the concept of privacy by design (EDPS, 2007).

In object-level tagging, the low price of the tag is essential for massive deployment. To keep tag costs down, ways may be found by research and development to shift security and privacy behind the tag, either to another part of the RFID system (readers-backend) or to personal devices for tag management.

The future will bring further automation. The 'Internet of things' is a vision of a global network where not only computers but also billions of items tagged with RFID can communicate. This, together with sensor networks where RFID-type communication will also play an important role, will become part of a pervasive intelligent environment, called Ambient Intelligence (Daskala, Maghiros 2007). If security is not properly elaborated before it happens, huge amounts of data continuously collected in such an environment will be beyond control.

There is a need to complement technical security with legal measures and their enforcement, and to promote best practices by industry. Technical solutions alone will not be sufficient to protect against illegal retention and abuse of personal data or function creep. Moreover, the deployment of RFID and its benefits may be limited not only by real threats but also by the concerns of potential users, resulting from their lack of awareness. Awareness and trust should be created simultaneously with the development of appropriate measures to counter real threats.

5. Acknowledgments

This chapter is based on research done within the EU-funded project "Study on RFID Technologies: Emerging Issues, Challenges and Policy Options", led by JRC-IPTS.

The author would like to thank Ioannis Maghiros from the European Commission, DG Joint Research Centre - IPTS for many helpful comments and suggestions and Patricia Farrer from DG JRC - IPTS for help with preparation of the manuscript.

The views expressed in this article are those of the author, and in no way represent the European Commission's official position.

6. References

Albrecht K., McIntyre L. (2005). *Spychips. How major corporations and government plan to track your every move with RFID.* Nelson Current 2005.

Alien Technology (2005). *EPCglobal Class 1 Gen 2 RFID Specification.* Alien Technology. Whitepaper 2005

Atkinson, R. (2006). RFID - There's Nothing To Fear Except Fear Itself. *Opening Remarks at the 16th Annual Computers, Freedom and Privacy Conference, 4 May 2006, Washington DC.*

Avoine, G. (2004). Privacy Issues in RFID Banknote Protection Schemes. *International Conference on Smart Card Research and Advanced Applications - Cardis, August 2004*

Avoine, G., Kalach K. & Quisquater J.J. (2008). Passport: Securing International Contacts with Contactless Chips. Financial Cryptography, January 2008, LNCS, Springer-Verlag.

Bar-El, H. (2003). Introduction to Side Channel Attacks. Whitepaper, Discretix 2003. Available at: http://www.discretix.com/wp.shtml

Bono et al. (2005). Security Analysis of a Cryptographically-Enabled RFID Device. *14th USENIX Security Symposium, pages 1--16. USENIX, 2005.* Available at: http://www.usenix.org/events/sec05/tech/bono/bono.pdf

Bose, I. & Pal, R. (2005). Auto-ID: managing anything, anywhere, anytime in the supply chain *Communications of the ACM, vol. 48, no. 8, August 2005, pp. 100-106*

Carluccio, D.; Lemke K. & Paar C. (2005). Electromagnetic side channel analysis of a contactless smart card: first results. *Ecrypt Workshop, July 2005.* Available at: http://www.iaik.tu-graz.ac.at/research/krypto/events/RFID-SlidesandProceedings/Proceedings-WSonRFIDandLWCrypto.zip

Courtois, N.T; Nohl K. & O'Neil S. (2008). Algebraic Attacks on the Crypto-1 Stream Cipher in MiFare Classic and Oyster Cards. *Cryptology ePrint Archive: Report 2008/166,* available at: http://eprint.iacr.org/2008/166.pdf

Damgard, I. & Ostergaard, M. (2006). RFID Security: Tradeoffs between Security and Efficiency *IACR eprint, July 2006*

Daskala, B., & Maghiros, I. (2007). Digital Territories. Towards the Protection of Public and Private Space in a Digital and Ambient Intelligence Environment. JRC-IPTS report EUR 22765 EN. Available at: http://ftp.jrc.es/EURdoc/eur22765en.pdf

De Koning Gans, G.; Hoepman, J.H. and Garcia, F.D. (2008). A Practical Attack on the MIFARE Classic. *Proceding of the 8th Smart Card Research and Advanced Applications, CARDIS 2008*

EDPS (2007). *Opinion of the European Data Protection Supervisor on the communication from the Commission to the European Parliament, the Council, the European Economic and Social Committee and the Committee of the Regions on 'RFID in Europe: steps towards a policy framework'*. 20 December 2007.

EPCglobal (2004). *EPC™ Radio-Frequency Identity Protocols Class-1 Generation-2 UHF RFID. Protocol for Communications at 860 MHz - 960 MHz. Version 1.0.9.* EPCglobal Inc, 2004.

Fishkin, K.P. and Roy, S. (2003). Enhancing RFID Privacy via Antenna Energy Analysis. Tech. memo IRS-TR-03-012, Intel Research Seattle, 2003.

Garfinkel, S.; Juels, A. & Pappu, R. (2005). RFID Privacy: An Overview of Problems and Proposed Solutions. *IEEE Security and Privacy, May-June 2005*

Garfinkel, S. & Rosenberg, B. (2005). RFID: Applications, Security, and Privacy *Addison-Wesley Professional, July 2005*

Gemalto (2007). *Moving to the Second Generation of Electronic Passports.* Gemalto white paper, July 2007. Available at: http://www.gemalto.com/brochures/download/2nd_generation_passport.pdf

Graafstra, A. (2006). RFID Toys. *Wiley, 2006*

Halamka, J.; Juels, A.; Stubblefield, A. & Westhues, J. (2006) The Security Implications of VeriChip Cloning. *Journal of the American Medical Informatics Associacion, Vol. 13, Issue 6; Nov/Dec, 2006*

Hancke, G. (2005) A Practical Relay Attack on ISO 14443 Proximity Cards. *Manuscript, February 2005*

Hancke, G. (2006) Practical Attacks on Proximity Identification Systems (Short Paper). *IEEE Symposium on Security and Privacy, May 2006*

Hancke, G. & Kuhn, M. (2005). An RFID distance bounding protocol. *IEEE SecureComm 2005, 5-9 September 2005, Athens, Greece*

Hansche, S.; Berti, J. & Hare, C. (2004). Official (ISC)2 guide to the CISSP exam *Auerbach Publications*

Harrop, P.; Das, R. & Holland, G. *Item Level RFID 2008-2018.* IdTechEx Report, 2008.

Henrici, D. (2008). *RFID Security and Privacy. Concepts, protocols and architectures.* Springer-Verlag 2008.

Heydt-Benjamin, T.; Chae, H.J.; Defend B. & Fu K. (2006). Privacy for Public Transportation *Workshop on Privacy Enhancing Technologies - PET, June 2006*

Hoepman, J.H. et al. (2006) Crossing Borders: Security and Privacy Issues of the European e-Passport. *Advances in Information and Computer Security, volume 4266 of LNCS,* pp. 152-167. Springer Berlin / Heidelberg, 2006

Juels, A. (2004). Minimalist Cryptography for Low-Cost RFID Tags. *International Conference on Security in Communication Networks - SCN, September 2004*

Juels, A. (2005). Attack on a Cryptographic RFID Device. RFID Journal, 28 Feb. 2005. Available at: http://www.rfidjournal.com/article/articleview/1415/1/39/

Juels, A. (2006). RFID Security and Privacy: A research Survey. *IEEE Journal on Selected Areas in Communication.* No 24 Vol 2, pp. 381--394, February 2006.

Juels, A.; Molnar, D. & Wagner D. (2005). Security and Privacy Issues in E-passports. *SecureComm, September 2005*

Juels, A. ; Rivest, R. & Szydlo, M. (2003). The Blocker Tag: Selective Blocking of RFID Tags for Consumer Privacy. *Conference on Computer and Communications Security - ACM CCS, October 2003*

Karjoth, G. & Moskowitz, P. (2005). Disabling RFID Tags with Visible Confirmation: Clipped Tags Are Silenced. *Workshop on Privacy in the Electronic Society - WPES, November 2005*

Kefauver, B. (2007). ePassports: The Secure Solution. *ICAO MRTD Report, Vol. 2. No. 2, pp. 4-10. International Civil Aviation Organization, 2007*

Kfir, Z. & Wool, A. (2005). Picking virtual pockets using relay attacks on contactless smartcard systems. *SecureComm, September 2005.*

Kirschenbaum, I. & Wool, A. (2006). How to Build a Low-Cost, Extended-Range RFID Skimmer. *IACR eprint, February 2006*

Lewan, T. (2007). Chip Implants Lined to Animal Tumors. Associated Press, 8 September 2007

Maghiros, I.; Rotter, P. & van Lieshout, M. (editors): RFID Technologies: Emerging Issues, Challenges and Policy Options. *EUR Technical Report, EC DG-JRC, IPTS, 2007.*

Marburger, A; Coon, J.; Fleck, K.; Kremer, T. (2005). Verichip. Implantable RFID for the Health Industry *Unpublished document*

Nohl, K.; Plötz, H. (2007). *Mifare - Little security despite obscurity.* Presentation on the 24th Congress of the Chaos Computer Club in Berlin, December 2007. Available at: http://events.ccc.de/congress/2007/Fahrplan/events/2378.en.html

Rankl, W. & Effing, W. (2004). *Smart Card Handbook.* John Wiley & Sons Ltd, 2004.

Reid, J., et al. (2006). Detecting Relay Attacks with Timing Based Protocols. *Proceedings of the 2nd ACM symposium on Information, computer and communications security, Singapore 2007, pp. 204-213*

Rieback, M.; Crispo, B. & Tanenbaum, A. (2005). RFID Guardian: A Battery-Powered Mobile Device for RFID Privacy Management. *Australasian Conference on Information Security and Privacy - ACISP, July 2005*

Rieback, M.; Crispo, B. & Tanenbaum, A. (2006). Is Your Cat Infected with a Computer Virus? *Pervasive Computing and Communications - PerCom 2006, March 2006.*

Rotter, P. (2008): A Methodological Framework for the Assessment of Security and Privacy Risk for RFID Systems. In: *IEEE Pervasive Computing, Vol. 7, No. 2, April/June 2008, pp. 70-77.*

Rotter, P.; Daskala, B. & Compañó, R. (2008). RFID implants: opportunities and challenges for identifying people. *IEEE Technology and Society Magazine,* Volume 27, Issue 2, Summer 2008, pp. 24 - 32

Snijder, M. (2007). Security & Privacy in Large Scale Biometric Systems. Report based on an experts meeting held in Brussels on 25 September 2006. EC, Joint Research Centre, IPTS. Available at: http://is.jrc.es/documents/SecurityPrivacyFinalReport.pdf

Thornton et al. (2006). *RFID Security.* Syngress Publishing, Inc., 2006.

Weis, S.; Sarma, S.; Rivest, R. & Engels, D. (2004). Security and privacy aspects of low-cost radio frequency identification systems. International Conference on Security in Pervasive Computing, March 2003, also published as LNCS Vol.2802, pp. 201-212, 2004.

Wortham, J. (2007). How To: Disable Your Passport's RFID Chip," Wired, vol. 15, no. 1, 2007. Available at: www.wired.com/wired/archive/15.01/start.html?pg=9.

Wustenberg, W. (2007). Effective Carcinogenicity Assessment of Permanent Implantable Medical Devices: Lessons from 60 years of Research Comparing Rodents with Other Species. 27 September 2007, available at: http://www.verichipcorp.com/files/RodentSarcomagenesis092807Wustenberg.pdf

Vaudenay, S. & Vuagnoux, M. (2007). About machine-readable travel documents. *Journal of Physics: Conference Series, Volume 77, Issue 1, 2007*

14

The Study of RFID Authentication Protocols and Security of Some Popular RFID Tags

Hung-Yu Chien
Department of Information Management, National Chi Nan University, Taiwan, R.O.C.

1. Introduction

A radio frequency identification (RFID) system consists of three components: radio frequency (RF) tags (or transponders), RF readers (or transceivers), and a backend server. Tag readers *inquire* tags of their contents by broadcasting an RF signal, without physical contact, at a rate of several hundred tags per second and from a range of several meters. The advancements of Silicon manufacturing also result in great cost reduction for RFID tags compared to barcodes, not to mention that the tags can carry more data and are more resistant to dust and twisting. Thanks to these excellent features, the world has seen many RFID systems already put to use by manufacturers and businesses of all kinds of goods for supply management and inventory control and such; in addition, many public facilities and parking lots have also brought in RFID systems to help them offer faster, easier and more user-friendly services. As a matter of fact, potential applications are everywhere [57]. Such features as great convenience, low cost, and wide applicability will soon make RFID systems the most pervasive microchips in history [57].

However, the wide distribution of RFID systems into modern society may very much likely get the security of both businesses and consumers exposed to threats and risks. For example, businesses may have malicious competitors on the market that collect unprotected RFIDs to gather information illegally, spread false tags to provide wrong information, or even launch denial of service (DOS) attacks against them. On the other hand, as a consumer, it is naturally preferred that the information of the purchase of RFID-tagged products be kept private from outsiders; however, a tag reader at a fixed location can read the content of an un-protected tag, tracing the RFID-tagged product or/and even identifying the person carrying the tagged product. Correlating data collected from multiple tag readers such as their locations and so on can also possibly be used to spy on an individual and track down his/her social interactions. Besides passive eavesdropping and tracking, a thief might use counterfeit tags to fool automated checkout or security systems into accepting wrong information like price, proof of presence or other information.

RFID authentication protocols

To protect the private information on the RFID tags, some special devices (such as a blocker tag [26]) can be used here to deter the reader from accessing the tags, or tag authenticates the reader before its access. An RFID authentication protocol is a cryptographic protocol that allows a reader and a tag to authenticate each other, and the protocol is especially suitable for cases where resource-limited RFID tags are involved. In fact, although there are high-cost RFID tags like [25] available on the market that can support conventional symmetric key computations or even public key computations, the mainstream tags targeted at the majority of consumers are low-cost and can only support simple computations and very limited storage [50]. For example, for such tags as Gen 2 [16, 58] or iso 15693, conventional authentication protocols that require symmetric key computations or even public key computations are not applicable. Therefore, most of the efforts both the businesses concerned and the academic community have made so far are focused on the research and development of low-cost tags with higher security levels. Therefore, the topic of the next section is authentication protocols that are designed for low-cost RFIDs. Please also note that since well-designed conventional cryptographic protocols can be effectively implemented on resource-abundant backend servers and readers, it is usually assumed that the channels between backend servers and readers are secure; however, now that the focus is on RFID authentication protocols, this study has to assume that the channel between tags and readers is insecure. Figure 1 shows the components of an RFID system.

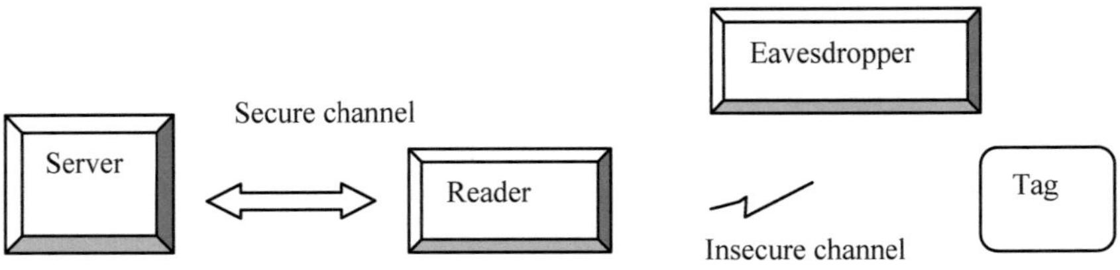

Fig. 1. Components of RFID systems

In addition, there are two special situations where the authentication of RFID tags is required to be done on extra conditions. To begin with, yoking proof protocols like [4, 7, 23, 24, 48, 53, 60] require the proof of simultaneous presence of two (or more) tags, and RFID distance bounding protocols like [5, 39, 56], on the other hand, not only authenticate the tags but also ensure that the authenticated tags are within a pre-assumed distance from the verifiers (the readers) so that the system is immune to message relay attacks like those brought up by [56]. In the following paragraphs, we shall briefly introduce yoking proof protocols and RFID distance bounding protocols. For detailed information, please refer to [4, 5, 7, 23, 24, 39, 48, 53, 56, 60].

Yoking proof

In 2004, Juels introduced an interesting RFID yoking proof protocol [23], which allows a verifier to prove the simultaneous presence of two tags in the communication range of a specific reader. Juels proposed several possible yoking proof protocol applications [23]. Let us take one example. Suppose a hard disk manufacturer wishes to ship each hard disk with its information leaflet. In such a case, each hard disk and each leaflet can be labeled with a different tag so that the yoking protocol can be applied to prove the simultaneous presence of the tagged products before shipping. In fact, the yoking proof protocol is a variant of the cryptographic authentication protocol, and it additionally requires the evidence of the simultaneous presence of two tags (or more tags).

RFID distance bounding protocols

Due to the short communication range, an authenticated RFID tag is deemed to be in proximity by its verifier (for example, an RFID reader), and the security of many RFID applications depends on this proximity assumption. However, this belief of proximity could be maliciously manipulated and thus become misleading when relay attacks like [56] are launched. For example, the access control system of a building would allow the access only when an authenticated tag is in the proximity. However, a specific kind of relay attack named the mafia attack, introduced by Desmedt [14], could cheat the system where an attacker sets up a rogue tag (say $\hat{A}$) and a rogue reader (say $\hat{B}$) sitting between the real reader and the real tag, and $\hat{A}$ and $\hat{B}$ cooperatively relay the messages between the real tag and the real reader so that the real reader wrongly believes that the tag is in its proximity (but it is not). A distance bounding protocol is a cryptographic mechanism that can prevent relay attacks from working. It is executed by a tag A and a reader B, and the tag A can convince the reader B of A's identity and A's physical proximity to B.

2. RFID authentication protocols

An RFID authentication protocol provides mutual authentication between the reader and the tag, and should resist potential security threats and attacks like the replay attack, man-in-the middle attack, etc. In addition to mutual authentication, anonymity and forward secrecy are also desirable properties for RFIDs. The point of ensuring the system's anonymity is to protect the privacy of the tags' identities such that un-authorized readers cannot identify or track a specific tag. Forward secrecy property, on the other hand, aims to protect the past communications where a tag is involved even if we assume that an attacker may have the power to compromise the tag some time later [50].

Just like tags of variant kinds currently available on the market, RFID authentication protocols can be quite different from one another, and the differences may come from the distinct resources required or the varied mechanisms adopted. Accordingly, we can classify these protocols and specify the features each kind has. Following the classification brought up by [52], for example, a protocol can be either a single-round design or a multi-round system. The former allows the reader and the tag to authenticate each other after a single round of operation of the protocol, while the latter has to run multiple rounds to do the job. Generally speaking, a single round protocol is more efficient than a multi-round protocol in terms of the number of interactions. Another classification, proposed by Chien [11], is based on the resources demanded by the protocols. This classification is very practical, because as we said earlier, on the market there are varieties of tags, of which most are resource-limited, and the resources required by these protocols can be very different. Under such circumstances, of course we will have a better view of the whole market if we classify the protocols and tags according to what kinds of resources are required. A third classification is based on the kind of cryptographic approach adopted, for the approach decides how well the protocol performs. Section 2.1 classifies the protocols as either single-round methods or multi-round methods, reviews the protocols and discusses the security properties. In Section 2.2, according to the required resources, we classify the protocols into four classes and introduce their corresponding applications. Finally, based on the cryptographic approaches, Section 2.3 classifies the protocols and discusses their performance.

2.1 RFID authentication protocols

Some single-round protocols are introduced in Section 2.1.1~2.1.6, while multi-round protocols are introduced in Section 2.1.7. Even though tags' data and keys are stored in the backend server in most of the cases, we do not differentiate the role of backend sever and the reader to simplify the description in the following sections. The notations used are introduced as follows.

r, r_T, r_R : l-bit random numbers.

ID_T, ID_R : the identity of tag T, the identity of reader R.

k_i : the secret key shared between tag T_i and the reader R.

$h()$, $g()$: secure one-way hash function; $h()$, $g()$: $\{0,1\}^* \rightarrow \{0,1\}^l$.

$CRC()$: cyclic redundancy code.

$f()$: a pseudo random number generator (PRNG function).

2.1.1 Weis et al.'s schemes

Weis et al. proposed a series of RFID authentication protocols [63, 64], and we review their hash-based access control protocol and the randomized access control.

Hash-based access control: Each hash-enabled tag T_i in this design will have a portion of memory reserved for a temporary $metaID_i$ and will operate in either a locked state or an unlocked state. Initially, a tag owner stores the hash of a random key, $metaID_i \leftarrow h(k_i)$, in the tag through either the RF channel or a physical contact to lock the tag. The owner also stores both the key and the $metaID_i$ in a backend server. Upon receipt of a $metaID_i$ value, the tag enters its locked state, and responds to all queries with only its $metaID_i$ and offers no other functionality. To unlock a tag, the owner inquires the tag, looks up the appropriate key in the back-end database and finally transmits the key to the tag. The tag hashes the received key and compares it to the stored $metaID_i$. If the values match, the tag unlocks itself and offers its full functionality to any nearby readers. The protocol is depicted in Figure 2.

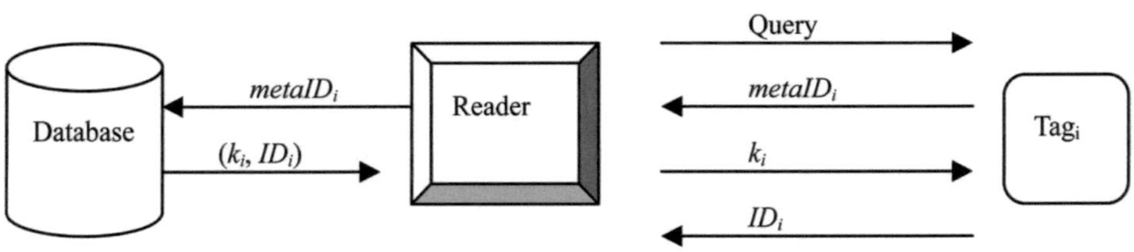

Fig. 2. Weis et al.'s Hash-based scheme: unlocking protocol

Randomized access control: In the previous scheme, a tag always responds with its $metaID_i$ to the queries, which allows any party to track an individual. So, Weis et al. proposed their randomized access control schemes where a tag will not respond predictably to queries by unauthorized users, but must still identifiable by only legitimate readers. The randomized access control schemes require tags equipped with a random number generator, in addition to the one-way hash function. Upon receiving a query from the reader, a tag responds with the values $(r, h(ID_i \| r))$, where r is a randomly chosen number. A legitimate reader identifies one of its tags by performing a brute-force search of its known IDs, hashing each of them concatenated with r until it finds a match. This mode is only feasible for owners of a relatively small number of tags. The protocol is depicted in Fig. 3.

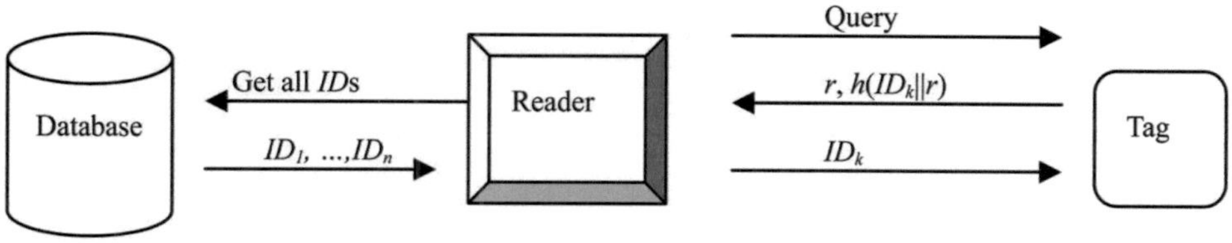

Fig. 3. Weis et al.'s randomized access control

Weakness of the hash-based scheme: In Figure 2, the reader broadcasts the tag's key in the forward channel. Since the signal in forward channel is strong enough for an adversary to monitor the transmission without being detected, this will allow an adversary easily eavesdrop the key and spoof a legal reader later.

Weaknesses of the random-access scheme: The Random-access scheme was designed to protect the *metaID* in the hash-based scheme to avoid individual tracking. However, it has poor scalability: it cannot support a large volume of tags because it has to perform the brute-force search to find a matched ID. It also gives the adversary (who resides in the range of the backward channel) a very high probability to find the matched tag, since he also searches only a small database of possible IDs. What makes it worse: the legal reader will broadcast the matched ID in the forward channel. So, an adversary might record the eavesdropped data ($r, h(ID_k \| r)$) and then easily spoofs the tags later.

2.1.2 Ohkubo et al.'s scheme [43]

The reader and each tag T_x initially shares a distinct hash seed s_{1_x}. T_x updates $s_{i+1\_x} = h(s_{i\_x})$ for $i \geq 1$ and responds with $a_{i\_x} = g(s_{i\_x})$ in the *i*-th authentication, where $h()/g()$ are two different hash functions. The reader can follow the hashing chains to authenticate the tag. The protocol is depicted in Fig. 4.

This scheme provides only one-way authentication of the tag, but it owns the forward secrecy property; that is, even assuming a tag is compromised some day in the future, the past communications from the same tag can not be traced. However, Ohkubo et al.'s original version cannot resist the replay attack [1]- a simple replay of old message can cheat the reader into accepting a forged tag. The scheme has the poor scalability problem [2, 3] - the computational cost to identify a tag is O(nm), where n is the number of potential tags and m is the maximum length of the hash chain. Avoine et al. [1] discussed the techniques to conquer the replay attack, and Avoine et al. [1, 2] also proposed their improvements to reduce the time complexity at the cost of extra memory.

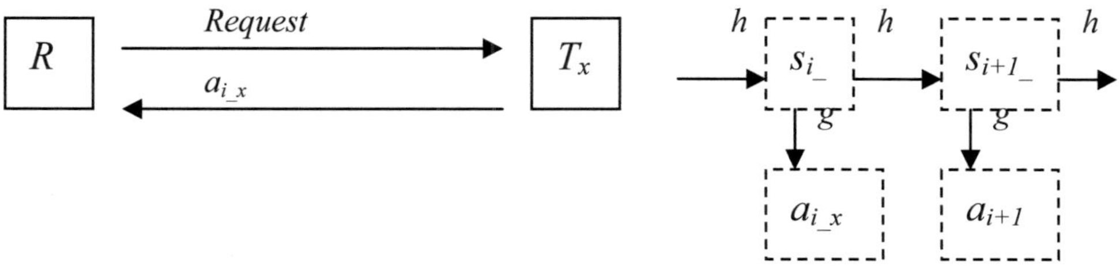

Fig. 4. Ohkubo et al.'s scheme

2.1.3 Karthikeyan-Nesterenko's scheme [28]

Karthikeyanand and Nesterenko, based on simple XOR operation, $\oplus$, and matrix operation, designed an efficient tag identification and reader authentication scheme. Initially, two matrices M_1 and M_2^{-1} are stored on each tag, and two matrices M_2 and M_1^{-1} are stored on the reader, where all the matrices are of size $p \times p$, and M_1^{-1} and M_2^{-1} are the inverses of M_1 and M_2 respectively. The tag and the reader also store a key K which is a vector of size q, where $q=rp$. That is, K can be represented as $K=[K_1, K_2, \ldots, K_r]$, where $K_i, i=1,2\ldots,r$ are vectors of size p. As a slight abuse of notation, the notation $X=KM$, where K is a vector of size q and M is a $p \times p$ matrix, denotes a component-wise multiplication of K and M. That is, $X=[X_1, \ldots, X_r]=[K_1M, \ldots, K_rM]$.

When the reader inquires a tag, the tag computes $X = KM_1$, and sends back X to the reader. The reader then forwards the message to the backend server, where the server will search its database to find a match. If it can find a match, then the tag is identified, and the server performs the following operations to authenticate itself to the tag and renew the key. The server first computes $Y = (K_1 \oplus K_2 \oplus \ldots \oplus K_r)M_2$, randomly selects a vector X_{new} of size q, computes $K_{new} = X_{new}M_1^{-1}$ and $Z = K_{new}M_2$, and finally sends (Y, Z) to the reader, which forwards (Y, Z) to the tag. Upon receiving the response from the reader, the tag verifies whether the equation $YM_2^{-1} \stackrel{?}{=} (K_1 \oplus K_2 \oplus \ldots \oplus K_r)$ holds; if so, the tag updates the key as $K_{new} = ZM_2^{-1}$. The scheme is depicted in Fig. 5.

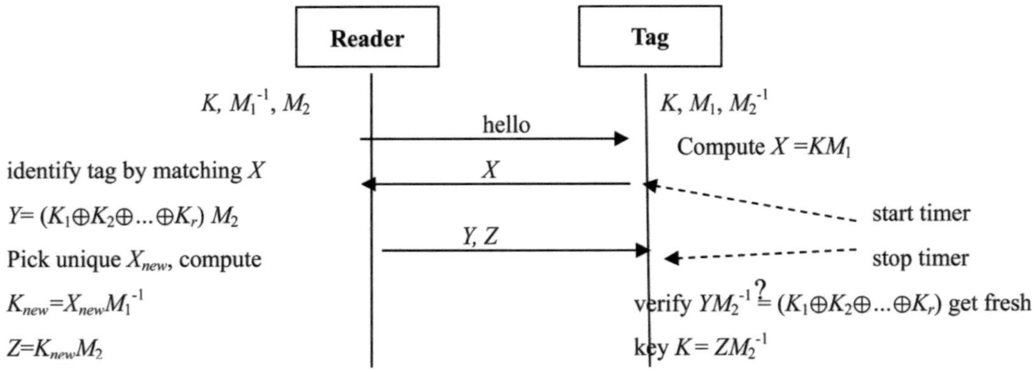

Fig. 5. Karthikeyan-Nesterenko's scheme

Weaknesses of Karthikeyan-Nesterenko's scheme

The scheme cannot resist the following attacks and threats- Denial of Services attack (DOS), replay attack and individual tracing.

In Karthikeyan-Nesterenko's scheme, the tag does not authenticate the received value Z when updating the key. Therefore, an attacker can replace the transmitted Z with an old one $\bar{Z}$ or any random value Z^* without being noticed; Upon receiving a valid Y and the fake Z^*, the tag will authenticate the Y successfully and then will update the key as $K^* = M_2^{-1} \cdot Z^*$. So, the legitimate reader and the tag cannot authenticate each other any more since the key is wrongly updated.

If the attacker replaces the Z with an old one $\bar{Z}$ (assuming $\bar{Y}$ and $\bar{Z}$ are previously sent in the ith legal session) in the above mentioned attack, then the attacker can replay the $\bar{Y}$ in the next session to cheat the tag in wrongly accepting the request and access the tag accordingly. He can even record the transmitted data from several sessions, and then launches the above attack several times. This will allow the attacker to trace the tag. Therefore, the anonymity property is violated.

2.1.4 Duc et al.'s scheme [15]

Duc et al.'s scheme was designed for improving the security of EPCglobal Class-1 Generation-2 tag (which is called Gen-2 for short later). Initially, each tag and the backend server share the tag's EPC code (the identity of the tag), the tag's *access* PIN, and an initial key K_0 (this key will be updated after each successful authentication, and K_i denotes the key after *i*th authentication). The steps of (*i*+1)th authentication are described as follows, where "Reader → tag: *M*" denotes the reader sends the tag a message *M*.

1. Reader → tag: *Query request.*
2. Tag → reader → server: M_1, r, C.

The tag selects a random number r, computes $M_1 = CRC(EPC \| r) \oplus K_i$ and $C = CRC(M_1 \oplus r)$, and sends back (M_1, r, C) to the reader, where the reader will forward (M_1, r, C) to the backend server.

3. Server → reader: the tag's info or "*failure*".

For each tuple (EPC, K_i) in its database, the server verifies whether the equations $M_1 \oplus K_i \stackrel{?}{=} CRC(EPC \| r)$ and $C \stackrel{?}{=} CRC(M_1 \oplus r)$ hold. If it can find a match, then the tag is successfully identified and authenticated, and the server will forward the tag's information to the reader and proceed to the next step; otherwise, it stops the process with failure.

4. Server → Reader → tag: M_2

To authenticate itself to the tag and update the information on the tag, the server computes $M_2 = CRC(EPC \| PIN \| r) \oplus K_i$ and sends M_2 to the tag through the reader. Upon receiving M_2, the tag uses its local values to verify the received M_2. If the verification succeeds, the tag will accept the "end session" command in the next step.

5. Reader → tag: "end session"
   Reader → server: "end session".

- Upon receiving the "end session" command, both the server and the tag update their shared key as $K_{i+1} = f(K_i)$.

The weaknesses

Duc et al.'s scheme cannot resist the DOS attack against tags and readers, cannot detect the disguise of tags, and cannot provide forward secrecy.

(1) In the last step of Duc et al.'s scheme, the reader sends the "end session" commands to both the tag and the backend server to update the key. If one of the "end session" commands is intercepted, then the shared key between the tag and the server will be out of synchronization. Thus, the tag and the reader cannot authenticate each other any more. The DOS attack succeeds. (2) If it is the "end session" command to the server is intercepted, then the server will hold the old key; therefore, a counterfeit tag can replay the old data (M_1, *r*, *C*) to disguise as a legitimate tag. So, the scheme fails to detect a disguised tag. (3) The scheme cannot provide forward secrecy. Suppose a tag is compromised, then the attacker would get the values (EPC, PIN, K_i) of the tag ; So, from the eavesdropped data (M_1, M_2, *r*) of the

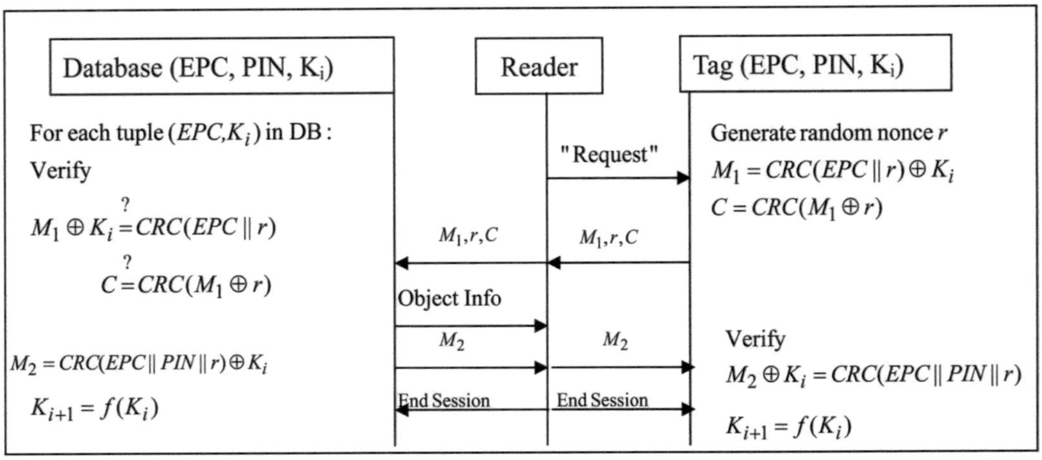

Fig. 6. Due et al. scheme

past communications, the attacker can verify whether a communication comes from the same tag by performing the following checking. For each eavesdropped communication (M_1, M_2, r), he computes $M_1 \oplus M_2$ to derive the value $CRC(EPC \oplus r) \oplus CRC(EPC \| PIN \| r)$, and then, using the compromised values (EPC, PIN, K_i) and the eavesdropped r, he can do the same computation to verify whether it came from the same tag. So, the past communications of a compromised tag can be traced.

2.1.5 Peris-Lopez et al.'s protocols [45-47]

Peris-Lopez et al. proposed a series of ultra-lightweight RFID authentication protocols [45-47] which were designed for very low-cost tags. Their schemes were very efficient: they require about 300 gates only and involve only simple bitwise operations. We review the LAMP protocol [45], which is one of Peris-Lopez et al.'s ultra-lightweight protocols.

LMAP involves only simple bitwise operations- bitwise XOR ($\oplus$), bitwise AND ($\wedge$), bitwise OR ($\vee$), and addition mod 2^m (+). The random number generator is only required on the reader. To protect the anonymity of tags, they adopt the technique of pseudonyms (*IDS*s), which is 96-bit length and is updated per successful authentication. Each tag shares an *IDS* and four keys (called *K1*, *K2*, *K3*, and *K4*, each with 96 bits) with readers, and they update the *IDS* and the keys after successful authentication. It needs 480 bits of rewritable memory and 96 bits for static identification number (*ID*).

The protocols consist of three stages- tag identification phase, mutual authentication phase, and pseudonym updating and key updating phase. In the following, ID_i denotes the static identification of Tag$_i$, IDS_i^n denotes the pseudonym of Tag$_i$ at the n-th run, and $K1_i^n / K2_i^n / K3_i^n / K4_i^n$ denote the four keys of Tag$_i$ at the n-th run. LMAP is depicted in Fig. 7.

Tag identification: Initially, the reader sends "*hello*" to probe Tag$_i$, which responds with its current IDS_i^n.

Mutual authentication phase: the reader uses IDS_i^n to find the corresponding four keys in its database, via the help of the backend server. It then randomly selects two integers *n1* and

$n2$, and computes the values A, B, and C (the calculation equations are specified in Fig. 7). From $A||B||C$, Tag$_i$ first extracts $n1$ from A, and then verifies the value of B. If the verification succeeds, then it extracts $n2$ from C, and computes the response value D. Upon receiving D, the reader verify the data D to authenticate the tag.

Pseudonym updating and key updating: After the reader and the tag authenticated each other, they update their local pseudonym and keys as specified in Fig. 7.

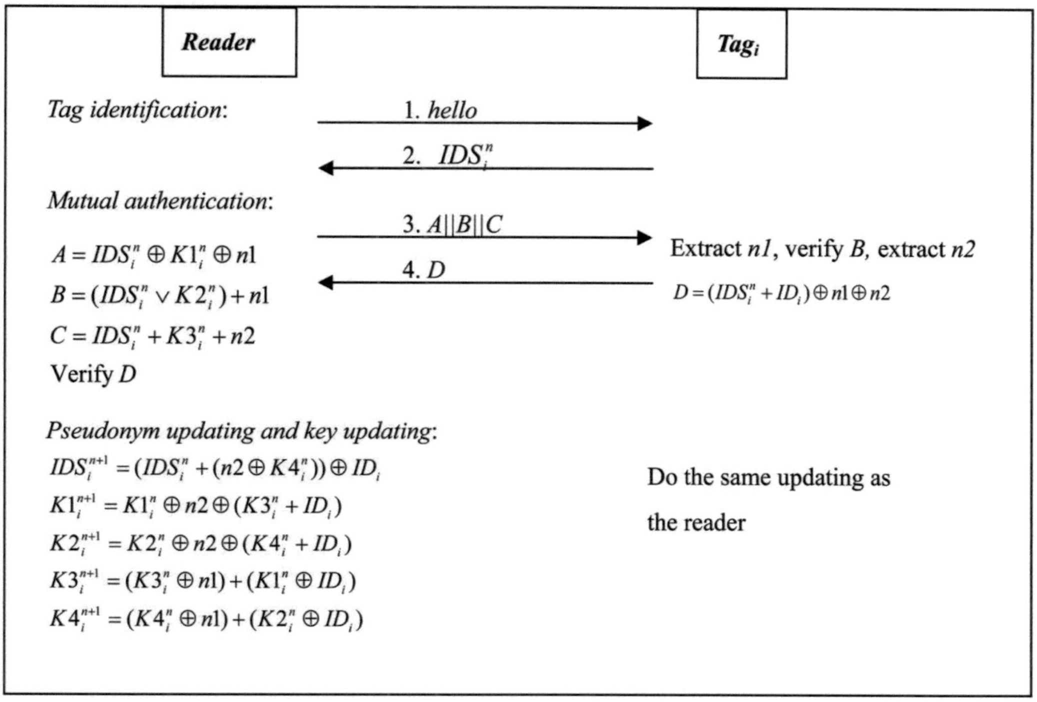

Fig. 7. LMAP

The weaknesses

The authentication of reader and tag in LMAP depends on the synchronization of pseudonym and keys. However, it is very easy to de-synchronize these values by intercepting the data in Step 4. In addition to the DOS attack, one can fully disclose the secrets of tags as follows.

We assume that an attacker can intercept, modify, and replay message between reader and tag in a reasonable time, and there is a completion message to indicate the completion of successful authentication. The attack scenario consists of five phases, but our attack is much more efficient than Li-Wang's work [34]. The whole scenario is depicted in Fig. 8.

In the attack scenario in Fig. 8, we omit the superscript n and the subscript i of pseudonym and of keys without causing ambiguity, since we are attacking the same tag within a successful session. In Phase 1, an attacker impersonates a reader and acquires the current IDS of a tag, and then the attacker (now impersonating the tag) uses the IDS to get a valid message $A||B||C$ from the reader in Phase 2.

In Phase 3, the attacker iteratively inverts the j-th (for $1 \leq j \leq 96$) bit of A, modifies B, and sends $A_j' \| B_j' \| C$ to the tag. From the tag's response (which is either a message D or an error message), the attacker can derive the *j*-th bit of *n1*. After deriving the value of *n1*, it further derives the values of *K1* and *K2* from *A*, *B*, *IDS* and *n1*. The detail of deriving the *j*-th bit is as follows. Let A_j' denotes the value by inverting the *j*-th bit of *A*. If the tag receives A_j', then it will derive n_1', which is equal to either $n1+2^{j-1}$ or $n1-2^{j-1}$, and each of the cases is with probability 1/2. So, the attacker can assume $n_1'=n_1+2^{j-1}$, computes $B_j'=B+2^{i-1}$, and sends $A_j' \| B_j' \| C$ to the tag. After receiving $A_j' \| B_j' \| C$, the tag extracts n_1' from A_j', verifies B_j', and then responds with either a message D_j or an error message. If a proper D_j is returned, the attacker can conclude that $n_1'=n_1+2^{j-1}$ and $n1[j]=0$ ($n1[j]$ denotes the *j*-th bit of *n1*); otherwise, it concludes that $n1[j]=1$. With this technique, the attacker launches 96 runs to derive all the bits of n_1, and then solves the values of *K1* and *K2* accordingly. Now the rest is to derive the values of *n2*, *K3*, *K4* and *ID*.

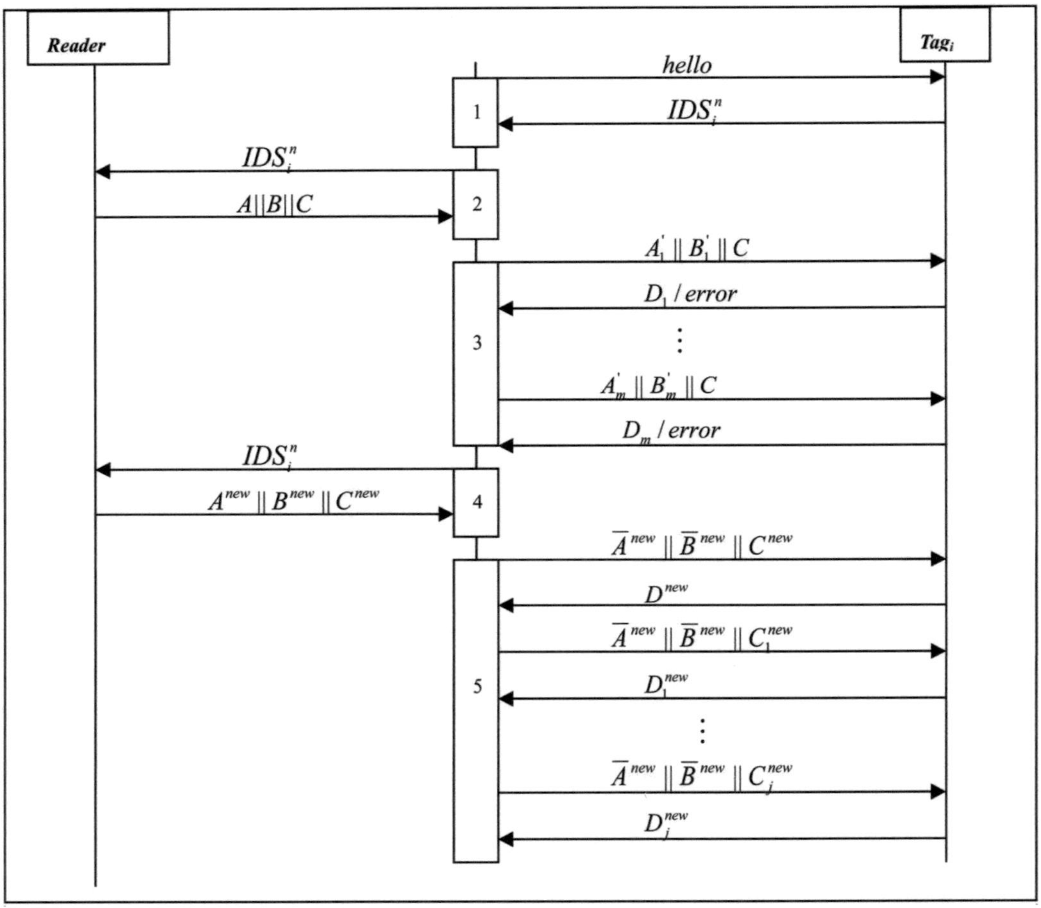

Fig. 8. Full-disclosure attack on LMAP

In Phase 4, the attacker impersonates the tag to the reader to get a new response $A^{new} \| B^{new} \| C^{new}$.

In phase 5, since the values of *IDS*, *K1*, and *K2* are already known, the attacker first sets $\overline{n1}^{new}$ =0 to have $\overline{A}^{new} = IDS \oplus K1$ and $\overline{B}^{new} = IDS \vee K2$, and sends $\overline{A}^{new} \| \overline{B}^{new} \| C^{new}$. So, the tag will respond with $D^{new} = (IDS + ID) \oplus n2$. Next, the attacker sets $C_1^{new} = C^{new} + 1$, and sends $\overline{A}^{new} \| \overline{B}^{new} \| C_1^{new}$ to the tag, which will extract $n2+1$ and will respond with $D_1^{new} = (IDS + ID) \oplus (n2+1)$. Now we have the equation $n2 \oplus (n2+1) = D^{new} \oplus D_1^{new}$. The possible values of $D^{new} \oplus D_1^{new}$ are summarized in Table 1. From Table 1, we can see that (1) if $D^{new} \oplus D_1^{new}$ has the form 0…01, then $n2[1]$=0; (2) if $D^{new} \oplus D_1^{new}$ has i+1 1s on the right and the rest are 0s, then $n2$ has i 1s from the right and followed by a zero. So, two simple interactions with the tag, the attacker can determine i+1 ($i \in [0,95]$) bits of $n2$. Following that, the attacker sets $C_{i+2}^{new} = C^{new} + 2^{i+1}$, and sends $\overline{A}^{new} \| \overline{B}^{new} \| C_{i+2}^{new}$. After getting the response D_{i+2}^{new}, the attacker computes $D^{new} \oplus D_{i+2}^{new}$ to determine the next few bits. It repeats this process until all the 96 bits of $n2$ are solved. This phase takes 2 interactions in the best case and 96 interactions in the worst case. After deriving $n2$, the attacker can further solve $K3$ and ID from the data C and D. With two successive pseudonyms IDS_i^n and IDS_i^{n+1}, the attacker further derives $K4$.

If $n2[1]$=0 (that is, $n2$ has the form xxxx….x0)*	then $n2 \oplus (n2+1) = 000...01$
If $n2[1]$=1 and $n2$ has the form xxx01…1 (that is, $n2$ has i 1s from the right followed by a 0)	then $n2 \oplus (n2+1) = 0...01...1$ (that is, $n2 \oplus (n2+1)$ has i+1 1s on the right and the rest are 0s)

*x denote the bit value is either 0 or 1.

Table 1. The possible values of $D^{new} \oplus D_1^{new}$

For more details of weaknesses of Peris-Lopez et al.'s ultra-lightweight protocols [45-47], one can refer to [13, 33-35].

2.1.6 Chien's SASI protocol [11]

Chien's SASI was designed for very low-cost RFID tags. Each tag has a static identification (*ID*), and pre-shares a pseudonym (*IDS*) and two keys *K1*/*K2* with the backend server. The length of each of *ID*/*IDS*/*K1*/*K2* is 96 bits. To resist the possible de-synchronization attack, each tag actually keeps two entries of (*IDS*, *K1*, *K2*): one is for the *old* values and the other is for the *potential next* values. The protocol consist of three stages- tag identification phase, mutual authentication phase, and pseudonym updating and key updating phase. In each protocol instance, the reader may probe the tag twice or once in the tag identification phase, depending on the tag's *IDS* is found or not. The reader first sends "*hello*" message to the tag, and the tag will respond with its *potential* next *IDS*. The reader uses the tag's response *IDS* to find a matched entry in the database, and goes to the mutual authentication phase if a matched entry is found; otherwise, it probes again and the tag responds with its old *IDS*. In the mutual authentication phase, the reader and the tag authenticate each other, and they respectively update their local pseudonym and the keys after successful authentication. After successful authentication, the tag stores the matched values to the entry ($IDS_{old} \| K1_{old} \| K2_{old}$) and stores the updated values to the entry ($IDS_{next} \| K1_{next} \| K2_{next}$). The random number generator is required on the reader only, and the tags only involve simple bit-wise operations like bitwise XOR ($\oplus$), bitwise OR ($\vee$), bitwise AND ($\wedge$), addition mod

2^m (+), and left rotate ($Rot(x,y)$). $Rot(x,y)$ is defined to left rotate the value of x with y bits. The protocol procedures are described as follows.

Tag identification: Initially, the reader sends "*hello*" to the tag, which first responds with its potential next *IDS*. If the reader could find a matched entry in the database, it steps into the mutual authentication phase; otherwise, it probes again and the tag responds with its old *IDS*.

Mutual authentication phase: the reader uses *IDS* to find a matched record in the database. It could be the potential next *IDS* or the old *IDS* of the tag. It then uses the matched values and two generated random integers *n1* and *n2* to compute the values *A*, *B*, and *C* (the calculation equations are specified in Fig. 9). From $A||B||C$, the tag first extracts *n1* from *A*, extracts $n2$ from *B*, computes $\bar{K}1$ and $\bar{K}2$ and then verifies the value of *C*. If the verification succeeds, then it computes the response value *D*. Upon receiving *D*, the reader uses its local values to verify *D*.

Pseudonym updating and key updating: After the reader and the tag authenticated each other, they update their local pseudonym and keys as specified in Fig. 9. The scheme also provides confirmation of the synchronization values ($\bar{K}1, \bar{K}2$) when the reader and the tag successfully authenticate each other.

The weaknesses

Sun et al. [62] had noticed that SASI is still vulnerable to DOS attacks. One attack scenario is described as follows. Assume that there is a synchronized tag T_x in which ($IDS_{next}, K1_{next}, K2_{next}$) equals to ($IDS_1, K1_1, K2_1$) stored in the database. Now, suppose the reader probes the tag, and sends out (A', B', C'), which is eavesdropped by the attacker. At the end of the protocol, the attacker interrupts the message *D* so that the reader will not update its variables. However, the tag will update its variables as follows: a) ($IDS_{old}, K1_{old}, K2_{old}$)=($IDS_1, K1_1, K2_1$), b) ($IDS_{next}, K1_{next}, K2_{next}$) =($IDS_2, K1_2, K2_2$).

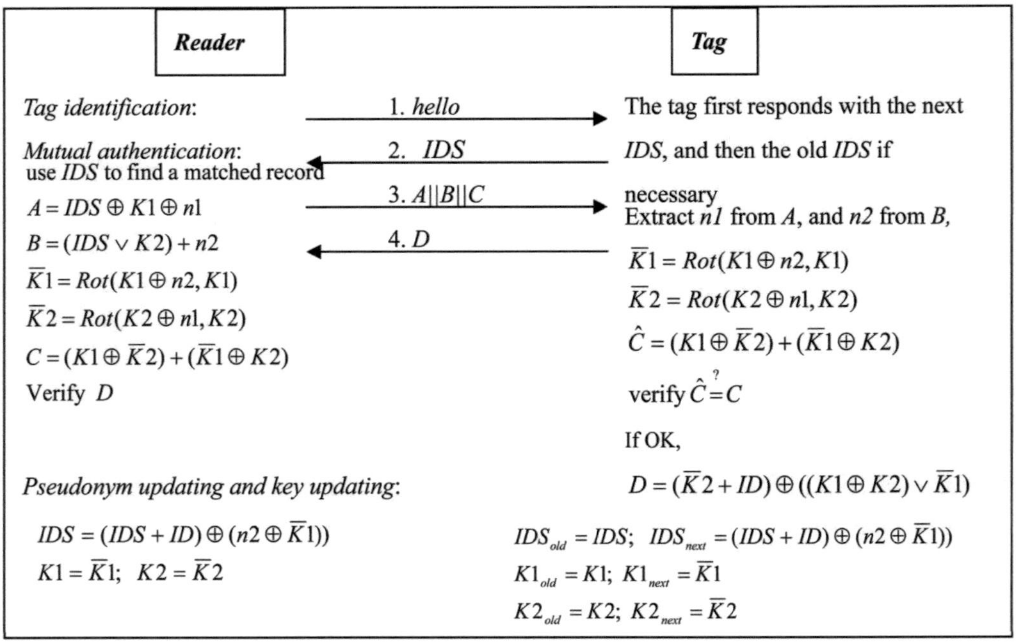

Fig. 9. SASI protocol

Next, the attacker allows the reader and the tag to run the protocol again without intervening. Because IDS_2 is not found in the database, both the reader and the tag use IDS_1 to complete the authentication. Thus, the database will update its variable list to ($IDS_3, K1_3, K2_3$), but the tag would own the values ($IDS_1, K1_1, K2_1$) and ($IDS_3, K1_3, K2_3$).

Finally, the attacker imitates as a valid reader to probe the tag. The tag first replies $IDS_{next} = IDS_3$, the attacker ignores this reply, which triggers the tag to reply IDS_1. The attacker now replays the recorded message (A', B', C'), which is valid and the tag would update its values a) ($IDS_{old}, K1_{old}, K2_{old}$)=($IDS_1, K1_1, K2_1$), b) ($IDS_{next}, K1_{next}, K2_{next}$) =($IDS_2, K1_2, K2_2$). Now, the genuine reader and the tag are out-of-synchronization. Sun et al. had shown another attack scenario, and other researchers like [8, 35, 49] had further shown passive attack to disclose the secrets of tags.

2.1.7 Multi-round authentication protocols- HB$^+$ [27]

The HB protocol [20, 21], proposed by Hopper and Blum, is a multi-round protocol and is based on the hardness of the LPN (Learning parity with noise) problem. However, the HB protocol is only secure to passive attacks, and successive improvements like [6, 18, 29, 40, 51] tried, but in vein, to protect from active attacks. In the following, we introduce the LPN problem and HB$^+$ protocol [27]. Interested readers are referred to [6, 18, 29, 40, 51] for other HB-related works.

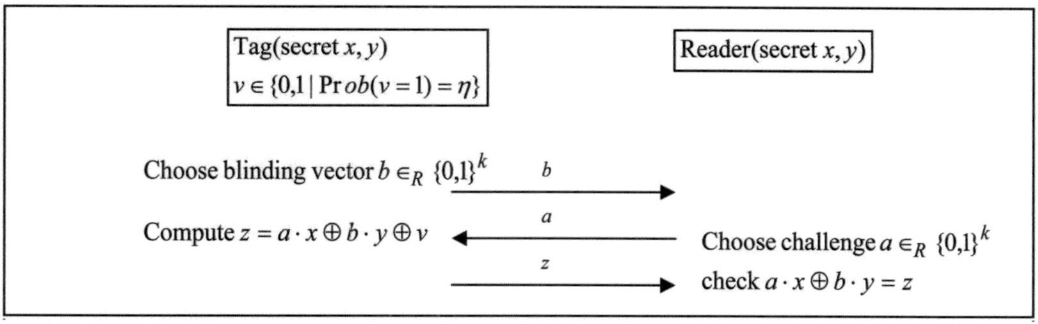

Fig. 10. One round of HB$^+$

The LPN problem: The LPN problem with security parameters q, k, η, with $\eta \in [0, \frac{1}{2}]$ is defined as: given a random $q \times k$ binary matrix A, a random k-bit vector x, a vector v such that $|v| \le \eta q$, and the product $z = A \cdot x \oplus v$, find a k-bit vector x' such that $|A \cdot x' \oplus z| \le \eta q$.

HB$^+$: Juels and Weis [27] tried to improve the HB protocol to resist active attacks. There are two k-bits secrets x, y between the reader and the tag. The protocol is composed of q rounds, one of which is depicted in Fig. 10. The tag is successfully authenticated if the check fails at most $q\eta$ times.

Gibert et al. [18] had shown a man-in-the-middle attack on HB$^+$. In their model, they assume that an attacker can learn whether an authentication procedure succeeds or not. One attack scenario is depicted in Fig. 11. The attack consists of two phases. First, the attacker replaces the challenge a sent by the reader with $a' = a \oplus \delta$ in all q rounds of the authentication

process, where δ is a k-bit constant vector. If the authentication succeeds, she can conclude that $\delta \cdot x = 0$ with high probability; otherwise, $\delta \cdot x = 1$ with high probability. The attacker can set only one bit of δ on each time, and repeats the process k times to reveal all the bits of x. In the second phase, the attacker impersonates the tag and sends a well chosen blinding vector b to the reader. After that, she responds to the reader challenge a with $z = x \cdot a$. If the authentication succeeds, she learns that $y \cdot b = 0$ with high probability; otherwise, she concludes that $y \cdot b = 1$ with high probability. After manipulating the bits of b and repeating the process k times, she can learn all the bits of y.
Even though several successive variants of HB+ have been proposed [6, 9, 18, 29, 40, 51], none of them can resist all possible active attacks, and all the variants of HB series did not consider the anonymity and forward secrecy property.

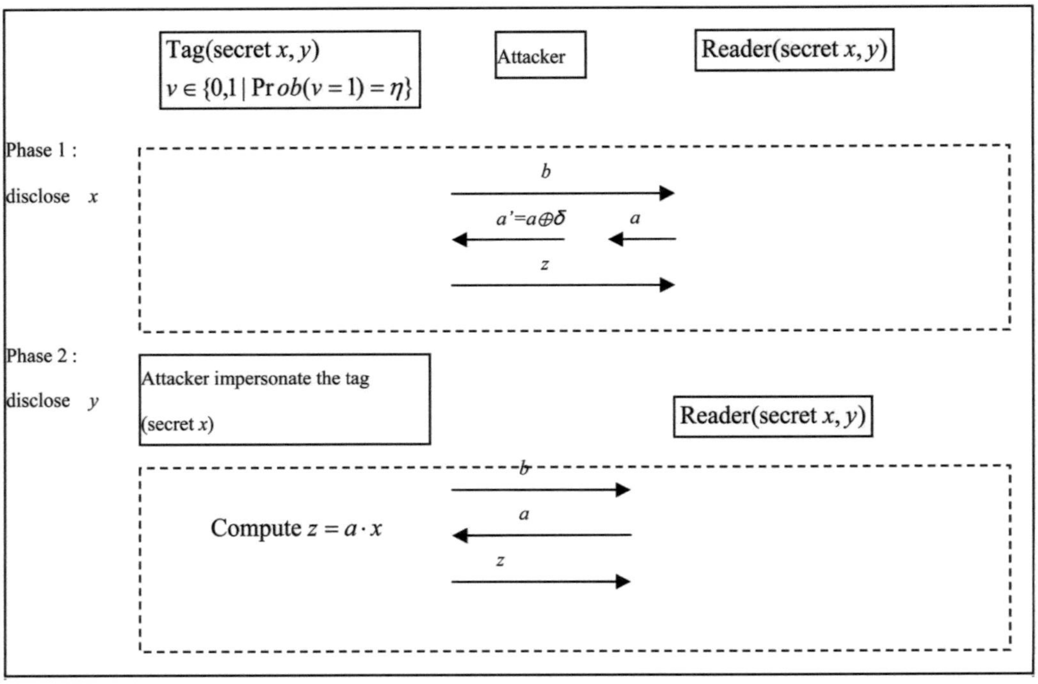

Fig. 11. Attack on HB+

2.2 The resources-based classification of RFID authentication protocols

In Section 2.1, we review and discuss several RFID authentication protocols without discussing their required resources. Actually, there are various RFID tags on the market, and the capacities of these tags are quite varying: some can support public key computations, and some can only support simple bit-wise operations. However, to have large market penetration, the cost of RFID tag plays an important factor, and most of the tags targeted for consumer market would be low-cost or even very low-cost. Even though most of the RFID authentication protocols introduced in Section 2.1 are targeted for such kind of tags, the required resources of these protocols are quite varying.
Based on the required capacity on tags, we roughly classify the RFID authentication protocols into four classes. The first class called "*full-fledged* class" refers to those protocols

(like the schemes [25]) that demand the support of conventional cryptographic functions like symmetric encryption, cryptographic one-way function or even the public key algorithms. One of the main applications of these full-fledged protocols is E-passport and credit card.

The second class called *"simple"* is for those protocols (like the schemes [10, 19, 38. 43, 59, 63, 64, 66, 67]) that should support random number generator and one-way hashing function on tags

The third class called *"lightweight"* protocols refers to those protocols [6, 9, 11, 12, 15, 18, 20, 21, 22, 27-29, 40, 51] that require random number generator and simple functions like Cyclic Redundancy Code (CRC) checksum but not hash function. The EPC Gen 2 tag [17] supports both random number generator and CRC function.

The fourth class called *"ultra-lightweight"* refers to the protocols [8, 11, 13, 33-35, 45-47, 49, 62] that only involve simple bit-wise operations (like XOR, AND, OR, etc) on tags. Peris-Lopez et al. first proposed a series of *ultra-lightweight* authentication protocols [45-47], where the tags involve only simple bit-wise operations like XOR, AND, OR and addition mod 2^m. These schemes are very efficient, and they only require about 300 gates. Unfortunately, Li-Wang [34] and Li-Deng [33] reported the de-synchronization attack and the full-disclosure attack on these protocols, and Chien and Hwang [13] further pointed out the weakness of Li-Wang's improved scheme. The security weaknesses of SASI protocols are explored in [8, 35, 49, 62]. In addition to design ultra-light authentication protocols, other researcher like [41, 65] focused on designing lightweight hash function or encryption functions.

2.3 The classification based on cryptographic approaches

Contrary to the authentications in conventional applications where anonymity and un-traceability are usually not necessary properties, anonymity and un-traceability are desirable properties in many RFID applications. Therefore, this section discusses those RFID authentication protocols that consider anonymity and un-traceability, and those protocols like [6, 18, 29, 40, 51, 63, 64] that do not consider or do not well protect anonymity and un-traceability are excluded from the following discussion. Based on the technique a RFID authentication protocol uses to identify a tag while protecting the anonymity, we may classify anonymous RFID authentication protocols into the following different approaches. In describing these approaches, we focus on the techniques to identify tags while preserving the anonymity, without covering the details of the protocols.

Simple challenge-response approach. In this approach, each tag T_i shares a distinct key k_i with the server S/ the reader R. When the reader R probes a tag T_i by sending a random value N_R as a challenge, T_i responds with $h(k_i, N_R)$, where $h()$ denotes a secure one-way function or some function that can output commitment on its inputs while protecting the un-disclosed input k_i. Upon receiving the response $h(k_i, N_R)$, the server computes $h(k_j, N_R)$ for each potential tag T_j in its database to see whether there is a matched tag. This approach allows the server to identify a tag without disclosing the identity to eavesdroppers. Each tag just keeps one secret key, but the server needs to perform the computation for each potential tag to identify the tag. So, the tag's storage space is O(1) but the computational cost for identifying a tag is O(n), where n the number of possible tags. The previous schemes like [12, 15, 28, 30, 38, 44, 59, 66, 67] adopt this approach.

Tree-walk approach. In this approach, the tags are organized as a tree, where each leaf node in the tree denotes one tag and each edge in the tree is associated with a key. Fig. 12(b) shows one simple example. In the example, tag T_1 owns the key $K1$ and $K3$, and tag T_2 owns the keys $K1$ and $K4$. When a reader probes T_2 by sending a challenge N_R, T_2 responds with $\{h(K_1,N_R)$, s} on which the server can perform depth-first-search to identify the tag. This approach requires O($\log n$) key space on each tag and demands O($\log n$) computational cost to identify a tag. The key space requirement is a serious burden on low-cost tags. One more serious weakness of this approach is that once a tag is compromised, other tags that share the same keys on the same key paths could be partially traced. The more the number of keys one tag T_i shares with the compromised tag T_j, the more the tag T_i could be identified and traced. The schemes like [31, 32] adopt this approach.

Hash chains approach. One distinguished work of this approach is Ohkubo et al.'s scheme [43]. In this approach, the server and each tag T_x shares a distinct hash seed s_{1_x} initially. T_x updates $s_{i+1\_x} = h(s_{i\_x})$ for $i \geq 1$ and responds with $a_{i\_x} = g(s_{i\_x})$ for each query request, where $h()/g()$ are two different hash functions. This approach owns the forward secrecy property; that is, even assuming a tag is compromised one day in the future, the past communications from the same tag can not be traced. However, Ohkubo et al.'s original version cannot resist the replay attack [1], and has the poor scalability problem [2, 3] - the computational cost to identify a tag is O(nm), where n is the number of potential tags and m is the maximum length of the hash chain. Avoine et al. [1] discussed the techniques to conquer the replay attack, and Avoine et al. also [1, 2] also proposed their improvements to reduce the time complexity at the cost of extra memory.

Varying Pseudonym (VP) approach. In this approach like [11, 19,45-47], each tag synchronizes its varying identifier and its internal state with the server. Please notice that, even though some challenge-response-based schemes like [2, 12, 15] also synchronize the state between the tags and the server, these schemes do not send a varying pseudonym to facilitate the server perform fast identification; we, therefore, do not count them in this VP approach. The varying identifier is called pseudonym in [11, 45-47], and is called metaID in [19, 31, 32]. Here, we all refer to them the pseudonyms. Upon receiving a challenge request, a tag responds with the current pseudonym and the commitment on the challenge and the secret internal state. Based on the commitment, the server can verify the tag. During the authentication, the tag and the server respectively update their pseudonyms and their internal state. In this approach, the pseudonym not only protects the anonymity of the tag but also facilitates the server to identify the tag in its database with O(1) computational complexity, because the server can directly use the pseudonym to locate the corresponding entry in its database and perform necessary computations for this matched entry only. Further more, each tag only needs constant quantity of internal values- O(1) key storage. It is these excellent features that make it quite attractive than the other approaches. However, due to the synchronization requirement, the VP-based schemes are prone to the de-synchronization attacks (or the denial of service attacks) [8, 13, 33-35, 49, 62], if adversaries can manipulate the communications such that the tag and the server are out of synchronization. Fig. 12 depicts the main ideas of these approaches.

3. Security analysis of the mifare ultralight card and OV-chipkaart

In Section 2, we have examined several RFID authentication protocols published in the literature. In this section and the next, we shall examine the security of some popular tags on the market. Section 3 will discuss the Mifare Ultralight card [36], and Section 4 will cover the EPC Class 1 Generation 2 card [16, 17].

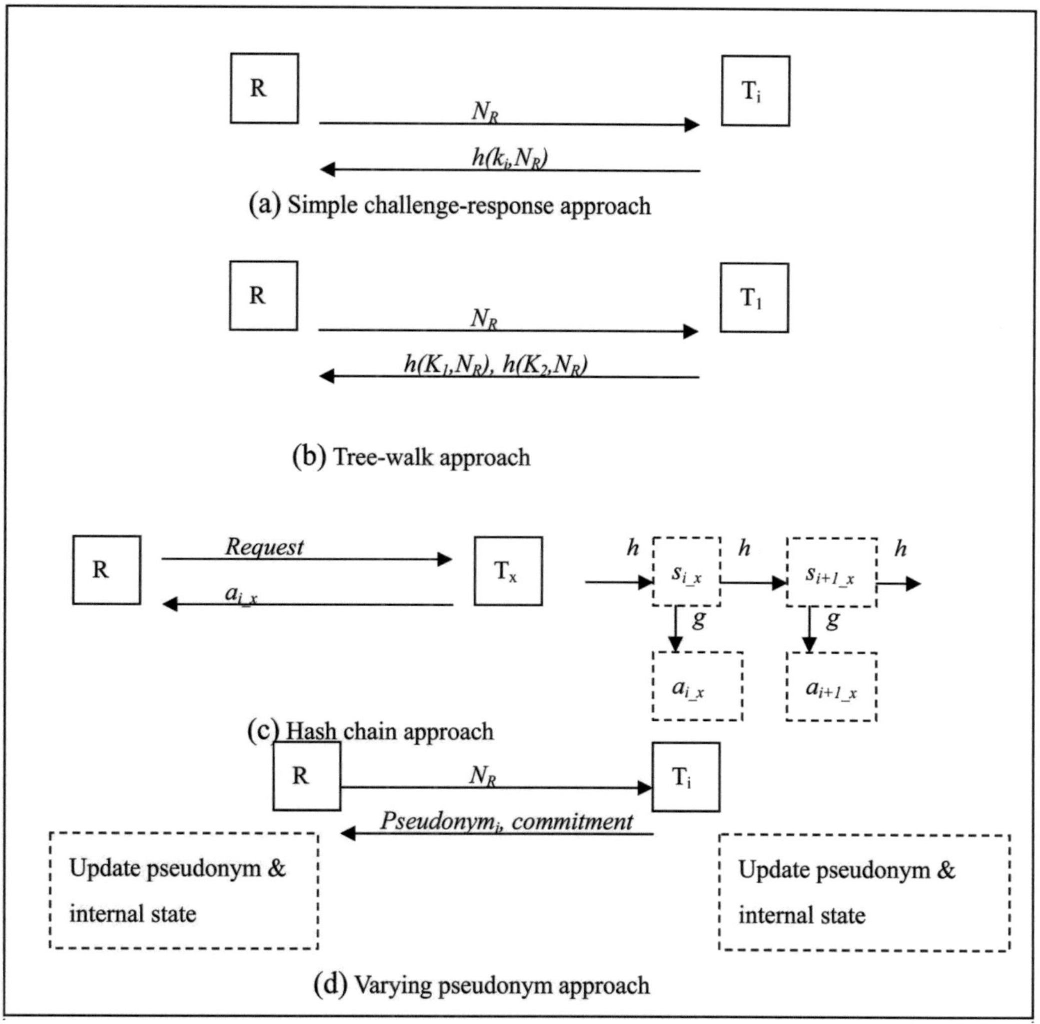

Fig. 12. Approaches to protect RFID tag identity

Mifare is a trademark of contactless RFID products and technologies developed by NXP Semiconductors [42]. Mifare cards have been widely used in many countries, and some of the cards feature a high level of security. However, Mifare Ultralight [42], one of Mifare card series, is focused on supporting faster applications at a cheaper cost. Thus, there is not any security mechanism implemented on the Mifare Ultralight chip. All privileges are fully accessible by anyone on the memory block. Section 3.1 right below will introduce Mifare card series, followed by Section 3.2 that deals with the memory organization of Mifare Ultralight. Then, Section 3.3 will take the OV-chipkaart in the Netherlands, which runs on the basis of the Mifare Ultralight card, as an example to discuss the security weaknesses and possible threats.

Feature	Mifare Ultralight	Mifare Mini	Mifare 1k	Mifare 4k
Security		Classic Mifare crypto algorithm		
Memory capacity	64 byte	320 byte	1000 byte	4000 byte
Acc. Standard	ISO/IEC 14443A			
Key applications	• Limited-use tickets in public transportation. e.g.: disposable OV-chipkaart in Netherlands • Event ticketing e.g.:entrance tickets of Great Wall	• Access control • Loyalty card	• Public transportation e.g.: Easy Card in Taipei • Access management • Event ticketing • Gaming • Identity	• Advanced public transportation e.g. : personal and anonymous OV-chipkaart in Netherlands • Access management • e-Commerce • e-Business • Gaming • Identity

Table 1. Mifare series products' features and key applications (I) [36]

3.1 Introduction

Mifare is a trademark of contactless RFID products and technologies developed by a subsidiary company of Philips—the NXP Semiconductors [42]. With more than 1 billion smart card ICs and 5 million reader components sold around the world, Mifare is also the industry standard on the contactless RFID market, holding an 80% market share by value [36]. The applications of Mifare cards include automatic fare collection systems, inventory systems and household security systems, etc., and among them automatic fare collection systems are the most successful. Products of NXP Semiconductors include Mifare Ultralight, Mifare Mini, Mifare 1k, Mifare 4k, Mifare Plus, Mifare DESFire and Smart MX. Table 1 and Table 2 show the features and key applications of Mifare cards.

Feature	Mifare Plus	Mifare DESFire	Smart MX
Security	AES/DES	3DES	3DES/RSA/ECC
Memory capacity	2000/4000 byte	4000 byte	20000 ~ 144000 byte
Acc. Standard	ISO/IEC 14443A		
Key applications	• Public transportation • Access management • Event ticketing • Gaming • Identity	• Advanced public transportation • Access management • Event ticketing • e-Government • Identity	• Advanced public transportation e.g. : e-Passport in Singapore • e-Government • Banking / Finance • Mobile communication

Table 2. Mifare series products' features and key applications (II) [36]

Mifare Ultralight, one popular card of Mifare series, is non-reloadable and usage-limited. It can be used as one way tickets, tourist weekend passes, zone-based tickets, and multiple trip tickets such as 10-ride tickets. Once the value recorded in the card IC has run out, the card becomes invalid and no further access is allowed. Therefore, it is a very good replacement for the traditional paper ticket. The Ultralight card can be easily integrated into the existing Mifare Classic applications without any additional cost and any additional equipment. The public transport payment system between Amsterdam and Rotterdam has used the OV-chipkaart [61] as access control tokens, and the disposable OV-chipkaart, one of the three types of OV-chipkaart, is a personal, anonymous, and as the name indicates, disposable ticket, and it is based on the Mifare Ultralight card design.

Mifare Ultralight IC [42] is a contactless RFID card which confirms to ISO/IEC14443A standard. The following list gives an overview of Mifare Ultralight:

- Operating distance: Up to 100mm (depending on antenna geometry)
- Operating frequency: 13.56MHz
- Fast data transfer: 106kbit/s
- No battery needed (energy is offered by reader)
- True anti-collision
- 7 byte serial number (Unique Identification, UID)

Each Mifare Ultralight card IC has its own UID. The anti-collision function can use the UID to distinguish a large number of cards in the communication field simultaneously, and then select one card per transaction. To support faster applications at cheaper costs, there is no security mechanism implemented in Mifare Ultralight; for instance, there is neither 3-way authentication support nor key storage on the IC chip.

3.2 Mifare ultralight memory organization

Mifare Ultralight contains an EEPROM of 512 bits, which are organized into 16 pages with 4 bytes each. The first 80 bits are reserved for manufacturer data and UID. The following 16 bits are used for the read-only locking mechanism, and the next 32 bits are the OTP area (One Time Programmable Area). The final 384 bits are user programmable read/write memory. Fig. 13 shows the Mifare Ultralight memory organization. In short, Mifare Ultralight divides the whole memory into four parts: (1) UID/serial number, (2) lock bytes, (3) OTP bytes and (4) data pages or user area.

UID/serial number: It contains a 7-byte UID, two check bytes and a 1 byte of internal data. It is stored in page 0, page 1 and the first 2 bytes of page 2 (Fig. 13). These bytes are write-protected after the IC manufacturer has programmed the card. The UID consists of the first three bytes of page 0 (SN0, SN1 and SN2) and the four bytes of page 1 (SN4, SN5, SN6 and SN7). Each card has a unique serial number so no collision happens. Each UID should be unique and should be unpredictable, and this way the UID can be used to prevent card forgery. The two check bytes, BCC0 and BCC1 defined as $CT \oplus SN0 \oplus SN1 \oplus SN2$ and $SN3 \oplus SN4 \oplus SN5 \oplus SN6$ respectively, follow the regulation of ISO/IEC 14443-3. The CT field, which has a value of 0x88, is the cascade tag, and is used for compatibility with the Mifare Classic 4 byte UID. In addition, SN0 is reserved for the manufacturer ID, according to ISO/IEC 14443-3 and ISO/IEC 7816-6 AMD.1.

Byte Number	0	1	2	3	Page
Serial Number	SN0	SN1	SN2	BCC0	0
Serial Number	SN3	SN4	SN5	SN6	1
Internal / Lock	BCC1	Internal	Lock0	Lock1	2
OTP	OTP0	OTP1	OTP2	OTP3	3
Data read/write	Data0	Data1	Data2	Data3	4
Data read/write	Data4	Data5	Data6	Data7	5
Data read/write	Data8	Data9	Data10	Data11	6
Data read/write	Data12	Data13	Data14	Data15	7
Data read/write	Data16	Data17	Data18	Data19	8
Data read/write	Data20	Data21	Data22	Data23	9
Data read/write	Data24	Data25	Data26	Data27	10
Data read/write	Data28	Data29	Data30	Data31	11
Data read/write	Data32	Data33	Data34	Data35	12
Data read/write	Data36	Data37	Data38	Data39	13
Data read/write	Data40	Data41	Data42	Data43	14
Data read/write	Data44	Data45	Data46	Data47	15

UID/serial number — One Time Programmable — Lock Bytes — User data area

Fig. 13. Mifare Ultralight memory organization [42]

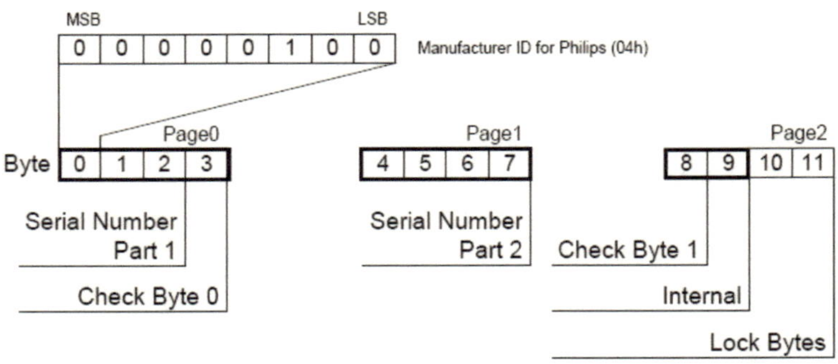

Fig. 14. Mifare Ultralight UID/Serial number[42]

(2) Lock Bytes: This part includes the last two bytes of page 2, which hold the field-programmable read-only mechanism (see Fig. 15). The lock bytes include 3 block-locking bits (BL) and 13 locking bits (L). The initial value is set to "0". The locking bits correspond to the pages from page 3~15 respectively, and when a locking bit is set to "1," it indicates that the corresponding page is locked. The 3 block-locking bits of lock byte 0 on page 2 are the

locks of locking bits, of which each bit handles pages 10~15, pages 4~9 and page 3, respectively. Once the block-locking bit is set, the corresponding lock bits are locked. This process is irreversible; if a bit is set to"1", then it cannot be altered to "0." For example, assume a block-locking bit, here let's take BL9-4 for example, is set to "1", then the corresponding memory area L9 to L4 cannot be changed.

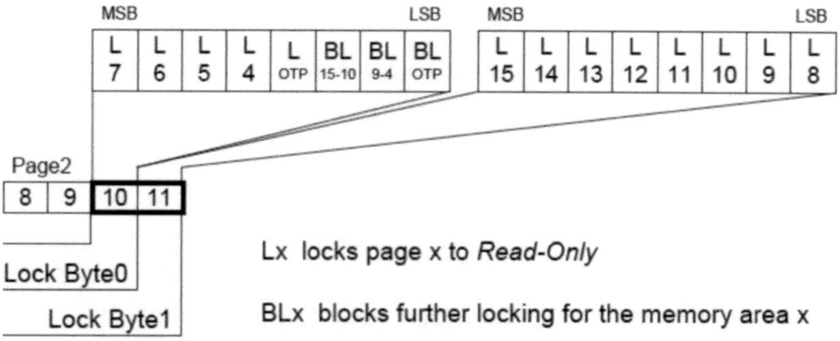

Fig. 15. Mifare Ultralight Lock Bytes [42]

(3) OTP Bytes: All the 32 bits of page 3 are the OTP bytes. The initial value of each bit is "0", and it can only be set to "1" once, and the process is irreversible. The writing process of OTP byes is as follows. The current contents of the OTP bytes are bit-wise "or-ed" with the new bytes, and the result becomes the new contents of the OTP bytes (see Fig. 16). The OTP area can be used as a one-time counter with a maximum value of 32 (by setting every bit to 0). Please see Fig 17. For example, if a card has four rides available, then the value in the OTP bytes will be 1111 1111 1111 1111 1111 1111 1111 0000. An OTP with three rides available is 1111 1111 1111 1111 1111 1111 1111 0001. An OTP with no ride available is 1111 1111 1111 1111 1111 1111 1111 1111.

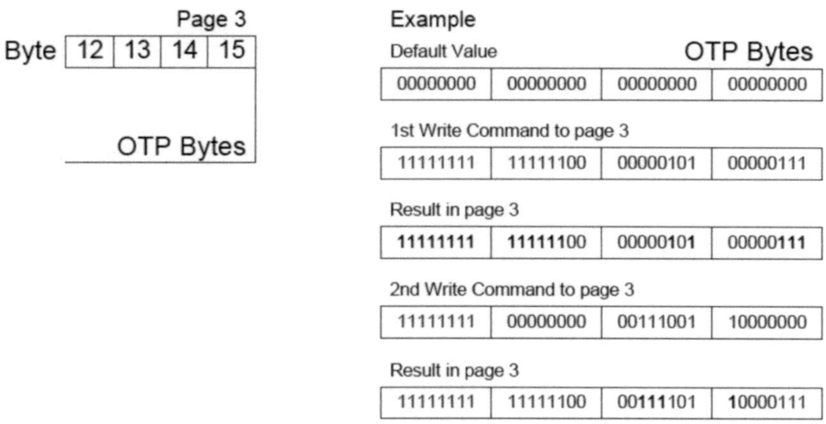

Fig. 16. OTP Bytes and an example [42]

(4) Data Pages: Pages 4~15 are user programmable read/write memory. These 384 bits are fully accessible to anyone. The values of these data pages are pre-set to "0".
In the next section, we will examine the security of Netherlands' OV-chipkaart system, which is based on that of the Mifare Ultralight card. The following discussion is based on the report [61].

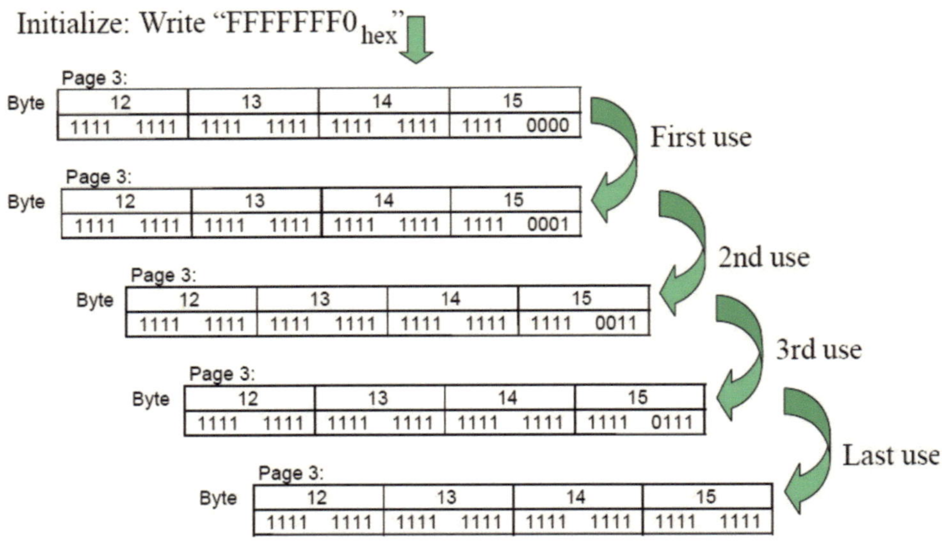

Fig. 17. Example—a four-ride ticket [42]

3.3 The security weaknesses of OV-chipkaart on mifare ultralight

The OV-chipkaart in use by the transport payment systems of Amsterdam and Rotterdam in the Netherlands is based on RFID technologies. The OV-chipkaart provides three types of tickets: personal tickets, anonymous tickets and disposable tickets. The personal ticket and the anonymous ticket are based on Mifare Classic 4k, and the disposable ticket is based on Mifare Ultralight. Readers and Mifare Classic cards together can make the three-pass authentication process a reality, and then the transferred data get encrypted by using NXP-developing crypto algorithms. However, Mifare Ultralight neither provides three-pass authentication nor data encryption. The data on Mifare Ultralight are fully accessible to anyone, and the communication is not done under the protection of encryption. According to the report [61] released by University of Amsterdam, the disposable OV-chipkaart based on Mifare Ultralight has the three weaknesses as follows: (1) failure of defense mechanism, (2) repeated check-out and (3) free travel problem. In the discussion below, we shall focus on the Mifare-Ultralight-based OV-chipkaart, and refer to it as the OV-chipkaart for short.
In the OV-chipkaart, the data pages of the card are configured into three sectors of 4 pages each: the first sector (slot 1) consists of pages 4~ 7, the second sector (slot 2) contains pages 8~11, and the last sector consists of pages 12~15. The first two sectors are used to store the check-in and check-out transactions, and the last sector is used to store the product information such as the number of available rides and the expiry date. Data on the first two

pages of slot 1 and slot 2 are stored in plain text while the last two pages are encrypted. The former records the transaction serial number, the type of transaction, the date and the time, and the latter contains the stations where the transactions took place.

There are three highlights in the lifetime of a disposable OV-chipkaart ticket: (1) getting purchased from a vending machine, (2) checking in when arriving at the station, and finally (3) checking out when leaving.

The legitimate access to the memory of a disposable OV-chipkaart ticket happens this way and in this order. (1) After a disposable ticket is purchased, a purchase transaction will be written into slot 2, and the lock bytes will be set as 0x00F0. Pages 12~15 are locked as read-only and cannot be changed anymore. (2) When arriving at a station, the user uses the disposable ticket to check-in. During the check-in process, the OTP bytes are set to indicate the number of remaining rides. The check-in transaction is written into slot 1. If the value in the OTP bytes is 0xFFFF, the system will refuse the card since it has no more rides available. (3) When the traveler arrives at the destination station and checks out, a new check-out transaction is written into slot 2, and overwrites the purchase transaction. When the user uses the card again, a new check-in transaction is written into slot 1, overwriting the old check-in transaction, and the new check-out transaction is written into slot 2 and overwrites the old check-out transaction. This process is repeated until no more rides are available.

To prevent an attacker from using an invalid card to check-in, the OV-chipkaart offers a defense mechanism. When someone attempts to use an invalid card to check-in, the defense mechanism gets triggered the moment the reader sets the lock bytes to 0xF8FF. Pages 3~15 enter a read-only state and no more access is allowed. Now this card is considered as permanently invalid.

The University-of-Amsterdam report [61] reveals that irregular usage behaviors disclose three vulnerabilities of OV-chipkaart: (1) failure of defense mechanism, (2) repeated check-out and (3) free travel. The report also proposes their countermeasures and improvements as follows.

(1.a) Failure of defense mechanism: Upon purchasing a disposable OV-chipkaart, the attacker sets all three block-locking bits to 1 to lock the locking bits from L3 through L15. Thus, when a reader detects an invalid card and fails to change locking bits, the defense mechanism fails.

(1.b) The solution: Adding a new checking function into the OV-chipkaart reader so if any card that is detected to have any block-locking bits set, that card is regarded as invalid and is refused.

(2.a) Check-out repetition problem: Because check-in transactions and check-out transactions are stored in user programmable read/write memory, one can readily reverse the state of transaction as follows: the attacker backs up the purchase transaction after purchasing a ticket, and then uses this back up data to overwrite the check-out transaction. This vulnerability would allow a bunch of people to use the same card to check out many times.

(2.b) The solution: In the OV-chipkaart system, the OTP counter is checked only for check-in, and is ignored when the ticket user checks out. Therefore, one proposed solution is to check the OTP for both check-in and check-out. Due to the feature of irreversibility of the OTP counter, the system can also keep record of the number of check-out transactions to prevent repeated check-out.

(3.a) Free travel problem: Because the transactions are stored in user read/write memory, an attacker can back up pages 4~11, follow normal check-in and check-out processes, and then write back the back-up data to the memory. This modified card can be used again and again until it expires.
(3.b) The solution: The weakness resulting in the free travel problem is that the system would not reject a modified card with manipulated pages 4~11, even though the OPT counter shows that no more ride there is available. Therefore, the software for check-in and the OTP counter should both be redesigned carefully, getting every record of check-in and check-out transaction verified.
Because Mifare Ultralight neither supports authentication nor provides encryption, the security of the applications depends on the well-designed memory configuration and the software semantics. For example, one can adopt 3DES to encrypt the data before writing them back to the card, and can add the MAC (Message Authentication Code) to ensure the integrity of the stored data.
In addition to the weaknesses of Mifare Ultralight cards, the authors of [54, 55] have also found some forgery problems that might bug Mifare Classic cards. These reports further arouse people's attention on the security of RFIDs.

4. Memory access and weaknesses of EPC Class1 Gen2

EPC Class1 Generation 2 (Gen 2 for short) is one of the tag standards proposed by EPCglobal [16], which is believed to have great influence on the RFID consumer markets, especially the logistics management part. In a typical application, a reader first probes a Gen 2 tag, and then uses the information on the tag to further acquire the information of the labeled product. However, Gen 2 tags have limited computation capacity, and the communication protocols between readers and tags are not secure. These weaknesses result in privacy violation, tracking problems and cloning problems. This section discusses the memory configuration and the weaknesses of Gen 2 tags.

4.1 Introduction

During the 1990s, the Auto-ID Center at MIT [37] was established, and it soon developed the Electronic Product Code (EPC for short). Extended from the structure of barcode, EPC contains more information about an object than barcode. In Oct. 2003, Universal Code Council (UCC) and Electronic Article Numbering (EAN) purchased the technology of EPC from the Auto-ID Center. Then they founded a nonprofit organization EPCglobal [16] to advance the standards of EPC and EPC network and to enhance the efficiency of the supply chain operation.
EPC Class 1 Generation 2 (Gen 2 for short) [17], a passive tag proposed by EPCglobal, is designed to be applied universally in supply chains, and now it is ISO 18000-6C standard. Gen 2 tags use no battery, and they are powered by the electricity the readers provide. It communicates at UHF band-860~960MHz with the communication range stretching from 2m to 10m. However, Gen 2 tags are easily interfered by metal or liquids, resulting in poor quality communication. Currently, Gen 2 tags are employed in a number of applications such as logistics management, luggage management, etc. In the future, the price of the Gen 2 tag will be around $0.05~0.1 [58], and many related devices are being developed [57]. In the next sub-section, we shall introduce the memory organization of the Gen 2 tag.

4.2 Gen 2 tag memory configuration

Gen 2 tag memory is logically divided into four banks: user memory (bank 11), TID memory (bank 10), EPC memory (bank 01) and reserved memory (bank 00). Each bank may have zero or more memory words (1 word = 16 bits). Fig. 18 shows the Gen 2 tag logical memory.

- User memory (bank 11)

This bank allows users to organize and store user data.

- TID memory (bank 10)

This bank stores a unique tag ID which is 8 bits (00_h~07_h) of ISO/IEC 15963 class-identifier value - either $E0_h$ or $E2_h$. If the class-identifier value is $E0_h$, then the memory location 08_h~$0F_h$ contains an 8-bit manufacturer identifier, and 10_h~$3F_h$ contains a 48-bit tag serial number which is assigned by the tag manufacturer.

- EPC memory (bank 01)

This bank consists of three parts: the Electronic Product Code (EPC), 16-bit Protocol-Control (PC) bits and a 16-bit cyclic redundancy code (CRC-16). The Electronic Product Code contains the header, general manager number, object class, and serial number. It is unique and can be extended to meet the requirements of different industries. Currently, the EPC can only be encoded either the 64-bit way or the 96-bit way, but the 256-bit method is on the way. Users, depending on the requirements, can choose their own way of encoding. The Protocol-Control (PC) bits record the lengths of PC and EPC, and the structure of EPC, whether it follows the EPCglobal™ Tag Data Standards or ISO/IEC 15961.

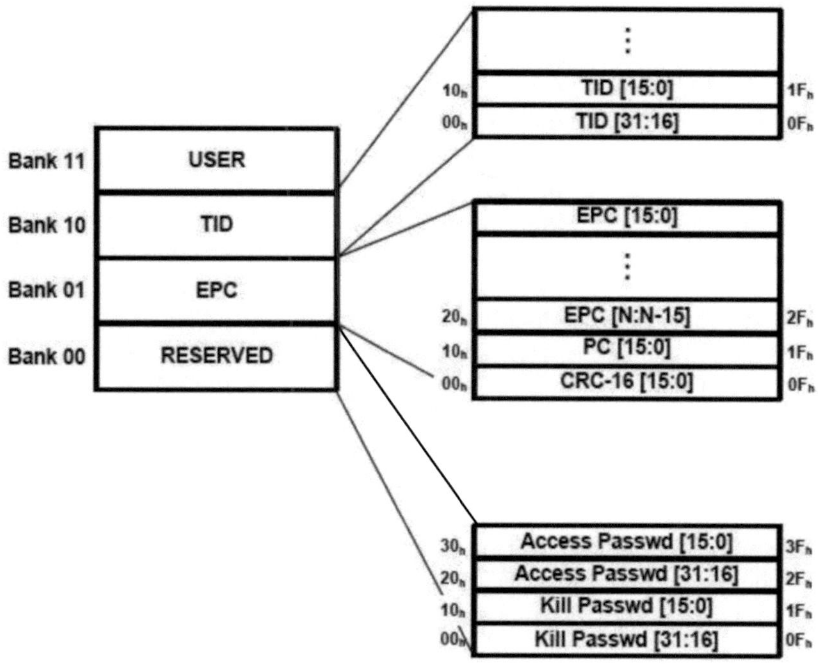

Fig. 18. Gen 2 tag logical memory [17]

- Reserved memory (bank 00)

This bank stores a 32-bit kill password and a 32-bit access password.

- **Kill password**: A reader may pre-set a kill password to a tag. When the reader later issues the same kill password to the tag, the tag is permanently disabled. Once the reader executes the above operation, the tag is permanently silent. This function is used to disable the access to protect the privacy of the consumer when a product is purchased.
- **Access password**: A reader may pre-set an access password to a tag. When the reader later issues the same access password to the tag, the reader would lock one memory bank such as EPC memory, TID memory or User memory to prevent an un-authorized reader from writing data into the above-mentioned memory banks. Or the reader would lock a kill password or an access password to prevent an un-authorized reader from modifying the passwords.

The Gen 2 tag supports a 16-bit pseudo-random number generator (PRNG), a slot counter and the bitwise XOR operation. A 16-bit cyclic redundancy code (CRC) is used to protect the integrity of the communications. A 32-bit kill password is used to disable the access to protect the privacy of the consumer when a product is purchased, and a 32-bit access password is used to authenticate the readers so that no illegal writing will happen.

In the communications between readers and tags, each tag generates at least two random numbers—RN16 and handle. The tag uses the RN16 to identify whether a reader is the authenticated object, and the reader uses one or more handles to execute such operations as writing, locking, killing, etc. The communication protocols are to be introduced in the next section.

4.3 Gen 2 communication protocols

The communication protocols between readers and tags consist of three parts. Firstly, a reader picks a number of tags as the communication objects. Secondly, the reader obtains the EPC codes of the tags. After that, the reader will execute some specific operations such as reading, writing, locking, and killing. The process where the reader obtains the EPC code is shown in Fig. 19, and the steps are as follows.

Step 1. After the reader picks a number of tags, the reader issues a *query* command (which contains a 5-bit CRC) to start the communication between the reader and the tags.

Step 2. Upon receiving the *query* command, each of the tags generates a random number *RN16* and computes a slot value. Then the reader issues some *queryRep* commands to the tags to make the tags' slot values to descend progressively. When the slot value of one of the tags hits zero, the tag responds to the reader with its *RN16,* and communicates with the reader.

Step 3. When receiving a random number *RN16*, the reader sends the same *RN16* back to the tag as a reply. This way, the validity of the received *RN16* can be checked, and the reader authenticated.

Step 4. If the verification succeeds, the tag transmits *EPC* and *PC* to the reader.

Step 5. After obtaining the tag's *EPC,* the reader executes operations such as reading, writing, locking and killing. Before executing the operations, the reader needs to issue a *Req_RN* command with the *RN*16 attached to it to make the tag respond with a new random number—the *handle.*

Step 6. The tag generates a new random number *handle* and responds to the reader with it.

Step 7. The reader executes such operations as reading, writing, locking and killing commands (the access commands in Fig. 19 that contain *handle*). If the reader executes a writing command to write data into the tag, the written data will be divided into successive words, and then each word is XORed with *handle* when it is transmitted to the tag. If the reader locks or kills the tag, the reader needs to input the access password or the kill password respectively. The password is divided into successive words, and then each word is XORed with *handle* when it is transmitted to the tag.

Step 8. The tag verifies the received *handle*. If the verification succeeds, the tag executes the specified operation (reading, writing, locking or killing), and then responds to the reader with *handle*. If the tag receives a reading operation, the tag responds with extra *Data* that the reader requests to read.

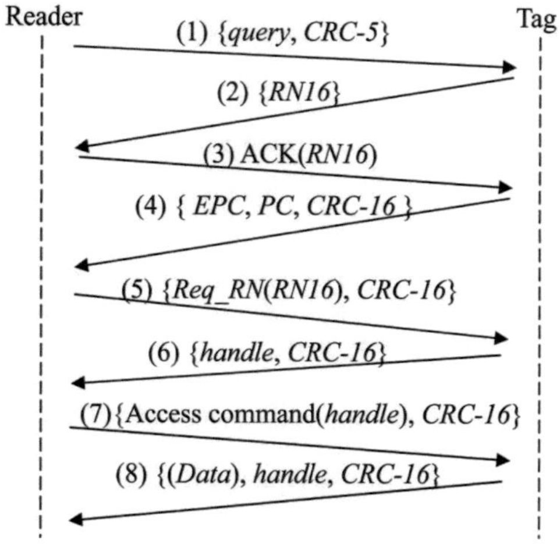

Fig. 19. The Gen 2 tag communication protocol

4.4 The weaknesses

Duc et al. [15] revealed that an attacker can easily derive the passwords from the communications because these passwords are only XORed with a transmitted random number. If an attacker obtains the random number *handle* in Step 6, then in Step 7 the tag's *passwords* could leak out, and in turn the attacker could access the contents in Step 8. Furthermore, please also notice that the Gen 2 tag always responds to any requests with its EPC code; therefore, it is not suitable for applications concerning anonymity or tracking.

In this chapter, we have introduced RFID authentication protocols, have examined their security weaknesses, have classified these protocols into several categories and have discussed the features of these categories. We also introduce Mifare Ultra-light card and EPC gen 2, two popular tags on the market, and introduce the security of the two tags.

5. Reference

[1] G. Avoine, E. Dysli, and P. Oechslin, "Reducing time complexity in RFID systems," The 12th Annual Workshop on Selected Areas in Cryptography (SAC), LNCS 3897, pp. 291-306, Springer, 2006.

[2] G. Avoine, and P. Oechslin, "A scalable and provably secure hash-based RFID protocol," IEEE PerCom, 2005, pp. 110-114, 2005.

[3] G. Avoine, and P. Oechslin, "RFID traceability: a multi-layer problem," in Financial Cryptography 2005, LNCS 3570, pp. 125-140, Springer, 2005.

[4] L. Bolotnyy, and G. Robins, "Generalized yoking-proofs for a group of RFID tags", in Proc. of MOBIQUITOUS 2006, July 2006.
[5] S. Brands and D. Chaum, "Distance-bounding protocols", In Advances in Cryptology EUROCRYPT'93, LNCS 765, pp. 344–359. Springer-Verlag, 1993.
[6] J. Bringer, H. Chabanne and E. Dottax, "HB++: A Lightweight Authentication Protocol Secure against Some Attacks," IEEE International Conference on Pervasive Service, Workshop on Security, Privacy and Trust in Pervasive and Ubiquitous Computing - SecPerU, 2006.
[7] M. Burmester, B. de Medeiros, and R. Motta, "Provably secure grouping-proof for RFID tags", Cryptology ePrint Archive, Report 2007/407, 2007.
[8] T. Cao, E. Bertino, H. Lei, "Security Analysis of the SASI Protocol", IEEE Transactions on Dependable and Secure Computing, 2008. http://doi.ieeecomputersociety.org /10.1109/TDSC.2008.32.
[9] H. Chabanne and G. Fumaroli, "Noisy Cryptographic Protocols for Low-Cost RFID Tags", IEEE Trans. on Information Theory 52(8), Aug. 2006.
[10] H. Y. Chien, "Secure Access Control Schemes for RFID Systems with Anonymity", *accepted and to be printed in proceedings of 2006 International Workshop on Future Mobile and Ubiquitous Information Technologies* (*FMUIT'06*), May, Japan.
[11] H. Y. Chien, "SASI: A New Ultra-Lightweight RFID Authentication Protocol Providing Strong Authentication and Strong Integrity," IEEE Transactions on Dependable and Secure Computing 4(4), October, 2007.
[12] H.-Y. Chien, and C.-H. Chen, "Mutual Authentication Protocol for RFID Conforming to EPC Class 1 Generation 2 Standards," Computers Standards & Interfaces 29(2), pp 254-259, 2007.
[13] H.-Y. Chien, C.-W. Huang, "Security of Ultra-Lightweight RFID Authentication Protocols and Its Improvements," ACM Operating System Reviews 41(2), pp. 83-86, 2007.
[14] Y. Desmedt, "Major security problems with the'unforgeable' (Feige)-Fiat-Shamir proofs of identity and how to overcome them", in SecuriCom '88, pp. 15–17, SEDEP Paris, France, 1988.
[15] D. N. Duc, J. Park, H. Lee and K. Kim (2006), "Enhancing Security of EPCglobal Gen-2 RFID Tag against Traceability and Cloning," *The 2006 Symposium on Cryptography and Information Security.*
[16] EPCglobal, http://www.epcglobalinc.org/ .
[17] EPCglobal Inc., UHF Class 1 Gen 2 Standard Version 1.1.0, 2007, http://www.epcglobalinc.org/standards/uhfc1g2/ (2007/10/20).
[18] H. Gilbert, M. J.B. Robshaw, and Y. Seurin, "HB#: Increasing the Security and E ciency of HB+", IACR eprint, eprint.iacr.org/2008/028.pdf.
[19] A. D. Henrici, and P. MÄuller (2004), "Hash-based Enhancement of Location Privacy for Radio-Frequency Identification Devices using Varying Identifiers," *In the Proceedings of PerSec'04 at IEEE PerCom*, pp.149-153.
[20] N. Hopper and M. Blum. A Secure Human-Computer Authentication Scheme. Technical Re-port CMU-CS-00-139, Carnegie Mellon University, 2000.
[21] N. Hopper and M. Blum. Secure Human Identification Protocols. Adv. in Cryptology | Asia-crypt 2001, LNCS 2248, pp. 52-66, 2001.
[22] A. Juels (2005), "Strengthening EPC Tag against Cloning," *To Appear in the Procceedings of WiSe '05.*
[23] A. Juels, "Yoking proofs for RFID tag", in: Proc. of the Second IEEE Annual Conference on Pervasive Computing and Communications Workshops, pp. 138-142, DC, USA, 2004.
[24] A. Juels, "Generalized 'yoking-proofs' for a group of RFID tags", In MOBIQUITOUS 2006, 2006.

[25] A. Juels, D. Molner, and D. Wagner, "Security and Privacy Issues in E-passports", RSA Laboratories, and UC-Berkeley.
[26] A. Juels, R. Rivest, and M. Szydlo, "The Blocker Tag: Selective Blocking of RFID tags for Consumer Privacy," 8th ACM Conf. Computer and Comm. Security, V. Atluri, ed., ACM Press, 2003, pp. 103–111.
[27] A. Juels and S. Weis. Authenticating Pervasive Devices with Human Protocols. Adv. in Cryptology | Crypto 2005, LNCS vol. 3621, Springer-Verlag, pp. 293-308, 2005.
[28] S. Karthikeyan, M. Nesterenko (2005), "RFID security without extensive cryptography," *Proceedings of the 3rd ACM workshop on Security of ad hoc and sensor networks*, pp. 63-67.
[29] J. Katz and J.-S. Shin. Parallel and Concurrent Security of the HB and HB+ Protocols. Adv. in Cryptology — Eurocrypt 2006.
[30] J. Kim, D. Choi, I. Kim, and H. Kim, "Product authentication service of consumer's mobile RFID device," ISEC'06, pp. 1-6, 2006.
[31] S. M. Lee, Y. J. Hwang, D. H. Lee, and J. I. Lim, "Efficient Authentication for Low-Cost RFID Systems," International Conference on Computational Science and its Applications - ICCSA 2005, May 2005.
[32] Y. K. Lee and I. Verbauwhede, "Secure and Low-cost RFID Authentication Protocols," Adaptive Wireless Networks - AWiN, November 2005.
[33] T. Li, and R. H. Deng, "Vulnerability Analysis of EMAP-An Efficient RFID Mutual Authentication Protocol," The Second International Conference on Availability, Reliability and Security (AReS 2007), Vienna, 2007.
[34] T. Li, and G. Wang, "Security Analysis of Two Ultra-Lightweight RFID Authentication Protocols", IFIP SEC 2007, May 2007.
[35] T. Li, G. Wang, R. H. Deng, "Security Analysis on a Family of Ultra-lightweight RFID Authentication Protocols", Journal of Software 3(3), Mar. 2008.
[36] MifareNet, http://mifare.net/.
[37] MIT Auto-ID center. http://www.autoidcenter.org/.
[38] D. Molnar and D. Wagner (2004), "Privacy and security in library RFID: Issues, practices, and architectures," *Conference on Computer and Communications Security – CCS'04*, pp. 210–219.
[39] J. Munilla and A. Peinado, "Distance bounding protocols for RFID enhanced by using void-challenges and analysis in noisy channel", Wireless Communications and Mobile Computing 2008, DOI:10.1002/wcm.590.
[40] J. Munilla, and A. Peinado, "HB-MP: a further step in the HB-family of lightweight authentication protocols," Computer Networks, doi:10.1016/ j.comnet.2007.01.011, 2007.
[41] NTRU. GenuID. http://www.ntru.com/products/genuid.htm.
[42] NXP Semiconductors, "Mifare Ultralight features and hints"http://www.nxp.com/acrobat /other/identification/M073120.pdf.
[43] M. Ohkubo, K. Suzuki, and S. Kinoshita, "Cryptographic approach to 'Privacy-friendly' tag," in RFID Privacy workshop, MIT, USA, 2003.
[44] K. Osaka, T. Takagi, K. Yamazaki, and O. Takahashi, "An efficient and secure RFID security method with ownership transfer," Proc. of International Conference on Computational Intelligence and Security 2006, pp. 1090-1095, LNCS 9743, Springer, 2006.
[45] P. Peris-Lopez, J. C. Hernandez-Castro, J. M. Estevez-Tapiador, and A. Ribagorda, " LMAP: A Real Lightweight Mutual Authentication Protocol for Low-cost RFID tags", in: Proc. of 2nd Workshop on RFID Security, July 2006.
[46] P. Peris-Lopez, J. C. Hernandez-Castro, J. M. Estevez-Tapiador, and A. Ribagorda, "M2AP: A Minimalist Mutual-Authentication Protocol for Low-cost RFID Tags", in: Proc. of International Conference on Ubiquitous Intelligence and Computing UIC'06, LNCS 4159, pp. 912-923, Springer, 2006.

[47] P. Peris-Lopez, J. C. Hernandez-Castro, J. M. Estevez- Tapiador, and A. Ribagorda, "EMAP: An Efficient Mutual Authentication Protocol for Low-cost RFID Tags," in: OTM Federated Conferences and Workshop: IS Workshop, November 2006.
[48] P. Peris Lopes, J. Hernandez-Castro, J.M. Estevez-Tapiador, and A. Ribagorda, "Solving the Simultaneous Scanning problem Anonymously: Clumping proofs for RFID Tags", in Proc. of SecPerU'07, 2007.
[49] R. C.-W. Phan, "Cryptanalysis of a New Ultralightweight RFID Authentication Protocol - SASI", IEEE Transactions on Dependable and Secure Computing, 19 June 2008. http://doi.ieeecomputersociety.org/10.1109/ TDSC. 2008.33.
[50] T. Phillips, T. Karygiannis, R. Kuhn (2005), "Security Standards for the RFID Market," IEEE Security & Privacy, Vol. 3, No. 6, pp. 85-89.
[51] S. Piramuthu, "HB and Related Lightweight Authentication Protocols for Secure RFID Tag/Reader Authentication," CollECTeR Europe Conference, June 2006.
[52] S. Piramuthu, "Protocols for RFID tag/reader authentication", Decision Support Systems 43(3), pp. 897-914, 2007.
[53] S. Piramuthu, "On existence proofs for multiple RFID tags", in: Proc. of IEEE International Conference on Pervasive Services, Workshop on Security, Privacy and Trust in Pervasive and Ubiquitous Computing- SecPerU 2006, France, June 2006. IEEE.
[54] H. Plötz and K. Nohl, " Mifare little security, despite obscurity", 24th Chaos Communication Congress, http://events.ccc.de/congress/2007/Fahrplan/events/2378.en.html.
[55] The Digital Security research group of the Radboud University Nijmegen, "Security Flaw in Mifare Classic", http://www.ru.nl/ds/research/rfid/.
[56] J. Reid, J.M. Gonzlez, T. Tang, B. Senadji, "Detecting Relay Attacks with Timing-Based Protocols. http://eprints.qut.edu.au/archive/00003264/.
[57] RFID Journal, http://www.rfidjournal.com/.
[58] RFID Journal. Gillette to Purchase 500 Million EPC Tags. http://www.rfidjournal.com, January 2003.
[59] K. Rhee, J. Kwak, S. Kim, and D. Won (2005), "Challenge-response based RFID authentication protocol for distributed database environment," *International Conference on Security in Pervasive Computing – SPC 2005*, pp. 70–84.
[60] J. Saito and K. Sakurai, "Grouping proof for RFID tags", in: Proc, of AINA 2005, Taiwan, March 2005. IEEE.
[61] P. Siekerman, M. v. d. Schee, "Security Evaluation of the disposable OV-chipkaart", System and Network Engineering, University of Amsterdam, http://staff.science.uva.nl /~delaat/sne-2006-2007/p41/Report.pdf
[62] H. M. Sun, W. C. Ting, and K. H. Wang, "On the security of Chien's Ultra-lightweight RFID authentication protocol", IACR eprint, eprint.iacr.org/2008/083.pdf.
[63] S. A. Weis, (2003), "Security and Privacy in Radio-Frequency Identification Devices," *Masters Thesis MIT*.
[64] S. A. Weis, S. E. Sarma, R. L. Rivest, and D. W. Engels, "Security and Privacy Aspects of Low-Cost Radio Frequency Identification Systems," In the Proceedings of the First Security in Pervasive Computing, LNCS, Vol. 2802, pp.201-212, 2003.
[65] D. J. Wheeler and R. M. Needham, "TEA, a tiny encryption algorithm", technical report, Computer Laboratory, University of Cambridge, 1995.
[66] J. Yang, J. Park, H. Lee, K. Ren, and K. Kim. (2005), "Mutual authentication protocol for low-cost RFID," *Handout of the Ecrypt Workshop on RFID and Lightweight Crypto.*
[67] J. Yang, K. Ren and K. Kim (2005), "Security and privacy on authentication protocol for low-cost radio," *The 2005 Symposium on Cryptography and Information Security.*

15

A Secure Mutual Authentication Protocol for Low-cost RFID System

N.W. Lo, Tzu-Li Yang and Kuo-Hui Yeh
National Taiwan University of Science and Technology
Taiwan, R.O.C.

1. Introduction

With extended data storage space and advanced wireless transmission capability, Radio Frequency IDentification (RFID) is rapidly deployed to replace barcode position in our daily lives and considered as the next generation identification technology in ubiquitous communication environment. The most important key factor of RFID technology is to enable systems with the ability to automatically identify labeled objects without the constraint of line of sight. RFID technology is a well known AIDC (Automatic Identification and Data Capture) technology to provide the benefits including contactless read, long transmission range and transaction time saving (Garfinkel & Rosenberg, 2005). Most of innovative applications designed for RFID system can be divided into following classes such as asset management, tracking, authenticity verification, matching, process control, access control, automated payment and supply chain management (Karygiannis et al., 2007).

In spite that the adoption of RFID technology becomes popular in a board range of applications, the cost of a RFID tag is still too expensive to be fully adopted by logistic and retailer industries. Even though from the logistic and retailer industries point of view, to label RFID tags on all sale items is still cost-prohibitive under the current price of a passive RFID tag. Nevertheless, the convenience of RFID technology still has a great attraction for inventory management. For example, in 2005, Wal-Mart which is the biggest retailer in America declared a new policy to force its top 500 suppliers to adopt RFID technology for inventory management; otherwise, Wal-Mart will deny new transaction contracts from those who do not comply this new policy. Because of this policy, all top 500 suppliers start to apply RFID tags onto their merchandises by spending and absorbing extra RFID cost. In contrary, the introduction of RFID technology can provide great benefits for Wal-Mart to control logistic process accurately, replenish empty stock efficiently and lower space requirement for goods storage.

Although the widespread use of RFID technology makes human life better than past, the security invasion and user privacy disclosure are still concerned by individuals and organizations. For example, in 2006, Metro AG which is the biggest supermarket chain store in Germany used the RFID technology to not only automatically manage production and stock but also help customers search their target items quickly. Metro AG gave VIP cards to the top 10% customers and based on the historical shopping behaviors of a VIP customer to recommend products nearby the customer's current location. However, Metro AG did not notify VIP customers that the VIP card is embedded with RFID. Three months later, a VIP member curiously disassembled his card and recognized the RFID secret of the VIP card. About ten thousand members' location privacy is at risk of disclosure because the unique customer number stored in each VIP card can be easily read by a malicious stalker using a handheld RFID reader.

As we mentioned above, the RFID technology faces serious security threats and privacy concern (Juels et al., 2005; Weis, 2003). Wireless communication and cost-down consideration on RFID systems are the two main factors that cause these security threats. In RFID operation environment, a passive RFID tag must be powered and triggered by a broadcast signal through the forward channel from a RFID reader, and the reader receives the response from the tag via the backscatter channel. An adversary may capture transmitted messages between reader and tag easily with wireless eavesdropping device. Furthermore, an adversary can utilize the captured messages to invoke other attacks such as object tracking, tag compromise and tag impersonation. In short, the concerns on information security and privacy protection will impede the future development of RFID technology. In order to secure data integrity, data confidentiality, non-repudiation, and availability of a RFID system, a straight forward thought is to apply existing authentication protocols on wireless networks. However, due to the nature of restricted computation ability and limited memory storage of a low-cost passive RFID tag, it is difficult to implement a secure or robust RFID system with powerful cryptographic operations such as RSA, DES, and AES (Datasheet Helion Technology, 2005) as existing authentication protocols did.

In the past five years, many researchers had proposed ideas to protect data security and user privacy (Weis et al., 2003; Lo & Yeh, 2007) on RFID systems. These researches use powerful cryptographic operations (Feldhofer et al., 2004; Kumar & Paar, 2006) such as symmetric key encryption, public key infrastructure and one-way hash function to prevent information leakage. Although those operations can provide strong protection to defend against malicious attacks, low-cost RFID tags with highly constrained resource are not able to carry out expensive cryptographic primitives to perform strong authentication. In fact, a passive tag can only contain 5K - 10K gates; on the contrary, a cryptographic primitive requires 250 - 3K gates. Hence, powerful encryptions are hardly possible to be built in a passive tag in the near future. In order to comply with the resource constraint, a few new authentication protocols with lightweight encryptions (Peris-Lopez et al., 2006; Chien, 2007; Yu et al., 2007; Juels, 2005) are invented to fit the physical limitation of a passive tag. However, those proposed schemes cannot provide enough security level in general; more specifically, they cannot prevent all major or general attacks such as eavesdropping, tracking, replay attack and Denial of Service, and preserve the forward secrecy of tagged object at the same time. Therefore, in order to successfully defend against those security threats, we propose a new secure mutual authentication protocol for low-cost RFID systems, named as SMAP-LRS, to achieve higher security level and be compatible with the hardware restriction of passive RFID tag at the same time. The design of SMAP-LRS protocol adopts simple cryptographic operations to comply with existing RFID standards. In addition, a bit flag mechanism is introduced in our scheme to resolve the Denial of Service attack and save the memory space for protocol implementation at backend server.

The rest of this chapter is organized as follows. Section 2 reviews previous work on RFID authentication protocol. Next, we propose a new authentication scheme for low-cost RFID system in section 3. The security analysis of our scheme is presented in section 4. Finally, we summarize our conclusion in section 5.

2. Related work

In recent years, the vast literatures have addressed the security and privacy concerns on the use of RFID tags. Based on the type of encryption primitive used on RFID system, we classify RFID authentication protocols into four classes. The first class of RFID authentication protocol is hash-based. Most of those schemes only use hash function for data encryption. In 2003, Weis et al. (Weis et al., 2003) proposed a new authentication protocol for RFID system using hash function to achieve data security and user privacy. In their hash-based access control mechanism, the tag does not change its identification in authentication sessions. An adversary can easily trace his target RFID object by eavesdropping the same ID transmitted through air interface. Ohkubo et al. (Ohkubo et al., 2003) developed a secure authentication protocol based on hash chain mechanism. This scheme provides indistinguishability and forward security. Through their scheme, a RFID tag can generate a responding message whose content is indistinguishable from truly random value to achieve indistinguishability. At the same time, the property of forward security is preserved because even if an adversary gathers information from transmitted messages during authentication sessions and the secret data stored in a compromised tag, the adversary still cannot derive the secret information of the tag before it is compromised. However, this scheme cannot resist replay attack. Henrici & Müller (Henrici & Müller, 2004) proposed a novel authentication which is based on hash function to provide anonymity and location privacy by updating tag identification in each session. Nevertheless, the tag always responds reader query with the same hashed value of identification before the tag successfully updates its current identification at the end of authentication session. This security flaw allows an attacker to track a specific tag by eavesdropping.

The second class of RFID authentication protocol utilizes hash function and random-number generator. Weis et al. also proposed another authentication protocol in their paper (Weis et al., 2003) by using randomized access control and hash function. The advanced scheme certainly provides stronger anonymity property than the previous hash-based scheme they derived. However, the backend server does not update the database information at all after authentication. An adversary can eavesdrop the transmitted messages between a reader and tags, as well as injecting arbitrary messages into the communication channel. In other words, the adversary can impersonate the original tags and send arbitrary message to backend server until the next authentication session. An and Oh (An & Oh, 2005) developed a new authentication protocol which is based on hash function and random number generator. Although authors claimed that their scheme provide data security in different databases, this scheme cannot prevent replay attack and tag tracking. Rhee et al. (Rhee et al., 2005) proposed a challenge-response protocol for authentication to enhance the anonymity and resist replay attack via hash function and pseudo-random number generator. Unfortunately this scheme cannot efficiently support forward secrecy when it encounters adversary attacks. Once the tag is compromised, the adversary can derive or identify the past transmitted messages through revealed secret information from the tag. Kim et al. (Kim et al., 2006) proposed a new scheme which generates stream blocks to update the shared secret information between tag and backend server in an authentication process. Their scheme supports tag anonymity and relay attack resistance. However, the identification of tag can be calculated by using XOR operation with the transmitted message consisting of E_{ID} and random value R2′; the adversary can use the specific characteristic to track a tag virtually anywhere. A new authentication protocol which is based on AES encryption primitive is designed by Feldhofer et al. (Feldhofer et al., 2004). Although the scheme reaches the strongest level of security requirement, it is not suitable for systems using low-cost RFID tags since the computing capability of a passive tag at present cannot support such large computation workload as the AES encryption process requires.

The third class of RFID authentication protocol adopts lightweight encryption primitive. Those schemes utilize the common bit-wise arithmetic operations to perform data encryption task. By doing so, both the low-cost requirement and security robustness for a passive RFID tag can be achieved simultaneously. In 2006, Peris-Lopez et al. (Peris-Lopez et al., 2006) proposed a series of authentication protocols which involve simple bit-wise operations such as AND, OR, XOR and addition mod 2^m. These schemes are very cost-effective and attractive to RFID systems with resource-constrained tags. Nevertheless, Li et al. (Li & Wang, 2007; Li & Deng, 2007; Li, 2008) showed that there are two vulnerabilites, de-synchronization and full-disclosure attack, in these schemes proposed by Peris-Lopez et al. However, Li-Wang's enhancement scheme still cannot successfully remedy these two security weaknesses as shown by Chien and Hwang (Chien & Huang, 2007). In 2007, Chien (Chien, 2007) proposed a new lightweight authentication protocol and corrected the drawback of Peris-Lopez's schemes by applying bit-rotation function. Even though Chien claimed his scheme can provide more robust security features than Peris-Lopez's schemes, the Chien's scheme still is vulnerable in subtle situations. For example, if the *IDS* value of Chien's scheme does not update in a period of time, the tag sent the same *IDS* response to reader might be tracked by adversary.

The forth class of RFID authentication protocol complies with the EPCglobal standard. Sarma et al. (Sarma & Engels, 2003) developed a mutual authentication scheme using pseudo-random number generator only. Although the scheme meets the implementation requirements of the EPCglobal standard, it suffers the problem of tag identification disclosure. Chien and Chen (Chien & Chen, 2007) proposed an enhanced EPCglobal complied authentication protocol. However, Lo and Yeh (Lo & Yeh, 2007) showed that Chien and Chen's scheme cannot provide forward security and suffer heavy computation workload at the backend server. Correspondingly, Lo and Yeh proposed a new authentication scheme to improve user privacy and data security.

3. Proposed SMAP-LRS protocol

As we mentioned above, the research in the past does not guarantee enough security for RFID system; previously proposed schemes only prevent a few specific types of security attacks. To implement encryption module in a passive RFID tag still requires lots of gates and space. In consequence, the cost of tag becomes more expensive and the tag needs more power to drive. Strong encryption operations, as more computing time required, might also delay tag response time. Most of passive tags cannot afford the resource demand from strong encryption primitive at present. The EPCglobal Class1 Gen2 tag standard only defines CRC function and pseudo-random number generator for tag to operate. Although some lightweight encryption primitives for RFID tags are introduced and claim that they are adaptive to the resource constraint of RFID tag (Duc et al., 2006; Juels, 2005; Karthikeyan & Nesterenko, 2005), most of them have not demonstrated that these schemes can really work on passive tags to achieve security requirement. Poschmann et al. (Poschmann et al., 2007; Poschmann et al., 2006) had proposed a new hash function requiring less number of gates to supply the need of lightweight encryption primitives for RFID authentication. Although this method seems to be lightweight enough to fit in a low-cost RFID tag, the security strength of this hash function still remains as an open question. In the following, we introduce a newly designed authentication protocol, which uses simple bit-wise arithmetic operations such as AND, OR, XOR and ROT (bit rotation) to achieve the security and privacy requirements of low-cost RFID system.

3.1 System assumption

We assume that tag is vulnerable to be compromised. When the tag was compromised, the secret information of tag which contains shared symmetric key and tag identification can be retrieved by adversary. The system assumption of our scheme is described below. Our protocol has three main components: tag, reader and the backend server. Tags are passive tags, reader is the equipment to collect data from tags, and the backend server is to analyze the collected data. The communication channel between tag and reader are classified into two categories, forward channel and backscatter channel. The backscatter channel is namely as back channel and reverse channel. The communication channel between reader and backend database is a well protected and trusted system, so that transmitted message cannot be violated or eavesdropped by adversary. In other word, it cannot get any secret information from backend server. Each tag contains four filed data including ID, T_{key}, t and *flag*. ID is the identification of RFID tag. According to EPC global standard, the length of tag identification can be 64bits, 96bits and 128bits and 256bits. Accordingly, we assume a reasonable length of tag identification is 96 bits. Sometimes, it has the probability of $1/2^{96}$ to generate the same identification because the length of tag identification has only 96 bits. Many researchers also provide complete solution for tag collision (Shih et al., 2006; Lee et al., 2004). Hence, we think that tag collision is almost impossible happened for RFID tag. T_{key} is the shared secret information in RFID tags as well as an encryption key. t is the counter value represented as total query times. The database includes two data, ID and T_{key}. We assume the length of T_{key} and t is the 96 bits as ID. Finally, we present the system notation in the following. Note that the flag mechanism design at backend server is used for solving *DoS* attack.

- S: random generator number is generated by reader for each session.
- *flag*: the value is used to indicate the tag is normal state(*flag*=0) or exceptional state(*flag*=1).
- i : the i th session
- ID_i, ID_i': the identification of tag at tag and backend server.
- ID_{iL}, ID_{iL}': the left half of tag identification at tag and backend server.
- ID_{iR}, ID_{iR}': the right half of tag identification at tag and backend server.
- T_{key} , T_{key}': the secret symmetric key of tag at tag and backend server.
- T_{keyL}, T_{keyL}': the left half of secret symmetric key of tag at tag and backend server.
- T_{keyR}, T_{keyR}': the right half of secret symmetric key of tag at tag and backend server.
- t: a counter value of tag, when flag is one, it generates a value to encrypt the message.
- M_1, M_2, M_3, M_4, M_1', M_2', M_3' and M_4': the encrypted message at tag and backend server.
- K_1, K_2, K_1' and K_2': the symmetric secret keys of tag which update for each session at tag and backend server.
- R, R': the certificated message at tag and backend server.
- R_L, R_L': the left half of certificated message R at tag and backend server.
- R_R, R_R': the right half of certificated message R of tag at tag and backend server.

- ID_{i+1}, $ID_{i+1'}$: the updated identification of tag at tag and backend server.
- ID_x: the identification of tag in any session
- $\oplus$: XOR
- $/\backslash$: AND
- $\backslash/$: OR
- $\|$: Concatenation
- +: ADD
- $Rot(x, y)$: left rotate the value of x with y bits

3.2 Mutual authentication protocol

In this section, we propose a new mutual authentication protocol namely SMAP-LRS. SMAP-LRS is based on two conditions, the first one is normal state (flag is zero) and second one is exceptional state (flag is one). After the authentication is successfully completed, the protocol switches to normal state and the flag of tag will be changed from one to zero.

The proposed scheme consists of two different conditions based on previous authentication session is safely terminated (*flag* = 0) or not (*flag* =1). The condition of normal state is illustrated as Fig. 1.

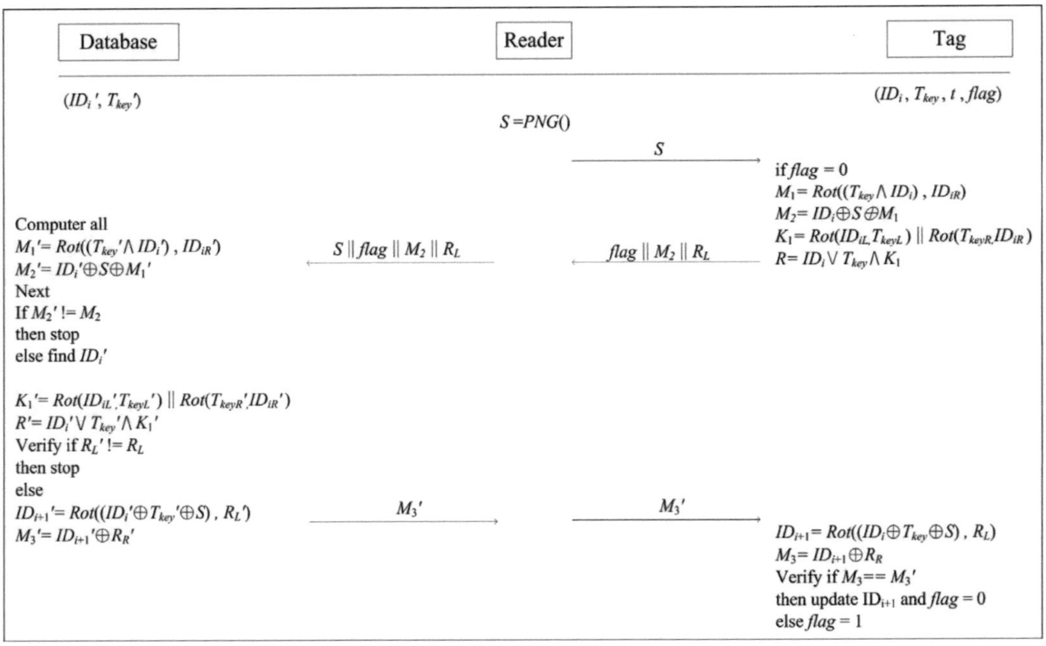

Fig. 1. The normal state of mutual authentication protocol

Condition 1: previous authentication session is safely terminated (*flag* = 0)

Step1: *Reader → Tag: Query*

The reader generates random number S and sends it as a query command to tag.

Step2: *Tag → Reader: flag*, M_2, R_L

When tag receives the query S from reader, it checks the flag state to decide the protocol is normal state. First, tag computes $M_1 = Rot((T_{key} /\backslash ID_i), ID_{iR})$ and response value

$M_2=ID_i \oplus S \oplus M_1$ which protect *ID* to avoid from eavesdropping. Second, tag computes T_{keyL}, T_{keyR} and $K_1=Rot(ID_{iL}, T_{keyL}) \| Rot(T_{keyR}, ID_{iR})$ to generate certificated message $R=ID_i \vee T_{key} \wedge K_1$. The certificated message *R* will be used to authenticate the tag and reader. Finally, the tag will send these response value *flag*, M_2, R_L to reader.

Step3: *Reader → Backend Server: S, flag,* M_2, R_L

After the reader receives the response from tags, it appends the number *S* and forwards to backend server.

Step4: *Backend Server → Reader:* M_3'

When backend server receives the authentication request (*flag*, M_2, R_L, *S*) from reader, server computers all $M_1'=Rot((T_{key}' \wedge ID_i'), ID_{iR}')$. Next, the server reuses M_1' to creates the $M_2'=ID_i' \oplus S \oplus M_1'$ to verify the M_2. If M_2' is the same as M_2, it finds the corresponding record form the database. Otherwise, it terminates the authentication immediately.
After retrieving the value of relative field in the corresponding record, the server computes the $K_1'=Rot(ID_{iL}', T_{keyL}') \| Rot(T_{keyR}', ID_{iR}')$. Next, the backend server keeps to create the certificated message $R'=ID_i' \vee T_{key}' \wedge K_1'$. The server uses the left half of certificated message R', called R_L' to verify whether R_L' is equal the R_L or not. This verification process can ensure the data integrity; otherwise it will terminate the process and respond anything. In order to avoid the tracking attack, the server updates the identification of tag $ID_{i+1}=Rot((ID_i \oplus T_{key} \oplus S), R_L)$ for each session. With new identification, the server can calculates the certificated message $M_3'=ID_{i+1}' \oplus R_R$ and transmits it to tag though reader.

Step5: *Reader → Tag :* M_3'

When tag receives M_3', it computes the new identification of tag and uses the updated identification of tag ID_{i+1} to generate the certificated message M_3. If the M_3 is equal to M_3', the tag updates the old identification *ID* with new identification ID_{i+1}. Until the process is successful finished, the tag also resets the flag value to zero.
When the authentication between tag and reader is not completely finished, the flag value will be changed from zero to one. For example, when the authentication is proceeding, once tag does not receive any response from original reader in a period time or the response is invalid, the tag which still receives the query from reader may change its condition to exceptional state. The condition of exceptional state is illustrated as Fig. 2.

Condition 2: previous authentication session is not safely terminated (*flag* = 1)

Step1: *Reader → Tag: Query*

The reader generates random number *S* and sends it as a query command to tag.

Step2: *Tag → Reader: flag,* M_2, M_3, R_L

When tag receives the query again and not terminates safely, it means that it is an exceptional state. So, the tag will calculate the $t = (t+2^t+T_{keyL}) \text{ } mod \text{ } length \text{ } (ID_i)$ value by using T_{key} and mod function. By using *t* value, the tag generates the another identification, namely as $M_1=Rot(ID_i, t)$ and computes the $M_2=S \oplus T_{key} \oplus M_1$ with *S* and T_{key}. In order to use the *t* value to resolve the M_2, we must send the *t* value to the backend server. The only way is to protect *t* value by using T_{key} and M_1. Thus, the $M_3=(T_{key} \wedge M_1) \oplus t$ is a ciphertext to protect the *t* value. At the same time, the tag computes the $K_1=Rot(T_{keyL}, T_{keyR}+t) \| Rot(T_{keyR}, T_{keyL}-t)$ to generate the message $R=T_{key} \vee M_1 \wedge K_1$. The certificated message *R* value will be utilized to conform whether the tag is legal or not. Finally, the tag responds *flag*, M_2, M_3 and R_L to reader.

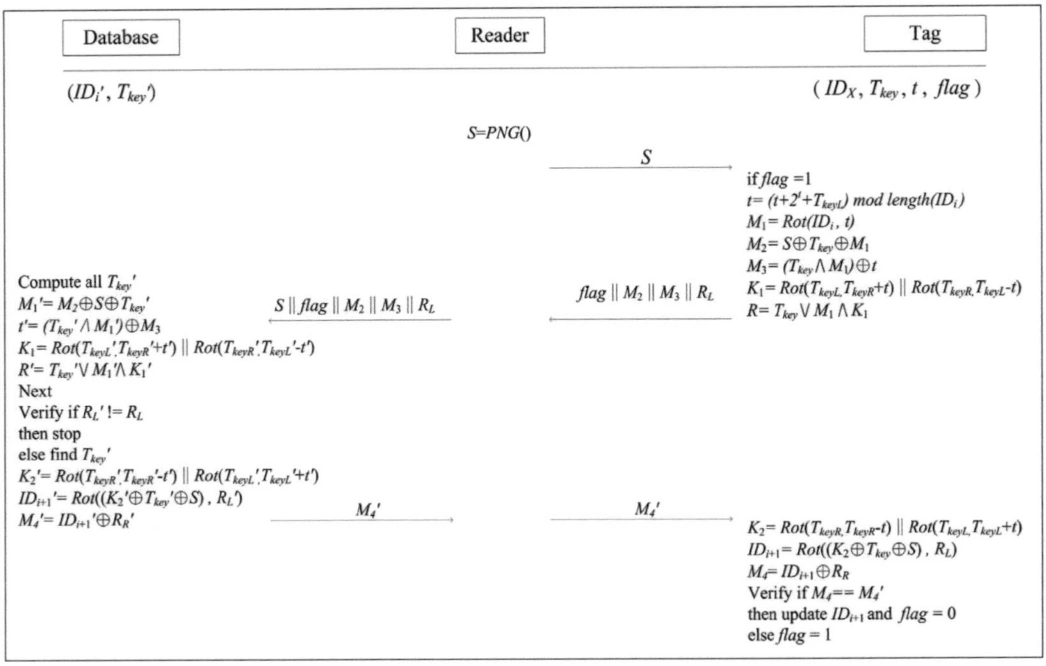

Fig. 2. The exceptional state of mutual authentication protocol

Step3: Reader → *Backend Server: S, flag,* M_2, M_3, R_L

When reader receives the response from tag, it appends S and forwards to the backend server.

Step4: *Backend Server→ Reader:* M_4'

When backend server collects a round of message from reader, it retrieves the $M_1'=M_2'\oplus S\oplus T_{key}'$ by using S, T_{key}' and M_2'. M_2' value is the same as M_2 which sends from tag. then, the backend server decrypts the M_3 with T_{key}' and M_1' to obtain the $t'=(T_{key}' \wedge M_1')\oplus M_3$ value. By using t' value, we can calculate $K_1= Rot(T_{keyL},T_{keyR}+t') \| Rot(T_{keyR},T_{keyL}-t')$ to generate the certificated message $R'=T_{key} \vee M_1' \wedge K_1'$. Next, backend server verifies whether the R_L' is equal to R_L or not. If the pair of values is not match, the authentication process will be terminated immediately. Otherwise, it means that the backend server can identify correctly the corresponding tuple of database. Finally, it computes the $K_2'=Rot(T_{keyR}', T_{keyR}'-t') \| Rot(T_{keyL}', T_{keyL}'+t')$ with T_{keyR}', T_{keyL}'. By using the updated identification of tag $ID_{i+1}'=Rot((K_2'\oplus T_{key}\oplus S), R_L')$ and the right half of R' to create the certificated message $M_4'=ID_{i+1}'\oplus R_R$, the certificated message M_4' provides a proof for tag to verify the reality of reader.

Step5: *Reader → Tag :* M_4'

while the tag receives the message M_4' from backend server, it calculates the new tag identification $ID_{i+1}=Rot((K_2\oplus T_{key}\oplus S), R_L)$. By using the right half of R and ID_{i+1}, the backend server can create the certificated message $M_4 =ID_{i+1}\oplus R_R$ to compare whether the M_4' is equal to M_4 or not. if M_4' is the same as M_4, the identification of tag will change to ID_{i+1} and reset the flag to zero.

4. Security and performance analysis

For the sake of clarity, the aim of this section is to analyze our authentication scheme and compare it with related literature based on following security and performance criterions. First of all, we explain that how to ensure that the protocol is well protected. We illustrate each security analysis in section 4.1. Secondly, we have a comparison for our scheme in storage, operation and communication in section 4.2.

4.1 Security analysis

In this section, we conduct security analysis to proposed authentication scheme.

- Data security

The transmitted message between tag and reader is a ciphertext by using AND, OR, XOR and ROT function. The encrypted message for each session is encrypted by random-generated one time valid numbers to perform beneficial computation. Even if the ciphertext can be modified or eavesdropped, the transmitted messages which provide the security robustness of meaningful data will not be compromised. So we believe that the transmitted message is secure enough to ensure the confidentiality of the transmitted data.

- Anonymity

For each tag, the information of tag is changed dynamically in each session. Even if the authentication process between tag and reader is failure, the tag still has its mechanism to keep the responded message different. In normal state, the transmitted messages are encrypted by different *S* and *ID*. In exceptional state, the transmitted message still keeps being changed by using updated *t* value. Generally speaking, no matter the authentication is success or not, the tag will modify its own data in every session. Hence, the attacker cannot find consistent clues of each tag response to track a specific tag easily.

- Replay attack resistance

SAMP-RLS is a challenge-response protocol using pseudo-random number to prevent replay attack. The message M_1, M_2 and M_3 are refreshing by using *S* and *ID* in each section. Hence, the malicious attack cannot reuse the original message to pass the authentication.

- Denial of Service resistance

As we noted above, *DoS* attack have two different definition. By using a flag mechanism, our scheme allows the tag with constant secret key can still be authentication by backend server and re-synchronize its data with databases. Additionally, comparing other schema against Dos attack, our schema can replace dual tuple of secret information values (*new* and *old*) to save lots of storage space in backend server.

- Forward security

If the adversary collects a series of past transmitted messages and get the secret information of tag in a period. The adversary infers transmitted messages to obtain previous relationship of data. Because the identification *(ID)* of tag is dynamically changed for each session, the adversary is unable to obtain the previous data by using the current secret information of tag and have no co relationship between messages transmitted in consecutive session. The adversary cannot generate new identification and track further recorder. However, if the adversary try to compromise tag to know all data stored in, the attacker still could not trace back the trajectory of compromised tag in our scheme.

- Mutual authentication

SAMP-RLS provides both tag to reader and reader to tag authentications. The R_L is the certificated code to verify the tag. On the contrary, the R_R is the certificated code to verify the reader. Hence, our scheme indeed reaches the aim of mutual authentication.

Introducing the security analysis in our scheme provides the well protection for command attacks. A simple comparison of recent authentication protocols is listed in Table 1. We compare the similar operations of authentication protocols such as EMAP, M2AP, LAMP, SASI, etc.

According to the Table 1 above, our scheme use simple operation to secure message to achieve the requirement of security. It also provides strong security against all kinds of command attacks.

	SMAP-LRS	EMAP	M2AP	LAMP	SASI
Data security	Y	N	N	N	Y
Anonymity	Y	N	N	N	N
Replay attack resistant	Y	N	N	N	Y
DoS resistant	Y	N	N	N	Y
Forward security	Y	N	N	N	Y
Mutual authentication	Y	N	N	N	Y

Table 1. Comparison of other simple operation scheme

4.2 Performance analysis

Our protocol also compares the performance analysis, including storage, operation and communication. In our research, we know that the memory space of our scheme decrease 5*L* of storage and 0.5*L* of communication for the SASI mechanism which is the most low-cost scheme currently. Hence, our scheme reduced about fifty percent of memory space is less than other scheme at present.

In our scheme, we assume that the lengths of the identification or key are 96 bit as *L* bits. First, storage is separated into two parts, one is the memory of tag and the other is the memory of database. The database memory of our scheme contains *ID* and T_{key} are 2*L* bits. Because the memory space of *flag* is one bit, the tag memory of our scheme contained *ID*, T_{key}, *t* and *flag* are about 3*L* bits. Second, the recent papers in designing the authentication protocol usually use hash, Pseudo-random number generator and CRC to protection their protocol. However, our scheme only uses simple operations that fit the requirement of passive tag such as AND, XOR, OR and Rot function. Hence, we believe that simple operation can ensure not only security requirement but also low-cost demanded, especially for EPC global standard. Third, the communication between reader and tag also should be considered because the energy of passive tag comes from reader. The length of message decides the consumption of energy to transmit range. It is an important factor to dispatch the power energy and control the communication. The total communications of our scheme including *flag*, M_2, M_3' and R_L is 2.5*L* bits when our scheme is a normal state. Even if our scheme is exceptional state, the communication of our scheme including *flag*, M_2, M_3, M_4' and R_L is only 3.5*L* bits. We believe that our communication is less 0.5*L* than SASI at least. We list a comparison summary of various schemes in Table 2. We also count the number of simple operation in detail to compare with other low cost authentication protocols in Table 3.

	Memory storage		Operation	Communication
	Tag	Backend Server		
EMAP (Peris-Lopez et al., 2006)	6L	6L	⊕, /\ , \/	5L
M2AP (Peris-Lopez et al., 2006)	6L	6L	⊕, /\ , \/ , +	5L
LMAP (Peris-Lopez et al., 2006)	6L	6L	⊕, /\ , \/ , +	4L
SASI (Chien, 2007)	4L	7L	⊕, /\ , \/ , + , Rot	4L
SMAP-LRS	3L	2L	⊕, /\ , \/ , Rot, mod	3.5L

Table 2. The comparison of required memory, operation and communication

	LMAP		M2AP		EMAP		SASI		SMAP-LRS		SMAP-LRS	
Authentication state									Flag = 0		Flag =1	
	T	R+B	T	R+B	T	R+B	T	R+B	T	R+B	T	R+B
AND	0	0	1	1	2	2	0	0	2	2	2	2
OR	0	1	1	1	1	1	2	2	1	1	1	1
XOR	2	2	1	2	6	5	6	6	3	3	4	4
ADD	1	3	1	2	0	0	3	3	0	0	0	0
ROT	0	0	0	0	0	0	0	0	1	1	1	1
Update state									Flag = 0		Flag = 1	
AND	0	0	0	0	0	0	0	0	0	0	0	0
OR	0	0	0	0	0	0	0	0	0	0	0	0
XOR	10	10	10	10	10	10	4	4	2	2	2	2
ADD	5	5	5	5	0	0	1	1	0	0	0	0
ROT	0	0	0	0	0	0	2	2	3	3	5	5
Total	18	21	19	21	19	18	18	18	12	12	15	15

Table 3. The counter of simple operation

5. Conclusion

In this chapter, we present a secure mutual authentication protocol for low-cost resource-constrained RFID tag system under insecure wireless communication environment. The introduction of three security-enhanced designs in our scheme provides a more robust RFID authentication process. First, a flag state mechanism is proposed to prevent DoS attack and reduce the data storage space at the backend server by eliminating the need of storing dual tuples in database. Second, simple operations such as AND, XOR, OR, bit addition (mod 2^m) and bit rotation function are introduced to be compatible with EPCglobal Class1 Gen2 standard and to fit in the computation limitation of resource-constrained tag. Third, the proposed scheme SAMP-RLS provides data security to defend against major security threats such as replay attack and eavesdropping. In addition, SAMP-RLS possesses privacy protection features such as anonymity and forward secrecy. In terms of resource utilization, the required memory space of our scheme for a RFID system decreases about 45% to 50% in comparison with other existing mutual authentication protocols. In summary, our mutual authentication protocol offers data security enhancement, privacy protection ability and better resource utilization in comparison with other RFID authentication protocols.

6. Acknowledgments

The authors gratefully acknowledge the support from TWISC projects sponsored by the National Science Council, Taiwan, under the Grants No NSC 96-2219-E-001-001 and NSC 96-2219-E-011-008.

7. References

An, Y. & Oh, S. (2005). RFID System for User's Privacy Protection, In 2005 Asia-Pacific Conference on Communications, pp. 516-519.

Chien, H. (2007). SASI: A New Ultralightweight RFID Authentication Protocol Providing Strong Authentication and Strong Integrity, *IEEE Transactions on Dependable and Secure Computing*, vol. 4, pp. 337–340.

Chien, H.Y. & Chen, C.H. (2007). Mutual Authentication Protocol for RFID Conforming to EPC Class 1 Generation 2 Standard, *Computer Standards & Interfaces*, Vol. 29, Issue 2, pp. 254-259.

Chien, H.Y. & Huang, C.W. (2007). Security of ultra-lightweight RFID authentication protocols and its improvements, in *ACM SIGOPS Operating Systems Review* Vol. 41 New York, NY, USA.

Datasheet Helion Technology. (2005). MD5, SHA-1, SHA-256 hash core for Asic, http://www.heliontech.com .

Duc, D.N.; Park, J.; Lee, H. & Kim, K. (2006). Enhancing Security of EPCglobal Gen-2 RFID Tag against Traceability and Cloning, Proceedings of the 2006 Symposium on Cryptography and Information Security.

Feldhofer, M.; Dominikus, S. & Wolkerstorfer, J. (2004). Strong authentication for RFID systems using the AES algorithm, Workshop on Cryptographic Hardware and Embedded Systems–CHES, vol. 3156, pp. 357–370.

Garfinkel, S. & Rosenberg, B. (2005). *RFID: Applications, Security, and Privacy*, Addison-Wesley Professional.

Henrici, D. & Müller, P. (2004). Hash-based enhancement of location privacy for radio-frequency identification devices using varying identifiers, in Proceedings of the Second IEEE Annual Conference on Pervasive Computing and Communications Workshops, Orlando, Florida, pp. 149-153.

Juels, A. (2005). Strengthening EPC tags against cloning, in Proceedings of the 4th ACM workshop on Wireless Security, pp. 67-76.

Juels, A.; Molnar, D. & Wagner, D. (2005). Security and privacy issues in e-passports, in *IEEE Secure Comm.* Vol. 5.

Karthikeyan, S. & Nesterenko, M. (2005). RFID security without extensive cryptography, in Proceedings of the 3rd ACM workshop on Security of ad hoc and sensor networks, ACM, pp. 63-67.

Karygiannis, T.; Eydt, B.; Barber, G. & Bunn, L. (2007). *Guidelines for Securing Radio Frequency Identification (RFID) Systems,* in National Institute of Standards and Technology, April.

Kim, H.W.; Lim, S.Y. & Lee, H. J. (2006). Symmetric Encryption in RFID Authentication Protocol for Strong Location Privacy and Forward-Security, in Proceedings of the 2006 International Conference on Hybrid Information Technology Vol. 02, pp. 718-723.

Kumar, S. & Paar, C. (2006). Are standards compliant elliptic curve cryptosystems feasible on RFID, in Proceedings of Workshop on RFID Security, Austria, July.

Lee, J.; Kwon, T.; Choi, Y.; Das, S.K. & Kim, K. (2004). Analysis of RFID anti-collision algorithms using smart antennas, in Proceedings of the 2nd International Conference on Embedded Networked Sensor Systems, Baltimore, pp. 265-266.

Li, T. & Deng, R.H. (2007). Vulnerability Analysis of EMAP-An Efficient RFID Mutual Authentication Protocol, in the Proceedings of the Second International Conference on Availability, Reliability and Security-AReS, pp. 10-13.

Li, T. & Wang, G. (2007). Security Analysis of Two Ultra-Lightweight RFID Authentication Protocols, IFIP SEC.

Li, T. (2008). Security Analysis on a Family of Ultra-lightweight RFID Authentication Protocols, *JOURNAL OF SOFTWARE,* vol. 3, p. 1.

Lo, N.W. & Yeh, K.H. (2007). An Efficient Mutual Authentication Scheme for EPCglobal Class-1 Generation-2 RFID System, in the 2nd International Workshop on Trustworthiness, Reliability and services in Ubiquitous and Sensor networks, TRUST. Vol. 7, LNCS.

Ohkubo, M.; Suzuki, K. & Kinoshita, S. (2003). Cryptographic approach to "privacy-friendly" tags, in RFID Privacy Workshop, MIT, MA, USA, pp. 624-654.

Peris-Lopez, P.; Hernandez-Castro, J.C.; Estevez-Tapiador, J.M. & Ribagorda, A. (2006). EMAP: An Efficient Mutual Authentication Protocol for Low-cost RFID Tags, OTM Federated Conferences and Workshop, IS Workshop.

Peris-Lopez, P.; Hernandez-Castro, J.C.; Estevez-Tapiador, J.M. & Ribagorda, A. (2006). LMAP: A Real Lightweight Mutual Authentication Protocol for Low-cost RFID tags, in Proc. of 2nd Workshop on RFID Security.

Peris-Lopez, P.; Hernandez-Castro, J.C.; Estevez-Tapiador, J.M. & Ribagorda, A. (2006). M2AP: A Minimalist Mutual-Authentication Protocol for Low-cost RFID Tags, in Proc. of International Conference on Ubiquitous Intelligence and Computing UIC'06, LNCS 4159, pp. 912-923.

Poschmann, A.; Leander, G.; Schramm, K. & Paar, C. (2006). A Family of Light-Weight Block Ciphers Based on DES Suited for RFID Applications, in Workshop on RFID Security–RFIDSec. Vol. 6.

Poschmann, A.; Leander, G.; Schramm, K. & Paar, C. (2007). New Light-Weight Crypto Algorithms for RFID, in Proceedings of The IEEE International Symposium on Circuits and Systems, ISCAS.

Rhee, K.; Kwak, J.; Kim, S. & Won, D. (2005). Challenge-response based RFID authentication protocol for distributed database environment, in International Conference on Security in Pervasive Computing–SPC. Vol. 3450, pp. 70–84.

Sarma, S.E. & Engels, D.W. (2003). On the Future of RFID Tags and Protocols, in white paper, Auto-ID Center, Massachusetts Institute of Technology.

Shih, D.H.; Sun, P.L.; Yen, D.C. & Huang, S.M. (2006). Taxonomy and survey of RFID anti-collision protocols, *Computer Communications*, Vol. 29, pp. 2150-2166, Elsevier.

Weis, S.A. (2003). Security and Privacy in Radio-Frequency Identification Devices, Massachusetts Institute of Technology.

Weis, S.A.; Sarma, S.E.; Rivest, R.L. & Engels, D.W. (2003). Security and Privacy Aspects of Low-Cost Radio Frequency Identification Systems, in Security in Pervasive Computing, pp. 201–212.

Yu, S.; Ren, K. & Lou, W. (2007). A Privacy-preserving Lightweight Authentication Protocol for Low-Cost RFID Tags, in IEEE Military Communications Conference, MILCOM, pp. 1-7.

16

Privacy Enhancing Techniques on RFID systems[1]

Masataka Suzuki[1] and Kazukuni Kobara[2]
[1]*Bank of Japan*
[2]*National Institute of Advanced Industrial Science and Technology*
Japan

1. Introduction

An RFID system is a tracking and tracing system, and is useful for the management of various items and animals in a supply chain, animal husbandry and so on. According to a Japanese investigation firm, the number of RFID tags in Japan will increase rapidly from 51 million in 2007 to 1.7 billion in 2012 (Yano Research Institute, 2008).

In RFID systems, RFID tags, which have unique IDs, are attached to items, and RFID readers confirm whether something is there and identify what it is by obtaining its ID. It is, however, pointed out that exploiting RFID systems could lead to some privacy issues. One issue is that someone may know what you have by getting the IDs of your items. Another one is that someone may know when and where you were by recording the time and the place at which the IDs were obtained. Many kinds of countermeasures against these issues have been proposed. Some of them have been implemented in RFID products.

This chapter explains the privacy issues concerning RFID systems, their countermeasures and finally compares them from the security point of view.

2. An RFID system and its privacy issues

2.1 A basic RFID system

At first, we explain a basic RFID system in which an RFID tag, hereafter called *a Tag*, emits a plaintext of its ID to *a Reader*. The RFID system consists of the Tags, the Reader and *a Server*. The Server assigns a unique ID to each Tag preliminarily (Fig. 1-1)). This task may be done by manufacturers when shipping. The Server records the IDs and their corresponding information to its database (Fig. 1-2)). In the phase of reading the ID of the Tag, the Reader sends an ID-query to a Tag (Fig. 1-3)) and receives the ID as a response from the Tag (Fig. 1-4)). The Reader forwards the ID to the Server (Fig. 1-5)), and the Server looks up its corresponding information in the database (Fig. 1-6)).

---

[1] Views expressed in this chapter are those of authors and do not necessarily reflect the official views of the Bank of Japan and National Institute of Advanced Industrial Science and Technology.

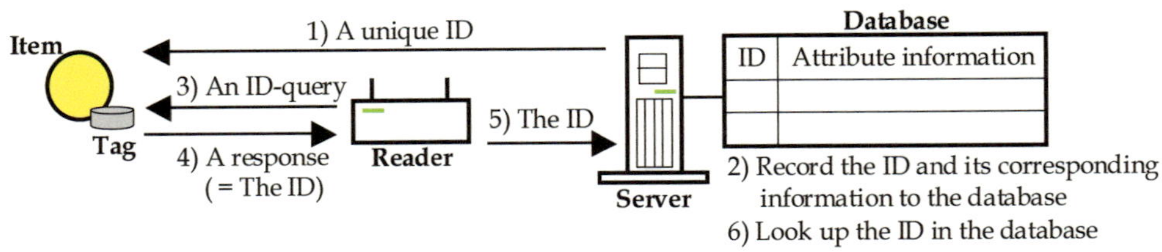

Fig. 1. A basic RFID system

2.2 Privacy issues on the RFID systems

It is worried that usage of RFID systems could lead to privacy violations in the following two senses. One issue is that someone, hereafter called *an Adversary*, may know what you have by obtaining the IDs of your items if the Adversary knows the relationship between the IDs and their corresponding information. The IDs can be obtaining by eavesdropping on the communications between legitimate Readers and Tags or by the Adversary sending an ID-query to your Tags. The Adversary may guess the corresponding information of a new ID from known IDs if the ID format is "an identifier of a company, an identifier of a product, an identifier of an individual product." We call this issue *belongings privacy* in this chapter.

Another issue is that the Adversary may know where you have been. Suppose you go around in a city with Tag-attached items. And the Adversary is supposed to locate many Readers in various places in the city, e.g. a hospital, a supermarket or an apartment, in order to collect IDs from people who pass near the Readers. For example, your ID, to be accurate, an ID of your item, is contained in two sets of IDs. The sets consist of IDs which were collected at the hospital and at your apartment, respectively. The Adversary may guess you are sick by confirming that your ID is contained in the two sets. That is, the Adversary knows where you have been by confirming the link of collected IDs, in other words, whether the IDs are emitted by the same Tag or not. This issue is called *location privacy*.

2.3 Approaches of countermeasures against privacy issues

Countermeasures against belongings privacy are to prevent an Adversary from obtaining IDs themselves or the relationship between IDs and items. The countermeasures against obtaining IDs are 1) to conceal the existance of Tags by preventing Tags from emitting any IDs and signals, 2) to prevent the Adversary from obtaining IDs by generating jamming and 3) to record not plaintexts of IDs but ciphertexts of the IDs in the Tags. The countermeasures against obtaining the relationships are 4) to make it difficult to guess the product of a new ID from the known relationship and 5) to employ strict access control on a database which records the relationship.

Countermeasures against location privacy are countermeasures 1) and 2) above because they prevent the Adversary from obtaining IDs. Countermeasure 3) above is not effective against location privacy. The Adversary can confirm the link of ciphertexts emitted from Tags by regarding the ciphertexts as new identifiers of the Tags if the ciphertexts are static. Therefore, for solving the location privacy, we need to make it difficult for the Adversary not only to extract IDs from the ciphertexts but also to confirm the link of ciphertexts. We call this countermeasure 6). Of course the Server must resolve the IDs from the recieved ciphertexts.

The countermeasures 4) and 5) above are not specific ones for RFID systems but can be used in general for information systems which use IDs and databases. Therefore we may refer to operational countermeasures for such kinds of systems. Countermeasures 1) and 2) are effective against both issues. In addition, countermeasure 6) is more sophisticated than countermeasure 3). Consequently, we focus on countermeasures against location privacy, i.e. countermeasures 1), 2) and 6), hereafter. Fig. 2 shows the relation between the three countermeasures.

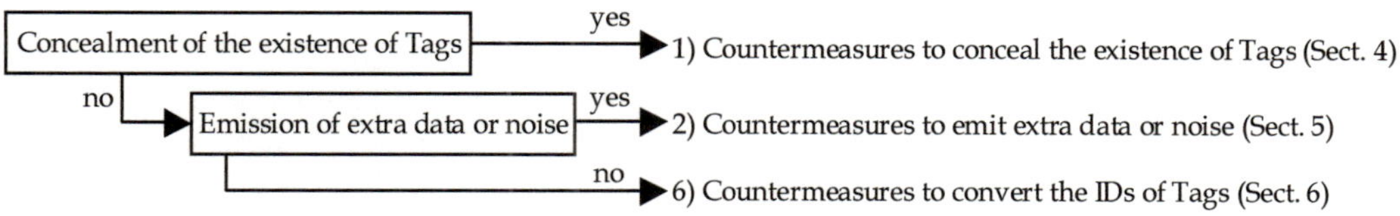

Fig. 2. Approaches of the countermeasures against location privacy

3. Assumptions of an adversary

We assume the following conditions for an Adversary:

- The Adversary cannot eavesdrop upon communications between a legitimate Server and legitimate Tags (Fig. 3-1)) because the ID assignment process is done in a secure area.
- The Adversary cannot obtain IDs and the corresponding information from the Server's database because the database is appropriately managed.
- The Adversary can eavesdrop upon the communication between a legitimate Reader and legitimate Tags (Fig. 3-3), 4)) because they communicate through a public channel.
- The Adversary can send ID-queries to the legitimate Tags (Fig. 3-3'), 4')).
- The Adversary cannot eavesdrop the communication between the legitimate Reader and the Server (Fig. 3-5)) because they communicate through a secure channel, e.g. Virtual Private Network.
- The Adversary can extract secret information, e.g. an ID and a cryptographic key, from the Tag.

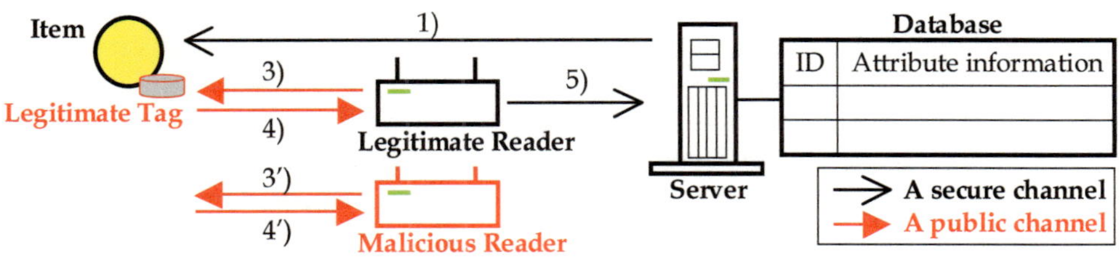

Fig. 3. The secure and public channels in the RFID system

Fig. 3 shows the secure channels and public ones in the RFID system. In order to make descriptions simple after section 4, we assume the Reader has the Server's resources, i.e. the computation power, the memory and the database. In other words, we describe the Server's task as the Reader's one.

4. Countermeasures to conceal the existence of Tags

This section introduces countermeasures which prevent Tags from emitting responses and signals against ID-queries. These countermeasures can conceal the existence of a Tag though we cannot receive the services of the RFID system.

4.1 Destroying / detaching Tag

This countermeasure is to destroy or to detach a Tag from an item. After destroying or detaching, we cannot permanently use the Tag. The means of destruction are, for example, to cut the antenna of the Tag by scissors, to burn off the logical circuits in the Tag by a high voltage electrical current and so on.

4.2 Faraday cage

This countermeasure is to wrap a Tag with some material, e.g. foil, which intercepts electromagnetic waves, i.e. ID-queries, in order to prevent the Tag from emitting responses. It is, however, difficult to apply the countermeasure to some items, such as large things, pets and domestic animals.

4.3 Kill command

This countermeasure is to implement a specific command on a Tag, which prevents the Tag from emitting responses. The Tag does not respond permanently after the Tag executes the command. Compared with the destroying/detaching of the Tag, this countermeasure disables the Tag only by executing the command. Therefore in order to protect against its misuse, we need to authenticate the Reader which directs the Tag to execute the command.
The command is implemented as *a Kill command* in a Tag, conforming to EPCglobal specification. The Tag has an authentication mechanism in which the Tag verifies a password sent by a Reader. The bit lengths of the password are 8 bits in EPCglobal Class 1 in frequency range of 860 MHz - 930 MHz (Auto-ID Center, 2002), 24 bits in Class 0 in that of 900 MHz (Auto-ID Center, 2003a) and Class 1 in that of 13.56 MHz (Auto-ID Center, 2003b), and 32 bits in Class 1 Generation 2 (ISO 18000-6 Type C, EPCglobal Inc., 2005), respectively. However, the number of the variation of the password is 256 at most in the case of an 8-bit password, it is not secure from the viewpoint of cryptography, that is, the Adversary may cause the Tag to execute the command.

4.4 Access password schemes

This countermeasure, hereafter called *an access password scheme,* is to respond an ID of a Tag when the Tag receives a correct password from a Reader. The countermeasure is relatively easy to implement and can conceal the existence of the Tag from an Adversary. However, the Adversary can obtain the password and the ID by eavesdropping upon the communication between the legitimate Reader and the Tag. The countermeasure is adopted in EPCglobal Class 1 Generation 2 (ISO 18000-6 Type C) and the bit length of password is 32 (EPCglobal Inc., 2005).

4.5 Hash Lock scheme

This countermeasure, called *Hash Lock scheme,* involves a Tag authenticating a Reader before sending its ID as a response (Weis et al., 2003). This scheme is executed with the following procedures:

1. A Reader generates a password for each Tag and calculates a hash value for each password. The Reader assigns a unique ID and the hash value to each Tag. In addition, the Reader stores the IDs, the corresponding hash values and passwords in the database.
2. Upon receiving an ID-query from the Reader, the Tag sends its hash value to the Reader.

3. The Reader looks up the corresponding password in the database and sends the password to the Tag.
4. The Tag calculates the hash value of the password and compares it with the stored one. The Tag sends its ID if it matches.

The advantage of Hash Lock scheme over the access password scheme is that a lot of time is required in order to guess the password from the secret information in the Tag. An Adversary must analyze the hash value in the Hash Lock scheme but need not analyze it in the access password scheme. The Adversary, however, can obtain the password against both schemes only by eavesdropping upon communications between a legitimate Reader and the Tags. The Adversary can confirm the links between the responses, i.e. the hash values, because the responses are static in Hash Lock scheme.

4.6 Change of operation modes

This countermeasure is to flexibly select whether to conceal the existence of the Tag or not by changing the operation modes of the Tag. Many schemes related to this countermeasure are proposed. We introduce two of these schemes in this section: EPCglobal Class 1 in frequency range of 860 MHz - 930 MHz (Auto-ID Center, 2002) and *LKI scheme* (Liu et al., 2004).

In EPCglobal Class 1 at 860 MHz - 930 MHz, the operation mode in which the Tag sends its ID is called *an Active mode* and the operation mode in which the Tag dose not emit its ID and signal is called *a Quiet mode*. And two commands are also implemented. The command, called *a Talk command*, changes the Tag to Active mode. Another one, called *a Quiet command*, changes the Tag to the Quiet mode. This scheme should keep supplying electric power to the Tag in order to maintain the Tag in each mode. Therefore, an Adversary may notice the existence of the Tag by detecting the supply source. Moreover, it needs an authentication mechanism that prevents the Adversary from executing the commands.

In LKI scheme, the operation mode in which the Tag does not emit its ID and signal is called *a Silent mode*. LKI scheme assumes that the Tag has a non-volatile memory to record the operation mode. Then the Tag maintains its operation mode without electric power. This scheme also needs the authentication mechanism. According to Liu et al., the password-based authentication may be enough if the password is managed appropriately and legitimate Readers pay appropriate attention to eavesdropping upon communication of authentication (Liu et al., 2004).

5. Countermeasures to emit extra data or noise

This section introduces countermeasures using extra devices which emit extra data or noise. The countermeasures can prevent an Adversary from obtaining a Tag's ID even if the Tag emits its ID as it is. The countermeasures cannot conceal the existence of the Tag, or to be accurate the extra device, but require no changes or only small changes to the Tag.

5.1 Jamming

An extra device in this countermeasure emits jamming to prevent a Reader from obtaining a Tag's ID. This countermeasure requires no changes in the Tag. Moreover, a legitimate Reader cannot obtain the ID if the device emits jamming. Its disadvantages are follows:

- Some countries restrict the emission of jamming. Moreover, the emission of electromagnetic waves is restricted in some areas, e.g. hospitals. This countermeasure cannot be employed in such situations.

- This countermeasure may prevent Readers in other RFID systems from communicating with Tags.
- An Adversary may be able to trace the Tag by continuously observing the source of the jamming.

5.2 Blocker Tag

Readers use anti-collision protocols for obtaining multi Tags' IDs sequentially in RFID systems. An extra device, called *a Blocker Tag*, emits many dummy IDs in order to obstruct the execution of *a Binary Tree protocol* which is one of the anti-collision protocols (Juels et al., 2003). We introduce a Blocker Tag.

At first, we explain the mechanism of the Binary Tree protocol. Suppose each Tag has a 2-bit ID and there are two Tags, whose IDs are 00 and 10 respectively, in the range where a Reader can communicate with them. The Reader using the protocol obtains the IDs with the following procedures (Fig. 4):

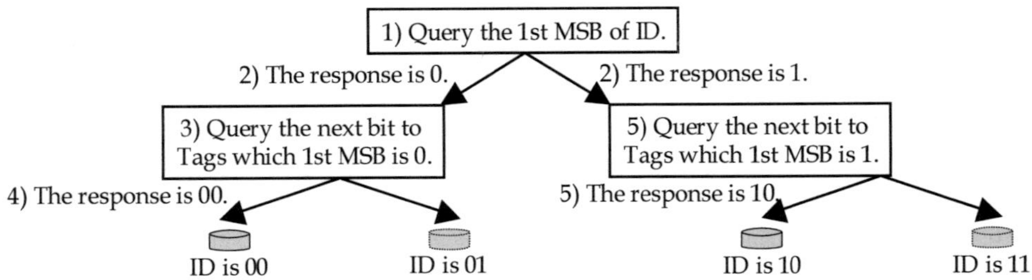

Fig. 4. Binary Tree protocol

1. The Reader asks Tags their most significant bits (MSBs) of their IDs.
2. Each Tag sends 0 or 1, and the Reader detects MSBs of the Tags in the area where the Reader can communicate.
3. The Reader asks the Tags for the next bit of the IDs whose MSBs are 0.
4. The Tag sends 00 and the Reader detects the Tag with 00.
5. The Reader asks the Tags for the next bit of the IDs whose MSBs are 1 in the same way. And it also detects the Tag with 10.

A Blocker Tag behaves as if there were every ID in the range. That is, the Blocker Tag emits "00 and 01" and "10 and 11" in the 4th and 5th steps of the above, respectively. Therefore, the Reader obtains all of the IDs, i.e. 00, 01, 10 and 11. The bit length is only 2 bits in the above case, but the length is longer, e.g. 128 bits, in practical systems. Then the Reader fails to obtain the all of the IDs.

This scheme needs no change to the Tag. Moreover, the Blocker Tag is not more sever than jamming from the viewpoint of the restriction of the emission of electromagnetic waves because the Blocker Tag reacts only when receiving queries[2]. The disadvantages of the Blocker Tag are a) a user must carry it, b) it may obstruct Readers communicating with Tags in other RFID systems, c) an Adversary with a smart Reader, which can accurately identify the location of a Tag sending its ID, may obtain the ID.

---

[2] The restriction of the emission is not completely solved because the Blocker Tag reacts when receiving the queries.

A more sophisticated Blocker Tag, called *a Selective/Partial Blocker Tag,* is proposed (Juels et al., 2003). A Selective/Partial Blocker Tag obstructs the read of only pre-defined IDs, e.g. the MSB of the IDs is 1, though the original Blocker Tag obstructs the read of every ID.

6. Countermeasures to convert the IDs of Tags

This section introduces countermeasures which make it difficult for an Adversary to confirm the link of responses sent by Tags. Of course legitimate Readers can identify their IDs from the responses. Receiving ID-queries, the Tags always send the response. Therefore the countermeasures cannot conceal the existence of the Tags.

6.1 Randomized Hash Lock scheme

Randomized Hash Lock scheme assumes Tags are implemented with a pseudo-random generator and a one-way hash function H()[3] (Weis et al., 2003). The Tags emit the hash value of their IDs and random numbers. The security of this scheme is based on the difficulty of inversion of the one-way hash function. A concrete procedure is as follows:

1. Upon receiving an ID-query, a Tag generates random number r and calculates hash value h of its ID and r, i.e. $h = \mathrm{H}(\mathrm{ID} || r)$, where $\mathrm{ID} || r$ denotes concatenation between ID and r. And the Tag sends h and r as its response to the Reader.
2. Upon receiving h and r, the Reader calculates the hash value $\mathrm{H}(x || r)$, where x denotes each ID of the Tags managed by the Reader. And the Reader exhaustively searches for x such that $\mathrm{H}(x || r)$ matches h. The Reader regards the corresponding x as the Tag's ID if it matches.

It is difficult for an Adversary to identify the Tag's ID by comparing the responses mutually because the responses sent by the Tag change at each ID-query. However, this scheme may have two problems: a) the Adversary may be able to identify the ID from the response by exhaustively searching, like the Reader, if the number of candidate IDs is small, b) the computational complexity for the Reader identifying the ID is proportional to the product of the following two factors: the number of the Tags that the Reader manages and the number of the Tags in the area where the Reader can communicate.

6.2 Symmetric Key Cryptography based schemes

It is possible to solve problems a) and b) of Randomized Hash Lock scheme if Tags are implemented with a pseudo-random generator, a symmetric key encryption function and a non-volatile memory. The scheme with the symmetric key cryptography, hereafter called *an SKC-based scheme,* is as follows:

1. A Reader preliminarily records symmetric key k, which is common to the RFID system, to each Tag.
2. Upon receiving an ID-query, the Tag generates random number r and encrypts r and its ID with k and SE(), where SE() denotes a symmetric key encryption function. And it sends ciphertext $c = \mathrm{SE}(k, r || \mathrm{ID})$ as its response to the Reader.
3. Upon receiving c, the Reader obtains $r || \mathrm{ID}$ by decrypting c with k and extracts the ID.

---

[3] A one-way function transforms an arbitrary length bit string into a fixed-length one. It is easy to calculate its output from the bit string and is difficult to calculate the string from the output.

The responses sent by the same Tag are different because the Tag generates random numbers at each ID query. Moreover, an Adversary needs to break the symmetric key cryptography for extracting IDs from the responses. Therefore, it is difficult for the Adversary to guess the IDs and to confirm the links between the responses if the symmetric key cryptography adopted is secure.

6.3 Hash Chain scheme

An Adversary may record ID-queries, the corresponding responses and dates/places when recording, and store them for a long time. The Adversary will have the opportunity to confirm the link of the responses if the secret information, e.g. a secret key, is leaked in the future. Therefore, a new security feature, called *forward security*, is proposed. In this feature, it is difficult for the Adversary who obtains the leaked secret information to confirm the link of the responses. We introduce *Hash Chain scheme* which is a typical scheme to fulfill the role of this feature (Ohkubo et al., 2003). This scheme assumes that Tags are implemented with two one-way functions H() and G(). Its procedure is as follows:

1. A Reader preliminarily assigns a different key to each Tag and stores each ID and the corresponding key in the Reader's database. We describe the initial key of Tag-*i* as $k_{i,0}$ ($1 \le i \le n$), where n is the number of Tags managed by the Reader.
2. Upon receiving an ID-query, Tag-*i* calculates hash value $h_{i,0} = H(k_{i,0})$ and sends $h_{i,0}$ as its response to the Reader. And Tag-*i* updates $k_{i,0}$ with G() and replaces $k_{i,0}$ by $k_{i,1} = G(k_{i,0})$. The key of Tag-*i* is updated when receiving the queries.
3. Upon receiving the response h, the Reader calculates $h_{i,t+j} = H(G^j(k_{i,t}))$ and searches for i and j such that $h = h_{i,t+j}$, where $G^j() = G(G^{j-1}())$, $0 \le j \le s$. s denotes a range where the Reader searches for the hash value. And $k_{i,t}$ denotes the Tag-*i*'s key recorded in the database at that time.
4. The Reader considers the sender of the response as Tag-*i* if matching. The key of Tag-*i* is $k_{i,t+j}$ at this time. The Reader replace $k_{i,t+j}$ in the database by $k_{i,t+j+1} = G(k_{i,t+j})$.

The security of this scheme is based on the difficulty of inversion of the hash functions. It is difficult for the Adversary to guess the former keys from the key obtained by the Adversary at a certain time because the keys are updated with G(). Then the scheme satisfies forward security if G() is sufficiently secure. Moreover, it is also difficult for the Adversary to confirm the link of the responses because the responses are generated by calculating the hash value of the keys with H(). However, the computational complexity for the Reader identifying the ID is proportional to the product of the following three factors: the number of the Tags managed by the Reader, the number of the Tags in the area where the Reader can communicate, and the number of the key updates which are not comprehended by the Reader. Moreover, the Tag updates its key even if a malicious Reader sends an ID-query to the Tag. The Tag's key goes out of the range in which the Reader searches if the malicious Reader sends ID-queries s times. That is, the legitimate Reader cannot identify the ID in this case.

6.4 Public Key Cryptography based schemes

We can construct a scheme which contains the following two features if Tags are implemented with a pseudo-random generator and a public key encrypting function: a) the scheme satisfies forward security, b) the Reader need not search for IDs exhaustively. The procedure of the scheme, hereafter called *a PKC-based scheme*, is as follows:

1. The Reader preliminarily writes its public key and a Tag's ID into the Tag.
2. Upon receiving an ID-query, the Tag generates a random number and encrypts the number and its ID with the public key. And the Tag sends the ciphertext as its response to the Reader.
3. Upon receiving the ciphertext, the Reader decrypts it with the Reader's secret key and extracts the ID.

An Adversary needs a Tag's ID, the Reader's public key, the responses and the random numbers used for generating the responses in order to confirm the link between the responses. The Adversary can obtain the ID and the public key by analysing the Tag, and can obtain the responses by eavesdropping upon the communication between the Reader and the Tag. The Adversary, however, cannot obtain the random numbers because the numbers are deleted when the responses are generated. Therefore, this scheme satisfies forward security if the pseudo-random generator adopted is secure. Moreover, the Reader need not search for IDs exhaustively because the Reader can obtain the IDs only by decrypting the responses. However, general public key cryptosystems (PKCs), e.g. RSA and elliptic curve cryptography, are not suitable for low performance RFID tags because of the computational complexities of such cryptosystems. Then, Suzuki et al. focus on Niederreiter PKC which is a lightweight PKC and is suitable for Tags because its encryption can be performed only with exclusive-OR in the parallel processing (Niederreiter, 1986). In addition, Suzuki et al. propose a new scheme in which the PKC is optimised suitably for the Tag (Suzuki et al., 2006).

6.5 Re-encryption schemes

Some PKCs in a specific class can update ciphertexts only with their public keys, the ciphertexts and random numbers. Of course the plaintexts of the updated ciphertexts are the same as the plaintexts of the original ciphertexts. ElGamal PKC is known as one of such PKCs (ElGamal, 1985).

A scheme, called *a Re-encryption scheme*, using such a PKC has been proposed (Juels & Pappu, 2003). This scheme assumes a Tag is implemented with a pseudo-random generator and the PKC. A Reader preliminarily writes a ciphertext of a Tag's ID and the public key of the Reader into the Tag. Upon receiving an ID-query, the Tag sends its ciphertext as its response and updates the ciphertext. The Reader can identify the ID in the same way as a Reader in a PKC-based scheme does.

The advantage of the Re-encryption scheme over the PKC-based scheme is that it does not store the plaintext of the ID in the Tag. On the other hand, the Tag cannot flexibly execute the reactions which correspond with its ID because the Tag does not know its ID. Moreover, the computational complexity of the updating is not low because the complexity is equal to that of encryption with ElGamal PKC. As a result, Juels and Pappu proposed a scheme in which the Reader updates the Tag's ciphertext and writes the ciphertext in the Tag in order to reduce the computational complexity of the Tag (Juels & Pappu, 2003). However, the Tag in the scheme needs to authenticate the Reader in order to prevent a malicious Reader from forging it.

7. Comparisons

This section compares the above schemes from the viewpoints of four security features: 1) concealment existence of Tags from an Adversary, 2) secrecy of IDs, which is the feature that the Adversary cannot identify the IDs, 3) unlinkability, which is the feature that the Adversary cannot confirm the link between responses, 4) foward security. Table 1 shows the results of the comparisons.

"o" and "x" in Table 1 denote that the countermeasure satisfies the corresponding feature and that the countermeasure does not, respectively. The schemes with "o" concering feature 1) are some of the schemes based on the approach of not emitting the ID and signals. The majority of schemes introduced in this chapter can be represented as the mark of "o" concering features 2) and 3). The schemes with "o" concering feature 4) are a part of the schemes based on the approach of converting the Tag's ID.

The Adversary may confirm the link by sending ID-queries frequently and by tracing the source of the responses continuously if the schemes cannot conceal the existence of the Tag. It is preferable to adopt the schemes with "o" concering feature 1) for protecting against this attack. On the other hand, it may be preferable to adopt the schemes with "o" concerning feature 4) if the attack is not assumed. For example, Hash Chain scheme, PKC-based schemes and Re-encryption schemes correspond to such schemes. However, these schemes assume a Tag whose performance is middle or more.

Countermeasures \ Security features		Concealment of existence of Tags	Secrecy of IDs	Unlinkability	Forward security
No countermeasures		x	x	x	x
Destroying/detaching Tags	(Sect. 4.1)	o	o	o	x
Faraday cage	(Sect. 4.2)	o	o	o	x
Kill command	(Sect. 4.3)	o	o	o	x
Access password schemes	(Sect. 4.4)	x *a	x *a	x *a	x
Hash Lock scheme	(Sect. 4.5)	x *a	x *a	x *a	x
EPCglobal Class 1 (Quiet mode)	(Sect. 4.6)	x *b	o	o	x
LKI scheme (Silent mode)	(Sect. 4.6)	o	o	o	x
Jamming	(Sect. 5.1)	x	o	o	x
Blocker Tag	(Sect. 5.2)	x	o	o	x
Randomized Hash Lock scheme	(Sect. 6.1)	x	o	o	x
SKC-based schemes	(Sect. 6.2)	x	o	o	x
Hash Chain scheme	(Sect. 6.3)	x	o	o	o
PKC-based schemes	(Sect. 6.4)	x	o	o	o
Re-encryption schemes	(Sect. 6.5)	x	o	o	o

Table 1. Security features of each countermeasure. Grey cells show the negative features. "*a" denotes the fact that the Adversary can obtain IDs and passwords, if the Adversary eavesdrops upon communications between a legitimate Reader and the Tags. After obtaining them, the Adversary can detect the Tags, can identify the IDs and can confirm the link. "*b" denotes the fact that the Adversary may notice the existence of the Tags due to detecting the power sources.

8. Conclusions

We explained two privacy issues on RFID systems in this chapter. One is an adversary may know the items you have and the other is your locations might be traced by linking RFID responses. In addition, we explained known approaches against these issues with concrete schemes. Finally, we compared them from the viewpoint of the four security features. Some of the schemes, which do not require heavy burden on tags, have already been implemented in some current RFID products. The other schemes, however, require certain technical break through to reduce the cost for implementing them and leave themes to study.

9. References

Auto-ID Center. (2002). 860 MHz - 930 MHz Class I Radio-Frequency Identification Tag Radio Frequency & Logical Communication Interface Specification Candidate Recommendation, Version 1.0.1

Auto-ID Center. (2003a). Draft protocol specification for a 900 MHz Class 0 Radio Frequency Identification Tag

Auto-ID Center. (2003b). 13.56 MHz ISM Band Class 1 Radio Frequency Identification Tag Interface Specification: Candidate Recommendation, Version 1.0.0

ElGamal, T. (1985). A public key cryptosystem and signature scheme based on discrete logarithms, *IEEE Transactions on Information Theory*, IT-31, pp. 469-472

EPCglobal Inc. (2005). EPC Radio-Frequency Identity Protocols Class-1 Generation-2 UHF RFID Protocol for Communications at 860 MHz - 960 MHz Version 1.1.0

Juels, A., Rivest, R., & Szydlo, M. (2003). The Blocker Tag: Selective Blocking of RFID Tags for Consumer Privacy, *the 10th ACM conference on Computer and Communications Security (CCS 2003)*, USA, October 2003

Juels, A. & Pappu, R. (2003). Squealing Euros: Privacy protection in RFID-enabled banknotes, *Financial Cryptography 2003*, January 2003

Liu, D., Kobara, K. & Imai, H. (2004). Pretty-Simple Privacy Enhanced RFID and Its Application, *The Seventh International Symposium on Wireless Personal Multimedia Communications (WPMC 2004)*, Italy, September 2004

Niederreiter, N. (1986). Knapsack-type Cryptosystems and Algebraic Coding Theory, *Problems of Control and Information Theory*, Vol. 15, No. 2, pp. 159-166

Ohkubo, M., Suzuki, K. & Kinoshita, S. (2003). Cryptographic Approach to a 'Privacy Friendly' Tags, *RFID Privacy Workshop*, USA, November 2003

Suzuki, M., Kobara, K. & Imai, H. (2006). Privacy Enhanced and Light Weight RFID System without Tag Synchronization and Exhaustive Search, *IEEE International Conference on Systems, Man, and Cybernetics (SMC 2006)*, Taipei, October 2006

Weis, S. A. (2003). Security and Privacy in Radio-frequency Identification Devices. *Master's thesis of Department of Electrical Engineering and Computer Science, Massachusetts Institute of Technology*

Weis, S. A., Sarma, S. E., Rivest, R. L. & Engels, D. W. (2003). Security and Privacy Aspects of Low-Cost Radio Frequency Identification Systems, *In First International Conference on Security in Pervasive Computing*, Germany, March 2003

Yano Research Institute. (2008). Result of the survey on RF-ID market. Research Express (in Japanese)

17

An Improved Forward Secrecy Protocol for Next Generation EPCGlogal Tag

L.M. Cheng, C.W. So and L.L. Cheng
City University of Hong Kong
Hong Kong

1. Introduction

Radio Frequency Identification (RFID) (Landt, 2001) is a prevalent technology that replaces barcode technology and it will be massively applied in both consumer and commercial products as the trend predicts. However, the computation power and memory of RFID including the EPCGlobal Gen-1 and Gen-2 RFID tags are restricted. These made the implementation of well-known cryptographic algorithms, both computational and memory intensive, on the tags not possible.

Although various cryptographic privacy enhancing technologies for RFID have been proposed, they include Hash-lock approaches, Digital Signature approaches, Encryption approaches, Time Stamping approaches, Pseudonyms approaches and Challenge-response approaches, the EPCGlobal tags continue to operate in a limited security protection.

The Hash-lock approach (Weis, 2003) is based on locking the tag using a hash of the key on the tag, where the key is stored in a back-end server. This approach assumes that tags can not be operated securely in a long isolated environment. This approach can be used for authentication, by matching the right hash of key. The cloning resistance is weak and enhanced techniques providing better privacy protection and scalability have been derived (An, 2005; Nohara, 2005; Wang, 2007).

Digital signatures approaches (Juels, 2003; Bono, 2005; de Dormale, 2005) provide better tracing and forgery resistance of RFID during authentication process. The approaches use a PKI encryption technique to avoid static identifiers and information to be read by others. Authentication is performed by verifying data on the tag is signed using a valid public-key digital signature to check the valididty of authentication.

Encryption approaches (Golle, 2004; Feldhofer, 2004; Ranasinghe, 2004; Ateniese, 2005) are similar to digital signature approaches except simplified private key standard or propriertay cryptographic algorithm is used.

Time stamping approaches (Glidden, 2004; Molnar ,2005; Tsudik, 2006; Ith, 2007) are the most popular approaches which provide a dynamic matching of time information that help avoiding replay attacks.

Pseudonyms approaches (Juels, 2004; Juel, 2006; Molnar, 2005; Avoine, 2005) is very similar to time stamping approaches except the dynamic information is scheduled from a pre-defined list of pseudo-random data called pseudonynms.

Challenge-response approaches (Ree , 2005; Dimitriou, 2006; Duc, 2006; Chien, 2007) are the most secured techniques developed from multi-pass authentication process to provide a wide range of security and privacy protection.

In actual situation, EPCGlogal Gen 1 (Garcia-Alfaro et al, 2008) tag implementation uses protocols that only require RFID readers to use the tags' unique serial numbers to identify the tags. Tags with the same ID will certainly confuse the reader. Gen 2 rectifies this problem by allowing reader to read tags even if two or more tags have the exact IDs. The unique sequence of communication makes anti-collision algorithm more robust and reduce the possibility of interference from other tags when a reader is talking to a certain tag.

In order to handle reading multiple tags reliably and securely, a number of security techniques have been deployed in Gen 2 (EPC, 2005; Roberti, 2005). They are Session Concept, Dense Readers Conditions, Enhanced Secured Protocols, Ghost Reads Improvements and Covered Coding. However, there are still some privacy concerns in both user and application levels left in the standards and a security loophole in defining and managing the key 'random number' that require attention in order to avoid possible privacy violation or information leakage by eavesdropping on the communication channels.

1.1 Fake tag ID (ghost reads) problems

Fake Tag ID (Ghost Reads) is the major reading problems with Gen 1 tag. Noise or glitches are a hurdle in adopting this valuable technology. As confirmed by the report from RFID Alliance Lab (Deavours, 2005), the Class 0 ghost read rate is about 1.3 per 1,000. These ghost reads can create havoc in many solutions. Gen 2 protocol has an edge on that and it comes with a very vigilant solution to tackle this problem using a 'Query' Concept. The 'Query Concept' establishes a mutual authentication flow to allow secure exchange data and to eliminate fake tag ID problem.

Strict timing constraints in Tag-Reader communications create an illusion of full duplex link. In fact, the communication is still operated in a half-duplex mode. Tag will not talk when it is listening to the reader commands but the timing constraints make sure that tag must response to reader command within a preset time. If the tag fails to response within the preset time, the task will be terminated and the entire process has to be started from beginning.

1.2 Password protection and effective randomness

A secure communication channel is essential for data transmitted over an air interface. In Gen 1 Class 1, an 8-bit password was used with a 'kill' command to safeguard the data. However, this 8 bit password is not secure and eases to break because of just 256 possible values. In Gen 1 Class 0, a 24-bit password is used which gives a better data protection against eavesdropping. Gen 2 with even better safeguard uses 32 bit password while offering 4 billion possible values and makes the brutally search for the correct password difficult and thus provides a very high level of secure communication that EPC tags never had before (Roberti, 2005).

To strengthen the password requirement in communications, a random number is used in Gen 2 to scramble the data commuted. Tags will generate and use a 16-bit pseudo-random number generator (PRNG) throughout the communication link session. Because the PRNG is close to truly random, the communication link is ensured to be safe. For example, having a tag population of up to 10,000, the probability that any two or more tags simultaneously generate the same sequence of RN16s (16-bit random number) is less than 0.1%. The 32 bit password protection in Gen 2 is further enhanced by using "cover coding" while EXORING the data with random numbers to mess up the data (EPC, 2005) during transmission, a matching random number is needed to recover the transmitted data at the receiver end.

1.3 Security problems and possible attacks

With increasing mobility requirements, RFID readers are integrated in a handheld device or even in mobile phones. The low-cost tags are likely the factor for widespread adoption of the technology, deployment on such massive scale has created new threats to user and application privacy due to the powerful tracking capability of the tags (Luo et al, 2005). All UHF standards do provide a security mechanism for reading user memory but any reader can read the tag ID on fly. Security check must be imposed on the tag before transmitting the ID and a mechanism should be defined to recognize the trusted reader to address privacy concerns.

The transmission protocol (EPC, 2005) defines the mechanism to exchange instructions and data between the reader and the tag, in both directions. It is based on the concept of "interrogator (reader) talks first" and it simply means that every tag compliant to UHF standards will always answer to reader's query with its identification (ID) at very first hand. This makes the RFID technology susceptible and any intruder reader can track a tag. The attacker can obtain concrete product information associated with EPC/UID code. This product information is usually provided in the public network. Although current UHF protocols have 'kill' command/option which makes tag presently dead and can be executed before moving the tag in the hands of end-users but it is not the solution for most applications. Some applications require permanent tag tracking, for example, tags associated with objects for security purposes, personal identification systems, vehicle tracking system etc.

Another tag security issue relates to the scenario. Since the communication between a tag and a reader is by radio means, anyone can access the tag and obtain its output, i.e. attackers can eavesdrop on the communication channel between tags and readers, which is a cause of consumers' apprehension. Therefore, the authentication scheme employed in RFID must be able to protect the data passing between the tag and the reader, i.e. the scheme itself should have some kind of encryption capability (EPC, 2005).

Gen 2 provides a good mechanism for securing the data communication between the tag and reader. The exchange of cover-coding is first initiated by a random number request, i.e. RN16, from the tag. If a lower secured mechanism or plaintext only is used, eavesdrop on the communication channel will break the entire security process of the cover-coding. The generation and management of this 'random number' are vital for ensuring the security and integrity of the system but its size should be reconsidered and time of command to response should be restricted with precise values. So that, random number and time for command to response should be directly proportional. Although the random RN16 secures the communication link but its 16 bit size still makes it susceptible as generating or searching 65536 combinations is very easy with ordinary processors. The duration of command to response time makes it more vulnerable which means that reader A would start querying the tag but reader B (an intruder) can jump in the communication link with fake random numbers.

1.4 Possible solutions

Duc et al (Duc et al, 2006) proposed schemes for enhancing security of EPCglobal Gen-2 RFID tag against Traceability and Cloning. It enhances the weaknesses of Rhee (Rhee, 2005), Juels (Juels, 2006), and Dimitriou (Dimitrios, 2006) schemes, which are either not conform to EPCGlogal standard or unable to resist the privacy or/and DoS attack.

Duc et al's authentication relies on the synchronized session key between the tag, *T*, and the server, *S*, an adversary can initiate replay attack, man-in-the-middle attack and brute force attack that will cause DoS in the RFID system. If any one of the "end session" command was intercepted, the shared session key between *T* and *S* will be out of synchronization. As a result, *T* cannot be authenticated anymore. Thus, Duc et al.'s protocol is not able to resist the DoS attack, and it does not provide forward secrecy to the RFID system. Chien (Chien, 2007) provides an enhnacement to Duc's approaches by introducing a pair of old (previous) and new Session Keys, and a pair of old and new random number to avoid DoS attack but it cannot resist Man-in-the-Middle attack caused by a spoofed reader.
We start our discussion in Section 1 with a short introduction and Section 2 we present Duc et al's scheme as an enhancement to both Juels' and Dimittriou's schemes, and we elaborate how Duc's technique will fail. In Section 3, we will give a scenario of all possible threats in RFID environment. In Section 4, we will propose a new security protocol to close these security loopholes and the corresponding simulation results in Section 5. The security analysis of the newly proposed scheme will be given in Section 6. We will conclude new security RFID solutions in Section 7.

2. Duc et al's scheme review

Duc et al. proposed a communication scheme (Duc et al, 2006) to protect user privacy for RFID system. The scheme based on a synchronous session key between tags and back-end database server to authenticate each other. This mutual authenticate scheme takes the advantages of the hash properties of CRC function and a PRNG that are supported by EPCglobal Class-1 Gen-2 tags. The underlying idea is by using the same PRNG with the same seed at both tag and back-end database to generate the same session key on both side. To prevent tag send static message before update of the session key, a random number is added in the authentication process. Data will be encrypted by performing logic operation ⊕ with the session key before transmission. Session key will be updated after each successful authentication. The following paragraphs will briefly explain the protocol flow.

2.1 Symbol notations

T	-	RFID Tag
R	-	RFID Reader
S	-	Backend Database Server
r	-	Pseudo-Random Number Generated by Tag's PRNG
CRC(:)	-	CRC Function
PRNG(:)	-	PRNG Function
K_i	-	Session Key for i^{th} Session
A	-	Adversary

2.2 Initialization of tags and back-end database server

Initially during the manufacturing time, the tag has assembled with its EPC and the necessary parameters for the PRNG. A random seed number for PRNG and PIN is chosen and then stored into both *T*'s memory and *S* entry corresponding to the matching EPC. This is very important that each EPC must exactly match with its PRNG seed number and PIN, otherwise the tag can not be authenticated by the back-end server.

2.3 Communication channel between *R* and *S*

The scheme assumes that *R* is communicating with *S* in a secure channel, both *R* and *S* are able to perform standard cryptography authentication. *S* can send the EPC and data to *R* in an encrypted form. *S* can even depend on the privilege of *R*, to determent what kind of information can send to the reader.

Protocol flow

Figure 1 shows the protocol flow of Duc et al's scheme. The flow is illustrated as below:

Step 1. First of all, *R* sends a query request to *T*

Step 2. *T* generates a nonce *r* and form the message $M_{1T} = CRC(EPC \| r) \oplus K_i$ and $C = CRC(M_{1T} \oplus r)$. CRC in M_{1T} actually is acting like a hash function while in C is functioned as an error detection function. M_{1T}, C and *r* will then be sent to R.

Step 3. R forwards M_{1T} , C and r to S.

Step 4. For each tuples in *S*, it generates a message M_1 in the same way as M_{1T} in *T* until a match where $M_{1T} = M_1$ is found. If matched, *T* is successfully identified and authenticated. *S* forwards *T*'s information to *R*. If match failed, *S* will send a tag reject message to *T* via *R*. Information on *T* will be updated if *R* is authenticated to *T* with the generation of M_2. *S* uses the matched tuple's EPC, PIN and K_i to generate the message M_2, where $M_2 = CRC(EPC \| PIN \| r) \oplus K_i$. Finally, *S* will send the corresponding object data and M_2 to *T* via *R*.

Step 5. *T* generates a message M_{2T} to verify M_2 from *R*. *T* uses its EPC, PIN, *r* and K_i to generate the message M_{2T} same as M_2. If M_{2T} matches M_2, data exchange is XORING data with the session key K_i to encrypt or decrypt. However, if M_{2T} does not match M_2, R is rejected and the session ends immediately. When data exchange is completed, *R* sends an "end session" message to both *S* and *T*. Both *S* and *T* updates the session key where $K_{i+1} = PRNG(K_i)$ and wait till a new session starts.

3. Possible attacks and vulnerabilities on Duc et al.'s scheme

Duc et al.'s protocol is not able to resist the DoS attack and it can not provide forward secrecy to the RFID system. Since this authentication reply on the synchronized session key between *T* and *S*, an adversary can initiate replay attack, man-in-the-middle attack and brute force attack and causes DoS in the RFID system. If any one of the "end session" command was intercepted, the shared session key between *T* and *S* will be out of synchronization. As a result, *T* can not be authenticated anymore. The above DoS attacks actually are based on this vulnerability, aiming at intercepting the delivery of the "end session" command sent from *R* to *T*.

3.1 Replay attack

An adversary can use a spoofed *R* to send a query request to tags, then record the replay messages M_{1T} and nonce *r* from *T*. Recorded message will replay with a session started with an authorized *R*, finally *S* will update its session key while *T*'s session key will remain unchanged. As the session key is out of synchronization between *T* and *S*, therefore *T* can not be authenticated anymore. This is one of the high level threats to the RFID system, as the replay attack can perform on a large numbers of *T* at a time.

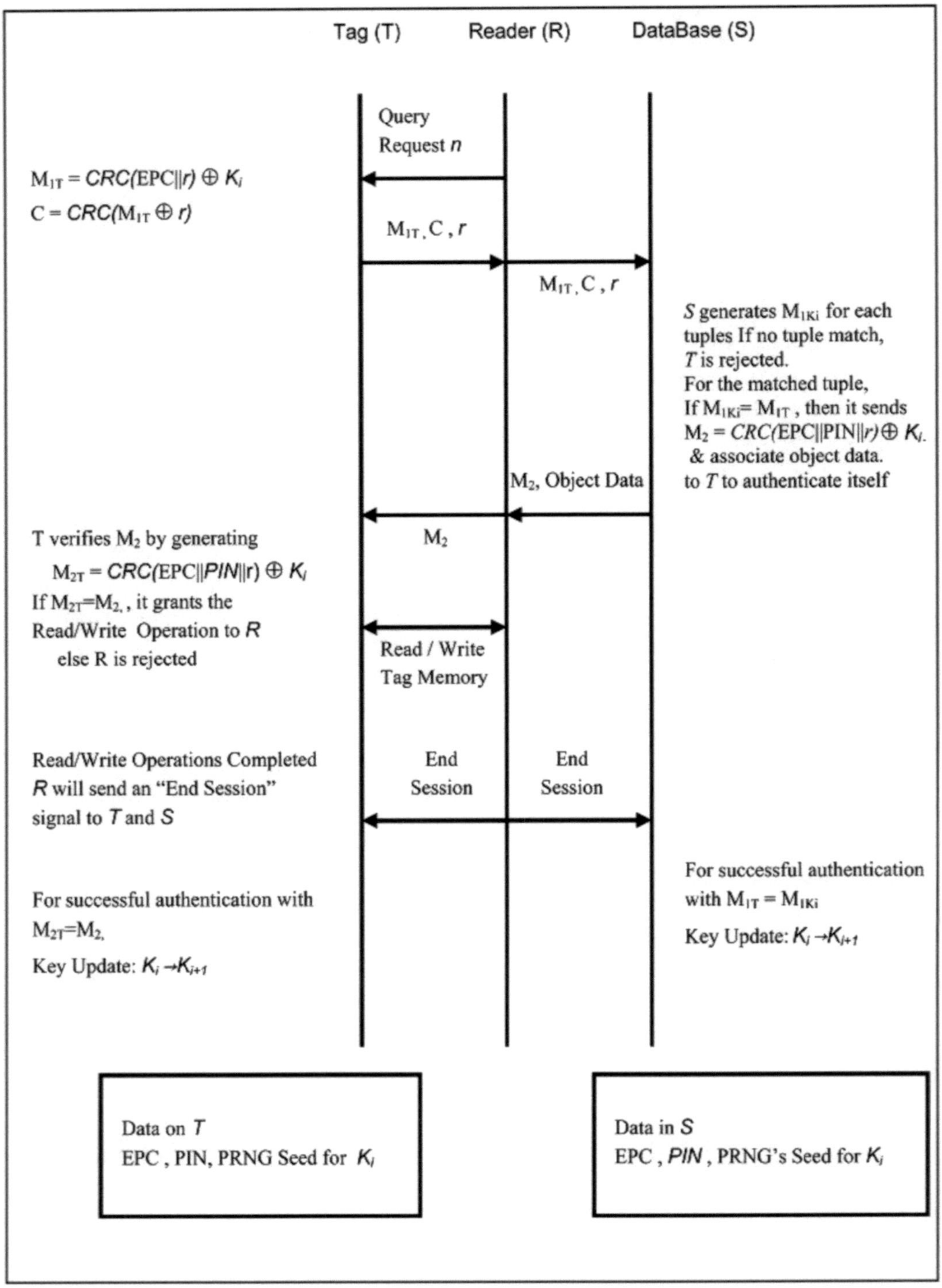

Fig. 1. Duc et al's Protocol

3.2 Man-in-the-middle attack

Man-in-the-middle attack is very similar to replay attack, an adversary acts like a hub to store and forward messages between *R* and *T*. However, an adversary will intercept the command "end session" from *R* to *T*, to make the session key out of synchronization. Man-in-the-middle attack is a high level threat to the RFID system too, as it can also perform on a large numbers of tag at a time.

3.3 Brute force attack

Since tag-to-reader authentication relies on the correspondence between nonce *r*, EPC and session key. It is important to take note that, before the update of the session K_i in a successful authentication, the session key will remain unchanged, while the EPC is always a constant. Therefore, the variance of M_{1T} is basically determined by *r*. An adversary can take this property to initiate a brute force attack on the message M_{1T}. A random message M is chosen, then an adversary can send along with different *r* in each session until a reply of M_2 from reader. As the length of *r* is 16 bits only, the maximum trial times for *r* in a particular M is only 65536. The probabilities that the random message finds a match in *S* is mainly depends on the number of tuples exist in *S*. This is a very dangerous attack to the whole system, as the message length of M_{1T} is 16 bits only, an adversary can send all the combination of M_{1T} and *r* to *R*, it only cost 2^{32} trial times to match all the tuples exists in the database.

3.4 Forward secrecy

If the tag is compromised, an adversary can obtain the EPC, PIN, K_i . From the eavesdropped communication data, we can trace the past communication record between *T* and *R* by computing the respective M_{1T} and M_2 with the obtained parameters. For instance, an adversary can take $M_{1T} \oplus M_2$ from the past communication, that can eliminate the session key and remain only the $CRC\ (EPC \oplus r) \oplus CRC\ (EPC \| PIN \| r)$. Then we may use the obtained parameter from *T* and generate with *r* to trace the past communication of *T* from the eavesdropped past communication data.

4. Proposed new protocol

With the understanding of the possible attacks and vulnerabilities in Duc et al.'s security scheme, a new security scheme that improves the security performance for RFID system is proposed.

The major differences between the proposed scheme and Duc et al.'s scheme are the additional random number challenge from the reader, and the database will keep the old session key for each tag, update the access PIN after each successful authentication and acknowledgement of M_2 from *T*. The flow of the proposed protocol is further explained below.

The tag T was manufactured and assembled with its corresponding EPC with preset parameter for the pseudo random number generator PRNG. A random seed number for PRNG and PIN was chosen and was stored into both *T*'s memory and backend data server S with entry corresponding to the matching EPC. The database will store the session key K_{i-1} and PIN_{i-1} after the first authentication. The communication between Reader *R* and server *S* was through a secure channel of which cryptographic algorithm can be used in authentication and for the object data exchange. The protocol below can also provide secured communications between *R* and *T* even for an insecure wireless channel.

4.1 Proposed protocol flow

The protocol flow for proposed scheme is shown in Figure 2. The protocol sequences are as follows.

Step 1. *R* generates a 16-bit random number *n* by its Pseudo Random Number Generator (PRNG) and sends it together with Query Request message to *T*.

Step 2. *T* generates a 16-bit random number *r* by its PRNG and the message $M_{1T} = CRC\,(EPC \| n \| r) \oplus K_i$ and the error checksum code $C = CRC\,(M_{1T} \oplus n \| r)$ and sends M_{1T}, C and *r* to *R*. [N.B. ⊕ is an exclusive OR function]

Step 3. *R* checks $C = CRC\,(M_{1T} \oplus n \| r)$ and detects any transmission error in the channel and *R* forwards M_{1T}, C , *r* and *n* to *S* if no error was found, otherwise, the tag is rejected.

Step 4. *S* generates $M_{1Ki} = CRC\,(EPC \| n \| r) \oplus K_i$ and $M_{1K(i-1)} = CRC\,(EPC \| n \| r) \oplus K_{i-1}$ for each tuples in *S*.

Step 5. If no tuple matches for $M_{1Ki} = M_{1T}$ or $M_{1T} = M_{1K(i-1)}$, the tag is rejected.

Step 6. If $M_{1T} = M_{1K(i-1)}$, it reveals that the session key is out of synchronization. The following steps are then executed.

Out of Synchronization Flow:

Step 6.1 *S* generates $M_2 = M_{2K(i-1)} = CRC\,(EPC \| PIN_{i-1} \| n \| r) \oplus K_{i-1}$ and sends it to *T* via *R*.

Step 6.2 *S* then informs *R* to send the "end session" command to *T*

Step 6.3 T updates its K_i and PIN_i after receiving the "end session" command, *S* continues to keep both K_i and PIN_i and K_{i-1} and PIN_{i-1} unchanged. In this session, *R* will not perform any read and write operation to *T* .

Step 6.4 Finally, *R* will re-initiate a new session with *T* using an updated session key.

Step 7. If the tuple is matched where $M_{1T} = M_{1Ki}$, *S* generates $M_2 = M_{2Ki} = CRC\,(EPC \| PIN_i \| n \| r) \oplus K_i$. *S* sends M_2 and the associated object data to *R*. *R* then forwards only M_2 to *T*.

Step 8. *T* verifies M_2 by computing $M_{2T} = CRC\,(EPC \| PIN_i \| n \| r) \oplus K_i$, if $M_{2T} = M_2$, i.e. *R* is authenticated, reading and writing *T*'s memory is granted to *R*; otherwise, the request from *R* is rejected.

(Note: Data exchange between *T* and *R* is encrypted and decrypted by exclusive OR operation ⊕ with the session key K_i.)

Step 9. When *R* has finished the reading and writing operation to *T*, *R* sends an "end session" command to both *R* and *S* to trigger the key update process. Both *T* and *S* will update K_i and PIN_i using $K_i = PRNG(K_{i-1})$ and $PIN_i = PRNG(PIN_{i-1})$. A session is completed at this stage.

5. Simulation

In order to find out the average appearing time of M_{1T} for a given tag, a simulation programme was built using VB.net and MySQL data base to test both Duc's and our proposed protocols. The number of occurrences of M_{1T} in each successful key update session was measured and used for comparison. The tests was based on 65,535 random entries using a PRNG satisfied Gen-2 requirements with a repeated period of 59,092.

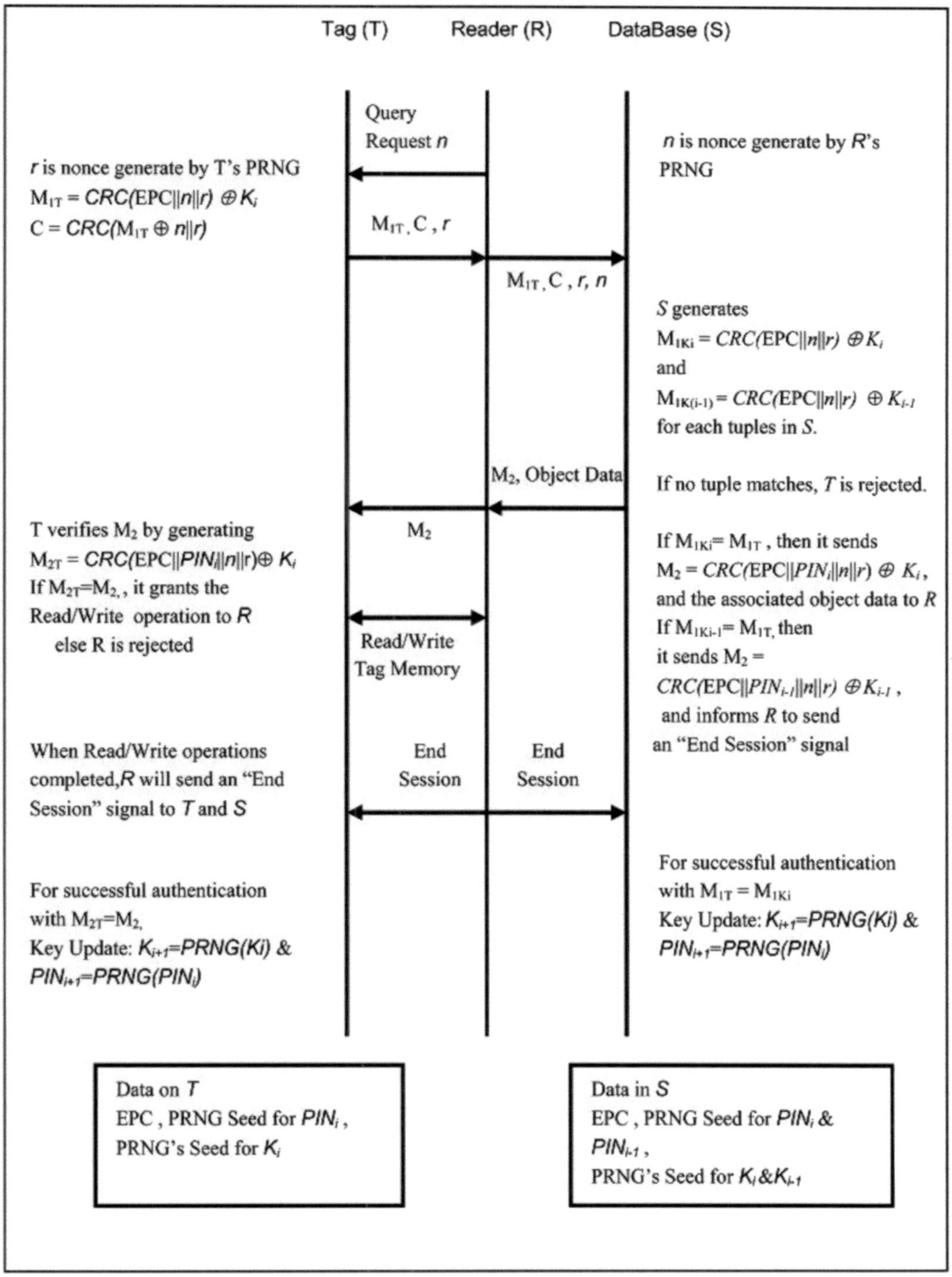

Fig. 2. New Proprosed Protocol

5.1 Pseudo-random number generation

There are four essential parameters for each tag which includes a 96-bit EPC, 32-bit PIN, 16-bit K_i and PRNG's parameters. EPC , PIN_i and seed for K_i are randomly chosen from a PRNG. The random number should satisfy the Gen-2 tag requirements.
A popular class of PRNG, linear congruential generator in the form of $X_{i+1} = (aX_i+c) \mod m$ was selected; where a, c and m defined the PRNG parmeters, X_i was the seed. In our simulation, a = 61979, c = 0 and m = 59093 were used. A repeating period of 59092 was obtained and satisfied the requirement of Gen-2 tag.

5.2 Protocol simulation programme

Figure 3 shows the layout of the simulation programme layout. The main screen in the centre simulates the protocol flow. The grid view on the left simulates the population of tags, while the right hand side grid view simulatse tag's information tuples maintained in the database. On the right hand corner, the programme performs simulation for M_{1T} out of a given trial times before a successful session key update for a selected tag in the tag's grid view. In order to find out how the CRC function affect the average appearing times out of a given trial, the programme can simulate the generation of M_{1T} for both CRC-16-CCITT used in current Gen-2 tag and CRC-32.

5.3 Duc et al's protocol under man-in-the-middle-attack

An adversary appears in between the tag and the reader, emulating a store and forward hubs. It forwards query request from the reader and then sends M_{1T} , C , r to the reader received from tag, and forward M_2 to tag like an ordinary authentication process. However, after the mutual authentication, it blocks the "end session" command send from reader. As a result, the tag can not be authenticated anymore. Since the tag remains its session key and PIN unchanged while back-end database server updates them with PRNG. The simulation programme shows the tag is rejected in the next authentication. This Man-In-The-Middle Attack simulation shows that the Duc et al's proposed RFID protocol can collapse.

5.4 Simulation on average appearing time M_{1T} and effect of CRC

A tag was randomly chosen to loop recursively to generate 65535 trials for M_{1T} in the simulation programme. The main difference of M_{1T} between the two protocols was the introduction of an additional random number challenge n generated from reader .
The simulation found that M_{1T} was equalled to T_D =1.5148, T_M =1.5312 and T_M =1.5234 for Duc et al and for the new proposed schemes using n=16 bit and n=32 bit respectively, where T_D was the Average Appearing Time out of 65535 Trial for Duc et al.'s Protocol, and T_M was the Average Appearing Time out of 65535 Trial for the Proposed Protocol.

5.5 Simulation result analysis

The simulation result reveals that the period of CRC output in M_{1T} does not follow the period of r. Since r is the only changing parameters in the M_{1T} throughout the trial, it is expected that T_D should approximate equal to r's period. The period of r is found to be 59092 time, therefore the average appearing times out of 65535 trial should be around 1.1. However, T_D is found to be around 1.5 which has a significant difference from r's period.

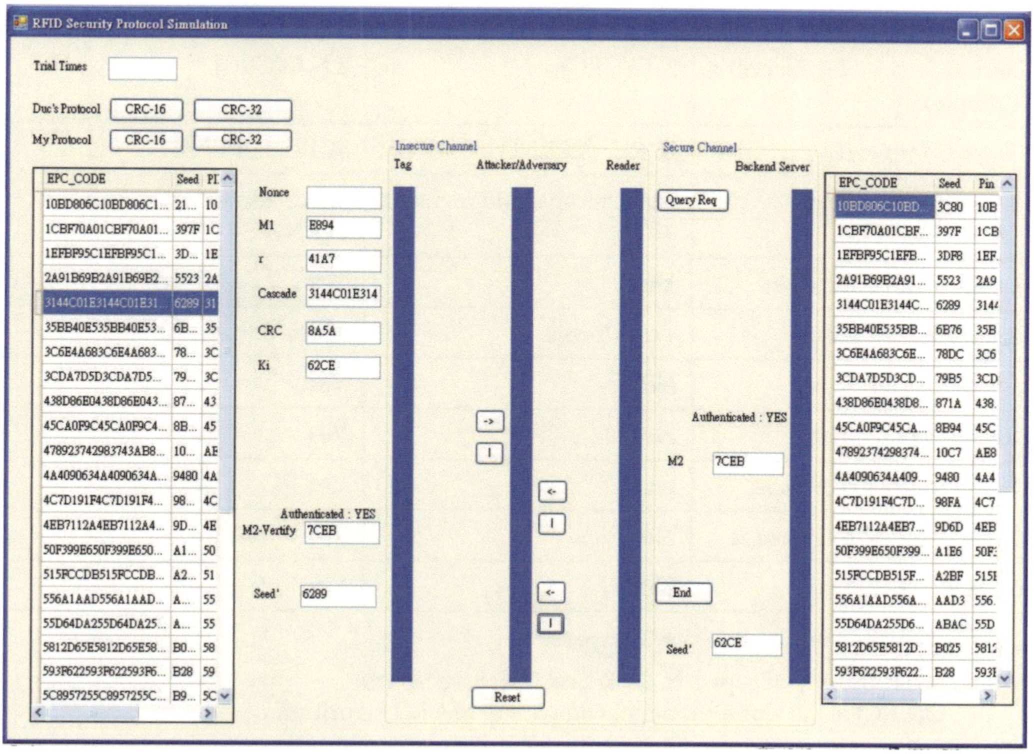

Fig. 3. Protocol Simulation Programme

The simulation results also show that T_M and T_D behaved the same and thus an addition of random number challenge will not reduce T_M, although the combined number from *n* and *r* is found to have no repetition out of the 65535 trials.

Further tests on the effect of CRC length on T_D and T_M was conducted using 32-bit CRC, the results T_D =1.000168 and T_M =1.000015 were obtained using n=16 bit respectively. This concludes that CRC affects the behavior of M_{1T} but not the n.

Since both T_D and T_M have dropped from around 1.5 to around 1 by using 16-bit CRC and 32-bit CRC. It can be concluded that the repetitive of PRNG is caused by the CRC bit size used. Since a 32-bit CRC is not available in the existing RFID standards, its vulernability against replay attack is weak and thus needs enhancement in the next/new generation tags.

6. Security & complexity analysis

6.1 Security analysis

Table 1 provides a security performance comparison of the Duc et al and this proposed schemes related to tag anonymity, data privacy, mutual authentication, forward secrecy, key attack, DoS attack and replay attack.

Tag Anonymity

Tag will never emit static ID, a new random number is chosen from Reader and Tag in each session to ensure tag anonymity.

	Duc et al.'s Protocol	Proposed Protocol
Backend Server's Complexity	N O(CRC)	2 N O(CRC)
Tag's Complexity	2CRC + 2PRNG	3CRC + 3PRNG
Reader's Complexity	Send, receive and forward	Send, receive and forward + 1PRNG
Reader Authentication	Yes	Yes
Tag Authentication	Two Phrase	Three Phrase
Spoof Reader Attack	No	Yes
Resist to Dos Attack	No	Yes
Resist to Replay Attack	No	Yes
M_{1T} Collision in Database	No	Yes
Forward Secrecy	No	Yes

Table 1. Security and Complexity Comparison

Where, N is Number of tuples in Back-End Database Server
O(CRC) is the Computational complexity of CRC algorithm.

Data Privacy

Tag never sends any plain text data through insecure channel, data is always encrypted by a session key with nonce. Reader can use cryptography algorithm to exchange data between back-end database server. Therefore, data privacy is protected.

Mutual Authentication

The new protocol performs both tag-to-reader and reader-to-tag authentication. Database authenticates the tag by verifying the message M_{1T}. Tag verifies M_2 generated by database. This mutual authentication scheme ensures data exchange will be granted to authenticated parties only.

Forward Secrecy

Even if the tag is compromised at some time later, as the PIN and session key is updated after each successful authentication, an adversary can not trace and track the compromised tag from the past eavesdropped communication data. Therefore, the forward secrecy is protected.

Key Attack

The shared secret session keys are chosen randomly for each tag and they are different from each other. Exposure for a single key will therefore not expose other's tags secret information.

DoS Attack

The database will maintain six values including the old session key and old PIN for each tag. Even though the tag is out of synchronization with the database, it can still communicate with the database, by performing a session key and PIN update process to synchronize with database. Although it may increase the communication cost, it can ensure that M_{1T} will not subject to any replay attack.

Replay Attack

The random number challenge from the reader can effectively prevent replay attack from the spoofed tag. The generation of M_{1T} has involved the random number from reader, therefore an adversary can not replay M_{1T} from an eavesdropped communication between spoofed reader and tag.

6.2 Complexity analysis

The proposed security scheme complexity is studied according to its computation, storage requirement and authentication phrase.

Computation Complexity

To communicate with the tag, the reader requires only a PRNG and cryptography algorithm to authenticate to allow the transfer of data between reader and back-end database server. The requirements are feasible in the current generation reader. In the authentication process, reader actually acts like a store and forward hub between back-end database server and the tag with the computation complexity are mainly handled by the back-end database server.

The authentication between the tag and reader two CRCs and one PRNG to generate the message M_{1T}. The Reader authentication process requires one CRC and one XOR operation to verify M_2. The key and PIN update process requires two PRNGs. A total of three CRCs and three PRNGs being used in the whole authentication protocol.

The database generates both M_{1Ki} and M_{1Ki-1} for each tuple, so the computation complexity is equal to 2N (2CRC + PRNG), where N is number of tuples in database.

Storage Requirement

For the tag, it is required to store 3 parameter, i.e. EPC , PIN_i and K_i . For the database, the storage requirement is the same as Duc's scheme and it is required to store five values for each tag. In addition, it requires to store the tag's EPC, PIN_i, PIN_{i-1} and PRNG's seed for K_i, K_{i-1}.

6.3 Authentication phrases

The proposed security scheme is a three-phrase mutual authentication protocol.

Phrase one: Random number challenge from reader.

Phrase Two: Tag generates M_{1T} to authenticate itself to reader.

Phrase Three: Back-end database server generates M_2 which included tag's access PIN to authenticate itself to tag, in order to grant the read and write right to reader.

7. Conclusions

In this chapter, we have evaluated Duc et al.'s security schemes under different attacks and pointed out its vulnerabilities, including DoS attacks, forward secrecy weakness and reader-

to-tag authentication collision weakness. To overcomes these weaknesses, a new protocol is proposed. In our new proposed scheme, it distributes the authentication computational complexity or loading to the back-end database server and the reader and keeping the complexity in the tag unchanged. The scheme also conform to existing EPCglobal Gen-2 specification.
The simulation results conclude that the average appearing time of M_{1T} is affected by the CRC function but not only the random number input. This unwanted repetitiveness can be avoided by using a CRC-32 function instead, so hopefully it will be implemented in the next available generation of RFID tag in the near future.

8. References

An, Younghwa; Oh, Soohyun (2005). RFID System for User's Privacy Protection, *Asia-Pacific Conference on Communications*, pp. 516- 519.

Ateniese, G.; Camenisch, J.; de Medeiros, B. (2005). Untraceable RFID tags via insubvertible encryption, Proceedings of the 12th ACM conference on Computer and *Communications Security*, pp. 92 – 101.

Avoine, G.; Oechslin, P. (2005). A scalable and provably secure hash based RFID protocol, *Third IEEE International Conference on Pervasive Computing and Communications Workshops (PERCOMW'05)*, pp. 110-114.

Bono, S.; Green, M.; Stubblefield, A.; Juels, A.; Rubin, A. (2005). Security Analysis of a Cryptographically-Enabled RFID Device, 14th USENIX Security Symposium.

Chien, H.Y.; Che-Hao Chen, C.H. (2007). Mutual authentication protocol for RFID conforming to EPC Class 1 Generation 2 standards, *Computer Standards & Interfaces Volume 29, Issue 2*, pp. 254-259

de Dormale G. M.; Ambroise, R; Bol, D; Quisquater, J. J. (2005). Low-Cost Elliptic Curve Digital Signature Coprocessor for Smart Cards, Proceedings of the IEEE 17th International Conference on Application-specific Systems, Architectures and Processors (ASAP'06), pp. 347-353.

Deavours, D.D; Ramakrishnan, K.M; Syed, A. (2005). Technical Report ITTC-FY2006-TR-40980-01, October 2005

Dimitrios, T. (2006). A Secure and Efficient RFID Protocol that could make Big Brother(partially) Obsolete, *Proc. Intern. Conf. on Pervasive Computing and Communications, PerCom2006*, Pisa, Italy.

Duc, D.N.; Park, J.; Lee, H.; Kim, K .(2006). Enhancing Security of EPCglobal Gen-2 RFIDTag against Traceability and Cloning, *SCIS 2006*, Hiroshima, Japan.

EPC (2005). EPC™ Radio-Frequency Identity Protocols Class-1 Generation-2 UHF RFID, EPCglobal Inc., January 2005.

Feldhofer, M.; Dominikus, S.; Wolkerstorfer J. (2004). Strong authentication for RFID systems using the AES algorithm; Workshop on Cryptographic Hardware and Embedded Systems–CHES,*Lecture Notes in Computer Science, Vol.* 3156, pp. 357-370.

Garcia-Alfaro, Joaquin; Barbeau, Michel; Kranakis, Evangelos (2008). Analysis of Threats to the Security of EPC Networks, *Communication Networks and Services Research.*

Glidden, R.; Bockorick, C.; Cooper, S.; Diorio, C.; Dressler, D.; Gutnik, V.; Hagen, C.; Hara, D.; Hass, T.; Humes, T.; Hyde, J.; Oliver, R.; Onen, O.; Pesavento, A.; Sundstrom, K.; Thomas, M. (2004). Design of ultra-low-cost UHF RFID tags for supply chain applications, IEEE Communications Magazine, *Volume: 42, Issue 8*, pp 140-151.

Golle, P.; Jakobsson, M.; Juels, A.; Syverson, P. (2004).Universal Re-encryption for Mixnets, *Topics in Cryptology - CT-RSA 2004*, The Cryptographers' Track at the RSA Conference, Lecture Notes in Computer Science, Volume 2964, pp. 1988.

Ith, P.; Oyama, Y.; Inomata, A.; Okamoto E. (2007). Implementation of ID-based signature in RFID system; 2007. APCC 2007. Asia-Pacific Conference on Communications,,pp. 233-236.

Juels, A.(2004). Minimalist cryptography for low-cost RFID tag, Conference on Security in Communication Networks - SCN'04, LNCS, Amalfi, Italia, September (2004) Springer-Verlag

Juels, A. (2006). RFID Security and Privacy: a Research Survey, *IEEE Journal on Selected Areas in Communications*, 24 (2), pp. 381-284.

Juels, A.; Pappu., R. (2003). Squealing Euros: Privacy Protection in RFID-Enabled Banknotes, Financial Cryptography - FC'03. Lecture Notes in Computer Science, Volume 2742, Springer-Verlag, Le Gosier, Guadeloupe, FrenchWest Indies, pp. 103–121.

Landt, J. Shrouds of Time, The history of RFID, *AIM Publication*, Ver. 1.0, October 1, 2001. *Conference (cnsr)*, pp. 67-74.

Luo, Zongwei; Chan, T.; Li, J.S. (2005). A lightweight mutual authentication protocol for RFID networks, *IEEE International Conference on e-Business Engineering*, pp. 620 - 625.

Molnar, D.; Soppera, A.; Wagner, D. (2005). A scalable, delegatable, pseudonym protocol enabling ownership transfer of RFID tags, *Handout of the Ecrypt Workshop on RFID and Lightweight Crypto, July.*

Nohara, Y.; Inoue, S.; Baba, K.; Yasuura, H. (2005). Quantitative evaluation of unlinkable ID matching schemes, *Proceedings of the 2005 ACM workshop on Privacy in the electronic society*, pp. 55-60.

Ranasinghe, D.; Engels, D.; Cole, P. (2004). Security and Privacy: Modest Proposals for Low-Cost RFID Systems, *Auto-ID Labs Research Workshop.*

Rfid (2006): http://www.rfidalliancelab.org/

Rhee, K.; Kwak, J.; Kim, S.; Won D. (2005). Challenge-Response Based RFID Authentication Protocol for Distributed Database Environment, Security and Pursvarive Computing, Lecture Notes in Computer Science Volume 3450, pp. 70-84.

Roberti, M. (2005). Understanding the EPC Gen 2 Protocol, *RFID Journal Special Report.*

Tsudik, G. (2006). Yet another trivial RFID authentication protocol, International Conference on Pervasive Computing and Communications - PerCom 2006, Pisa, Italy, March

Weis, S.; Sarma, S.; Rivest, R.; Engels, D. (2003). Security and privacy aspects of low-cost radio frequency identification systems, *International Conference on Security in Pervasive Computing -SPC 2003, Lecture Notes in Computer Science, Vol. 2802. Springer-Verlag, Berlin Heidelberg New York*, pp. 454-469.

Wang, Xiao-hua; Zhou, Xiao-guang; Sun, Bai-sheng (2007). An Improved Security Solution of RFID system, International Conference on Wireless Communications, Networking and Mobile Computing, pp. 2081-2084.

18

RFID System Integration Design with Existing Websites via EPCglobal-like Architecture for Expensive Material Handling

Shing Tenqchen[1,2,3], Chui- Yu Chiu[2] and Saad Laraqui[3]
[1]*Dept of Electrical Engineering, National Taiwan University,*
[2]*Dept of Industrial Engineering and Management, National Taipei University of Technology,*
[1]*ChungHwa Telecom Labs.,*
[3]*Dept of Management & Technology, University of Maryland University College,*
[1,2]*Taiwan*
[3]*USA*

1. Introduction

Radio Frequency Identification System (RFID) Technologies have been widely applied to many fields for many years (Glover & Bhatt (2006)). In the past, due to the high prices of reader/tag, and many complex existing material management softwares in operating unit, there are not many successful driving projects working on constructing an integrated service for expensive materials handling in this area. On the end of 2006, the design of middleware using RFID reader and tag to collect traffic information implemented on urban-bus for intelligent transportation system has been successfully deployed in Taipei city for a trail system (Tenqchen *et. al.* (2007)). However, the quick response for the material ID, the price is lower down to certain acceptable region in reader/tag, and more successful stories in RFID related applications have motivated the birth of integrated service for expensive material handling. The driving force can be seen from Figure 1-1 to see the interior and exterior forces. Especially, the stock manager can easily investigate the total amount of important controlled equipments via the hit of website over different operating systems and existing material management systems.

In this following, we will handle the problems of the following:

- manual counting all stocks in warehouses;
- one can not allocate and transfer the wanted material in a short time; and
- it is hard to find the information of mapping the amounts of materials while the materials are in the installing/disassembling equipments.

Thus, the objectives of the design of website are:

- Check stock materials easily by e-information (e-info);
- Provide the mapping for material and customer;
- Provide stocking e-info for operating organization; and
- Provide the history of expensive materials.

The system architecture for this kind of system is shown in Figure 1-2. The dotted red block is called the middleware to handle the read/write of RFID tags and readers. The RFIDIS is the website for us to design such that one can monitor the status of expensive equipments on line. The Material Acquisition and Support Information System (MASIS) and Spare-Parts Administration System (SPAS) may be the existing material management software systems in the website.

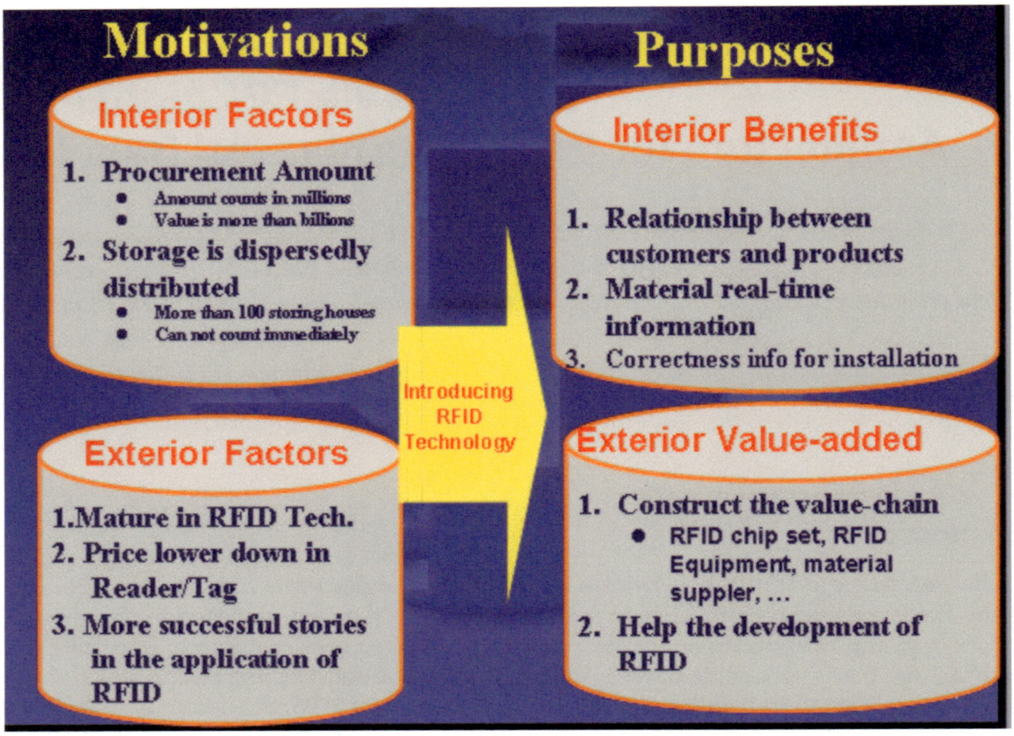

Fig. 1-1. Motivations and Purposes for RFID

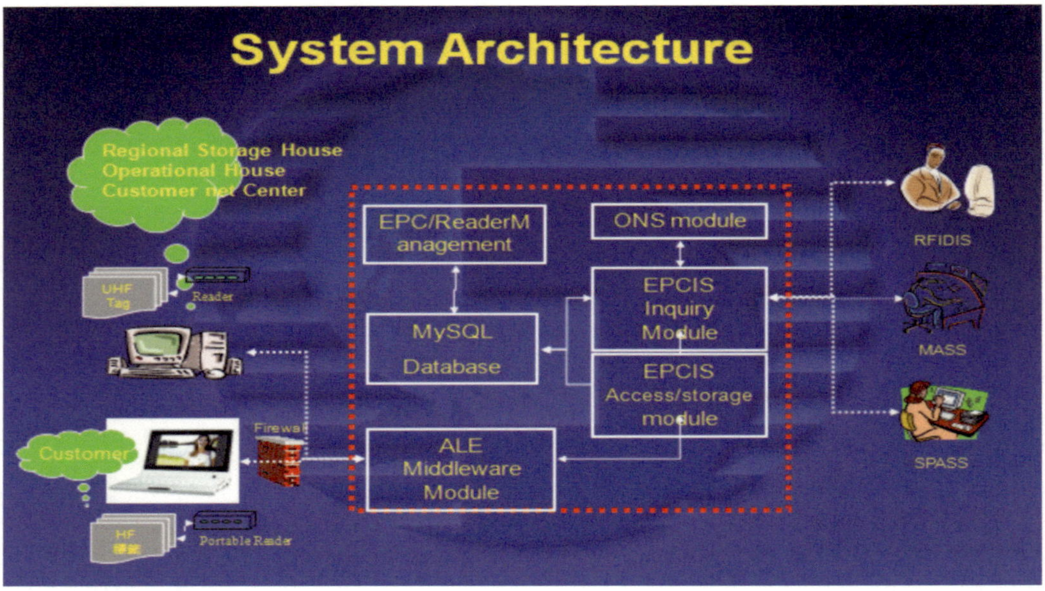

Fig. 1-2. System Architecture

The most elegant benefits of handling expensive materials using RFID technology can be described in four main folds as shown in Figure 1-3:

- Reducing the stock level;
- Increasing the efficiency of managements; and
- The historical display for material in item level or case level; and the realization of e-info for e-management.

In section 2, we describe the system requirements of the existing technology plan for RFID. This section of the rest will include some of the historical reviews in this area. What are the previous solutions that they ever did? What are the possible solutions in the future? What are the approaches of our methods and the logical architecture? The standardization of EPCglobal in this area will be included. In section 3, we will discuss the commercial RFID middleware. The RFID middleware platform, key elements of the EPCIS, a typical ONE query, and centralized deployment structure will be included in this section. The other structures please refer to Glover and Bhatt (2006). In section 4, we will focus the interface methodology of the system architecture of RFID with the existence systems like MASIS, SPAS and others. The sequence of logic operations is important to the message right from the enquiry of website on-line checking. In section 5, the system integration test is given to verify the logic sequences of our RFID system's results according to Table 4-1.

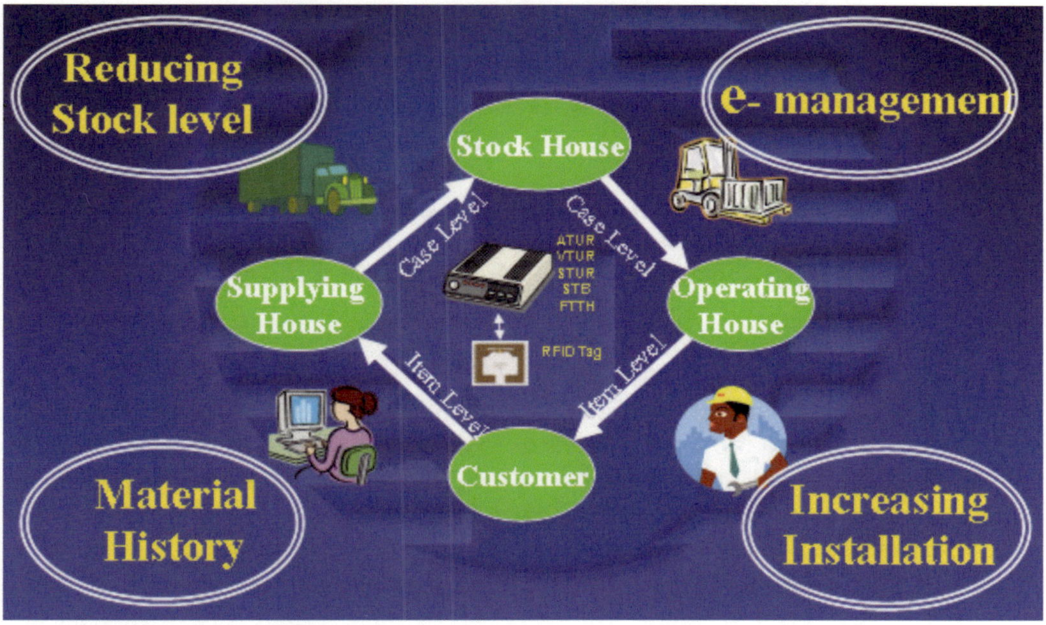

Fig. 1-3. Benefits

2. The system requirements

The ADSL Transceiver Unit Remote (ATUR), Symmetrical Transceiver Unit Remote (STUR), VDSL Transceiver Unit Remote (VTUR), Fiber to the Building (FTTB), Set-Top-Box (STB), and multi-functions telephones are served as examples for expensive materials as shown in Figure 2-1. The total volumes of expensive devices are generally larger than millions. They deserve our attentions to look at the material management problems. However, in most cases, the stock houses always have an existing material handling software. Thus, the integration of existing software like MASIS and SPAS (at least, but not limited to) in existing environments is a MUST condition for entire program.

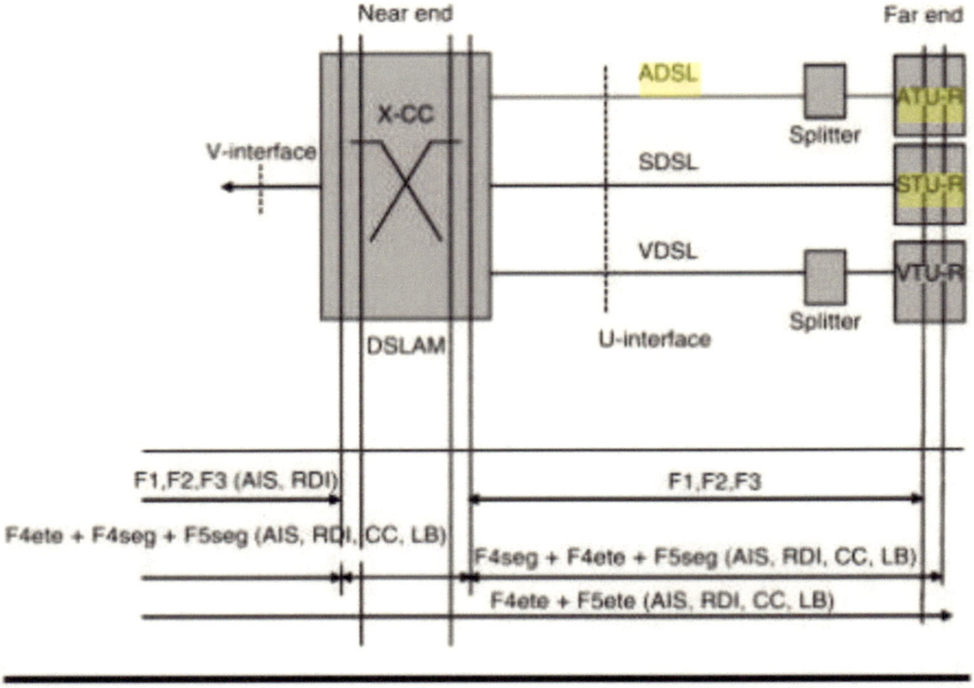

Fig. 2-1. General System architecture and Operators, administration, and maintenance (OAM) information flows. Source: Golden, Dedieu, and Jacobsen (2007), p. 448.

2.1 Primary typical system architecture

The primary typical system architecture of RFID components found in a retail store as shown in Figure 2-2 (modified from Finkenzeller & Waddington (2003)). In the bottom-left corner of the diagram, it is a set of RFID tags with antennas that represent the tagged merchandise. The readers may read tags hundreds or even thousands of times per minute, but most of these reads will not be interesting to our applications. The readers must be configured and managed to know how to work together to cover blind spots a reader should fail. The box marked RFID middleware represents one or more software modules that handle these responsibilities. The box marked edge applications represents any enterprise applications that have components running inside the store – for instance, Point of Sale (POS) system components. The box marked RFIS information service represents a mechanism to store RFID events and related data at the edge. This is because RFID information is stored at various points in the infrastructure: at the edges, within the data center, and with business partners.

RFID readers, also called *interrogators*, are used to recognize the presence of nearby RFID tags. An RFID reader transmits RF energy through one or more antennas. An antenna in a nearby tag picks up this energy, and the tag then converts it into electrical energy via induction. This electrical energy is sufficient to power the semiconductor chip attached to the tag antenna, which stores the tag's identity. The tag then sends the identity back to the reader by raising and lowering the resistance of the antenna in one kind of Morse code. This is only one scenario, and different tags can work in slightly different ways, but this is typical of the way readers and tags interact.

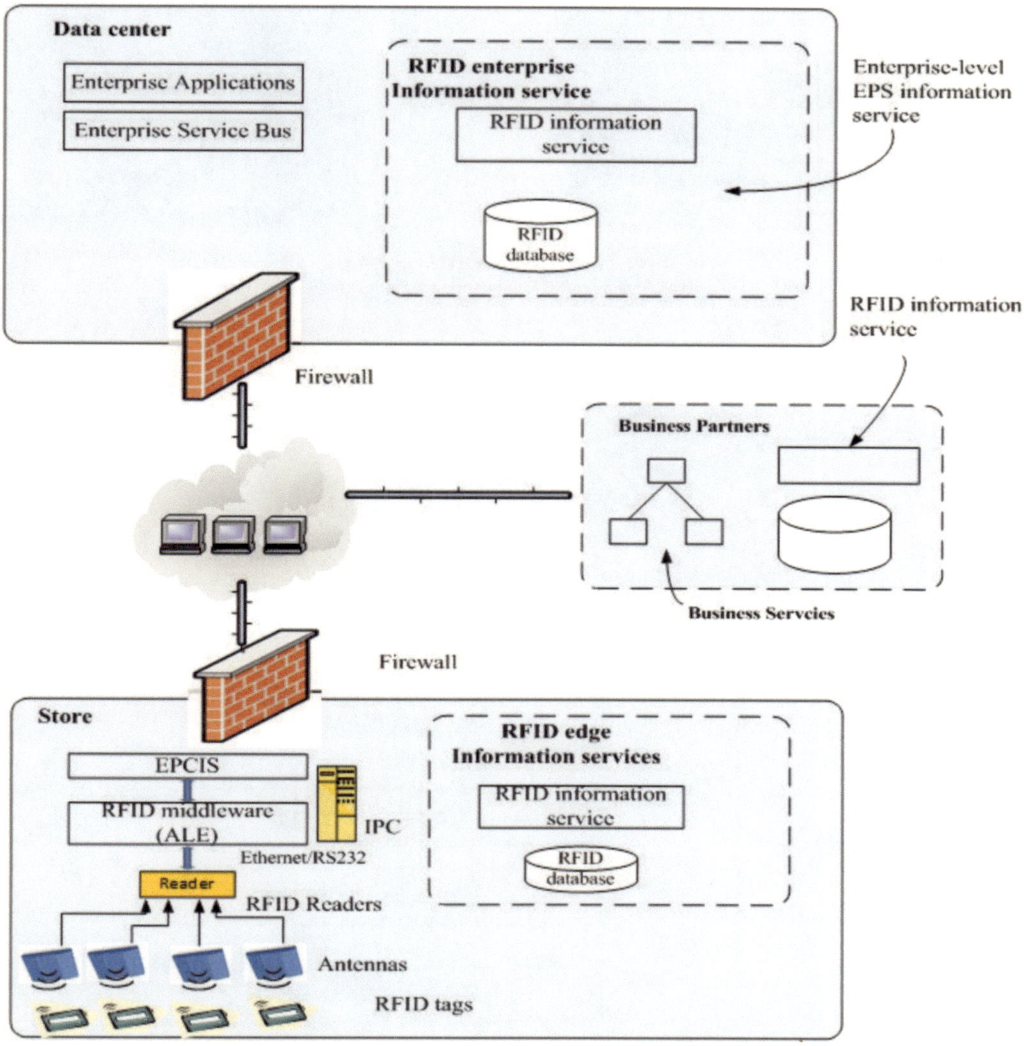

Fig. 2-2. RFID System Architecture

The implementation of RFID system architecture can be simulated as an RFID-based Management Platform as shown in Figure 2-3. It is only a trial system to verify the designed websites have a good communicational records with existing material management softwares like MASIS, SPAS, TOPS, and STARS, etc.

Many readers come in many shapes and sizes and can be found in stationary, as well as portable, handled varieties. Figure 2-4 shows how a reader fits between tags, antennas, materials and the components of a reader.

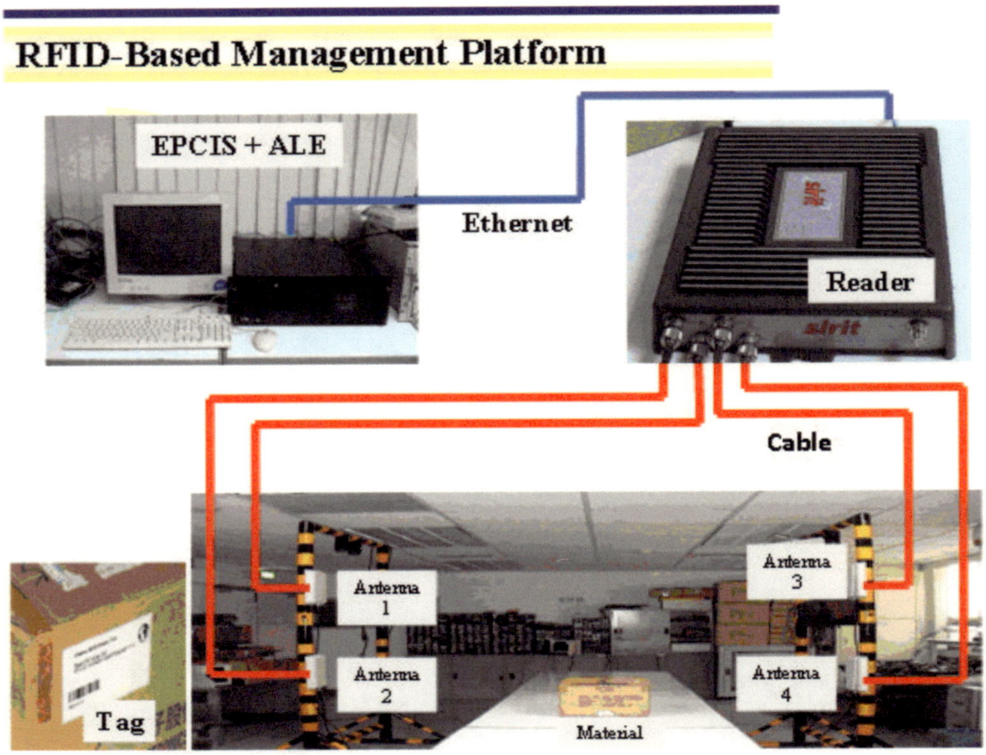

Fig. 2-3. the implementation of RFID-Based Management Platform

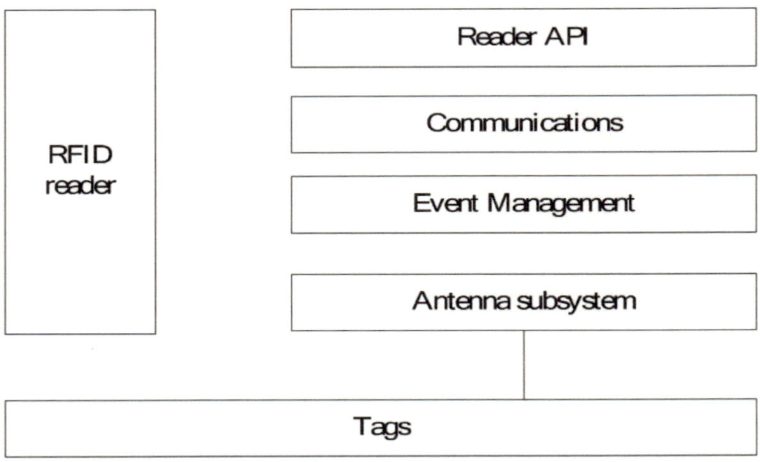

Fig. 2-4. Parts of a reader [Source: Glover and Bhatt (2006)]

1. *Reader API*

Reader API is the application programming interface that allows programs to register for and capture RFID tag read events. It also provides capabilities to configure, monitor, and manage the reader.

2. *Communications*
Readers are edge devices, and like any other RFID devices; they are connected to the overall edge network. The communications component handles the networking functions.
3. *Event management*
When a reader sees a tag, we call this an *observation*. An observation that differs from previous observations is called an event. The analysis of observations is called event filtering. Event managements defines what kinds of observations are considered events and determines which events are interesting enough to merit being either put in a report or sent immediately to an external application on the network.
4. *Antenna subsystem*
The antenna subsystem consists of one or more antennas and the supporting interfaces and logic that enable RFID readers to interrogate tags.
5. *RFID middleware*
There are three primary motivations for using RFID middleware: providing connectivity with readers (via the reader adapter), processing raw RFID observations for consumption by applications (via the event manager), and providing an application-level interface to manage readers and capture filtered RFID events. The event-processing middleware is introduced between the readers and the applications. This approach is suitable for small-scale deployments using the capabilities provided by application integration products.
Several dozen types of RFID readers are available in the marketplace today, and each has its own proprietary interface. It would be impractical to expect application developers to learn different types of reader interfaces. Reader interfaces, as well as data access and device management capabilities, differ widely. One should try to use middleware that shields you from having to learn the idiosyncrasies of individual readers. The adapter layer encapsulates the proprietary reader interfaces so that they don't need to come in contact with the application developers.
We should filter out any such spurious observations to avoid sending a flood of inaccurate observations to the inventory control system. Filter 2-5 illustrates a procedure for filtering and smoothing system devised to address typical scenarios experienced in retail stores.

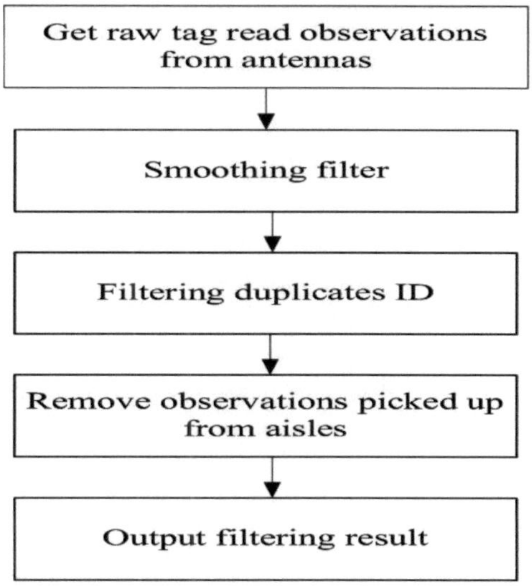

Fig. 2-5. Procedure of Filtering the Events

2.2 Logical architecture

A typical RFID-enabled distributor or retailer with several hundred or more stores will have hundreds, if not thousands, of readers. Each of these readers will be chirping away several times a second to read the RFID tags around them. As shown in Figure 2-6, the raw observations from RFID readers and sensors lack application-level context. More instance, an order management application would want to know when the in-store inventory for a particular item drops below its threshold. As you can imagine, an order management system wouldn't be the least bit interested in knowing whether RFID readers are employed in tracking the items in the stores, let alone how many readers there are per store and in what configuration. The process of making them more meaningful for enterprise applications is called *event filtering*. The data mechanism of Reader and Tag shown in Figure 2-6 contains the reader subsystem and event subsystem. The component that provides the event filtering functions is called the *event manager*.

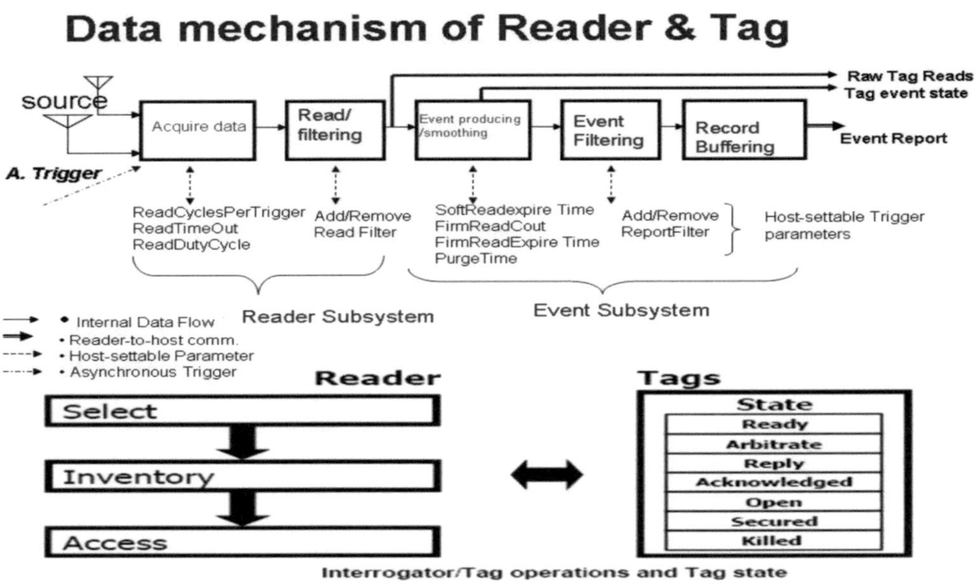

Fig. 2-6. The Data mechanism of reader and tag

Figure 2-7 is from the event volume to application relevance through different layers of an RFID system. Figure 2-8 shows a conceptual model of the RFID middleware. The RFID middleware receives raw observations from one or more data sources. A data source can be any sensor that collects data about the physical world, like an RFID reader or temperature sensor. After receiving observations from the readers, the event manager component of the middleware aggregates, transforms, or filters them to prepare them for consumption by applications. In addition to making the RFID observations more relevant to applications, the event manager helps reduce the sheer volume of data that the applications must process. The RFID middleware as shown in Figure 2-8 can support reader discovery, provisioning, monitoring, and management: provide data collection, translation, aggregation, and filtering. The grouping mechanisms are to support service-oriented interfaces using standards such as JSP, J2EE, .NET, EJB, and web services; and offer remote provisioning, monitoring, and management capabilities.

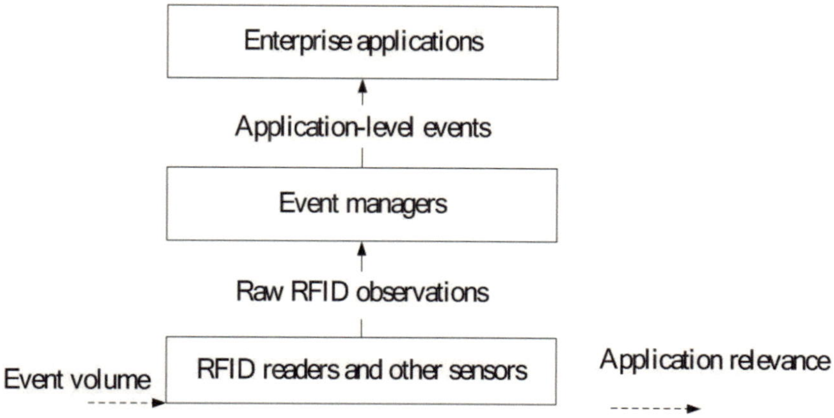

Fig. 2-7. Event volume and relevance through different layers of an RFID system

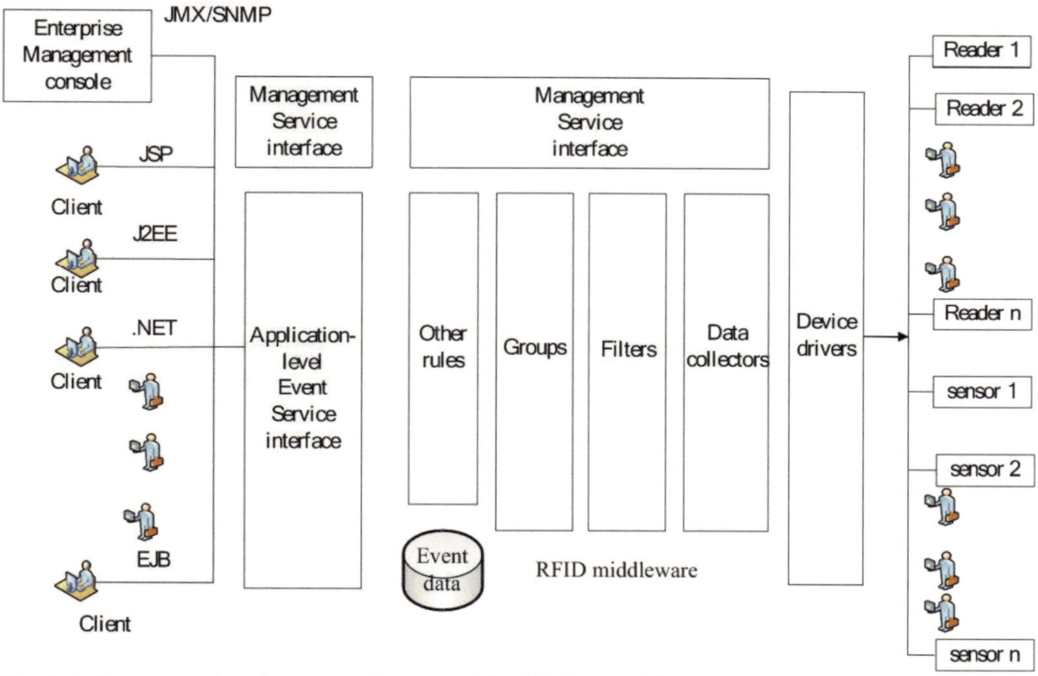

Fig. 2-8. Conceptual architecture for a modified RFID middleware product

3. Commercial RFID middleware

A report in ALE-speak is the output from an event cycle, which is returned as an ECReport instance. A report specification, represented as ECReportSpec, provides filtering, grouping and other data processing instructions. Figure 3-1 shows the primary data elements.

The EPCglobal's Application Level Events (ALE) specification defines a reader-neutral interface for receiving events from RFID readers and filtering and grouping them. The remainder of this chapter provides an overview of the ALE 1.0 specification as published by EPCglobal. We'll describe this specification in considerable detail so that you can familiar yourself with its key concepts and API, but you should be aware that several vendors' implementations of the ALE specification are available in the market, each providing its own extensions and benefits.

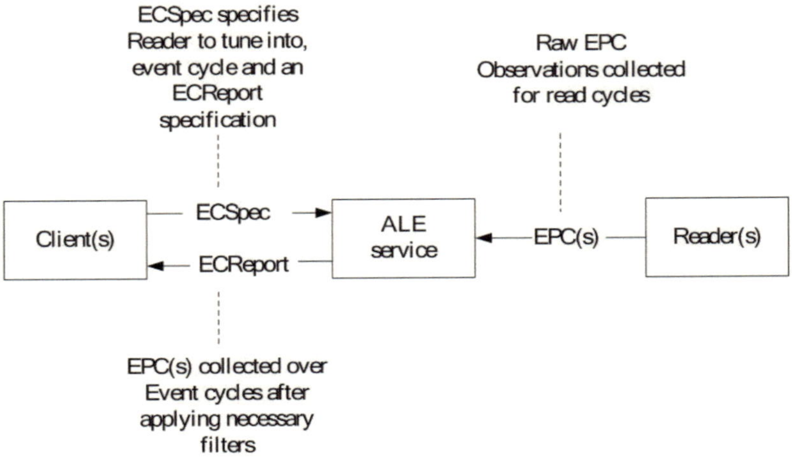

Fig. 3-1. Primary data elements [Source: Glover & Bhatt (2006)]

3.1 RFID middleware platform

The main components of the Edge Server offering include the filtering and collection engine (also known as the ALE Engine) and the device management agent. The RFID Tag aware edge server interfaces to a wide variety of popular readers and printers, as well as various sensor inputs that are used as triggers to the reader control. The edge server implements the EPCglobal ALE API and includes extensions for tag writing and other capabilities not yet covered by the standard. Figure 3-2 shows the RFID Tag Aware Middleware Platform.

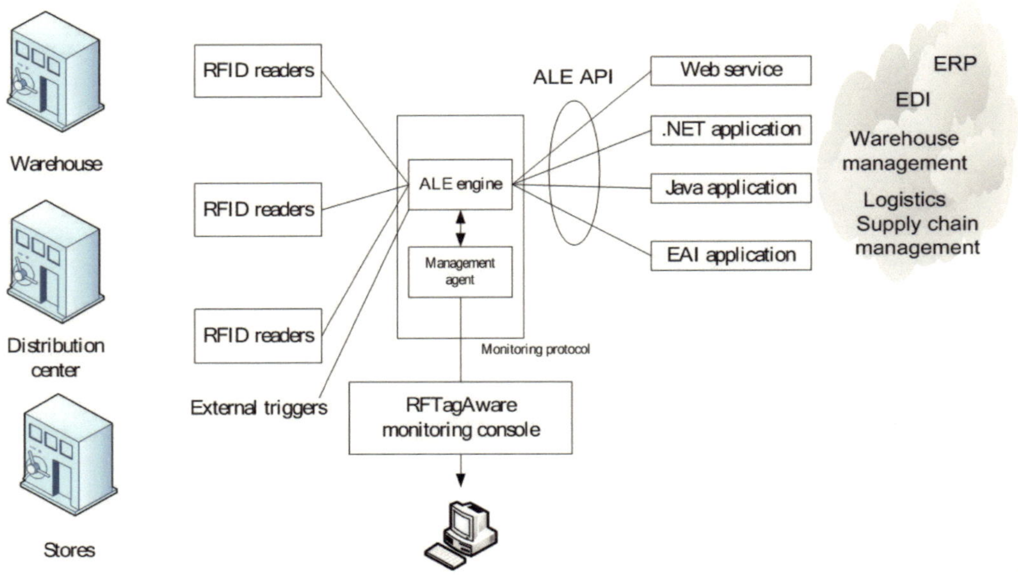

Fig. 3-2. RFID Tag Aware Middleware Platform [Source: Glover & Bhatt (2006)]

3.2 Key elements of the EPCIS

The EPCIS defines a standard interface for capturing and sharing EPC-related data. It should be noted that the EPCIS focuses only on the service interface and semantics of EPC-related data, such as location information that gets registered as products move through supply chain. Vendors are provided the flexibility to compete on implementations and add-on functionality. EPC observations are captured using the EPC Capture interface, and they are queried using the EPC Query interface. Additionally, the EPCIS provides a common model for location information and other important data. The EPCIS standards were envisioned for use with events within an organization, where they may be subscribed to or stored and queried, or between companies, where the same operations will be able to take place. The EPCglobal's reader protocol interface insulates the higher layers from knowing what RF protocols and reader makes and models are in use, the EPCIS insulates enterprise system from having to know the details of how individual steps in a business process are carried out. The EPCIS-level data differs from lower layers in the EPC Network Architecture as shown in Figure 3-3 because it incorporates semantic information about the business process in which EPC data is collected and provides historical observations.

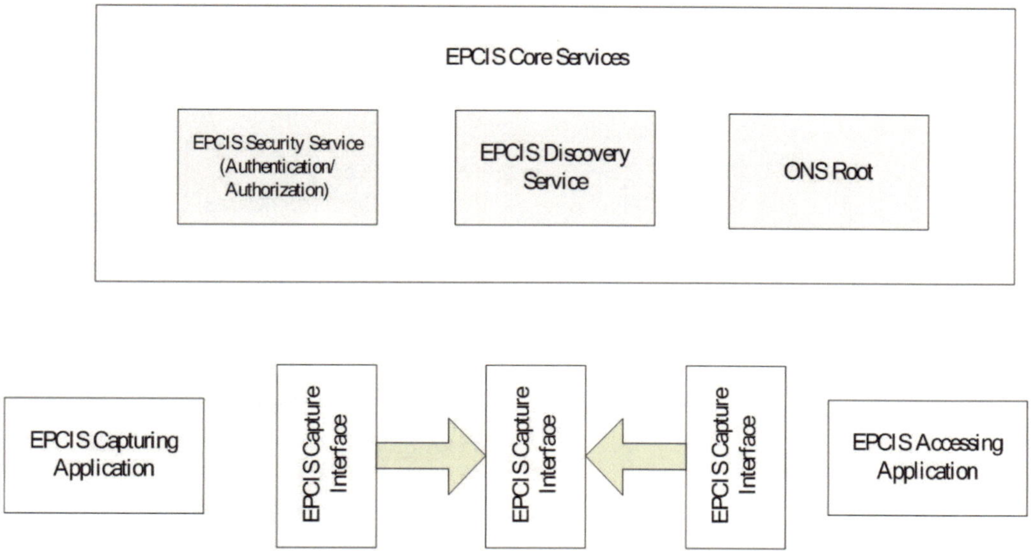

Fig. 3-3. Key Elements of the EPCIS [Source: Glover & Bhatt (2006)]

3.3 A typical ONS query

The integrated EPCglobal system architecture proposed by Amerio etc (2007) EPCglobal division has been published in RFID standard. There are five sections as shown in Figure 3-4 to deal with the EPCglobal architecture framework. The first deals with tag emit EPC data and emit detected data. The second is the reader captures events and filtering. The third is Savant server deals with data report, manages readers, and makes an advanced filtering. The fourth is EPCglobal subscriber server deals with cross-enterprise elements (EPC data exchange and EPC object exchange), and the last deal with intra-enterprise elements (EPC Infrastructure).

There is a static ONS that let EPC transfer to the wanted EPC data. Another is the dynamic ONS is to provide founded EPC flow control for tracking. The enterprise's trading needs the action of EPC Access Registry.

3.4 Centralized deployment

In the 1980s and 1990s, microcomputers exploded out of the data centers, onto desktops, and into server closets enabling end users in new ways and at the same time bringing new challenges to Information Technology (IT) operations staff. Over the pass few years, most organizations have worked steadily toward a return to more centralized management of computing devices by moving servers out of market shift server closets and into climate-controlled, secure data centers. The advantages of this configuration include more efficient physical access to systems by IT operations staff and increased reliability due to the protection data centers offer from power and temperature fluctuations, dust, and vibration. Improved physical security is an additional benefit of moving all of these servers to a central local within the restricted area. The centralized approach of edge deployment options is shown in Figure 3-6.

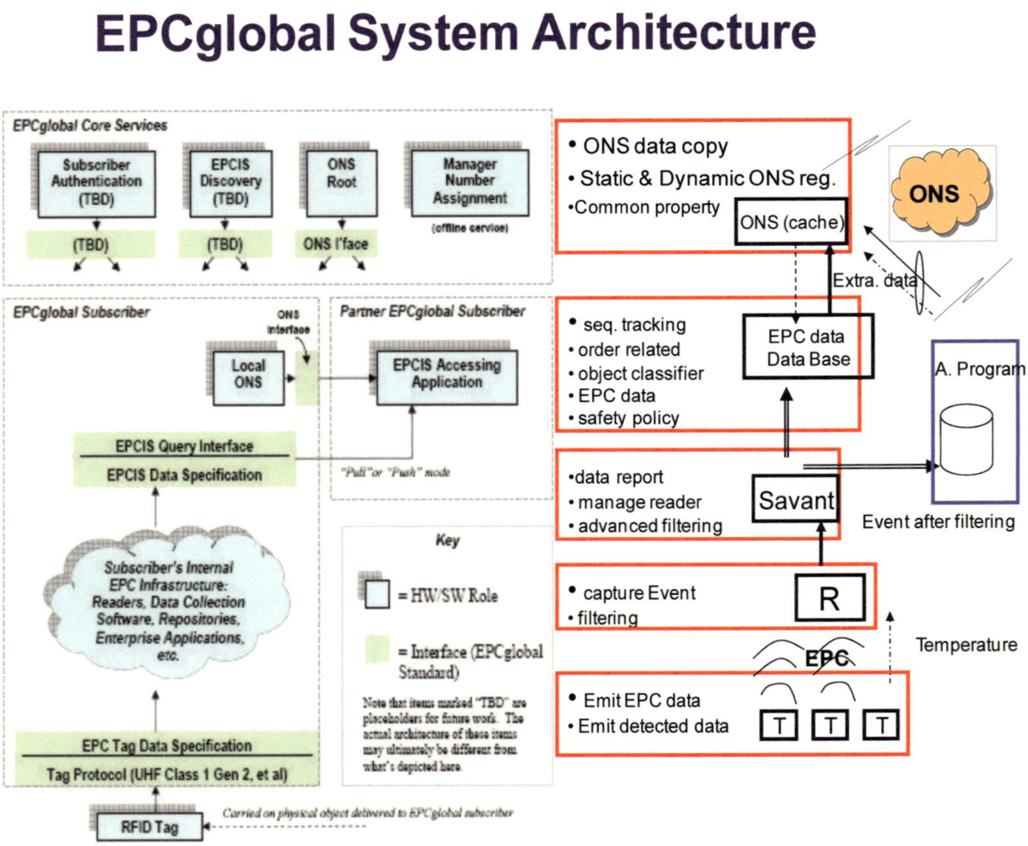

Fig. 3-4. A typical EPCglobal System Architecture

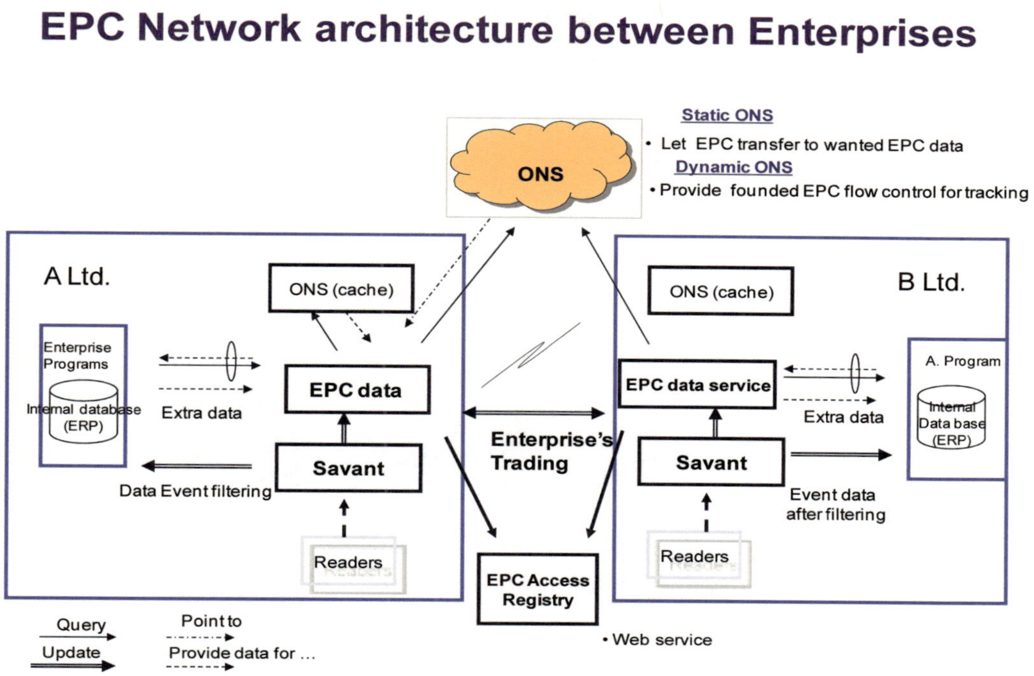

Fig. 3-5. The EPCglobal Network architecture between enterprises

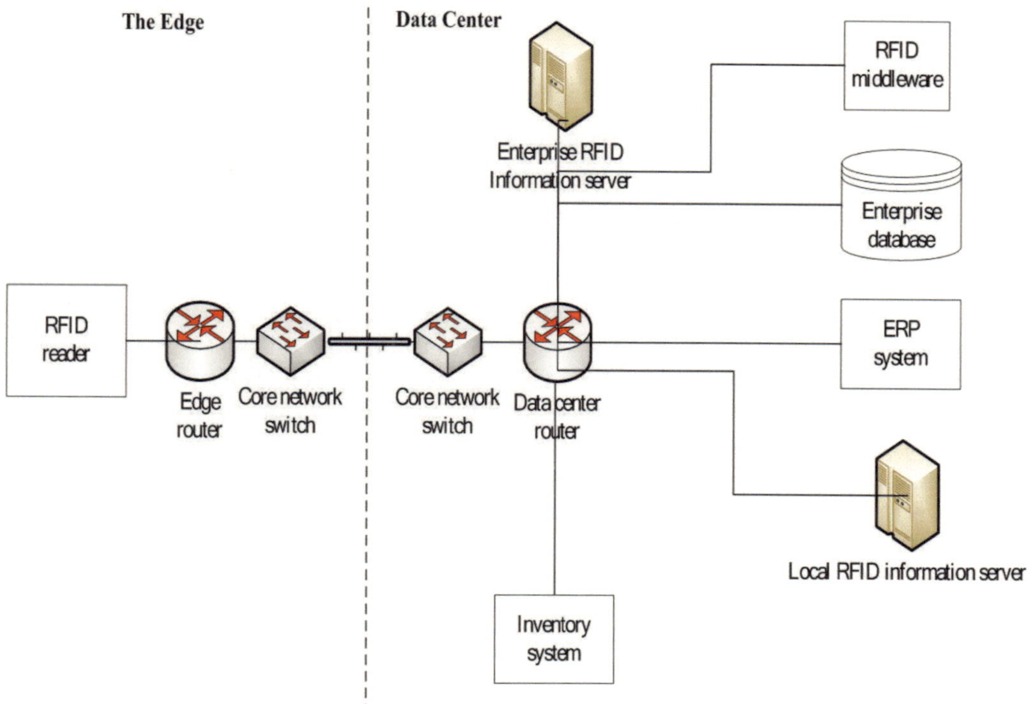

Fig. 3-6. The centralized approach

4. Interface methodology of the system architecture of RFID with the existence of MASIS, SPAS and others

According to the protocol of RFID-aid material information system (RamMIS) and other material-related management system (MASIS, SAPS and others), the lifecycle of material management can be classified into three roles: applicant, material management person, and deploy person. The RFID materials has been classified with an extensive closer look on the relationship of existing material management systems as shown in Figure 4-1. In this section, there is focused on the first item called MASIS for an example of demonstration.

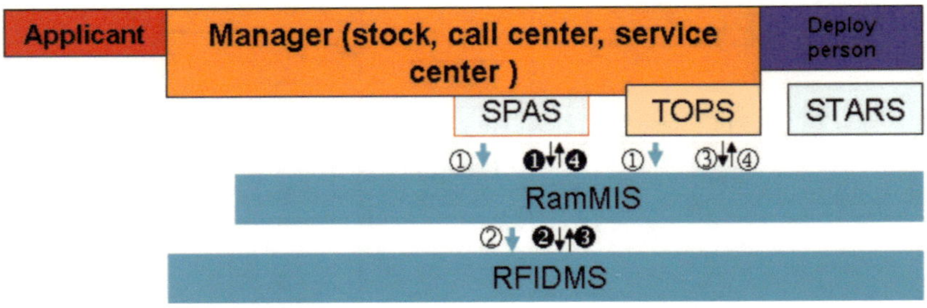

Fig. 4-1. RFID material Management Information System (RamMIS)

Mainly, the electronic Purchasing Information System (ePIS) is the top system for purchasing any possible items for one company. The systems are that Material Acquisition and Support Information System (MASIS), Spare-parts Administration System (SPAS), Top Service System (TOPS), and Survey of Testing And Repairing System (STARS),....
The RamMIS will be designed to be WEB-based with the tools of (Java Server Pages (JSP), Servlets, Java 2 Platform Enterprise Edition (J2EE), and Signed Applets). The J2EE applications will be used for material flow control with Enterprise JavaBeans (EJB). It is the server-side component architecture Java and provides the interface system's info and query response (XML over HTTP).
Thus, the RFID Monitoring System (RFIDMS) is suggested to provide an EPCglobal-like architecture framework, which provides the platform of frontend software/hardware including tag, reader, and middleware with AP-based gateway and signed applets for other interface systems. At the backend, the softwares shall include ALE, EPCIS database, and EPCIS query interface. The J2EE applications can include receiving the raw data from the feedback of Middleware Servlets, the query of web services, and the alarm message of monitoring system. The EPCglobal-like architecture framework as shown in Figure 4-2 is mainly divided into two parts. The top one is the RamMIS itself. The second is the RFID monitoring system (RFIDMS). The signed-applets are designed for receiving/applying at stock house, and for other information interface systems. It contains the parts of EPCIS access interface, access/application interface (ALE) software, filtering/collecting interface, filtering & collecting (RFID middleware), and reader protocols. The blue color representative represents the interface for EPCglobal standard and brown color represents the hardware/software. Thus, there are a lot of design work can be done with different requirements for material handling.
The tag, reader, and middleware are the front-end hardware/softwares. They need AP-based gateway PC and signed applets are used to provide the other interface systems to show their instant information. The ALE, EPCIS database, and EPCIS inquiry interface are the back-end softwares. J2EE applications mainly provide the following functions:

- To receive and process the Servlets of raw data from middleware software.
- To provide the inquiry response of web services from general purpose equipment.
- To provide the maintenance report for alarm message, etc.

The EPCglobal-like architecture framework is shown in Figure 4-2. This framework provides the needs of design items for each different faces on the website.

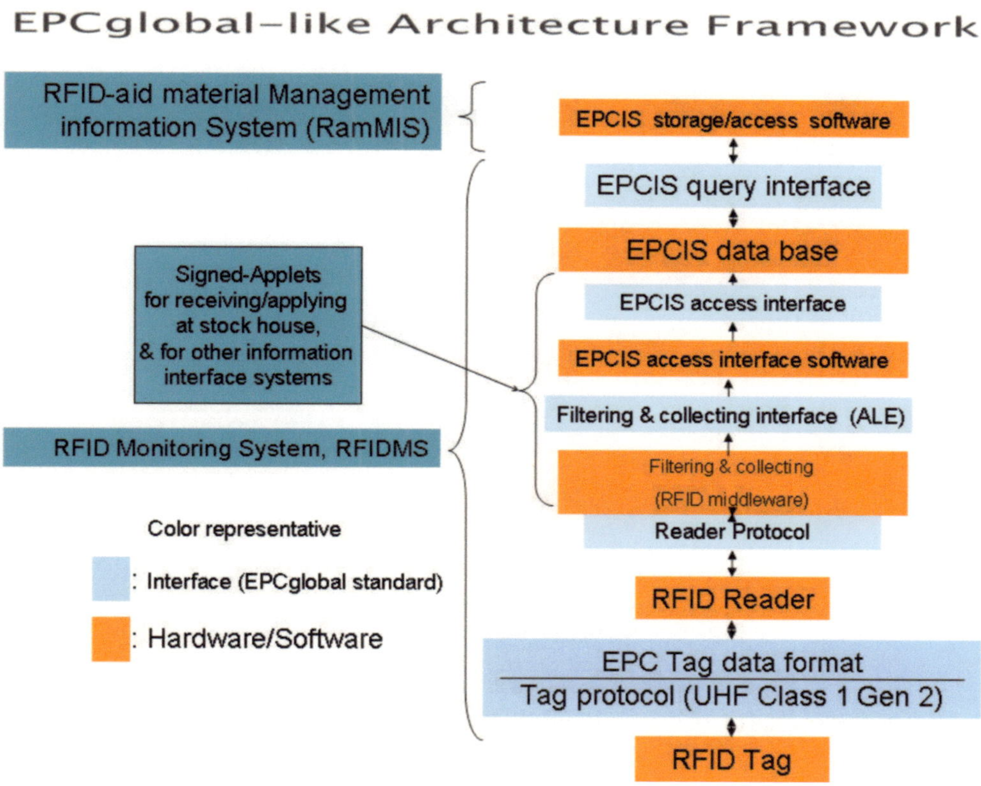

Fig. 4-2. EPCglobal-like Architecture Framework

The RFID-aid material management information system (RamMIS) is designed for RFID serving system. The signed-Applets for receiving/applying at stock house and for other information serving systems could include the EPCIS access interface, filtering and collecting interface, and filtering and collecting (RFID middleware).

4.1 Timing of message negotiating and usage methods

In the following, we will discuss the timing of message negotiating and then the usage methods.

Timing of Message negotiating

a. The tag of RFID will have an access signal to the system before purchasing equipment (connecting with MASIS)
b. confirm with income/outcome material (connecting with MASIS)
c. inform after outcome material (connecting with SPAS)
d. acquisition on material after change (connecting with SPAS)
e. synchronization with Assembly & removal batches (connecting with SPAS)

f. material status on-line query (connecting with SPAS, STARS, and TOPS …)
g. Query for basic material information (connecting with STARS, TOPS …).

RamMIS will connect with all the existing systems with http's packet by the content of XML. The detail is demonstrated as following section. The Functional Structure of RamMIS Website can be designed as Figure 4-3.

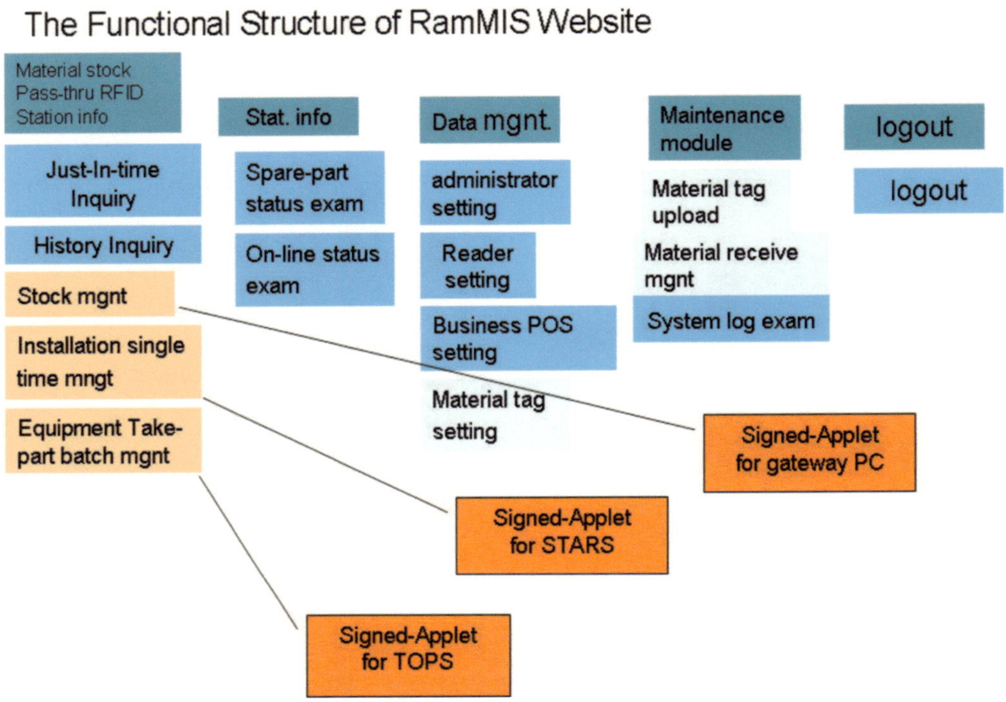

Fig. 4-3. The Functional Structure of RamMIS Website.

4.2 Usage methods for message negotiating

A. [RBN*, RBR*] XML over HTTP POST

*: Please refer to Table 4-1 for more detail explanations.

MASIS needs to inform the RFID platform when, what, and where to have RFID elements. RamMIS responds the processing result whether it successes or not. If it is wrong, RamMIS shall response the reasons.

B. [RCQ*, RCR*] xml over Http

RFID related gateway PC needs to wait for the lists of monitoring tag. RFID platform is immediately response the reader's results.

C. [RBN*, RBR*] XML over http

MASIS needs to inform RFID platform when, what, and where to have RFID elements. RamMIS responds the processing result whether it successes or not. RamMIS responds the processing result whether it successes or not. If it is wrong, RamMIS shall response the reasons.

D. [RCQ*, RCR*] XML over HTTP

RFID related gateway PC needs to wait for the lists of monitoring tags. RFID platform shall respond MASIS what the reading results of materials.

E. [CBS*, CBR*, RSQ*, RSR*] XML over HTTP

SPAS needs to ask the RFID platform to ask what kinds of materials, when will be here recently, and where the place will be? The batch response will follow the process of from the sequence of SPAS > RamMIS > SPAS.

F. Signed applet, [RSQ*, RSR*, PIQ*, PIR*, PIP*] XML over HTTP

STARS may consider the platform of RFID and enquiry the basic materials by tag ID. SPAS will require informing the RFID platform of information by tag ID.

G. Signed applet, [RSQ, RSR, PIQ, PIR, PIP] XML over HTTP

TOPS or SPAS will response the batch information by RFID platform. TOPS will require executing the batch information. TOPS will require the basic information by tag ID from RFID platform. SPAS will require the basic information by tag ID from RFID platform. The relationship of websites for different management systems is shown in Figure 4-4.

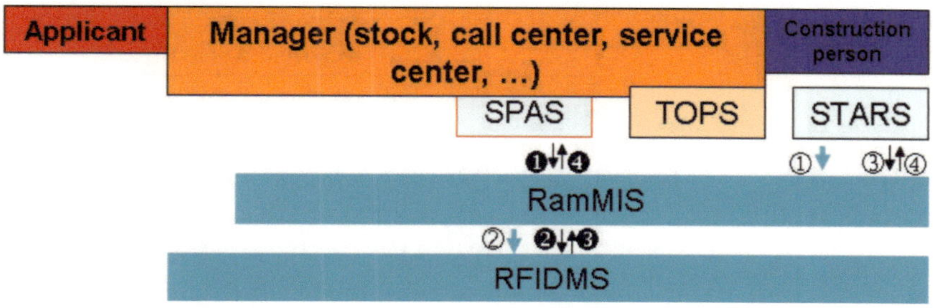

Fig. 4-4. The relationship of websites for different management systems

The material monitoring flow of RFID Monitoring System (RFIDMS) is shown in Figure 4-5.

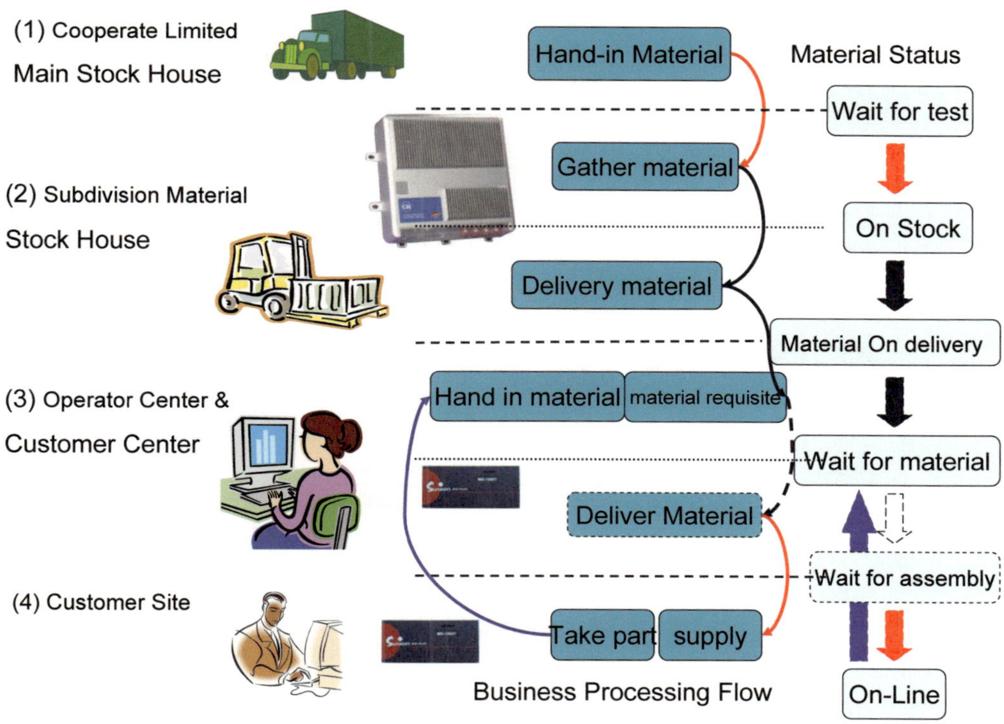

Fig. 4-5. RFID materials monitoring flow

The sequence is mainly followed from the main stock house, to subdivision house, operator center and customer center, and customer's site. The material status begins from wait for test, on stack, material on delivery, wait for material, and wait for assembly, etc. One can check the status from on-line website such that he/she may understand how long he/she needs to wait. The service center can make arrangement for the installation person to customers to install their equipments. Thus, it is a good service for both sides. The managers also can know how many items left for future installation. If he/she wants to meet the requirements of installation, managers can easily plan the purchase orders according to the life cycles of equipments for the future.
The abbreviations of the mentioned symbols will be listed as the following table 4-1.

Seq.	Module	Abbr.	Message function
1	Inform before materials get in	**RBN**	MASIS delivers material input (/apply material) message to RamMIS.
		RBR	RamMIS delivers back the processing results to MASIS.
2	Acknowledge before materials get in	**RCQ**	MASIS delivers materials get in (output) inquiry to RamMIS.
		RCR	RamMIS delivers back materials get-in message to MASIS.
3	Inform after materials leave	**OAN**	RamMIS delivers out the message of materials leaving SPAS.
		OAR	SPAS delivers back the processing result to RamMIS.
4	Inform material change	**ICN**	RamMIS delivers out the message of material exchange SPAS.
		ICR	SPAS delivers back the processing results to RamMIS.
5	Install, tease, & move (ITM) batch synchronously	**CBS**	SPAS delivers out the batch message of RamMIS
		CBR	RamMIS deliver back the processing results to SPAS.
6	Material instant condition enquiry	**RSQ**	Client system delivers out the instant condition enquiry message to RamMIS.
		RSR	RamMIS deliver back the processing results to Client system.
7	Material basic information enquiry	**PIQ**	Client system delivers the message of basic information to RamMIS.
		PIR	RamMIS delivers back the enquiry result to Client system
		PIP	RamMIS actively response basic material info to Client.

Table 4-1.

5. System integration test

5.1 Material gets in information

1. Applicant login the website of RamMIS and feds in the file of RFID tag's information.
2. Before MASIS system gets in the materials, it needs to send **RBN** message to RamMIS system.
3. RamMIS system receives the notice, proceeds to relative records to monitor, and responds **RBR** message to MASIS system.

Fig. 5-1. RFID material hand-in notice

5.2 material get-in acknowledgement

1. Material manager needs to login RamMIS system first and execute the operation of RFID gateway reading.
2. Material manager needs to login MASIS system and make sure the reception is OK, then, MASIS needs to send **RCQ** message to RamMIS.
3. RamMIS system receives the message of acknowledgement and then transfer to the platform of RFIDMS.
4. The platform of RFIDMS receives the message back to RamMIS system, and then RamMIS transfers the **RCR** message to MASIS system.

It is shown as Figure 5-2.

Fig. 5-2. RFID material get-in acknowledgement

5.3 Notice before apply material

1. Before MASIS system applies for materials, it shall send **RBN** message to RamMIS.
2. When RamMIS system receives the notice, it needs to relative records to ne monitored, and respond to **RBR** message to MASIS.

It is shown as Figure 5-3.

5.4 Acknowledge when materials leave the stock house

1. Manager needs to login RamMIS system to execute RFID gateway reading procedure.
2. Manager again login MASIS system and make sure the acknowledgement is OK when materials leave the stock house.
3. RamMIS system receives the out source message and ask again to RFIDMS platform.
4. RFIDMS platform will send the message back to RamMIS system, then, RamMIS system replies **RCR** message to MASIS system.

It is shown as Figure 5-4.

5.5 Inform when material gets out

1. After RamMIS system the material gets out, it needs to send **OAN** message to SPAS system.
2. After SPAS system receives a notice, SPAS needs to precede the relative records and send **OQR** message back to RamMIS system.

It is shown as Figure 5-5.

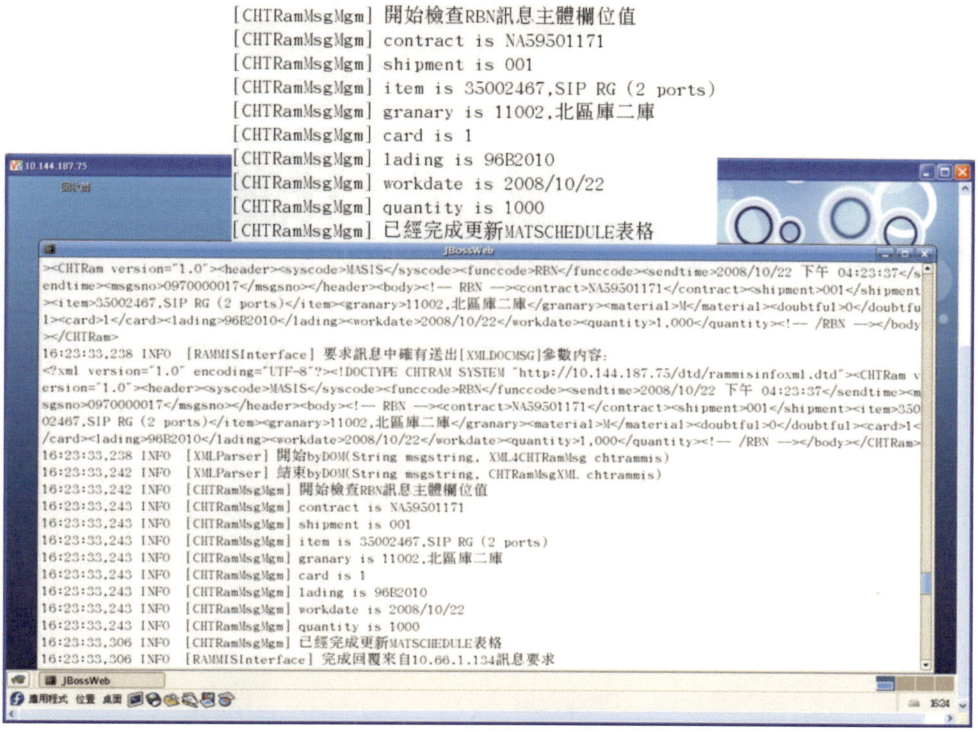

Fig. 5-3. Notice before apply material

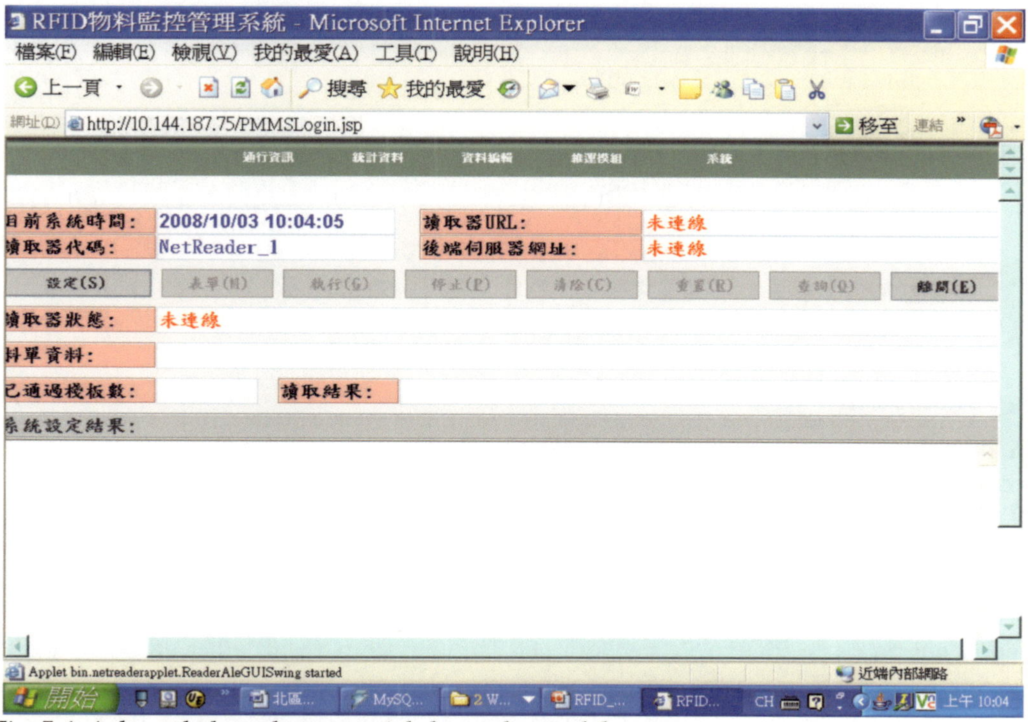

Fig. 5-4. Acknowledge when materials leave the stock house

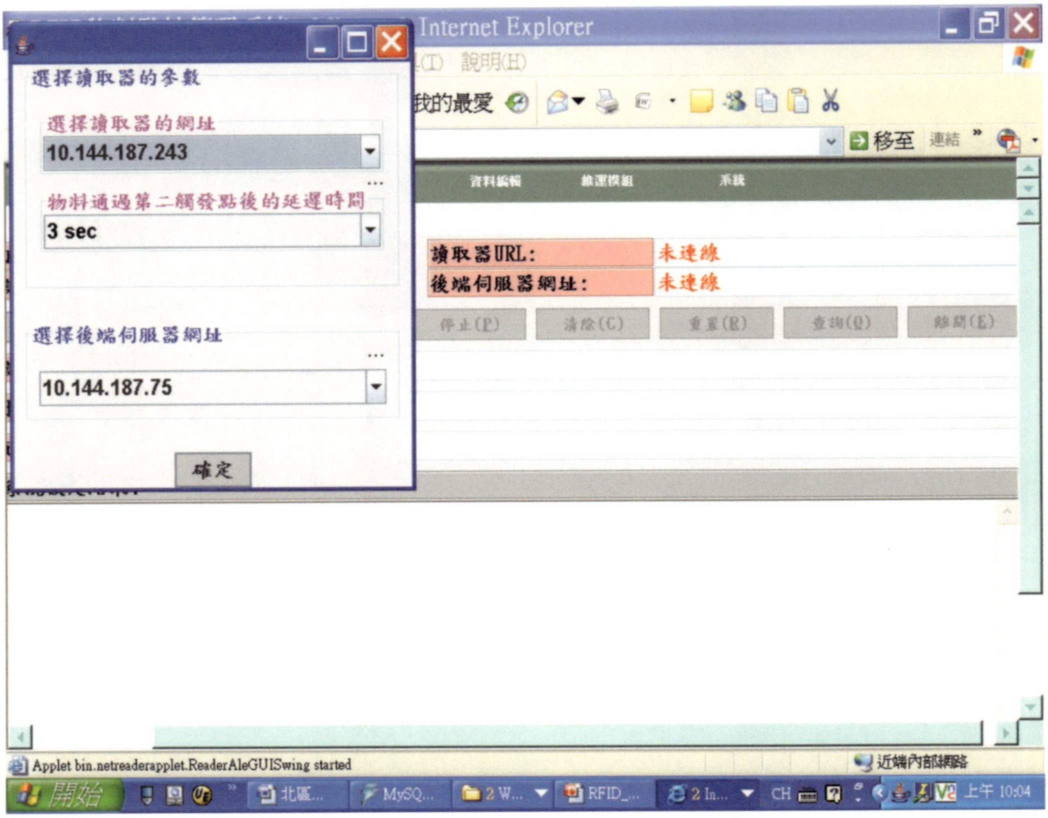

Fig. 5-5. Inform when material gets out

5.6 Information notice for material change

1. After materials get out, if the users login RamMIS system, the total numbers of RFID's tag will be changes. RamMIS system will send the message to **ICN** to SPAS system.
2. SPAS system receives the notice and execute the recording the relative work of managements, and send **ICR** message to RamMIS system.

It is shown in Figure 5-6.

6. Conclusion

In this chapter, we propose an example of handling expensive materials using RFID technological approach on an open platform environment and follow the standardization of EPCglobal Gen II. We discuss the integration case for the case of centralized deployment.
We also discuss the general cases in the future. We hope to have a good reference site for your design. The high unit price and big volume materials have an urgent request to have a clear request on input/output information needed by operating units, which is not dependent on a specific mobile network and is interoperable with other ad hoc material operating systems, like some existing softwares. One can design an interface based on integrated database for existing material management systems. It can develop and

implement a case level and item level management for expensive materials based on RFID platform on line.

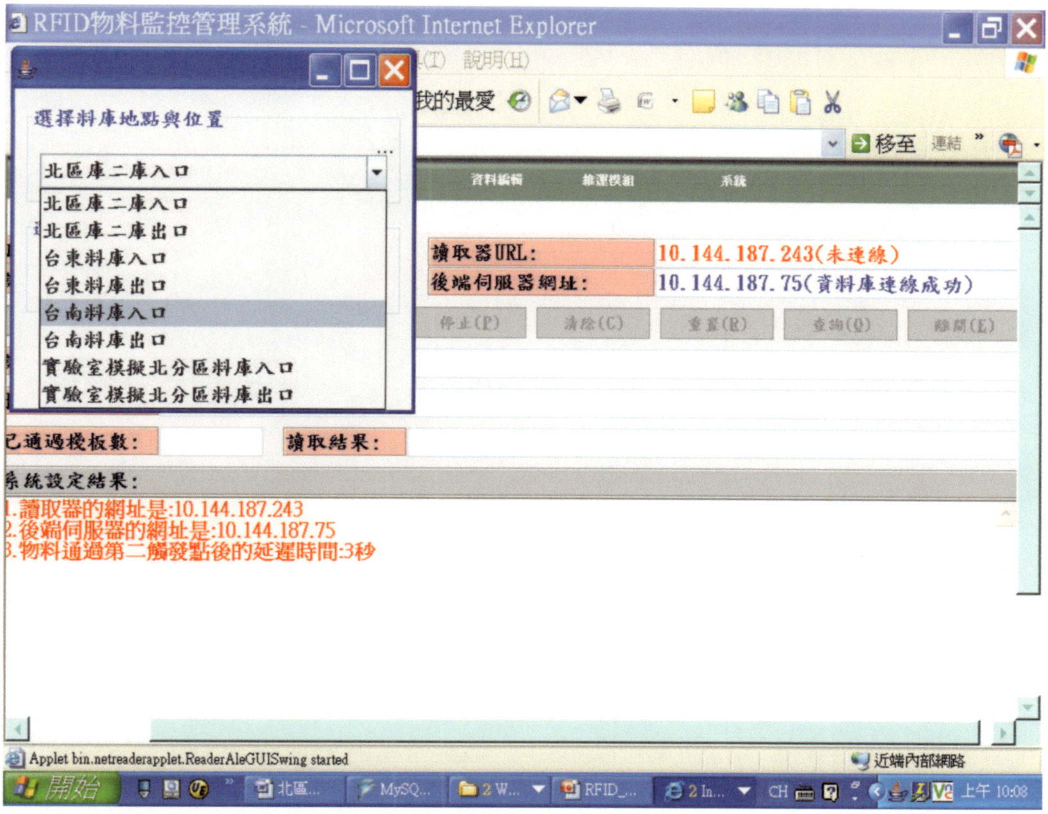

Fig. 5-6. Information notice for material change

7. References

Amerio, F. etc. (2007). The EPCglobal Architecture Framework, p.27, http://www.epcglobalinc.org/standards/architecture/architecture_1_2-framework-20070910.pdf.

Finkenzeller, K. and Waddington (2003). R. RFID Handbook, *John Wiley* p61-159.

Golden, P. Dedieu, H., and Jacobsen, K.S. (2007/10). *Implementation and Applications of DSL Technology*. CRC Press, Auerbach Publication. P.448.

Glover, B. and Bhatt, H. *RFID Essentials* (2006). O'Reilly, Media, Inc.

Tenqchen, S., Y.-K. Huang, H.-H. Huang, F.-S. Chang, K.-Y. Chen, Y.-K. Tu, C.-H. Wang, Y.-C., Lee, C.-H. Lee, S.-L. Tung, P.-C. Chi, "Design of Middleware Using RFID Reader and Tag to Collect Traffic Information Implemented on Urban-bus for Intelligent Transportation System Application," *Proceedings of 14TH World Congress on ITS 2007, Oct.9-12.*

Tenqchen, S. Y.-K. Huang, C.-H. Lee, W.-S. Feng, C.-K. Wang "Design of Middleware with EPC global by Using RFID Reader and Tag to Collect Traffic Information

Implemented on Urban-bus," *Proceeding of International Conference on Signal Processing and Communication Systems*, Australia, Gold Coast, 17-19 December 2007.

19

RFID Product Authentication in EPCglobal Network

Tieyan Li[1] and Wei He[2]
[1]Institute for Infocomm Research
[2]Singapore Institute of Manufacturing Technology
Singapore

1. Introduction

Estimated by the International Chamber of Commerce (ICC) in 2006, nearly 5-7% of the global world trade is in counterfeit goods, with the counterfeit market being worth approximately US$600 billion annually. Existing technical countermeasures, such as holograms, smart cards, biometric markers and inks, represent a flexible portfolio of solutions against some counterfeiting behaviors. Recently, RFID was reportedly used in product authentication solutions to achieve a higher degree of automation when checking the authenticity of a product. For example, Euro banknotes are attached with RFID chips to com- bat counterfeiting by European Central Bank. The United States Food and Drug Administration (US FDA) has issued a report that endorses RFID as a tool to combat counterfeiting of pharmaceuticals. So far, these RFID-based solutions seem pretty promising [28]. With wide adoption of RFID technology witnessed in various industries, the future of RFID for product authentication purpose looks optimistic.

The main objective of a product authentication solution is to distinguish a genuine product from a fake one. The basic concept of applying RFID to product authentication lies in its original function of *identification*. Imagine a scenario in the future, in which every object will be attached with an RFID tag that contains a unique number belonging to the object. Once the tag is interrogated, the unique object number is emitted and interpreted by the back-end system to identify the object. If, for instance, all the unique object numbers are stored in a database, we can then check the database to verify the identity of an object. Unfortunately, identification alone is insufficient for solving the anti-counterfeiting problem. Problems exist in such a straightforward solution. For example, the unique object number can be eavesdropped and copied onto blank tags to produce clones, and the database would not be able to distinguish a legitimate tag from a cloned tag containing the same object number. There are many other ways to attack such a simplified identification system. For example, in a "tag removal and reapply" attack, counterfeiter can remove a tag from an authentic product, perform reverse engineering on the tag to extract out key attributes, and replicate these attributes onto blank tags.

In fact, product authentication has stronger requirements on security and needs a more complex system to implement. RFID-based product authentication solutions leverage on the benefits provided by the RFID tags and the back-end information system within the RFID-enabled production and distribution flow. RFID tags can have certain security functions implemented in them, which raises the barrier for counterfeiting them. Furthermore, a counterfeiter would now need to counterfeit both the product and the tag, which raises his costs for counterfeiting. The back-end information system assists in drawing and maintaining real-time profile over the movements and activities of goods, thereby facilitating fast tracking of the goods. Essentially, a simplified product authentication system could consist of the following components - the object that is to be protected, the RFID tag that is attached onto the object, the RFID reader and the back-end system. Fig. 1 depicts the components in a generic RFID-enabled product authentication system.

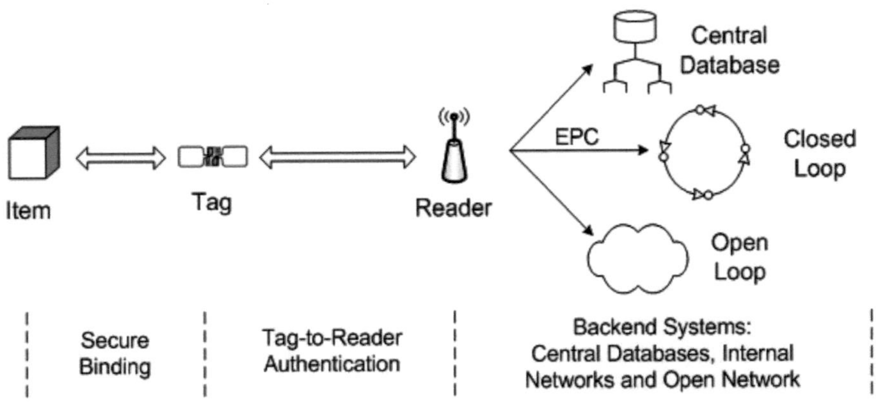

Fig. 1. RFID Product Authentication System.

Traditional product authentication methods rely on optical technologies such as watermarks, holograms and micro-printing to authenticate and verify goods. Other more advanced methods include the use of biological, chemical, or even nano-technologies (*e.g.*, using DNA markers, nano-level material characteristics, *etc.*). RFID technology, with the use of RFID tags that are attached to goods, opens up a new way to authenticate products. Like optical solutions, RFID technology authenticates the information stored on an external object (the RFID tag) rather than the product itself. If the RFID tag is authenticated, we claim that the product is authenticated too. To ensure the effectiveness of such a solution, the RFID tag needs to be securely bound to the product. Some secure binding mechanisms that are used in RFID systems will be discussed in greater detail in Section 5.

The authentication of an RFID tag is carried out through interactions with an RFID reader. RFID tag-to-reader authentication protocols resemble much of the existing two party authentication protocols based on challenge-response. In fact, a large number of research works conform to this principle and rest on symmetric or public key cryptographic primitives. We summarize these solutions in section 6. Unfortunately, these solutions do not provide a practical solution in realistic product authentication scenarios. This is because most RFID tags (for example, those being used on fast moving consumer goods) are too cheap to incorporate even lightweight cryptographic primitives. Currently, there exists a gap between what needs to be implemented for a substantial level of security on the tag and what could be realistically supported on the tag. Achieving proper authentication with low-cost RFID tags is still very challenging.

Besides the secure binding of an RFID tag to an object and the authentication between an RFID tag and a reader in the end system, another area that needs to be considered for a more complete product authentication solution is that of the back-end system. In a supply chain, as the goods are moved from one part of the world to another, many different activities can be taking place at each intermediate point. In fact, each intermediate point could potentially represent a point of vulnerability, where counterfeiting behavior might exist. Hence, in addition to checking at the end points, checks may need to be conducted at each intermediate point as well. This requires a systematic back-end support that connects itself to all the intermediate points. The simplest back-end system is a single standalone database that records up-to-date information on the goods by collecting data at each intermediate point. A verifier can then check the database for the details and/or status (*e.g.*, ID, some stored secret, current location, history, *etc.*) of a particular product, and based on this knowledge, determine the authenticity of the product. With a powerful database, there is a high chance that even a perfectly cloned tag can be detected. However, collecting and collating all relevant information into one single database is rather ambitious and unlikely to be scalable. How to disseminate these information into decentralized locations is very much desirable in both closed loop solutions and open loop solutions.

Product authentication solutions may be customized for different product distribution scenarios by considering hybrids involving the closed loop and open loop solutions. For example, an e-pedigree solution for combating counterfeit drugs is promoted and piloted as a major anti-counterfeiting effort of the US FDA. The potential high risk of drug misuse and increasing market of counterfeit drugs are the main drivers of this countermeasure. In general, for a product authentication solution to be feasible, the cost of implementing the solution must be lower than the losses suffered due to counterfeiting activities. Moreover, the cost of breaking the system should be high in order to provide a substantial barrier against counterfeiting behavior. Hence, when customizing a product authentication solution, we need to consider the cost-effectiveness of the customizations. Challenges arise when we face dynamic and complex application environments, such that each of them requires a different security level. In such cases, it would be difficult to design an optimal solution that fits all the requirements.

The rest of this chapter is organized as two parts: Part 1 introduces the security issues and countermeasures with RFID systems, which includes Section 2-the common threats that are faced by RFID systems; Section 3-the security and privacy issues with RFID systems; and Section 4-the countermeasures. Part 2 presents various RFID product authentication solutions including the secure binding of an RFID tag to the target object in Section 5; RFID authentication protocols in Section 6; and some network level solutions in Section 7 and 8. Finally, we conclude the chapter with some remarks.

PART 1: RFID SECURITY ISSUES AND COUNTERMEASURES

2. Common threats against RFID systems

The proliferation of RFID tags implies that RFID enabled systems might suffer from unintended risks. For example, unauthorized data collection, where attackers gather illicit information by either actively issuing queries to tags or passively eavesdropping on existing tag-reader communications. RFID threats refer to malicious user abuse in RFID context and are categorized as *Gather*, *Mimic*, and *Denial of Service (DoS)* [2]. *Gather* threats include *Skimming*, *Eavesdropping* and *Data tampering*; *Mimic* threats include *Spoofing*, *Cloning* and

Malicious code; Denial of Service threats include *Killing, Jamming* and *Shielding*. The details of these threats are explained as follows:

- *Skimming* data is the unauthorized access of reading of tag data. Data is read directly from the tag without the knowledge or acknowledgement of the tag holder.
- *Eavesdropping* is unauthorized listening/intercepting, through the use of radio receiving equipment, of an authorized transmission to monitor or record data between the tag and reader for the purpose(s) of: collecting raw transmissions to determine communications protocols and/or encryption; collecting the tag's data, or determining traffic patterns.
- *Data tampering* is unauthorized erasing of data to render the tag useless or changing of the data.
- *Spoofing* is defined as duplicating tag data and transmitting it to a reader. Data acquired from a tag is transmitted to a reader to mimic a legitimate source.
- *Cloning* is defined as duplicating data of one tag to another tag. Data acquired from a tag is written to an equivalent tag. A cloned tag is indistinguishable from its original tag.
- *Malicious code* insertion of a executable code/virus to corrupt the enterprise systems is hypothetically possible given a tag with sufficient memory and range.
- *Denial of Service* occurs when multiple tags or specially-designed tags are used to overwhelm a reader's capacity to differentiate tags, rendering the system inoperative. E.g., A blocker tag [19] is a kind of denial of service that confuses the interrogators so that they are unable to identify the individual tags.
- *Killing* of a tag (electronic or mechanical) is an operational threat in that the physical or electronic destruction of the tag deprives downstream users of the tag data.
- *Jamming* is the use of an electronic device to disrupt the reader's function.
- *Shielding* is the use of mechanical means to prevent reading of a tag.

Utilizing a combination of above threats, more serious attacks can be launched on RFID systems including unwanted location tracking of people and objects (by correlating RFID tag sightings from different RFID readers). Beyond these threats, RFID tags suffer from a variety of subtle attacks such as physical invasive attack, where an adversary physically compromises the inlay of an RFID tag and reads the memory for any information; and side channel attack, where an adversary uses timing analysis, power analysis or electro-magnetic analysis (e.g., [24]) to extract tag information. The design of RFID product authentication solutions shall consider appropriate countermeasures to defend against all possible threats.

3. RFID security and privacy issues

3.1 RFID security issues

In traditional IT systems, security means to prevent unauthorized reading and changing of data in the systems. RFID security means protecting the data on the tag, the data transmitted between the tag and reader, and even the data on the reader, to ensure it is accurate and safe from unauthorized access. RFID systems must employ mechanisms to achieve one or more of the security objectives such as confidentiality, integrity, availability, authentication and access control, to alleviate various security concerns. In the following, we describe the security objectives in details and show that meeting these security objectives eliminates the security threats posed by inherent weaknesses in low cost RFID systems.

Confidentiality involves a mechanism to keep information from all but those that are authorized to see it. In an RFID system, sensitive data such as a secret key needs to be kept confidential either when it is stored on tag or reader, or transferred between a reader and a tag.
Integrity ensures that information has not been altered by unauthorized or unknown means. Alteration in an RFID context may involve the capture, substitution, or deletion or insertion of information and the retransmission of that altered information to a reader or a tag.
Availability in RFID systems is important since readers need to be ready to detect tags that may enter their reading range at certain intervals of time. RFID systems meeting the availability criteria will ensure that there are services in place to thwart a DoS attack.
Authentication The objective of authentication in RFID context can be expressed as authenticating the devices involved (the tags and the reader) or in a supply chain application where the tags are used to label products, as product authentication. The objectives of tag and reader authentication and product authentication are discussed below.

- Tag/Reader Authentication: In RFID context, authentication simplifies to the proofs of the claimed identity of a tag or a reader. Authentication is an important RFID security measure for preventing counterfeiting behaviors. In some applications where perhaps the tag is an integral part of the tagged object, authentication of the tag may be adequate to guarantee the authenticity of the object to which it is associated.
- Product Authentication: In certain use cases where tags are placed as an external label to a high value item, authentication of the tag is not sufficient to guarantee the authenticity of the product to which the tag is attached. Since these tagged goods are subject to some specific attacks such as the "remove and reapply" attack. Hence, product authentication refers to the establishment of the authenticity of a product by the secure binding of the identity of a tag and the legitimacy of the product with an irrefutable link between the product and the tag that can be verified by a third party.

Access Control implies a mechanism by which a tag or a reader grants access or revokes the right to access some data or perform some operation in the interaction between RFID readers and tags. Generally tags will require access control mechanisms to prevent unauthorized access to tag contents.
To achieve these security objectives, RFID systems require solid implementations of appropriate security mechanisms. While security cannot be solely accomplished by these mechanisms, we stress that proper legislation, procedural techniques and enforcement of laws are also required.

3.2 RFID privacy issues

Compared with security properties, privacy is not easily defined, as many different interpretations can be found under a variety of real situations. It is not possible to enumerate every scenario in which RFID technology may potentially compromise personal privacy, because those scenarios depend on the application of RFID technology and on the personal information involved. However, most such scenarios have a common root cause stemming from the potential to automatically associate human identification information with object identification information. The objectives of a privacy preserving RFID system include anonymity and untraceability as explained below.
Anonymity is probably the concealment of the identity of a particular person involved in some processes, such as the purchasing of an item, visiting to a doctor or a cash transaction.

In RFID context, mitigating the problem of anonymity will involve the prevention of associating an EPC of an item with a particular individual. As the EPC can be used to obtain information regarding a particular process and that information may be associated with a particular person.

Untraceability is defined as a means by which the ability of other parties to learn or track the location of people, based on information obtained from RFID tags in possession of that person, is prevented. Hence, providing untraceability would need to involve the prevention of other parties from obtaining RFID tag data without the tag owners' consent; and/or the prevention of associating an EPC of an item with a particular individual; and/or preventing tags from emitting any kind of a unique identification information; etc.

Note that existing barcode system may have many of the same privacy risks, as the barcode can be read and cloned easily. However, RFID deployments present more potential vulnerabilities for those operations to be performed over the air and apparently obtrusive on an immense scale. It is good to know that privacy is a multi-dimensional issue involving many aspects. The successful implementation of privacy objectives above will not only require security mechanisms but will also require the formulation of public policies, legislation and the enforcement of the law by the relevant law enforcement agencies. Public policy is a vital aspect because the security mechanisms used to ensure privacy are most effective when implemented in conjunction with a well-defined policy. In fact, there are existing privacy polices that can be applied directly in RFID systems. They may however need to be clarified, refined or amended to cover aspects specific to RFID Systems.

4. Countermeasures

Toward these RFID security and privacy issues, many countermeasures have been proposed. To our knowledge, a couple of hundreds of research articles addressing RFID security and privacy problems have been published (refer to [17] for a literature survey). Countermeasures can be categorized from basic to sophisticated. In general, the more sophisticated the countermeasures, the more expensive the tag. Furthermore, not all countermeasures are applicable to all threats. No single countermeasure is 100% effective in all situations. Combinations of countermeasures can be used to improve RFID security. The countermeasures are categorized into 4 classes as follows.

4.1 Physical protections

RFID deployments have some practical limitations, which can be considered as effective protection mechanisms. Firstly, the tag-to-reader channel is assumed to be private, since the backscatter channel from the tag to the reader has a relatively shorter range (*e.g.*, several centimeters) than that of the forward channel. The low power of the backscatter channel relates to the fact that while the reader-to-tag communication can be eavesdropped from a long way away, it is only possible to eavesdrop on the tag-to-reader channel if the person is close to a legitimate reader. Thus, an attacker, not within the range, cannot get reply from the tag. In the case of the "clipped tag", the range can be further reduced by tearing off part of the tag's antenna. Alternatively, one can use Faraday cages or other shielding mechanisms to protect a tag within certain (safe operation) range.

Secondly, one can permanently deactivate a tag with physical tag removal or destruction. For example, one can use a momentary switch, electrical, or physical add-on to alter the readability of a tag. Thirdly, a level of security is provided by wafer programming, in which the True Write-Once-Read-Many (WORM) tags are programmed at the fabrication facility with a unique code that cannot be changed. For instance, wafer programming of a WORM device at the IC foundry prevents data from being inadvertently or clandestinely altered later in the supply chain. ISO/IEC 15963 [1] defines a unique tag identification (Tag ID) encoded by the I.C. manufacturer. A Tag ID shall be serialized in accordance with the standard to uniquely identify the chip and then locked by the I.C. manufacturer. The Tag ID can be used to authenticate that the chip is the original and not a copy, but only if one assumes that an attacker cannot obtain a tag in the unlocked state and program his own unique ID. In other words, all chip manufacturers have to agree to lock such memory at manufacture time - if any one chip manufacturer sells a tag in which this memory is unlocked, this countermeasure will not be effective.

Last but not least, the likely detection of physical presence of an attacker, who tries to hide between a legitimate reader and a tag in an active session, can defend some obvious man-in-the-middle attack. And technically, it is not easy to intercept a message and modify the message over the air in real-time without being detected, because of shared bearing medium plus the error detection codes that the protocols employ. This could make the possibility of launching active man-in-the-middle attacks low.

4.2 Access controls

Proper access control mechanisms can prevent the tags from certain unauthorized accesses. As one example, memory lock is typically used to disable the write/rewrite function on the tag or a given block of memory, and prevent unauthorized users from deleting or changing data or inserting unexpected data. In another example, the EPC UHF Gen2 specification defines a *Kill* command, which will totally disable a tag once issued. Another command, *Access*, is also defined to allow for either read or write operations to tag mem- ory after presenting a correct "Access Password".

To provide privacy protection on tags' identifiers, a cloaking mechanism can be used to alter the transmitted EPC code to a different encoded code, thereby obfuscating the identity of the item to which the tag is attached. In the research field, one widely adopted assumption is that tags can support a one-way hash function, which incurs a family of researches on hash based ID variation protocols. For example, the very first one is the hash-lock scheme [29], which is improved with a randomized hash-lock scheme [33]. These are extended to a class of hash chain model [25] by embedding some hash functions in a tag. By changing the IDs or pseudonyms of a tag each time being queried, the *untraceability* property of the tag is protected.

4.3 Cryptographic countermeasures

Above we assume that the RFID tags can support some cryptographic primitives such as hash function. Traditional security systems rely on cryptographic solutions to achieve the security properties like confidentiality (by using encryption) or integrity (using authentication code). If an RFID tag can support cryptographic primitives like traditional security devices, we can just apply existing security solutions to solve the security problems with RFID tags. However, to implement symmetric ciphers, or even asymmetric ciphers on a low-cost RFID tag is still too heavy, because of the extreme resource constrains on those tags. A fair comparison in terms of power consumption, chip area, and clock cycles on the implementations of some standardized cryptographic algorithms (*e.g.*, SHA-256, SHA-1, MD5, AES-128, and ECC-192) on passive RFID tags is presented in [14].

The primary goal of implementing a cryptographic primitive in an RFID tag is to achieve (mutual) authentication of the tag and reader, as in contrary to the common sense (of applying encryption first). The objective of the authentication protocol is for the RFID reader to verify whether a tag knows a secret key. The reader first sends a challenge to the tag. The tag uses the challenge and its secret key as the inputs to some cryptographic function and computes a result. The response will then be checked by the reader, since the reader shares the same secret with the tag. More details of privacy preserving authentication protocols proposed so far are given in Section 6.

4.4 Active devices

To protect the wireless channels between the tag and the reader, we can alternatively choose some active countermeasures by using active tags or proxy devices. For instance, a `blocker' tag is proposed in [19] as a device that simulates RFID tags during tree-walking singulation. The blocker tag works by responding to singulation queries of a reader such that the reader is led to traverse the entire tree or a sub-tree. This way, the presence of actual tags that are to be protected is hidden from unauthorized readers.

In [26], a "selective RFID jamming" mechanism is proposed, in which a battery-powered mobile device is used to selectively transmit jamming signals to block responses from tags. The mobile device holds an access control list (ACL), which specifies the queries that may be allowed from readers. Based on the ACL, the device checks whether a query sent from a reader should be allowed. When a disallowed query is encountered, the device blocks off the tag response to the query by transmitting a jamming signal. Hence, unauthorized reading of a tag can be prevented.

Similarly, an "RFID Enhancer Proxy" (REP) is proposed in [27], which is a high power proxy device that can acquire the identity of RFID tags. Tags that have their identities acquired by the REP will remain in dormant mode until their identities are released back to them. The REP will then take part in the singulation process on their behalf. For security, the REP is equipped with the capability to authenticate readers to ensure that private information is only communicated to authorized readers.

With active countermeasures, we can alleviate some of the security and privacy problems encountered in RFID systems. However, non-trivial cost will be put on building such devices with comprehensive security functionalities.

PART 2: RFID PRODUCT AUTHENTICATION SOLUTIONS

5. Secure binding between tag and object

An RFID-enabled product authentication system typically authenticates the RFID tag attached to the product, instead of the product itself. Hence, the authenticity of the product can only be ensured if the RFID tag is securely bound to the product and is not tampered with. There are generally two categories of secure binding - physical binding and electronic binding.

Physical binding refers to the use of physical means (which may involve the use of mechanical or chemical mechanisms) to pack the RFID tag with the product tightly so that the binding is either impossible to be tampered with (tamper-resistant) or leaves clear evidence when the it has been tampered with (tamper-evident). An example of such binding is the electronic seal used to guarantee the integrity of containers [21]. Secure physical binding is used to defend against attacks based on removal and re-attachment of RFID tags.

Electronic binding refers to methods in which the unique fingerprint of a product is stored on the RFID tag. During authentication, an authentication device would be used to re-generate the fingerprint and compare it with the value stored on the RFID tag. The fingerprint is typically signed by the manufacturer of the product and can be verified by the authentication device. The digital signature guarantees the authenticity of the product, but not the authenticity of the tag, since the fingerprint, together with its signature, can be skimmed and copied onto other tags. It is possible that the cloned tag not only contains a part of authentic information, but also some other misleading information about this product. Thus, it is natural to bind the RFID tag with the product using methods proposed in [22] (the secure binding of object unique feature on tags) and [23] (the integration of tags on machine readable documents).

In [22], the authors proposed a method of secure binding that is achieved by signing on the unique features of the product, as well as that of the attached tag. For the tag, the Tag (or Transponder) IDentification number (TID) was used as the unique feature. The TID is essentially a globally assigned unique number that is programmed onto the tag by the chip manufacturer and set to a "locked" state. One cannot easily "unlock" the state and change the TID, although dedicated attackers might break it with some invasive attacks. The EPC is another globally assigned unique number for a specific product, but it is written by the product manufacturer and can be erased and overwritten with another EPC so that the tag can be re-used. In short, it is easy to clone the EPC, but difficult to clone the TID [4]. Hence, we consider the TID to be a good authenticator of an RFID tag that can be used to tighten the binding proposed in [22].

Here, we stress that there is no "absolute security". All security measures can very likely be broken given the time and resources. Nonetheless, for a product authentication solution to provide "good enough security", it should guarantee cost-effectiveness in preventing and detecting massive counterfeits in a timely manner. For the products that require very high level of security, strict security design techniques should be used and stringent tests and analysis should be carried out on those techniques before they can be put to deployment.

6. Tag-to-reader authentication

The RFID security research community has been paying a lot of attention on RFID authentication. Over several years, a large number of privacy-enhanced authentication protocols have been proposed in the literature. We focus our attention on tags that come with the capability to store some secret values, and we categorize these tags into three different classes based on the resources available on them - namely Crypto-tag, Light-tag and Gen2-tag. Crypto-tags support classic cryptographic primitives and hence, traditional authentication schemes can be applied here. Light-tags can not perform cryptographic functions, but can conduct bitwise operations such as XOR. Gen2-tags conforming to the EPC Class 1 Generation 2 specification [9], which can only perform 16 or 32 bits bitwise operations and are embedded with 16-bit PRNG and CRC functions.

6.1 Authentication with classic cryptographic primitives

The objective of such an authentication protocol is for the RFID reader to verify whether a Crypto-tag knows some secret key that is shared between the reader and the tag. The reader first sends a challenge to the tag. The tag uses the challenge and its secret key as inputs to some cryptographic function and computes a result, which is returned to the reader as a response to the challenge. The response will then be checked by the reader for verification. If the reader needs to authenticate a lot of tags, it has to store the IDs and secrets of all these tags, which is not scalable.

With regards to Crypto-tags, one widely-adopted assumption is that these tags can support a one-way hash function. The very first approach of using hash function was the hash-lock scheme, proposed by Sarma *et al.* [29]. Following that, a lot of RFID authentication protocols based on hash functions have been proposed. Besides these hash-based solutions, there were other solutions that require a Pseudo-Random Function (PRF) on a tag or make use of symmetric ciphers instead of hash functions. Another work [18] even assumed the use of public key cryptographic primitives, in which tags update their IDs with a re-encryption scheme. Although public key cryptography can reduce the key management overload, it is still too heavy to be implemented on medium-cost Crypto-tags.

Promisingly, there are some ongoing research efforts that lead to ultra-lightweight cipher designs. For example, the block cipher PRESENT-80 [6] features a compact implementation of only 1, 570 Gate Equivalents. Comparable lightweight stream ciphers, like Grain, has about 1, 300 Gate Equivalents [16]. More efficient hardware/software stream cipher designs are proposed and evaluated (currently within the ECRYPT project) for minimal footprint hardware implementation even in low-cost RFID tags.

6.2 Authentication with lightweight primitives

Light-tags are restricted to a much lower gate count (less than hundreds of GEs) than Crypto-tags for the implementation of security features. Some authentication schemes that do not rely on assumptions on classic cryptographic primitives have been proposed so that they can be supported on low-cost tags.

HB family of Authentication Protocols. In 2005, Weis *et al.* introduced the Hopper and Blum Protocol (HB) under the RFID setting [32]. The protocol can achieve sound security and can be implemented with extremely less circuits. Subsequently, Juels and Weis proposed a lightweight authentication protocol (HB^+) in [20]. The security of both the HB and HB^+ protocols are based on the Learning Parity with Noise (LPN) problem, whose hardness over random instances remains as an open question. However, Gilbert *et al.* showed that HB^+ is not secure against a simple man-in-the-middle attack [15]. To defend against such active attacks, Bringer *et al.* extended the protocols to HB^{++} protocol [5]. Later on, the HB family of protocols is enriched by several other complementary designs. But the protocols are still not mature enough to be applied in practical due to inherent security and performance pitfalls.

Ultra-Lightweight RFID Authentication Protocols. In 2003, Vajda and Buttyan presented a set of extremely-lightweight challenge-response authentication protocols [31] that are suitable for authenticating tags, but their protocols can be broken by a powerful adversary as was shown in [7]. Besides this, there are a number of approaches employing existing or self-designed mathematical primitives to build ultra-lightweight mutual authentication protocols for low-cost RFID tags. Unfortunately, almost all such light-weight protocols are being attacked in one way or another, and their practical deployment could be at risk unless strict security analysis is conducted beforehand.

6.3 Authentication with Gen2 functions

Some approaches, conforming to EPC Gen2 specifications [9] that rely solely on the specified functions like 16-bit CRC and PRNG, have also been proposed. For instance, in 2006, Duc *et al.*'s authentication protocols used 16- bit PRNG, CRC and XOR operations to replace the 128-bit strong cryptographic PRNG and MAC functions [8]. But the tradeoff of the replacement is the reduced (perhaps better than nothing) security. Thus far, all of the authentication protocols based on Gen2 functions are vulnerable even under a weak security model. Obviously, Gen2 tags provide almost no security at this moment, but the security issues are being investigated and improved in the next generation (Gen3) specification. With the fast development of lightweight cryptographic research and semiconductor technologies, we are optimistic on expecting lower-cost and stronger-security RFID tags being massively produced in the near future.

6.4 Further discussion

A secure tag-to-reader authentication scheme might enable a completely of- fline RFID product authentication solution. Suppose an RFID-tagged product is dispatched by the manufacturer, distributed along the supply chain, and finally comes under possession by an end user. The end user would verify the product with a standalone authentication device, which means that the end user can only rely on this device to check the authenticity of the product. In this case, the verifier scans the tag to obtain its ID and takes part in a challenge-response authentication protocol to prove that the tag owns some shared secret. As long as the secret on the tag is not disclosed, the authentic- ity of the tag is guaranteed. This can resist certain copycat attacks where all data except the secret of the tag is cloned on another tag.

In such a solution, the requirements on the authentication device are high. The device integrates a combination of functions including cryptographic algorithms, physical feature extraction functions and huge memory to store the relevant information for all tags. The end system is expensive due to the cost of tag, the binding and the authentication device. However, the system/network overhead is rather low in this ubiquitous setting.

7. Legacy product authentication solution

Above we described an extreme case of authentication involving an authentication device in the end system that can authenticate the tag and product without any online support. Such an ideal solution requires a secure binding between the RFID tag and the product (Section 5) and the tags must be capable of taking part in an authentication protocol (for example, the Crypto-tags in Section 6). The high cost of such an end system limits its application to supporting high-value products only. For ordinary products, a more economical anti-counterfeiting solution would have to be used and the cost-effectiveness of the solution has to be weighed carefully. To support high-volume usage, the item-level tags for ordinary products would have to be extremely low-cost and thus, it is unlikely that there would be sufficient resources to support security features.

Even when Crypto-tags are used, these tags could still be compromised by side channel attacks [24]. Hence, under some circumstances, there might be a need to rely on a back-end system for stronger authentication. This gives rise to the other extreme case, where a central database dominantly grasps all product information. Suppose the centralized database maintains all the product's activities during its life cycle, it can check the history of the product to see whether the information in the online request is logically sound (*e.g.*, a drug that is mandated to be sold only in US should not be available in South Africa). The result of this check is sent back to the verifier. In this case, the cost of the end system is relatively low, while the cost of maintaining the centralized back-end database is extremely high. The collection and analysis of the status information of a tag is not likely to be easy. Moreover, defining the granularity of the information collected is also important. Other challenging issues include the sensitivity and/or privacy of the data that is to be shared and the requirement for protecting against a single point of failure. Hence, this is another impractical solution since it does not scale well and potentially suffer from Distributed Denial of Service (DDoS) attacks and result in a single point of failure. Next, we shall study some distributed product authentication solutions that are practical, economical and reasonably scalable.

One good example of such a distributed solution is the existing **E- pedigree** solution in pharmaceutical supply chain. Initially promoted by the US FDA, the E-pedigree specification was then ratified by EPCglobal in the beginning of year 2007 [12]. The purpose of the new standard is to provide the pharmaceutical supply chain partners with a common format on collecting pedigree information and building their pedigree software platforms. The standard comprises instructions on how supply chain partners can create an E-pedigree, update information on it and digitally sign it. Many companies are accelerating their initiatives towards integrating E-pedigree pilots into existing legacy supply chain systems to enhance product integrity and further protect patient safety.

An E-pedigree system consists of all partners involved in the distribution of a medicine, including its manufacturer, the wholesaler, the retailer, and the pharmacy or any other entities administering or dispensing the medicine. These partners form a limited distributed system and establish some business relationship between each other (Public Key Infrastructure, or PKI, is typically assumed in this scenario for establishing entity trust relationships). As the medicine goes through the distribution path, it forms a growing certified chain of custody while each participant contributes to the E-pedigree. Here, we briefly describe how an E-pedigree system works:

1. 1. A medicine is produced by a manufacturer and attached with a unique RFID tag. The manufacturer starts to build the initial E-pedigree with the medicine's serial number, transaction information and other product- related information. Then, the E-pedigree is digitally signed with the pub- lic key of the manufacturer. The E-pedigree, together with the digital certificate of the manufacturer, is ready to be sent to a downstream partner (typically before the medicine is shipped out).
2. 2. On receiving the E-pedigree, the downstream partner first authenticates the E-pedigree by verifying the digital signature with the public key in the certificate. If the verification is successful, the partner continues to match the information on the E-pedigree with that on the product (assuming the medicine has been shipped in at this moment). A successful match completes the verification procedure. If the medicine is going to be shipped out, the partner needs to update the E-pedigree with its own information and signs on the renewed E-pedigree. Once again, the updated E-pedigree, together with the certificate of the partner, is ready to be transmitted to the next downstream partner in advance.
3. 3. The same procedure is repeated by every participant in the distribution path until the medicine reaches its destination. The procedures described above actually represent a typical (aggregated) document authentication flow. It does not really need to involve an RFID tag, except that a tag's ID is recorded in the E-pedigree for an additional match. In fact, the tag can be made more useful by strengthening its binding with the E-pedigree.

As in Section 4.1, we know that the tag could be a good authenticator due to its fabricated TID. A manufacturer can combine the tag's unique feature with the initial E-pedigree tightly by signing on them together and storing the signature on the tag (take for example TI's electronic marking scheme [30]). Then, the signature can be verified by the forthcoming partners. Under a secure access infrastructure, this piece of additional information can even be encrypted to ensure its confidentiality. The strong binding between the tag and the E-pedigree provides another layer of security.

The E-pedigree solution has been adopted rapidly in pharmaceutical supply chains since it is a natural extension of the legacy IT systems. Many of them already have existing internal, closed-loop RFID systems. Although the solution is promising, its success in real world applications will depend more on the non-technical issues such as privacy protection and legal agreements among multiple partners.

8. Product authentication with EPCglobal network

An E-pedigree solution revolves product authentication around a number of supply chain participants. However, it is more desirable that individual products can be tracked throughout the global supply chain to realize the greatest benefits of RFID technology. This inspires a globally available service - an **EPCglobal network** that offers another huge opportunity to obtain services from an open and standard interface (via Internet). As an essential part of the new supply chain management system, the emerging network enables real-time visibility of all products throughout the supply chain, improves efficiency in inventory control and reduces occurrences of product loss. Thus, EPCglobal network will provide a more open and efficient infrastructure for product authentication solutions.

8.1 EPCglobal network architecture

EPCglobal network resembles the Internet, but constructs an overlay of the Internet architecture. EPCglobal network architecture is shown in Fig. 2.

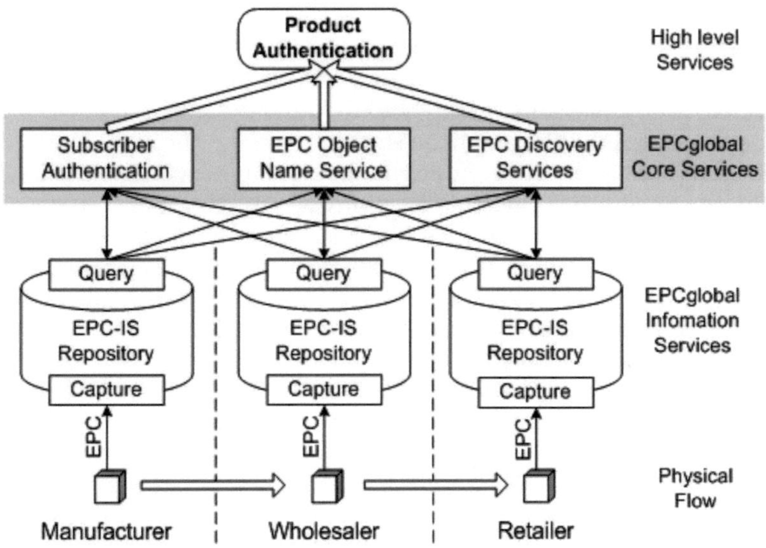

Fig. 2. EPCglobal Network Architecture.

The EPCglobal network [10] employs Electronic Product Code (EPC) to allow companies to track individual product through the global supply chain. The network provides (near) real-time tracking and product life cycle monitoring that make business processes more efficient. To realize these benefits, the EPCglobal committee specifies a standard framework to regulate the tracking, security and collaboration between different supply chain partners. The EPCglobal network manages RFID information through a number of core services: *Object Name Service (ONS)*, *EPC Information Services (EPC-IS)*, and *EPC Discovery Services (EPC-DS)*. Of which, the EPCglobal Architecture Framework identifies three possible ways to locate the informative service according to a specific EPC of an object:

- A party may use the Object Name Service (ONS) to locate the EPC-IS service of the EPCglobal Subscriber who commissioned the EPC of the object.
- A party may know in advance exactly where to find the information by means of being given the network address of the other party's EPC-IS service as part of a business agreement.
- A party may use Discovery Service (EPC-DS) to locate the EPC-IS services of trading partners that have information about the object, including partners other than the one who commissioned the EPC of the object.

Briefly, these core services can be described as below:

ONS: With an EPC that uniquely identifies a single product unit, one can query the ONS to look up the address of the product manufacturer's EPC-IS. Thus, ONS can be thought of as a lookup service that takes an EPC as input, and produces as output the address (in the form of a Uniform Resource Locator, or URL) of an EPC-IS repository designated and implemented by the EPC Manager of the EPC. This is similar to the Domain Name Service (DNS) on Internet, which matches the internet domain names to the IP addresses. From the EPC Manager's EPC-IS repository, one can then obtain detailed product information relating to the EPC.

EPC-IS: EPC-IS regulates the specification for supply chain partners to share EPC-related data. It controls the storage and retrieval of detailed product information on individual product units. It provides a standard data model to enable track and trace, product authentication, diversion detection, and other use cases involving supply chain partners across multiple industries. EPC-IS defines a capture interface and a query interface to obtain and share business event information. In fact, EPC-IS is the foundation for increasing visibility, accuracy, and automation throughout the supply chain.

EPC-DS: The product information might be stored not only at the manufacturer's site, but at different sites along the supply chain (for example the ship-in and ship-out information of a product might be stored at intermediate locations where the product transits). This raises the question of how a trading partner identifies and locates all of the other parties who may have relevant EPC-IS data. The EPC-DS provides the lookup service to all these fragmented sources of information. It serves as a search engine for the EPCglobal Network with restricted access, where subscribers can query it with an EPC to obtain a list of EPC-ISs that they can query directly for more detailed information.

EPC-SAS: Additionally, EPCglobal specifies the Certificate Profile [11] for building Subscriber Authentication Service (EPC-SAS). The authentication of entities (subscribers, services, physical devices) operating within the EPCglobal network serves as the foundation of any security function incorporated into the network. It is expected, however, that the X.509 authentication framework will be widely employed within the EPCglobal network. To ensure broad interoperability and rapid deployment while ensuring secure usage, the specification defines a profile of X.509 certificate issuance and usage by entities in the

EPCglobal network, which are based upon two Internet standards, defined in the IETF's PKIX Working Group, that have been well implemented, deployed and tested in many existing environments.

Integrating these core services, EPCglobal network can provide product life cycle visibility and traceability, which are the foundations of advanced product authentication solutions.

8.2 EPCglobal network threats and mitigations

As being composed by millions of individual RFID systems, the EPCglobal network will be much more complex than any standalone system. It may suffer many new or even more serious security threats coming from internal systems or existing Internet.We identify some of these threats in this section and point out possible mitigations.

ONS Security

ONS is similar to the Domain Name Service (DNS) on Internet, which matches the internet domain names to the IP addresses. An ONS server can be considered as a DNS server (typically, ONS can share the same server with DNS). Therefore, the security threats related to DNS server are also applicable to ONS. Security Threats such as file corruption, unauthorized updates, ONS cache poisoning, IP address spoofing, packet interception, query prediction and all threats from client to server or from server to server, are all to be addressed [13].

We shall take a similar way on protecting ONS as we did on securing DNS (e.g., using DNS Security Extensions-DNSSEC [3]). To protect ONS data, we need to provide security properties like confidentiality, origin authentication and data integrity, by using a brunch of security measures like key exchange protocols, digital signature and mutual authentication schemes. On safeguarding the systems, typically firewalls and intrusion detection systems are to be installed. Also, some good security practices such as secure backing-up of the files, applying proper read and write permissions; defining access control lists, are to be applied. However, if ONS server is to be implemented together with DNS server, privacy issues will arise and have to be investigated.

EPC-IS Security

EPC-IS repository can be considered as both a database server serving internal enterprise applications and a web server serving Internet requests. Thus, it suffers all threats coming from internal or outside. Some of them are traditional database threats like SQL injection, viruses, insider attacks; some are intrusion, worms, DoS attacks from Internet.

To protect the EPC-IS repository, we consider the whole set of system level security measures: authentication, authorization, access control and auditing. We rely on security tools like ant-virus softwares and hardwares, firewalls and intrusion detection systems, and backup mechanisms. The database SQL injection attack can be typically prevented by checking for buffer overflows, validating and sanitizing input data before passing it to SQL Query, disabling web script execution by outside sources, setting up appropriate access rights and enforcing access control policy, etc.

EPC-DS Security

EPC-DS provides visibility in the supply chain for all parties who have a right to know. The discovery of where data resides, the actual exchange of data, and the security policies governing these activities are all related. Of which, authentication and authorization are intimately connected with discovery. For example, merely discovering that one party in a supply chain has information about a particular EPC may or may not be privileged information subject to data authorization policies. Serving as a search engine for the EPCglobal Net- work with restricted access, EPC-DS server suffers both new threats from the subscribers (insider attacks) and common threats from the existing network infrastructure (similar to above threats on ONS and EPC-IS).

On the one hand, the EPC-DS infrastructure must define elegant access control and authentication policies to securely manage the information to be discovered and shared. Note that EPC-DS specification is still an on-going effort. On the other hand, all system level security measures (introduced above) have to be applied to protect EPC-DS servers.

EPC-SAS Security

EPC-SAS can be considered as a Trusted Third Party (TTP) to provide (web-based) information service. It shall have a higher level of security compared with above services. All subscribers' registration information are to be protected well and kept in secure storage. It is essential to authorize and authenticate legitimate subscribers based on their credentials and provide only the appropriate data that is relevant to them. Besides, EPC-SAS servers suffer all above common threats like intrusions, trojans, injections, and is especially sensitive to DoS attacks.

EPC-SAS depends on many traditional public key cryptographic mechanisms to authenticate the identity of an EPCglobal subscriber and issue credential to the subscriber. Then, without other prior arrangement, a subscriber can authenticate itself to any EPCglobal services providers and use those network services. Additionally, all system level security protection mechanisms (as introduced in EPC-IS security) have to be applied here. Specially, a distributed architecture is expected to avoid the central point of failure caused by DoS attacks.

8.3 EPCglobal network product authentication service

The ongoing efforts of the committee also includes the establishment of some specific business cases such as brand protection, product authentication and chain of custody. These use cases could utilize a combination of the core services described above. For example, the *EPC Product Authentication Service* (EPC-PAS), once regulated, might provide an all-in-one interface for the entities within a supply chain to authenticate a product.

While the EPC-PAS solution is very much desirable, it is not easy to regulate and could potentially encounter many obstacles when put under real operations. One of the major challenges in the design is the privacy of partners along the supply chain. There can be issues with regards to how much information a partner would want to keep with itself instead of sharing them with other partners; and how to define the minimal level of authentication-relevant information that should be shared. If there is insufficient information available on product visibility, then one cannot make a good judgement on the authenticity of a product.

In addition, the solution provided by EPC-PAS faces other limitations. Firstly, only authorized personnel can access the service, which is in conflict with our expectation towards a public service where everyone can authenticate a tagged product in hand. Secondly, even if the service is not provided to all, but to a group of subscribers, there could exist several desired service levels for different groups (*e.g.*, for ordinary users or for supply chain partners). Under such circumstances, how to define the privacy levels for different groups in a dynamic deployment setting would be a big issue. Thirdly, we need to think of how to prevent these services from abuse for malicious purposes, such as the tracking of a particular person. In addition, there is also a lack of practical experience on handling such a huge information system. Beyond that, there are also other issues like the likelihood of social acceptance and legislative support.

9. Conclusion

In this chapter, we presented a high level view on RFID-based product authentication solutions. Firstly, we exposed the threats that might be launched on RFID systems. We also investigate the security and privacy issues with the RFID systems and reviewed some countermeasures against the threats. As a promiscuous and ubiquitous technology, RFID presents unique security features and requirements. Assessing RFID's security and privacy risks requires a case-by-case analysis, due to the diversity of possible RFID deployments. The risk evaluation depends on the type of RFID used, the information stored on the chip, and the context in which the implementation is deployed. Accordingly, taking effective and balanced security measures to mitigate the risk is necessary to avoid jeopardizing RFID's usability. We stress that the success of RFID relies on all kinds of factors like professional devotion, social acceptance and legislative support.

Secondly, we pay attention on the security requirements and potential mitigations with EPCglobal network. With EPCglobal network, RFID not only acts as an additional authenticator for authenticating a product, but also provides an easy way to share a product's information throughout the global supply chain. Although the solutions are not perfect at this moment (and is unlikely to be in the near future), they look promising with the potential to act against massive counterfeits. The heartening thing is that the product authentication solutions are being piloted and deployed at many companies. With these precious experiences gained, implementors should be equipped with better knowledge and be in a better position to design optimal security solutions in their fight against counterfeiters.

10. References

[1] Information technology - Radio frequency identification for item management - Unique identification for RF tags. ISO/IEC 15963:2004.

[2] Information technology - Radio frequency identification for item management - Implementation guidelines - Part 4: RFID guideline on tag data security. ISO/IEC PDTR 24729-4:2008.

[3] R. Arends, R. Austein, M. Larson, D. Massey, and S. Rose, DNS security introduction and requirements, *Request for Comments - RFC 4033*, March 2005.

[4] AIM Global Analysis: Counterfeit Tags, Jun. 2005.

[5] J. Bringer, H. Chabanne, and E. Dottax. *HB++*: a Lightweight Authentication Protocol Secure against Some Attacks. In: *Proc. of SecPerU'06*, pp. 28-33. IEEE Computer Society Press, 2006.

[6] A. Bogdanov, L.R. Knudsen, G. Leander, C. Paar, A. Poschmann, M.J.B. Robshaw, Y. Seurin, and C. Vikkelsoe PRESENT: An Ultra-Lightweight Block Cipher. *Cryptographic Hardware and Embedded Systems - CHES 2007*, Vienna, Austria, Sept. 2007.

[7] B. Defend, K. Fu, and A. Juels. Cryptanalysis of two lightweight RFID authentication schemes. In *Fourth IEEE International Workshop on Pervasive Computing and Communication Security (PerSec) Workshop*, March 2007.

[8] D.N. Duc, J. Park, H. Lee, K. Kim, Enhancing security of EPCglobal GEN-2 RFID tag against traceability and cloning, In *The 2006 Symposium on Cryptography and Information Security*, 2006.

[9] EPCglobal Inc., EPC Radio-Frequency Identity Protocols Class-1 Generation-2 UHF RFID Protocol for Communications at 860MHz-960MHz Version 1.1.0, EPCglobal Standards, Oct. 2007.

[10] EPCglobal Inc., Architecture Framework Standard v1.0 http: //www.epcglobalinc.org/ standards/architecture/Architecture 1 0-StandardApproved-20050701.pdf

[11] EPCglobal Inc., EPCglobal Certificate Profile v1.0.1 http://www.epcglobalinc.org/ standards/cert/cert 1 0 1-standard-20080514.pdf

[12] EPCglobal Inc., Pedigree Standard v1.0 http://www.epcglobalinc.org/standards/pedigree/Pedigree 1 0-StandardRatified-20070105.pdf
[13] B. Fabian, O. Gunther, and S. Spiekermann. Security Analysis of the Object Name Service for RFID. In: *Proc. of SecPerU'05*, IEEE Computer Society Press, 2005.
[14] M. Feldhofer, J. Wolkerstorfer. Strong Crypto for RFID Tags-a Comparison of Low-Power Hardware Implementations, In: *IEEE International Symposium on Circuits and Systems (ISCAS 2007)*, pp.1839-1842, New Orleans, USA, May 27-30, 2007.
[15] H. Gilbert, M. Bobshaw, H. Silbert, An Active Attack against HB^+-A Probable Secure Lightweight Authentication Protocol, *Cryptology ePrint Archive, Report 2005/237*, 2007.
[16] T. Good, W. Chelton, and M. Benaissa. Hardware Results for Selected Stream Cipher Candidates. In *SASC 2007*, February 2007.
[17] A. Juels. RFID Security and Privacy: A Research Survey. *IEEE Journal on Selected Areas in Communications*, 24(2): 381-394, Feb. 2006.
[18] A. Juels and R. Pappu. Squealing euros: Privacy protection in RFID-enabled banknotes. In: *Proc. of FC'03*, LNCS 2742, pp. 103-121. Springer-Verlag, 2003.
[19] A. Juels, R. L. Rivest, and M. Szydlo, The Blocker Tag: Selective Blocking of RFID Tags for Consumer Privacy, in *Proc. of ACM Conference on Computer and Communications Security (ACM CCS) '03*, pp. 103-111, Oct 2003.
[20] A. Juels and S. Weis. Authenticating pervasive devices with human protocols. In: *Proc. of CRYPTO'05*, LNCS 3126, pp. 293-308. Springer-Verlag, 2005.
[21] R. Johnston, Tamper-Indicating Seals, *American Scientist*, Nov-Dec 2005.
[22] Z. Nochta, T. Staake, E. Fleisch, Product Specific Security Features Based on RFID Technology. In *Proceedings of the International Symposium on Applica- tions and the Internet* Workshops. IEEE Computer Society, 2006.
[23] M. Lehtonen, T. Staake, F. Michahelles, E. Fleisch, Strengthening the Security of Machine Readable Documents by Combining RFID and Optical Memory Devices. In *Conference on Ambient Intelligence Developments - AmID,* Sophia-Antipolis, France, September 2006.
[24] Y. Oren and A. Shamir. Remote Password Extraction from RFID Tags. In: *IEEE Transactions on Computers*, 56(9):1292-1296, 2007.
[25] M. Ohkubo, K. Suzuki, and S. Kinoshita. "Cryptographic approach to privacy- friendly tags." In: *Proc. of RFID Privacy Workshop*, 2003.
[26] M. R. Rieback, B. Crispo, and A. S. Tanenbaum, Keep on Blockin' in the Free World: Personal Access Control for Low-Cost RFID Tags, in *Proc. of the 13th International Workshop on Security Protocols*, Apr 2005.
[27] A. Juels, P. Syverson, and D. Bailey, High-Power Proxies for Enhancing RFID Privacy and Utility, in *Proc. of the 5th Workshop on Privacy Enhancing Technologies (PET '05)*, 2005.
[28] T. Staake, F. Thiesse, E. Fleisch, Extending the EPC Network-The Potential of RFID in Anti-Counterfeiting. In *Proceedings of the 2005 ACM symposium on Applied computing*, pp. 1607-1612. New York (NY): ACM Press.
[29] S. Sarma, S. Weis, and D. Engels. RFID systems and security and privacy implications. In: *Proc. of CHES'02*, LNCS 2523, pp. 454-469. Springer-Verlag, 2003.
[30] Texas Instruments and VeriSign Inc.: Securing the pharmaceutical supply chain with RFID and public-key infrastructure technologies. *Whitepaper*, 2005.
[31] I. Vajda and L. Buttyan. Lightweight authentication protocols for low-cost RFID tags. In: *Proc. of UBICOMP'03*, 2003.
[32] S. Weis. Security parallels between people and pervasive devices. In: *Proc. of PERSEC'05*, pp. 105-109. IEEE Computer Society Press, 2005.
[33] S. Weis, S. Sarma, R. Rivest, and D. Engels. Security and privacy aspects of low-cost radio frequency identification systems. In: *Proc. of 1st Int. Conf. on Security in Pervasive Computing*, LNCS 2802, pp. 201-212. Springer-Verlag, 2003.

20

RFID Information Value Chain and ETRI RFID Ecosystem: Value-added Environment Linking Physical and Virtual Worlds

Tae-Su Cheong[1] and Yong-Jun Lee[2]
[1]School of Industrial and Systems Engineering, Georgia Institute of Technology,
[1,2]Electronics and Telecommunications Research Institute,
[1]USA
[1,2]Republic of Korea

1. Introduction

Since MIT Auto-ID lab envisioned the concept of "a networked physical world" with tagged objects (Sarma et al., 2001), RFID technology has gained a lot of significant attentions from academia and industries, and its research on hardware issues as well as software platform has been actively studied so far. In fact, RFID technology has shown itself to be a promising technology to keep track of the items in real time and further enhance operational efficiency. As the RFID technology is being spread and applied to real world systems, the research focus is changed accordingly from hardware to business application areas and relevant software systems. Having considered the significance of software and business integration for (ideally) automated system, it is indispensable to have intelligent software platform designed for RFID technology to deal with large amounts of data and complex business contents in order to create *value* in the sense of business performance efficiency.

This chapter is divided into two folds: First, we introduce the concept of *RFID Information Value Chain* (RFID IVC), which reflects the evolution of RFID data semantics and investigate how RFID software systems should be organized in order to support such value chain. In this part, we focus on the information evolution activities converting raw RFID data to useful information which may even be used to lead to automated business process execution and further to knowledge to support decision making. We examine the elements and relevant activities on the information value chain.

Second, *ETRI RFID Ecosystem* which materializes RFID Information Value Chain is presented from architectural point of view. ETRI RFID Ecosystem consists of RFID middleware performing raw RFID data processing in the primitive level, rule engine generating business semantic events and orchestration engine coordinating automated business process. In this part, we explain how the software systems are implemented in association with RFID Information Value Chain and elaborate each individuals of ecosystem in the implementation perspective. We remark that the usual concerns of scalability, reliability, distributed management and interoperability are as relevant to RFID middleware as they are in other middleware domain. In addition, the extensibility for readers as well as user-defined functions, and global RFID standard compatibility such as EPCglobal and

ISO/IEC standards must be taken in consideration for better market positioning. Therefore, we analyze the requirements and highlight several architectural considerations in order to build a distributed, reliable and standard-compatible RFID software system. We present our efforts aimed at the RFID software platform which is distributed, reliable and global RFID standard-compatible.

2. EPC network and RFID middleware: representative reference model for RFID software platform

The EPC Network is a networked infrastructure for gathering, sharing and accessing EPC-related information about physical movement of each EPC-tagged items as it passes through supply chain. It was proposed and developed by Auto-ID Center (now, Auto-ID Labs[1]) at MIT and is currently maintained by EPCglobal[2]. It comprises of the following main components: (i) Electronic Product Code (EPC), an item identifier, (ii) RFID reader and tags, (iii) RFID middleware (*Filtering & Collection*; Its interface is named as Application Level Events (ALE)) (EPCglobal, 2005b; EPCglobal, 2008a) located in-between RFID readers and business applications and performing filtering and aggregation of EPCs, (iv) EPC Information Services (EPCIS; whose interfaces are *EPCIS Capture Interface* and *EPCIS Query Interface*) (EPCglobal, 2007b) which is a set of interfaces and repository for enabling EPC-related data sharing within or across the enterprises, and (v) Object Name Service (ONS) (EPCglobal, 2008b) and Discovery Services, authoritative directories of information sources or EPCISs associated with EPCs.

When it comes to the RFID middleware, the researches in EPCglobal have been continued with somewhat gradual progresses. In the initial specifications from Auto-ID Center (Oat Systems et al., 2002), the software called 'Savant' was to perform data routing operations for data capturing, monitoring, and transmission, and the specification focused on the functional analysis for RFID middleware software system.

In the later version of the specification about Savant from EPCglobal (Clark et al., 2003), a matter of concern moved from the specific processing features into a flexible container with the generalized external interfaces for outer service applications. Savant became a container of processing modules for specific features and may be customized to meet the needs for applications. Finally, in the latest version about the middleware, now called 'Application Level Events (ALE)' (EPCglobal, 2005b; EPCglobal, 2008a), it specified the external interfaces only for defining the control and delivery of the filtered and collected tag read data, and then outer applications have access to ALE in order to obtain the tag read data of interest.

We briefly presented EPC network architecture and one of its main software component, RFID middleware. We remark that most of currently available middleware software systems in the market follow EPC network architecture spanning not only middleware itself but also EPCIS (Sun Microsystems Inc., 2006; IBM, 2006; Oracle, 2008; Floerkemeier, 2007). However, EPC network architecture is mainly focused on supply chain management-oriented mechanism and associated information traceability. In this chapter, we view the RFID-based software architecture from a different viewpoint as the RFID data semantics evolution and propose RFID software framework supporting such the data transformation process as well as RFID-based automated business process execution in the end.

---

[1] http://www.autoidlabs.org/

[2] http://www.epcglobalinc.org/home

3. RFID information value chain

In this section, we discuss how to manage the information flow from RFID tag sensing to the delivery to RFID applications. In general, RFID system has its advantage on the automatic identification of individual objects without human intervention. Moreover, the structure of EPC Network is designed to support the information exchange among trading partners over the supply chain. However, in order to create value by utilizing RFID infrastructure, the following process must go through: information that is filtered and aggregated from RFID raw data is well-used for the enterprise applications as valuable resources, and later significant knowledge is derived from the accumulated information. In other words, low-value raw RFID data is transformed into useful information and further proper guidance for management and this implies that such transformation activities are in a value-added process. Therefore, we define this value-added data transformation process as '*RFID Information Value Chain* (RFID IVC)' and it pursues the maximization of the benefits from the RFID deployment. Figure 1 shows RFID IVC along with the RFID solution components and the correspondent activities for each step are described and, here after, we briefly present the key activities per each stage corresponding with RFID IVC.

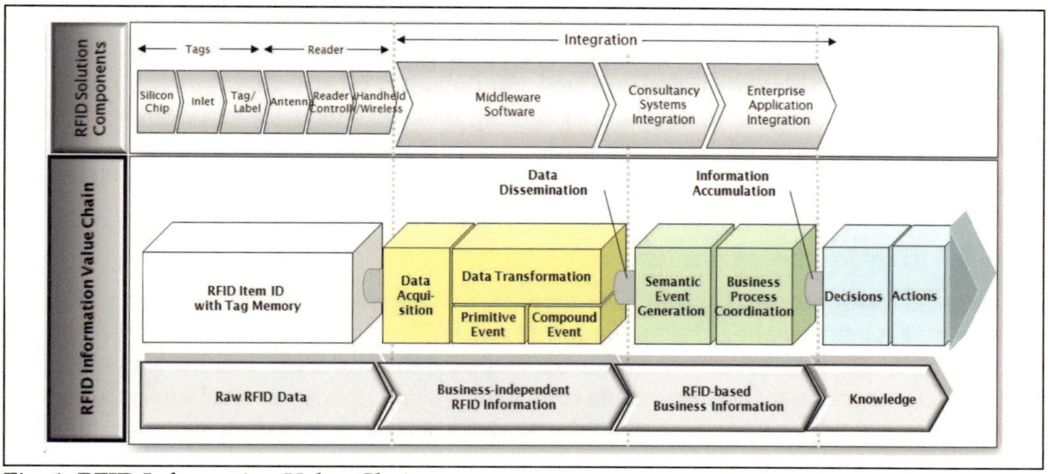

Fig. 1. RFID Information Value Chain.

3.1 Data acquisition

This stage is the first step to gather the RFID data from various types of RFID readers and other types of object identification devices like barcode system. At the present, there exist different kinds of RFID tag-reader protocols including EPCglobal Class 0, Class 1, and Class 1 Generation 2 tag-reader protocols (EPCglobal, 2005a), and ISO/IEC 18000-x series (ISO/IEC JT1/SC31/WG4, 2008). Moreover, the market-available reader systems use the different I/O interfaces. In addition, the heterogeneity among RFID reader systems must be dealt with in order to support the seamless data in-flow.

3.2 Data transformation

In general, RFID system generates a vast amount of data in the low level and such data right from RFID reader has little direct value to the enterprise applications. Hence, it is desirable to provide mechanism that the captured RFID raw data is transformed into the correponding information in the business context so that such mechanism can endow the RFID captures with a significant value in the business sense. This stage, data transformation, primarily involves either filtering or aggregation of raw RFID data. Here, the meaning of 'filtering' is that not all data may be necessary to the applications, so signification data items are identified and the remainder discarded. With the consideration of the characteristics of RFID, the tag data captured by RFID readers have the technical limits as follows :

First, RFID readers cannot guarantee 100% accuracy of tag reading at the present because of interference, limited bandwidth, collision in the dense readers environment and high sensitivity by the surrounding environment (Floerkemeier, 2007). Therefore, it is necessary to resolve the physical limits and the smoothing process can be considered as one method (EPCglobal, 2006).

Second, RFID reader is physically non-contact to communicate with tags and the number of RFID tag data flowed from a RFID reader ranges from 10s of tag data per second up to more than 100s a second. This leads to bring the big burden to the host system which is responsible to process all the data, so the appropriate scheme for data volume reduction is required.

Third, as pointed out at the second, an RFID reader can read multiple RFID tags simultaneously and, sometimes, tags are unintentionally sensed from an area beyond one intended to be monitored by the reader due to many reasons including the reflection of radio wave. Therefore, among the captured tag data, not all the data are required to the applications that consume RFID data, so the tag data in which the applications are not interested should be filtered out.

This stage is responsible to handle the problems above and two levels of RFID event processing are considered: primitive and compound event processing.

Primitive Event Processing (Purification)

The data emerging directly from RFID readers may be regarded as a stream of records of the form (r, n, g, u, t), denoting the antenna n of the reader r reads tag g at time t (with tag memory u). In this stage, it is responsible for mapping the low-level data stream having the limited information to more manageable form suitable for application-level interactions.

An objective on this sub-stage is the volume reduction of data stream by discarding the redundant tag data. In general, when a tag appears present to a particular RFID reader for many read cycles, this generates a lot of data. Moreover, the tag might not appear every read cycle although the tag stays in the read range. As the methods to help overcome these problems, the event smoothing process is introduced (EPCglobal, 2006).

The meaning of 'smoothing' here is to pass the tag read only when something of interest happens such as when a tag is first captured by the reader or when the tag is no longer present. Figure 2 presents the state transition diagram used in the smoothing process.

As seen in Figure 2, there are three states per each tag - Unknown, Peeked, Captured - and four different events - eventPeeked, eventDisappered, eventCaptured and eventExpired, and the tag read can be passed to next stage only if the state transition between two adjacent states occurs in suitable situation. Initial state of a tag is 'Unknown' and, when a tag appears for the first time in the read range, the event 'eventPeeked' is generated and the current tag state is moved onto the state 'Peeked'. The event 'eventCaptured' is generated when the tag is seen for a certain period and the event 'eventExpired' occurs if current state of the tag is 'Captured' and has not seen for a while. The event 'eventDisappered' is generated when the tag hasn't seen for a time without subsequently generating 'eventCaptured' event. If a tag generates 'eventPeeked' and then 'eventDisappered', it means that the tag enters the read range briefly and it can be regarded as a candidate for an inadvertent tag read.

What we can gain from this primitive event processing sub-stage is as follows: (a) reduction of data volume by reporting only if the state transition occurs, (b) establishment of the decision basis which is used to distinguish tags that remain for time interval enough to be regarded as tags that the reader should monitor, as compared to tags that enter the read range only too shortly or are reported unwillingly.

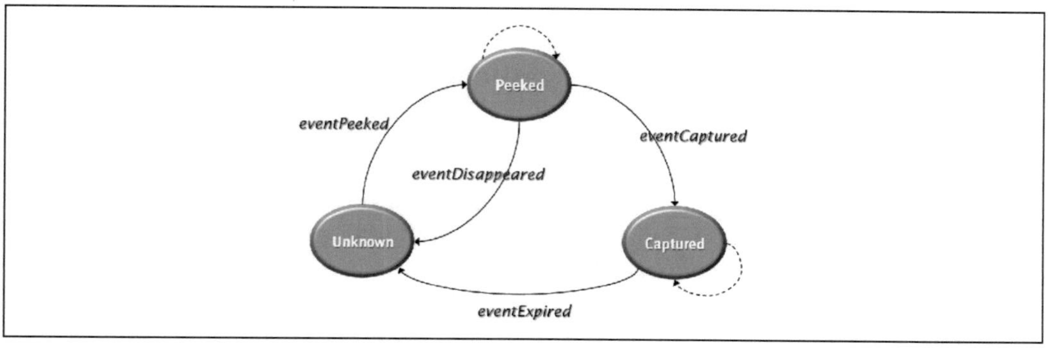

Fig. 2. Primitive Event Generation.

Compound Event Processing (Refinement)

The primitive event processing sub-stage mainly focuses on the duplicated data volume reduction. On the other hand, this stage - compound event processing - concentrates on the selection of tag data that are meaningful to the business domain (but, not dependent on business domain) while having the limited information delivered from RFID reader. There are several literatures dealing with complex event processing by utilizing existing complex event process techniquies (Ku et al., 2007; Dutta et al., 2007) and most of them describe the complex event with ECA rule-type algebraic notations and the corresponding algorithm for interpretation and exeuction. We also remark that Application Level Events (EPCglobal 2005, 2008) specification can falls into this category.

Data from readers follow the feature of record (*r, n, g, u, t)* and, as passing through the primitive event processing sub-stage, the event type *e* - which itself can be regarded as significant event information - is added as an additional field in the record as like the record (*r, n, g, u, t, e*) in exchange for the elimination of redundant data. At this stage, the information meaningful to business application has to be derived from the restricted information.

Basically, the event filtering can be applied to each field of the record or the combination of fields. For example, an application may want to select the tag reads which are captured by specific readers and another application may be interested in the tag reads that their IDs start with the specific pattern. Here, we have five categories to apply the filtering operation - reader ID, antenna number, tag ID, timestamp, event type and even tag memory (if possible) - and those are the basic filtering schemes. The basic filtering schemes based on the following fields – 'reader ID', 'antenna number', 'tag ID', 'event type' and 'tag memory' – are applied like this: the filtering operations applied to the associated field on the event record (*r, n, g, u, t, e*) forward the records that only fit within specific range or pattern; on the other hand, the basic filtering scheme based on the field 'timestamp' can be utilized as duration-based filtering since RFID data are highly temporal. The timestamp-based filtering scheme usually combines with other basic filtering schemes to support more advanced filtering. In addition to the basic filtering schemes, many other filtering schemes are considered as follows:

Association In the case that several readers are associated with the same tag reads, it needs to allow a group of readers to be filtered for duplicate RFID tags so that the application which consumes the tag read seems as if the tag read is originated from the single reader. To provide the association, it needs to combine the event record (r, n, g, u, t, e) with the relationship (r, l) providing the logical location l of reader r. EPCglobal Application Level Event (ALE) specification handles association with `Logical Reader'.

Aggregation & Containments some cases, a number of tags is sensed by readers at the same time, then the group of sensed tags itself seems significant event to the applications. For example, suppose that readers located in the entrance of warehouse capture a number of cases and one pallet containing them within a short time interval. In order to derive the meaningful information to the applications, the sensed tag list including both cases and a pallet should be processed as a batch. That is, this batch filtering scheme clusters tag reads into groups of distinct tags based on when they have been observed during the given interval.

Read-Range In/Out For the situation of warehousing of goods or taking them out of the warehouse, reporting in-field and out-of-field events when tags move in and out of the read range is more significant to the corresponding business applications. This situation can be handled by taking advantage of the 'event type'-based basic filtering scheme; that is, use the event type 'eventCaptured' and 'event Expired' respectively among the fields of the record (r, n, g, u, t, e).

In general, the filtering schemes which applications want to apply vary from application to application and, we believe that the schemes discussed above are considered as the commonly used filtering schemes (for more detailed complex event process, see ALE specification or, Ku et al.(2007) and Dutta et al.(2007)).

3.3 Data dissemination

A critical task of any information system is to deliver the right information to the right people or applications at the right time. The dissemination activity involves moving the filtered and aggregated information from RFID readers to enterprise applications or other business integration applications. The purpose of data dissemination activities is to determine who needs what information and to deliver it to them on time.

We note that RFID-enabled business applications tend to have event-driven nature and, in order to meet this requirement, it must be required to provide the following types of data dissemination models at the same time; *the push model* (or *query/response model*) that RFID data captured by RFID readers keep flowing upward to the backend applications over pre-defined event pipeline, and *the pull model* (or *publish/subscribe model*) that applications which are interested in handling RFID data subscribe to the middleware system with additional information such as notification cycle, reader set which they are interested in listening to and so on, and the middleware plays a role of dispatching messages to the subscribers asynchronously.

3.4 Semantic event generation

At the second stage of 'data transformation', filtering and aggregation methods of raw RFID data are demonstrated. However, it may not be sufficient for the filtered data to become the meaningful information to enterprise applications. At this stage, the filtered data are combined with different sources residing in the legacy system or external sources such as purchased data services so that semantic events which are significant in the business domain are generated. Here, the term "semantic event" means that, as it says, RFID data capturing events merely indicate raw observation and taking business actions based only on the event reports are somewhat limited; therefore, the raw observation events need to be combined with additional business context information in order to construct business events (here, we call 'semantic event') upon which legacy applications can play with. Furthermore, the reason why this stage is required is to enable sophisticated RFID-based data processing. As domain-specific information is integrated with RFID tag data, content-based filtering and routing become possible by mapping and combining tag data with corresponding object information and applying basic filtering scheme on the combined data.

In other way, Event-Condition-Action (or ECA) rules (Palmer, 2006) can be applied to generate the semantic events. When a primitive tag data are delivered from the previous stage as an event, the rule set associated with the event is evaluated and then appropriated actions are taken. For example, ECA rule mechanism can be useful when predefined action is taken by the comparison of the captured tag list with the scheduled information with little human intervention. If the sensed tag list mismatches the schedule information, another action to detect the problem can be taken. We may regard such the inference process as the semantic event generation process making use of RFID tag data and additional information stored in backend systems and supporting the conversion from raw RFID data to actionable information.

3.5 Business process coordination

Main objective in the stage is to enable business processes and solutions to leverage the real-time data captured by RFID infrastructure. The key benefit of RFID technology is automatic identification of individual objects coupled with automatic data capture. The employment of low-levels of process automation toward the process automation and efficiency improvement ultimately leads to the high return in terms of efficiency and cost reduction.

3.6 Decision / actions

One of the desired advantages adopting RFID technology is real-time information gathering, exchange and real-time item visibility. It means that real-time decision making could be realizable. For example, real-time inventory monitoring suggests optimum reorder points based on usage and improves inventory accuracy. Another purpose of this stage is to provide guidance for action to decision maker based on the accumulated information and ultimately produce the knowledge. As a lot of RFID data and related production information are accumulated, it is possible to elicit the valuable knowledge from them - for example, RFID data warehousing (Gonzalez et al., 2006). There are many methods to produce guidance for decision maker and further knowledge as following:

Views of current or historical information. This is the simple approach and the modeling usually consists of aggregation, summarization and filtering.

Forecast. This requires using a methodology like statistical regression based on the current and historical information.

Recommendation of the best and alternative decisions. To find the best recommendation, an optimization model searches among various alternatives and decides the best. Finding a reorder point in inventory problem through various optimization techniques is an example.

Inference through data mining. This is the process to elicit knowledge by searching for the pattern hidden within accumulated information.

4. ETRI RFID ecosystem

ETRI RFID Ecosystem is an RFID software platform that supports not only the presented capabilities that RFID middleware platform must provide but also the activities occurred on RFID IVC, and ultimately provides the seamless environment spanning from the edge of the enterprise network to the enterprise systems.

Figure 3 presents ETRI RFID Ecosystem in accordance with RFID IVC. ETRI RFID Ecosystem is a multi-layered middleware platform in Java environment. The first layer - RFID Event Management System (REMS[3]) - deals with primitive and compound event processing in order to obtain purified and refined RFID event while having less business context. The second layer - Real-time Business Process Triggering System (RBPTS) - is responsible for generating the semantic business event by utilizing refined RFID event. On the top layer, Orchestration Engine (OE) supports the autonomous business process execution. The top layer deals with the generation of invaluable knowledge. Additionally, Tagged Object Information Repository manages the tagged object information and makes them available to whatever the information are required for the purpose of interoperability and exchange within or among enterprises. It is designed to offer the seamless environment extending from RFID hardware infrastructure to backend software systems, and support the RFID IVC. In this section, we introduce the architectural considerations for RFID software platform implementation (mainly, RFID middleware implementation), and then, the functional features and the architecture of each individual that constitutes ETRI RFID Ecosystem is discussed in the following.

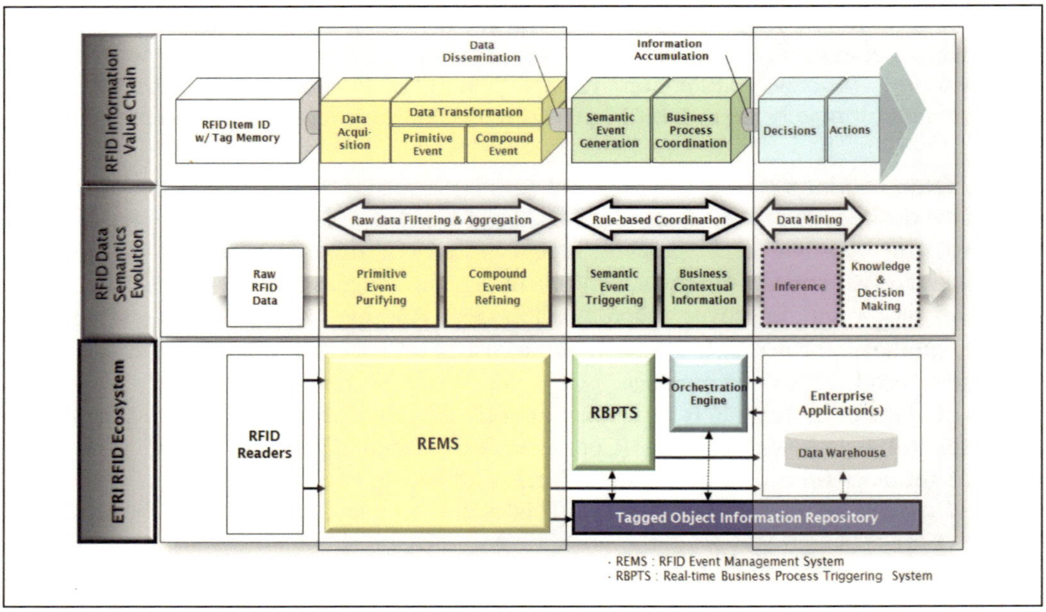

Fig. 3. ETRI RFID Ecosystem and its Correspondence with RFID IVC.

[3] ETRI REMS/EPCIS (including REMS) earned EPCglobal software certification (ALE/EPCIS). http://www.epcglobalinc.org/certification/sw_cert/

4.1 Architectural considerations for RFID middleware platform

We discuss the several architectural issues and our concerns toward REMS, RFID middleware since how well RFID middleware can perfom must be a decisive factor for overall system performance of ETRI RFID Ecosystem. We here deliver our approach to meet those requirements. There are several literatures which deal with concerns on RFID software requirements (Luo et al, 2006; Floerkemeier et al., 2007). Especially, Luo et al. (2006) proposed the following requirements for RFID middleware benchmarking: Streaming, Reactivity (event triggering features) and Integration. In the next, we present how ETRI RFID Ecosystem absorbs those requirements.

Component/Service-oriented Architecture

In general, many applications adopt the component or service-oriented frameworks - for example, Spring Framework[4] - in order to enhance the system flexibility and reusability. Among various component-based software architectures, we have chosen Avalon framework (Loritsch, 2001; SourceForge Inc., 2008a) to implement the RFID middleware servers. Using Avalon, it is straightforward to have the components of each server interact, to instantiate different instances of the components, and to reuse code while having light-weight and minimal features comparing with other application containers.
An Avalon applications, then, is composed of the Avalon infrastructure, the specified components, and a 'container' that reads the configuration files and starts the process running. The container reads the configuration files, loads the specified implementation classes, and invokes the Avalon interfaces in order. In this implementation, we use Phoenix (SourceForge Inc., 2008c) as a container.

Distributed System Architecture

The main role of RFID middleware is to filter and aggregate a lot amount of RFID tag events coming from RFID readers. If the situation of the item-level RFID tags attachment on individual items become realized in the near future, the single server which even equips with high computational capabilities may not control a great amount of RFID tag reads flowing into the system. Therefore, it is unlikely to handle a lot of data by a single server.
In this context, the term 'distributed' has somewhat different meaning from general sense. For business applications in general, enterprise-scale business applications are distributed and operated over the physically separated hardware in order to achieve the load balance and increase scalability. In this sense, it is applicable to the RFID middleware software as well. However, it has more than that: most of RFID readers can communicate with only one system at a time. Therefore, if a reader is deployed into a RFID software, then no other software except it can capture the tag reads by the reader. It implies that an application which wants to utilize the tag reads by specific RFID reader must cooperate with the software which have the connection with the reader.
In this implementation, we adopt the registry-based federated architecture. We name the software system that offers the resource locator service as 'Service Broker'. As shown in Figure 4, Service Broker acts as a 'federator' and other subsystem including our edgewares like 'Reader Management Subsystem' and 'Event Management Subsystem' are 'federatees'. When a subsystem connects into the network, it starts to send the predefined heartbeat messages including the server name and IP over the network. When the Service Broker

[4] http://www.springframework.org/

receives the message, it sends the acknowledge message back to the caller. If it is allowed to join the federation, it sends the registration information including the server name, the access information, and server type and so on, and asks for the download of common resources like RFID reader adapters if necessary.

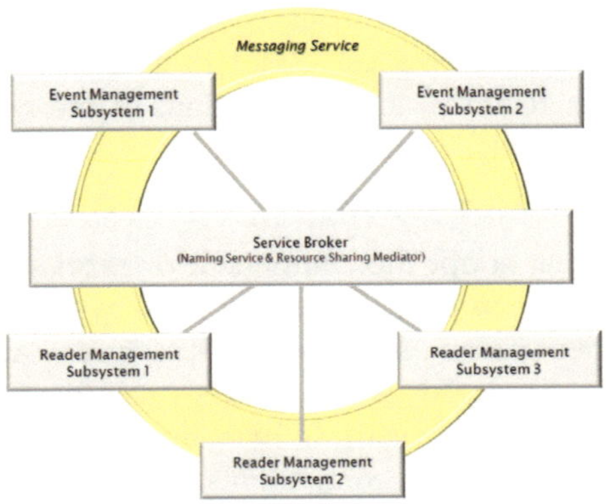

Fig. 4. Building Blocks for REMS, as a distributed system.

Reliability (Fault-Tolerant System Operation)

To ensure the reliable RFID data delivery to the desirable destinations, it is critical to eliminate any single points of failure over all the layers. For example, Sun Java System RFID Software (Gupta et al., 2004; Sun Microsystems Inc., 2006) adopts Jini technology (Sun Microsystems Inc., 2008) for this purpose. When we consider our distributed RFID middleware federation, there are three types of system failures that this RFID middleware platform must take into account.

Failure of RFID readers The failure of any single RFID reader may link directly with the loss of significant data in the business context. RFID middleware system only recognizes and tracks items within range of readers and the failure of a reader may result in the data missing for the items. To cope with this situation, our subsystem periodically checks the aliveness of the connection with readers. If a reader stays unconnected in a pre-specified period of time, the subsystem sends the failure notification by e-mail to the administrator in order for him to examine the reader and periodically keeps trying to reconnect to the reader until establishing the connection between the reader and the subsystem again.

Failure of the Individual Middleware Subsystem Each distributed subsystem may face the system down because of several reasons - for example, the increase of system overhead caused by the tremendous amount of data influx and failure of managing the system overflow. To avoid the subsystem failure, we adopt a simple solution: that is for the service broker acting as a control center to perform active monitoring, which consist of having individual subsystems periodically send keep-alive messages to inform the service broker of their aliveness. Thus, service broker always has an image about the health of their federation over the network. If it has not received the keep-alive information from a subsystem for a timeout, fault-detector module performs reachability test to the subsystem for conformation. It is certain that it has a problem that it generates the amount of traffic; however, an appropriate timeout period can be mediated with the consideration of the trade-off between alleviation of network traffic and responsiveness of failure occurrence.

When the service broker detects a subsystem's failure, it sends the failure notification by e-mail and then it packages all the operational resources related to the failed one and feeds them to the temporal running server in order to take over all the responsibilities which the failed subsystem has taken care of until then. At the startup stage of the service broker, the temporal subsystem also starts up for this case. This approach can be possible because the service broker keeps track of all the changes happened in the federation - that is, we adopt the centralized meta-information sharing in order to cope with such failure.

Failure of Service Broker It is important to guarantee the stable running of the service broker because it governs all the distributed individual subsystems as a control center. In order for the safe operation, the service broker operates as a dual mode and the secondary service broker maintains the redundant information with the primary one by periodically replicating the information the primary manages so that it makes sure the continuous and reliable operation of the service broker.

Various Passive/Active RFID Readers & External Sensors Support and their Management

RFID middleware should provide the means to handle the heterogeneity of RFID readers in terms of the vendors and versions. Most RFID middleware software systems can connect to the RFID devices via the reader adapters which play a role of managing communication in a standardized way between the reader and the middleware. In the implementation, eclipse-based adapter development toolkit is developed. Reader adapter programmers can write java codes for the reader adapter which they want to provide through this middleware and perform the Ant-based build process to generate the Jar package. The adaptor is designed to be compatible with EPCglobal RFID Reader Managment Protocol (EPCglobal, 2007a).

In addition, it is necessary to look at what custom configuration settings you may need to tune on the reader that you choose. Currently, many middleware vendors support varying levels of configuration on the RFID readers; however, some are limited in the amount of control, which means that you are not able to control key settings such as antenna power or antenna cycling. This may lead the users to manually configure the readers outside the middleware if a tunable parameter is not supported. This manual tuning process may make it difficult to manage the readers.

In order to alleviate such difficulties, we devise XML-based configuration for setting tunable parameters for each reader as shown in Figure 5. The adapter programmers can decide the scope of tunable parameters which are willing to be opened by simply exposing parameters in the XML documents like Figure 5 (b).

Lastly, for the applications which do not necessarily require continous RFID reading (Floerkemeier et al., 2007), it is preferable to have a mechnism to initiate tag reading by external sensors. At this time, the reader adaptor can also cover the registration of external sensors in the limited manner enough that the sensor-triggered RFID reader activation is achievable.

Global RFID Standards Compatibility

There are several RFID-related global standards which RFID middleware must concern: (a) interface between RFID tags and readers, (b) interface between RFID readers and host applications and (c) information exchange interface between RFID middleware and RFID applications.

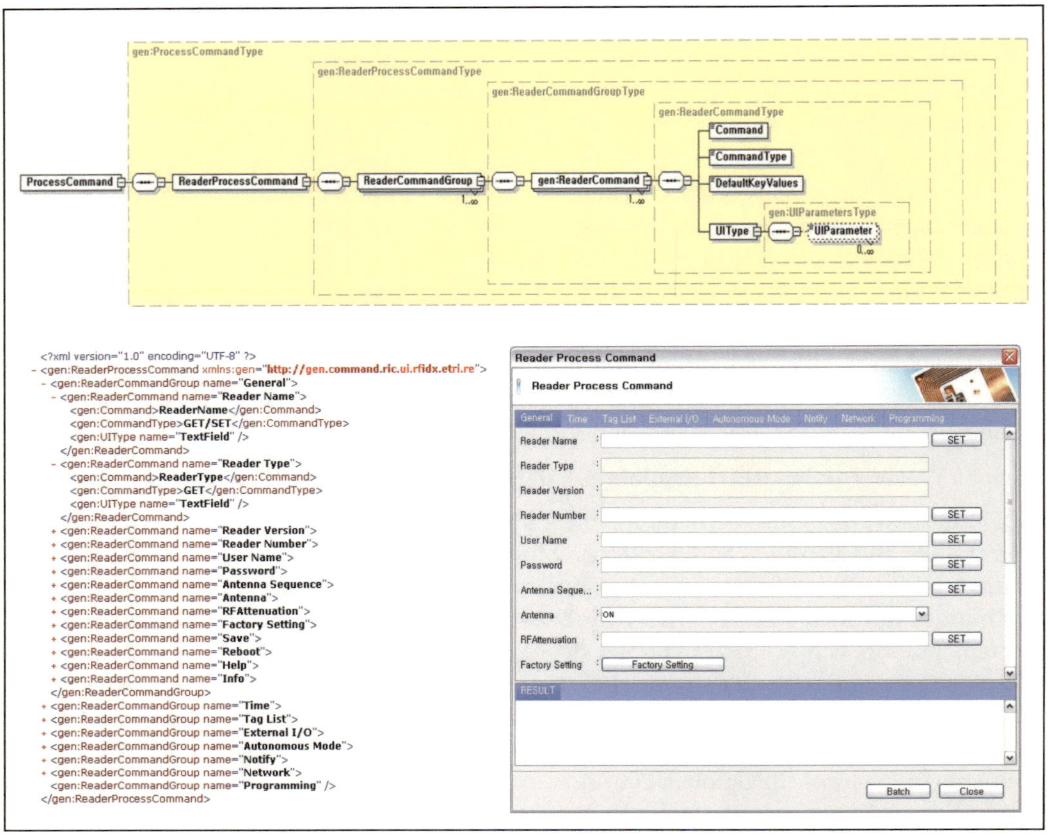

Fig. 5. Configuration for vendor-specific tunable parameters and commands: (a) XML schema for configuring tunable parameters for reader setting and commands (upper), (b) Example for Alien RFID reader[5] which is an instance of (a) (lower left), and (c) user interface rendering from (b) (lower right).

Tag/Reader Interface (Air Interface Protocol): EPCglobal UHF RFID Class 0, Class 1 Gen 1, Class 1 Gen 2 Protocol and ISO/IEC 18000 series

RFID Tag/Reader interface defines the physical interactions between RFID tags and readers. EPCglobal ratified the EPCglobal Class 1 Gen 2 protocol in 2004 and there are ISO 18000 series as a de jure air interface protocol standards in RFID system.

This interface has little concern with the RFID middleware implementation; however, it is required to consider the tag memory structures described in those standard specifications. The memory structure on the tags decides what types of information can flow in. Therefore, this tag memory structure affects the reader/host protocol design, which subsequently has an influence on the middleware implementation

Reader/Host Interface: EPCglobal Reader Protocol, Reader Management Protocol and ISO/IEC 15961, 15962

Reader/Host interface defines a set of commands for reader control, configuration and management, tag reading and writing. Each RFID reader vendor defines its own

[5] http://www.alientechnology.com

reader/host protocol and some vendors provide libraries to help programmers to develop RFID applications using their RFID readers in more convenient way. As part of resolving the diversities of reader/host protocol varying from vendors, the demand for developing general reader/host protocol leads to the development of EPCglobal Reader Protocol, and ISO/IEC 15961 and 15962. But, two protocols have different operations to have access to tag data due to the difference of tag memory structure and data organization in it. However, they define the general features which most RFID reader vendors also take into consideration, so most vendor-specific reader/host protocols can converge on either protocol.

When it comes to the middleware implementation, it is important to support as many market-available RFID readers as it can. Most middleware vendors provide toolkits to develop so-called 'reader adapter', which is a driver-like pieces that interfaces to actual RFID reader and provides a unified interface for RFID middleware to have access to readers. We also take the same approach to support various types of readers and devise the vendor-neutral reader/host API set in order to have access to those readers in a seamless way. The two global standard specifications play a critical role of deriving the common API set for the reader adapter while satisfying the diversities of vendor-specific protocols.

Middleware/Application Interface: EPCglobal Application Level Events

Application Level Events (ALE) is a software specification for the filtering and collection of RFID data being defined and ratified by EPCglobal. It defines a globally accepted method of filtering and collecting RFID information and it is the most representative standard interface between RFID middleware and external applications. As a result, this standard is expected to improve the interoperability between systems as it becomes widely accepted, so it is necessary to develop the middleware software that complies with this EPCglobal mandates, ALE. We fully implement ALE 1.0 (EPCglobal, 2005b) and currently extend it to meet ALE 1.1 (EPCglobal, 2008a) while reflecting ISO active tag features.

Application Integration

The ability to integrate RFID into the legacy systems or existing ones is absolutely critical to deliver the sensed tag events to the right applications in the right time. The simple approach is just to dispatch the captured events from readers to a series of applications at the low level; whereas, some form of enterprise application integration (EAI) is needed to get the full value from RFID events. Many major EAI solution providers like 'Tibco' try to integrate their solutions with RFID middleware and release to the market (Tibco Software Inc., 2006).

In particular, it is necessary to support various types of application integration methods including *push, pull* and *publish/subscribe* for application-level RFID information capture. For this, we take the following approach: our middleware is equipped with several standards-based adapters required to ensure connectivity to backend applications. In addition, it is necessary to allow application developers to register their own adapters to send the filtered RFID events to their legacy applications. In this implementation, there are six different protocols in order to let legacy applications receive the notification of RFID event messages - that is, HTTP, TCP/IP, JMS, File and Web Service (SOAP/HTTP). Users can subscribe to the middleware by entering URIs with specifying the protocol. Table 1 shows how to specify URI for each protocol.

<table>
<tr><th>Protocol</th><th>URI Template</th><th>Example</th></tr>
<tr><td>HTTP</td><td>http://<ip>:<port>/<web page>?pollingInterval=<millisecond></td><td>http://129.254.238.16:8080/reports_log.jsp?pollingInterval=50000</td></tr>
<tr><td>TCP/IP</td><td>tcp://<ip>:<port></td><td>tcp://129.254.238.16:7000</td></tr>
<tr><td>JMS</td><td>jms://<queue | topic>/<JMS Connection Factory>/<queue | topic name>?jndiInitialContextFactory=<Java class URI for JMS context factory}&jmsProviderURL=<URL of JMS Provider></td><td>jms://topic/JmsTopicConnectionFactory/triggerTopic?jndiInitialContextFactory=org.exolab.jms.jndi.InitialContextFactory&jmsProviderURL=rmi://localhost:1099</td></tr>
<tr><td>File</td><td>file:///<directory>:/${SpecName}_${yyyy MMddhhmmss}.xml</td><td>file:///c:/sample_${SpecName}_${yyyyMMddhhmm}.xml</td></tr>
<tr><td>Web Service</td><td>axis://<end-point address>?optQName=<Operation QName>&optServiceName=<Service Name></td><td>axis://localhost:8080/ECReportsService.asmx?optQName=escape('http://www.etri.re.kr')&optServiceName=OperationProcess</td></tr>
</table>

Table 1. URI definitions and their examples for RFID event subscription

Moreover, application programmers can develop their own event dispatch modules inheriting from designated Java interface we suggest and deploy them into the system in order to ensure the application-dependent protocol-based communication between the RFID middleware and their legacy applications.

4.2 ETRI RFID event management system (ETRI REMS)

RFID middleware is a software system that manages data communication between RFID readers and enterprise applications. In this section, we present the layered RFID middleware while considering architectural design aspects discussed in the previous section. We divide the RFID middleware into three layers – that is, 'device monitoring and management layer', 'data management layer' and 'business integration layer'. As given in Figure 6, the ETRI RFID middleware covers lower two layers and consists of two subsystems: Reader Device Abstraction & Management Subsystem for device monitoring, Event Management Subsystem for RFID data management and delivery. Also, there exists Service Broker for offering the name lookup service and resource sharing. The functional features and internal component architecture of each subsystem are described in the following.

Reader Device Abstraction and Management Subsystem (RMS)

The primary roles of RMS are to support the seamless integration between the middleware software and various kinds of RFID readers, and to monitor and manage the deployed readers. In order to handle the heterogeneity of readers and reader/host interface protocols, we abstract the reader/host interface APIs with help of the existing global reader/host interface standards mentioned in Section 4.1 and offer the eclipse[6]-based development toolkit for 'reader adapter'. Single reader adapter is developed per each vendor and version

[6] http://www.eclipse.org/

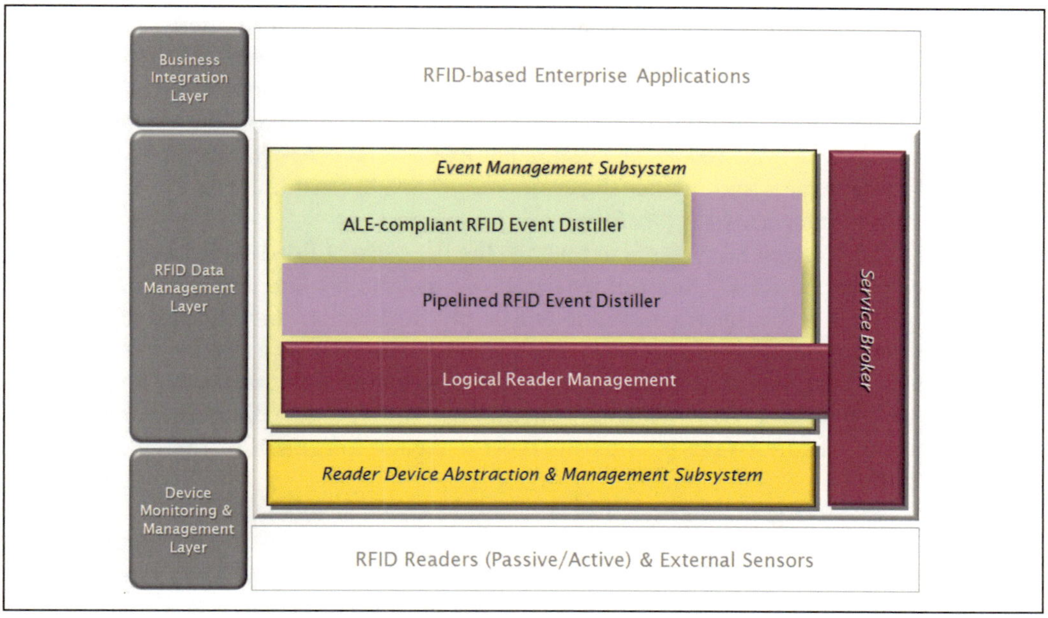

Fig. 6. Layered conceptual architecture of ETRI REMS.

if the vendor does not guarantee the backward compatibility of reader/host interface. Moreover, reader adapter developers take a responsibility of organizing the tunable reader-specific parameters by editing XML file shown in Figure 5 and implementing the proper execution codes for them. Those activities improve the extensibility for newly-introduced RFID readers at this middleware system.

In Figure 7(a), we show the component architecture of RMS using Avalon and the dependencies among the components. All the components are deployed and controlled by Phoenix. As shown in the Figure 7(a), there are three major components: Connection Manager, Monitor and Reader Agent Manager with Reader Agent.

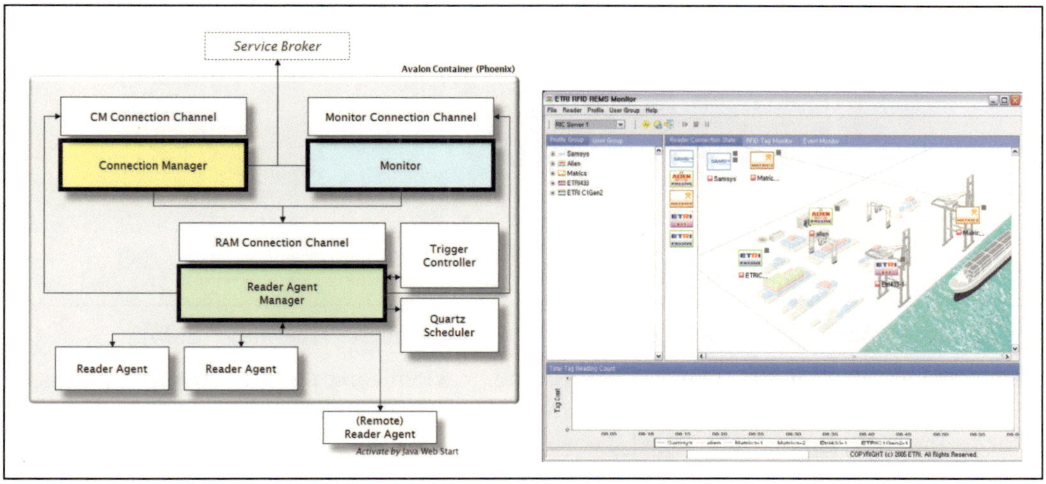

Fig. 7. Reader Device Abstraction & Management Subsystem: (a) component architecture (left) and (b) main user interface for reader configuration and monitoring (right).

The task of Connection Manager is to manage all the context information related to the deployed RFID readers and issue commands for reader management. All the commands exposed at the user interface are sent to this component. Then, this component validates and dispatches them to appropriate other components. The Monitor is responsible for monitoring the events occurred in the system. For example, it keeps track of the aliveness of individual active readers and notifies the erroneous events to the administrator or the reader management user interface in Figure 7(b).

The task of Reader Agent Manager is to manage the life cycle of Reader Agents representing the actively connected RFID reader or external sensors, and act like a container for Reader Agents. In general, a Reader Agent binds with a physical reader and handles all activities related with corresponding reader such as sending commands to a reader to control the reader and receiving tag data. Besides the reader-specific tunable parameter setting, each Reader Agent provides the following operations: (a) connect/disconnect reader, (b) suspend/resume reading tags (c) modify Reader Agent information, (d) delete Reader Agent and so on.

In addition, Reader Agent Manager has a dependency with Scheduler, Quartz, which is java-based open source scheduler. Basically, Reader Agent operates in a polling manner as a default operation mode in order to capture tag data in a reading zone; however, our implementation allows it to operate in the on-demand mode or user-specified schedule-based mode. For the latter case, the administrator specifies the Unix crontab-like expression with duration and Quartz scheduler awakes the Reader Agent based on the crontab expression and let it collect tag reads for the duration.

RFID Event Management Subsystem (EMS)

EMS is the core system in this middleware platform which filters data extracted from the RFID readers, aggregates the information and routes the data to the RFID-enabled applications (see Figure 8).

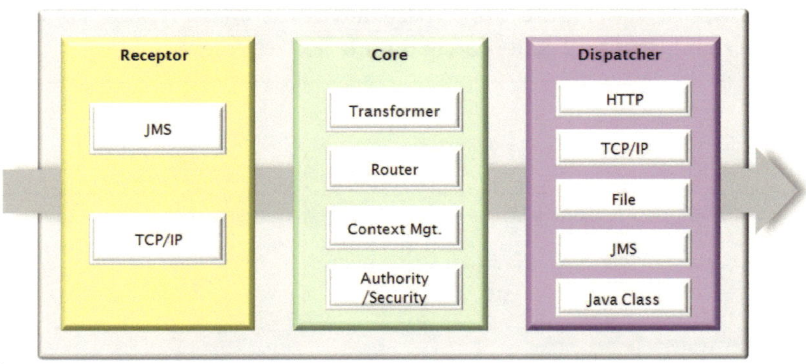

Fig. 8. Conceptual Representation of RFID Event Management Subsystem (EMS).

EMS enables ALE-based event queries and customizable event stream processing operations. Such the processes allow raw RFID data to be transformed into business information that can be leveraged by RFID-enabled external applications. In order to support reliable event processing and better performance, EMS adopts pipeline architecture as a basis for data processing. A pipeline consists of a set of primitive task processors called 'valves' and 'chain's connecting two valves. Pipelines categorize the influx data and process those categories with a set of primitive tasks. By following the 'divide-and-conquer' like approach, it is expected to increase overall throughput and the average speed for high-volume data processing. Moreover, the XML-based event description named 'ECSpec' in ALE specification can be expressed in a pipeline way, so we stack ALE-specific processing modules over the pipeline-based processing modules in order to support ALE API as shown in Figure 9(a).

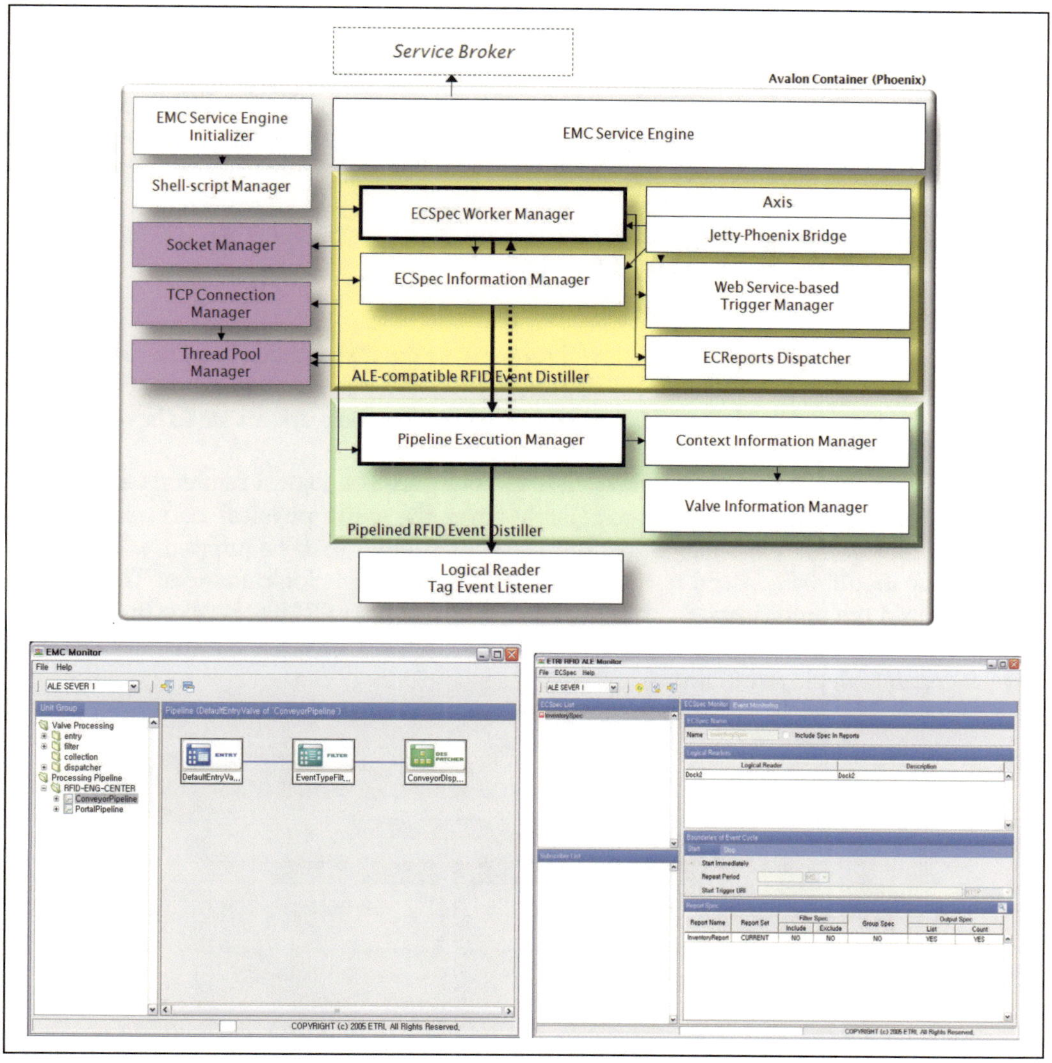

Fig. 9. Event Management Subsystem: (a) component architecture using Avalon (upper), (b) pipeline designer (lower left) and (c) ALE ECSpec designer and monitor (lower right).

Figure 9(a) shows the component architecture of EMS using Avalon and the dependencies among the components. As stated above, the building blocks for pipeline support reside in the bottom layer, followed by ALE-specific components. The major components are Valve Information Manager, Pipeline Execution Manager and ECSpec Worker Manager. Valve Information Manager is responsible for managing built-in or custom event processors.

Application programmers can create custom event processing valves as well as event dispatchers which pre-process filtered RFID data prior to propagating the information to external applications and use them while building up the event stream processing pipeline.
The task of Pipeline Execution Manager is to control the execution of pipelines and exception handling. We provide the user interface to define a pipeline instance as shown in Figure 9(b).
ECSpec Worker Manager takes a role of managing individual ECSpec Worker per each ECSpec given by outers. When the request for ECSpec is received, the ECSpec Worker Manager parses the request described in XML, transforms it into a pipeline and then asks Pipeline Execution Manager to execute the pipeline. The pipeline instance is linked with an ECSpec Worker internally so that the result of event processing by the pipeline is at first delivered to the ECSpec Worker. The role of ECSpec Worker is to manage the information of subscribers to the related ECSpec and handles the pipeline executions. Figure 9(c) shows the administration user interface for defining and monitoring ECSpec.
Lastly, we deploy the lightweight web server, Jetty (Mort Bay Consulting, 2008), with Axis in order to support the Web Service ALE API.

Service Broker

The Service Broker plays a role of a control center over the distributed network discussed in Section 4.1. It keeps track of the all the distributed subsystems and provides the name look-up service for them - especially, our user interface uses this look-up service to have access to individual subsystem.
Another major role of this service broker is to configure the logical readers. Generally, a single reader is not often sufficient to reliably cover the entire physical area relevant to a business process. For example, a loading dock may have to be equipped with several readers that should be exposed to client applications as a single logical reader. This enables EMSs to avoid the modifications from the change of readers in RMSs. Service broker offers methods to configure the logical readers and the relations between a logical reader and physical readers deployed to RMSs.

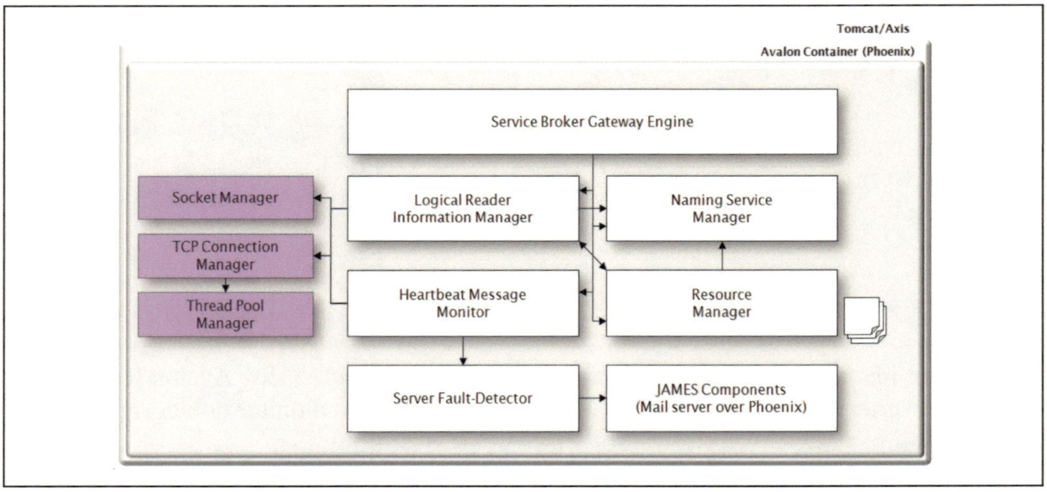

Fig. 10. Component Architecture of Service Broker.

Figure 10 shows the component architecture of Service Broker. We develop the Java servlet pseudo-container for Avalon components in order to reuse components we already developed and keep consistency with other systems. As major components, Naming Service Manager maintains distributed system configuration information such as access information of all RMSs and EMSs. Heartbeat Message Monitor and Fault-Detector keep track of the aliveness of all subsystems and take reactions whenever any failure of them is captured. The task of Logical Reader Information Manager is to maintain the configuration of logical readers and the relations between logical readers and physical readers.

4.3 Real-time Business Process Triggering System (RBPTS)

Real-time Business Process Triggering System (RBPTS) is built on a rule-based inference engine which provides mechanism to extract semantic and business-context information from tag read events through ECA-type rules. Such semantic information is derived from domain-specific knowledge provided by domain experts or business collaboration partners and embedded in rule definitions. The semantic events - as discussed in Section 3.4 - are produced by associating the RFID primitive events with the domain-specific information residing in legacy system. This system receives a continuous stream of filtered and unfiltered RFID data from RFID middleware or RFID readers and produces the RFID-triggered business event by using set of rules. The produced semantic event is utilized for the query to execute collection of rules to perform various predefined actions ranging from one-time actions such as DB operation, the notification, alerts, actuator operations, or actions that involve the long term business process actions which require interoperation with workflow systems such as ebXML[7] engine and BPEL[8]-based workflow engine. The development of RBPTS is driven by the requirement of flexible way of incorporating RFID data with business applications; that is, to convert the data from lower RFID middleware layers to actionable semantic information for the upper layers.

In order to achieve the goal and be suitable for RFID environment, the rule engine of RBPTS adopts the backward chaining inference mechanism. As the physical RFID readers involve the specific business goals - for example, gate open/close, inventory check and so on - and the business actions triggered by the collected data fall into small number of categories, it is expected that possible conclusions can be chosen at the time that a set of tag data is collected by specified readers. The domain experts define a set of rules which are described as the 'If-condition(s)-then-action' pattern. The event message delivered by REMS includes the 'action' indicator called 'query' to be proved, so the inference process starts with a conclusion with the help of 'query'. The rule engine searches for the rule set which has the action clause that matches the action which the event message includes and then evaluate the associated condition clauses. The condition is described as not just a simple form like value matching but also complicated form like a predefined java class or access to database located in the legacy system.

Figure 11 shows the RBPTS components and the internal message flow. We note that we revised the open source java class library, MANDARAX (SourceForge Inc., 2008b), in order to implement ECA-type rule engine, and XML Schema for ECA rule definition is newly defined as presented in Figure 12.

---

[7] http://ebxml.org
[8] http://www.w3.org

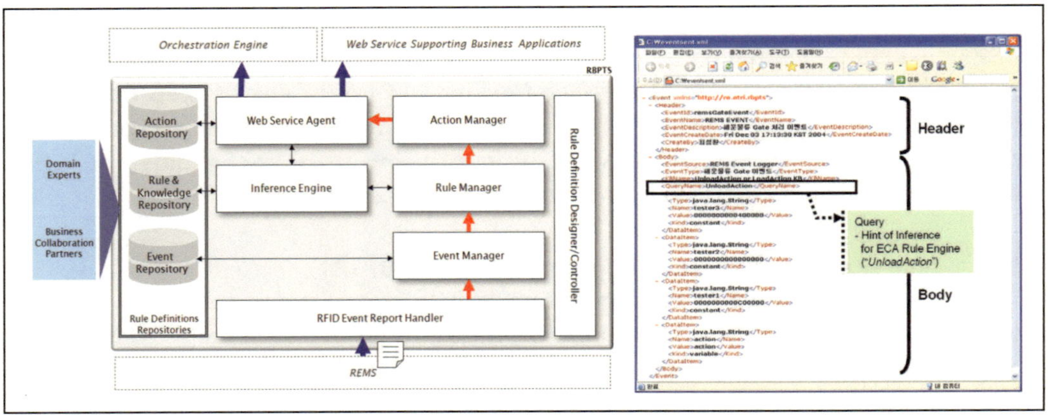

Fig. 11. Architecture of Real-time Business Process Triggering System and Sample Event XML Message over SOAP/HTTP delivered from REMS.

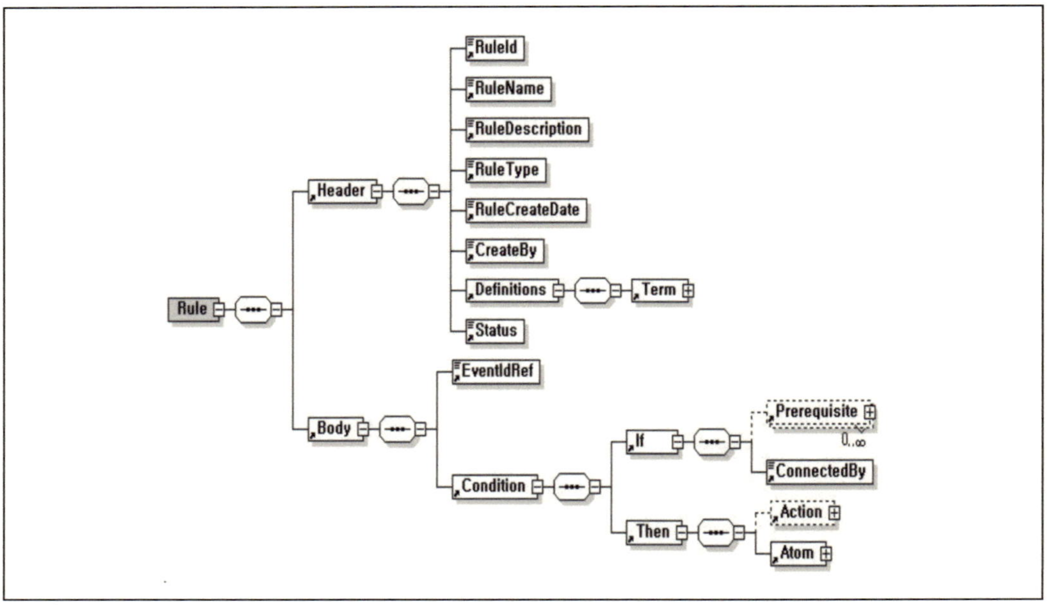

Fig. 12. XML Schema of Rule Definition.

REMS accumulates RFID tag data over intervals of time, filters to eliminate duplicate tag data and the tag data that are not of interest, and then reports in the XML/SOAP message form which follows the input format of RBPTS (see Figure 11). The RBPTS-specific event XML messages are generated by custom message dispatcher registered in REMS. RFID Event Report Handler accepts the SOAP message and then passes it to Event Manager in turn. Event Manager unmarshals the event message, checks whether the message is valid by looking up the event registry and checking its register status. Afterward, Event Manager reorganizes the valid event message into sort of event query message that is used for the next step - inference process - and delivers it to Rule Manager. Rule Manager inquires for the rule set associated with the event query and constitutes all the matters that are essential for the reasoning: database drivers that have access to the legacy database, repository information and so on. Rule Manager feeds all the prepared materials into the Inference Engine and then this evaluates the conditions for each rule and generates the result set. This is during the process that business semantics are disclosed. Types of conditions span from simple forms including direct comparison to more complex ones such as inquiring to external applications. Based on the result set, Rule Manager organizes the action execution list and passes the list to Action Manager. Action Manager searches for the web service for each action execution and configures the information for the web service call. Action Manager asks for Web Service Agent to call the dynamic web service and records the execution result on the log database. RBPTS supports the application triggering via web service only.

In addition, RBPTS provides web-based user interface for rule design to model RFID event, related business rules and the detailed actions which in result provide more flexible way to adapt to the rapidly changing business environment.

4.4 Orchestration Engine (OE)

To build a concrete RFID middleware platform, it is desirable to orchestrate RFID-based end-to-end processes that associate with multiple applications or legacy systems so that it ultimately provides the RFID-enabled process automation environment.

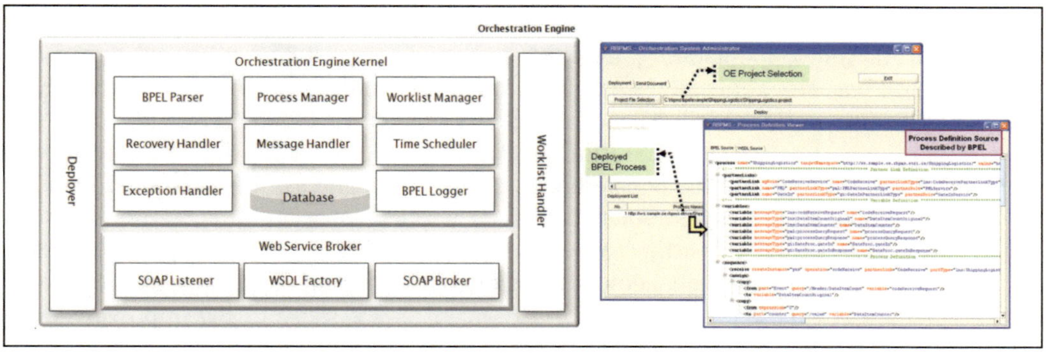

Fig. 13. Architecture of Orchestration Engine (OE) and GUI Window of OE Administrator.

For this purpose, the business process/workflow coordination engine called 'Orchestration Engine' is developed and the system architecture is shown in Figure 13. Currently, the most recent answer to the integration challenge is the Service Oriented Architecture (SOA) and the Web Service technologies. We suppose that we can access different functionalities of different legacy and other developed applications in a standard way through Web Services. Under the environment that all applications expose the functionalities via Web Services, we develop a business process definition and execution engine that provides a way to compose the Web Service-exposed functionalities in J2EE framework. Mostly, the business processes defined by Orchestration Engine are triggered by RFID-related events including not only the primitive event as the output of data transformation on the RFID IVC but also the semantic event generated by RBPTS.

In this implementation, we adopt BPEL (Business Process Execution Language for Web Services) (Andrews et al., 2003), an XML-based industry standard for business process management, as the definition language of business processes. BPEL builds on top of XML and Web Services, and BPEL process specifies the exact order in which participating Web

Services should be invoked. As the typical scenario we develop under the ETRI RFID Ecosystem, a BPEL business process receives a SOAP request from RBPTS when the raw observation by REMS are dispatched to RBPTS with XML event message and the rule instance, which is invoked by the message and contains the action clause of calling the BPEL process, is evaluated as true. Then, new instance is started and managed by Process Manager. It calls the external Web Services specified in the BPEL definition - for example, invoking EPCIS Web Service in order to get the prices of products and then invoking calculator Web Service to sum up the prices of items which are checked out - and returns the results when the process instance is done.

7. Conclusion

One of major issues in RFID software platform is how to (i) handle vast amount of RFID data efficiently and (ii) transform raw RFID tags to more valuable forms so that RFID-driven business applications can create value and achieve their ideal goals - automated business execution and integration. In this chapter, we introduce the concept of RFID Information Value Chain (RFID IVC), sequential activities for RFID data semantics evolution and value creation, and then discuss the several architectural considerations when we developed our RFID software platform called 'ETRI RFID Ecosystem ', whose middleware has features of component-oriented architecture, distributed and reliable operations, support of various types of RFID devices including external sensors, compatibility with existing RFID standards and business applications integration. We believe that we address the majority of the RFID software implementation concerns which other products should regard.

In addition, we demonstrate the RFID software platform with the internal service component architecture and the dependencies among components. It is presented how the discussed architectural considerations have been applied in order to construct such the RFID middleware. In fact, this artifact was tested in the field of several RFID pilot projects in South Korea including yard management project in harbour (Kim et al., 2006) and it showed the competitive advantages of ETRI RFID Ecosystem in the sense of the improvement in operational efficiency rather than partial deployment as desired.

8. References

Andrews, T., Curbera, F., Dholakia, H., Goland, Y., Klein, J. Leymann, F., Liu, K., Roller, D., Smith, D., Thatte, S., Trickovic, I. & Weerawaran, S. (2003). Business Process Execution Language for Web Services version 1.1. http://download.boulder.ibm.com/ibmdl/pub/software/dw/specs/ws-bpel/ws-bpel.pdf

Clark, S., Traub, K., Anarkat, D. & Osinski, T. (2003). Auto-ID Savant Specification 1.0, *MIT Auto-ID Center working paper*

Dutta, K., Ramamritham, K., Karthik, B. & Laddhad, K. (2007). Real-time Event Handling in an RFID Middleware System, *Proceedings of Workshop on Databases in Networked Information Systems,* pp. 232-251, ISBN 978-3-540-75511-1, Japan, October 2007, Springer

EPCglobal (2005a). EPC Radio-Frequency Identity Protocols Class-1 Generation-2 UHF RFID Protocol for Communications at 860 MHz - 960 MHz Version 1.0.9. http://www.epcglobalinc.org

EPCglobal (2005b). The Application Level Events (ALE) Specification, Version 1.0. http://www.epcglobalinc.org

EPCglobal (2006). Reader Protocol Standard, Version 1.1. http://www.epcglobalinc.org

EPCglobal (2007a). Reader Management 1.0.1. http://www.epcglobalinc.org

EPCglobal (2007b). EPC Information Services (EPCIS) Version 1.0, Specification. http://www.epcglobalinc.org

EPCglobal (2008a). The Application Level Events (ALE) Specification, Version 1.1. http://www.epclobalinc.org

EPCglobal (2008b). EPCglobal Object Name Service (ONS) 1.0.1. http://www.epclobalinc.org

Floerkemeier, C., Roduner, C. & Lampe, M. (2007). RFID Application Development With the Accada Middleware Platform. *IEEE Systems Journal*, Vol. 1, No. 2, December 2007, pp. 82-94

Gonzalez, H., Han, J., Li, X. & Klabjan, D. (2006). Warehousing and Analyzing Massive RFID Data Sets, *Proceedings of the 22nd International Conference on Data Engineering*, pp. 83-92, ISBN 0-7695-2570-9, Atlanta, USA, April 2006, IEEE Computer Society, USA

Gupta, A. & Srivastava, M. (2004). Developing Auto-ID Solutions using Sun Java System RFID Software. http://java.sun.com/developer/technicalArticles/Ecommerce/rfid/sjsrfid/RFID.html

IBM. (2006). *IBM WebSphere RFID Handbook: A Solution Guide.* http://www.redbooks.ibm.com/redbooks/pdfs/sg247147.pdf

ISO/IEC JT1/SC31/WG4. http://usnet03.uc-council.org/sc31/sc31_wg5.cfm

Kim, Y., Yoo, J. & Park, N. (2006). RFID Based Business Process Automation for Harbor Operations in Container Depot, *Proceedings of Industrial & Manufacturing Engineering Graduate Research Symposium at Wayne State University*, pp. 213-226, Detroit, USA, April 2006

Ku, T., Zhu, Y. & Hu, K. (2007). Semantics-based Complex Event Processing for RFID Data Streams, *Proceedings of First International Symposium on Data, Privacy and E-Commerce*, pp. 32-34, ISBN 978-0-7695-3016-1, Chungdu, China, November 2007, IEEE Computer Society, USA

Loritsch, B. (2001). Developing with Apache Avalon. *Apache Software Foundation*

Luo, Z., Wong, E., Cheung, S.C., Lionel, M. & Chan, W.K. (2006). RFID Middleware Benchmarking, *Proceedings of the 3rd RFID Academic Convocation*, Shanghai, China, October 2006

Mort Bay Consulting. (2008). *Jetty – Java HTTP Servlet Server.* http://www.mortbay.org/jetty/.

Oracle. (2008). *Oracle RFID and Sensor-based Services.* http://www.oracle.com/technologies/rfid/index.html

Oat Systems & MIT Auto-ID Center. (2002). The Savant Version 0.1 Alpha, *MIT Auto-ID Center Technical Manual*

Palmer, M. (2006). RFID and Complex Event Processing, *RFID Today,* http://www.rfidtoday.co.uk/articles/objectsore_rfid.html

Sarma, S.; Brock, D.L. & Ashton, K. (2001). The networked physical world proposals for engineering the next generation computing, commerce & automatic-identification. *MIT Auto-ID Center white paper*

Sun Microsystems Inc. (2006). *Sun Java System RFID Software 3.0 Developer's Guide.* http://docs.sun.com/source/819-4686/ale-web-services.html

Sun Microsystems Inc. (2008). *Jini Network Technology.* http://www.sun.com/software/jini/

SourceForge Inc. (2008a). *Avalon – Framework branch.* http://freshmeat.net/projects/avalon/?branch_id=18263

SourceForge Inc. (2008b). *Mandarax Project.* http://mandarax.sourceforge.net/

Tibco Software Inc. (2006). RFID Solution. http://www.tibco.com/solutions/biztech/rfid.jsp

21

Enhancing the Interactivity of Learning-Guide Systems with RFID

Yo-Ping Huang[1], Yueh-Tsun Chang[2], Wei-Po Chuang[2] and Frode Eika Sandnes[3]
[1]*Department of Electrical Engineering, National Taipei University of Technology,*
[2]*Department of Computer Science and Engineering, Tatung University,*
[3]*Faculty of Engineering, Oslo University College,*
[1,2]*Taiwan*
[3]*Norway*

1. Introduction

Many countries, including Taiwan, have over the last few years been actively promoting digital archives programs. Recent advances in information science and computer technology has opened up novel new means in which the general public can enjoy and be educated on historical and cultural relics that are important parts of a country's national heritage. New technologies has led to a growing number of extensive digital databases for artifacts such as pictures and visual art, writings, records, and other cultural objects. Researchers have therefore started to explore new ways in which these digital databases can be associated with the actual relics such that the knowledge and understanding of these objects can become more accessible to the general public.

Currently, many exhibition centers employ professional guides that explain the objects on display, answer questions and provide guided tours. Although such services are effective, pedagogic and promoting social interaction, it is limited by the human resources available and is therefore usually offered during peak hours and for groups with a minimum number of participants. Exhibition centers and museums therefore often provide self-service prerecorded audio guides to increase accessibility. Visitors carry portable audio players, on loan from the museum, and listen to the prerecorded guide through headsets as they walk through the exhibition. Such guides usually provide sufficient information about the objects on display. However, the static nature of the prerecorded guided tours means that the visitors need to view the exhibitions in a particular order and this restriction leaves little room for visitor participation and interaction [3]. Furthermore, visitors have different expectations and interests, and some visitors may be too impatient to complete their tours. Some museums employ digital audio devices that allow the users to enter a set of digits matching digits displayed next to an artifact. This allows users to manually control the playback order. More advanced audio guides have been proposed, such as Sotto Voce [21], which promotes social interaction where visitors eavesdrop others' personal audio guide. This study targets indoor guides, as outdoor guides pose different challenges [20, 23].

Radio Frequency Identification (RFID) is a wireless communication technology that has been successfully applied to various fields such as transportation, distribution, supply chain, telemedicine, etc. RFID technology was used during World War II to identify airplanes as friend or foe, but was then forgotten for many years. Because the electronic tags have been expensive, RFID technology has not until recently become widely embraced. In 2003, Wal-Mart, the leader of retail business in United States started using RFID technology and incorporated an extensive RFID system in their storehouse and circulation. This event was picked up by others resulting in the RFID market rising rapidly and catching wide attention. The price of electronic tags is decreasing, and the RFID technology is now widely applicable and about to become acceptable for all kinds of fields.

RFID systems can be classified as non-contact and automatic identification technology that consists of two components - RFID readers (also called interrogators) and electronic tags (also called transponders). Unlike traditional bar-code system, RFID systems can carry dynamic as well as static data. According to different kinds of electronic tags, the functionality and memory size vary. Traditional bar-codes are reliant on an unobstructed label face of the codes because they need to be scanned to identify the digital data, and it takes a while to identify the traditional bar-codes. On the other hand, with RFID systems, tags can be hidden and there is no need to pay attention to the label face. As long as the tags are in the range of the radio wave that the RFID reader sends out, the information in the tags can be accessed and identified, and they can be identified rapidly through the radio wave. If there is more than one item, traditional bar-code must be scanned sequentially. In contrast, multiple tags can be read simultaneously.

We propose a novel interactive mobile guiding environment, where PDA's and both UHF and HF RFID technologies [6] are used to overcome the static nature of prerecorded guides and the code-entry effort of older interactive digital audio guides. In addition, the interactive learning infrastructure includes a wireless network and the system makes use of data mining and information retrieval technology [4]. The combination of RFID and wireless connectivity also allow remembering tools [22] to be realized such that visitors can review their visit remotely via the web at a later time. The proposed system allows an art museum to provide all its related artifact data or questionnaires to the visitors by equipping each artifact with an electronic tag. The long-distance RFID reader allows art galleries to promote exhibitions and attract visitors to the "hidden treasures" of less popular areas [8]. The system also recommends viewing routes for visitors. These recommendations are obtained by performing Apriori-like collaborative filtering on the individual viewing records [13, 14].

Our system is different to previous RFID-based guide systems [16, 17, 24] as two separate RFID systems that employ the HF and the UHF frequency bands respectively are used in parallel. The HF RFID system is used to increase personalization service and to attract visitors to specific artifacts [9]. The visitors do not have to wait long for the system to respond once they approach the target artifact as the HF tag instantaneously communicates with the system such that the related information is provided to the visitors without intervention [5, 11]. The long-distance UHF RFID reader, positioned at a strategic location within the museum, reads the UHF tag attached to the PDA of the visitors and is used to promote less-known artifacts.

The HF frequency band RFID system is based on ISO 15693 passive electronic tags, RFID readers and middleware. The passive ISO 15693 tags are not reliant on power. Consequently, there is a limited transmission distance, which makes them readable within a given radius of an artifact. Furthermore, passive tags are especially suitable for exhibitions because of their ubiquitous attributes - namely, compact size, long life, no need for maintenance, and low cost [12].

Visitors' PDAs are equipped with electronic EPC class 1 UHF tags. UHF frequency band RFID readers are installed at strategic locations throughout the exhibition venue. These are typically locations that receive fewer visitors and need to be actively promoted. Information read from an EPC class 1 electronic tag, attached to a visitor PDA, is immediately transmitted to the server middleware. Next, the middleware determines which visitor the tag belongs to and then delivers promotional information about the given exhibition via the wireless network to the visitors' PDAs.

2. Exhibition recommendation

The exhibition centers generally group artworks according to themes or type, such as photographs or Chinese ink-water paintings, in different sections. Occasionally, artworks are put together for a particular artist for his/her commemoration or to present a retrospective angle of the artist.

Data mining is an emerging technology that extracts useful information or patterns from large databases, data warehouses or other storage repositories. Data mining is also a field bringing together techniques from machine learning, pattern recognition, statistics, databases, and visualization, etc.

2.1 Mining association rules

Association analysis involves finding association rules from data stored in databases. Association rules relate the associations of attribute-value frequently appearing in the given dataset. A traditional example of association rule mining is the market basket analysis. The process finds associations from grocery selection of a customer in the basket and then discovers the purchase behavior of customer.

Normally, two steps are required to extract the association rules from databases.

1. Find the sets of large itemsets: The first step is to find all the itemsets and supports of these itemsets must be larger than a predefined minimum support (threshold). The minimum support is usually defined by domain experts and is case dependent.
2. Generate strong association rules from large itmesets: By applying both support and confidence values to the large itemsets, strong association rules can be solicited. For example, the rule "item_C$\Rightarrow$item_D [*support* = 25%, *confidence* = 50%]" means that 25% of transactions show that both item_C and item_D were bought together, and there are 50% of chance that if people bought item_C then they will buy item_D too.

The Apriori algorithm [14] can be applied to get large inter-itemsets. However, the processing cost of the first two iterations (i.e., obtaining L_1 and L_2, representing large 1-itemset and large 2-itemset, respectively) dominates the total mining cost. The reason is that a low minimum support induces a very large L_1, which in turn results in a huge number of itemsets in C_2 (i.e., candidate 2-itemset). This problem is even more severe when mining inter-transaction association rules. Tung et al. proposed the FITI algorithm that first finds the frequent intra-transaction itemsets and then generates the inter-transaction itemsets from the frequent intra-transaction itemsets. The work reduces the number of candidates and therefore improves the mining efficiency [25]. Lu et al. employed EH-Apriori to reduce

the number of candidate inter-itemsets [26]. Pei et al. [27] employed a projection scheme in the PrefixSpan algorithm where the customer sequences (transactions) were projected into overlapping sets called projected databases such that all the customer sequences in each set have the same prefix that corresponds to a frequent sequence (itemset). The underlying principle of the PrefixSpan is that, instead of projecting sequence databases by considering all the possible occurrences of frequent subsequences, the projection is only based on the frequent prefixes. This holds because any frequent subsequence can always be found by growing a frequent prefix. The advantages of the PrefixSpan algorithm are (1) no candidate sequence needs to be generated; and (2) the projected databases shrink [27].
The visitors' viewing history is continuously recorded as visitors browse the exhibition using the mobile guiding system. Information stored in the database includes timestamps for the various viewing events, the particular contents viewed, and the details browsed for particular artifacts [14]. Once a substantial amount of data has been collected a data mining algorithm is used to discover association rules between artifacts and the artists. These association rules are subsequently used to suggest useful exhibition-related information to future visitors. For example, assume the system generates an association rule as follows:

If "**Little Flying Phoenix** (Sculpture, Yang Yuyu)" and "**Suckling Lamb** (Sculpture, Ju Ming)," then "**Field Laboring** (Print, Yang Yuyu)."

Then, if a visitor inquires information about the 'Little Flying Phoenix' by Yang Yuyu and the 'Suckling Lamb' by Ju Ming, the system automatically recommends the information about the 'Field Laboring' by Yang Yuyu to the visitor as shown in Figure 1(a).

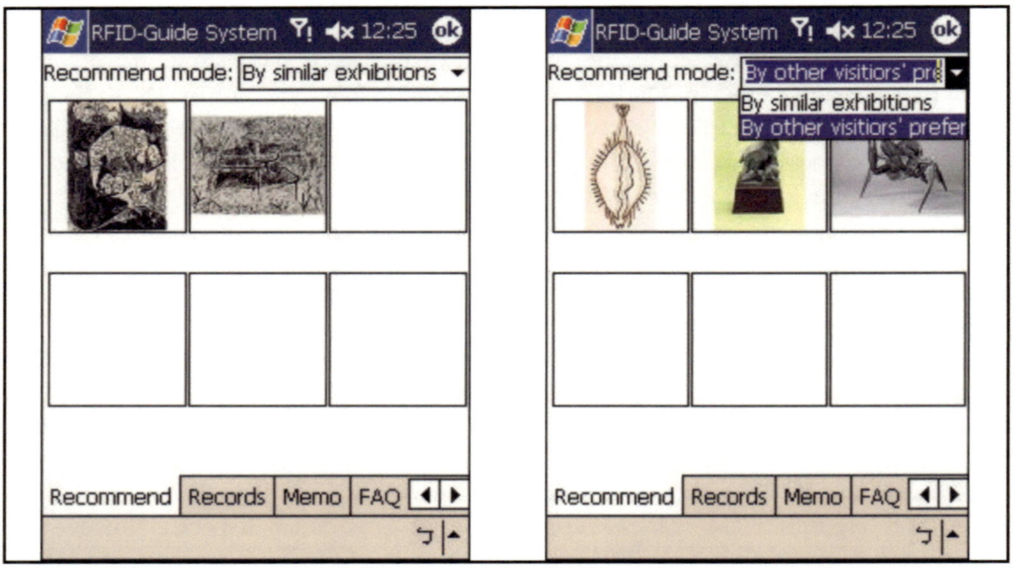

(a) The user can select different recommendation modes. (b) The other recommendation mode.

Fig. 1. Exhibition recommendation.

2.2 Collaborative filtering

Information about visitors is used to perform collaborative filtering [2]. Recommendations are created according to groups of visitors with similar and related interests and preference.

Item-based filtering is a strategy where the connection between items is identified according to visitors' selections, preferences, and browsing patterns. This technique is commonly employed on e-commerce web sites where, for example, warehouse retailers collect information about products purchased by customers. Next, the connection between purchased products and customers' purchasing habits are then estimated. Finally, the warehouse retailers can recommend additional and relevant products when the customer purchases certain items.

	Big Pot	Square Dish	Revolve	Soul-box II	Moon Bowl
Visitor A	1	1	0	0	1
Visitor B	0	1	0	1	1
Visitor C	0	0	1	0	0
Visitor D	1	1	0	0	0
Visitor E	1	0	0	1	0

Table 1. Collaborative filtering example.

Table 1 shows an example of collaborative filtering where artifacts viewed by a visitor are assigned 1 and artifacts that are not visited are assigned 0. In this example, the viewing patterns of visitor *A* and visitor *D* have the highest similarity. Therefore, artifacts viewed by visitor *A* that has not yet been viewed by visitor *D* should be recommended to visitor *D*. The similarity between visitors and artifacts, and the most suitable information of the exhibition for recommendation can be calculated as follows [15] where the similarity of viewing patterns between two visitors *a* and *b* is denoted by *sim(a,b)*:

$$sim(a,b) = corr_{ab} = \frac{\sum_{j=1}^{N}(p_{aj}-\bar{p}_a)(p_{bj}-\bar{p}_b)}{\sqrt{\sum_{j=1}^{N}(p_{aj}-\bar{p}_a)^2}\sqrt{\sum_{j=1}^{N}(p_{bj}-\bar{p}_b)^2}}, \tag{1}$$

N is the total number of exhibitions, p_{aj} is visitor *a*'s rating on the exhibition *j*, and $\bar{p}_a$ is visitor *a*'s average rating on all exhibitions. Let $H = \{h_1, h_2, \ldots, h_m\}$ be the set of visited exhibitions by visitor *c*. The query likeness score *QLS(c,j)* of visitor *c* on a new exhibition *j* is determined by [15]:

$$QLS(c,j) = \frac{\sum_{i \in H}(p_{ij}-\bar{p}_i)\times sim(c,i)}{\sum_{i \in H} sim(c,i)}. \tag{2}$$

When viewing artifacts in one section, visitors may miss artworks of interest on display in other locations of the museum or artwork currently not on display at all. Museums usually have limited real-estate to display artifacts, and therefore have to rotate the artifacts on display during different exhibitions. The recommendation system therefore emphasizes exhibitions from different sections that have been viewed by other visitors during previous exhibitions. These exhibitions may be in different styles but from the same creator or exhibitions with the same style but in a different category [7]. Artifacts not on display can therefore be viewed on the PDA via the recommendation system as shown in Figure 1.

The system also provides a directory map to help visitors easily navigate to the recommended item when recommended artifacts are on display in different locations to where the visitors presently are (see Figure 2).

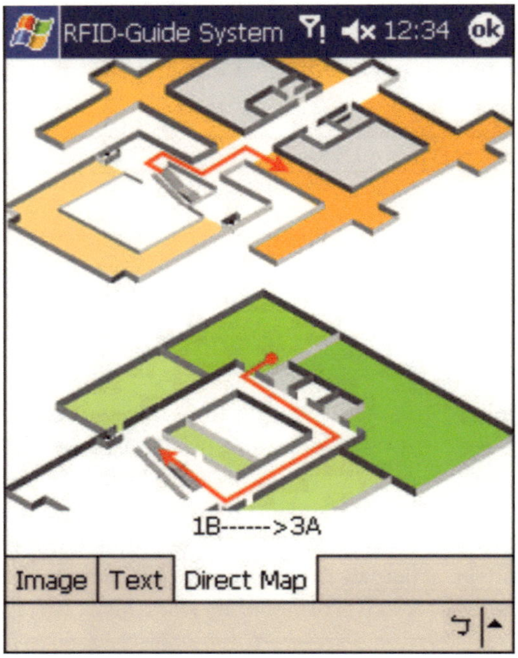

Fig. 2. Exhibition directory map.

3. System implementation

The architecture of the mobile guide system is illustrated in Figure 3. The mobile guide infrastructure comprises PDA clients, RFID readers, a wireless network and EPC class 1 electronic tags.

Each exhibition sign in the exhibition center is fitted with a unique ISO 15693 passive electronic tag. These tags are read by the PDA RFID readers allowing the artifact to be identified. Detailed content associated with the identified item is then downloaded via the wireless network and presented on the PDA together with a FAQ or a questionnaire. In addition, the electronic tag read by the RFID also identifies the whereabouts of the visitor allowing their movements to be tracked [1]. The back-end server is managed by exhibition centre personnel who have to update the database each time an exhibition is changed. They also provide answers to questions posted by visitors.

Traditional prerecorded guides require the visitors to follow prewritten scripts and fixed routes around the items on display. This is problematic for visitors who are unable to keep up with the explanations, visitors who wish to deviate from the tour and explore the exhibition on their own, and sometimes, visitors are unable to find the exhibitions because they are unfamiliar with the floor plan, misunderstand the guides, or the museum is so crowded such that physical movement and vision is restricted. The user-friendly interface helps overcome these limitations and difficulties as it provides a more convenient environment for the visitors to browse the artworks and exhibitions.

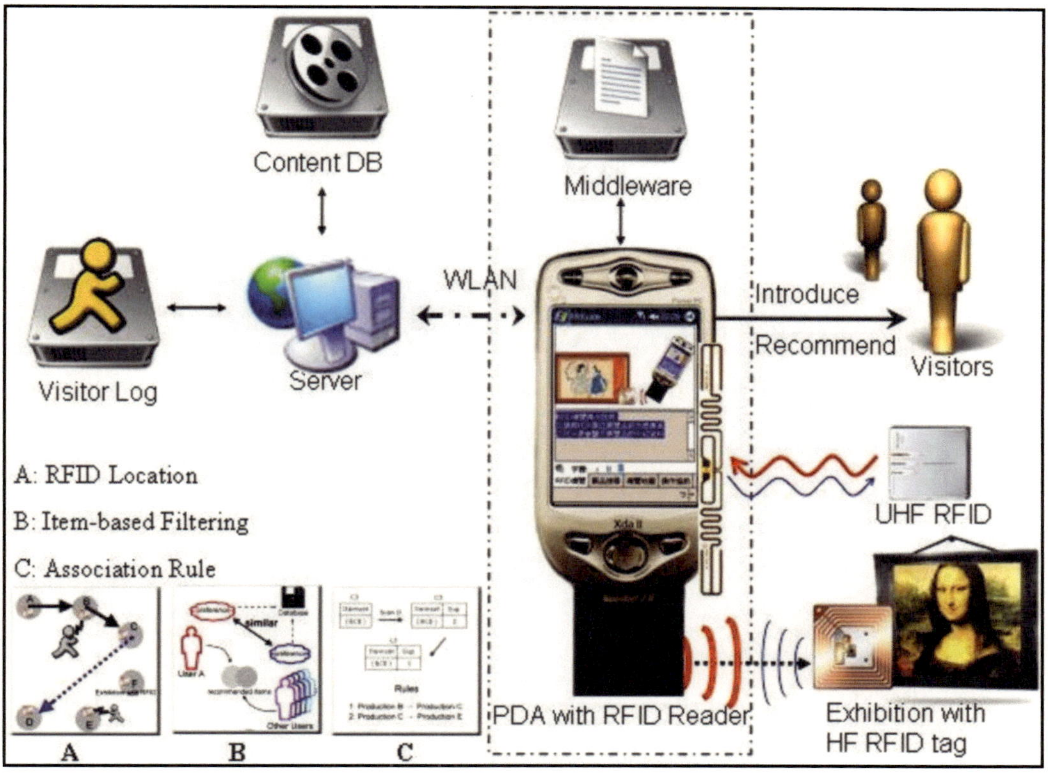

Fig. 3. The architecture of the RFID guide system.

3.1 RFID guiding

The digital content is prepared by the exhibition center staff. The multimedia contents include textual explanations, digitally recorded audio, graphical illustrations and short video clips. Each entry in the content database is linked to the identifier of the encased electronic RFID tag that is attached to the sign of the physical artifact on display together with RFID guiding marks that help visitors know where to point the RFID reader.

Visitors simply move the RFID equipped PDA close to the RFID guiding mark to read the discernment code of the electronic tag. The middleware identifies the content associated with the discernment code which then can be displayed in the PDA as shown in Figure 4. Visitors do not need to input textual queries. Text input is particularly difficult, error-prone and time-consuming on handheld devices [18, 19]. Further examples from an art gallery in Taipei county are shown in Figures 5-8.

Museum may employ multiple RFID-tags and guide marks around popular artifacts that attract many simultaneous visitors. For instance, RFID-tags could be placed on both sides and beneath a wall mounted piece, or in all four directions, or more, around some items on a pedestal.

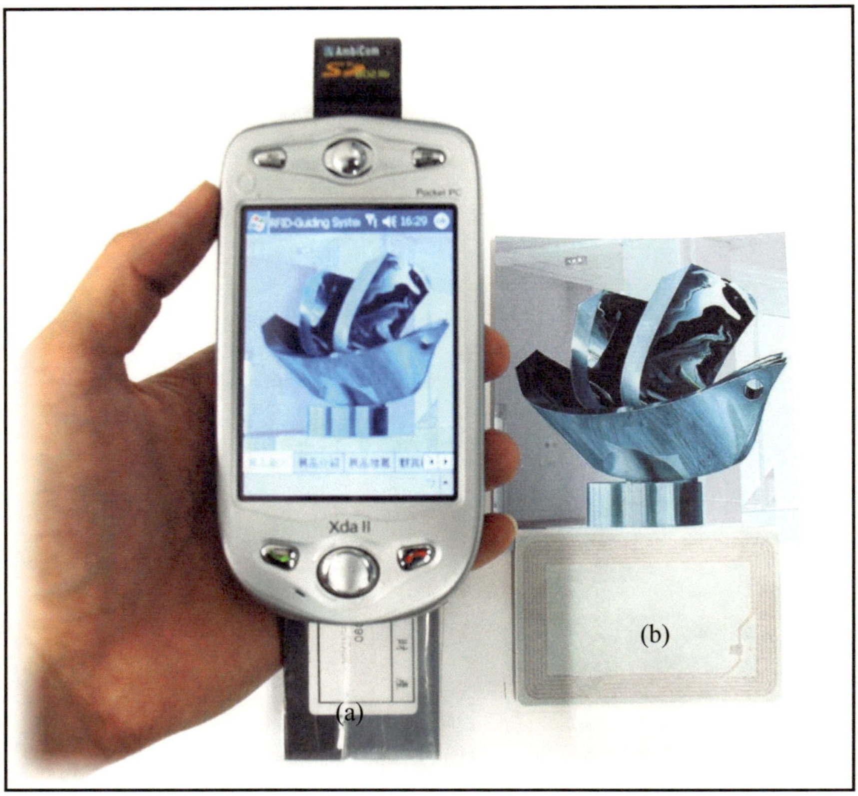

(a) HF RFID reader. (b) ISO 15693 passive electronic tag.

Fig. 4. The RFID-based guide system.

Fig. 5. A painting with a RFID-enhanced information plate.

Fig. 6. The visitor moved the handheld computer close to the information plate. The RFID-reader picks up the ID of the painting.

Fig. 7. Information about the painting is downloaded from a server via the wireless network.

Fig. 8. The visitor browses the downloaded information.

3.2 Discovering "Hidden Treasures" with UHF RFID

The mobile guiding system is designed to help visitors explore the exhibitions freely according to their personal interests and preferences. The system allows visitors' viewing patterns to be recorded and these records can subsequently be analyzed by the exhibition staff such that future exhibitions can be improved. The layout of the exhibition can be changed by placing popular and famous objects at strategic locations such that visitors have to walk past and discover lesser known objects, in a similar manner to which merchandise are strategically placed in modern supermarkets. However, we also exploit UHF RFID to actively attract visitors to less popular objects.

Imagine for instance a painting on display in the exhibition depicting two people touching hands on a shadowy background. Uninformed visitors may only catch a brief glimpse of the painting, judge the painting unimportant and then move on. However with the UHF RFID installed, the system could for instance raise the visitors' attention by delivering the message "What is it in their hands?" as they walk past the painting.

When visitors are in the vicinity of a less popular artifact with an UHF RFID reader installed, the UHF RFID reader identifies the UHF electronic tag attached to the PDA device carried by the visitor and transmits the information to the back-end server. The server identifies the wireless network IP address of the PDA with this particular tag. Then the message can be sent to the PDA and reported to the visitor as shown in Figure 9. Consequently, visitors are more likely to discover that the two people in the painting are holding a dandelion in their hands and to realize that the painting is conveying the happiness associated with the arrival of spring.

3.3 Multimedia streaming

Multimedia presentations, such as video and audio, are presented to the visitors on their PDAs (see Figure 10). However, most PDAs have limited storage capacity and the multimedia content is therefore hosted on the server, instantly encoded and streamed over the wireless network before it is decoded and presented to the visitors on their PDAs. This also simplifies maintenance of the system as only the server needs to be updated and not each of the individual PDAs.

In addition to admiring the beauty of the artworks, the visitors may also gain insight into additional relevant information through the presentations, such as the complexity of the creation process. Hopefully, this makes learning more entertaining and fun.

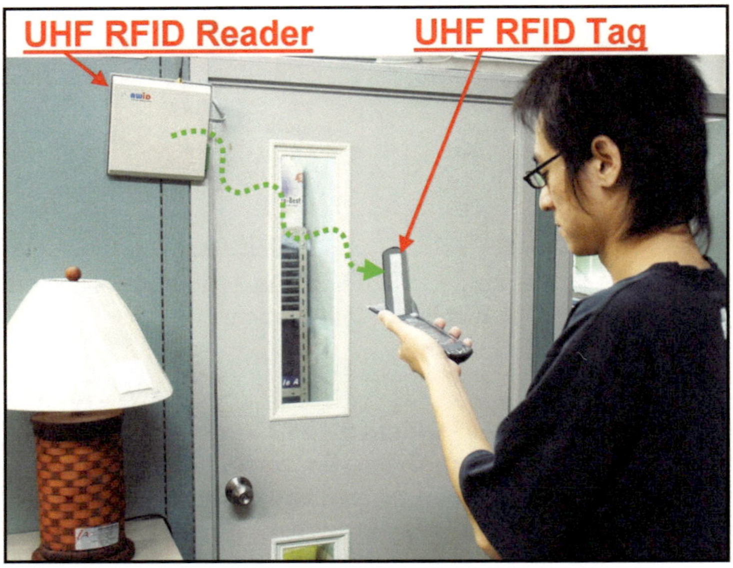

Fig. 9. RFID promotion.

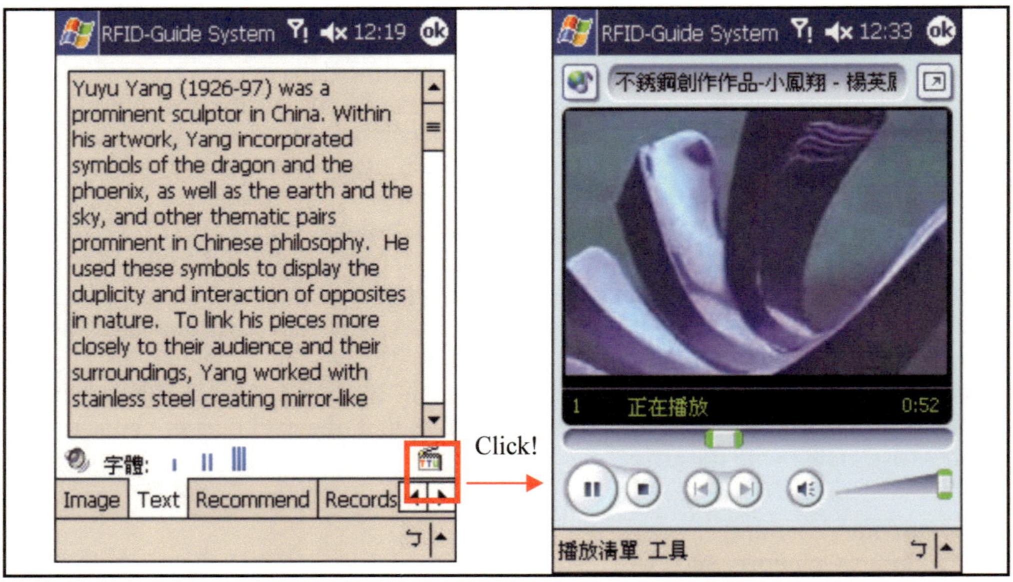

(a) The video streaming function. (b) Playing the video.

Fig. 10. Multimedia streaming.

4. Evaluation and analysis

An evaluation experiment was set up to assess the effectiveness of the mobile guide. The purpose was to examine the stability of the system, to acquire users' opinions about the system and to identify potential problems. HF electronic tags were attached to the back of photographs of historical and cultural relics. After having completed the interactive guided tour the participants were asked to complete a questionnaire. In addition, the proposed system was compared to existing guide systems.

4.1 Results of the experiments

A total of 133 questionnaires were collected. The results of the six-part questionnaire can be summarized as follows. (1) 82.7% of the users think that the system is interesting. (2) 78.1% of the users enjoy learning using the mobile guiding system. (3) 75.9% of the users believe that it is suitable to use such a system in exhibition centers. (4) All users agree that it is uncomplicated to manage this system and do not require any prior learning for operation. (5) 84.2% of the users liked the recommendation function. (6) Some parents suggested including artifact-related games for children to promote fun learning.

Other observations include: (1) The RFID guiding function can correctly read and respond to 100% of the information delivered from the exhibitions. (2) There was a system delay of approximately one second after an RFID electronic tag was read. (3) The RFID recommendation system yields a success rate of 94.7% in recommending interesting exhibitions to other visitors. (4) Without the multimedia streaming technology, it takes on average three seconds to download a two-minute video before it can be played. The multimedia streaming technology efficiently partitioned the video such that the perceived downloading delay diminished.

The ubiquitous nature of RFID technology allows RFID tags to be successfully read although they are hidden out of sight. Consequently, the visual appearance of artifacts on display are not affected or disturbed as they are with conventional signs and posters [11]. There is an interaction distance of approximately 15 centimeters between the HF RFID tag and the reader and this was found to be a reasonable distance. If the distance is too large, the reader may simultaneously detect multiple tags and the system will be unable to resolve which artifact the user is actually browsing.

However, the reading distance of the UHF frequency band RFID is much larger. If there are two, or more, visitors within the interaction range, there will be a collision between the UHF electronic tags. Furthermore, during the experiment we found that the UHF frequency band RFID is affected by moisture. Most visitors put the UHF RFID tickets in their pockets. These tickets did not operate accurately when they were too close to the human body and were affected by human moisture. The UHF electronic tag was; therefore, attached to the PDA instead. This has greatly improved the operational accuracy of the UHF subsystem, but it also helps reduce costs as the UHF electronic tags can be reused.

4.2 Guide system comparison

Most guide systems can be categorized as being either designated guide systems or active guide systems. A designated guide system includes stationary components such as exhibition labels and information kiosks. Exhibition labels contain simple and general exhibition-related textual information. On the other hand, an information kiosk is a powerful multimedia workstation which can be placed in each section to provide interactive responses to exhibition inquiries. Although, a kiosk is a powerful information resource it somewhat is inconvenient, as users must move away from the artifact in order to use the kiosk.

Active guide systems include professional guides, brochures, prerecorded audio, and digital mobile guide systems. Professional guides are individuals that have been trained by the exhibition center to escort visitors around the exhibition while explaining the details of the artifacts on display. Due to limited budgets and few staff professional guided tours are often only provided during certain peak hours or need to be pre-booked. Brochures and booklets that contain information about the exhibition are often distributed to visitors when they purchase tickets or enter the exhibition centers. Prerecorded audio guides force the visitors to view the exhibitions in a particular order. Digital mobile guide systems, such as digital voice players or PDAs, provide enhanced interaction to users. Visitors can obtain exhibition-related content by entering the artifact name or ID number into the system as they move to each exhibition on display.

Table 2 presents a comparison of the different strategies discussed herein according to five key objectives, variable cost, and the simplicity of updating the information.

	RFID-based guide system	Assigned professional guides	Pre-recorded audio	Brochures
Response instantaneity	Very high	High	Lowest	Low (Self-searching)
Dynamic interaction	High	Very high	Lowest	Low
User-friendly interface	High	Very high	Lowest	Low
Abundant in content	Very high	High	Low	Very low
Information gathering	Very high	Low	Lowest	Very low
Fixed costs (training, etc)	High	Very high	Low	Very low
Cost for each service	Very low	Very high	Low	High
Update simplicity	Very high	Very low	Low	High

Table 2. Comparisons of guide systems.

5. Conclusions

An integrated interactive RFID guide system based on information retrieval, association rules, and personalized recommendation that assists visitors browsing an exhibition centers was presented. Visitors retrieve exhibition-related multimedia information by using RFID equipped PDAs. Recorded visitor viewing patterns can be subsequently analyzed and used to improve the exhibition to achieve more effective learning.

The database needs to be maintained such that it accurately reflects the physical artifacts on display at any given time. In addition, the question-and-answer function did not offer suitable answers to all questions. Thus, there are two directions for further improvements:

Tailored content: Users have different levels of knowledge and the contents of the exhibitions should be designed accordingly. Thus, contents tailored for different visitor groups such as children, young adults and senior visitors could greatly help improve the overall user experience.

On-line oracle: A computer system is no substitute for a real human expert. One can therefore use the wireless infrastructure to connect visitors to a real guide which can offer visitors an immediate remote on-line one-to-one guiding service.

6. Acknowledgment

This work is supported by National Science Council, Taiwan under Grants NSC94-2745-E-036-001-URD, NSC94-2745-E-036-002-URD and NSC94-2213-E-036-021.

7. References

[1] P. Bahl and V.N. Padmanabhan, "RADAR: An in-building RF-based user location and tracking system," *Proc. of IEEE Computer & Communications Societies*, Tel-Aviv, Israel, vol. 2, pp.775-784, Mar. 2000.

[2] D. Billsus and M.J. Pazzani, "Learning collaborative information filters," *Proc. of Int. Conf. on Machine Learning*, Madison, WI, USA, pp.46-54, July 1998.

[3] M. Derntl and K.A. Hummel, "Modeling context-aware e-learning scenarios," *Proc. of the 3rd IEEE Int. Conf. on Pervasive Computing and Communications Workshop*, Kauai Island, HI, USA, pp.337-342, Mar. 2005.

[4] K. Facer, R. Joiner, D. Stanton, J. Reid, R. Hull and D. Kirk, "Savannah: mobile gaming and learning," *Journal of Computer Assisted Learning*, vol. 20, no. 4, pp.399-409, Sept. 2004.

[5] J. Hightower, R. Want and G. Borriello, "SpotON: an indoor 3D location sensing technology based on RF signal strength," *UW CSE 00-02-02, Department of Computer Science, University of Washington,* Seattle, WA, USA, Feb. 2000.

[6] Y.-P. Huang and T. Tsai, "A fuzzy semantic approach to retrieving bird information using handheld devices," *IEEE Intelligent Systems,* vol. 20, no. 1, pp.16-23, Jan./Feb. 2005.

[7] R.J. Mooney and L. Roy, "Content-based book recommending using learning for text categorization," *Proc. of the 5th ACM Conf. on Digital Libraries*, San Antonio, TX, USA, pp.195-204, June 2000.

[8] L.M. Ni, Y. Liu., Y.-C. Lau and A.P. Patil, "LANDMARC: indoor location sensing using active RFID," *Proc. of the IEEE Conference of Pervasive Computing and Communications,* Vienna, Austria, vol. 10, no. 6, pp.701-710, Nov. 2004.

[9] S. Poslad, H. Laamanen, R. Malaka, A. Nick, P. Buckle and A. Zipf, "CRUMPET: Creation of user-friendly mobile services personalized for tourism," *Proc. of the 2nd Int. Conf. on 3G Mobile Communication Technologies*, London, UK, pp.28-32, Mar. 2001.

[10] R. Baeza-Yates and B. Ribeiro-Neto, *Modern Information Retrieval*, Addison-Wesley, Reading, MA, USA, 1999.

[11] R. Want, "Enabling ubiquitous sensing with RFID," *IEEE Computer*, vol. 37, no. 4, pp.84-86, Apr. 2004.
[12] R. Weinstein, "RFID: a technical overview and its application to the enterprise," *IT Professional*, vol. 7, issue 3, pp.27-33, May-June 2005.
[13] S. Zadrozny and J. Kacprzyk, "An extended fuzzy Boolean model of information retrieval revisited," *Proc. of The 14th IEEE Int. Conf. on Fuzzy Systems*, Reno, Nevada, USA, pp.1020-1025, May 2005.
[14] J. Han and M. Kamber, *Data Mining: Concepts and Techniques*, Morgan Kaufmann Publishers, San Francisco, CA, USA, 2001.
[15] C.Y. Kim, J.K. Lee, Y.H. Cho and D.H. Kim, "VISCORS: A visual-content recommender for the mobile web," *IEEE Intelligent Systems*, pp.32-39, Nov./Dec. 2004.
[16] S. His and H. Fait, "RFID enhances visitors' museum experience at the exploratorium," *Communications of the ACM*, vol. 48, no. 9, pp.60-65, Sept. 2005.
[17] H. Hlavacs, R. Gelies, D. Blossey and B. Klein, "A ubiquitous and interactive zoo guide system," *Lecture Notes in Computer Science*, vol. 3814, pp.235-239, Nov./Dec. 2005.
[18] F.E. Sandnes and Y.-P. Huang, "Chord level error correction for portable Braille devices," *Electronics Letters*, vol. 42, no. 2, pp.82-83, Jan. 2006.
[19] F.E. Sandnes and Y.-P. Huang, "Chording with spatial mnemonics: automatic error correction for eyes-free text entry," *Journal of Information Science and Engineering*, vol. 22, no. 5, pp.1015-1031, Sept. 2006.
[20] K. Cheverst, K. Mitchell and N. Davies, "The role of adaptive hypermedia in a context-aware tourist GUIDE," *Communications of the ACM*, vol. 45, no. 5, pp.47-51, May 2002.
[21] P.M. Aoki, R.E. Grinter, A. Hurst, M.H. Szymanski, J.D. Thornton and A. Woodruff, "Sotto voce: exploring the interplay of conversation and mobile audio spaces," *Proc. of SIGCHI Conf. on Human Factors in Computing Systems*, Minneapolis, MN, USA, pp.431-438, Apr. 2002.
[22] M. Fleck, M. Frid, T. Kindberg, E. O'Brien-Strain, R. Rajani and M. Spasojevic, "From informing to remembering: deploying a ubiquitous systems in an interactive science museum," *IEEE Pervasive Computing*, vol. 1, no. 2, pp.13-21, Apr.-June, 2002.
[23] K. Cheverst, N. Davies, K. Mitchell, A. Friday and C. Efstratiou, "Developing a context-aware electronic tourist guide: some issues and experiences," *Proc. of SIGCHI Conf. on Human Factors in Computing Systems,* The Hague, The Netherlands, pp.17-24, Apr. 2000.
[24] D. Raptis, N. Tselios and N. Avouris, "Context-based design of mobile applications for museums: a survey of existing practices," *Proc. of 7th Int. Conf. on Human Computer Interaction with Mobile Devices & Services*, Salzburg, Austria, pp.153-160, Sept. 2005.
[25] A.K.H. Tung, H. Lu, J. Han and L. Feng, "Efficient mining of intertransaction association rules," *IEEE Trans. on Knowledge and Data Engineering*, vol. 15, no. 1, pp.43-56, Jan./Feb. 2003.
[26] H. Lu, J. Han, and L. Feng, "Stock movement and n-dimensional intertransaction association rules," *Proc. SIGMOD Workshop Research Issues on Data Mining and Knowledge Discovery*, vol. 12, pp.1-7, 1998.
[27] J. Pei, J. Han, B. Mortazavi-Asl, J. Wang, H. Pinto, Q. Chen, U. Dayal and M.C. Hsu, "Mining sequential patterns by pattern-growth: the PrefixSpan approach," *IEEE Trans. on Knowledge and Data Engineering*, vol. 16, no. 11, pp.1424-1440, Nov. 2004.

22

Object Recognition Using a 3D RFID System

Se-gon Roh and Hyouk Ryeol Choi
School of Mechanical Engineering, Sungkyunkwan University, Korea

1. Introduction

We cannot think about something if that something does not exist. If something exists, we can understand or imagine about it no matter what it may be. Understanding and imagination in this way fully depend on recognition. A human being's object recognition is executed in real-time by identification intelligence, which is developed from experience, learning, and presumption. On the other hand, a robot's object recognition is executed by sensation, perception, and identification. Sensation means the response to the stimulus and intensity of the object; for example, a vision system captures images obtained from a CCD camera (Weiss et al. 2001). Perception implies the estimation or acquirement of the object geometry, of which invariants are extracted from the two-dimensional luminance data. Identification matches and determines the object from a database based on the representations of the extracted geometry. These processes need to compute enormous data, so that real-time process is almost impossible. In addition, matching uncertainty is immanent in this recognition because the robot has much difficulty in identifying the existence of an object.

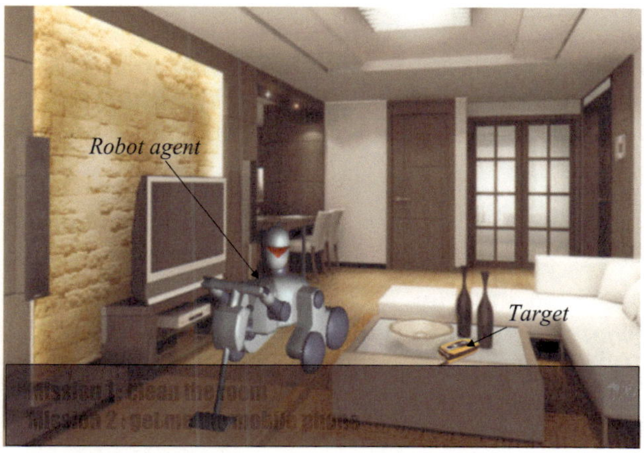

Fig. 1. Recognition for executing a task.

Suppose that a robot agent is commanded to clean the room as illustrated in Fig. 1. The robot has sensors such as ultrasonic, vision, and laser range finder. When executing the cleaning mission, it receives another order, that is, to bring the commander his/her mobile phone. Executing mission needs the classification of physical objects, which is to be kept or cleaned. At the same time, it should find the mobile phone. However, sensors that should recognize objects for cleaning are beyond their computational capacity. Thus, the robot stops cleaning, and then begins to find the mobile phone. First, it scans all objects using vision, sonar, etc., and then will try to compare the objects with the target mobile phone. It, however, cannot find the target despite every effort because the sensors cannot scan the target object, which is hidden by bottles and a dish. Consequently, the robot cannot even confirm the existence of the target in the room and it will conclude that the target is in another room. RFID is an attractive technology to supplement the limitation of robot faculty. The basic but powerful function of this technology is to identify the existence of the object. In the same mission, there is another robot with the RFID system and all objects have built-in tags. This tag gives to the robot the information about the property or characteristic of the object. To clean the room, the robot easily chooses the objects, which should be removed or kept. To execute the second mission, the robot searches the ID lists of objects, which have been obtained while cleaning. Through this list, the robot has already knows that there is the target mobile phone in the room. To find the target, this robot moves to the position where the target has been detected, and then scans nearby objects using its sensors. Hence, the robot can complete its task more easily after object identification. The robot, however, does not know where the target is because the sensors cannot detect the target, which is hidden by other objects, and the mission is not completed. Obviously, the target is there, but something is often thought not to exist if it is not detected. Unfortunately, the RFID system also cannot present the solution for object recognition, because it is not enough to identify and confirm the existence of the object. In order to overcome this limitation, the authors have developed the advanced RFID system based on 3D tag. The proposed RFID system can identify the object, and estimate the object's position and orientation. Because of these characteristics, the robot with the automated systems can recognize objects easily and rapidly. Naturally, this recognition mechanism can also simplify other robot processes such as localization, navigation, and manipulation. The authors have developed the algorithm and application of such processes based on the proposed RFID system. In this chapter, we mainly focus on the fundamental principle and algorithm of this system. In Section 2, the basic idea of this system is addressed. Sections 3 and 4 describe the structure of the system and the 3D tag, which characterize the system. The algorithms for estimating the position and orientation of the target object are given in Section 5, and experimental results are briefly presented in Section 6.

2. Problem statement and idea

An RFID system has been used for the artificial landmark to obtain the geographical information for navigation and localization of the mobile robot (Yamano et al, 2004; Kulyukin et al, 2004; Ni et al, 2003; Kubitz et al, 1997; Tsukiyama, 2002; Hahnel et al, 2004; Ruff & Hession, 2001). Using this system, several researchers have developed the application of RFID system to support object recognition and manipulation (Boukraa & Ando, 2002; Mae et al, 2000; Chong et al, 2004). Their studies are very useful and practical, in that, the RFID system supplements the limitation of the robot's capability. Most of previous RFID systems have omni-directional read range. One RFID tag is embedded into one object, and this tag is detected by the antenna of the RFID system. This system informs the robot of the existence of an object. To recognize objects, however, a robot considerably depends on other sensors though it is equipped with the RFID system. For example, let us suppose the missions described above. There are objects with new type tags called a 3D tag. This tag provides the robot with the position and orientation of the object. Since the orientations of objects and tags are the same, the robot can easily estimate the pose of the object when it detects the tags. To manipulate the target mobile phone, the robot directly approaches the target without the unnecessary scanning and sensing of other objects. The robot already knows the position and orientation of the target and some other objects (bottles and dish around the target) as shown in Fig. 2(a).

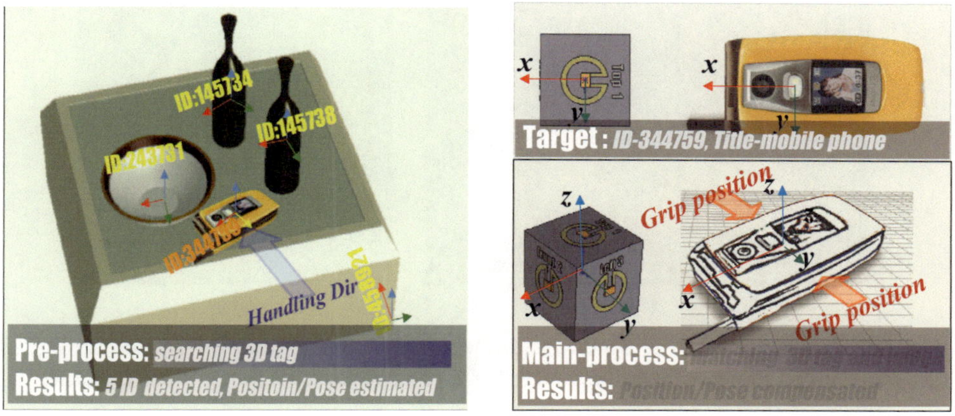

(a) (b)

Fig. 2. The proposed Idea. (a) Preliminary process for object recognition. (b) Main process for object recognition.

The robot determines the location and direction for moving so that it can handle the target without obstructing by obstacles. The robot has estimated the position and pose of the object from the information provided by the 3D tag, and already known the characteristics of the object such as the shape and size. Thus, it is simple to compare and match the real target and the model. The robot begins to manipulate the target considering the corrected position and orientation, which are obtained from the match process, and deliberating the shape, weight, material, and size of the target. For the materialization of this concept, the 3D tag and the 3D RFID system were developed.

3. Characteristics of the 3D RFID system

3.1 Recognition module

The DRP I (Dynamically Reconfigurable Personal Robot I) which is equipped with the RFID system is modularized, as shown in Fig. 3(a) (Roh et al, 2004). It is composed of four modules. Each module is functionally distributed and reconfigured. The proposed RFID system is a part of the module called a recognition module. The major function of this module together with the sensor module is to recognize, and judge from the existing state of objects. For more detail recognition, this module has also a vision and voice recognition system. These systems are physically synchronized as shown in Fig. 3(b). They simultaneously try to scan the object for the purpose of recognition. Especially, with the RFID system, the robot can identify and confirm targets easily, so that other recognition systems of the recognition module and the sensor module can perform their functions quickly.

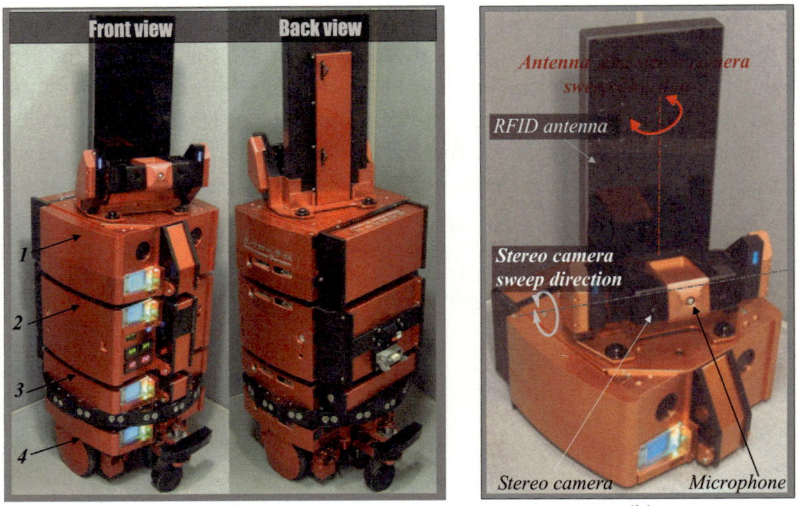

(a) (b)

Fig. 3. 3D RFID-based robot. (a) Modularization of robot. 1, 2, 3, and 4 represent recognition, arbiter, sensor, and mobile module, respectively. (b) Structure of recognition module.

3.2 Sensing range of the 3D RFID system

The proposed RFID system is composed of an antenna and reader to detect the 3D tag (Roh et al, 2006). The antenna can be swept by the actuator in Fig. 3(b), and it has the uni-directional read range as shown in Fig. 4. These features are used for estimating the position and orientation of the 3D tag. The 3D tag is composed of several passive tags, which have the dipole antennas as shown in Fig. 4(b). It has no self-power source, so it has to obtain its required power through electromagnetic induction, especially magnetic field.

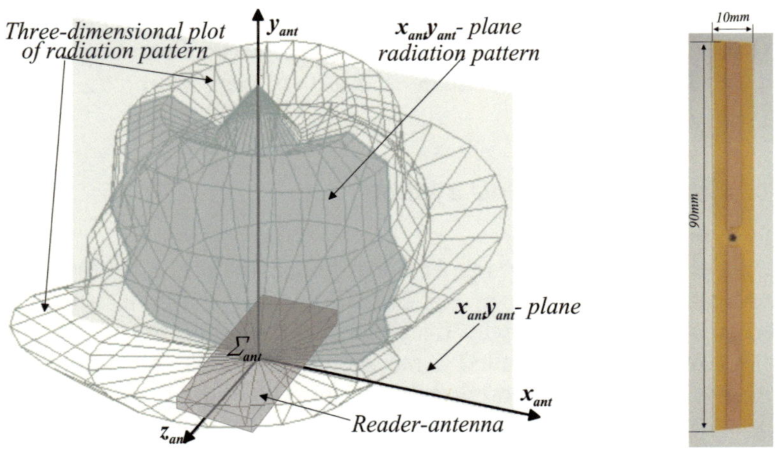

(a) (b)

Fig. 4. RFID antenna and tag. (a) Simulated sensing range of the RFID antenna. (b)Tag with a small dipole antenna

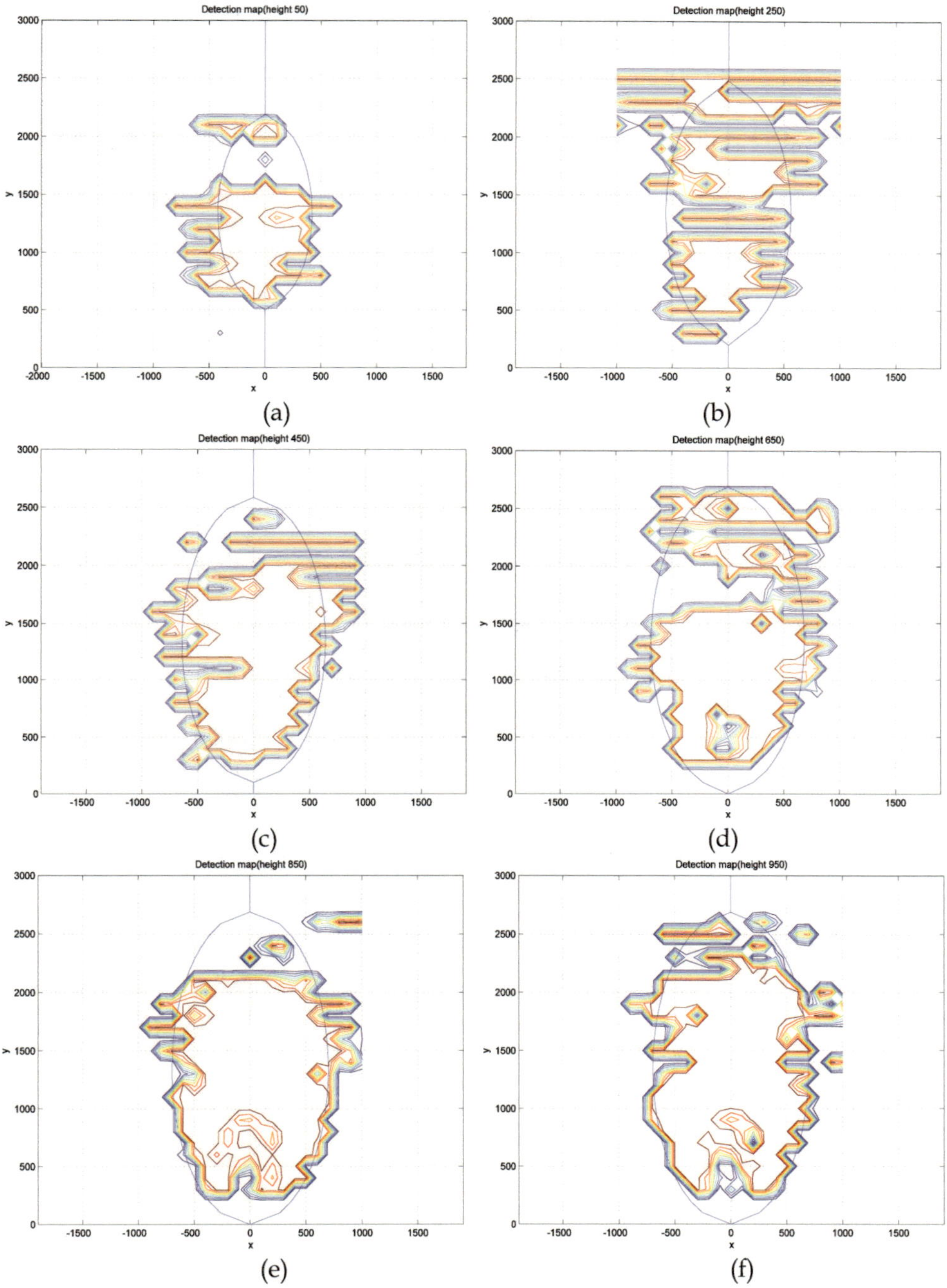

Fig. 5. Read range pattern of the RFID antenna according to the height of the tag. (a) z=50mm. (b) z=250mm. (c) z=450mm. (d) z=650mm. (e) z=850mm. (f). z=950mmmm.

To induce sufficient operating voltage, the tag must be placed within the range of the detectable angle and distance in the magnetic field of the electromagnetic wave from the antenna. If the tag is out of the range of detectable angle, it cannot be sensed by the reader even if it is placed within the detectable distance. The specific detectable angle and distance of the RFID system can provide the robot with more definite information. The authors measured the sensing range of the RFID antenna to the tag when the 3D RFID system was combined into the DRP I. To measure the range with respect to height and distance, the test space of (2000*mm*× 3000*mm*×1000*mm*) and 6-axis manipulator were used. In this experiment, the concept of a detection rate was introduced. The detection rate in this chapter means how many times an RFID-reader detects and counts a single tag per 1 second; the detection rate is 100% if the reader detects and counts the tag 20 times per 1 second. Fig. 5 shows sensing range pattern of the reader-antenna with respect to the tag height, depending on the detection rate. The experimental sensing range formed the geometry similar to the simulated sensing range. Based on the experimental results, the detectable range of our system was modeled as an ellipsoid as shown in Fig. 6.

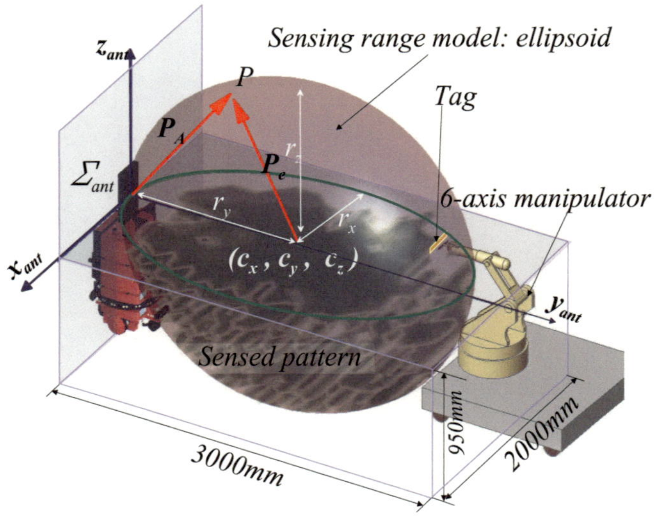

Fig. 6. Sensing model.

The equation of the ellipsoid is Eq.1. The reference frame is denoted by Σ_e. (c_x,c_y,c_z), which is the center of the ellipsoid and r_x, r_y, and r_z are the radii of the ellipsoid, respectively.

$$\frac{(x_e-c_x)^2}{r_x^{\,2}}+\frac{(y_e-c_y)^2}{r_y^{\,2}}+\frac{(z_e-c_z)^2}{r_z^{\,2}}=1, \tag{1}$$

where $c_x=0$, $c_y=1350$, $c_z=0$, $r_x=700$, $r_y=1350$, and $r_z=1000$. This equation can be rewritten suitable for the sweepable antenna. In spherical coordinates, Eq. (1) can be rewritten by

$$\mathbf{P}_e(\varphi_e,\phi_e)=\begin{bmatrix} |\mathbf{P}_{x_e}|\cdot\cos\varphi_e\sin\phi_e \\ |\mathbf{P}_{y_e}|\cdot\sin\varphi_e\sin\phi_e \\ |\mathbf{P}_{z_e}|\cdot\cos\phi_e \end{bmatrix}, \tag{2}$$

where $\mathbf{p}_e$ is the position vector inside the ellipsoid. $\mathbf{p}_{x_e}, \mathbf{p}_{y_e}$, and $\mathbf{p}_{z_e}$ are x_e, y_e and z_e vectors of $\mathbf{p}_e$, respectively ($|\mathbf{p}_{x_e}| \le r_x$, $|\mathbf{p}_{y_e}| \le r_y$, $|\mathbf{p}_{z_e}| \le r_z$). φ_e is the azimuthal angle in the $x_e y_e$ plane from the x_e-axis, and φ_e is the polar angle from the z_e-axis ($0 \le \varphi_e \le 2\pi$, $0 \le \phi_e \le \pi$). The position vector $\mathbf{p}_A$ inside the ellipsoid, of which the reference frame is Σ_{ant}, can be written by

$$\mathbf{P}_A = \mathbf{C}_A + \mathbf{P}_e(\varphi_e, \phi_e), \tag{3}$$

where $\mathbf{C}_A$ is the vector from Σ_{ant} to Σ_e.

3.3 Detectable tag orientation

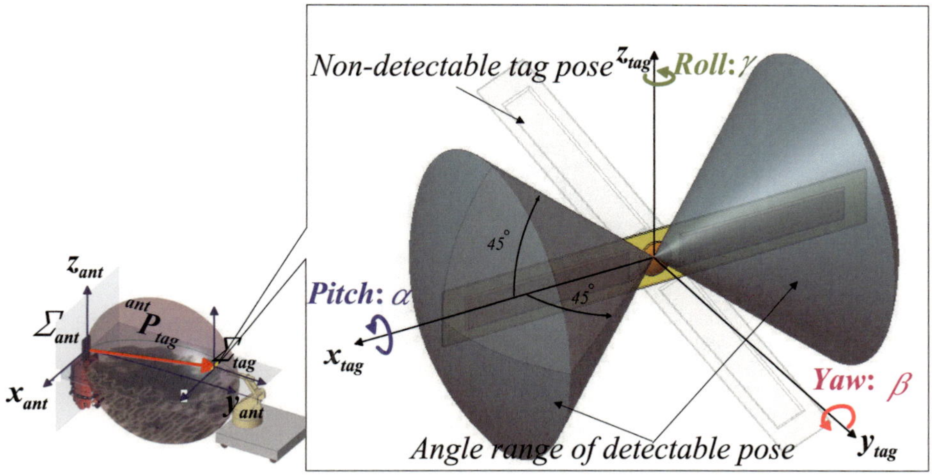

Fig. 7. Detectable tag orientation.

The detection rate of the tag changes with the angle between the tag in the sensing range and the antenna. Let us assume that the relation between the coordinate frames Σ_{ant} and Σ_{tag} is given by the vector ${}^{ant}\mathbf{P}_{tag}$ without rotation in Fig. 7 (Σ_{ant} and Σ_{tag} are the coordinate frames of the antenna and the tag, respectively). Each of the detection rate in the tag rotating around x_{tag}, y_{tag} and z_{tag} is measured as plotted in Fig. 8, so that the detectable angle range of tag is obtained as follows:

$$\begin{aligned}&[-180^\circ \le \alpha \le 180^\circ],\\&[-45^\circ \le \beta \le 45^\circ, 135^\circ \le \beta \le 225^\circ],\\&[-45^\circ \le \gamma \le 45^\circ, 135^\circ \le \gamma \le 225^\circ],\end{aligned} \tag{4}$$

where α, β and γ are the angles of pitch, roll, and yaw, respectively. The results mean that the detection rate can estimate the pose of a tag within the limits. Furthermore, the orientation of the object with the built-in tags can be estimated if the change of the detection rate in accordance with the axes can be properly combined. The 3D tag is developed based on this characteristic.

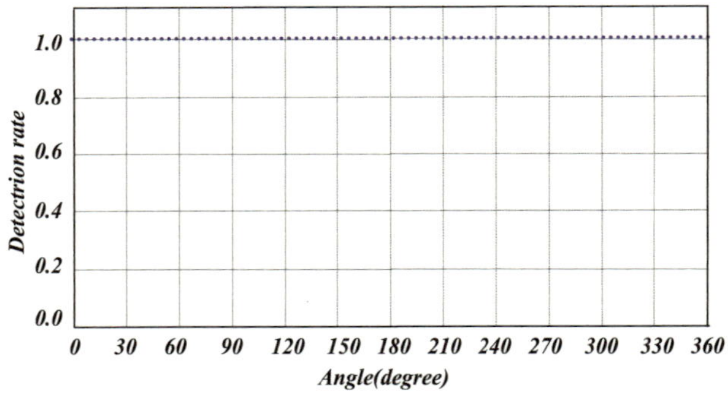

(a)

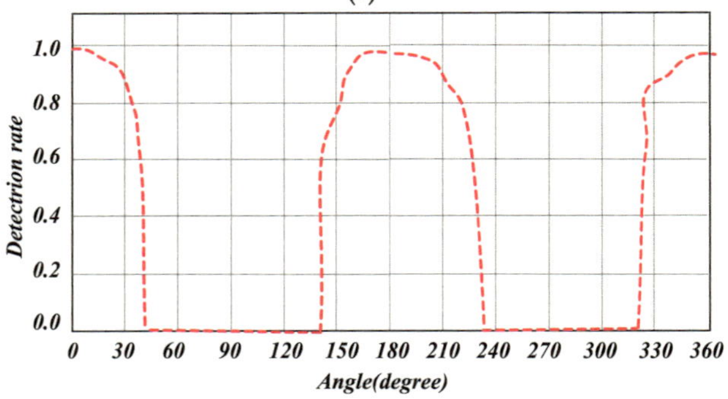

(b)

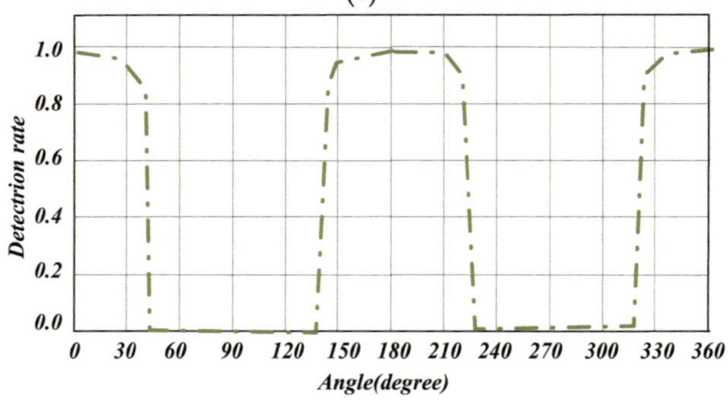

(c)

Fig. 8. Detectable angle range of a tag. (a) Pitch. (b) Yaw. (c) Roll.

4. 3D tag: union tag

The 3D tag is comprised of several tags called tag units. As illustrated in Fig. 9, the tag units are attached to the six edges of a cube. The surface of the cube is covered with the radio shielding material. This shield limits the pitch angle for the detection of the unit tag, while it does not limit the roll and yaw angles. As shown in Fig. 9(b), the antenna, which can detect one tag unit U_{TF} of six units, should be placed within the angle range of 180°; the other tag units have the same ranges 180° similarly to U_{TF} . This detection of U_{TF} means that the *y*-axis of Σ_{ant} (hereinafter referred to antenna direction) faces the side of the top or front or both; the letters *T* and *F* of U_{TF} denote top and front. Similarly, U_{FL} , U_{LT} , U_{DB} , U_{BR} , and U_{RD} correspond to the front or left, left or top, down or back, back or right, and right or down, respectively, as shown in Fig. 9(c).

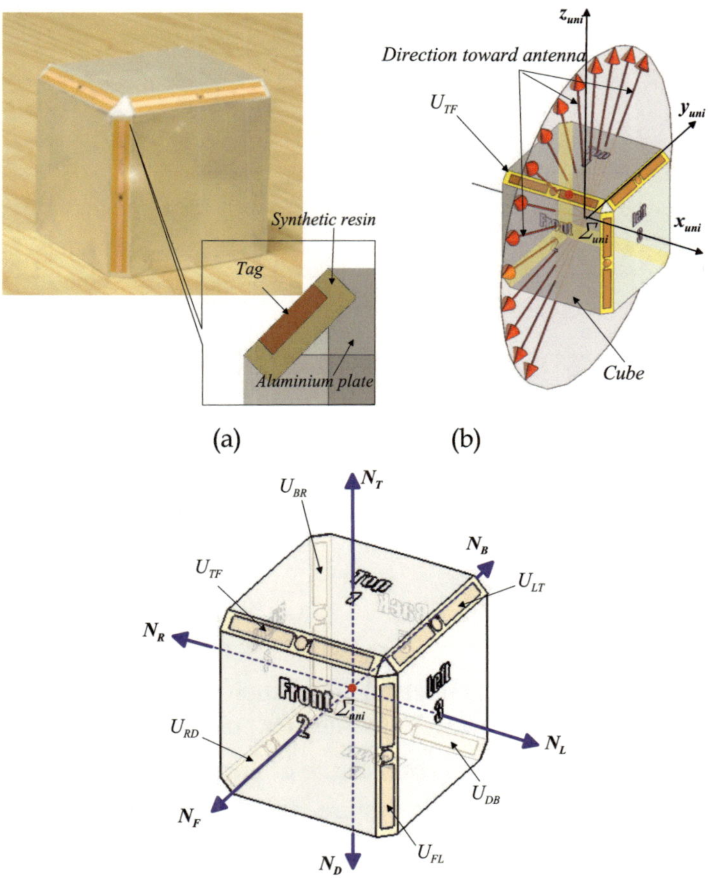

(c)

Fig. 9. Union tag. (a) Prototype of a union tag. (b) Antenna direction capable of detecting U_{TF}. (c) Notation of a union tag.

The proposed 3D tag composed of the tag units is named a union tag, and its detection angle range is characterized by the tag units. With this detectable angle range, the orientation of the union tag can be estimated, and thus the orientation of the object fitted with the union tag can be estimated because the union tag is aligned with the orientation of the object. As shown in Fig. 10(a), an ordinary object has four postures per one face when the face is rotated in 90-degree increments. In this case, there are 24 kinds of poses of the object .The classification like this is useful in that a human being frequently understands the pose of an object with the base of 90-degree such as top, bottom, front, back, left and right. On the other hand, the pose of the object with the built-in union tag is divided into twenty-four classes from a different standpoint. For the tag unit U_{TF}, four kinds of object pose allows U_{TF} of the union tag to be detected, as shown in Fig. 10(b).

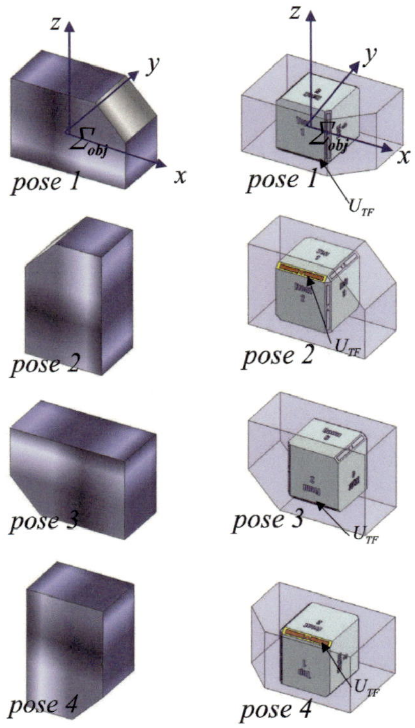

(a) (b)

Detected $U_{\{code\ A\ code\ B\}}$		Determined pose	
Step 1	Step 2	Front side	Left side
U_{TF}	U_{FL}	N_T	N_L
U_{TF}	U_{LT}	N_F	N_L
U_{TF}	U_{RD}	N_F	N_R
U_{TF}	U_{BR}	N_T	N_R
U_{FL}	U_{TF}	N_L	N_T
U_{FL}	U_{LT}	N_F	N_T
U_{FL}	U_{RD}	N_F	N_D
U_{FL}	U_{DB}	N_L	N_D
U_{LT}	U_{TF}	N_L	N_F
U_{LT}	U_{FL}	N_T	N_F
U_{LT}	U_{BR}	N_T	N_B
U_{LT}	U_{DB}	N_L	N_B
U_{RD}	U_{TF}	N_R	N_F
U_{RD}	U_{FL}	N_D	N_F
U_{RD}	U_{BR}	N_D	N_B
U_{RD}	U_{DB}	N_R	N_B
U_{BR}	U_{TF}	N_R	N_T
U_{BR}	U_{LT}	N_B	N_T
U_{BR}	U_{RD}	N_B	N_D
U_{BR}	U_{DB}	N_R	N_D
U_{DB}	U_{FL}	N_D	N_L
U_{DB}	U_{LT}	N_B	N_L
U_{DB}	U_{RD}	N_B	N_R
U_{DB}	U_{BR}	N_D	N_R

(c)

Translated No.		Translated No.	
u_{i1} u_{i2}	u_{j1} u_{j2}	n_i	n_j
1 2	2 3	1	3
1 2	3 1	2	3
1 2	4 6	2	4
1 2	5 4	1	4
2 3	1 2	3	1
2 3	3 1	2	1
2 3	4 6	2	6
2 3	6 5	3	6
3 1	1 2	3	2
3 1	2 3	1	2
3 1	5 4	1	5
3 1	6 5	3	5
4 6	1 2	4	2
4 6	2 3	6	2
4 6	5 4	6	5
4 6	6 5	4	5
5 4	1 2	4	1
5 4	3 1	5	1
5 4	4 6	5	6
5 4	6 5	4	6
6 5	2 3	6	3
6 5	3 1	5	3
6 5	4 6	5	4
6 5	5 4	6	4

(d)

Fig. 10. Pose relation of 3D tag and object. (a) Orientation of normal object. (b) Object with built-in 3D tag. (c) Pose estimation. (d) Translation for pose estimation.

This detection of U_{TF} means the detection of one of the object's two faces: the front side or top side of the object. Owing to this feature, the orientation of the object cannot be determined by the detection of one tag unit. Two tag units at the least should be detected so that one of the twenty-four poses can be determined. The most effective method is to detect two neighboring tag units. Let us assume that the antenna of the robot detects the object from $-y_{uni}$ axis direction when the object is fixed, as shown in Fig. 9(b), and then it detects from x_{uni} axis direction. 3D RFID system reads U_{TF} and U_{LT} in order. This means that 3D RFID system detects the front side denoted by N_F, and then the left side by N_L. N_F and N_L also denote the direction vectors, which are normal to the faces of the front and left, respectively (the subscripts F and L denote the front and left). The detection case by this order is only one. There are a total of 24 cases like this, as shown in Fig. 10(c) and each case determines the pose of the object. To generalize these cases, the authors developed an algorithm. We defined that U_i is the first detected tag unit, and U_j the second where $U_i=(i1,i2)$ and $U_j=(j1,j2)$. T, F, L, R, B, and D are denoted by 1, 2, 3, 4, 5 and 6, respectively. $[n_i,n_j]$, which means the first and second face detected, is shown in Fig. 10(d). The equations for the relations of these numbers are found through observation, and the relations can be written by

$$[n_i, n_j] = \begin{cases} [\min(U_i)\min(U_j)] & \text{if } S_{MM} \le 7 \text{ or } S_{MM} \ge 9 \\ [\max(U_i)\max(U_j)] & \text{if } S_{mm} = 7 \\ [\max(U_i)\min(U_j)] & \text{if } S_{mM} = 7 \\ [\min(U_i)\max(U_j)] & \text{if } S_{Mm} = 7 \end{cases} . \quad (5)$$

where S_{MM} , S_{mm} , S_{mM} ,and S_{Mm} are as shown below.

$$\begin{aligned} S_{MM} &= \max(U_i) + \max(U_j) \\ S_{mm} &= \min(U_i) + \min(U_j) \\ S_{mM} &= \min(U_i) + \max(U_j) \\ S_{Mm} &= \max(U_i) + \min(U_j) \end{aligned} . \quad (6)$$

Using Eqs. (5) and (6), the orientation of the object with the built-in 3D tag can be determined.

5. Position and orientation estimation of object

5.1 Position estimation

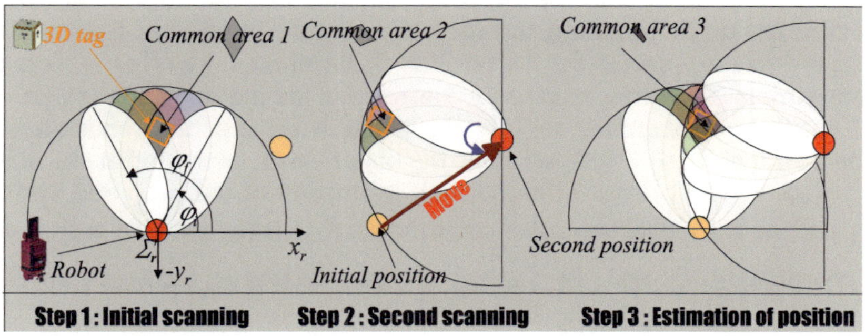

(a)

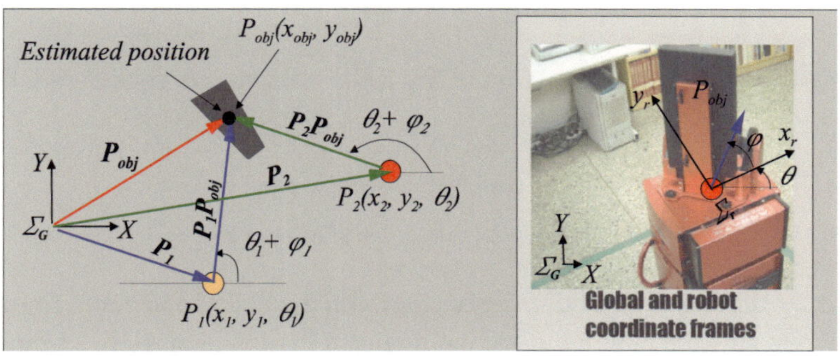

(b)

Fig. 11. Algorithm for position estimation. (a) Procedure for position estimation.
(b) Relations between positions of target and robot.

To complete a mission for manipulating a target, the robot has to know the position of the target. Fig. 11 shows how the 3D RFID system is used. The proposed system can rotate the RFID antenna to scan for finding objects, and it can estimate the position of the 3D tag. As shown in Fig. 11(a), the robot rotates the antenna from the right-hand to the left-hand. The 3D tag is initially detected when the direction angle of the antenna is φ_i, and the tag is detected until the angle is φ_f. Thus, it can be said that the tag will be placed in the common area 1 through this scanning procedure. In the next step, the robot moves to the second position, and then repeats the procedure. From steps 1 and 2, the position of the 3D tag is one point in the common area 3. Object detection by scanning has been used by the other robots, which have the sensors with a detection range. However, the results obtained from scanning of these sensors differ from the results of the 3D RFID system. For example, the main purpose of the procedure using ultrasonic sensors is to give the robot not the position but geometry data of the environments such as walls and obstacles (the position is already determined because the distance and direction of the object have been measured when detected), or the procedure is performed to compensate the error data when the sensor does not detect the object because the object is a sharp edgy shape, etc. The ultrasonic sensor or laser range finder is very useful for sensing unspecified objects such as mapping environments or avoiding obstacles, but these sensors cannot find and sense the specified target because the sensor cannot identify target. On the other hand, the robot with the 3D RFID system can identify the target even if the scanning procedure by the 3D RFID system cannot exactly determine the position of the target because the RFID antenna itself does not have capability to read distance between the antenna and the tag. Thus, this scanning should be used not for determining, but for estimating the position of the target. The steps in Fig. 11(a) are the pre-process for the position determination by other sensors capable of distance measuring. The other sensors of the robot with the 3D RFID system can easily detect the object and determine rapidly the position because the robot already knows roughly the position of the object, whether the object exists or not, what the object is, and how it is posed. Fig. 11(b) shows the position estimation of the proposed system. In this figure, Σ_G and Σ_r are global and robot coordinates. $\mathbf{P}_{obj}$ is the position vector of the object in global coordinates. $\mathbf{P}_1$ and $\mathbf{P}_2$ are the position vectors of the robot at $\mathbf{P}_1$ and $\mathbf{P}_2$, respectively. $\mathbf{P}_{obj}$ can be written by

$$\mathbf{P}_{obj} = \mathbf{P}_1 + \mathbf{P}_1\mathbf{P}_{obj} \ . \tag{7}$$

When the robot's pose and the direction of the RFID antenna are considered, Eq. (7) can be rewritten by

$$\mathbf{P}_{obj} = \begin{bmatrix} x_1 + |\mathbf{P}_1\mathbf{P}_{obj}| \cdot \cos(\theta_1 + \varphi_1) \\ y_1 + |\mathbf{P}_1\mathbf{P}_{obj}| \cdot \sin(\theta_1 + \varphi_1) \end{bmatrix}, \tag{8}$$

where θ_1 and θ_2 are the angular differences between the global and robot frames at $\mathbf{P}_1$ and $\mathbf{P}_2$. φ_1, which means the antenna direction angle, is the angle of $\mathbf{P}_1\mathbf{P}_{obj}$ from the x_r-axis, and φ_2 is the angle of $\mathbf{P}_2\mathbf{P}_{obj}$ from the x_r-axis. $|\mathbf{P}_1\mathbf{P}_{obj}|$ is rewritten by

$$|\mathbf{P}_1\mathbf{P}_{obj}| = |\mathbf{P}_1\mathbf{P}_2| \cdot \frac{\sin(\theta_2 + \varphi_2 - \psi)}{\sin(\theta_2 + \varphi_2 - \theta_1 - \varphi_1)}, \tag{9}$$

where $\psi = \tan^{-1}\frac{y_2 - y_1}{x_2 - x_1}$ and φ_1 and φ_2 are obtained from $(\varphi_i + \varphi_f)/2$ at $\mathbf{P}_1$ and $\mathbf{P}_2$, respectively. Thus, we finally obtain

$$\mathbf{P}_{obj} = \begin{bmatrix} x_1 + |\mathbf{P}_1\mathbf{P}_2| \cdot \frac{\sin(\theta_2 + \varphi_2 - \psi)}{\sin(\theta_2 + \varphi_2 - \theta_1 - \varphi_1)} \cdot \cos(\theta_1 + \varphi_1) \\ y_1 + |\mathbf{P}_1\mathbf{P}_2| \cdot \frac{\sin(\theta_2 + \varphi_2 - \psi)}{\sin(\theta_2 + \varphi_2 - \theta_1 - \varphi_1)} \cdot \sin(\theta_1 + \varphi_1) \end{bmatrix}. \tag{10}$$

Eq. (10) is obtained from the relations of the steps 1 and 2. Theoretically, the position can be estimated with only step 1 because the common area 1 is determined with the two ellipsoids computed. The steps 1 and 2, however, are required to execute the orientation estimation, which will be presented in the next section. Since the orientation estimation is always accompanied with the position estimation and the position estimation by the two steps is more accurate, the position estimation using these steps is reasonable.

5.2 Orientation estimation

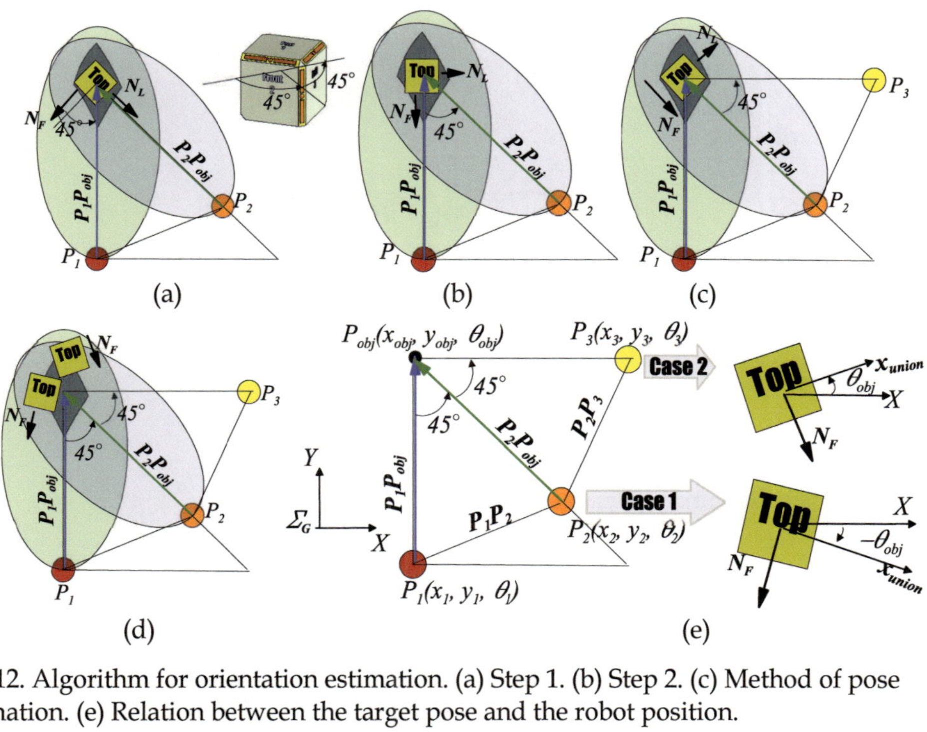

Fig. 12. Algorithm for orientation estimation. (a) Step 1. (b) Step 2. (c) Method of pose estimation. (e) Relation between the target pose and the robot position.

The proposed method for estimating the orientation of the object is based on the characteristic of the 3D tag: the 3D tag's detectable angle is limited. To define the pose of the target, two sides of the 3D tag should be detected as mentioned in the section 4. If the robot detects the front side at the initial position P_1, and then detects the left side at the second position P_2, the 3D tag poses as shown from Fig. 12(a) to Fig. 12(c). To obtain the pose of the target, the robot must determine the second position. If the second position is not chosen properly, the left side cannot be detected because the antenna has a limited direction and sensing range for detecting the 3D tag. The determination of the second position depends on the detectable angle range of the tag unit of the 3D tag; the maximum is 45° and the minimum -45° from Eq. (4). When the tag poses as shown in Fig. 12(a), the angle between the vectors N_F denoting the front side and $-\mathbf{P}_1\mathbf{P}_{obj}$ should be less than 45° to detect the front side of the 3D tag. In case of Fig. 12(b), N_L and $-\mathbf{P}_2\mathbf{P}_{obj}$ should be more than 45° to detect the left side. Thus, it is reasonable that the angle between $\mathbf{P}_1\mathbf{P}_{obj}$ and $\mathbf{P}_2\mathbf{P}_{obj}$ is 45°. In addition, when the robot is placed at the second position P_2, the target should be within the sensing range of the antenna. Hence, the distance between the target position P_{obj} and P_2 should be chosen properly. This distance is substituted with that between the common area 1 and P_1. In the previous section, the common area 1 is used to obtain φ_1, but the position of the area is not considered. In other words, this common area is used as a guide for the choice of P_2 to estimate the orientation even if the area is not used for the position estimation. Though the robot moves to P_2, the left side of the 3D tag cannot be detected as shown in Fig. 12(c). In such a case, the robot moves to P_3 and detects the left side. As shown in Fig. 12(e), the positions of P_1, P_2, P_3, and $\mathrm{P}_{\mathrm{obj}}$ are solved as follows.

$\mathbf{P}_2$ can be written by

$$\mathbf{P}_2 = \mathbf{P}_1 + \mathbf{P}_1\mathbf{P}_2, \tag{11}$$

where $\mathbf{P}_1\mathbf{P}_2$ is written by

$$\mathbf{P}_1\mathbf{P}_2 = \begin{bmatrix} \sqrt{2-\sqrt{2}}\,|\mathbf{P}_1\mathbf{P}_{obj}|\cdot\cos 22.5^\circ \\ \sqrt{2-\sqrt{2}}\,|\mathbf{P}_1\mathbf{P}_{obj}|\cdot\sin 22.5^\circ \end{bmatrix} = \begin{bmatrix} \dfrac{|\mathbf{P}_1\mathbf{P}_{obj}|}{\sqrt{2}} \\ 1-\dfrac{|\mathbf{P}_1\mathbf{P}_{obj}|}{\sqrt{2}} \end{bmatrix}. \tag{12}$$

Thus, $\mathbf{P}_2$ becomes Eq. (13).

$$\mathbf{P}_2 = \begin{bmatrix} x_1 + \dfrac{|\mathbf{P}_1\mathbf{P}_{obj}|}{\sqrt{2}} \\ y_1 + 1 - \dfrac{|\mathbf{P}_1\mathbf{P}_{obj}|}{\sqrt{2}} \end{bmatrix}. \tag{13}$$

$\mathbf{P}_3$ can be written by Eq. (14)

$$\mathbf{P}_3 = \mathbf{P}_1 + \mathbf{P}_1\mathbf{P}_3, \tag{14}$$

where $\mathbf{P}_1\mathbf{P}_3$ is given by

$$\mathbf{P}_1\mathbf{P}_3 = \begin{bmatrix} |\mathbf{P}_1\mathbf{P}_{obj}| \\ |\mathbf{P}_1\mathbf{P}_{obj}| \end{bmatrix}. \tag{15}$$

Thus, $\mathbf{P}_3$ becomes Eq. (16).

$$\mathbf{P}_3 = \begin{bmatrix} x_1 + |\mathbf{P}_3\mathbf{P}_{obj}| \\ y_1 + |\mathbf{P}_3\mathbf{P}_{obj}| \end{bmatrix}. \tag{16}$$

Consequently, we finally obtain

$$\mathbf{P}_{obj} = \begin{bmatrix} x_1 + |\mathbf{P}_1\mathbf{P}_2| \cdot \frac{\sin(\theta_1 + \varphi_1 + 22.5^\circ)}{\sqrt{2}} \cdot \cos(\theta_1 + \varphi_1) \\ y_1 + |\mathbf{P}_1\mathbf{P}_2| \cdot \frac{\sin(\theta_1 + \varphi_1 + 22.5^\circ)}{\sqrt{2}} \cdot \sin(\theta_1 + \varphi_1) \\ \theta_{obj} \end{bmatrix}, \tag{17}$$

where $|\mathbf{P}_1\mathbf{P}_2| = \sec 22.5^\circ |x_2 - x_1|$. The orientation θ_{obj} of the target can be obtained by the following method. If the robot detects the front side of the 3D tag at the position P_1 and detects the left side of the 3D tag at P_2, as shown in the case 1 of Fig. 12(e), the orientation θ_{obj} of the target can be written by

$$-45^\circ \le \theta_{obj} \le 0^\circ. \tag{18}$$

If the robot detects the front side of the 3D tag at the position *P*1, does not detect the left side of the 3D tag at P_2, and then detects the left side of the 3D tag at P_3 , as shown in the case 2 of Fig. 12(e), the orientation θ_{obj} of the target can be written by

$$0^\circ \le \theta_{obj} \le 45^\circ. \tag{19}$$

This result is satisfied wherever the target is placed in the common area 1 and how it is posed, as shown in Fig. 12(d).

6. Experiments

Figure 13 shows the procedure of the experiment based on the proposed algorithm. In the test space, the target is placed at the random position P_{target} (1040, 900, 0). The target is the box with the built-in 3D tag. The robot is initially located on P_r (500, 500, 0) in the global map. Using only the 3D RFID system, the robot DRP I searches and detects the target, and then, the robot estimates the position and orientation of the target. This experiment is repeated several times. Figure 14 shows the results of the position estimation. The results of the orientation estimation are always regular ($0^\circ \le \theta_{obj} \le 45^\circ$). Currently, the orientation of the object by using the 3D RFID system is estimated when the face is rotated in 45°

increments. This is because the detection rate cannot be matched to the change of continuous orientation of the object, as show in Figs. 8(b) and 8(c). If the RFID tag and antenna suitable for the stable detection rate should be developed, the object's orientation estimation using the 3D tag can be extended to various automation systems.

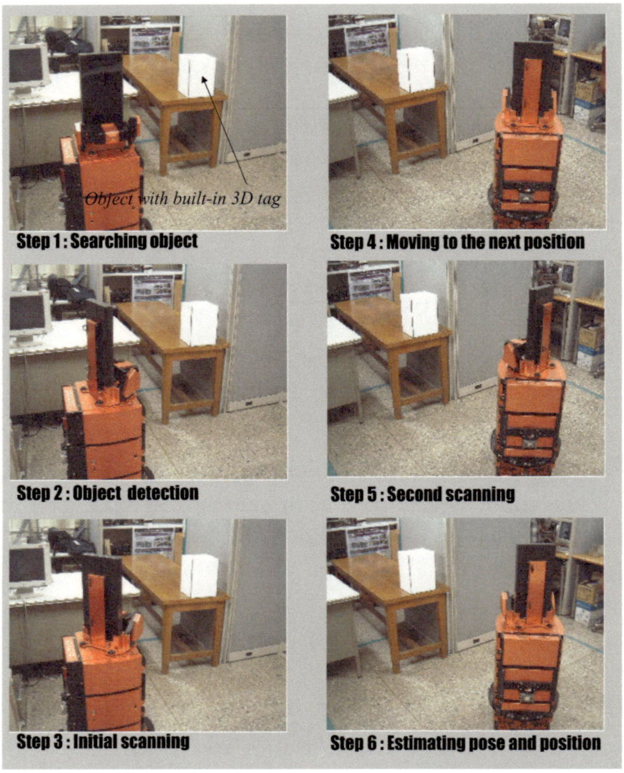

Fig. 13. Experiment of DRP 1.

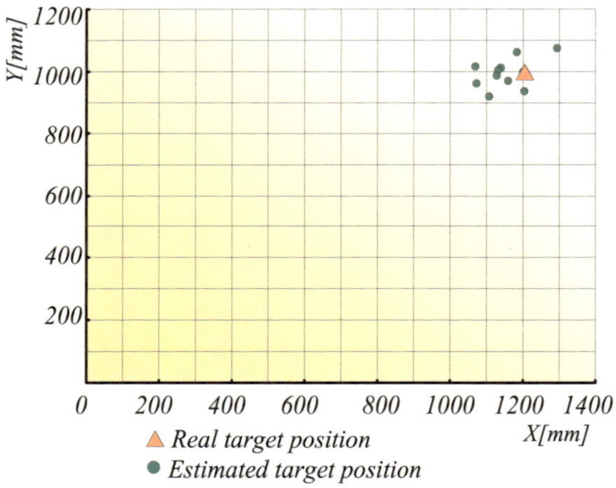

Fig. 14. Location estimation by only the RFID system.

7. Conclusions

Up to now, object recognition in robotics has been typically done by vision, ultrasonic sensors, laser ranger finders etc. Recently, RFID has emerged as a promising technology that can strengthen object recognition. In this chapter, the 3D RFID system and the 3D tag were presented. The proposed RFID system can determine if an object as well as other tags exists, and also can estimate the orientation and position of the object. This feature considerably reduces the dependence of the robot on other sensors such as vision systems required for object recognition. With the fast growth of RFID technology, the field of robotics will benefit greatly, and our research will support the investigation of its use in practical applications.

8. Acknowledgment

This work was supported by the Ministry of Knowledge Economy (MKE) and Korea Industrial Technology Foundation (KOTEF) through the Human Resource Training Project for Strategic Technology.

9. References

Weiss, I. & Ray, M. (2001). Model-Based Recognition of 3D Objects from Single Images, *IEEE Trans. Pattern Analysis, Machine Intelligence,* vol. 23, pp. 116-128.

Yamano, K.; Tanaka, K.; Hirayama, M.; Kondo, E.; Kimuro, Y. & Matsumoto, M. (2004). Self-localization of Mobile Robots with RFID System by using Support Vector Machine, in *Proc. IEEE/RSJ Int. Conf. Intelligent Robots, Systems,* pp. 3756-3761.

Kulyukin, V.; Gharpure, C.; Nicholson, J. & Pavithran, S. (2004). RFID in Robot-Assisted Indoor Navigation for the Visually Impaired, in *Proc. IEEE/RSJ Int. Conf. Intelligent Robots, Systems,* pp. 1979-1984.

Ni, L. M.; Liu, Y.; Lau, Y. C. & Patil, A. P. (2003). LANDMARC: Indoor Location Sensing Using Active RFID, in *Proc. IEEE Int. Conf. Pervasive Computing and Communications,* pp. 407-415.

Kubitz, O.; Berger, M. O.; Perlick, M. & Dumoulin, R. (1997). Application of radio frequency identification devices to support navigation of autonomous mobile robots, in *Proc. IEEE Int. Conf. Vehicular Technology,* vol. 1, pp. 126-130.

Tsukiyama, T. (2002). Global Navigation System with RFID Tags, in *Proc. SPIE,* vol. 4573, pp. 256-264.

Hahnel, D.; Burgard, W.; Fox, D.; Fishkin, K. & Philipose, M. (2004). Mapping and Localization with RFID Technology, in *Proc.IEEE Int. Conf. Robotics, Automation,* vol. 1, pp. 1015-1020.

Ruff, T. M. & Hession-Kunz, D. (2001). Application of radio-frequency identification systems to collision avoidance in metal/nonmetal mines, *IEEE Trans. Industry Applications,* vol. 37, pp. 112-116.

Boukraa, M. & Ando, S. (2002). Tag-based vision: assisting 3D scene analysis with radio-frequency tags, in *Proc. IEEE Int.Conf. Information Fusion,* pp. 412-418.

Mae, Y.; Umetani, T.; Arai, T. & Inoue, E. (2000). Object recognition using appearance models accumulated into environment, in *Proc. IEEE Int. Conf. Pattern Recognition,* vol. 4, pp. 845-848.

Chong, N. Y.; Hongu, H.; Miyazaki, M.; Takemura, K.; Ohara, K.; Ohba, K.; Hirai, S. & Tanie, K. (2004). Robots on Self-Organizing Knowledge Networks, in *Proc. IEEE Int. Conf. Robotics, Automation*, pp. 3494-3499.

Roh, S. G.; Park, K. H.; Yang, K. W.; Park, J. H.; Kim, H. S.; Lee, H. G. & Choi, H. R. (2004). Development of Dynamically Reconfigurable Personal Robot, in *Proc. IEEE Int. Conf. Robotics, Automation*, pp. 4023-4028.

Roh, S. G.; Lee, Y. H. & Choi, H. R. (2006). in *Proc. IEEE/RSJ Int. Conf. Intelligent Robots and Systems,* pp. 5725-5730.

Roh, S. G. & Choi, H. R. (2009). 3D Tag-based RFID System for Recognition of Object, *IEEE Trans. Automation Science and Engineering,* (in press).

23

RFID System Architecture Reconsidered

Dirk Henrici, Aneta Kabzeva and Paul Müller
University of Kaiserslautern
Germany

1. Introduction

The RFID technology is already of high commercial relevance. It breaks into new application areas, and new markets are emerging. RFID becomes more and more an indispensable part of our everyday life. However, the technology also introduces security and privacy problems. Despite of the numerous research efforts, no satisfactory solutions for these issues have yet been found and widely implemented. For this reason, there are many people who take fright at RFID.

Today's RFID system architecture is carried over from the architecture used in other auto-id systems, chiefly optical barcode systems. As RFID introduces new functionalities and privacy risks, this classic architecture is no longer appropriate. For instance, the classic architecture fails to provide location privacy and self-determination for the affected users while being scalable and open. In this chapter, the problem is explained, the limitations in extending the classic architecture are outlined, and important aspects of a new architecture are sketched.

In the remainder of this first subchapter, an overview of the security and privacy goals and the main concepts for reaching them is provided. The requirements that RFID systems should fulfill are outlined in a separate section. The second subchapter introduces into the current RFID system architecture and the general direction of RFID security and privacy research. Subchapter 3 shows the practical deficiencies of the current architecture and illustrates, using an example, why incremental improvements and extensions lack to provide satisfactory solutions. Finally, considerations on how a completely new RFID architecture might look like are performed.

1.1 Security and privacy goals

There are five high-level issues of great importance for the RFID system security and the users' privacy [Henrici, D. (2008)]. Fig. 1 shows an overview on them. The remainder of this section provides a more detailed explanation of each goal.

Maintain data security: In many cases, RFID systems operate with privacy sensitive data that shall not become public. Such data may be some product information or even personal information. Security mechanisms for the prevention of illegal access to such data are one of the main challenges for RFID systems.

Prevent counterfeiting: With the change-over from barcodes to RFID, a better prevention of product counterfeiting is desired. Plagiarism in not only an economic issue but, e.g. in the case of drugs, can also be a mortal danger.

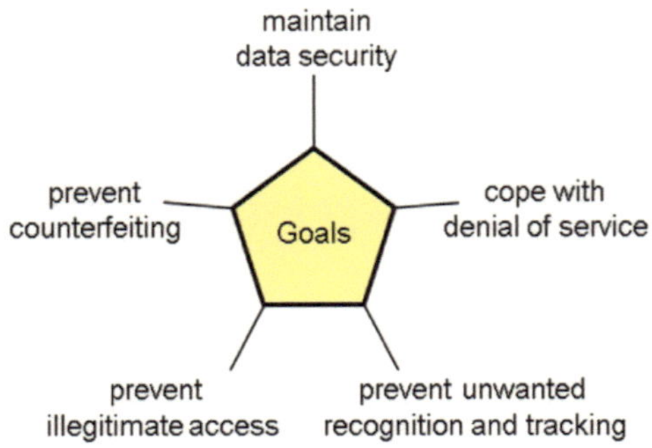

Fig. 1. Security and privacy goals for RFID systems [Henrici, D. (2008)]

Prevent illegitimate access: Reading an RFID tag, a reader creates a "read event" that is processed in the backend systems. It should be ensured that valid events cannot be generated by spoofed devices. This prevents, for example, that an attacker can fabricate events indicating a false location of an RFID tag. In general, illegitimate access to system components shall be effectively prevented.

Cope with denial of service: It is easy to put parts of an RFID system out of order. The reading of many kinds of RFID tags can be interfered with tinfoil, for example. Since a prevention of all denial of service attacks is not feasible, RFID systems should at least provide means for detection and recovery. RFID protocols should not introduce additional denial of service vulnerabilities.

Prevent unwanted recognition and tracking: Recognition and tracking of objects belong to the main purposes of RFID systems. Logistics applications rely on such functionality. However, from a privacy perspective, this is not always desired, especially when persons are involved. For this reason, a mechanism that allows people to decide when their RFID tags can be used for recognition/tracking and by whom is reasonable.

1.2 Reaching the security and privacy goals

In the previous section, five high-level security and privacy goals were introduced. This section explains the general approaches for reaching them.

Maintain data security: Data can be stored directly on RFID tags or in databases in backend systems. For achieving data security, it makes sense to store as few data as possible directly on tags. This means that an RFID tag should contain just a unique identifier which acts as a reference to other data stored in backend systems.

This approach offers many advantages. The RFID tags need little storage which keeps their cost low. Furthermore, potentially privacy sensitive data are not transferred between tags and readers on an insecure medium, and costly measures for protecting these data are not required. Instead, secure and flexible access control methods can be applied for accessing the data in the backend. There are also some other advantages like flexibility, interoperability, and increased speed of reading. Due to all the advantages, storing only identifiers directly on the tags and all other data in backend systems is the approach used in most RFID research. We assume the approach as basis in the following.

Prevent counterfeiting: Using the "track & trace" approach, it is possible to implement plausibility checks. By keeping a detailed object history and by storing the intended movement of items, it is possible to detect unexpected object locations that indicate fraud.

However, keeping such an object history requires a dense network of readers and standards for data exchange. Using "track & trace" with its extensive product history is also add odds with privacy requirements and does not allow for "real" security.
For preventing counterfeiting effectively without requiring such an extensive data collection, an authentication mechanism is an inevitable requirement. The authenticity of RFID tags should become verifiable to prevent tag cloning.
Prevent illegitimate access: Also for achieving this goal, the authentication ability of RFID tags is essential. If only data of authenticated tags is processed, attackers cannot enter invalid data into the RFID system. Note that there is a class of attacks called "relay attacks" [Kfir, Z. & Wool, A. (2005)] that cannot be prevented by tag authentication alone. Considerations regarding this class of attacks are beyond the scope of this text.
Cope with denial of service: Denial of service attacks cannot be fully prevented in practice. For instance, tags can be permanently destroyed by mechanical, chemical or electromagnetic means. Temporary denial of service can be performed by shielding the tags or transmitting disturbing noise. One can only try to implement mechanisms to detect such actions, provide means for sanctioning, and implement processes for recovery. For security and privacy researchers who implement new concepts and protocols for RFID communication, it is important that no additional means for denial of service attacks are introduced with the new solutions.
Prevent unwanted recognition and tracking: It is important that outsiders, i.e. potentially unwanted readers, are not able to abuse the data stored on RFID tags for unwanted recognition and tracking. Arbitrary static data, e.g. an identifier or even encrypted data, acts as a means for recognition and tracking. Even constellations of tags with different amounts of data can be used for tracking purposes.
For preventing unwanted recognition and tracking, no static data may be stored on tags. This means that a periodic change of the tag identifiers is required. The idea behind this concept is that only authorized parties can link the changing identifiers to the tag identity and the data stored in backend systems. Illegitimate parties can no longer distinguish whether two tag identifiers obtained at different times belong to the same tag or not.
Altogether, for reaching the five presented goals, three core functionalities are required for RFID tags: (unique) identification, authentication, and modification (regular changes of tag identifiers). As important constraint, implementing these functionalities should not introduce additional possibilities for denial of service attacks. *Identification* is the basic functionality required for RFID systems to operate. *Authentication* mechanisms prevents tag cloning (and therewith counterfeiting) and illegitimate access. *Modification,* i.e. a regular change of the tag identifiers, prevents unwanted recognition and tracking. RFID researchers implement these functionalities in different ways and propose various schemes.

1.3 Solution requirements

In addition to fulfilling functional requirements, RFID systems should fulfill many non-functional criteria, i.e. provide quality or have certain qualities. This section defines a set of requirements that can be used for the evaluation of an RFID system and its core components. The requirements are mostly also usable for evaluating RFID communication protocols.
Security and Privacy: The importance of security and privacy for RFID systems and the complexity of their achievement have been discussed at the beginning of the chapter. The

previous section presented the properties a system should have in order to fulfill these requirements.

Resources: Since the amount of resources needed for the realization of RFID solutions have an effect on the costs, it is important to keep the required resources as low as possible. The cost factor determines the economic incentive of the technology. As RFID tags have to be produced in oodles to be applied on everyday objects, the tag cost and therewith the available tag resources are the most limiting aspects.

Performance: The performance of RFID systems can be measured on the time needed to read a bunch of tags, i.e. to get all the relevant data. The result depends on many factors like the bandwidth of the physical communication channel, the amount of data to be transferred, the number of message roundtrips, the time for retrieving data from backend systems, the use of caching mechanisms, etc. Performance can be improved by keeping message size and number of messages small and by using caching and delegation mechanisms.

Scalability: This is an important quality for systems intended for inter-organizational or even worldwide use. The design of a system has to allow in best case an unlimited extension of users, data, and devices. There should not be any bottlenecks in the system. In practice, this requirement is fulfilled by distributing load without requiring central systems for control.

Reliability and availability: RFID systems often become part of business processes. Like with most information systems, companies and people start to rely on the operation of the technology. Failures and errors disturb business process and can thus become very costly. Therefore, RFID systems should always be available and operate reliably.

Usability: RFID systems should be calm, just like any ubiquitous computing system [Weiser, M. (1991)]. This means that these systems should not require the user's attention if possible. The RFID systems have to work for the user and ease his everyday life without disturbing or bothering him. User interactions should be required as seldom as possible, and the technology should not require a special behaviour from the user, e.g. waiting until an operation is completed.

Sustainability: In some application areas, RFID tags have a long life span. For example, RFID tags are used to identify university inventory, firm inventory, and library books. This needs consideration when implementing mechanisms that rely on cryptographic primitives. What is secure today is often no longer secure some years ahead. RFID systems have to keep up with the times. This means that new tags should be able to use up to date cryptographic primitives while older tags still use less secure ones.

RFID tags usually implement primitives in hardware that cannot be updated. Replacing such tags may be economically infeasible. It should be possible that such tags remain in operation. As they use less secure primitives, these primitives might get broken with feasible effort at some time. The impact of such a security breach should be as limited as possible, e.g. the affected tag should still be identifiable and only lose some privacy features.

Universality: In order to use RFID tags all over the world and in different applications, the tags have to be designed in a generic and application independent manner. Having the same kinds of tags reduces costs due to mass production. Having the same level of security and privacy protection everywhere relieves the users from the burden to pay attention to the level which is implemented with a particular tag. Of course, it should be possible to hold arbitrary application specific data in an RFID system and to use different cryptographic primitives. This means that a high level of flexibility should be provided with only a small number of different kinds of RFID tags.

Scope: The scope of RFID application areas varies from local to global and from intra-organizational to inter-organizational. Ideally, RFID system architectures and therewith RFID systems are able to operate on a global, inter-organizational scope.
Practicability: Many proposals regarding RFID security and privacy are of academic nature. Some proposals even only work in theory but inherently fail in practice (e.g. some protection schemes require ideal synchronization of data transmissions and/or do not consider the behavior of the physical layer). Practicability is thus a crucial requirement.
Practicability also needs to be considered with respect to the already mentioned *usability* and other requirements: Low costs, fast processing of the data, minimal user involvement, and secure handling of data are some of the problems that are, at least indirectly, linked to RFID's practicability.

2. Current RFID system architecture

This subchapter describes the common RFID system architecture. It shows that the architecture is based on the one known from other kinds of auto-id systems, like optical barcode systems. Afterwards, the general direction of security and privacy research is outlined. Overall, the subchapter shows a "mainstream direction" of RFID architecture and RFID security and privacy research.

2.1 Barcode systems as the guide

A typical barcode system is depicted in figure 2. This example shows the operation of a cash register in a retail market.

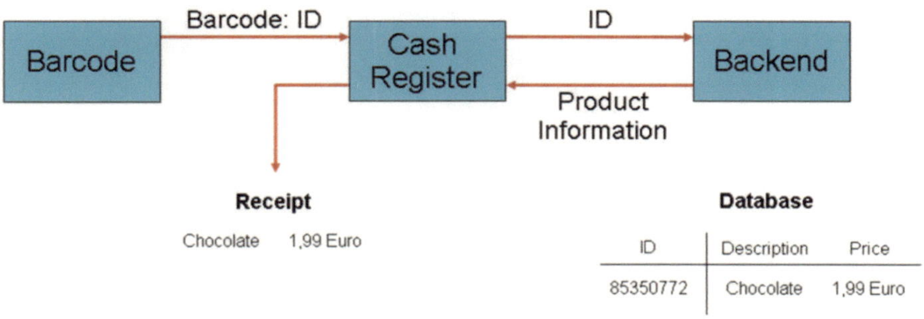

Fig. 2. Barcode system example

Optical one-dimensional barcodes provide product information via bars with different width and space between them. To read the information for a given product, the barcode of the product is captured using a barcode reader, sometimes also called "scanner". In the example, the reader is directly connected to the cash register. In a supermarket, the barcode contains an identifier for the manufacturer and an identifier for the product type. These identifiers do not provide the required information like the product price to the cash register. Thus, the scanned information is transferred to the backend of the system. The backend has a database with item information and retrieves the database record associated with the given barcode data. The database record includes information like a product description and the current product price. The product information is then transferred back to the cash register. Usually, the backend also keeps a product inventory, i.e. information on the number of items available. This information is updated when triggered by the cash register. The cash register now has all required product information to print the customer receipt.

RFID systems have corresponding system components: RFID tags, RFID readers, and backend systems. Low-cost passive RFID tags are intended as an alternative to optical barcodes: The tags can store a certain amount of data, e.g. 128 bit. For the point-of-sale, the data forms an identifier that is structured into manufacturer, product type, and a serial number. The serial number part makes the identifier unique.

Apart from the additional serial number, RFID systems for the point-of-sale are fully compatible to barcode systems. The process that has been shown in figure 2 remains the same in such cases. Using RFID labels as a replacement for optical barcodes is therefore a simple task. Only if additional possibilities that result from the serial number shall be used, vendors have to provide new software.

Based on the similarities between barcode systems and RFID systems, figure 3 shows an "Auto-ID Triangle". The barcode or the RFID tag is the "media" on which some data, i.e. the "content", is stored. The data has a meaning which is defined by numbering schemes. The media is read using readers which generate read events that are processed. Such a "data processing" can have arbitrary forms. Read data, usually containing an identifier, can be linked to additional data that is stored in databases. The mentioned read events can be filtered, data can be linked to other data, or whatever else.

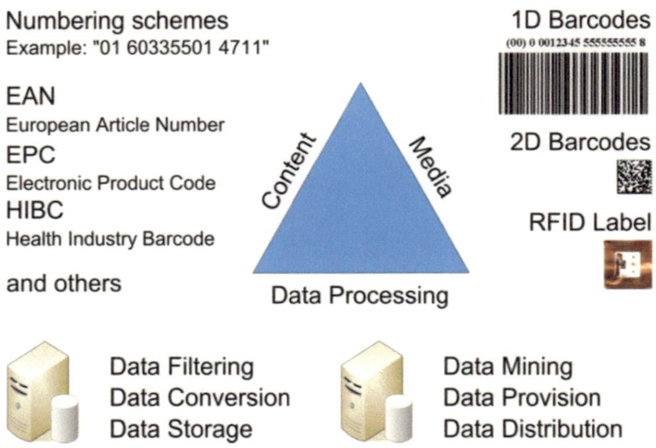

Fig. 3. Auto-ID Triangle

Note that RFID tags can have additional functionalities like sensors or smartcard-like functionality. Such additional functionality breaks the analogy between barcodes and RFID. It is thus not covered by the Auto-ID Triangle in figure 3.

Interestingly, the major limitations of auto-id systems today are within the backend systems, i.e. the data processing. This means that efficiency and productivity could be much higher if there were better backend systems. For instance, if products are processed by different companies, often each company applies its own barcode to the product instead of using already present barcodes. This is often not necessary if the reader and the backend are flexible enough. RFID tags or 2D-barcodes with their means for unique identification can make reuse much simpler because there is no possibility for identifier collisions any more.

Data sharing between companies is another aspect with room for improvement. In many business processes, there are still many manual steps that could be avoided using

appropriate information technology enabling inter-organizational data sharing. For instance, if bills had a unique identifier that could be read by an auto-id system, one could retrieve payment information automatically instead of entering the required data.
The Object Name Service (ONS) [EPCglobal (2008)] maps unique identifiers (here: Electronic Product Codes, EPCs) to additional data in an application independent manner. It operates quite similarly to the domain name system. However, it is only a first building block. For flexible and efficient inter-organizational data sharing, a lot more functionality and open standards are required.

2.2 Security and privacy research

There is an essential difference between barcode systems and RFID systems. RFID systems use electromagnetic fields or waves as data transport medium. Since this medium is easy to tap, the communication channel between RFID tags and reader needs to be secured (ref. to fig. 4). In contrast, the optical channel in barcode systems requires a line-of-sight. A barcode on an item in a bag cannot be read, whereas this is possible with RFID tags.

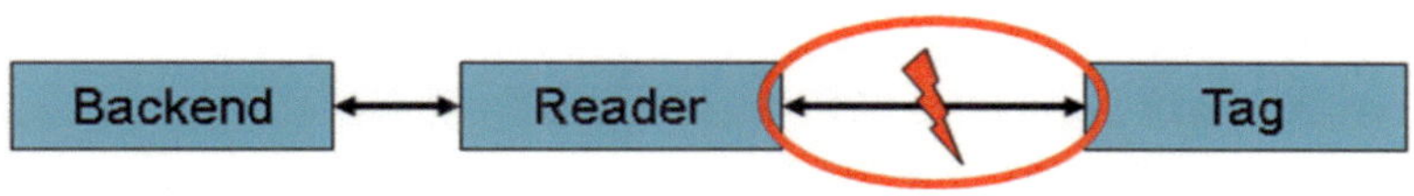

Fig. 4. RFID system with vulnerable communication channel

Research in RFID systems is thus concerned with the definition of communication protocols that secure the vulnerable channel between tags and readers. A lot of researchers concentrate on methods that not only provide security features but also assure privacy protection in RFID systems.
The achievement of the stated security and privacy goals (see subchapters 1.1 and 1.2) requires three basic functionalities. Protocols need to implement identification, authentication, and modification. This means that an RFID tag must be able to identify itself, to authenticate itself, and to change its identifier regularly. Many researchers propose protocols that implement these three basic functionalities in different ways, to a different extent, and in different quality. First protocols for RFID communication have been presented in 2002 [Sarma et al., (2002)]. However, the research continued in this direction, and still today (2008) new communication protocols are introduced on major conferences (e.g. [Henrici, D. & Müller P. (2008)]).
A presentation, comparison, or evaluation of the published approaches can be found, for example, in [Lehtonen et al., (2006)] with a focus on authentication and for more general approaches in [Avoine, G. (2005)]. Avoine also maintains a website (see http://www.avoine.net/rfid/) with links to publications.
Protocols have to consider the amount of resources consumed by the RFID tags. This amount has to be kept as low as possible to assure cost-effective production of a large number of tags. For this reason, the implementation of the three functionalities needs to be kept as simple as possible. This leads to avoidance of symmetrical and asymmetrical cryptographic techniques and the exclusive application of one-way hash functions in many proposals. Most requirements stated in subchapter 1.3 also apply for RFID communication protocols.

3. Reconsidering the current architecture

The common RFID architecture is based on the one known from barcode systems. Extensions are added when required. Researchers are looking for solutions that secure the communications channel between tags and readers. Regular identifier changes shall protect against unauthorized creation of movement profiles.
In this subchapter, some major deficiencies of the current architecture are presented. Afterwards it is reasoned why extensions to the architecture are not sufficient to address all these deficiencies and thus to provide a satisfying solution.

3.1 Deficiencies of the current architecture

There are several reasons why the described architecture and the proposed security mechanisms are inappropriate for today's demand. Some important deficiencies are the following:

1. There are no explicit possibilities for infrastructure sharing.
2. The proposed lightweight security protocols supporting identifier changes are not practical.
3. Most lightweight security protocols do not support delegation.
4. Tag bearers are not a part of the RFID architecture.
5. Some solution requirements like sustainability are not considered adequately.

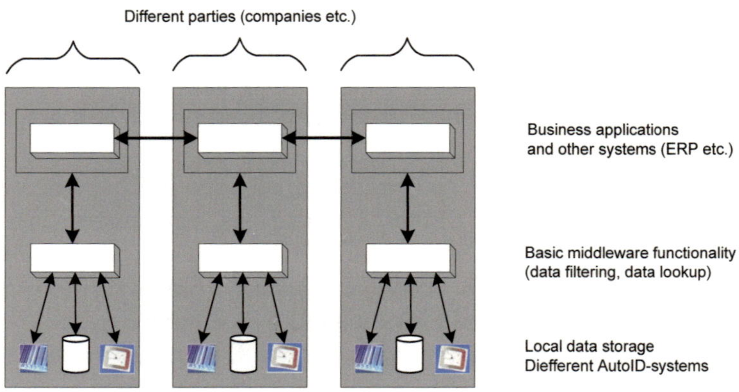

Fig. 5. Today's inter-organizational data sharing

There are no explicit possibilities for infrastructure sharing.
Each company has to build its own RFID infrastructure, i.e. install its own readers. There is no standardized way to make use of readers installed by other companies.
In practice, if a company wants to obtain read events of subcontractors, e.g. for providing item locations to customers in logistics applications, such data sharing needs to be implemented on the application layer (see figure 5). Standards for data formats evolve, but there are not enough defined interfaces and procedures so that enabling data sharing is a costly task.
The proposed lightweight security protocols supporting identifier changes are not practical.
For providing location privacy, tag identifiers have to be changed regularly. However, frequent identifier changes are difficult to provide without requiring the tags to interact with backend systems. Besides the challenge to implement a protocol with good characteristics,

regular identifier changes put high load on the infrastructure. Even without such additional load, the data produced by RFID events is a burden for networks and the backend systems.

Most lightweight security protocols do not support delegation.

Today's business processes often take place among different companies. This means that companies no longer act independently from others but have, for instance, subcontractors involved in a business process. Therefore, delegation becomes an important feature (see Molnar et al., (2005)).

To provide location privacy, identifier changes are performed. This means that subcontractors can no longer identify a tag without communication with the tag owner. For efficiency reasons, the possibility to delegate the ability to identify a tag is required.

Tag bearers are not a part of the RFID architecture.

Tag bearers are the users that carry items with RFID tags. Obviously, the privacy of such tag bearers is in danger (data security, location privacy). Nevertheless, the current architecture and therewith the RFID protocols do not consider the tag bearers at all. Instead, they regard the owner of an RFID tag (which is usually not the user carrying the tag) as trusted. For instance, if a user carries a subway ticket which might be abused for unwanted recognition and tracking, the user needs to trust the transport company which is the owner of the RFID tag on the ticket. This is not wanted. The tag bearer needs explicit consideration in the RFID architecture.

Some solution requirements like sustainability are not considered adequately.

The main research topics in the area of RFID in the past years have concerned mainly security, privacy, and resource consumption requirements. Other requirements like sustainability and scope have hardly been examined. However, they have immense significance for the quality of RFID systems. RFID systems that adhere to all the requirements stated in subchapter 1.3 are required.

3.2 Limitations in patching the current architecture

The question that arises is to what extent the existing architecture and methods can be improved to address all the deficiencies and to meet all quality requirements. Detailed analyses show some contradictory features. Using an example, it is shown in the following that just extending the current architecture does not lead to satisfactory solutions. The aim is to prove that a completely new concept, i.e. a "clean-slate approach", is required to address all the issues.

There is a conflict between systems of inter-organizational scope and the protection of privacy. This example shows the limitation of patching the current architecture. The trade-off between the two requirements is explained in the following.

As already stated, regular changes of the RFID tag identifiers are a requirement for privacy protection. Using regular identifier changes, location privacy can be provided. The idea is that the current tag identifier does not provide any information regarding the tagged item and that due to the regular identifier changes it is no longer possible to use the identifier for unwanted recognition and tracking.

However, this method poses a restriction for inter-organizational systems. Only the organization administrating the tag can also identify it because it keeps track of the identifier changes in a backend database. Only this organization can link the current tag identifier to associated data. For every other organization trying to read the tag and wanting to obtain associated tag data, the identifier appears to be just a random number. This is necessary because the reader might also be illegitimate or even an attacker. In consequence, the identifier also appears random for legitimate readers, e.g. a subcontractor. A subcontractor cannot even find out who is the owner of the tag (see figure 6). The subcontractor might work with a huge number of other companies and now does not get the information which company is responsible. Requiring contacting all possible owners is not a scalable approach and puts too much load on the infrastructure and the backend.

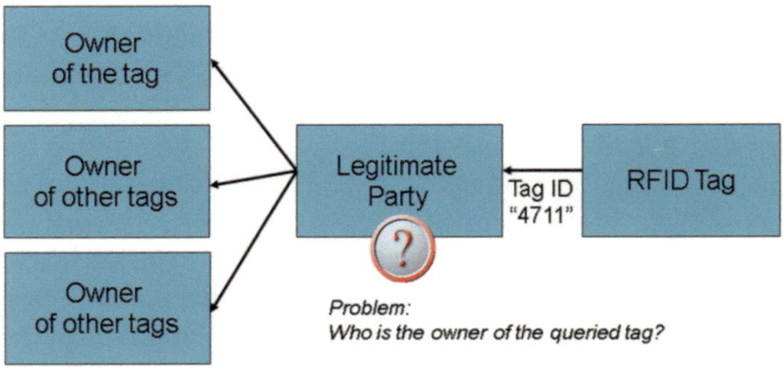

Fig. 6. Non-interpretable identifiers in inter-organizational systems

The information which organization/company is responsible for the tag may not be stored on the tag and be provided to the reader since that data can be misused for recognition and tracking purposes (at least by considering constellations). Consequently, a regular change of tag identifiers helps protecting privacy but causes problems in inter-organizational RFID systems. This means that the privacy requirement and the scope requirement run into conflict.

Theoretically, this conflict is solvable [Henrici, D. (2008)]. One possible solution to the problem can be provided by the use of pseudonymization infrastructures. They are employed for example in anonymous remailers and operate on the shared secret principle. In practice, such solutions solve one problem but bring new problems in other areas. For instance, the mentioned pseudonymization infrastructures solve the conflict between privacy and scope but require a lot of resources, lead to low performance, have scalability issues and are costly to implement.

The described conflict between privacy and scope and the possible architecture extension by using pseudonymous infrastructures is just a single example. There are many issues like the described deficiencies that need to be addressed in practice. However, the many efforts made to improve and extend the existing architecture do not seem to be able to provide adequate solutions that can fulfill all quality requirements.

4. Creating a new RFID system architecture

Subchapter 3 showed that there are many deficiencies in the current RFID architecture that need to be addressed. However, the example in the previous section showed that it will not be possible to ever meet all specified requirements in a satisfactory manner only with the use of incremental improvements. The trade-off points between some of the quality attributes require lying out a fundamentally new direction. The goal of this subchapter is to provide some considerations on how such a new, "clean-slate" architecture might be created and how it could look like.

4.1 Starting points

The toughest problem in the current architecture is probably the conflict between providing location privacy and creating inter-organizational RFID systems. This conflict was described in subchapter 3.2.

Data security demands tag identifiers that appear random. However, random identifiers do not provide information regarding the tag owner. But this would be required in inter-organizational systems. Location privacy requires a periodic change of the tag identifiers. This leads to high infrastructure load and affects the scalability of the system negatively. Even if the issues can be solved in some way, this comes at high costs.

Does this mean that security and privacy are incompatible with the economic interests and the technical needs? It seems so, a least when the current architecture is considered. The only possible way out seems to be the creation of a redesigned architecture based on new concepts.

One starting point can be found by closer consideration of the nature of privacy. When do people consider their privacy protected and when not? Answering this question is not easy. The fulfillment of their privacy expectations is of great importance to people. Meeting these expectations requires a certain degree of privacy protection. Yet, the expectations and the resulting requirements are not explicitly definable. They are context-sensitive and person-specific. This makes the modeling of privacy requirements and their technical realization a difficult task.

However, there are ways to solve the conflict between privacy protection and scalability: Humans need no total privacy protection. They are social beings and are used to give away some private information under certain circumstances. It is only important to them that their expectations regarding privacy protection are fulfilled.

Another starting point is that not all of the privacy requirements need to be addressed by technical means. The identification and sanction of a violation is often easier to implement and fully sufficient in some cases. Moreover, incentive systems can be applied. If the effort for a successful attack is higher than the expected gain for the attacker, the attacker will not perform an attack. This means, if there is no incentive for an attack, it is no longer an interesting option for the attacker. Vice versa, there is a stimulus to behave system-conform, if there is some kind of reward for such behavior.

Thus, there are also non-technical ways to implement privacy protection, e.g. via legislative or economic means. Under consideration of this knowledge it is possible to define a system architecture that fulfills the quality requirements on RFID better than the existing one.

Please note that this does not mean that technical protection schemes are not necessary. Instead, they are essential. Legislation alone does not provide adequate protection. For instance, sending unsolicited email is prohibited. Nevertheless, mailboxes are full of such email. The reason is that there are no effective technical safeguards. The possibility to sanction misbehavior is missing, too. It is thus very important that not only a kind of "pseudo security" is provided but that the sum of technical, legislative, economic and social means for providing protection deliver an effective solution.

4.2 New concepts

This subchapter provides some new concepts and outlines a new RFID architecture. Some of the concepts can also be used in other auto-id systems like optical barcode systems.

Infrastructure sharing

One of the deficiencies stated in subchapter 3.1 is the lack of explicit possibilities for infrastructure sharing in the current architecture. Each company or organization needs to implement its own infrastructure of readers. Of course, companies can exchange data regarding read events, but such a data exchange needs to be implemented explicitly. It often takes place on the application layer and is costly to realize.

Creating a generic mechanism of infrastructure sharing is fairly simple and straightforward. First, each RFID tag needs to have an owner that shall be notified regarding the whereabouts of the tag. Second, there needs to be a policy that defines that each well-behaving reading party must contact the tag owner when a tag is read using one of the party's readers. This way, the tag owner gets a notification each time a tag is read. If the notification also includes additional information like the location of the reader, there is no difference any more whether an RFID reader is operated by the tag owner itself or by somebody else: The tag owner gets a read event each time a tag is queried.

This mechanism is very powerful. It allows companies to track items, e.g. in logistics, without requiring to define special interfaces for data exchange with each subcontractor and without operating a dense network of readers.

A tag only needs to store two pieces of information: One that enables the reading party to send a message to the tag owner (a kind of address of the tag owner). Another one is required to identify the tag uniquely. Both pieces of information together can form the tag identifier. If no other information is stored on the tags, the reading party has a strong incentive to contact the tag owner, perform the notification and request additional information regarding the tag. This is an example where a policy is enforced by an incentive.

Note that the Object Name Service (ONS) [EPCglobal (2008)] already defines a mechanism that could be used for notification purposes: Using the ONS, the reading party can contact the tag owner and get links to additional information regarding the read tag. This mechanism just needs to be adapted so that the ONS query contains information regarding the reading party.

An infrastructure sharing mechanism is highly desirable for getting the optimal results with a given amount of infrastructure hardware. It also saves costs. But if implemented as described, it also has a number of disadvantages: It puts a high load on the network infrastructure if each tag query results in a notification to the respective tag owner. It also seems bad for privacy at first sight as it makes independently operated readers to a powerful tracking device. At second sight, it only highlights a problem that is already present. If readers become networked, the tracking capability gets more powerful. But in the scenario presented here, there is also a big opportunity for privacy protection: We "just" need to solve the privacy problem once for the stated infrastructure sharing mechanism, and we get a generic solution for all users and all applications.

Tag bearers become a part of the RFID architecture

One further problem that was stated in the list of deficiencies of the current architecture is the non-consideration of people carrying RFID tags. A tag bearer is not considered in the architecture and does not have the freedom to decide when tags are read and what tag information is visible and to whom.

To solve this problem, the tag bearers need to be considered explicitly in the RFID architecture. Only this way, they get the ability to protect their privacy effectively. Figure 7 shows the integration of tag bearers into the architecture.

The procedure of reading a tag is as follows: The reading party, i.e. the party operating the reader, queries an RFID tag and obtains the tag identifier. Instead of the RFID tag, an optical barcode could be used, too. If the identifier does not provide valuable information to the reading party (which we presume), the reading party has to notify the tag bearer of the read event and request additional information regarding the read tag. The tag bearer can now decide whether to proceed or to notify the reading party that no data is provided for privacy reasons. In case that the tag bearer decides that the request for data is allowed, he/she forwards the request to the tag owner. The tag owner can decide whether to abort or to provide the requested data regarding the tag. Thus, if both, tag bearer and tag owner, agree, the reading party obtains the information required to identify an item as well as the requested additional item information.

Fig. 7. Reader, tag bearer and tag owner as separate entities

The presented procedure is a straight-forward extension to the mechanism described in the *infrastructure sharing* paragraph. But it results in a number of new issues that need to be addressed.

Of course, tag bearer and tag owners cannot decide manually whether to allow tag identification or not. This would be a burden to the users. Instead, they require the ability to define policies so that most of the requests can be answered automatically.

Another issue is anonymity. For example, it should be possible for a customer to read the price of a product in a supermarket with his/her mobile phone without revealing his/her identity. This requires a neutral third party in the communication path between the reading party and the tag bearer. Another requirement is the support of changing tag identifiers for the protection of location privacy.

Reducing load while providing location privacy

There are still a lot of issues to be addressed. One major problem is the conflict between the provision of location privacy using identifier changes and the scalability requirements of inter-organizational systems. Furthermore, the *infrastructure sharing* mechanism introduced at the beginning of this subchapter puts a high load on the infrastructure by read notifications. Even more load would be caused by frequent identifier changes.

To address all the open issues, a new architecture is required. It needs to provide location privacy and needs to reduce the load on the infrastructure significantly. Some starting points were discussed in subchapter 4.2. Additionally, we require effective mechanisms for caching and delegation.

On this account, the *ID-Zone Architecture* [Henrici, D. (2008)] was created. It does no longer use the barcode-system principle but can be applied for different kinds of auto-id systems, including RFID and barcode systems. The approach is compatible with existing backend solutions.

The basic idea of the ID-Zone Architecture is the creation of zones. The physical space is separated into disjoint (non-overlapping) zones. An example is shown in figure 8. Each zone matches an administrative competency, e.g. a shop or a library.

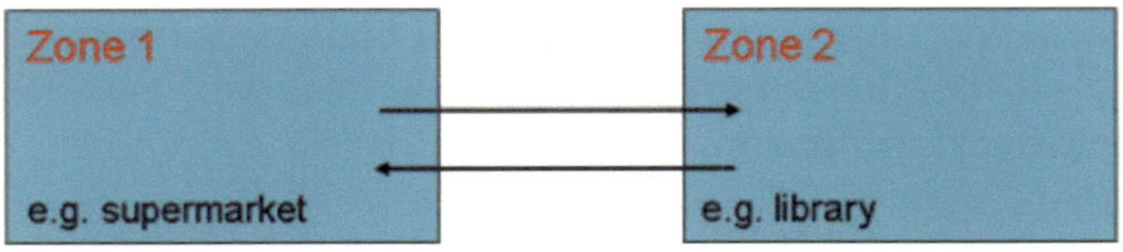

Fig. 8. Example for the zone concept

As long as an object with an RFID tag is within a zone, the tag identifier remains constant (there are some optional exceptions for preventing industrial espionage that are not discussed here). On the one hand, this approach considerably reduces the load on the infrastructure caused by identifier changes. On the other hand, the privacy protection is slightly restricted. But since recognition within an administrative zone is often desired and can also be realized via other methods (e.g. by personnel, video cameras), this is not a relevant limitation in practice.

If a tagged object leaves a zone, the tag on the object has to identify itself with another, new identifier in the new zone so that it is not possible for an outsider to recognize that it is the same tag, i.e. the same object. This is important for privacy protection since otherwise a movement profile could be created if different zones cooperate. If a tag returns to a previous zone, i.e. a zone where it already was some time before, it is for the same reason necessary to use a new identifier, different to the former one used previously in that zone. This behavior is shown in an example in figure 9: At first, a tag has the identifier "P123:456" and enters another zone. If it returns, it gets the identifier "P123:963".

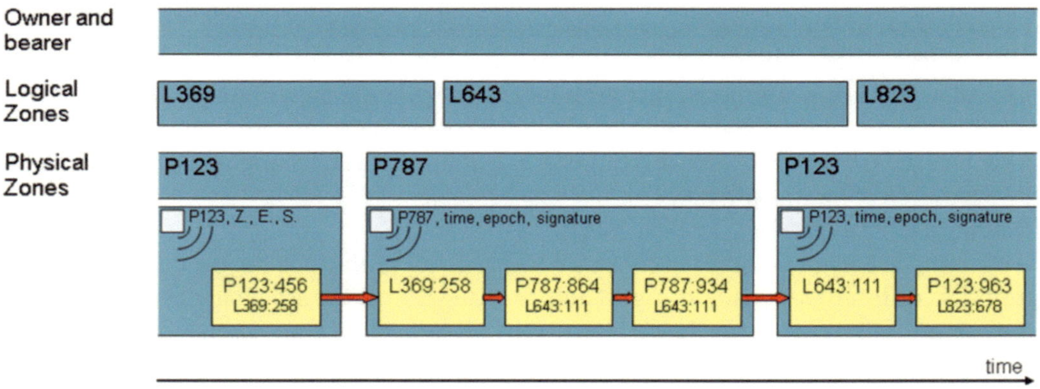

Fig. 9. Layering and example for the ID-Zone Architecture

This approach solves the conflict between privacy protection and scalability. The "logical zones" (refer to figure 9) provide anonymity for tag bearers. However, a number of new questions arise: How does a tag know from a reliable source in which zone it is? When (before leaving a zone, on entering a zone, on the first read in the new zone, etc.) and how should the change of the identifier take place?

The answer to these questions is already depicted in figure 8 but an explanation goes beyond the scope of this text. The implementation of the concepts of the ID-Zone

Architecture with appropriate efficient RFID protocols is also out of scope. More information is provided in [Henrici, D. (2008)] for the interested reader.
The message should be clear now: With new concepts and architectural changes, there are powerful possibilities for addressing the challenges stated in subchapters 1.1 and 1.3. In contrast, the current architecture is at its limits and incremental improvements are no longer sufficient.

5. Summary and research directions

The RFID technology will be an inevitable part of our everyday life in the foreseeable future. It offers various application spectrums that raise productivity, increase comfort, and open new markets. An important aspect is the consideration of the user requirements since an abandonment of the technology will not be possible for the individual. Foods, clothes, books, and many other goods will be tagged and identified with RFID tags. Therefore, designing concepts and methods that ensure security and privacy protection for systems of global scope is one of the main research goals in the field of RFID.
The challenge is great since not only technical and economical aspects have to be considered but also ethical and social ones. There need to be technical safeguards and the possibility of informational self-determination for the users. Nevertheless, the solutions have to be cheap and easy to realize. Moreover, a range of quality requirements like reliability, scalability, flexibility, openness, and sustainability have to be considered.
Besides security and privacy, there are many more research challenges. Systems need to support inter-organizational business processes. Also, the integration of people carrying the RFID tags ("tag bearers") into the system is important for providing information self-determination. In contrast, current solutions consider only the interests of the RFID tag owner (i.e. supermarkets, libraries, employers, countries, etc.). Yet, the users concerned with privacy and data protection problems are the ones who carry the tags (buyers, library users, employees, citizens, etc.). Another sophisticated problem is the provision of location privacy, i.e. protect people from unwanted recognition and tracking.
The goal of this book chapter was to advise the reader to the different challenges in the sphere of RFID systems and to point out the need of a new system architecture as a solution to the various functional and qualitative requirements.
New concepts help to address the issues. Examples are infrastructure sharing and the consideration of tag bearers as part of the RFID architecture. The sketched *ID-Zone Architecture* provides ideas on how to even resolve conflicts that seem unsolvable with the current architecture.
The presented concepts and the proposed architecture are surely no universal remedy, but they appear promising. In contrast, the current architecture appears to be at its limits so that incremental improvements will not be able to meet all the practical requirements.
The various challenges can only be addressed by heading into new directions. One of the new directions is the consideration of technical security measures in the economical, legislative, and social context. Many existing proposals concentrate on technical security measures and neglect the non-technical constraints and possibilities. Yet, non-technical measures can support the technical security mechanisms. Researchers need to start to explore the possibilities and limitations that new concepts and new architectures provide.

6. References

Avoine, G. (2005). Cryptography in Radio Frequency Identification and Fair Exchange Protocols, PhD thesis, EPFL, Lausanne, Switzerland, December 2005

EPCglobal (2008). Object Name Service (ONS) 1.0.1, Ratified Standard Specification with Approved, Fixed Errata, May 2008, available at http://www.epcglobalinc.org/-standards/ons/ons_1_0_1-standard-20080529.pdf

Henrici, D. (2008). *RFID Security and Privacy – Concepts, Protocols, and Architectures,* Springer, ISBN 978-3540790754, Berlin

Henrici, D. & Müller, P. (2008). Providing Security and Privacy in RFID Systems Using Triggered Hash Chains, *Proceedings of the Sixth Annual IEEE International Conference on Pervasive Computing and Communications – PerCom 2008*, Hongkong, March 2008, IEEE, Los Alamitos

Kfir, Z. & Wool, A. (2005). Picking virtual pockets using relay attacks on contactless smartcard systems, *Proceedings of SecureComm 2005*, Athens, Greece, September 2005, IEEE, Los Alamitos

Lehtonen, M.; Staake, T.; Michahelles, F. & Fleisch, E. (2006). From Identification to Authentication - A Review of RFID Product Authentication Techniques, *Workshop on RFID Security -- RFIDSec'06*, Ecrypt, Graz, Austria, July 2006

Molnar, D.; Soppera, A. & Wagner, D. (2005). A Scalable, Delegatable Pseudonym Protocol Enabling Ownership Transfer of RFID Tags, *Selected Areas in Cryptography – SAC 2005,* LNCS, vol. 3897, pp. 276–290, ISBN 978-3-540-33108-7, Kingston, Canada, August 2005, Springer, Berlin

Sarma, S.E.; Weis, S.A. & Engels D.W. (2002). RFID Systems and Security and Privacy Implications, *Cryptographic Hardware and Embedded Systems -- CHES 2002*, LNCS, vol. 2523, pp. 454-469, Redwood Shores, CA, USA, August 2002, Springer, Berlin

Weiser, M. (1991). The Computer for the 21st Century, Scientific American, vol. 265, no. 3, pp. 94 - 104, September 1991, Scientific American Inc., New York

24

Generalized "Yoking-Proofs" and Inter-Tag Communication

Leonid Bolotnyy and Gabriel Robins
Department of Computer Science, University of Virginia, USA

1. Introduction

Some radio-frequency identification (RFID) scenarios require a proof of action (e.g., that a group of objects tagged with RFID tags were identified simultaneously). For example, pharmaceutical distributors may want to prove that a bottle of medicine was sold together with its instructions leaflet [Juels, 2004]; manufacturers may want to prove that safety devices were sold together with a tool or that a number of matching parts were delivered simultaneously; banking centers or security stations may want to prove that several forms of ID were read simultaneously; meeting organizers may want to prove that a group of people were present together at a meeting, etc. Third-parties can verify the validity of such proofs. For examples above, the verifying third parties can be regulatory agencies, company headquarters, etc.

We seek to ensure that if a group of tags is not read nearly-simultaneously, an entity (RFID reader) will not be able to forge a proof that they were by constructing a valid forged proof. Inspired by this problem, Ari Juels developed a protocol that creates such a proof for a *pair* of RFID tags [Juels, 2004]. He left open for future research the problem of generalizing his protocol to three or more tags. We develop a methodology that generalizes yoking-proof protocols to arbitrarily large groups of tags. We also define an *anonymous yoking* problem and propose an efficient solution for it [Bolotnyy & Robins, 2006]. Finally, we show how these yoking protocols can be sped up.

2. Assumptions

We assume that RFID tags are passive and have limited computational capabilities. We require that tags be able to execute keyed hash functions and store state information such as a key, a counter, and some data computed during the protocol execution. These requirements can be satisfied in practice by Class-2 Generation-2 EPC tags [EPCglobal, 2006]. Since we assume that tags are passive, the tags cannot communicate directly with each other, but they can communicate with each other indirectly through the reader. For now, we also assume that an adversary cannot physically steal tags' secret information. Later we discuss how this requirement can be relaxed.

Our verifier is assumed to be a trusted and computationally powerful machine. The verifier is considered to be off-line in the sense that it does not have to verify the proof immediately after it is created, and it does not need to communicate with the tags during the proof construction. A reader communicating with the tags is assumed to be adversarial, and we want the protocol to be secure against a reader that attempts to create the yoking proof without reading all the tags within the required time bounds. The tags are assumed to be not impaired by an adversary, and tags do not collude with an adversary.

Replays of previously constructed valid proofs (replay attacks) may or may not be considered a threat, depending on the application. In our generalized yoking-proof protocol, we consider adversarial replay attacks to be a viable threat, and thus we design the protocol accordingly. To avoid replay attacks, the proof verifier stores some information about previous correct proofs. The verifier is not required to store this information if replays of valid proofs are not considered to be attacks. This will be elaborated upon in the discussion following the protocol specification.

We require that the tag accessed first by the reader be able to implement a timeout after a specific time period t has elapsed. Perhaps surprisingly, timeouts can be implemented on clock-less RFID tags. FCC regulations require the reader to change the communication frequency of a tag-reading protocol within 400ms. Changing the communication frequency in the middle of the protocol execution will likely result in loss of power on-board a tag and cause protocol termination. Therefore, the FCC regulation can serve as the clock for honest (law-obedient) readers. However, if the reader is malicious and violates these FCC regulations, a capacitor discharge rate on-board a tag can be used for protocol timing [Juels, 2004].

3. Basic protocol for a pair of tags

We now briefly describe Ari Juels' "yoking"[1] protocol for a pair of tags [Juels, 2004]. Assume that an RFID system contains n tags, which are denoted by $T_1, T_2, \ldots, T_n$. Each tag T_i is assigned a unique key x_i and a counter c_i both are d bits long. A tag has the ability to compute a keyed hash function and a standard message authentication code, which is likely to be implemented as a keyed hash function, such as HMAC, in order to simplify the circuit. A reader will read two tags and produce a proof P that both tags were read near-simultaneously, i.e. within t time units. The verifier V knows all key assignments to tags and will verify that the proof P is valid (not forged).

Let $f: \{0,1\}^d \times \{0,1\}^* \rightarrow \{0,1\}^d$ be a keyed hash function, and let $MAC: \{0,1\}^d \times \{0,1\}^* \rightarrow \{0,1\}^d$ denote a standard *message authentication code*. Let $f_x[m]$ and $MAC_x[m]$ denote computation of f and MAC respectively with a secret key x on input message m. The proof that tags T_A and T_B were scanned simultaneously is $P_{AB} = (A, B, c_A, c_B, m_{AB})$ where m_{AB} is defined in the protocol in Figure 1. To verify proof P_{AB}, the verifier computes $a' = (A, c_A, f_{x_A}[c_A])$, and $b' = (B, c_B, MAC_{x_B}[a', c_B])$, as well as $m'_{AB} = MAC_{x_A}[c_A, b']$, and then checks if $m_{AB} = m'_{AB}$. If t time units elapse, the first tag terminates the protocol, thus depriving the reader of the ability to construct a valid proof. Juels's protocol, as shown in Figure 1, has a minor, yet critical omission. Specifically, the counter value on the first tag is not incremented on timeout, allowing an adversary to violate the near-simultaneous object read requirement. Our group yoking protocol corrects this problem.

Juels also introduced a "Minimalist MAC", which may be implemented on lower-cost tags without encryption or hash function support. However, the protocol that he suggests for a

---

[1] The term "yoking" suggests the *joining together*, or the simultaneous presence of all the tags.

"yoking-proof" using a "Minimalist MAC" can only construct the proof once, which seems to diminish the original purpose of the protocol. If the reader is trusted, there is no need for such a protocol; on the other hand, if the reader is potentially untrusted and aborts the protocol after the first read, there will be no proof and no chance of ever re-creating a provably unforgeable proof in the future. Next, we discuss several works that unsuccessfully attempt to solve the generalized "yoking-proof" problem.

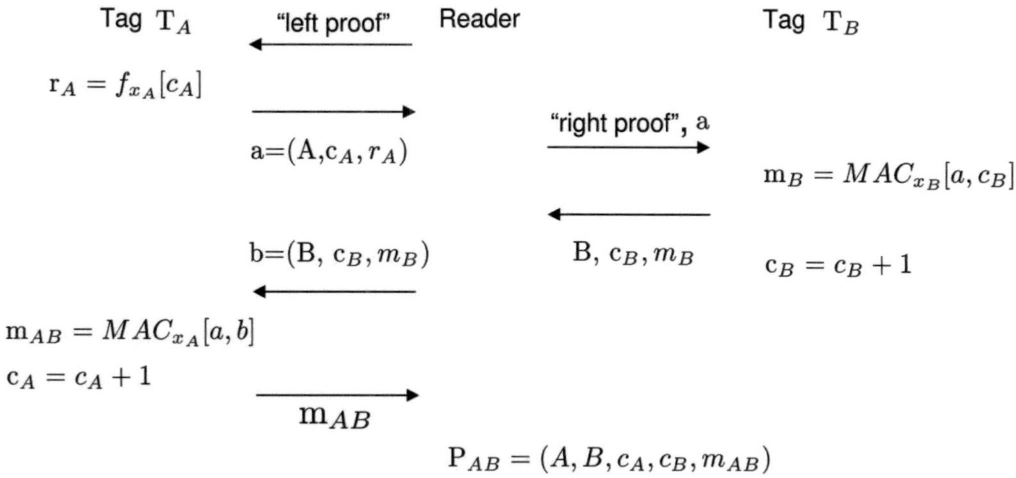

Fig. 1. A "yoking-proof" construction for a pair of RFID tags. The reader first communicates with tag T_A, then with tag T_B, and then with tag T_A again.

4. Related work

There have been several attempts to generalize the yoking-proof protocol. However, none of them solve the problem satisfactorily. Saito and Sakurai [Saito and Sakurai, 2005] apparently misunderstood Ari Juels' protocol for "strong MAC" and its modification for a "minimalist MAC". The authors presumed that in a minimalist "yoking-proof" each tag generates a random number for each proof, on which a keyed-MAC function is later applied, whereas Juels states that each tag is initialized with a one-time random number. Thus, the replay-attack the authors consider against the "minimalist-MAC" protocol is not applicable, since the "minimalist-MAC" protocol was designed for one-time use.

Saito and Sakurai suggest a group protocol which relies on time stamps provided by the back-end database [Saito & Sakurai, 2005]. In their scheme a reader receives a time stamp *TS* from the database, and it sends this timestamp to all the tags participating in the protocol. Each tag T_i computes $m_i = MAC_{x_i}[TS]$ and sends m_i back to the reader. In addition, the authors assume the existence of one powerful/leader tag among tags participating in the protocol. The reader sends all m_i to this leader tag, and the leader tag encrypts them together with the time stamp *TS* using encryption function *SK* keyed with a secret key *x*. Then, the leader tag sends the encryption result C_p to the reader, and the yoking proof is $P_n = (TS, C_p)$. Figure 2 shows their grouping protocol.

We discovered several flaws in Saito and Sakurai's solution. First, the assumption that one of the tags is more powerful than the others is not true in many practical scenarios. The second and main weakness of their protocol is that an untrusted reader can pick the time stamp *TS* as a future time stamp, and then use it on one tag and much later on another, thus violating the near-simultaneous read requirement. Even if the time stamp *TS* is encrypted, the reader can still separate tag accesses in time, since each tag access is independent of the others.

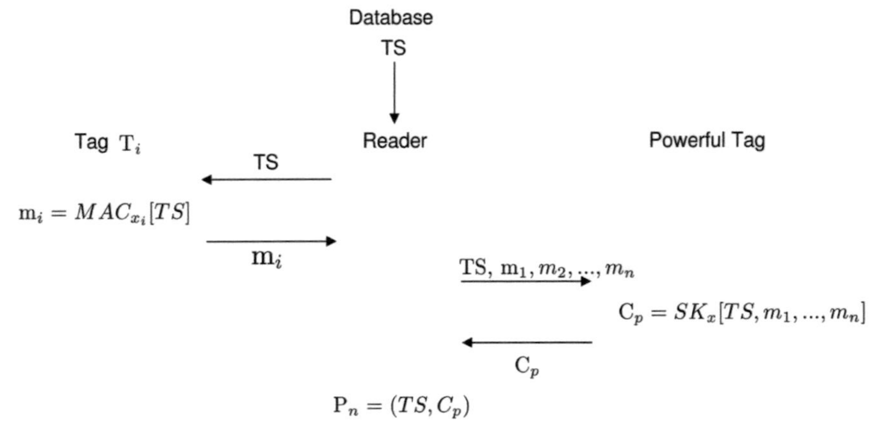

Fig. 2. Saito and Sakurai's protocol for a group of RFID tags.

Recently another paper [Piramuthu, 2006] independently observed the same problem with the "yoking-proofs" protocol of Saito and Sakurai [Saito & Sakurai, 2005]. Both papers [Saito & Sakurai, 2005] [Piramuthu, 2006] discuss a replay attack on Juels' one-time "yoking-proof". However, replay attacks are not an issue in Juels' Minimalist-MAC protocol [Juels, 2004] since the yoking-proof is a one-time proof. The one-time "yoking-proof" of Juels provides no security guarantees if the protocol is run more than once, and in fact, it is insecure in such re-run scenarios. The proposed "fix" of Piramuthu [Piramuthu, 2006] only works for a pair of tags, as opposed to an arbitrary number of tags. Moreover, the method of Piramuthu [Piramuthu, 2006] solves a different problem than the original problem formulation of Juels [Juels, 2004], since it relies on a random number the reader obtains from an on-line verifier, rather than accommodating an off-line verifier, as done in [Juels, 2004]. Our proposed solutions do not have these limitations.

Lastly, the work Peris-Lopez et al. [Peris-Lopez et al., 2007] repeats our observations of flaws in previous attempts to generalize yoking proofs. The authors [Peris-Lopez et al., 2007] offer their solution to our anonymous-yoking problem, discussed below, yet their solution does not satisfy the problem requirements. Specifically, it does not provide privacy since their protocol leaks the counter value on-board the tags. In addition, their solution is applicable only to 2 tags, requires an on-line verifier, and forces the reader to be trusted. It is worth mentioning that Peris-Lopez et al. [Peris-Lopez et al., 2007] state that our anonymous-yoking protocol takes $O(n^2)$, where n is the total number of tags in the system, to verify the yoking-proof for two tags. However, this bound is not tight since our anonymous-yoking protocol takes $\theta(k*n)$ where k is the number of tags participating in the yoking protocol and n is the total number of tags in the system. So, if the number of tags participating in the protocol is small, as is the case for most realistic scenarious, our protocol takes $\theta(n)$ time.

5. Our group yoking protocol

The idea of our generalized "yoking proof" for a group of tags is to construct a circular chain of mutually dependent MAC computations [Bolotnyy & Robins, 2006]. The purpose of the construction is to ensure that if an untrusted reader "breaks the chain" (i.e. does not read all the tags within t time units), it will neither be able to mount a replay attack nor create a proof that will be accepted as valid by the verifier.

Using the same notation and definitions as above, let $x_1, x_2, \ldots, x_n$ be the secrets and let $c_1, c_2, \ldots, c_n$ be the counters stored on tags $T_1, T_2, \ldots, T_n$, respectively. Tag secrets are shared with the verifier. In our protocol below, we assume that the reader generates a proof for tags $T_1, T_2, \ldots, T_k$ ($k \leq n$) and queries them in that order. (Tags are queried based on their IDs or on the random numbers that they generate.) The first tag computes $r_1 = f_{x1}[c_1]$ and sends $a_1 = (1, c_1, r_1)$ to the reader. The reader will then send a_1 to the second tag. The second tag will compute $r_2 = MAC_{x2}[c_2, a_1]$ and send $a_2 = (2, c_2, r_2)$ to the reader. The reader will then send a_2 to the third tag, which will in turn perform the same computation as the second tag, i.e. $r_3 = MAC_{x3}[c_3, a_2]$ and send $a_3 = (3, c_3, r_3)$ to the reader. The reader will then send a_3 to the fourth tag and so on, until the last tag k has computed r_k and sent $a_k = (k, c_k, r_k)$ to the reader.

The reader then sends a_k to the first tag, which computes $m = MAC_{x1}[a_1, a_k]$ (assuming that t time units have not yet elapsed since the initial tag access), and sends m to the reader. The reader R creates a proof $P_{1,2,\ldots,k} = (1, 2, \ldots, k, c_1, c_2, \ldots, c_k, m)$. The protocol is shown in Figure 3. The numbers in ellipses in the figure indicate the communication order. To verify the proof $P_{1,2,\ldots,k}$, the verifier V performs the same computations as the tags $1, 2, \ldots, k$, maintaining the order, and compares the proof P that it generates to the proof $P_{1,2,\ldots,k}$ that the reader provided. If the proofs match, the verifier outputs *success*; otherwise, it outputs *failure*. The pseudocode of algorithms for tag initialization, the reader, a tag, and the verifier can be found in the author's Ph.D. thesis [Bolotnyy, 2007].

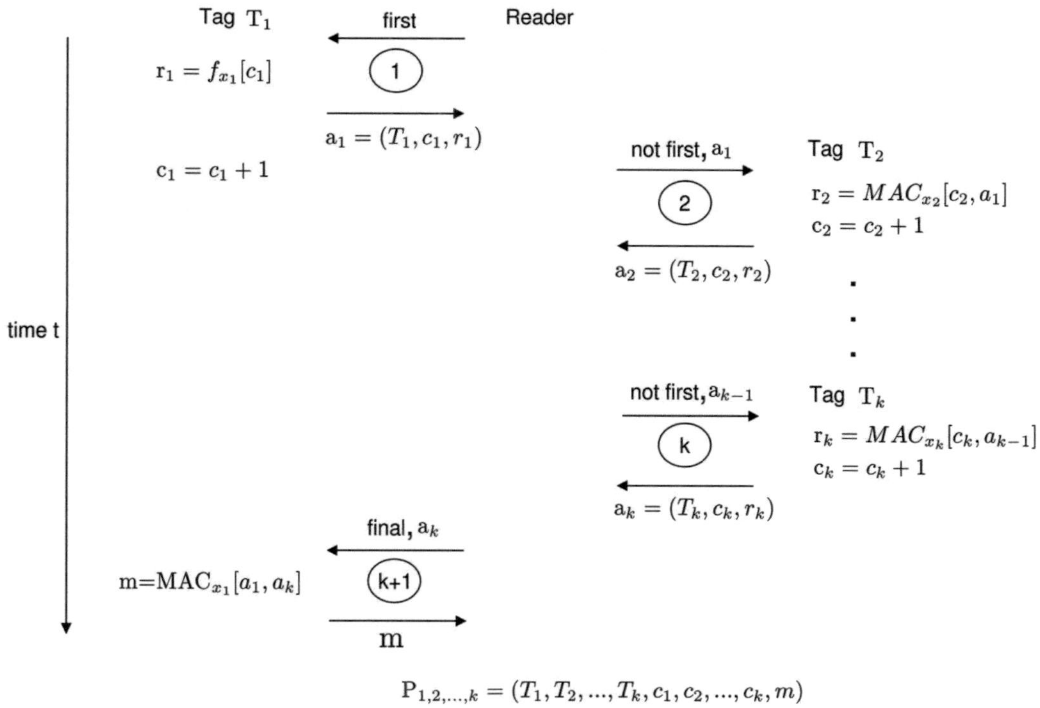

Fig. 3. Our "yoking-proof" protocol for a group of RFID tags.

Notice that each tag computes a MAC of a message that is a function of a MAC computed by the preceding tag in a yoking chain. This ensures that the reader has at most t time units to create the proof. To avoid replay attacks and allow temporal ordering of the proofs, each tag increments its counter immediately after it sends a_i to the reader. Also, observe that tags need not know how many tags participate in the "yoking-proof" protocol; instead, they only need to know when to timeout.

Note that if the first tag does not update its counter right after it sends its first message, a possibly malicious reader can create a proof P that will successfully pass through the verifier, without reading all the tags within the specified time bound t. In such a scenario, a proof can be forged as follows. The malicious reader can ask the first tag to compute a_1, then wait for t time units to elapse in order to cause T_1 to timeout, then send a_1 to T_2 to obtain a_2. Then, the reader will access T_1 for a dummy computation of a_1, and send a_2 to T_1 to obtain m, and construct a valid proof $P = (1, 2, c_1, c_2, m)$.

The basic yoking protocol of Juels, as shown in Figure 1, suffers from this problem unless the counter on the first tag is incremented on a timeout, but this is not specified in Juels's paper. The security of our scheme hinges on the probability δ that an adversary A is able to construct a "yoking proof" P_{AB}, which could fool the verifier V into reporting "success", without actually reading the IDs of all the tags involved in the protocol within t time units, as intended.

Next, we state the security property of our protocol. The proof of the theorem can be found in the author's Ph.D. thesis [Bolotnyy, 2007].

Theorem: Given random-oracle assumptions for f and *MAC* [Bellare and Rogaway, 1993], the success probability δ of an adversary A for a grouping protocol is bounded from above by 2^{-d} where d is the message length. □

Discussion: Our group-yoking protocol can be adapted to the "Minimalist MAC" protocol proposed by Juels. However, as discussed above, a one-time proof is not very useful given the problem assumption that the reader may be malicious.

To prevent an adversary from replaying old proofs, the verifier can store counter values of tags obtained from the latest verified-correct yoking proofs in which tags participated. A replay attack will use the counter value that is less than or equal to the last recorded counter for the tag. Storing tag counter values also allows for a temporal ordering of the yoking proofs, which may be desired. If there is more than one verifier in the system, the verifiers need to be able to share the tags' counter values to prevent replay attacks.

Having counters on-board tags allows for temporal ordering of the yoking proofs and guarantees that a keyed hash function will never be computed on the same point twice. However, such tag counters require tags to maintain a persistent state between several runs of the protocol. An alternative is to replace counter values with random numbers generated on-board the tags (the tags will need to be equipped with pseudo-random number generators and maintain secret seeds or have truly random number generators). However, if random numbers are used instead of counters, and a replay of past proofs is considered to be an attack, the verifier should store all previous yoking proofs and compare them to the new proof.

We made an assumption (see Section 2) that the reader cannot physically steal secrets from the tags. We can relax this assumption somewhat. Assuming that a malicious reader did not read one or more of the tags, and is pressed for a proof by the verifier, then the owner of the reader can try to find the "first tag" and physically steal the secret key from it, allowing the reader to complete the protocol. To prevent such an attack, each tag can update its secret key in a forward secure manner [Bellare & Yee, 2003]. For example, a one-way hash function can be used to update a tag's key at the time of a tag's counter update. The old key is then securely discarded. For this scheme to be practical, tags should maintain counters, instead of generating random numbers as timestamps, to allow the verifier to quickly determine the secret key that each tag used for computation.

In yoking proofs it is critical to rapidly establish communication with *all* the tags, otherwise the yoking-proof can not be created within the required time interval. This issue raises the object detectability problem [Bolotnyy & Robins, 2005] [Bolotnyy & Robins, 2007a]. Note, multiple *connected* tags, or *multi-tags*, can help address the object detection problem [Bolotnyy & Robins, 2005] [Bolotnyy, 2007].

6. Anonymous yoking

Observe that Juels's "yoking proof" protocol and our generalization do not hide the identities of tags - each tag sends its identifier and its counter value to the reader in the clear. Before the execution of the protocol, the reader is unaware of the identities of tags, and running the yoking protocol will reveal them to it. In some practical scenarios, we would like to preserve the identity of objects associated with the tags (i.e. not reveal tag IDs to untrusted readers). We therefore introduce a new problem formulation, called *anonymous yoking*, which in addition to the requirements of a "yoking-proof" problem, requires tags to preserve their privacy.

The protocol that we develop for anonymous yoking is very similar to the generalized "yoking proof" protocol discussed above, yet certain differences and details are important. Let $f: \{0, 1\}^d \times \{0, 1\}^* \rightarrow \{0, 1\}^d$ be a keyed hash function. Upon the reader's request, each tag will generate a random number r, and compute $a = f_x(r, value)$, where x is a secret key stored on a tag and *value* is an output of the previous tag in the chain, as in our generalized "yoking proof" scenario above. The first tag sets *value* equal to 0. Each tag will respond to the reader's request by sending (r, a) to it, and the first tag will close the chain. The detailed algorithms for the reader, a tag, and the verifier can be found in the author's Ph.D. thesis [Bolotnyy, 2007]. The proof of security is very similar to the one for the generalized protocol [Bolotnyy, 2007]. This anonymous yoking-proof protocol is privacy preserving in the Strong Privacy model of [Juels & Weis, 2006].

Note that the process of determining the tags' identifiers and verifying the proof are combined in the verification steps. The verifier will try to determine which secrets were used to compute each a_i given r_i. Since a_i is a function of a_{i-1} for all $2 \leq i \leq k$, the process of determining the secrets and verifying the proof coincide. To determine the tags' identifiers and to verify the protocol, it will take the verifier $O(k \cdot n)$ time where n is the total number of possible tags and k is the number of tags participating in the protocol. The running time can be further reduced to $O(k \cdot log(n))$ if the approach suggested by Molnar et al. [Molnar and Wagner, 2004] is used, namely arranging the tags at the leaves of a tree and associating a secret to each edge in the tree. Each tag stores all the secrets on the path from the root to the leaf where it is located. Instead of sending (r_i, a_i) to the reader, each tag will send r_i, $a_i^1 = f_{s_1}(r_i, a_{i-1}), \ldots, a_i^{\text{tree-depth}} = f_{s_{tree-depth}}(r_i, a_{i-1})$ where $s_1, \ldots, s_{\text{tree-depth}}$ are secrets from the root of the tree to the leaf where a tag is located. This algorithm modification will allow the verifier to determine a tag's identity in $O(log(n))$ time by finding the right path from the root to the leaf. However, this speedup in tag identification suggested in [Molnar & Wagner, 2004] comes at a cost, as leakage of secrets of one or more tags poses a privacy threat to other tags. The extent of the threat is analyzed in [Avoine et al., 2005].

Observe that anonymous yoking allows an adversary or even an innocent transient entity to inject an arbitrary ("yoking-proof" supporting) tag into the "yoking-proof" construction. If the verifier does not possess the injected tag's secret key, this action will prevent the verifier from verifying the complete proof, thus causing a denial of service attack.

7. Speeding up the yoking protocols

The run-time of the "yoking-proof" protocol is the sum of the time to perform encryption, the time to establish a reliable reader-to-tag communication channel, and the time to decode the reader requests and to encode the responses. We estimate that the establishment of the communication channel, and the decode and encode times take about 1ms per tag, assuming about 1,000 security-free tags can be identified per second. The majority of the protocol time will be spent on encryption operations. Therefore, we concentrate on encryption operations for the protocol run-time analysis. The tag that starts and closes the yoking chain performs 2 encryptions and the tags in the middle of the chain perform 1 encryption. Therefore, to yoke k tags requires $k + 1$ encryptions. The time to perform a single encryption depends on the frequency of a tag/reader communication, the communication distance, the communication standard, the power consumption of a tag, the cryptographic algorithm, and the length of the secret key.

For example, for high RFID communication frequency of 13.56MHz, ISO/EIC 18000 compliant standard, and 128-bit secret key AES implementations suggested by [Feldhofer et al., 2004] with 15μA current for AES module and 100KHz AES module clock frequency, one encryption takes about 10ms. So, under these conditions we can yoke ~ 36 tags and remain within the FCC required 400ms communication window ($36 \cdot 10ms + 36 \cdot 1ms < 400ms$). There may be applications where over 40 tags need to be yoked, or where a novel minimalist RFID encryption takes long computation times, or scenarios with real-time performance requirements (i.e., multiple yokings per second). For these types of applications, the "yoking-proof" protocol should be sped up.

The yoking-proof creation can be sped up by splitting the circular chain of dependent MACs into a group of arcs, where each arc consists of a sequence of dependent MACs, and where the adjacent arcs are inter-dependent. Each arc has a single element that plays the role of the "first" and the "last" tag. Let $ID_1, \ldots, ID_k$ be the tags' identifiers sorted by the tags ID, or by the random numbers generated on-board the tags for anonymous yoking. We split the sorted list of identifiers into the desired number of groups g (e.g., $ID_1, \ldots, ID_{i1}, ID_{i1+1}, \ldots, ID_{i2}, \ldots, ID_{i_g}, \ldots, ID_k$).

For example, ID_1 is the "first" and the "last" element of the first arc. It starts the chain of its group ($ID_1, \ldots, ID_{i1}$) as described in the generalized "yoking-proof" protocol, and it closes the chain of the group $ID_{i_g}, \ldots, ID_k$. In other words, the first element of each arc starts the chain of the arc and closes the chain of the preceding arc (see Figure 4).

Note that the protocol requires multiple readers or a single reader with multiple antennas. In addition, the protocol should incorporate a medium access control scheme that allows the reader(s) to communicate with more than one tag at a time to avoid/minimize tag response collisions. The time to create the yoking proof is the sum of the reader communication times with the tags belonging to the longest arc. Therefore, the overall speedup factor resulting from partitioning the tag set into arcs, can ideally approach the number of arcs.

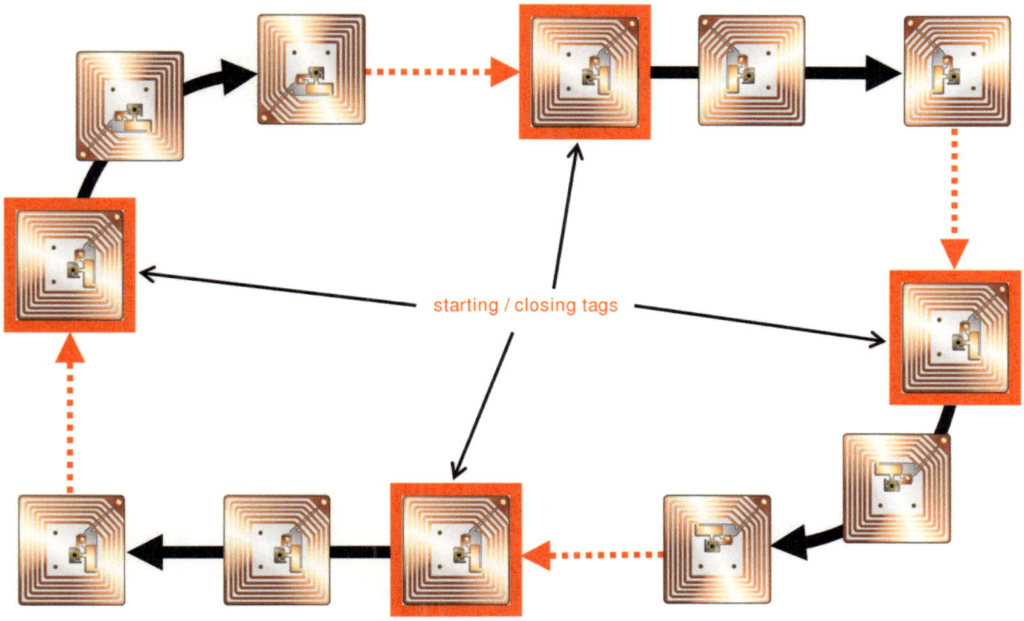

Fig. 4. Group "yoking-proof" protocol speedup. The circular chain is split into 4 arcs. Starting/closing tags are clearly marked and directions of the chain creation are shown with bold arrows. Dashed arrows represent closing operations with the source tag of the arrow proving the input to the destination tag.

The suggested speedup does not effect the security statement of the theorem for generalized yoking-proof protocol given above. The proof is also very similar and therefore omitted. The main difference in the proof is that there is more than one starting and closing operations.

7.1 Message authentication code implementations

A message authentication code (MAC) aboard a tag can be implemented using a standard cryptographic hash-based MAC (i.e., HMAC). Alternatively, Lamport's one-time signature scheme [Lamport, 1979] can be used, as noted in [Juels, 2004], where each bit position of the signature has two associated secrets - one for 0 and one for 1. The signature is an ordered sequence of secrets that correspond to 0 or 1 in each bit position. However, this scheme requires a prohibitably large memory on-board a tag. A more "minimalistic" approach, where each secret is just a single bit, was suggested by Juels [Juels, 2004]. To avoid simple forgeries, he suggests lengthening the message size, and making the message space sparse [Juels, 2004]. In our research [Bolotnyy & Robins, 2007b] [Bolotnyy, 2007] we showed how MACs can be implemented using physical hash functions (PUFs) that require an order-of-magnitude less hardware to implement than standard cryptographic hash functions. In addition, PUF-based MAC implementations allow messages to be signed more than once. Still, due to some restrictions of PUF-based MAC (limited message space or required tag presence during the signature verification), applicability of PUF-based MAC to yoking proofs may be limited.

The signatures of PUF-based MAC constructions can be long. To reduce the length of the signatures and to make them conform to the input length of each tag, a reader can apply a publicly known collision resistant one-way hash function to each MAC. The verifier will need to apply the same hash function to the signatures when verifying proofs.

8. Inter-tag communication

"Yoking-proofs" provide a good example of passive tags communicating with each other through the reader (Figure 5). In [Juels, 2004], inter-tag communication is performed as part of a larger protocol, and it did not receive enough attention as a general technique in its own right. We have found no literature that specifically discusses tag-to-tag communication between powerless tags. In the few sources where the term "tag-to-tag communication"[2] is mentioned, it refers to active (battery-powered) tags that can communicate with each other directly.

Fig. 5. Inter-Tag communication. Passive tags can communicate with each other through the reader.

In almost all RFID systems discussed in the literature, readers query RFID tags for their IDs or for some other data, and pass that data to the back-end server for processing. We envision a different paradigm of RFID systems where RFID tags, even passive or semi-passive ones, can communicate amongst themselves, using the readers as intermediaries. Such inter-tag communication capability can create additional heterogeneity in ubiquitous computing, where powerful wireless devices (e.g. readers) are able to initiate communication on their own, and communication-powerless devices (e.g. semi-passive tags, passive tags) can still communicate with their peers through powerful devices. Inter-tag communication creates opportunities for new RFID applications, but it also creates new security challenges (e.g., securing the communication channel between the tags even if "connecting" readers are untrusted, and maintaining the integrity of the transmitted data.)
Next, we give examples of applications where inter-tag communication can be beneficially utilized.

8.1 Example 1: battery-free sensing

In [Philipose et al., 2005] the authors provide evidence that battery-less tags can be used for sensing by harvesting their energy from the readers using RFID technology. Consequently,

[2] For example, [Brooke, 2005] contains the term "tag-to-tag communication" without explicitly referring to passive tags, the assumption being that the tags in question are active.

inter-tag communication can allow passive tags to share sensing data, and use it to enable future sensing activities, just like in wireless sensor networks.

8.2 Example 2: tags as mailboxes

We envision a scenario where off-line readers can exchange messages by using tags as "mailboxes". Here, one reader can send a message to another reader by writing a message onto an appropriate tag, and this message will later be retrieved by the recipient reader. Tags can also be used as proxies, permitting readers to send and receive data from tags that are outside their reading range. The data will propagate from one tag to another tag which is closer to the destination. Essentially, RFID tags can serve as powerless distributed storage devices. In this application of inter-tag communication, readers must verify the data that they read off the tags to ensure that it is not a virus (e.g., see attacks discussed by Rieback et al. [Rieback et al., 2006]).

8.3 Example 3: centralized authentication

The tag population can be partitioned into groups, with specific tags designated as *group leaders*. A group's leader is in charge of reader authentication and access control to data stored on-board all tags belonging to a group. An example of centralized authentication is a pallet tagged with a group leader tag, and individually-tagged items within the pallet serving as the group members. Individual tag access is authorized by a group leader (a powerful tag), which contains access policy information that can be updated over time. One of the benefits of such an approach is the centralization of group policy information, which allows for faster and more uniform policy updates. In addition, the tag complexity can be reduced by placing major functions on a single (group leader) tag.

Next, we describe the protocol for this centralized authentication scenario. Let k be the number of tags in the group, excluding the group leader. Let $f, h : \{0, 1\}^d \times \{0, 1\}^d \rightarrow \{0, 1\}^d$ be two pseudo-random functions. Let s be a secret key shared by the group leader and the reader, and for $i = 1, \ldots, k$ let s_i be the secret keys of tags T_i, known to the group leader. At the beginning of the protocol, the reader sends an access request to tag T_i and the tag responds with a *nonce* (i.e., a random "number used once"). The reader will then ask the group leader for an access certificate to tag T_i, providing it with the nonce received from the tag. The group leader will authenticate the reader using a challenge-response protocol and issue a certificate c based on the tag supplied nonce, if allowed by the access control policy. The tag will verify that the access certificate is valid and grant data access to the reader. The general form of the reader and group leader communication protocol is shown pictorially in Figure 6, and the algorithms for the reader, a tag, and the group leader are given in the author's Ph.D. thesis [Bolotnyy, 2007].

Tag data access control policy can be group-based or tag-based (i.e. the identity of a tag is not a secret, but access to the data stored on-board the tag is granted only to authorized readers). If the policy is tag-based, the reader will need to specify to the group leader the identity of the tag whose data it wants to access. Note that this scenario reduces the number of secrets that a reader needs to possess in order to collect information from the tags, since the reader only needs to share secrets with group leaders.

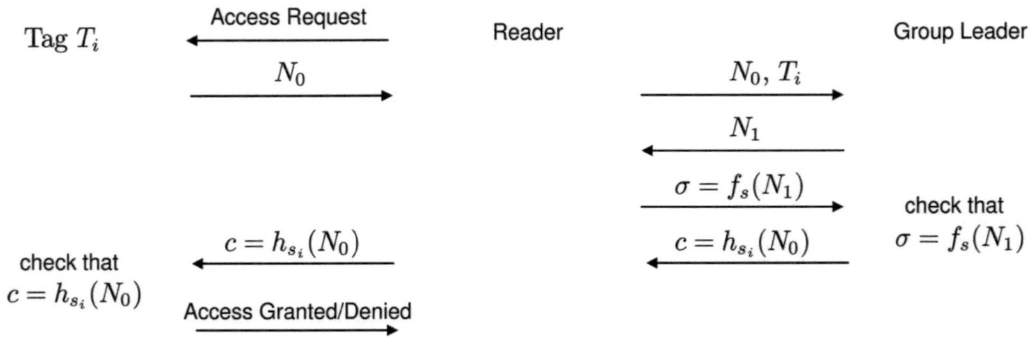

Fig. 6. Centralized reader authentication. The group leader issues a certificate to the reader, allowing access to tag T_i. Here *f* is the pseudo-random function that the reader and the group leader use for authentication, and *s* is the secret they share; *h* is the pseudo-random function that the group leader and the tags within a group use for authentication; and s_i is the secret that the group leader and tag T_i share.

8.4 Example 4: distributed access control

Inter-tag communication can also be used for distributed access control. A master tag grants the reader access to a resource only if all of its subordinate tags first authorize access to all dependent resources (i.e. all dependent *sub-resources* have to be acquired first, before the main resource itself is acquired). Examples of distributed access control include "safe deposit box opening" that requires multiple keys, and a system of doors where access to a door depends on successful access to a group of other doors.

Distributed Access Control Problem: A reader seeks to gain access to information, data or facility controlled by a master tag. To gain access to the master tag, the reader first needs to obtain recent authentication certificates from all subordinate tags, (i.e., all proofs must be generated within some time window *t*).

To solve this problem, the reader can perform an authentication algorithm with each subordinate tag based on a nonce provided by the master tag, and then present the collected authentication certificates to the master tag for verification. The master tag can time the reader, as in yoking-proofs, to verify the timely presentation of proofs.

However, if we also require the proofs to be verifiable at any time after they have been generated, the problem becomes more difficult. To provide such a proof, which joins together proofs from a number of tags, we can use a modified version of our anonymous "yoking-proof" solution described above. The reader will generate a proof, showing that it read all tags within time *t*, and then present this proof to the master tag. The "yoking-proof" will need to be augmented to allow for preliminary authentication, and timed from the start of the authentication rather than from the yoking construction. The master tag will then verify the validity of the "yoking-proof".

Another application of inter-tag communication is subordinate reader authentication / authorization. A group leader tag provides access to the reader or performs reader authentication if and only if the reader has successfully authenticated itself to other tags. To gain access to the main object/data, the reader needs to prove to the guardian tag that it has successfully authenticated itself to the subordinate tags. To do so, the reader authenticates itself to each subordinate tag and provides a proof to the group leader that it was successfully authenticated by all tags. Timing is important here as we do not want a reader to present an "old authentication" or an authentication it stole by eavesdropping on communication by another reader.

8.5 Example 5: location access control

Yet another application of inter-tag communication is location access control. For example, several readers want to gain access to a resource if all of them are present at specific locations. Each reader can prove its proximity to a corresponding tag using a distance-bounding protocol [Sastry et al., 2003] [Hancke & Kuhn, 2005], and the authentication result is then shared among the tags. Access to the resource is granted if the protocol completes successfully.

The above list of examples does not exhaust all possible applications of inter-tag communication, and future research can focus on finding new applications.

9. Conclusion

We reviewed the basic "yoking-proof" protocol for a pair of tags suggested by Juels and discovered a critical omission in his protocol. We also described weaknesses in attempts by other researchers to generalize the protocol to a yoking-proof problem. We designed a protocol that creates a proof that an arbitrarily large group of RFID tags are read within a given time bound. This protocol generalizes the basic "yoking" protocol by Juels. The yoking-proof is improbable to forge and it is verifiable off-line by a trusted verifier. We modified the security requirements of the yoking-proof problem, requiring the system to maintain privacy, and gave an algorithm for this new anonymous yoking problem formulation. We also described a way to speedup our "yoking-proof" protocols.

We briefly discussed viable low-cost message authentication code (MAC) implementations. MAC implementations relying on physical unclonable functions (PUFs) require less hardware resources than the known cryptographic implementations of keyed hash functions and therefore, may be applicable to some applications of yoking-proofs. We proposed a new paradigm of inter-tag communication between passive RFID tags where tags communicate with each other through the reader. We put forward a number of interesting applications of inter-tag communication such as battery-free sensing, and distributed access control, among others. We suggested another type of "yoking-proof" for distributed access control in RFID, and described possible solutions.

The cost of tags capable to implement our generalized yoking proof protocol is likely to be greater than the cost of a basic RFID tag that only replies to the reader with a constant tag identifier. The tag cost may be even higher if in addition to the hash function, a PUF circuit is added to the tag to provide physical protection of the tag key(s) and counter value. Also, due to protocol timing constraints and the time to compute a hash value onboard a tag, it may not be possible to join all the tags within the required time period.

In this case, it is interesting to consider the use of a logical clock instead of a physical clock. Moreover, the tag capacitor charge and discharge rate may be imprecise, resulting in variable tag timeout. Since the capacitor charging rate is based on the power supplied to the tag by the reader, an adversarial reader can reduce the power supply to the first tag to a minimum in order to extend the object yoking period. Some provisions onboard a tag can be added to prevent the tag from communicating with the reader until the tag capacitor charges fully.
Future research opportunities lie in the development of new applications of inter-tag communication, utilizing the ability of tags to communicate with each other through the reader. In addition, much future work is needed in the area of PUF-based security, which may allow for low-cost implementations of yoking-proofs protocols. Inter-tag communication and yoking-proofs in particular give good examples of RFID that go beyond simple tracking of objects. Looking broadly at possible RIFD applications opens previously unexplored possibilities, thus paving the way to ubiquitous RFID deployments.

10. Acknowledgement

This research was supported by grant CNS-0716635 from the U.S. National Science Foundation.

11. References

[Avoine et al., 2005] Avoine, G., Dysli, E., and Oechslin, P. (2005). Reducing time complexity in rfid systems. In Preneel, B. and Tavares, S., editors, *Selected Areas in Cryptography (SAC), Lecture Notes in Computer Science*, volume 3897, pages 291–306, New York. Springer-Verlag.

[Bellare and Rogaway, 1993] Bellare, M. and Rogaway, P. (1993). Random oracles are practical: A paradigm for designing efficient protocols. In *Proc. ACM Conference on Computer and Communications Security*, pages 62–73.

[Bellare and Yee, 2003] Bellare, M. and Yee, B. (2003). Forward-security in private-key cryptography. In Joye, M., editor, *Topics in Cryptology - CT-RSA2003, RSA Conference*, volume 2612, pages 1-18. Lecture Notes in Computer Science.

[Bolotnyy, 2007] Bolotnyy, L. (2007). New Directions in Reliability, Security and Privacy in Radio Frequency Identification Systems. Ph.D. thesis, University of Virginia.

[Bolotnyy and Robins, 2005] Bolotnyy, L. and Robins, G. (2005). Multi-tag radio frequency identification systems. In *Proc. IEEE Workshop on Automatic Identification Advanced Technologies (Auto-ID)*, pages 83–88.

[Bolotnyy and Robins, 2006] Bolotnyy, L. and Robins, G. (2006). Generalized 'yoking proofs' for a group of radio frequency identification tags. In *International Conference on Mobile and Ubiquitous Systems (Mobiquitous)*, San Jose, CA.

[Bolotnyy and Robins, 2007a] Bolotnyy, L. and Robins, G. (2007a). Multi-tag rfid systems. International Journal of Internet and Protocol Technology, Special issue on RFID: Technologies, Applications, and Trends, 2(3/4).

[Bolotnyy and Robins, 2007b] Bolotnyy, L. and Robins, G. (2007b). Physically unclonable function -based security and privacy in rfid systems. In *Proc. IEEE International Conference on Pervasive Computing and Communications (PerCom 2007)*, pages 211–218, New York.

[Brooke, 2005] Brooke, M. (2005). Common mistakes, uncommon best practices. http://www.rfidjournal.com/article/articleview/1483/1/15/.

[EPCglobal, 2006] EPCglobal (2006). Epc radio-frequency identity protocols class-1 generation-2 uhf rfid version 1.0.9.

[Feldhofer et al., 2004] Feldhofer, M., Dominikus, S., and Wolkerstorfer, J. (2004). Strong authentication for rfid systems using the aes algorithm. In Preneel, B. and Tavares, S., editors, *Workshop on Cryptographic Hardware and Embedded Systems (CHES 2004), Lecture Notes in Computer Science*, volume 3156, pages 357–370. Springer-Verlag.

[Hancke and Kuhn, 2005] Hancke, G. and Kuhn, M. (2005). An rfid distance bounding protocol. In First International Conference on Security and Privacy for Emerging Areas in Communications Networks (SECURECOMM), pages 67 – 73.

[Juels, 2004] Juels, A. (2004). 'yoking-proofs' for rfid tags. In Sandhu, R. and Thomas, R., editors, *International Workshop on Pervasive Computing and Communication Security*, pages 138–143, Orlando, FL, USA.

[Juels and Weis, 2006] Juels, A. and Weis, S. (2006). Defining strong privacy for rfid. Technical Report Report 2006/137, http://eprint.iacr.org/2006/137, Cryptology ePrint Archieve.

[Lamport, 1979] Lamport, L. (1979). Constructing digital signatures from a one way function. Technical Report Technical Report CSL-98, SRI International.

[Molnar and Wagner, 2004] Molnar, D. and Wagner, D. (2004). Privacy and security in library rfid issues, practices, and architecture. In *Proc. ACM Conference on Computer and Communications Security*, pages 210–219, Washington, DC, USA.

[Peris-Lopez et al., 2007] Peris-Lopez, P., Hernandez-Castro, J., Estevez-Tapiador, J., and Ribagorda, A. (2007). Solving the simultaneous scanning problem anonymously: Clumping proofs for rfid tags. In *Third International Workshop on Security, Privacy and Trust in Pervasive and Ubiquitous Computing (SECPerU)*, pages 55–60.

[Philipose et al., 2005] Philipose, M., Smith, J., Jiang, B., Mamishev, A., Roy, S., and Sundara-Rajan, K. (2005). Battery- free wireless identification and sensing. In *Pervasive Computing*.

[Piramuthu, 2006] Piramuthu, S. (2006). On existence proofs for multiple rfid tags. In In IEEE Intl. Conference on Security, Privacy and Trust in Pervasive and Ubiquitous Computing (SecPerU), Lyon, France.

[Rieback et al., 2006] Rieback, M., Crispo, B., and Tanenbaum, A. (2006). Is your cat infected with a computer virus? In *Fourth Annual IEEE International Conference on Pervasive Computing and Communications (PerCom)*.

[Saito and Sakurai, 2005] Saito, J. and Sakurai, K. (2005). Grouping proof for rfid tags. In *International Conference on Advanced Information Networking and Applications (AINA)*, volume 2, pages 621–624, Taiwan.

[Sastry et al., 2003] Sastry, N., Shankar, U., and Wagner, D. (2003). Secure verification of location claims. In *Proceedings of the ACM Workshop on Wireless Security*, pages 1 – 10, San Diego, CA.

25

Shared Tag RFID System for Multiple Application Objects

Ji-Yeon Kim[1], Jong-Jin Jung[1], Yun-Seok Chang[1] and Geun-Sik Jo[2]
[1] *Daejin University,*
[2] *Inha University*
Korea

1. Introduction

Recently, the information technology has evolved toward the ubiquitous environment accessible to the network everywhere and every time. The ubiquitous environment can provide easy access to the devices and make one's economical benefit also. The RFID system is an important core technology in ubiquitous environment. RFID system consists of contactless devices to communicate each other by radio frequency. It provides technologies of automatic object identification in invisible range, read/write function and adaptability against various circumstances. These advantages make RFID system to be applied in various fields and expect one of the big markets in the area of human life such as traffic card system, toll gate system, logistics, access control, etc. As RFID tags identify many different types of objects, it is going to increase the tags that have to be carry in individual life. But it is uneasy for a person to control many tags in a hand because traditional RFID systems have restriction that is one tag per each object and it is difficult to distinguish tags without some kind of effort. That is why a tag is used to store identifying information just for a single object in common RFID applications.

We propose a multiple objects tag structure which can be shared by many different applications. As a tag is used to identify only one type of object, it is expected to have many tags increasingly by people or things. If a tag can be shared by many RFID application objects, it will be more efficient to RFID users and will be helpful to the information integration as well as device sharing. So, we design a RFID tag structure which has many different identifiers. This tag can be used to access many different RFID applications.

We also propose an efficient authentication protocol adapted to the multiple objects tag structure. We consider robustness of the authentication protocol against various attacks in the proposed scheme. RFID system often makes serious violation of privacy and security caused by various attacks through the weak wireless interface. Eavesdropping, location tracking, spoofing, message losses or replay attack can threaten RFID components anywhere and anytime. To protect RFID system from these kinds of attacks, researchers have studied several schemes such as Faraday cage scheme (mCloak, 2003), blocker-tag scheme (Juels et al., 2003), hash lock scheme (Weis et al., 2003), randomized hash lock (Weis et al., 2003), hash chain (Ohkubo et al., 2003) and variable ID scheme (Saito & Sakurai, 2003), etc. However, these schemes have restrictions that each object is just corresponding to one tag. So, it is difficult to distinguish tags without additional cost. That is why a tag is used to store identifying information just for a single object in common RFID applications. Therefore, we design an authentication protocol to support multiple objects based RFID system. Especially, we focus on efficiency of the authentication procedure by considering security levels of objects stored in a tag. Multiple objects-based tag can be shared by various kinds of RFID applications. Therefore, we classify various RFID application objects shared in a tag into different groups by security levels and apply the appropriate authentication procedure to the object according to its security level. In this way, our proposed RFID authentication protocol is designed to adjust to the multiple objects tag structure and to operate differently according to the security level.

The proposed authentication protocol can guarantee for various attacks. We evaluated the efficiency and stability for the proposed scheme compared with common single tag RFID scheme through various experimental results. As the result, the proposed scheme maintained stable operation through the test of error rates and made reasonable results of computation time for authentication procedure in spite of its high security.

2. Basic requirements

The standardization organizations for RFID tags are JTC1(Joint Technical Committee 1) by ISO/IEC and EPC Global by GS1(Global Standard 1). Tag identifier is a number to distinguish a tag from others in communication with readers. International standard observes the structure of the number (TTAR-06.0013, 2003). Permanent unique ID named as chip ID or tag ID is masked within a tag by tag producer according to the ISO/IEC 15963 standard. Another identifier is item ID which is used to distinguish the tagged objects. Tag memory has the item ID under user's decision and standardization on item ID is still under discussing. Therefore, we would like to use the item ID as the name of object ID on multiple objects access. Fig. 1 shows the typical example of the multiple item IDs in a same tag.

RFID systems often make serious violation of privacy and security caused by various attacks through the weakness of their wireless interface. Vulnerabilities to eavesdropping, location tracking, spoofing, message losses or replay attack can threaten RFID components. These attacks may affect individual privacy and information security. Therefore, efficient security mechanism against attacks must be considered when RFID applications are designed.

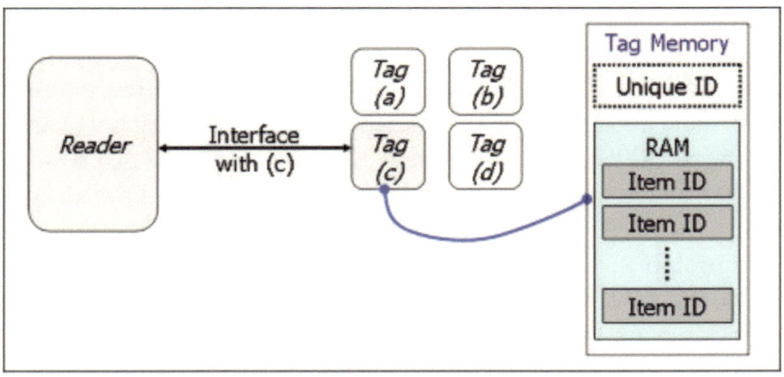

Fig. 1. Concept of tag identifier

Unprotected tags could be read by unauthorized readers. Attackers are used to eavesdrop on the general information stored in a tag such as serial number or code number from readers or tags though they can not understand the mean of the number. The attackers retransmit the eavesdropping data to the server and find out the critical information. With the eavesdropping data, the attackers also detect the tag location and analyze traffics. Therefore, RFID system has to be designed with data encryption and authentication protocol for preventing the attackers from replay of tapped data and location tracking. This reason makes the authentication protocol of RFID system may change the response of corresponding tag to the query of reader after each session. Spoofing is another type of attacks and it means that attacker joins in authentication protocol in the disguise of authorized tag or reader. To prevent the spoofing attack, RFID system has to control the authority of tag access in authentication protocol and must block the attacker' illegal information gathering. In addition to the above attacks, the attackers often intercept communications between reader and tag. This behavior causes loss of data which is important to authentication process. Therefore, the safe and reliable RFID systems should detect the interference coming from various kind of attack as many as it can.

3. Related works

Several papers have proposed the solution schemes against the attacks of privacy threat. There are two types of scheme: One is the physical scheme such as kill-command scheme, Faraday cage scheme and blocker-tag scheme. The other is the encryption scheme such as hash lock scheme, randomized hash lock, hash chain and variable ID scheme. In this study, we have focused on the encryption methods which are hot among researchers lately. The hash lock scheme is simple access control mechanism based on one-way hash function. It uses metaID to process authentication between reader and tag. The metaID is a temporary ID generated from one-way hash function with single key. Both the reader and tag store the metaID separately and match the key with metaID during the authentication process. A tag responds to all queries with only its metaID and decides whether the tag offers it or not. This scheme only requires a hash function on the tag and key management on the back-end database. So, it would be the best one of the cost-efficient solutions in the near future. Based on the difficulty of inverting a one-way function, this scheme also prevents unauthorized readers from reading tag contents. Maybe, spoofing attempts may be detected under this scheme but not prevented. The hash lock scheme can not prevent replay attack and location tracking either because the metaID has constant value. Randomized hash lock scheme improves the hash lock scheme and uses variable metaID. It can generate a different metaID with random number generator in every session. Therefore, a tag would not respond to queries from unauthorized readers. This scheme can prevent RFID tag from tracking but not replay attack and spoofing. Hash chain scheme uses two different hash functions to change the response message for reader. This scheme can prevent tag tracking and replay attack but still has weakness against spoofing. And actually, it is not practicable on account of their demand for circuit size and operation power because a tag has to keep two hash functions. Variable ID scheme changes tag ID by a random value in every session. For the replay attack, the scheme keeps transaction ID (TID) and last successful transaction (LST) in each session. However, this scheme may allow adversaries to track when LST was not updated on occurring message loss during the session.

4. RFID scheme for accessing multiple objects

4.1 Tag structure

Multiple objects tag is a new type of RFID tag structure that has multiple objects simultaneously for many RFID readers. As increasing the RFID system applies on individual life, people are expected to have several tags to identify the different types of objects. It causes serious problem on managing different tags in one's pocket and needs a new method to handle multiple objects in a simple way. But it is not easy to control many tags because traditional RFID systems can handle only one tag per each object. In ubiquitous computing environment, information integrating and device sharing have been increased on the various application areas and now we need an efficient solution to simplify the control and management method. Therefore, if there is a kind of method to share one tag for many RFID applications, it can be suitable for the ubiquitous computing environment in many ways. We propose a multiple objects tag structure which can be shared by different RFID applications. That is, the proposed tag is recognized by many different RFID readers. Fig. 2 shows the concept of multiple objects tag.

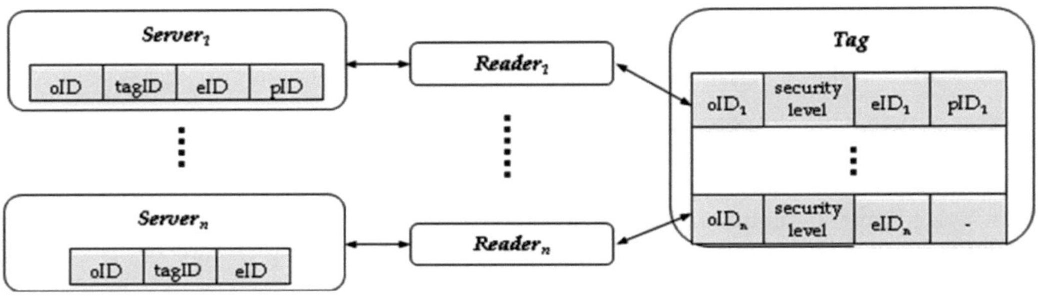

Fig. 2. Proposed multiple objects tag structure

As shown in Fig. 2, we need some types of IDs to design multiple objects tag. We define several types of data which are called tagID, oID, eID and pID. These IDs should be stored in tag memory and back-end database in server respectively as shown in Table 1.

RFID components	Stored data
Server	oID, tagID, eID, pID
Tag	oID, security level, eID, pID

Table 1. Data stored in server and tag

The meanings of each stored data are as follows:

- *tagID* is an unique identifier of a tag. It is used to distinguish tags each other in RFID system. In the proposed multiple objects tag, tagID is used as the key to create eID for distinguishing each distinct object.
- *oID*(object ID) is an identifier to distinguish different objects in multiple objects-based RFID system. In usual RFID systems, a key saved in a RFID tag is used to identify an object in a specific application. We define this key as oID. The multiple objects tag means that a single tag has more than an identifier for each distinct object with same *tagID*. This tag structure integrates different identifiers for different applications into data memory in a tag. The oID is stored in tag memory and server database

respectively. When a reader transmits a query to a specific tag, the query includes an oID.

- *eID*(encrypted ID) is an encrypted value to identify a certain object. In actual cases, multiple objects application causes security problem in multiple identifiers accesses and needs some appropriate method to tie up between identifier and object with security. To satisfying this specific need, we generate a new identifier with SEED encryption algorithm (IETF RFC 4269, 2003) using both tagID and oID. That is, the SEED algorithm takes both oID and tagID as input keys. The SEED is one of the famous block cipher algorithm developed by the KISA (Korean Information Security Agency) and broadly used throughout South Korean public and industry field. Since decryption of the SEED encryption value without input keys is very hard and expensive work, physical replication of tag is not worth for attacks. These characteristics of SEED provide a high level security and reliability in RFID systems. Therefore, we employ SEED algorithm on key encryption to identify a corresponding object. We call the encrypted identifier as *eID*. Different eIDs are generated by SEED algorithm using oIDs even if they have the same tagID. The numbers of eIDs are equal to the numbers of oIDs. These pairs of IDs are stored in a tag memory and back-end database respectively when the system is initialized. According to the object type, two or more eIDs can be stored in a tag. When a certain type of reader transmits query to a tag during wireless connection, the tag searches corresponding eID which is fit for the reader's object type.
- *pID*(partial ID) is a temporary version of eID which is generated by hash function embedded in a tag. It is a variant of eID made within a tag using random value transmitted from a reader. When a certain type of reader transmits query to a tag during wireless connection, the tag searches corresponding eID which is fit for the reader's object type. Then, the tag makes a variable ID with a random value which is transmitted from the reader when it queries to the tag. We call it *pID*. The pID is changed into different value at every session. The last successful pID is maintained in database and a tag. If the authentication process ends successfully, the tag and the server update the existing pID to prevent attack's threat. In the RFID system based on multiple objects tag, a tag or a server can perceive illegal adversaries' spoofing or replay attack by comparing its stored pID with the attacker's.

4.2 Authentication protocol using security level

To ensure the security and safety between tag and reader, we focused on encryption of IDs and authentication. As mentioned above, server and tags store the eIDs generated by SEED encryption algorithm respectively when RFID system initialize. The SEED algorithm provides a high level security and reliability in RFID systems. We also designed an authentication protocol by considering of robustness against privacy threats such as location tracking, spoofing, re-play and message loss. When we designed the protocol, we noticed the important characteristics about security of RFID applications. Multiple objects tag can be used on many kinds of RFID applications. These various applications are classified into two groups. One is that the security maintenance is very important such as financial system. The other is that the security maintenance is relatively less important. Therefore, we defined two types of security levels to represent strength of security for applications as high and low. In addition, we designed the authentication protocol which consists of high level procedure for high security level and low level procedure for low security level. As the result, we classify various RFID application objects into two groups with security level and apply the different authentication procedure to the groups. In this way, our proposed RFID authentication protocol is designed to adjust to the multiple objects tag structure and to operate differently according to the security level. The proposed authentication protocol includes the following steps.

1. The reader sends a query to the tag. The query includes a pair of values *(oID, RNo)*. *oID* represents the type of current RFID application object and *RNo* is a random value generated each session. *RNo* is composed of two parts of value in expression (a). *RNo* is not included in the query at the low level procedure.

$$RNo = R_{Forward\_no} \,||\, R_{Backward\_no} \quad \text{(a)}$$

2. The tag searches eID corresponding to the oID and check its security level. For the objects with high security level, it creates a pID for the eID using *RNo* by bit masking operation. If the tag memory has 128bits block size and holds the pair of (oID, eID) for an object in the same block, the size of eID can be *(128 – size of oID)* bits. We decide bi-directional masking points ($R_{Forward}$, $R_{Backward}$) for eID in the expressions (b) and (c).

$$R_{Forward} = R_{Forward\_no} \bmod (\text{sizeof eID} / 2) \quad \text{(b)}$$

$$R_{Backward} = R_{Backward\_no} \bmod (\text{sizeof eID} / 2) \quad \text{(c)}$$

$R_{Forward}$ is the starting point of forward masking in the eID and $R_{Backward}$ is the starting point of backward masking in the eID. The tag creates a pID through two steps. First step, it makes two 32 bits strings by bit masking operation using both $R_{Forward}$ and $R_{Backward}$. Next, it creates a 64 bits pID by concatenating two masking operation results. These operations are shown in Fig. 3. The creation of *pID* in this step is omitted in the low level procedure.

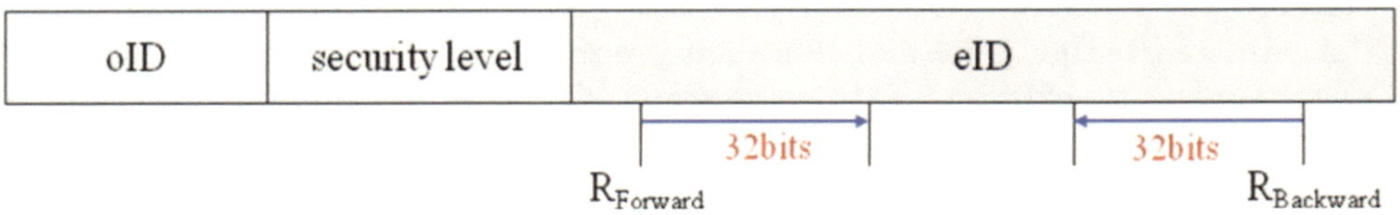

Fig. 3. Creation of pID

Since the tag maintains the pID generated through the last successful authentication process in its memory, the reader may be an illegal adversary in current session if pID in current session is equal to the stored value in tag memory. When the pID does not match to the one stored in tag memory, the tag recognizes the reader as legal and send the pID. If the illegal reader obtained the message at the step (1) in the former session, it would disguise as an authorized reader and retransmit the obtained message as replay attack. However, the tag can perceive replay attack by comparing stored pID with present one. In that case, the tag will not transmit the value to the illegal reader because the pID does not be changed. The pID is the most important key value in our scheme. As mentioned above, a temporal one-time key of pID can keep tag from spoofing attack and location tracking. The last successful pID can also prevent RFID components from illegal adversaries' spoofing or replay attack. Since the pID can be created by bit masking operation in a simple way, it can be easily processed in most of the low-cost RFID systems.

3. The tag transmits *pID* to the current reader for objects with high security level, whereas the tag transmits *eID* to the reader for objects with low security level. Therefore, the omission of step (2) can reduce computation time in the low level procedure.
4. After the reader receives the response message from the tag, it sends a message to the corresponding server. This message contains both *RNo* and *pID* in the high level procedure or *eID* in the low level procedure
5. Server retrieves proper encrypted ID with the *RNo* and partial ID pair in database. If there is a match, the server decrypts the encrypted ID by SEED decryption algorithm and resolves tag ID. At the end of this session, the server starts service to the reader.

The proposed authentication protocol is shown in Fig. 4.

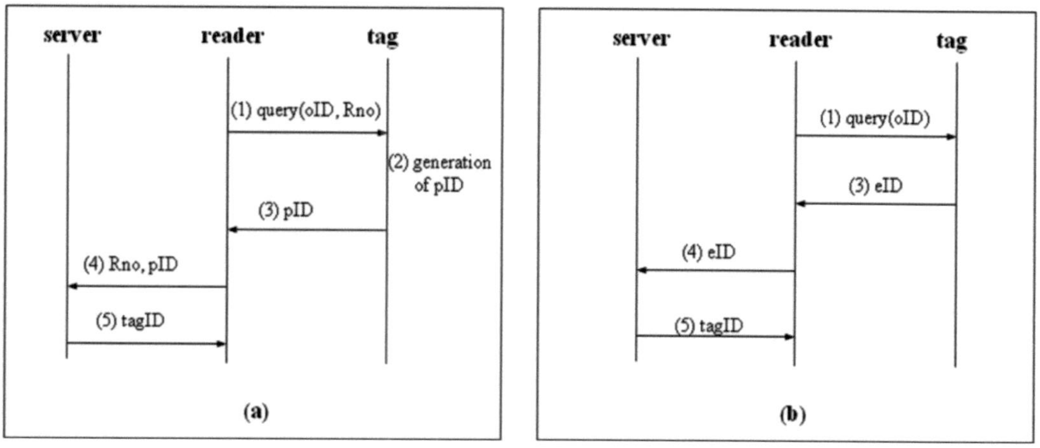

Fig. 4. The proposed authentication protocol: (a) high level procedure and (b) low level procedure

5. Evaluation

5.1 Evaluation of security

The proposed authentication protocol can keep its security for various kinds of attack. To prove this guarantee, we had evaluated the security of our scheme against the attacks as follows:

- Eavesdropping: The information tapped by attackers can be used for spoofing or location tracking. To solve this problem, a query message of reader should contain random value and then the tag should respond to the reader with variable value using this random value in the proposed scheme. In addition, this scheme will not let an attacker have any security information because the response message of variable pID is not complete value but the partial masking value of eID.
- Location tracking and replay attack: Since the tags respond differently for each query in our scheme, attackers can not catch the source of ID as well as the original value of ID. Therefore our scheme guarantees anonymity and supports blocking replay attacks.
- Spoofing: In our scheme, unauthorized readers have no way of joining in authorization process since readers should make a specified random value and transmit it to the tags. If an unauthorized reader uses the eavesdropped value again in current session, tags or server can be aware of abnormal status by comparing current pID with the stored one.

On the other hand, if an unauthorized tag catches a random value from the reader, it can not create the correct pID without eID either. In addition, though an unauthorized tag catches the pID by eavesdropping, the tag is unable to reuse it because the random value for pID is changed in every session.

- Message loss: The eID is non-volatile and fixed in the tag memory. Therefore, even if there is some data loss during the authentication process by transmission interference, the tag does not have to recover the data in the proposed scheme.

Table 2 shows the comparison of the proposed multiple objects tag scheme with the several existing schemes from the viewpoint of security.

	Location tracking	Spoofing	Replay attack	Message loss
Hash lock	x	x	x	x
Randomized hash lock	o	x	x	o
Hash chain	o	o	x	o
Variable ID	o	x	o	x
Proposed scheme	o	o	o	o

Table 2. Comparison in security (o: strong, x: weak)

5.2 Evaluation of efficiency

Almost existing schemes are not practicable on account of their demand for circuit size and computing power. Those schemes often require heavy operation in tag but the proposed scheme requires small operation. In the multiple objects RFID tag scheme, a tag computes bits masking once only for eID. This operation is simply compared with the other complicated hashing operations. On the other side, the server processes the tagID encryption with SEED algorithm for each tag in the proposed scheme when the system is initialized. The server also executes bit masking operations of n/2(n: numbers of ID in database) times on an average using message of (*RNo*, pID) to retrieve exact eID. If there is matched one, the server processes decryption of the retrieved eID once only by SEED algorithm. As the result, the proposed scheme guarantees more privacy and security than the existing schemes. Table 3 shows the comparison of the schemes from the viewpoint of efficiency.

	Tag	Reader	Server
Hash lock	hashing: 1	-	-
Randomized hash lock	randomizing: 1 hashing: 1	hashing: n/2	-
Hash chain	hashing: 2	-	hashing: (n/2)*i (i: update count)
Variable ID	hashing: 3	randomizing: 1	randomizing: 1 hashing: 3
Proposed scheme	bit masking: 1	randomizing: 1	bit masking: n/2

Table 3. Comparison in efficiency

5.3 Experimental results

We have simulated the proposed scheme and made experimental results about efficiency and stability for the scheme. To evaluate efficiency of the scheme, we have measured the computation time to perform one session in the proposed authentication procedures. One session means the stage to process the communication from server to tag via reader. To evaluate stability of the scheme, we have analyzed average error rates occurred during the authentication procedure. The experimental results were made under the environment as the reader of 13.56 MHz band, the tag with 2 Kbits of read/write memory and the communication standard of ISO15693. The experimental results are shown in Fig. 5.

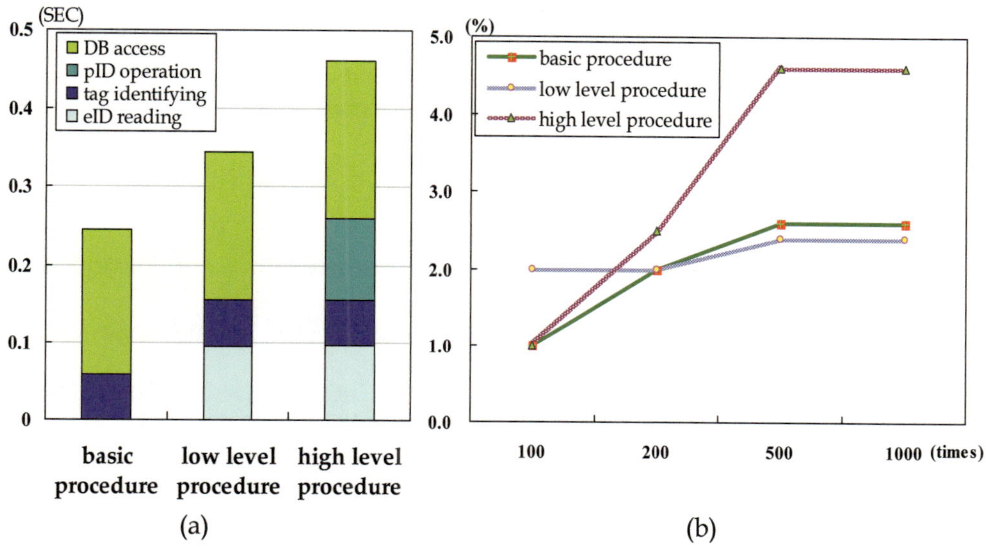

Fig. 5. Evaluation of the proposed scheme: (a) computation time and (b) error rate

Fig. 5(a) shows the comparison results of computation time values for three procedures: low level procedure, high level procedure and basic procedure of non-security level procedure in common RFID system. The values have been measured by 200 times repeatedly for different tags in the same condition. As shown in Fig. 5(a), the value of high level procedure is the largest in three cases but does not exceed double value of the basic procedure and the value of low level procedure is little more than the value of basic procedure. It means that the proposed scheme has reasonable computation time in spite of its strong security. In addition, we note that our scheme can save computation time of authentication procedure by using security levels because the scheme applies low level procedure to objects with low security level in multiple object tag. Fig. 5(b) shows the comparison results of average error rates for the above three procedures. As the result, all of the cases have similar error rates. The results show that the proposed multiple objects-based RFID system can be applied to areas of real environment.

6. Conclusion

We proposed a new RFID scheme including the multiple objects tag structure and the authentication protocol to give more privacy than existing schemes. The proposed multiple objects tag structure can maintain more than one object ID for different applications in a tag and allow applications to access them simultaneously. So, each application can share a tag on the multiple objects and it can result many tags in one tag. The proposed authentication protocol supports the proposed multiple objects tag structure and keeps the RFID components from various attacks without heavy system load. Especially, the protocol is designed to perform authentication procedure efficiently according to security level of object. We evaluated the security and efficiency of the proposed RFID scheme for several types of attacks. The evaluation results show that the proposed scheme has better performance in security and efficiency than existing schemes.

The multiple objects-based RFID scheme is just at the beginning in our study. We proposed basic concept on multiple objects tag structure in this study. For the deep research and implementation, RFID reader has to be physically redesigned to support the proposed multiple objects tag structure. We are going to design the RFID components fit with the proposed authentication protocol based on multiple objects tag with embedded SEED algorithm in further study.

7. References

Garfinkel, S. & Rosenberg, B. (2005). *RFID: Applications, Security and Privacy,* Addison Wesley, ISBN 0-321-29096-8

IETF RFC 4269 (2005). The SEED encryption algorithm, IETF(The Internet Engineering Task Force)

Juels, A.; Rivest, R. L. & Szydlo, M. (2003). The blocker tag: selective blocking of RFID tags for consumer privacy, *Proceedings of 10th ACM Conference on Computer and Communications Security(CCS 2003),* pp. 103-111, Washington DC., USA, October 2003

Kim, J.; Jung, J.; Ko, H.; Joe, S.; Lee, Y.; Chang, Y. & Lee, K. (2007). A design of authentication protocol for multi-key RFID tag, *Proceedings of APWeb/WAIM 2007,* pp. 644-653, Lecture Notes Computer Science 4537, Springer-Verlag

mCloak (2003). http://www.mobilecloak.com

Ohkubo, M.; Suzuki, K. & Kinoshita, S. (2003). Cryptographic approach to "Privacy-friendly" tags, *RFID Privacy Workshop @MIT,* November 15 2003

Saito, J. & Sakurai, K. (2004). Variable ID scheme of anonymity in RFID tags, *Proceedings of the 2004 Symposium on Cryptography and Information Security,* Vol. 1, pp. 713-718, Sendai, Japan, January 2004

TTAR-06.0013 (2006). Technical report on numbering an RFID tag, *TTA Technical Report,* TTA(Telecommunication Technology Association)

Weis, S.A.; Sarma, S.E.; Rivest, R.L. & Engels, D.W. (2003). Security and privacy aspects of low-cost radio frequency identification systems, *First International Conference on Security in Pervasive Computing,* pp.201-202, Lecture Notes in Computer Science 2802, Springer-Verlag

26

Object-Oriented Solutions for Information Storage on RFID Tags

Cristina Turcu, Remus Prodan, Marius Cerlinca and Tudor Cerlinca
Stefan cel Mare University of Suceava
Romania

1. Introduction

Already moving into the real world through a wide variety of applications, Radio Frequency Identification (RFID) technology uses radio waves to uniquely identify an entity (object, animal, or person). This data collection technology uses electronic tags to store identification data and other specific information, and a reader to read and write tags. A tag is a chip with an antenna. Tags fall into three categories: are active (battery- powered), passive (the reader signal is used for activation) or semi-passive (battery-assisted, activated by a signal from the reader). In certain tag types, the information on the tag is reprogrammable. There are existing and proposed RFID standards that deal with the air interface protocol (the way tags and readers communicate), data content (the way data is organized or formatted), conformance (ways to test whether products meet the standard) and applications (how standards are used on shipping labels, for example).
RFID solutions run at several frequencies:

- Low - from 125 KHz to 134 KHz (LF)
- High - 13.56MHz (HF)
- Ultra High - 860-960 MHz (UHF)
- Micro Wave - 2.45 GHz

The cost of simple RFID tags is likely to fall to roughly $0.05/unit in the next several years (Sarma, 2001), while tags as small as 0.4mm × 0.4mm, and thin enough to be embedded in paper are already commercially available (Takaragi et al., 2001). Such improvements in cost and size will ensure a rapid proliferation of RFID tags in many new areas.
The use of RFID is becoming more and more popular in industry, logistics, retail and other branches as an alternative to the barcode. In fact RFID tags are expected to replace conventional barcode labels due to their major benefits: high data storage capacity, read-write capability, read-speed rate, multiple entity identification, information updating, no-line-of-sight scanning, durability, and environmental resistance.
But how much data should be placed on RFID tags? There are two schools of thought:

1. as little as possible (just an ID);
2. more in support of efficiency and performance (e.g. item name, security data, etc.).

In the former case, the electronic product code (EPC) is a typical example. While the EPC standard continues to be adopted in various markets and employed in a wide range of applications (e.g. the retail supply chain), many RFID users are particularly interested in high-level functionality features to meet their own requirements. Using the EPC number as an identifier certainly provides benefits, but there are many applications that require additional memory on the tag in order to more fully meet the needs of many users (***, 2008). This paper considers the latter case and focuses on the general methods of data storage on a passive HF tag operating at 13.56 MHz. The International Organization for Standardization (ISO) has created standards that define how data is structured on the tag for specific applications. For example, ISO 11784 and 11785 describe the structure and the information content of the codes stored in the tag for RF identification of animals. But an ISO 11784 dedicated application allows only the identification of animals and cannot be used in other domains such as product identification, for instance. The current memory capacity for commercially available HF tags is typically 128 bytes, or 256 bytes. But standards-based ISO tags such as those operating at the high frequency (HF) can now provide for up to 8 Kilobits of memory. For example, the announcement by Hewlett Packard (HP) of its memory Spot technology provides for an RFID chip containing 4 Megabits of storage that can be written to and read multiple times (Greene, 2006). Anyway, by having over 128-bits of dedicated user memory, the tags allow the storage of additional item information and opens up new application areas. This available memory can be used to customize an application and allows users the flexibility that a standard EPC tag or ISO 11784 dedicated applications cannot fulfill.

Furthermore, as more applications make use of the same tag technology, memories of different capacities could become available. The proposed solution has been designed to address this particular aspect, so that tags with different memory capacities may be used in the same system.

The general rule with any memory-based system has always been that no amount of memory is ever sufficient. Invariably, the response to enlarging the memory capacity of a system is to increase the scope of the application so that it requires even more memory. But, there is a strong relationship between price and capacity, larger memory capacities directly increase the cost per tag and the price of tags with a larger storage capacity is rather high. On the other hand, the RFID application implementation costs must be as small as possible, so as the costs of the tags be low as well. Although tag prices have considerably lowered lately, the price of large-memory tags has remained fairly high because added capacity always pays off. Under the circumstances, solutions are sought in order to increase memory capacity on the limited space of small and average tags (at an affordable price). There are needed solutions that could be adapted to a variety of activity domains. We propose a solution based on templates defined according with any users' requirements. A template is used to describe the data format to be applied for writing/reading data into/from tags.

2. Data types

The memory space on tags is entirely dependent on the data type used to store information. Thus, our solution proposes the use of some fundamental data along with additional data type defined by the user. The list of fundamental data types includes eight data types that are presented in Table 1. The data types have variable length. For each data type, both the occupied space and value ranges are determined.

Let us consider the date type. As this data type allows us to store date values in the YYYY-MM-DD format, 4 bits will be employed for the data type and 15 bits for the representation of the date value.

NO.	DATA TYPE	AMOUNT OF STORAGE (BITS)		DESCRIPTION, RANGE
		TYPE	VALUE	
1	BIT	4	1	true/false or yes/no;
2	INT4	4	4	non-negative integer values between 0 and 15
3	INT8/CHAR	4	8	non-negative integer values between 0 and 255 / ASCII char
4	INT12	4	12	non-negative integer values between 0 and 4095
5	INT16	4	16	non-negative integer values between 0 and 65535
6	INT32/REAL	4	32	real values or integer values represented on 32 bits
7	DATE_TIME	4	26	date and time values between 1st of January 1969, 0:00 and 31st of December 2031, 23:59
8	DATE	4	15	date values between 1st of January 1969 and 31st of December 2031

Table 1. Fundamental data types

The date value will be represented in accordance with the following specifications:

14	13	12	11	10	9	8	7	6	5	4	3	2	1	0
century	year					MM				DD				

where the meanings 'century' and 'year' are indicated by the following formulas:

century = YYYY/100 – 19

year = YYYY-2000, if century = 1 (e.g. 2006)
2000-YYYY, if century = 0 (e.g. 1999)

The last two formulas may be simplified as follows:

$$\text{year} = (2 * \text{century} - 1) * (\text{YYYY} - 2000).$$

Different applications have different requirements and demand different data structures. Through this fundamental data type, new data type can be defined using list and class types. Using the list type, one is able to present different other kinds of data, such as strings (considered as list of characters), bits lists (which can be used in the case of true/false or yes/no operations), lists of integers, etc.

A user can also define a data type called *class* to represent a collection of different data (properties) and functions grouped together under a single name. The instance of class can be used as data field on a tag or as member object of another class. The proposed solution enables the definitions of each class to be encoded once at the beginning of the tag; the values of each data field of class type (instance/object) are encoded subsequently. But, four bits of memory should be used to encode the data type for each different data element and they prefix the corresponding values.

The class type is very useful especially if the information to be stored on the tag corresponds to certain logically grouped information types, and the information types in question are repeated with the same meaning on the same tag.

Defined data types are meant to compact the data for a more efficient encoding within the RFID tag memory and to organize the data in the memory. Consequently, a variety of data may be encoded and some of it may remain permanently locked. Furthermore, data types should support selective read/write operations, as well tag data updates.
The data fields and data encoding discussed later in this chapter only define, as example, a class for a particular application. But other classes may be easily defined, depending on users' needs.
Each data field should be defined by *data type*, *value* and several *associated attributes*. Thus, each data field can be associated with some attribute in the following categories:

- *empty*: indicating whether the associated field has no value;
- *read only*: data elements can be read, but not updated or erased;
- *normal*: read-write field;
- *codified*: on the RFID tag some data elements need to be represented by a value with certain meaning revealed by the template associated with the tag.

Users can define more than one new class and specify any access privileges (attributes).
The introduction of data types ensures the uniform coding of data, while the use of templates increases the flexibility of the data to be coded. Furthermore, fixed-length and variable-length fields may be easily handled simultaneously.

3. Script language

A user-defined class may contain data and function members. The user can define a function member as a subroutine (sequence of statements to perform an action) called *script*. This section includes the specification of the proposed script language and its instructions syntax. The defined data types and script language allow the data and the commands to be specified on a tag in a uniform way, irrespective of any particular application.
Our script language offers 16 distinct instructions presented in Table 2.
Using these instructions, users can easily define a script which can be compiled to byte-code on a PC or PDA. The code resulting from the compilation is small sized and can be stored on a tag. Every time the tag is read on a PC/PDA or even on a low-resources embedded device (station), the code can be interpreted and executed.

4. Data template

We described a solution where the possibility to define both the data type and the script has been taken into consideration. They may be easily tailored to the template which best suits the needs of a user for a given application domain. Other templates may be adopted at the choice of the user to meet any specific requirements.
A data tag is based on a specific template, which makes it unique not only within the particular domain of an application, but also among all other domains. A template:

- provides guidelines on how data shall be written on the tag;
- defines the desired data classes, based on specific requirements;
- specifies the data fields attributes;
- specifies the commands that are supported for defining the indispensable actions that must be executed at RF tag reading.

Since all data elements can be easily defined, users are free to use standard RFID tags and, depending on tag memory capacity, choose which data elements are most appropriate for a specific application.

INSTRUCTION	PARAMETERS	BINARY	OPCODE	BYTES	LENGTH
#DEFINE	BYTE	00000	0x00	1+1+1	3
	WORD	00001	0x01	1+1+2	4
INC	FIELD_NO	00010	0x02	1+1	2
	FIELD_NO_GIVEN_BY_CONSTANT	00010	0x02	1+1	2
DEC	FIELD_NO	00011	0x03	1+1	2
	FIELD_NO_GIVEN_BY_CONSTANT	00011	0x03	1+1	2
SETVAL	FIELD_NO, FIELD_NO	00100	0x04	1+1+1	3
	FIELD_NO, CONSTANT_VALUE	00100	0x04	1+1+1	3
	FIELD_NO_GIVEN_BY_CONSTANT, FIELD_NO	00100	0x04	1+1+1	3
	FIELD_NO_GIVEN_BY_CONSTANT, CONSTANT_VALUE	00100	0x04	1+1+1	3
IF	GATE==CONSTANT	00101	0x05	1+1	2
	Variable1 == Variable2	00101	0x05	1+1	2
	Vriable1 != Variable2	00110	0x06	1+1	2
IF	GATE != CONSTANT	00110	0x06	1+1	2
	FIELD_NO==CONSTANT	00111	0x07	1+1	2
	FIELD_NO!=CONSTANT	01000	0x08	1+1	2
	FIELD_NO_GIVEN_BY_CONSTANT == CONSTANT	00111	0x07	1+1	2
	FIELD_NO_GIVEN_BY_CONSTANT != CONSTANT	01000	0x08	1+1	2
EVENTS	SERVER_EVENT	01001	0x09	1+1	2
EVENTI	INTERNAL_EVENT	01010	0x0A	1+1	2
STOP		01111	0x0F	1	1
SCRIPT	SCRIPT_NO	10000	0x10	1+1	2
RETURN	-	10001	0x11	1+1	2
CONSTRUCTOR	SCRIPT_NO	10010	0x12	1+1	2
DESTRUCTOR	SCRIPT_NO	10011	0x13	1+1	2
GOTO	LABEL	10100	0x14	1+1	2
CALL	LABEL	10101	0x15	1+1	2
	SCRIPT_NO	10110	0x16	1+1	2
SETVAL_STRUCT	FIELD_NO0.FIELD_NO1. … VAR	10111	0x17	1+1+1+ …	2+…
SETVAL_STRING	FIELD_NO [INDEX] VAR	11000	0x18	1+1+1+1	4
GETVAL_STRUCT	VAR FIELD_NO0.FIELD_NO1. …	11001	0x19	1+1+1+ …	3+…
GETVAL_STRING	VAR FIELD_NO [INDEX]	11001	0x1A	1+1+1+1	4
SETVAL	FIELD_NO, VAR	11010	0x1B	1+1+1	3
GETVAL	FIELD_NO, VAR	11011	0x1C	1+1+1	3
ADD	Variable CONSTANT	11100	0x1D	1+1+1	3
IF	VAR1 == VAR2	11101	0x1E	1+1+1	3
	VAR1 != VAR2	11111	0x1F	1+1+1	3
//	COMMENT				

Table 2. Instruction set

Once the template has been created, users can specify the desired values for every defined field. On the other hand, the template is required for a complete understanding of the data tag in its entirety.
The logical tag content is divided into three sections as shown in Figure 1. Thus, the first logical section is a header which contains all the information regarding the physical and logical organization of the data on the tag. In this first section specific fields are included (for example, the length of the data tag) as well as information regarding the classes structure on which objects are instantiated in the tag data section. A class encapsulates the properties - data of the fundamental types (previously defined), and operations/methods of the class - described by scripts. A script will be considered as a static method that should be associated with a class rather than an object. Each class can vary in length and content (e.g. the number of data members, the type of data members, etc.). At the moment the implementation of all object-oriented programming principles (i.e. encapsulation, inheritance, polymorphism) is not an option because the tag memory size needed for this operation is considerably larger than the memory size available on current tags. The purpose of this header organized as a logical memory map is to provide extensibility for future, unanticipated data requirements.
The second section defined on the tag contains all the desired values. This section is associated with information encoded on the tag, and is made up of the data members and instances of the classes defined in the previous section. Whenever data must be encoded on the tag, the class notion defined in previously allows an optimization of occupied space.
The last section on the tag represents the map of the bits associated with each field or object on the tag. Within this map, there are 2 bits allocated for each field, and they are required for the encoding of the following states: normal (read-write), empty, read-only, codified (Figure 1).
Since the template structure is memorized directly on the tag, additional memory space is required for the definitions of classes and field types. This information is necessary in order to ensure the independence of tags. Furthermore, the classes and field types could be used by a low-resources embedded device (station) to interpret and modify the tags read.
When tag independence becomes an optional feature and when an application supports a large number of templates, there are other solutions to be sought. The solution we have chosen refers to the identification of each template through a unique number and the storage of this identification number associated with the template in the header section of the tag. The stations could memorize these templates and their associated identifiers. If the hardware resources do not support high memory usage, the template identified through the identification number on the current tag will be downloaded directly from a server. Obviously, this operation takes more time because it is necessary for the station to connect to the server and download the required template. In fact, much tag space is saved if the template identifier is memorized in the tag header.
If a tag template is unknown, its content cannot be interpreted and thus data privacy is ensured.
The evolution of the tag market and the demand for RFID-based applications will dictate the appropriate choice.

5. Security

The capacity of an RFID tag to be secured against unauthorized access, theft or damage is an issue to be considered. Some users may not wish to share the information and the data types stored on their RFID tags with their competitors. The data generated and used in our RFID system represent a valuable asset characterized by confidentiality, integrity and availability. We have devised a method to verify or authenticate whether the information read from a tag is genuine; taking into consideration the serial number of the tag, the solution proposed does not require any database to verify whether a tag is a copy or a fake.

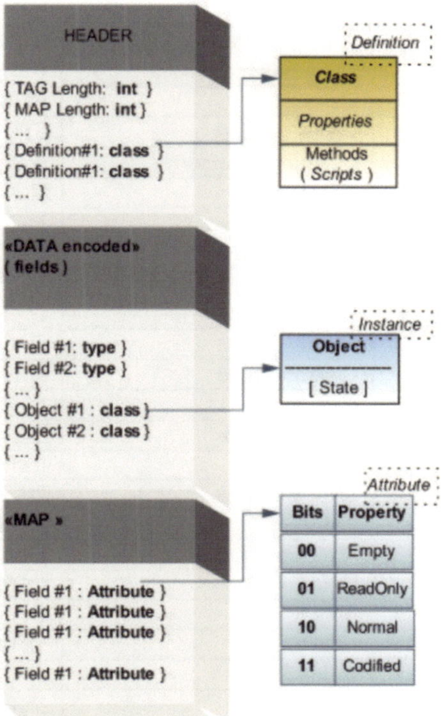

Fig. 1. The logical organization of a tag's content

One of the RFID privacy principles according to (BISG, 2004) consider that all businesses, organizations, libraries, educational institutions and non-profits that buy, sell, loan, or otherwise make available content to the public utilizing RFID technologies shall protect data by reasonable security safeguards against interpretation by any unauthorized third party. The template employed to define the data stored on the tag enables the reading of the tag content by authorized users only. Hence, it is impossible to identify the content of the tag without the corresponding template.

6. Case study

A selection of date type elements is presented below in order to help us illustrate how data encoding is performed and to estimate the amount of memory required to store this type of information. Within the adopted format, the length of these elements is either fixed or variable. Moreover, we have associated this format to a well-defined class depicted as a set of data and member functions. Data class members are stored sequentially in the tag memory. To illustrate the implementation of our solution, we will consider an RFID system to be used in an automotive service center. The list of fields to be stored on the tag should include:

Minimum data for car identification

		FIELD	DATA TYPE
Section 1	1	License number	STRING
	2	Brand	codified
	3	Car body series	STRING
	4	Engine type	STRING
	5	Energy source	codified
	6	Cylinders [cm3]	INTEGER
	7	Owner	STRING
	8	Phone number	STRING
	9	Insurance company	codified
	10	Color code	STRING
Section 2	1	Gate 1	INT5
	2	Date and time	DATE_TIME
	3	Gate 2	INT5
	4	Date and time	DATE_TIME
	5	Gate 3	INT5
	6	Date and time	DATE_TIME
	7	Gate 4	INT5
	8	Date and time	DATE_TIME
	9	Gate 5	INT5
	10	Date and time	DATE_TIME
	11	Gate 6	INT5
	12	Date and time	DATE_TIME
	13	Gate 7	INT4
	14	Gate 8	INT4
	15	Gate 9	INT4
	16	Gate 10	INT4
	17	Gate 11	INT4
	18	Gate 12	INT4
Section 3	1	Counselor	codified
	2	Date and time of car entry	DATE_TIME
1	1	Operation type	codified
	2	Recommended operation	codified
	3	Scheduled operation	DATE
	4	Operator identity	codified
	5	Execution date	DATE
	6	Execution status	3 bits
	...		
10	1	Operation type	codified
	2	Recommended operation	codified
	3	Scheduled operation	DATE
	4	Operator identity	codified
	5	Execution date	DATE
	6	Execution status	3 bits

Since it is evident that the information contained in section two is repeated six times, a class with two data members (i.e. gate and date) should be considered. This class may be used to store more information concerning the gate number and the entry date of any identified vehicle.

Similarly, since the information contained in section three is repeated ten times, the redundant information may be eliminated if one considers the introduction of a class that contains data members (i.e. properties) corresponding to various fields of interest, namely operation type, recommended operation, scheduled operation, operator identity, execution date, execution status. However, any user may feel free to adopt other classes, templates or scripts.

If we consider the classes defined above, then we realize that the tag memory capacity required for all proposed fields is 197 bytes. If a dedicated application is used, namely an application in which the meaning of the values stored on the tag is predefined, the required tag memory capacity would be 156 bytes. Therefore, the proposed solution is certain to be able to optimize the memory space on RFID tags and to offer a high generality degree.

Let us suppose, for instance, that every driver entering the service center is given an RFID tag as an entry or parking pass. The following script has been defined in order to screen entry to secure areas in the automotive service center:

Example

```
# DEFINE  BYTE  FIELD_ENTRIES             0x01
    //First field from tag (how many entries)
# DEFINE  BYTE FORBIDDEN_BARRIER1  0x02
    //2'nd field from the tag (one forbidden entry)
# DEFINE  BYTE OPEN_BARRIER               0x0E
    //constant in embedded system (open the barrier)
# DEFINE  BYTE BEEP_ALARM                 0x0D
    //also constant
IF (GATE == FORBIDDEN_BARRIER1)
        EVENTI (BEEP_ALARM)
    //beep if the car tries to enter in a forbidden area
IF (GATE == FORBIDDEN_BARRIER1)
        STOP
    //and then stop if forbidden door
IF (FIELD_ENTRIES == 0x00)
        STOP
    //no more entries allowed
DEC (FIELD_ENTRIES)
    //decrement the number of parking entries
EVENTI (OPEN_BARRIER)
    //open current barrier and let the car go inside parking area
STOP
```

The RFID tags used with this script should consider the fields presented in Table 3. The compilation process may be understood by following the information included in Table 4.

Let us consider only the values in the byte-code column for the IF instruction. The values 0x81 and 0x82 refer to the first defined and to the second defined field, respectively. This convention was adopted in order to help us distinguish between defined fields and integer values. For a maximum number of 127 of defined fields, the maximum integer number to be considered in an IF instruction is again 127.

DATA TYPE	NAME	LENGTH	EXPLICATIONS
BYTE	FIELD_ENTRIES	1	One byte that will be decremented every time the car enters to one allowed area of the parking.
BYTE	FORBIDDEN_BARRIER 1	1	One byte that contains the value of one restricted area. This value will never change.

Table 3. Allowed data types

INSTRUCTION	BYTECODES
#DEFINE BYTE FIELD_ENTRIES 0x01	0x00, 0x00, 0x01
#DEFINE BYTE FORBIDDEN_BARRIER1 0x02	0x00, 0x01, 0x02
#DEFINE BYTE OPEN_BARRIER 0x0E	0x00, 0x02, 0x0E
#DEFINE BYTE BEEP_ALARM 0x0D	0x00, 0x03, 0x0F
IF (GATE == FORBIDDEN_BARRIER1)	0x05, 0x81
EVENTI (BEEP_ALARM)	0x0A, 0x82
IF (GATE == FORBIDDEN_BARRIER1)	0x05, 0x81
STOP	0x0F
IF (FIELD_ENTRIES == 0x00)	0x07, 0x80, 0x00,
STOP	0x0F
DEC (FIELD_ENTRIES)	0x03, 0x80
EVENTI (OPEN_BARRIER)	0x0A, 0x83
STOP	0x0F

Table 4. Compiling script to byte-code

If we want higher values, then we must use the #DEFINE instruction.

Whenever an RFID-tagged car approaches a barrier with an RFID embedded system,

- the content of the tag is read;
- the script, if any, will be executed and tag values may be modified;
- the embedded device will decide upon opening the barrier, sending an event, running an internal event, etc.

We have implemented 2 versions of the *ScriptCompile()* function: one for the PC and another one for the PDA. We have also employed an *ExecuteScript()* function for every type of embedded device to be used. Our source code implementation of script language functions is compatible with ANSI C standard in order to be used with different compilers:

- MS Visual C++ and Borland C++ Builder at PC level;
- Embedded Visual C++ and .NET for Windows CE devices (PDA);
- Keil Compiler for 8051 compatible uC's;
- gcc for MicroBlaze soft processor.

The first version of the *ScriptCompile()* function source code for the PC was created using Microsoft VC++. The same source was successfully used with Borland C++ Builder. For Windows CE devices we have used either Embedded Visual C++, or EVC++ (for .NET programming).

Furthermore, the *ExecuteScript()* function was tested with Keil and gcc compilers. No malfunctions were reported at the level of the embedded system. Irrespective of the micro-controller used, the whole functioning was speedy and accurate.

7. Advantages and disadvantages

The presented specifications represent a flexible and multi-purpose solution which may be easily customized for a variety of purposes. Its major advantage, following the introduction of templates and class types, is that the amount of data on the tag may be increased upon request and there is no need to change the application. Secondly, depending on the application involved, users may select the relevant data to be encoded into the RFID tag memory.

Thirdly, the embedded devices support any type of microcontroller and no large memory capacity is required. As this solution proposes the implementation of processing logic on RFID tags, there is no need to modify the software of the embedded system to allow the same device to be used in different applications (e.g. security, parking, supply chain, etc.). Furthermore, this solution enables the provision of high-quality services with high additional values in numerous domains such as supply chain, security systems, or product tracking.

However, a series of disadvantages have been also identified:

- the execution of the script from the tag may take too long, depending on the length of the script, but also on the frequency of the processing device;
- if, through the commands of the script, a change in the value of a tag field is required, the processing time increases because several tag fields need to be written;
- after the script is stored on the tag, there is less storage space left for other interest information;
- the higher the amount of data on the tag, the longer the data reading time. Nevertheless new solutions have already been found: a high capacity, high-speed LSI for RFID tags complying with ISO/IEC15693.

8. Future developments

The most efficient encoding of tag information (e.g. the encoding of a character string presented as a field value) may be obtained by applying data compaction algorithms.

If codification and several security elements are considered, it is possible for the same value to be stored differently on two tags belonging to two different applications. In this way users may create personalized applications that will ensure a higher level of security. The implementation of such system requires a unique key for encryption/decryption, namely a combination between the tag ID and the key of some application. If the security standards are not met, there is no guarantee for data consistency when it is copied from one tag to another.

One major concern would be the design and development of a mechanism that will allow users to concatenate more tags (even with different classes) into a single one, without modifying the data class that already exists on the destination tag. This mechanism is suitable for the applications designed to assemble a small number of components into a final product. In this case, the tag associated with the final product will store information about the traceability of each component.

9. Conclusions

In this paper, we summarize our approach and our research. The discussion of different methods of data storage on a passive HF tag operating at 13.56 MHz has been the major focus of the present paper. The HF tags features that include different form factors, different memory sizes (over 128 bits) and different read ranges allow users to design customized RFID systems according to their specific application requirements. We described a novel solution for the structural optimization of information storage on RFID tags. The method proposed is designed to reduce the cost of RFID-based applications by increasing the memory capacity on limited space of small and average tags. The solution devised and presented in this paper ensures the introduction of used-defined types to memorize any kind of information. Users may develop their own templates to describe information formatting, content and specifications after defining a set of classes to represent their own data. Furthermore, they are entitled to define a script to determine what the code should be executed in certain conditions. Through the use of user-defined templates and scripts, the presented system can be easily adapted to meet future needs of the user just as well as it meets today's needs. The created templates are used in order to write some product tags with specific information. This feature enables reading of the tag content for authorized user only. Hence, it is impossible to identify the contents of the tag without the corresponding template. Thus, these specifications define the security mechanism that deals with anti-theft. The proposed solution ensures a flexible and intelligent handling of tag information processing. This solution is expected to allow the development of an RFID-based generalized system that can be easily implemented in various domains (such as supply chain, security systems or product tracking) without any modifications in the structural level of software applications.

8. References

BISG (2004), BISG Policy Statement POL-002, Radio Frequency Identification, Available at: http://www.bisg.org/docs/BISG_Policy_002.pdf

Greene, K. (2006). Wireless Wonder Chip, *MIT Technology Review*, July 2006, Available at: http://www.technologyreview.com/read_article.aspx?id=17182&ch=infotech&a=f

Sarma, S.E. (2001). Towards the five-cent tag, *Technical Report MIT-AUTOID-WH-006*, MIT Auto ID Center, Available at: http://www.autoidcenter.org.

Takaragi, K.; Usami, M.; Imura, R.; Itsuki, R. & Satoh, T. (2001). An ultra small individual recognition security chip, *IEEE Micro*, Vol. 21, No. 6, pp. 43–49, November 2001, ISSN 0272-1732.

*** (2008), Capturing the Value of EPC Gen 2 Custom Commands, *NXP Semiconductors and Sirit Inc*, Available at: http://www.rfidproductnews.com/whitepapers/files/Custom_Commands.pdf

27

Mobile Applications for RFID Based B2B Systems

Tudor-Ioan Cerlinca, Cornel Turcu, Valentin Popa and Felicia Giza
"Stefan cel Mare" University of Suceava
Romania

1. Introduction

Business-to-business or "B2B" is a term commonly used to describe the transaction of goods or services between businesses, as opposed to that between business and other groups, such as transactions between business and individual consumers (B2C) or business to public administration (B2G) transactions [Turcu et al., 2007]. Given today's general interest in RFID (Radio Frequency Identification) technology, B2B systems are expected to allow for considerable extensions and improvements. It is expected that RFID technology will enable the building of complete and complex B2B solutions in areas such as industry and commerce where mobility is a key factor. In fact, the overall success of any RFID_B2B (Radio Frequency Identification - Business to Business) system is highly dependent upon this factor. The RFID_B2B system's mobility resides in the use of multiple PDA devices and RFID readers connected to them.
This chapter presents the principles governing the design and development of a mobile application, as well as various aspects regarding its integration into a more complex RFID_B2B system. The main goal of such application is to extend the applicability of generalized RFID_B2B systems. Mobile applications are generally expected to handle large amount of data, to operate in stand-alone mode and to allow their easy integration into complex RFID_B2B systems. All these aspects will be detailed presented in this chapter. The chapter also proposes new solutions and ideas regarding the design and development of a secure and very fast method for the communication and synchronization between different B2B servers and mobile applications running on various mobile devices.

2. RFID_B2B mobile applications

2.1 RFID

RFID (Radio-Frequency Identification) technology has been considered one of today's "hottest" technologies due to its specialized capacity to track and trace objects in real time. RFID technology is classified as a wireless Automatic Identification and Data Capture (AIDC) technology that uses electronic tags to store identification data and other specific information, and a reader to read and write tags. Tags are small chips with antenna. They ca be active (battery powered), passive (uses the reader signal to be activated) or semi-passive (battery-assisted, activated by a signal from the reader). RFID technology currently allows to identify, locate, track and monitor each and every item (product, box, pallet, etc.) and to obtain continuous real-time information on these items from the factory, through shipping and warehousing, to the retail location [Finkenzeller, 2003]. Incorrect or outdated data used in invoices, bills of lading (a document from the carrier indicating the description of the goods being shipped) or purchase orders can result in product delivery errors and lost sales estimated at more than $50 billion annually [Lefebvre et al., 2006]. But RFID technology could prevent these costly data inaccuracies. Moreover, it is expected that RFID tags will replace conventional barcode labels due to their major benefits: high data storage capacity, read-write capability, read-speed rate, multiple entity identification, information updating, no line of sight scanning, durability, and environmental resistance [Turcu et al., 2006]. Also, it can be demonstrated that RFID enables more integrated and more collaborative business-to-business (B2B) ecommerce solutions.

2.2 RFID_B2B general presentation

The RFID_B2B system that integrates the mobile application is detailed described in [Turcu et al., 2007]; the generalized character of the system results from the fact that it can be easily implemented in various activity fields without any modifications in the structural level of software applications. Thus, the user can define the data format to be used for writing data into tags through an advanced template editor which allows user to establish necessary fields (e.g. acquisition date, location, current value) and their type (character, string, integer, real). [Turcu et al., 2007, Cerlinca et al., 2006]

The RFID_B2B system refers to the business relations in large enterprises, corporations and groups, as regards the control of the materials along their entire supply chain. The system proposes applying the RFID technology by using tags to identify materials and assemblies. Thus, based on the ID codes of the materials and assemblies, it is possible to control the content and the origin of any finite product, the content of assemblies and the origin of any constituent component, and so on, for each company which contributed to the creation of the finite product. By extending the system to the entire supply - chain - final producer, supplier, the manufacturer's suppliers, etc. - the customer can follow the course of materials included in the final product, up to the primary sources. In order to accomplish this, all the necessary tracking information will be comprised in the tags attached to the materials, assemblies and finite products. The RFID_B2B uses RFID technology by using passive 13.56 MHz tags for parts and finite products identification. The system also handles multiple PDA devices and PC servers and facilitates data sharing among these devices.

The general architecture of the RFID_B2B system is presented in figure 1 [Turcu et al., 2007]. Relating to this architecture we can note that the integrated RFID_ B2B system includes the following main components:

- one IBM-PC compatible computer which runs an OPC (OLE for Process Control) server with two main components: communication and data acquisition;
- one IBM-PC compatible computer which runs an OPC dedicated client. This computer can be the same as the first one;
- one network of different gates devices, each of them having attached an RFID reader, which provides local data processing;
- different PDA devices with RFID readers attached too;
- one IBM-PC compatible computer which runs a Local B2B server;
- one IBM-PC compatible computer which runs the Central B2B server.

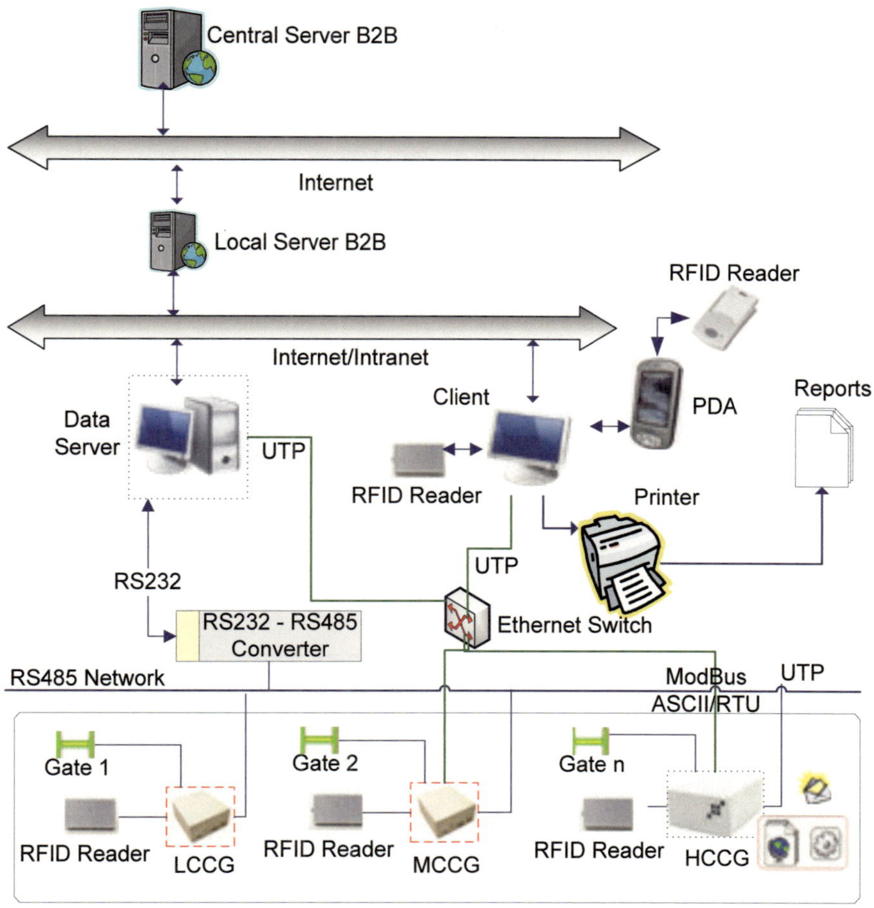

Fig. 1. The RFID_B2B system architecture

2.3 RFID_B2B mobile application facilities

Mobile applications present several interesting and complex challenges. Following our research, we have reached the conclusion that the software application that runs on such mobile devices and that is integrated into the complex RFID_B2B system should perform the following main functions [Cerlinca et al., 2008]:

- read and write RFID tags;
- work in stand-alone mode (independently of the main servers);
- store huge data;
- integrate and exchange information with complex RFID_B2B systems and other PDA mobile devices;
- ensure maximum security;
- employ a multi-user and user-friendly interface.

Within the mobile application integrated in the RFID_B2B system, the following main operations are facilitated:

- the bi-directional communication between the PC and PDA applications, allowing a total or a partial transfer of records within database tables from PC to PDA and in inverse order, for the update of the database from the PC and from the PDA;

- the communication between PDA and reader, enabling the reading of stored information into tags and the update of databases from the PDA, as well as the writing and updating of tags with database records and according to the settings performed by the user;
- enabling the system security, through system access control, as well as data encrypting from the database. These operations are performed both on the PC and the PDA;
- management of system registered users (users visualization, adding or deletion of certain users, profile modification, etc.);
- enabling the management of the database, which stores information related to tags;
- the configuration of the system in accordance with necessities;
- the visualization of the templates to be used for tag information storage;
- enabling the communication with other systems and allowing the connection with higher enterprise levels;
- the storage of all events and their administration (visualization, searching, filtering etc.);
- clock synchronization between the PDA and the PC;
- checking the connection state;
- basic file/directories based operations;

2.4 Specifications for implementation

There are many aspects to be discussed about the implementation of an RFID_B2B mobile application. However, this chapter will focus only on several most important aspects such as: data security, high degree of usage, communication/synchronization etc.

Needless to say, security is one of the most important aspects that should be taken into consideration when implementing an RFID_B2B system. Also, security is a major concern for mobile applications. Thus, wireless transmission, in a way, biases end-users to perceive mobile applications to be more vulnerable and unsecured. Our system provides several security enhancements and options to ensure the security of data and communication between applications:

- data encryption with the TripleDES algorithm for all important information such as user names, passwords, access rights, etc.;
- password-based access to all web services used for communication and synchronization between the PDA devices and the RFID_B2B systems;
- password-based access to the PDA's main application;
- support for different levels of access rights. This means that users are granted different rights to the application features. For example, some users will create new tags while others will only view the available database tags. The access rights are established at the PC level through a specialized application called User Management and transferred to the PDA through specialized web services.

Another important aspect we have focused upon in the implementation of the mobile application is the way in which a high level of generality can be provided. The application was designed in such a manner so that it can be used in different areas of activities. To insure the desired level of generality we took into consideration two important aspects. The first one is related to the use of tag templates to create specialized tags [Cerlinca et al., 2006, Cerlinca et al., 2008]. All templates are created at the PC level and then transferred to the PDA through specialized web services. The second aspect is related to the visual organization of the fields on a tag so that they can be read on the PDA display. Given our

experience in this respect [Cerlinca et al., 2006, Cerlinca et al., 2008], we consider that it is rather difficult to create/update a tag that has too many fields. Thus, the visual space on the PDA touch screen is far too small; the low display resolution and small display screen have inhibited information to be displayed completely and clearly. Also, it's difficult to manage information tag when the way that the template's fields were created and visual grouped may not correspond to the actual expectations of the user. That is why, an RIFD_B2B based application should allow users to define at the PC their own visual areas according to their needs and then group all tag fields. In general, each group will consist of several fields with the same purpose. All visual areas created at the PC level are then transferred to the PDA.
Let's suppose that a company is selling Desktop PCs. Each PC that is sold to a customer will need to have an RDIF tag attached. As we already mentioned, a tag is created by using a specific template. A minimal Desktop PC template will have the following fields:

Field	Type	Size	Description
Location	CHAR	30	company location – e.g. **Bucharest**
Name	CHAR	50	product name – e.g. **DC X-Line Home X2 4200+**
Code	INT8	1	product code – e.g. **102**
Processor	CHAR	30	processor type – e.g. **AM2 Athlon 64 X2 4200+ BOX**
Motherboard	INT8(CODED)	1	motherboard type ID – e.g. **1** (**nVidia nForce 630a/GeForce 7050PV**)
Memory	INT8(CODED)	1	memory capacity ID – e.g. **3** (**4GB**)
Video	INT16(CODED)	1	video card ID – e.g. **0** (**VGA GeForce 7050**)
HDD	INT16(CODED)	1	HDD type ID – e.g. **2** (**250GB SEAGATE Barracuda 7200 7200rpm/SATAII/8M**)
TAG_DATE	DATE	1	Tag creation date – e.g. **20/11/2007**
EXPIRATION_DATE	DATE	1	product expiration date– e.g. **20/11/2009**
PRICE	REAL	1	product price – e.g. **590.47**
SHOP_ASSISTANT	INT8(CODED)	1	seller ID – e.g. **2** (**John E.**)
CLIENT	CHAR	50	client name – e.g. **3** (**Peter A.**)
PAYMENT_TYPE	INT8(CODED)	1	payment type – e.g. **1** (**Credit card**)

Table 1. Desktop PC template

Note that all types in column 2 of table 1 are not built-in but application specific types. Figure 2 shows the tags editor window for the case when no visual group exists. Figures 3, 4 and 5 exemplify the visual organization of the fields on a tag.

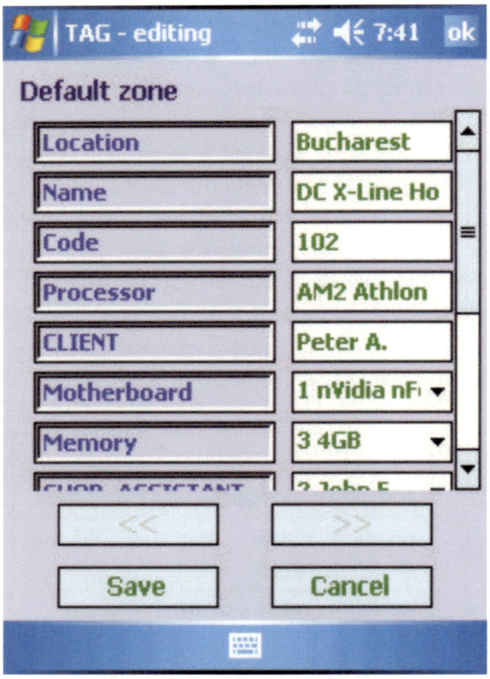

Fig. 2. Tags editor window. No visual groups

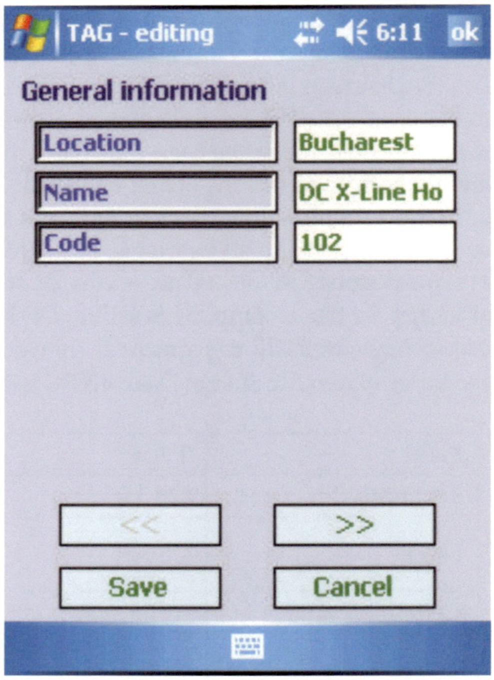

Fig. 3. Visual organization of tag fields. First group

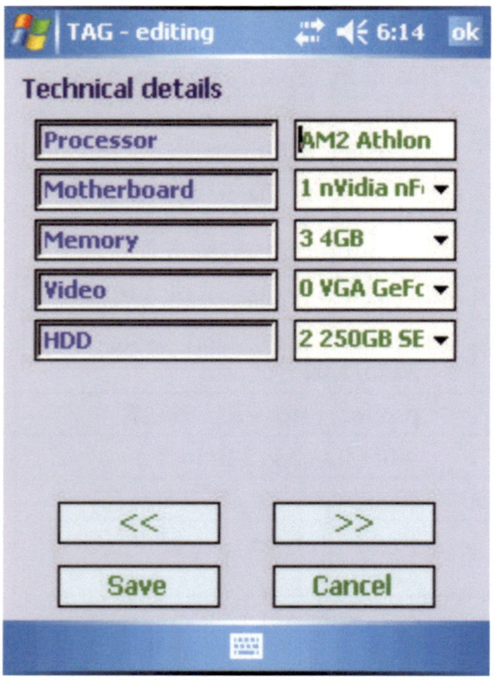

Fig. 4. Visual organization of tag fields. Second group

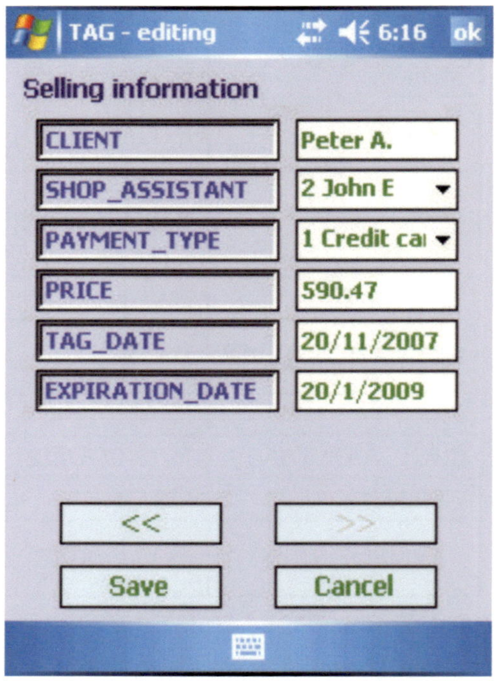

Fig. 5. Visual organization of tag fields. Third group

For the Desktop PC template, three visual groups were created. The mobile application also offers a 'visual groups' browser that allows user to browse through the visual groups and choose the one he wants to edit.

Perhaps the most important aspect in the design and development of RFID_B2B mobile applications is the one related to the communication and synchronization between mobile devices and different B2B servers and different mobile devices that run the same application. A powerful communication and synchronization component will provide the following facilities:

- the bi-directional communication between the PC and PDA applications, allowing a total or a partial transfer of records within database tables from PC to PDA and in inverse order, for the update of the database from the PC and from the PDA;
- support for multiple PDA devices/PC servers that could store different information about the same entities;
- support for intelligent updating of both PC and PDA databases;
- clock synchronization between the PDA and PC.

First of all, we should mention that a universal synchronization tool already exists and it is called Microsoft ActiveSync. In the following we will see what Microsoft ActiveSync is all about and why an RFID_B2B mobile application can't rely on it. Microsoft ActiveSync is a tool that allows users to create a synchronization relationship between a mobile device and a PC, by using a cable, cradle, Bluetooth, or infrared connection. Mainly, ActiveSync helps users to keep their information up-to-date on both mobile device and PC. If a change was made in one place, the next time when the user is synchronizing, the change will be automatically made to the corresponding information on the other computer. No matter where the user is viewing the information, he will know that it's up-to-date. ActiveSync can be used to synchronize Contacts, Calendar, E-mail, Tasks, Notes, Favorites and even files. When it comes to databases, ActiveSync can be successfully used to synchronize Pocket Access databases with Microsoft Access databases. But no RFID_B2B system can be built upon Pocket Access and Microsoft Access databases. Moreover, ActiveSync does not support record based synchronization. Only tables/databases synchronization is allowed. If we take into consideration the fact that different devices (PCs and PDAs) could store different information about the same entities and the database records must not be replaced but rather updated, then we can conclude that Microsoft ActiveSync is definitely not a viable solution for communication and synchronization in complex RFID_B2B systems.

In our RFID_B2B system, we used Sybase SQL Anywhere 10 for the PDA devices and Microsoft SQL Server 2005 for the PC Server. We consider Sybase SQL Anywhere 10 to be the best solution for PDA devices because:

- it is not just a database file but a real multi-user SQL server;
- supports stored procedures and user functions (using Watcom SQL, T-SQL, Java, or C/C++), triggers, referential integrity, row-level locking, replication (two technologies: SQL Remote, MobiLink), proxy tables (links to other databases), and events (both scheduled and in response to system events such as lack of free disk space) ;
- supports strong encryption of both database files and client-server communication.

The communication, which is a client-server process, is basically achieved through specialized password-based web services that are available on PC servers. While the RFID_B2B system supports the operation of several PC servers, the PDA must dynamically connect to any of these servers. This problem was solved at the PDA level by implementing a specialized software component capable of reading the description of any web service and then connecting to it.

All data transferred between the PDAs and the PC servers is first converted from the database format into the XML format. There is still one important detail to be mentioned here: the amount of data to be stored on both the PC and the PDA can be huge, hence data transfer may take longer than one might expect. Furthermore, there is a lot of important data shared by the PDA devices and the PC servers that needn't be overwritten but perhaps only updated. Our communication component is intended to perform an intelligent update of both the PDA and the PC databases, by transferring and updating only the new/modified data that is explicitly marked as being transferable. The communication component is also able to handle multiple PDA devices. Taking into consideration the fact that different PDA devices could store different information about the same tag, we can conclude that this is not an easy task. The solution to this problem implies:

- the design and development of an UID server that will give a unique identifier to each PDA/PC in the system. The UID server has to be capable to handle security problems also (e.g. no PDA/PC with pirated/cloned application will ever receive an ID);
- the design of a table that will contain all the possible states that a database record can get into (Transferable to PDA, Transferred to PDA, Transferable to PC, Transferred to PC, Removed from PDA, Removed from PC etc);
- the design of a table that will contain the current states of each and every database entity that is involved in the communication/synchronization process. Al long as the RFID_B2B system can have more than one PDA device, this table can contain multiple records for the same entity (one record for each device). It is obvious that the state of some entity can differ from one PDA to another. In order to avoid huge computation time and disk space wastage, we do not consider any database record as an entity. For example, the tag's fields are not entities; only the tag is an entity. The RFID_B2B database was designed in such a manner that each table which contains entities (tags, templates etc) has a field called ModificationDateTime. Each time an entity is modified the ModificationDateTime field will be automatically updated.
- the design and development of a mechanism that will continuously update the above mentioned table in order to reflect the most recent changes of database entities.

Clock synchronization is also a very important component in the process of communication/synchronization. A successful process of communication will take place only when the PDA's clock is correctly synchronized with the PC's one.

Let us consider a simple test test-case that will demonstrate the efficiency of this communication/synchronization method. Let's suppose that we have an RFID_B2B system with 2 PDA devices and only one PC server. The table that contains the current states of all database entities is called tblEntitiesStates. At the PC level, the user creates one template (e.g. Desktop PC) and two different tags (e.g. Tag_PC1 and Tag_PC2). At this time, the tblEntitiesStates table will not contain any information related to DesktopPC_Template, Tag_PC1 and Tag_PC2. Next, the user is connecting the PDA1 to a computer that has an Internet connection and synchronizes the PDA's clock with the PC Server's one. Then he initializes a database transfer process by issuing a specific command to the communication web service. In the first step, the web service located on the PC server checks the current state of the following entities: DesktopPC_Template, Tag_PC1 and Tag_PC2. No information could be found for PDA1, which means that these entities were never transferred to PDA1. At this point, the web service will perform the following tasks:

- builds an XML string with all the information that must be transferred;
- sends the XML string to the PDA1;
- adds new records in the tblEntitiesStates table with the following information: entity's ID, PDA's id and state's ID (1 - Transferred to PDA)

In the next step, the user is modifying Tag_PC2, first at the PC level and then at the PDA level. The user is modifying Tag_PC1 also, but only at the PC level. The database records from tblEntitiesStates that are referring to Tag_PC1 and Tag_PC2 will be automatically updated, in order to change the current state from 'Transferred to PDA' to 'Transferable to PDA'. The user initializes a new database transfer process. Tag_PC1 and Tag_PC2 were marked as being transferable to PDA, but only Tag_PC1 will be transferred to PDA1, because the Tag_PC2 on PDA1 is newer than the same tag on PC. Next, the user is connecting the PDA2 to a computer that has an Internet connection and synchronizes the PDA's clock with the PC Server's one. Then he initializes a database transfer process. At this step, the web-service checks the current state of the following entities: DesktopPC_Template, Tag_PC1 and Tag_PC2. No information could be found for PDA2, which means that these entities were never transferred to PDA2. In this case, all the information related to these entities will be transferred to PDA2. The web-service will perform the same three tasks described above.

Now, let's perform a more complicated test. The user is performing some modifications in the following order: Tag_PC2 at the PC level, Tag_PC1 at the PDA1 level, Tag_PC1 and Tag_PC2 at the PDA2 level, Tag_PC2 at PDA1 level and Tag_PC1 at the PC level. Then he is transferring the database from PDA1 to PC. Tag_PC1 will not be transferred to PC because information on PC is newer. Tag_PC2 will be transferred to PC, because the last modifications were made at the PDA1 level. The database records from tblEntitiesStates that are referring to Tag_PC2 will be updated as follows:

- for PDA1, the state will be changed from 'Transferred to PDA' to 'Transferred to PC';
- for PDA2, the state will be changed from 'Transferred to PDA' to 'Transferable to PDA'.

In the last step, the user is transferring the database from PC to PDA2. The state of Tag_PC1 entity for PDA2 is 'Transferable to PDA' but the tag was modified both on PC and PDA2 levels. The most recent data is the one from PC, in which case, the tag will be transferred to PDA2. As for the Tag_PC2, the state is also 'Transferable to PDA', because the tag was transferred from PDA1 to PC. The most recent data is the one from PC, in which case, the tag will be transferred to PDA2.

As it can be seen from the test-case, the proposed method of communication/ synchronization ensures that the information is up-to-date on all RFID_B2B devices (PDAs or PCs) and no information will be mistakenly updated or replaced.

Another important aspect in the development of an RFID_B2B mobile application is related to the tags management. An RFID_B2B mobile application should perform at least the following main operations: the creation of new tags (see figure 6) and the read/write of RFID tags. Perhaps one of the most important facilities that an RFID_B2B mobile application should provide is the ability to read/write RFID tags. Figure 7 present the application's window that allows users to select the database tags that will be psychically written on RFID tags. The RFID reader is connected to the PDA through SD port. The ability to read/write RFID tags was achieved through a specialized software component that is performing the following main tasks:

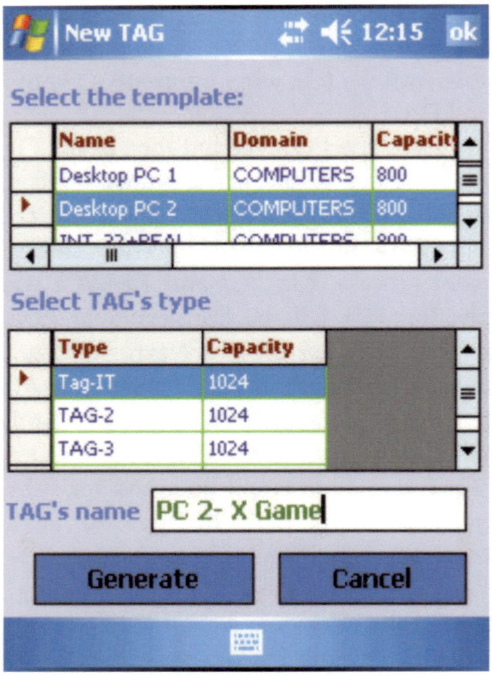

Fig. 6. Tags creation window

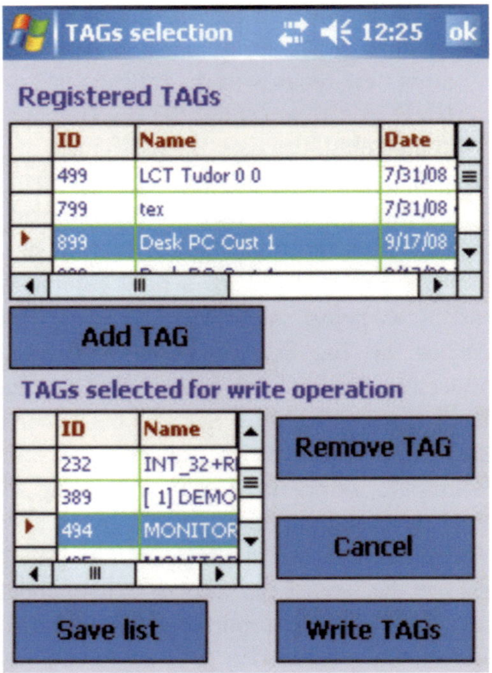

Fig. 7. Tags selection window

WRITE operation:
- establishing a connection with the RFID reader;
- getting the tag's data from the database;
- encoding the data to be written on the RFID tag;
- searching for an RFID tag in the proximity of the RFID reader;
- writing the encoded data to the RFID tag;
- closing the connection.

READ operation:
- establishing a connection with the RFID reader;
- searching for an RFID tag in the proximity of the RFID reader;
- reading all the data encoded in the RFID tag;
- decoding the data;
- updating the database;
- closing the connection.

Figure 8 presents the application's window that allows the reading/writing of RFID tags.

2.5 Advantages

The integration of the developed mobile application into the main RFID_B2B system has some considerable advantages:
- the PC-PDA communication component is fast and secure, allowing the use of several PDA devices within the same system and supporting an intelligent solution for updating data;
- the generality of our application: one application - multiple purposes;

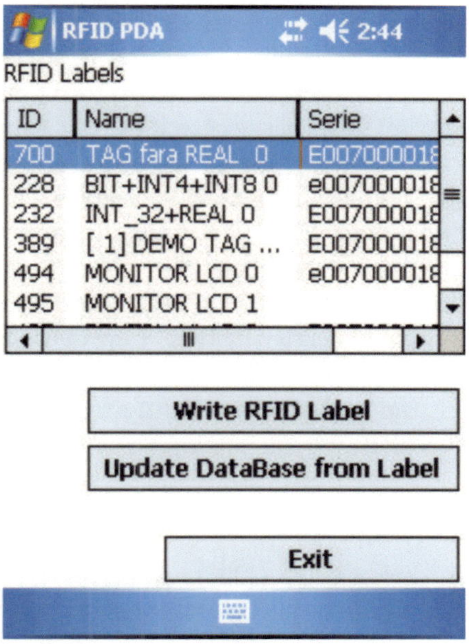

Fig. 8. RFID tags reading/writing

- the mobile application may be easily adjusted to users' requests, ensuring high performance and flexibility;
- the user graphics interface is simple to use and allows varied configurations depending on user preferences and necessities;
- the usage of the present system results in a considerable reduction of human errors;
- it promotes quality, security and ensures high-speed data processing.

Requiring no software modifications, the system is recommended for extremely varied activity fields.

3. Future directions for development

The following aspects might be taken into account as future directions for development:

- Application of agent technology, through the development of some intelligent agents, which allows the defining of the user's profile, simplifying, among others, the collecting of information and its filtering (considering the criteria chosen by users), etc;
- On-line processing of banking-financial-accounting transactions involved in B2B exchanges.

4. Conclusions

This chapter presents a PDA application that enables the development of complete and complex RFID_B2B solutions in industry and commerce. A tag will be attached to each material/ assembly, which will allow its identification based on an ID code. Thus, based on these ID codes attached to each product or assembly, it will be possible to check the constituents and origin of each finite product, the components of assemblies and the origin of the constituent components, and so on, for each company involved in the building process of the final product. By extending the system to the entire supply chain, the final

consumer will be able to track the origin of the materials included in the final product down to the primary sources.
The integration of the mobile application into the main RFID_B2B system has considerable advantages: the usage of several PDA devices within the same system; fast and secure PCPDA communication; a high degree of generality of the entire system; easily adjustment of mobile application to users' requests; a flexible user graphics interface; a high-speed and secure data processing; human errors reduction. Thus, the system is suitable for extremely varied activity fields without software modifications.
All in all, the implementation of our solutions has resulted in a high-performance multi-user mobile application, which can generate further improvements in RFID_B2B system. Companies have an excellent opportunity to improve their competitive advantage, using mobile technology to deliver important corporate information to employees, partners, and customers wherever they're located.

5. References

Cerlinca, M.; Graur, A. & Cerlinca, T. (2006). A Script Language for RFID Systems, *Proceedings of the Second European Conference on the Use of Modern Information and Communication Technologies ECUMICT* 2006, ISBN: 9-08082-552-2, March 2006, Ghent, Belgium.

Cerlinca, T.; Turcu, C. & Cerlinca, M. (2008). Integrating Mobile Applications into RFID Based B2B Systems, *22nd International Conference on Advanced Information Networking and Applications*, pp. 1341-1345, ISBN: 978-0-7695-3096-3, March 2008, Gino-wan, Okinawa, Japan.

Finkenzeller, K. (2003). *Fundamentals and Applications in Contactless Smart Cards and Identification, Second Edition,* John Wiley & Sons, ISBN: 978-0-470-84402-1.

Lefebvre, L.; Lefebvre, E. & Bendavid, E. (2006). RFID as an Enabler of B-to-B e-Commerce and Its Impact on Business Processes: A Pilot Study of a Supply Chain in the Retail Industry, *Proceedings of the 39th Annual Hawaii International Conference on System Sciences*, pp. 104a-104a, January 2006, Hawaii.

Turcu, C.; Turcu, O. & Popa, V. (2007). Integrating RFID Technologies in B2B Applications for Enterprise Supply Chain, *Proceedings of the 1-st RFID Eurasia Conference*, pp. 206-209, ISBN: 978-975-01566-0-1, September 2007, Istanbul, Turkey.

Turcu, C.; Prodan, R. & Popa, V. (2006). A Generalized Integrated RFID-Based System for the Identification and Traceability of Products and Subsets in Enterprises, *Proceedings of the Second European Conference on the Use of Modern Information and Communication Technologies ECUMICT* 2006, pp. 147-158, ISBN: 9-08082-552-2, March 2006, Ghent, Belgium.

28

Development of Consumer RFID Applications and Services

Yong-Woon KIM
ETRI
Republic of KOREA

1. Introduction

Basically RFID is a wireless communication technology within the L1 (Layer 1, the physical layer of the OSI 7-layer Reference Model) and L2 scopes between RFID tag and reader. An RFID tag works as a data storage and transmits stored data to an RFID reader via the wireless technology.

Such a basic communication scope can be extended by inter-networking of a series of networks to support business and consumer applications. Thus RFID has been adopted for enterprise business purposes in the form of Business-to-Business (B2B) RFID applications in retail, logistics, supply chain management, pharmaceutical industries, etc.

Nowadays an RFID reader is being equipped in a cell phone, which enables network-based consumer-purposes RFID applications called B2C (Business-to-Customer). Additionally a business partnership may integrate B2B with B2C applications into B2B2C applications. Consumer applications are provided as services to consumers. Such B2B, B2C and B2B2C applications must be based on network and communication among functional entities distributed in a closed enterprise network or an open, public network like Internet.

When RFID stayed in limited environments with limited purposes, development and standardization issues were not too many. But applying RFID to those new application areas requires consideration of L1 to L7 issues and expands the development and standardization scopes of RFID into higher layer scopes. Each RFID application type may produce new challenges with different characteristics and different requirements.

This chapter identifies a new business opportunity by integration of RFID with mobile telecommunication networks to enable consumer RFID applications and services, describes its development models in terms of communication model, functional configuration, message exchange procedure, and case studies, and then summarizes relevant activities of standardization in ISO/IEC, ITU-T and NFC Forum.

2. Integration with mobile telecommunication networks

This clause identifies a few problem statements about mobile telecommunication services and a new service model of the mobile RFID to mitigate them. Then it describes prospective business impacts of mobile RFID services.

2.1 Problem statements of mobile telecommunication services

Mobile telecommunication technologies will have a variety of problem statements to be tackled to provide better consumer services. A few of them RFID can mitigate are described.

Media break between off-line and on-line

The media break problem exists between off-line and on-line worlds. In the off-line world, everyone uses his signature to prove his legal actions. Under a seal-based operation system, a seal image shall be registered at a public certification center, for example, the government in Korea, and a seal certificate issued by the public certification center shall be attached to legal documents such as agreements and contracts. But instead digital certificates are used in the on-line world. So people should have two signatures in off-line and on-line, which means such off-line and on-line media cannot interwork each other. This is the media break problem.

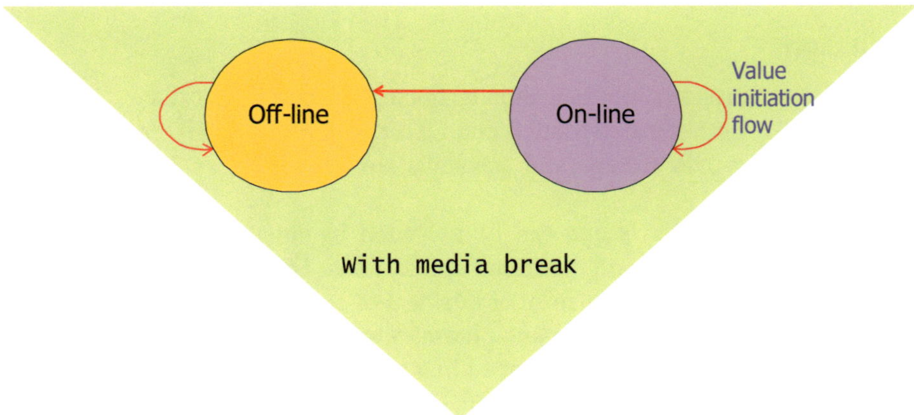

Fig. 1. View of the media break problem

In Figure 1, the off-line world can create values for itself in every off-line business process. For example, physical materials and corresponding products are manufactured; a group of the products are transported to wholesalers and retailers; and consumers buy them at shopping malls. This value chain occurs only in the off-line world. The on-line space also can create values for itself. For example, Google makes money through various on-line advertisement business models. The on-line world can affect the off-line value chain. For example, on-line shopping triggers off-line business processes, but not vice versa. The off-line space cannot initiate any value chain toward the on-line business space due to the media break problem.

Tree-based user interface

It may be said that mobile Internet services of mobile telecommunication service providers are supported via tree-based user interface.

The traditional way of information access was "tree" as shown in Figure 2. The traditional Gopher was based on the tree-based information hierarchy and was the top information access tool until the early 1990's before popularization of the Web.

Figure 3 shows the hyperlink-based information access. Since the Web was released in 1992 by Sir Tim Berners-Lee, its hyperlink-based access interface replaced the tree-based interface eventually.

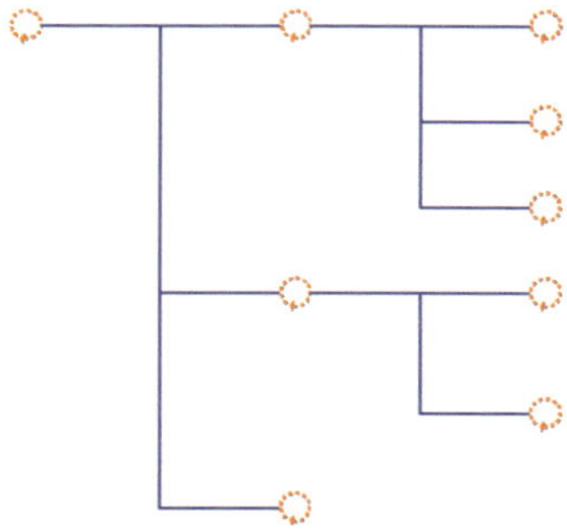

Fig. 2. Tree-based information access

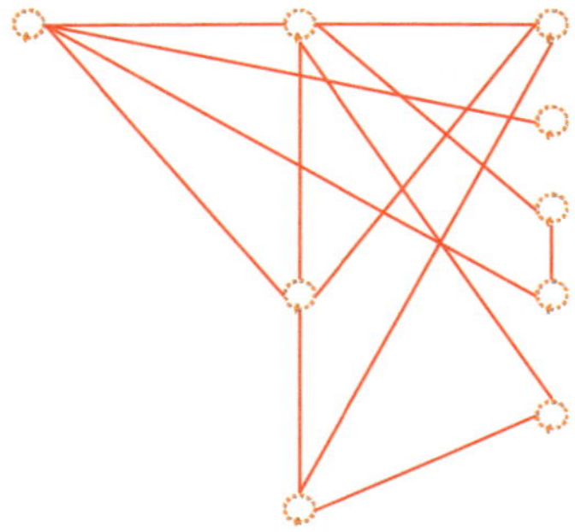

Fig. 3. Hyperlink-based information access

Internet services over mobile telecommunication networks are provided in various ways. The typical example is Wireless Application Protocol (WAP) which is a wireless-profiled Web service platform (WIKIPEDIA, 2008)(OMA, 2008). Other techniques also can provide Web-based information access services.

Even though such mobile Internet services are provided via hyperlink-based user interfaces, they look like working as tree-based user interfaces because cell phones cannot use mouse input interface. Usually consumers should follow links deeper and deeper to reach a final target even in hyperlink interface. They will make decisions to enjoy the target or not. Thus information service providers are struggling to take mobile Internet users to decision points for profitable contents and make them pay money for the contents. The easiest way to do this is to move the decision points up to the top level of the user interface in order to expose them to users more times. But this betterment must be limited due to too small screen size of cell phones.

Monopoly of mobile Internet service providers

Mobile Internet service providers are also called service portals because consumers have to navigate information contents of the service providers from a gate of information access. Consumers are usually locked within a service portal due to user interface problems of cell phone and feel a lot of difficulty in going out to other portals. The tree-based navigation structure limits users' options, that is, users can choose only provided items in the structure and cannot escape from the world made by a tree as illustrated in Figure 4. Theoretically they can go out to other contents and portals by inserting and executing URLs at a user interface of cell phones. But inserting alphabetical URLs in cell phones is difficult actually and almost every user stays within a service portal provided by a mobile telecommunication service provider.

Now the portal appears as power and contents providers must go to the portal to serve their contents to mobile Internet consumers. This kind of monopoly will cause a bad impact that contents providers cannot be revitalized much because such situation causes development of service contents restricted and controlled by service portals.

Fig. 4. Consumers locked in a service portal

2.2 Mobile RFID services

For a long time, RFID has stayed in B2B business fields such as transport and logistics, supply chain management, manufacturing and processing, and inventory control due to various problems such as still expensive RFID tag price, lack of 100% reading accuracy, limited operation conditions, etc.

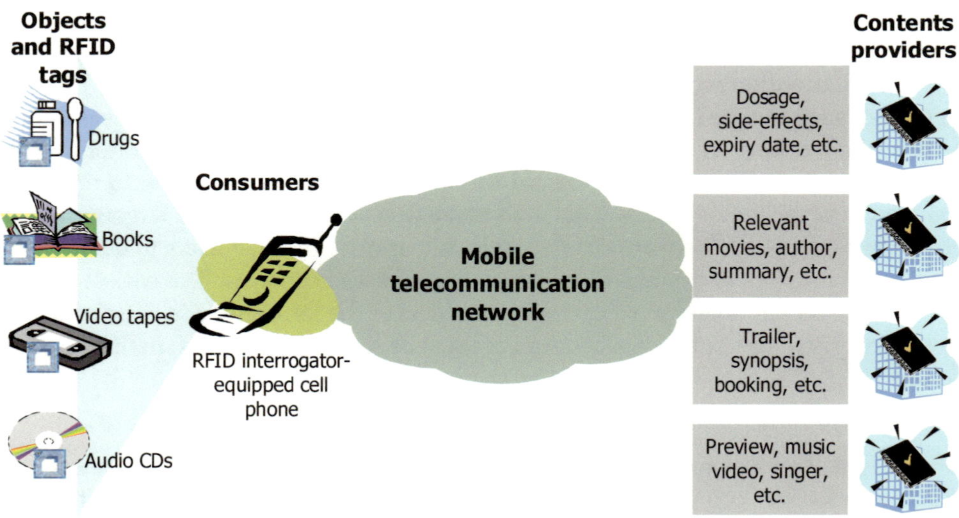

Fig. 5. Use cases of mobile RFID applications and services

RFID can be integrated with consumer applications for which a prominent terminal is cell phone. This type of RFID applications is called mobile RFID applications in a B2C manner where business parts provide information contents and consumers retrieve and enjoy them with their cell phones.

People may confuse the meaning of "mobile RFID." It doesn't mean that RFID is mobile and flowing. But it is a simply combined term of RFID and mobile telecommunication.

ITU-T described mobile RFID telecommunication services from a service point of view as follows: "RFID based mobile telecommunication services can be defined as services that provide information on objects equipped with an RFID tag over a telecommunication network. The RFID reader is installed in a mobile device such as a cell phone. The implementation of RFID in mobile telecommunications services would lead to a scenario where the tags are stationary and readers (that are integrated in the cell-phone) become mobile. There would also be applications where the mobile phone can be both, tag and reader at the same time (Seidler, 2005)." The information provided by an RFID tag on an object doesn't always depend on the object. For example, an RFID tag attached to a wine can give a restaurant name for advertisement purposes.

Figure 5 shows mobile RFID service cases. RFID tags are attached to physical objects; relevant information contents related to the objects are established at backend application systems; and, consumers execute a service operation by triggering an RFID reader-equipped cell phone to read an RFID tag and retrieving the information content associated with the RFID tag, i.e. its corresponding object.

2.3 Business impacts of mobile RFID services

Integration of RFID with mobile telecommunication services can mitigate the problems defined in clause 2.1 with the key feature of off-line hyperlink which can solve the media break problem, realize the hyperlink-base user interface acctually and enable easier navigation within a service portal and to other service portals.

Concept of off-line hyperlink

Merriam-Webster defines "hyperlink" as an electronic link providing direct access from one distinctively marked place in a hypertext document to another in the same or a different document (Merriam-Webster, 2008). It may be seen for easier understanding that a hypertext is an information text with one or more hyperlinks. A hyperlink shall have an address information to take its user to its target information. The Web technology represents the address information as URL (Uniform Resource Locator). Thus the hyperlink has a standardized format to incorporate URL. The following example shows a representation of URL into a corresponding hyperlink:

URL: *http://www.abcd.com/hyperlink*

Hyperlink: *<a href="http://www.abcd.com/hyperlink">hyperlink</a>*

This hyperlink exists logically as digital data and resides in the on-line world. Thus it may be called on-line hyperlink compared with the other case, off-line hyperlink.

An RFID tag is a tiny memory device which embeds some information such as identifier, price, name, manufacture date, shipping date, and manufacturer name of an object. But, the identifier of the object is the mandatory requirement and all the others are optional. ITU-T Y.2091 defines "identifier" as a series of digits, characters and symbols or any other form of data used to identify network element(s), function(s), network entity(ies), subscriber(s), user(s) providing services/applications, or other entities (e.g. physical or logical objects) (ITU-T Y.2091, 2006).

URL also is one of identifiers. An RFID tag may contain a URL for addressing information toward an information content associated with a physical object. That is, RFID tags which exist in the off-line and physical world contain a URL for certain information and then enable hyperlink access at the information. In other words, off-line and physical objects can trigger information users to access on-line information associated with the objects as shown in Figure 6. This is the off-line hyperlink or often called physical hyperlink. Even though, however, an RFID tag can contain a URL, it contains a shorter form of identifier usually in an alpha-numeric form instead of the URL because it has a small memory space and saving memory means saving money. In this case, a resolution from the identifier to the URL has to be provided.

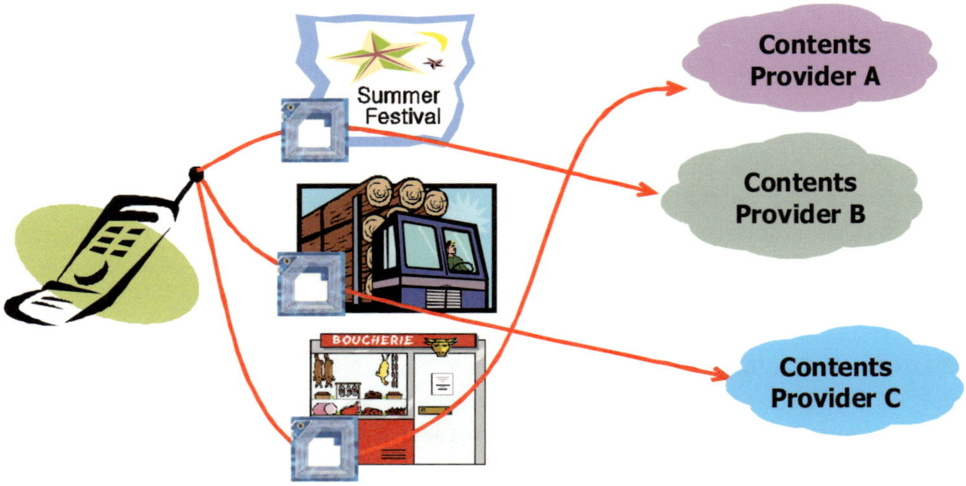

Fig. 6. Off-line hyperlink feature

Solving the media break problem

Clause 2.1 defined the media break problem with Figure 1. That is, there is a broken link between off-line and on-line worlds. The off-line hyperlink feature can solve this problem as depicted in Figure 7.

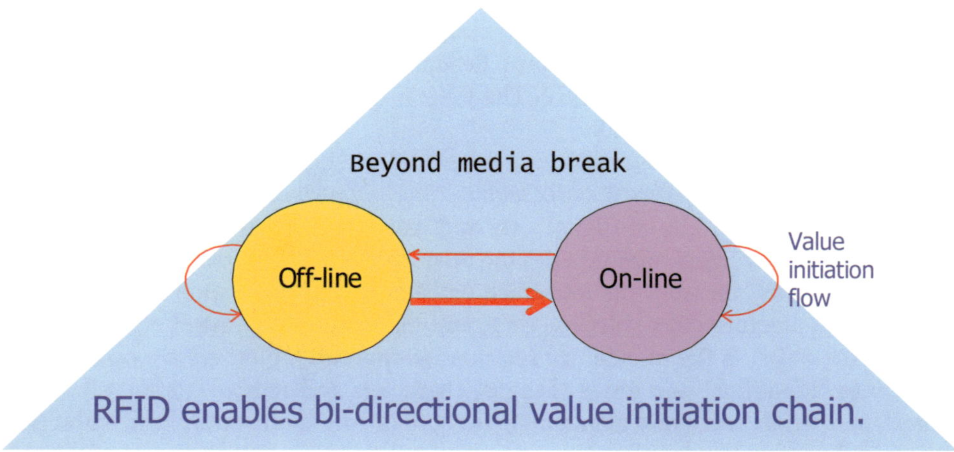

Fig. 7. Solution of the media break problem

An off-line and physical RFID tag contain an information pointer represented as URL or a corresponding alpha-numeric identifier which can take information users to an associated on-line content. That is, off-line objects such as movie posters enable mobile Internet users to access information contents associated with the movie posters and the new initiation line from off-line to on-line is realized as shown in Figure 7. So RFID can realize a bi-directional value initiation chain between on-line and off-line. This benefit makes much bigger business opportunities.

Realizing hyperlink-based user interface

Even though mobile Internet services have been developed in hyperlink-based user interface, their information contents look like being provided in tree-based user interface in which a decision to pay money for certain information occurs at the final step after a few navigation steps. This takes time, makes users feel user interface inconvenient and then causes them to give up their navigation. Thus how to take information users to their decision points to pay money is one of the key problems to be tackled for service providers.
The off-line hyperlink feature of RFID can realize hyperlink-based user interface actually as depicted in Figure 8. RFID can take information users directly to decision points to pay money. That is, cell phone users can enjoy hyperlink-based user interfaces with pressing keypad buttons just once or twice and a tremendous number of off-line objects will take them to on-line contents and services via the bi-directional value initiation flow.

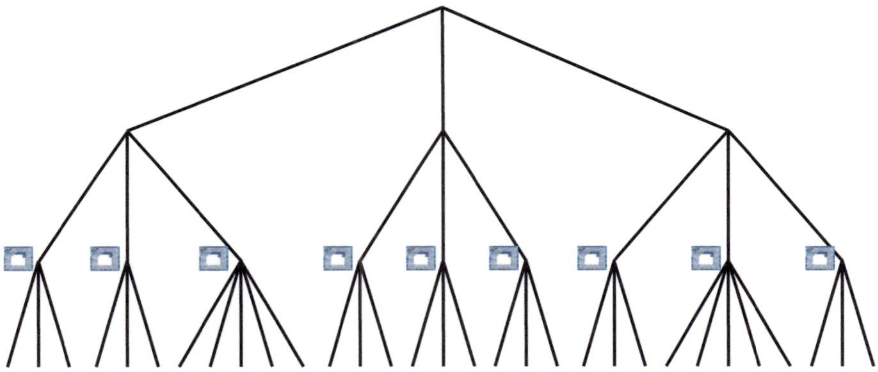

Fig. 8. Off-line hyperlink-based user interface

Enabling easier navigation to other service portals

Difficulty in inserting URL in a user interface of cell phones has caused users to stay within only a service portal. The off-line hyperlink solves this problem and enables information users to go directly to other service portals.

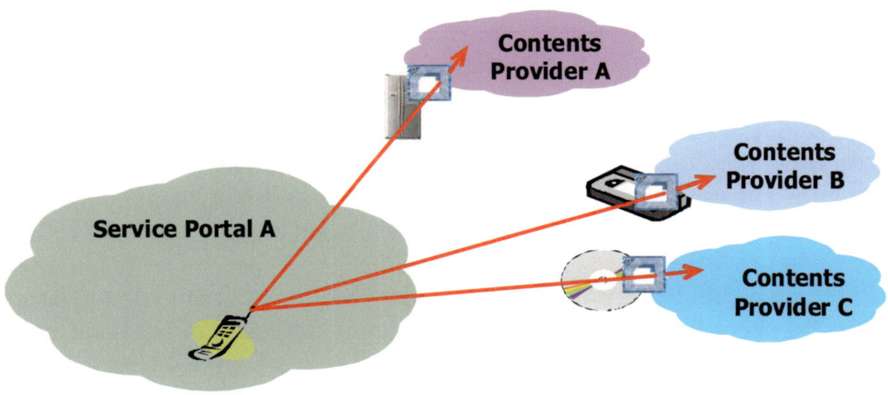

Fig. 9. Easier navigation to other service portals

As shown in Figure 9, anybody can build up an information contents portal at his/her application systems without negotiation or permission of other service portals. Each content provider in Figure 9 can provide its information contents as a service portal because the off-line hyperlink takes users directly to the information contents portal. RFID makes the power shift.

3. Development model of mobile RFID services

Mobile RFID services may be developed in various ways. This clause describes only a basic model and an extended model.

3.1 Basic and extended communication models

Figure 10 shows the case that an RFID tag is attached to a movie poster for a certain movie; an identifier for the movie is embedded in the RFID tag; an RFID reader is built in a cell phone; a contents provider system contains information contents associated with the movie; and, an identifier directory system has a location information for the information contents distinguished by the identifier where the location information may be represented as a URL.

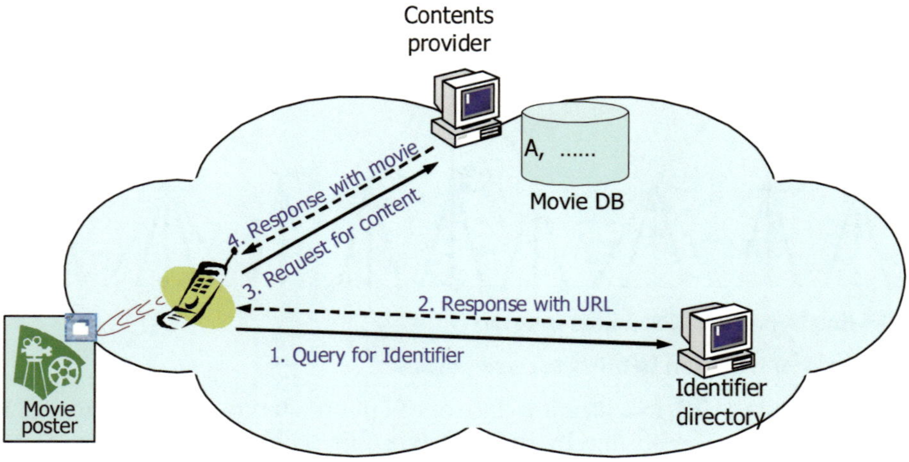

Fig. 10. Basic communication model

Figure 10 illustrates a basic communication model consisting of two end-to-end operations: identifier resolution (step 1 and 2) and content retrieval (step 3 and 4). This basic model is not a new one. Every client/server application works in the similar model: name resolution and content retrieval. For example, when a Web browser accesses a Web page via a URL, it has to get a network address, i.e. IP, resolved from a domain name embedded in a URL by consulting DNS and then it connects with a Web server system addressed by the IP to access the Web page. The former operation is the name resolution and the latter one is the content retrieval.

Mobile RFID has a different resolution target, identifier, not domain name. The identifier resolution is the operation that a client asks the identifier directory of resolving an identifier and receives a corresponding URL to the identifier. DNS is a typical example of protocol solutions for the identifier resolution. The content retrieval works through a generic Web access operation.

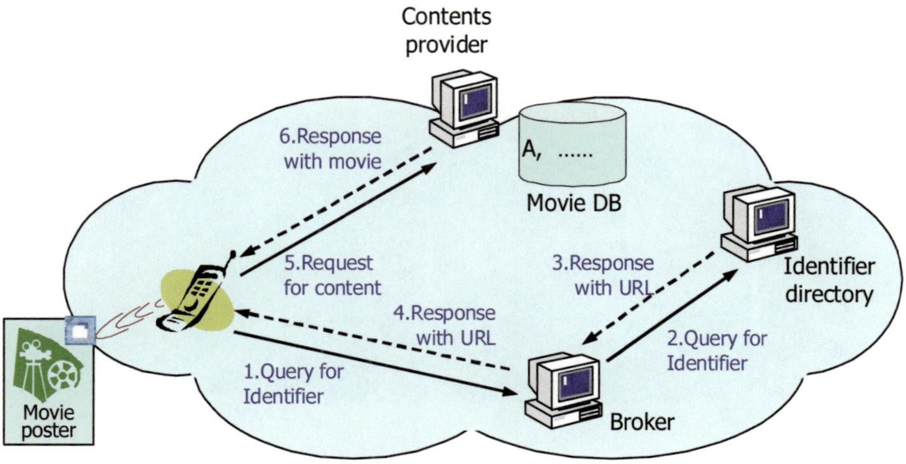

Fig. 11. Broker-based communication model

The basic model may be extended in various ways. One of them is to incorporate an intermediate broker between mobile RFID terminal and identifier directory as shown in Figure 11. It may simply relay identifier resolution messages or do additional functions such as user authentication and filtering for resolution requests.

3.2 Functional configuration of mobile RFID terminal

A mobile RFID terminal may be described in Figure 12 of which WAP browser is one of typical Web browsers in cell phone software environments; bearer service is mobile telecommunication service based on, for example, CDMA, GSM, UMTS, GPRS, etc.; device driver is a functional entity to control the RFID reader; middleware platform is an application execution and running environment like BREW, WIPI and J2ME; and mobile RFID applications run over the platform.

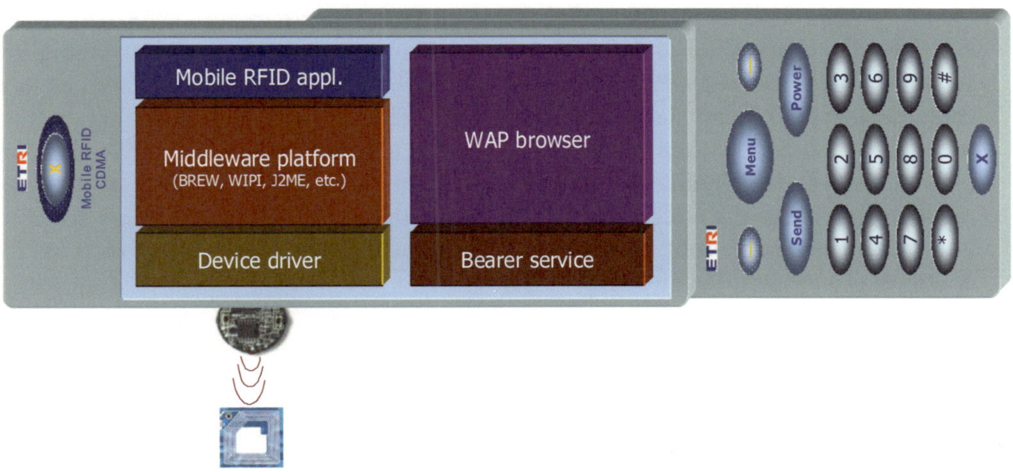

Fig. 12. Functional configuration of mobile RFID terminal

Figure 13 shows a more detailed functional configuration. Various functional entities may be needed to support mobile RFID applications at a mobile RFID terminal. Following functional entities are mandatory: interrogator control, user data processing, identifier processing and identifier resolution. All these functions are provided through proper APIs to mobile RFID applications.

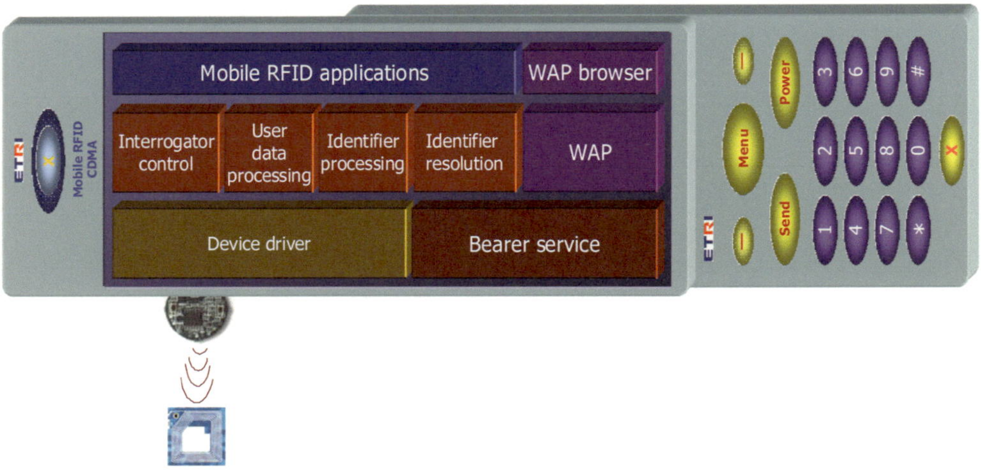

Fig. 13. Key functional entities of mobile RFID service platform

The interrogator control is a set of functions based on the device driver to access and control the RFID interrogator. The user data processing is a functional entity to process user data formatted in a standardized way and contained with an identifier in an RFID tag. For example, an RFID tag shall have an identifier for a product and may have additional information about the product such as name, price, manufacturer and expiry date. Such additional information is handled as the user data.

The identifier processing entity performs a set of functions with an identifier read by the interrogator control as follows:

- It learns what kind of the identifier is by using, for example, OID or EPC header. Learning what kind of the identifier is means getting its structure information. For example, if an identifier is learned as a telephone number assigned by the E.164 standard, the processing entity has to get the E.164 structure information from its local configuration data, internal database or an external directory service.
- It decodes raw binary data of the identifier into a standardized form with structure information of the identifier. For example, if the identifier is learned as an E.164 number, the processing entity coverts raw data of the identifier with the E.164 structure information into something like *1.2.345.6789* in case of using the dot notation for the delimiter of sub-identifier elements.

The identifier resolution is a function to resolve an identifier into associated information which means the information associated with an identifier, which results in maintaining mapping relationship(s) between the identifier and associated information. Example associated information instances are information content like audio, video, text and image, or another identifier like URL, URN, IP address and E.164 number. An identifier may have multiple associations. Example solutions to provide the identifier resolution are DNS, X.500, LDAP, etc.

Those functional entities can be described with an operation scenario as shown in Figure 14. The interrogator control reads tag data from an RFID tag. The tag data may consist of only an identifier or both identifier and additional application-specific data for which the identifier processing function manipulates the identifier and the user data processing function does the application-specific data.

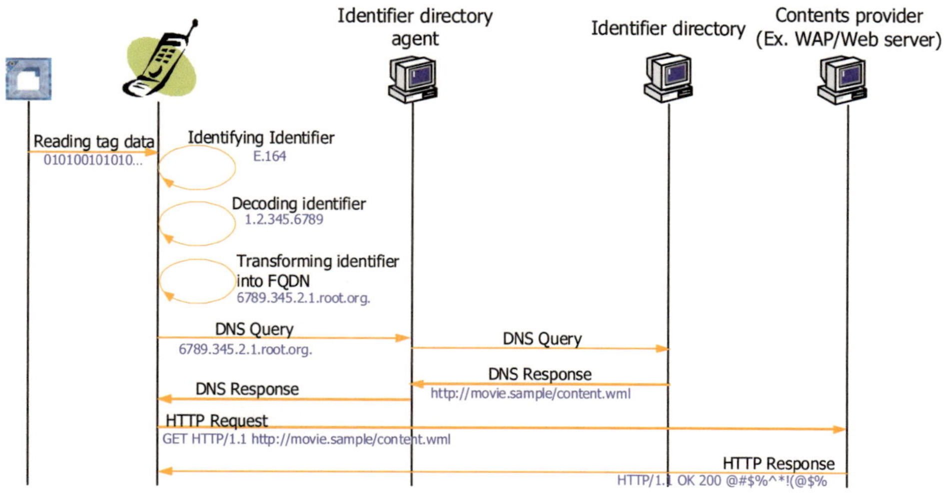

Fig. 14. Example operation scenario

The identifier processing function has to know identity of the identifier. ISO/IEC standards use OID to let users know the identity and EPC standards use EPC header. In the example, the identifier is learned as E.164. The identifier processing function should have E.164 structure information to manipulate the identifier. It converts raw binary data of the identifier into sub-identifier elements, 1, 2, 345 and 6789, which may be represented in various ways by standards. In case of using the dot notation, the identifier is denoted as *"1.2.345.6789."*

Then the identifier resolution process starts. Figure 14 shows the case of using DNS for identifier resolutions. An identifier resolution request is triggered by a DNS Query message which includes an FQDN-formatted identifier like *"6789.345.2.1.root.org."* according to the DNS protocol specification. A corresponding DNS Response message is replied, containing a URL registered with the identifier in the identifier directory.

Finally a Web browser accesses the contents provider through the URL information and fetches relevant information content.

3.3 Internal operation scenario

A service operation may work in a mobile RFID terminal as shown in Figure 15. The step 1 and 2 show that an application invokes the interrogator control function residing in the middleware platform in order to read an identifier from an RFID tag. An internal operation, invoking the identifier processing function, isn't depicted for a simplified presentation but performed actually before the step 3 and 4. Then the identifier represented in a standardized format like URN is returned to the mobile RFID application.

The application should have an address information for the content to access an information content identified by the identifier. It calls the identifier resolution entity to get the address information, URL. The step 3 and 4 shows an identifier resolution operation and a URL is returned to the application.

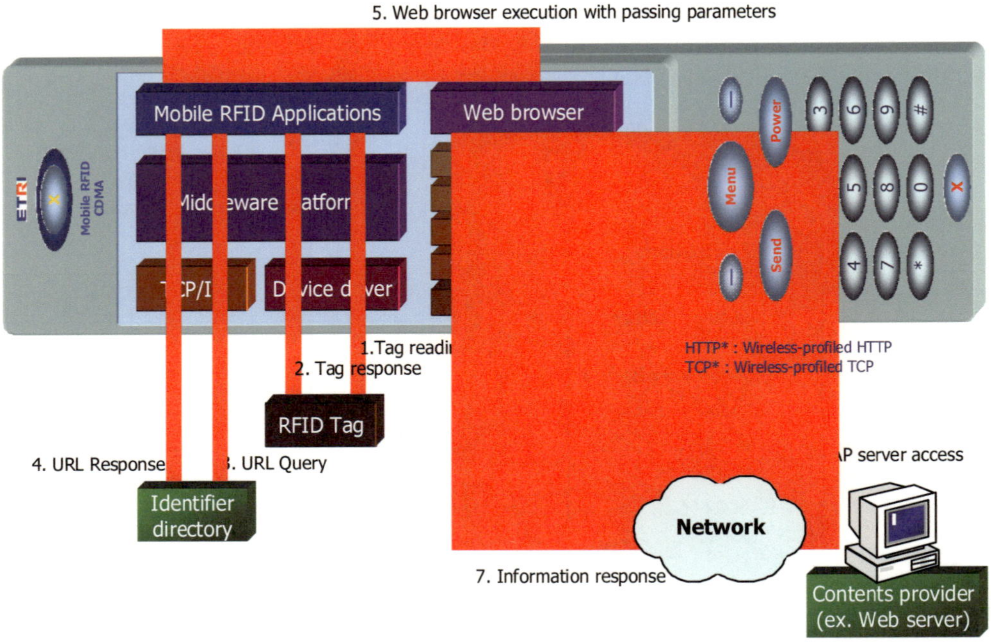

Fig. 15. Internal operation scenario

The mobile RFID application may access the information content directly via the URL or instead a web browser can do that. Since web browsers have rich information content processing features, they are better for information browsing of end-users than dedicated mobile RFID applications. Figure 15 shows the latter case and the mobile RFID application executes a web browser as shown in the step 5 with passing the URL in. Then the web browser takes over all remaining operations.

Figure 15 shows two types of web application platform, WAP 1.0 shown in the left side and WAP 2.0 shown in the right side. The application platform of conventional TCP/IP and web technologies is another alternative. The step 6 and 7 depict an ordinary web access process via the URL.

3.4 Case studies of mobile RFID development

In order to bring RFID to consumers, make RFID and related business opportunities bigger dramatically, and enable B2C RFID applications consequently, end users must have an RFID reader as well as a ubiquitous information terminal. Thus, RFID-equipped cell phones are the perfect candidate device to achieve such purposes.

Table 1 is a summary of mobile RFID development cases. Nokia developed an initial version of the mobile RFID in 2004. Then it developed NFC-based mobile RFID again. KDDI developed two types of mobile RFID with passive and active RFID solutions. Korea has decided to use the UHF band to use following advantages:

- Supporting both short and long read ranges by power control according to user and service requirements; and
- Avoiding duplicate RFID tag installations for both B2B and B2C applications by sharing a single tag. That is, a single RFID tag can be shared by different business domains, B2B and B2C and provide different contents according to applications.

	Frequency	Read range	Standard	Tag/reader features
Nokia's mobile RFID	13.56MHz	2~3cm	ISO/IEC 14443A	Separate
KDDI's mobile RFID with passive type	2.45GHz	~ 5cm		Separate
KDDI's mobile RFID with active type	315MHz	~ 10m		Separate
NFC	13.56MHz	~ 10cm	ISO/IEC 18092	Both
Korea's mobile RFID	908.55~ 913.95MHz	~ 80cm	ISO/IEC 18000-6C	Separate
uID Center's mobile RFID	13.56MHz	~ 5cm		Separate

Table 1. Summary of mobile RFID implementations

[Note] "Both" in the tag/reader features means an RFID device supports reader as well as tag features. Usually an RFID device works as either reader or tag but NFC supports both.

uID center (Kim & Koshizuka, 2006)

The Ubiquitous ID Center developed a network-based ID architecture called "Ubiquitous ID Architecture". The architecture defined "ucode" as an identification scheme to identify things and places. The ucode is stored on several kinds of tags such as RFID, barcode, 2D barcode, and sensor network tag embedded into several things and places in the real world. In the Ubiquitous ID Architecture, these tags that carry ucodes are called as ucode tags.

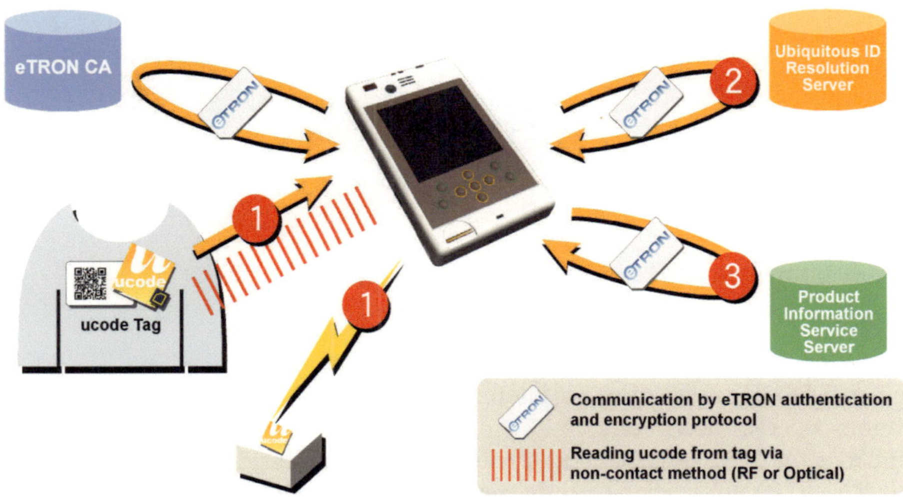

Figure 16. Basic configuration of Ubiquitous ID architecture (Source: uID Center, 2005)

In the basic configuration of the Ubiquitous ID Architecture, the ucode tag contains only a ucode. The information regarding things and places is stored in a database connected to networks and end-user's service browser retrieves the information from the database through networks. If the ucode tag has enough space of memory, it may contain the information too.

The terminal that reads information from the ucode tag is referred to as "Ubiquitous Communicator (UC)". In Figure 16, the UC reads a ucode from a ucode tag and gains an access privilege to the information service server depending on the obtained ucode (step 1). It gets an address information mapped with the ucode from the ubiquitous ID resolution server which maintains the correspondence between ucode and relevant service information (step 2). Then it retrieves product information from the product information service server via the address information (step 3).

Since the communication by the Ubiquitous ID Architecture is privacy-conscious and secure, it uses a public key cryptography system and requires a certificate authority for the system. In addition, in the case the thing attached with a ucode tag flows out to the public, the identification preventing communication shall be used on the non-contact communication interface of the ucode tag to prevent malicious persons from illegally reading out ucode information stored in the tag.

KDDI

Figure 17 illustrates KDDI's mobile RFID service architecture. A service broker is located. The multi-contact server seems to support both identifier resolution and information contents provisioning. An identifier read by a cell phone from an RFID tag is transmitted to the service broker which finally transmits resulting information back to the cell phone.

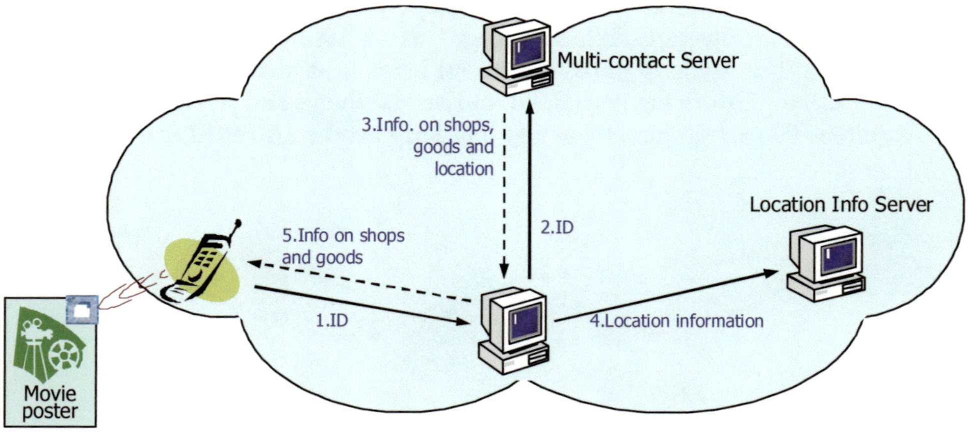

Fig. 17. KDDI's mobile RFID service architecture (Source: KDDI, 2006)

Nokia (Nokia, 2006)

Nokia developed a conceptual model of mobile RFID services based on ISO/IEC 14443A which is a 13.56MHz contactless air interface standard for smartcard applications. Its pilot applications worked successfully.

Then it chose the NFC technology and released the Nokia Field Fource solution in 2005 to support B2C RFID application/service models. Predesigned operations can be performed automatically by touching an RFID tag-equipped object with the NFC reader. Each tag contains a specific serial number which links to the initiation of a service, such as calling, messaging, browsing or recording data. Nokia developed Local Interactions Server (LI Server) which is a Web service that simplifies data capture, reporting, management, and communication with mobile workforces integrated into back-end application systems. That is, the LI server operates as a service broker as described in clause 3.1.

NFC

The Near Field Communication (NFC) supports three operating modes: Reader/Writer, Peer-to-Peer and Card emulation. The reader/writer mode works as a conventional RFID reader or tag. That is, an NFC device sometimes works as an RFID reader and sometimes works as an RFID tag, but it cannot support both modes simultaneously. Mobile RFID applications and services described in this chapter can be enabled by this operating mode.

The Peer-to-Peer mode enables two NFC devices to exchange data in the communication speed at 106, 212 or 424 kbit/s. When one device runs as a tag, the other device runs as a reader and then vice versa. So two devices exchange some data.

The card emulation mode supports a unified interface for various contactless smartcards to existing smartcard readers. It provides three RF communication modes: NFC FeliCa based on ISO/IEC 18092 and JIS6319-4; NFC Type A based on ISO/IEC 18092 and 14443A; and NFC Type B based on ISO/IEC 14443B, which can support ISO/IEC 14443, 15693 and Mi-Fare also. It is a solution to mitigate two problematic situations: a cell phone is usually equipped with a smartcard for various applicaitons and services and such smartcard cannot support various air interface protocols but do only a single air interface; and existing smartcard readers also have been deployed in such various ways and one smartcard reader cannot support other air interfaces. These situations restrict service coverage because one smartcard reader cannot communicate with other types of smartcards and vice versa. Thus, an NFC device with a smartcard in a cell phone can support a unified interface to almost every existing smartcard readers.

Korea

SKT and KTF of Korea developed mobile RFID service environments of which have been described in the clause 3.1, 3.2 and 3.3.

4. Standardization activities of mobile RFID technologies

Mobile RFID technologies are involved with a set of standardization issues and there are three relevant SDOs (Standards Development Organizations) as illustrated in Figure 18. Relevant bodies are ISO/IEC JTC 1/SC 31/WG 6; ITU-T SG 13, SG 16 and SG 17; and NFC Forum.

ISO/IEC JTC 1/SC 31 has developed AIDC (Automatic Identification and Data Capture) techniques which cover linear and 2-dimensional barcode symbologies as well as RFID. It consists of 6 working groups: WG 1 to WG 6. Among them, WG 6 is working on MIIM (Mobile Item Identification and Management) which covers mobile ORM (Optical Readable Media) as well as mobile RFID and additionally sensor interfaces specified by IEEE 1451. ITU-T is working mainly on network aspects of mobile RFID issues and NFC Forum is dealing with the whole parts of Figure 18. Even though SC 31/WG 6 and NFC Forum have identical work scopes, they have different architectures and technical views.

4.1 ISO/IEC JTC 1/SC 31/WG 6

After 1-year preliminary study of mobile item identification and management issues from March 2007, creation of WG 6 was approved in February 2008 by National Bodies of SC 31 and endorsed at the plenary meeting of SC 31 in June 2008. Its work scopes consist of sensor interfaces as well as RFID and ORM with mobile telephony.

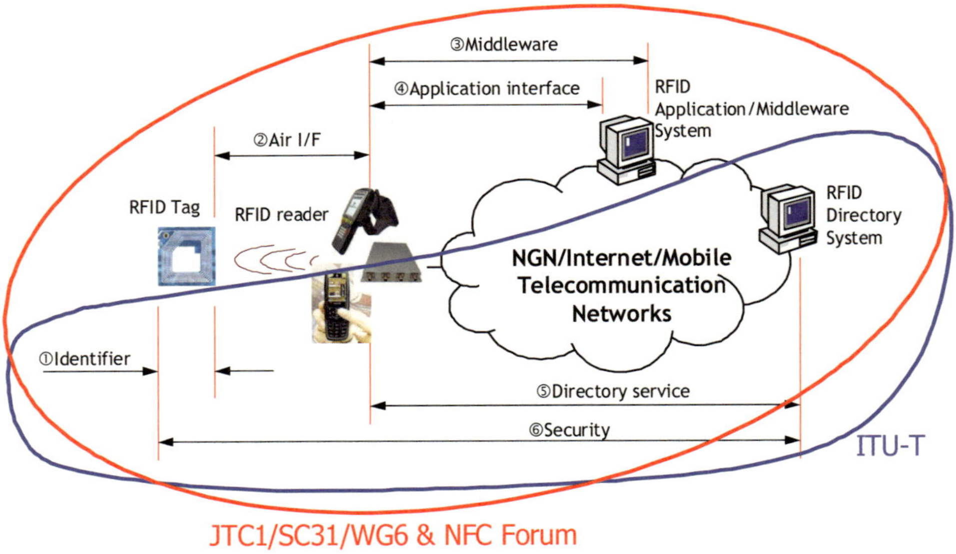

Fig. 18. Standardization scope of relevant SDOs

The mobile ORM is conceptually identical to the mobile RFID but there is the only exception that camera and barcode tag are used instead of RFID reader and tag for a cell phone to read an identifier from the barcode tag through the camera. After the identifier is captured, remaining operations are identical conceptually but technical implementations may be different.

The sensor interfaces have been considered by IEEE 1451 which has already published a few standards and is developing some standards. WG 6 is going to adopt the published standards of IEEE 1451 via the fast track standardization process of ISO/IEC JTC 1.

At its first meeting in April 2008, it made following resolutions to initiate relevant standards developments:

- Regarding mobile ORM,
  - Submission of a new work item proposal for *Implementation guidance for Optically Readable Media (ORM) reader and ORM displayed on Mobile equipment*
- Regarding mobile RFID,
  - Appointment of a project editor for ISO/IEC 29143, *Air Interface specification for Mobile RFID interrogator*
  - Submission of a new work item proposal for *Reference architecture for Mobile AIDC services*
  - Submission of a new work item proposal for *Mobile RFID interrogator device protocol*
  - Submission of a new work item proposal for *UII scheme and encoding format for Mobile AIDC services*
  - Submission of a new work item proposal for *Application data structure and encoding format for Mobile AIDC services*
  - Submission of a new work item proposal for *Consumer privacy-protection protocol for Mobile RFID services*

  - Submission of a new work item proposal for *Object Directory Service for Mobile AIDC services*
  - Submission of a new work item proposal for *Service broker for Mobile AIDC services*
  - Submission of a new work item proposal for *Mobile AIDC application programming interface*

  [Note] "Mobile AIDC" aims at supporting both mobile ORM and RFID.

- Regarding sensor interfaces,
  - Submission of IEEE 1451.0, *Smart Transducer Interface for Sensors and Actuators – Common Functions, Communication Protocols, and Transducer Electronic Data Sheet (TEDS) Formats,* for PSDO (ISO/IEEE Partnership Standards Development Organization) fast-track processing
  - Submission of IEEE 1451.1, *Smart Transducer Interface for Sensors and Actuators – Network Capable Application Processor (NCAP) Information Model,* for PSDO fast-track processing
  - Submission of IEEE 1451.2, *Smart Transducer Interface for Sensors and Actuators – Transducer to Microprocessor Communication Protocols and Transducer Electronic Data Sheet (TEDS) Formats,* for PSDO fast-track processing
  - Submission of IEEE 1451.5, *Smart Transducer Interface for Sensors and Actuators – Wireless Communication Protocols and Transducer Electronic Data Sheet (TEDS) Formats,* for PSDO fast-track processing

It is believed that some of them have been proposed to JTC 1/SC 31 for development of new standards and relevant balloting period for each proposal is being progressed. Standards development activities will be followed according to voting results.

4.2 ITU-T

ITU-T TSAG (Telecummunication Standardization Advisory Group) launched a study group called Correspondence Group to review network aspects of mobile RFID technologies in 2005. After its 1-year study with four deliverables of review of business models and brief service scenarios; relevant standardization issues; proposed ITU-T strategy for relevant standards development; and terms and definitions for the work, ITU-T TSAG recommended ITU-T Study Groups to initiate relevant standards development works in 2006. Then ITU-T SG 13, SG 16 and SG 17 started their works to enable tag-based identification applications and services over telecommunication networks or Internet.

ITU-T SG 13

SG 13 deals with NGN (Next Generation Networks) which is targetting the move from circuit switched to packet based networks and aiming at IP-based convergence networks with public switched telephone network (PSTN), digital subscriber line (DSL), cable television (CATV), wireless local area network (WLAN) and mobile networks (ITU-T SG 13, 2008).

It initiated two work items to provide tag-based identification applications and services over such evolved networks where tag-based identification was defined as the process of specifically identifying a physical or logical object from other physical or logical objects by using identifiers stored on an ID tag such as RFID tag or barcode tag, and tag-based identification applications and services are applications and services which use tag-based identification.

One of the two works was finished in September 2008 and as a result ITU-T Y.2213 was published. It dealt with requirements analysis from NGN points of view because this work is a starting point to develop technical specifications. It covered description and scope of tag-based identification applications and services with some example scenarios; high level service requirements of tag-based identification applications and services; and extended or new NGN capabilities based on the high level service requirements (ITU-T Y.2213, 2008).

The other one is ITU-T Y.idserv-arch. It is scheduled to be finished in 2009. It specifies functional requirements of the NGN architecture based on ITU-T Y.2213 to support tag-based identification applications and services; functional entities of the NGN for extended capabilities; functional architecture of tag-based identification applications and services in NGN; and analysis of ITU-T Y.2213 from architectural viewpoints (ITU-T Y.idserv-arch, 2008).

ITU-T SG 16

SG 16 deals with all aspects of multimedia standardization, including terminals, architecture, protocols, security, mobility, interworking and quality of service. It focuses its studies on conferencing systems, directory services, speech, audio and visual coding, PSTN modems and interfaces, facsimile terminals, ICT accessibility, etc. (ITU-T SG 16, 2008) Since it focuses on multimedia applications, services and systems at higher layers of the OSI Reference Model, its study doesn't depend on specific network technologies.

It has already published two Recommendations: F.771 and H.621. ITU-T F.771 specifies a high level functional model, a service description and requirements for multimedia information access triggered by tag-based identification. Its scope is limited to those applications and services that have both multimedia and tag-based characteristics (ITU-T F.771, 2008).

ITU-T H.621 defines the system architecture for the multimedia information access triggered by tag-based identification based on ITU-T F.771 and serves as a technical introduction to subsequent definition of detailed system components and protocols. The services treated by ITU-T H.621 provide the users with a new method to refer to the multimedia content without typing its address on a keyboard or inputting the name of objects about which relevant information is to be retrieved (ITU-T H.621, 2008).

SG 16 initiated two new work items in April 2008, H.IDscheme and H.IRP. The former one is dealing with identification schemes and incorporating two schemes proposed by Korea and Japan. Korea proposed an extensible code called xCode and Japan proposed its ucode described in the case study of uID center in clause 3.4. The latter one is dealing with resolution protocols from an identifier into an address information like URL and incorporating two solutions proposed by Korea and Japan. Korean solution includes how to incorporate the existing DNS infrastructure for identifier resolution and Japanese solution specifies a unique, dedicated technology for identifier resolution.

ITU-T SG 17

SG 17 deals with telecommunication security, ASN.1 language and X.500 directory matters.

It has developed two items: X.rfpg and X.1171 (a.k.a X.nidsec-1). ITU-T X.rfpg defines guidelines that provide guidance for RFID users and vendors (including service providers and manufacturers) to protect the personally identifiable information for privacy of individuals in the context of RFID technology. The guidelines can be applied to the cases where the RFID system might be used to infringe on individual privacy in such a way that personally identifiable information is recorded in an RFID tag and then collected, or the object information collected by means of the RFID is linked to a personally identifiable information; provided, however, that it does not be applied to such cases as the object information is collected and used without any risk of invasion of personally identifiable information and privacy (ITU-T X.rfpg, 2008).

ITU-T X.1171 was approved in September 2008 and specifies threats against PII (Personally Identifiable Information) and requirements for PII protection in a B2C-based environment of applications and services using tag-based identification.

4.3 NFC forum

The Near Field Communication (NFC) Forum is a non-profit industry association that promotes the use of NFC short-range wireless interaction in consumer electronics, mobile devices and PCs.

It published a set of standards which are positioned as showin in Figure 19 according to their functional relationships.

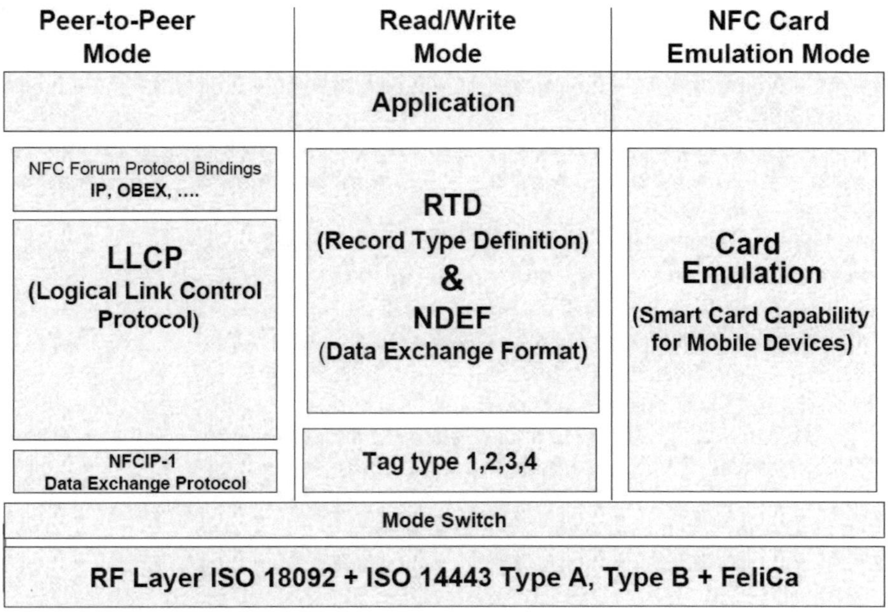

Fig. 19. Structure of NFC standards (Source: NFC Forum, 2007)

The NFC Data Exchange Format (NDEF) specifies a compact, common data format for NFC devices and NFC tags. The NFC Record Type Definition (RTD) specifies standard record types used in messages between NFC devices and between NFC devices and tags, and allows the use of Internet-standard media types. The tag type 1, 2, 3 and 4 specify, with a set of rules and guidelines, NFC device operation and management of a Type 1, 2, 3 or 4 tag. They also define the data mapping and how NFC device detects, reads, and writes NDEF data into the Type 1, 2, 3 or 4 tag platform in order to achieve and maintain interoperability. The Logical Link Control Protocol (LLCP) defines the protocol to manage the logical link between NFC devices (based on ISO/IEC 18092). The mode switch specification defines how to change the communication modes such as data transfer between devices, access at digital content, and creation of contactless transactions like mobile payment. (Romen, 2007)

5. References

ITU-T SG 13 (2008), Study Group 13 at a Glance, ITU-T, *http://www.itu.int/net/ITU-T/info/sg13.aspx*

ITU-T SG 16 (2008), Study Group 1 at a Glance, *ITU-T, http://www.itu.int/net/ITU-T/info/sg16.aspx*

ITU-T F.771 (2008), Service description and requirements for multimedia information access triggered by tag-based identification, *ITU-T Recommendation F.771 (2008)*

ITU-T H.621 (2008), Architecture of a system for multimedia information access triggered by tag-based identification, *ITU-T Recommendation H.621 (2008)*

ITU-T X.1171 (2008), Threats and Requirements for Protection of Personally Identifiable Information in Applications using Tag-based Identification, *ITU-T Recommendation X.1171 (2008)*

ITU-T X.rfpg (2008), Guideline on protection for personally identifiable information in RFID applications, *ITU-T draft Recommendation X.rfpg (2008)*

ITU-T Y.2091 (2006), Terms and Definitions for Next Generation Networks, *ITU-T Recommendation Y.2091 (2006)*

ITU-T Y.2213 (2008), NGN service requirements and capabilities for network aspects of applications and services using tag-based identification, *ITU-T Recommendation Y.2213 (2008)*

ITU-T Y.idserv-arch (2008), functional requirements and architecture of the NGN for applications and services using tag-based identification, *ITU-T draft Recommendation Y.idserv-arch (2008)*

Kim, Yong-Woon & Koshizuka, Noboru (2006), Review report of standardization issues on network aspects of identification including RFID, *ITU-T TSAG TD315, http://www.itu.int/md/T05-TSAG-060703-TD-GEN-0315/en*

Merriam-Webster (2008), hyperlink, *Merriam-Webster Online, http://www.merriam-webster.com/dictionary/hyperlink*

Nokia (2006), Nokia Field Force Solution, *Nokia, http://nfb.online.nokia.com/Page%20Content/Mobilize%20your%20Business/Knowledge%20Center/White%20Papers/WhitePaper_RadioFrequency.pdf | http://www.nearfield.org/2006/08/developing-nfc-applications | http://www.soc-eusai2005.net/documents/presentations/pres_6.pdf*

OMA (2008), Open Mobile Alliance, *http://www.openmobilealliance.org/*

Romen, Gerhard (2007), Near Field Communication Technology and the Road Ahead, *NFC Forum, http://mobile.hkwdc.org/nfc2007/presentations/NFCForum.pdf*

Seidler, Christoph (2005), RFID Opportunities for mobile telecommunication services, *ITU-T Technology Watch, http://www.itu.int/ITU-T/techwatch/docs/rfid.pdf*

WIKIPEDIA (2008), Wireless Application Protocol, *WIKIPEDIA, http://en.wikipedia.org/wiki/Wireless_Application_Protocol*

29

Efficient Outlier Detection in RFID Trails

Elio Masciari and Giuseppe M. Mazzeo
ICAR-CNR
Italy

1. Introduction

Radio Frequency Identification (RFID) applications are emerging as key components in object tracking and supply chain management systems. In the next future almost every major retailer will use RFID systems to track the shipment of products from suppliers to warehouses. In addition to providing insight into shipment and other supply chain process efficiencies, such data can be also valuable for determining product seasonality and other trends resulting in key information for the companies plans. Moreover, companies are already exploring more advanced uses for RFID. For example, tire manufacturers plan to embed RFID chips in tires to determine the tire deterioration. Many pharmaceutical companies are embedding RFID chips in drug containers to better track and avert the theft of highly controlled drugs. Airlines are considering RFID-enabling key onboard parts and supplies to optimize aircraft maintenance and airport gate preparation turnaround time.
Due to the streaming nature of RFID readings, large amounts of data are generated by these devices. In particular, RFID applications will generate a lot of so-called "thin" data, i.e. data pertaining to time and location. This phenomenon will be even more relevant when RFIDs are so cheap that every individual item can be tagged thus leaving a "trail" of data as it moves across different locations.
This scenario raises new challenges in effectively and efficiently exploiting such large amounts of data. In this paper we define a technique for detecting anomalous data in order to prevent problems related to inefficient shipments or fraudulent actions. To this end, we introduce a framework enabling users to control correct shipment of items. As an anomalous situation is detected, the system signals it in order to quickly recover the possible error. Moreover, while the trajectory is monitored we provide a validation step in order to take into account the natural "concept drift", i.e. deviation from the shipping plan due to modification of the service. In such a case, the signalled trajectory becomes new knowledge about the overall system. In order to check whether a trajectory has to be considered anomalous with respect to past ones, we adopt a distance-based approach. Specifically, we measure similarity between item routes by comparing their Discrete Fourier Transforms. We point out that the proposed approach is flexible and can be applied in different scenarios in order to monitor very different systems independently of their spatial extension. Preliminary experiments show the effectiveness of our approach.
This chapter is organized as follows. In Section 2 the problem and its application context are described. Section 3 introduces a measure of similarity between RFID series based on the Discrete Fourier Transform, which will be used to detect anomalies in object shipping as

described in Section 4. In Section 5 a technique for managing the concept drift related to the shipping plans is introduced. Section 6 presents some experimental results, showing the effectiveness of the adopted measure of similarity. In Section 7, some related work is presented. Finally, in Section 8 conclusion are drawn and some possible improvements to be developed in future works are proposed.

2. Problem statement

An RFID system consists of a set of *n* sources (i.e., tag readers), located at different positions, $R=\{r_1,\ldots, r_n\}$, producing *n* independent streams of data, representing tag readings. Each data stream can be viewed as a sequence of triplets ‹*rid, epc, ts*›, where: (i) $rid \in \{1,\ldots,n\}$ is the tag reader identifier (observe that it implicitly carries information about the spatial location of the reader), (ii) *epc* is the electronic product code, and (iii) *ts* is a *timestamp*. Basically, a reading ‹*rid, epc, ts*› denotes the fact that the item *epc* was at the location of the reader *tid* at time *ts*. The data streams produced by the sources are caught by a *RFID Data Stream Management System* (RfidDSMS), which combines the RFID readings into a unique data stream in order to support data analysis. We assume that *epcs* are gouped by classes and each class of *epcs* is assigned a shipping plan, which can be time-varying. As objects move across the distribution network, they are traced using the spatio-temporal information generated by RFID readers. The system architecture is shown in figure 1.

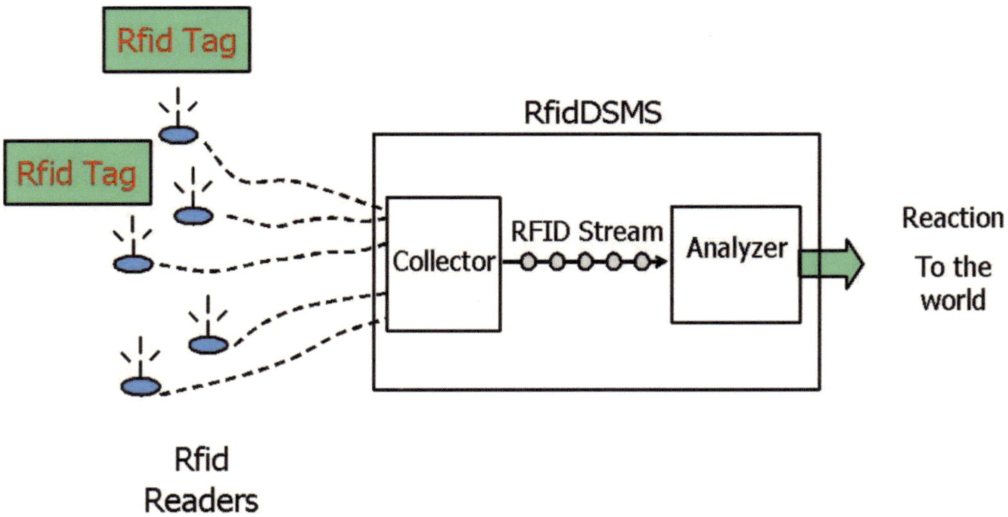

Fig. 1. RfidDSMS architecture

Our goal is to identify anomalies in the flow of the data. A possible anomaly can be an object that is planned to pass through a series of checkpoints but some check is missing, so it could be the case that someone modified the path with fraudulent intentions. Another relevant anomaly that would be interesting to be revealed is the continuous stay of an object at the same place, since it could be the case that the object is damaged or some shipment problem occurred. Thus, we can state the problem of detecting outliers in RFID data as follows.

Given an RFID system as described above, how is it possible to efficiently detect whether a new reading ‹rid, epc, ts› is representative of an anomaly in the shipping of epc?

In order to show how we tackle this problem, we introduce a running example to which we will refer through the rest of the paper. Consider the following sequence of readings being collected by the *RfidDSMS* assuming, for the sake of simplicity, that the system consists of 5 readers, r_1, r_2, r_3, r_4, r_5, 3 tagged items, o_1, o_2, o_3 (o_i represents the *epc* of the *i*-th item) and the initial time is 0. A possible sequence of readings could be:

$Seq = \{\langle r_1,o_1,0\rangle, \langle r_2,o_1,1\rangle, \langle r_1,o_2,1\rangle, \langle r_3,o_1,2\rangle, \langle r_2,o_2,2\rangle, \langle r_1,o_3,2\rangle, \langle r_4,o_1,3\rangle, \langle r_4,o_2,3\rangle, \langle r_2,o_3,3\rangle, \langle r_5,o_1,4\rangle, \langle r_5,o_2,4\rangle,\langle r_2,o_3,4\rangle, \langle r_2,o_3,5\rangle\}$.

In order to quickly identify the kind of anomalies described above, we need to define a technique for continuously comparing the streams generated by the readers with the original shipment plans. In order to accomplish this task we need to define a measure of similarity between two sequences. Intuitively, two sequences are said to have a similar structure if they correspond in the type of the readings they contain and in the order the readings appear. Observe that, even if it is easy to detect whether the structure of two streams is almost the same (i.e., the item has been scanned by the same set of readers), this information is rarely useful for our aims. Indeed, we would like to quantify the similarity between the structures of two streams, also emphasizing the differences that are most relevant. For instance, we consider similar two streams that exhibit the same features with different regularities, since this could be simply due to a shipment delay.
Much attention has been devoted to the problem of detecting similarity between time series using approaches such as time warping (Kruskall & Liberman, 1999). In this paper we propose a different approach, which is essentially based on the idea of associating each stream of readings related to an item with a time series representing its structure, and checking the structural similarity on the basis the Discrete Fourier Transform of its time series. As we will see, this approach is both efficient and effective.

3. A DFT-based similarity measure for RFID data

In this section we introduce a technique for encoding the input data stream into a time series and define a suitable similarity function that will be used to identify outlier data. First, we will describe how to encode RFID reading into time series representing the route of each item being monitored, and then we will introduce the metric for comparing the similarity of time series which will be adopted to identify outliers.
Given a sequence *Seq* of RFID readings, it is possible to obtain the route of a item x, that will be denoted as *Route(Seq, x)*, by extracting from *Seq* all the triplets with the *epc* value equal to that of x, and ordering these triplets by their ascending timestamp.
This operation yields an ordered set of triplets, from which the *epc* value can be disregarded (as it is equal to x for all the triplets), thus obtaining a sequence of pairs representing the association time-location of the item x that has been monitored.
These pairs can be represented as an array, where the cell indexes represent the timestamps, and the cell values represent the corresponding locations.
Observe that, for some timestamps the readings related to some items could not exist. In this case, we assume that the corresponding location for these timestamps is a *null* value.
From the sequence introduced in the previous section, we obtain the following routes

$Route(Seq, o_1)=[r_1, r_2, r_3, r_4, r_5, null]$
$Route(Seq, o_2)=[null, r_1, r_2, r_4, r_5, null]$
$Route(Seq, o_3)=[null, null, r_1, r_2, r_2, r_2]$

These arrays can be transformed into arrays of integers by assigning a positive integer identifier to each reader and assuming the *null* value to correspond to *0*.
For instance, by associating r_i with *i*, the routes of our running examples are those reported below:

Route(Seq, o_1)=[1, 2, 3, 4, 5, 0]
Route(Seq, o_2)=[0, 1, 2, 4, 5, 0]
Route(Seq, o_3)=[0, 0, 1, 2, 2, 2]

We call this encoding scheme of the readers *simple encoding*. We also introduce a more sophisticated encoding scheme, called *pairwise encoding*. This encoding scheme works as follows. Instead of assigning a unique identifier to each reader, as in the simple encoding, we assign a positive integer location to each different pair of readers. Thus, a function $E{:}R \cup \{null\} \times R \cup \{null\} \rightarrow N$ will represent the encoding function. For instance, in our running example, the encoding function E could be defined as in Fig.1.

	null	r_1	r_2	r_3	r_4	r_5
null	0	1	2	3	4	5
r_1	6	7	8	9	10	11
r_2	12	13	14	15	16	17
r_3	18	19	20	21	22	23
r_4	24	25	26	27	28	29
r_5	30	31	32	33	34	35

Fig. 1. Example of function adopted for the pairwise encoding scheme.

On the basis of the function *E* a vector which encodes the route $[p_1, p_2, \ldots, p_k]$, where $p_i \in R \cup \{null\}$, of an object *o*, is obtained by assigning $E(null, p_1)$ to the first element of *Route(Seq, o)*, and $E(p_{i-1}, p_i)$ to the *i-th* element ($1<i<k$) of *Route(Seq, o)*.
Thus, for our running example, the encoded routes are the following.

Route(Seq, o_1)= [1, 8, 15, 22, 29, 30]
Route(Seq, o_2)= [0, 1, 8, 16, 29, 30]
Route(Seq, o_3)= [0, 0, 1, 8, 14, 14]

In the following, we will omit *Seq* from *Route(Seq, x)*, thus writing *Route(x)*, when *Seq* can be implicitly assumed. Unfortunately, comparing two time series obtained by this encoding scheme can be very difficult, since time series could have different lengths and thus costly resizing and alignment operations, and stretching (or narrowing) would be necessary.
These drawbacks can be avoided if the signals are compared by examining their *Discrete Fourier Transforms* (Oppenheim & Shafer, 1999), which reveals much about the distribution and relevance of signal frequencies and can be computed incrementally as new readings arrive avoiding the problem of recomputing at each step the signal for each item.
Given a sequence of readings *Seq* and an item *x*, we denote as *DFT(Seq, x)* the Discrete Fourier Transform of the time series *Route(Seq, x)*. More formally, being r_x*=Route(Seq, x)* the array of size *N* representing the time series of item *x* extracted from the reading sequence *Seq*, *DFT(Seq, x)* is the array R_x of size *N* such that

$$R_x[k] = \sum_{t=0}^{N-1} r_x[t] \cdot e^{-\frac{2\pi i}{N} k \cdot t} \quad (k \in \{0, \ldots, N-1\})$$

In order to compare two routes, we consider the difference in the magnitude of the corresponding frequency components in their DFTs. This allows (i) to abstract from the length of the streams, and (ii) to know whether a given subsequence (representing for example a set of repeated readings) exhibits a certain regularity, no matter where it is located along the signal. Specifically, being x, y two items whose readings are in the stream *Seq*, r_x, r_y their routes and R_x, R_y the corresponding DFTs, we define the *DFT distance* ($dist_{DFT}(Seq, x, y)$) of the items as the sum of the squared difference of the magnitudes of the two signals, that is

$$dist_{DFT}(Seq, x, y) = \sum_{k=0}^{N-1} (|R_x[k]| - |R_y[k]|)^2$$

In the following, we will omit *Seq* from *DFT(Seq, x)* and $dist_{DFT}(Seq, x, y)$, thus writing *DFT(x)* and $dist_{DFT}(x, y)$, when *Seq* can be implicitly assumed. We will also adopt the notation $\mathrm{dist}_{\mathrm{DFT}}(\mathrm{p}, \mathrm{q})$ and $dist_{DFD}(P, Q)$ to represent the distance between two time series p and q whose DFTs are P and Q, respectively.

4. Outlier identification

In this section we define a strategy for identifying anomalies in the data flow, on the basis of the similarity measure for RFID time series defined in the previous section. In order to better understand this problem we describe two possible scenarios in our running example.

Consider three containers, whose epcs are o_1, o_2 and o_3, containing dangerous material that must be delivered through check points r_1, r_2, r_3, r_4, r_5 (in the given order). Assume that the RFID system reveals their routes to be *Route(Seq,* o_1*)*, *Route(Seq,* o_2*)*, and *Route(Seq,* o_3*)* reported in the previous section.

It easy to see that o_1 followed the correct route whereas o_2 and o_3 routes are affected by two different irregularities. Specifically, o_2 did not pass through r_3 and o_3 stayed too long at r_2. As regards o_2, two main explanations could be provided: (i) the original routing has been changed for shipment improvement, or (ii) someone changed the route with fraudulent intentions (e.g., in order to robber the content or to modify it). In our encoding strategy this case will produce two different signals exhibiting low similarity since the structure is different. As regards o_3, its sequence may occur either because (i) the container (or the content) is damaged so it cannon be shipped form r_2 until some recovery operation is performed or (ii) the shipment has been delayed. Depending on the anomaly detected, different recovery actions need to be performed. In our encoding strategy this situation will be reported in the frequency spectrum as a few components with amplitude much more larger than the others.

The two situations described above have an intuitive explanation but we point out that our encoding strategy enables detecting all the anomalies that cause the time series representing the routes to be different. Based on the above examples, we can now define our notion of outlier. Given the route r_x=*Route(x)* of an object x being traced whose planned sequence is represented by a time series r^*, and a threshold value T, we say that r_x is an outlier route if $dist_{DFT}(r_x,r^*)>T$. The threshold value can be chosen on the basis of the stream being monitored. Once defined our notion of outlier we can design an effective method for tracking objects. In particular, as objects enter our RFID environment, the readings are

collected by the *RfidDSMS*. At the *RfidDSMS* the signal corresponding to the correct shipping plan is stored (r^*) along with its *DFT* (R^*). As readings are generated, the *DFT* of the signal corresponding to the actual plan of the item x is computed (R_x) by using the previously described encoding strategy. If $dist_{DFT}(R^*,R_x)$ is larger than a predefined threshold T (suitable for the context being analyzed) a *signalling service* will notify the detected outlier in order to allow a proper recovery action.
In more details, as new readings arrives their contributions to the *DFT* of the proper *epc* is computed. This task can be accomplished very efficiently since the computation of the DFT can be performed incrementally without recomputing it from scratch.
If the computed distance between the original signal and the actual one is larger than T a *signalling service* will notify the detected anomaly, the location and the *epc* to be checked.

5. Managing changes in shipping plans

In the previous section it was assumed that the correct shipping plan for a given object is defined. Even though it is a realistic assumption, it is quite common that shipping plans change throughout time, for instance because the shipping has been optimized or because new requirements impose objects to pass through different check points. In order to manage this situation, it is necessary to enable the system to follow these changes. To this end, we propose a semi-automatic technique, in which the (i) system signals potential anomalies which can depend either on a shipping problem or on a planned shipping change and (ii) a human user check which of the two cases occurred taking the proper recovery action.
We assume that our system stores the DFT of the routes of the object currently in shipping. Furthermore, for each class of *epcs*, representing objects are shipped together, and for each location r_i, the *average routing* $R^*(epc\_class, r_i)$ is stored. Basically, $R^*(epc\_class, r_i)$ represents a weighted average of the partial DFTs of all the correct past routes followed by the objects belonging to the class *epc_class* arriving at r_i. These values are adopted for checking the validity of new routes, and they are continuously updated. Fig. 2 shows an algorithm which signals possible anomalies and maintains the average shipping plans up-to-date. The algorithm is invoked when a new triple ‹*l, o, t*› arrives (we recall that ‹*l, o, t*› denotes the fact that object o is at location l at time t) and works as follows. First, the DFT R_o of the route of object o is updated. The update requires time $O(N)$, where N is the number of locations traversed by the object. Then, the class e of epcs to which o belongs is retrieved. If a hash table is adopted to map the classes of epcs, this operation can be performed in constant time. Then, the DFT-based distance between R_o and the average DFT $R^*(e, l)$ of past correct routes

```
INPUT: a triplet ‹l, o, t›;
begin
        update_R(o, l, t);
        e ← epc_class(o);
        if dist(Ro, R*(e,l)) > T
                signal(‹l, o, t›);
        else
                R*(e, l) ← α * Ro + (1 - α) * R*(e, l);
end
```

Fig. 2. Algorithm for detecting wrong routes

is computed. If this distance is less than a given threshold T, $R^*(e, l)$ is updated, in order to take into account the last correct route. The update is performed by adopting a parameter α for weighting new valid routes. The computation of the distance, and the update of the average route can be performed in time O(N). If the distance is larger than the threshold T, the triplet is signalled as a possible anomaly. In this case, a human action is required. If the user considers valid the signalled route, then she will update $R^*(e, l)$ taking into account last route. The algorithm works in time $O(N)$.

6. Experimental evaluation

In this section, we present some experiments we performed to assess the effectiveness of the proposed approach in detecting outliers in RFID data streams. The direct result of each test is a similarity matrix representing the degree of similarity for each pair of streams. We analyzed about *104* streams, belonging to *4* classes of epcs: 1) `Tuna Fish`, whose readings are generated by *26* tagged containers storing *500* cans each, 2) `Tomato`, whose readings are generated by *23* tagged containers storing *400* cans each, 3) `Syrupy Peach`, whose readings are generated by *20* tagged containers storing *350* cans each, and 4) `Meat`, whose readings are generated by *35* tagged containers storing *600* cans each. For each pair of objects, the DFT-distance d between their routes is computed, and the similarity between the routes of the pair is assumed to be equal to $1/(d+1)$. We adopted both encoding schemes, simple and pairwise, thus obtaining two different *confusion matrices* depicted in Fig. 3. The confusion matrices are represented by images, where the colour of the pixel (i,j) represents the similarity between the routes of the i-th and j-th objects. More specifically, the higher the similarity, the darker the pixel colour. Note that the blocks on the diagonal of the matrix correspond to intra-class similarities, whereas the blocks outside the diagonal represent the inter-class similarities. A quantitative analysis of the results, averaged by classes of epcs, is reported in Tables 1 and 2.

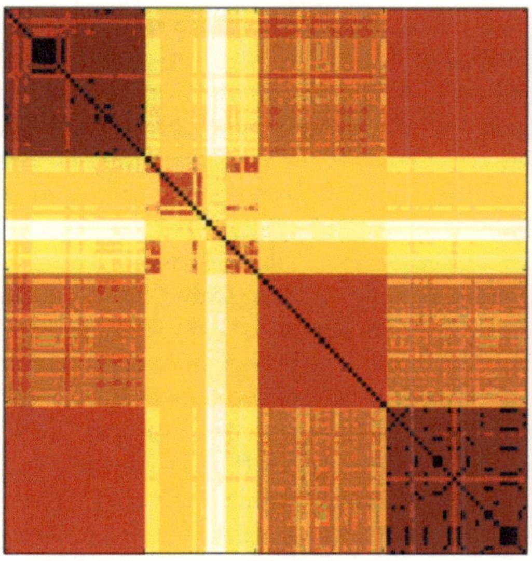
(a)

(b)

Fig. 3. Confusion matrix obtained using simple encoding (a) and pairwise encoding (b)

	Tuna Fish	Tomato	Syrupy Peach	Meat
Tuna Fish	0.9608	0.6039	0.7568	0.8094
Tomato	0.6039	0.8553	0.6545	0.6039
Syrupy Peach	0.7568	0.6545	0.8095	0.7265
Meat	0.8094	0.6039	0.7265	0.9436

Table 1. Similarity between routes of 4 different classes of items using simple encoding

	Tuna Fish	Tomato	Syrupy Peach	Meat
Tuna Fish	0.9824	0.7542	0.8043	0.7584
Tomato	0.7542	0.9225	0.6784	0.5264
Syrupy Peach	0.8043	0.6784	0.9545	0.6834
Meat	0.7584	0.5264	0.6834	0.9665

Table 2. Similarity between routes of 4 different classes of items using pairwise encoding

The quantitative results obtained reveal that our measure of similarity is effective. In fact, for all classes, the intra-class similarity values are sufficiently higher than the inter-class ones, allowing for separating all classes from one another and thus stating that the containers follows the predicted plan. Furthermore, the pairwise encoding provides better results than the simple one, as in the majority of cases, intra-similarity is higher and inter-similarity is lower. This is rather expected, as a new reading updates the DFT of current routes by taking into account not only the location where the reading was taken, but also the provenance of the object.

7. Related work

Sensor networks has emerged in recent year as a fruitful research field since it provide interesting challenges both for designing more efficient system and for data management. We disregard in our discussion the physical aspect and we shall focus on data management issues. Bonnet et al. (2001) describe the design guidelines for sensor database systems and recognize numerous advantages of the distributed approach over the warehousing approach, under the condition of having sensors with computational capability. In particular, they study the *in-network aggregation* and distributed query processing. The Cougar project (Yao & Gehrke, 2002) is focused on techniques for processing queries over sensor data. In this project there is a distributed sensor network environment and a central administration for sensors management. The *TinyDB*, proposed by Madden et al. (2002), is a distributed query processor that runs on each of the sensor nodes; this system is focused on optimizing data acquisition for long-running queries, being no data stored locally at the nodes. Yao and Gehrke (2003) extend the general guidelines for sensor databases proposed by Bonnet et al. (2001) focusing on the query processing issue, and propose the definition of a query layer that (i) improves the capabilities of a generic sensor network, and (ii) defines a declarative language for efficiently in-network query processing. Schlesinger and Lehner (2004) combine the Grid framework with sensor database systems. In their framework, each Grid node holds cached data of other Grid nodes using a *data replication scheme*. Furthermore, they propose a model describing inconsistencies in Grid organized sensor database systems, and a technique for finding optimal distributed query plans by deciding on which node of the grid a query should be evaluated. Concerning the specific case of RFID data management, it has been studied in (Gonzales et al., 2006). In particular the problem of defining an efficient warehousing model along with techniques for summarizing and indexing data are discussed. They introduce a model based on hierarchy of summary called *RFID-Cuboids* of the RFID data aggregated at different abstraction levels in order to allow different kinds of data analysis. Regarding time series comparison, a traditional approach known in literature is based on time warping (Yi et al., 1998), which mainly consists in considering every possible stretching and narrowing of the signals being compared, and choosing the best matching. However, time-warping-based approaches are quite expensive (quadratic in complexity) when dealing with long series. Furthermore, time warping cannot be incrementally computed as objects move throughout locations, differently from DFT, which can be efficiently updated.

8. Conclusions

In this chapter we addressed the problem of detecting outliers in RFID readings stream. The technique we have proposed is mainly based on the idea of representing an stream of readings as a time series. Thereby, the structural similarity between two series can be computed by exploiting the Discrete Fourier Transform (DFT) of the associated signals. Experimental results showed the effectiveness of our approach, with particular regard to some of the encoding schemes defined in the paper.

The current work is subject to further significant extensions. As a matter of fact, the structural similarity between routes can be refined exploiting additional information on the RFID stream such as the actual distance among the reader. In our current implementation, the reader encoding function does not take into account semantic similarities between tags being scanned (i.e., objects belonging to the same category). However, precision could be improved by exploiting tagged object similarity techniques, such as, e.g., the exploitation of suitable ontologies. FFT-based distance measures, different from the one introduced in the paper could be used. Indeed, the FFT transformation contains lots of information about the contents of the original stream, and different distance measures could be more appropriate to exhibit such information. A further possibility to improve the proposed encoding schemes is that of defining a different strategy for dealing with the stream of readings in particular we plan to use more robust methods (e.g., non parametric-ones) for determining outliers, like those introduced in (Subramaniam et al., 2006). However, eventually new implementation has to be carefully studied in order to avoid inefficiency that are common in a system that has to deal with huge amounts of data.

9. References

Bonnet, P.; Gehrke, J. & Seshadri, P. (2001). Towards Sensor Database Systems, *Proceedings of Second International Conference on Mobile Data Management (MDM 2001)*, ISBN: 978-3-540-41454-4, pp. 3-14, Hong Kong, China, January 2001, Springer-Verlag, London, UK

Gonzales, H.; Han, J.; Li, X. & Klabjan, D. (2006). Warehousing and Analyzing Massive RFID Data Sets, *Proceedings of the 22nd International Conference on Data Engineering (ICDE 2006)*, ISBN: 0-7695-2570-9, pp. 83, Atlanta, GA, USA, April 2006, IEEE Computer Society Washington, DC, USA

Kruskall, J., B. & Liberman, M. (1999). The Symmetric Time Warping Problem: From Continuous to Discrete, In: *Time Warps, String Edits and Macromolecules: The Theory and Practice of Sequence Comparison*, Kruskal, J.,B. & Sankoff, D. editors; pp. 125–161, Stanford: CSLI Publications, ISBN: 1575862174, Stanford

Madden, S., R.; Franklin, M. J.; Hellerstein, J., M. & Hong, W. (2003). TinyDB: An Acquisitional Query Processing System for Sensor Networks. *ACM Transactions on Database Systems (TODS)*, vol. 30, no. 1 (March 2005), pp. 122-173, ISSN: 0362-5915

Oppenheim, A., V. & Shafer R., W. (1999). *Discrete-Time Signal Processing,* Prentice Hall, ISBN: 978-0137549207

Schlesinger, L. & Lehner, W. (2004). Querying Asynchronously Updated Sensor Data Sets under Quantified Constraints, In *GeoSensor Networks*, vol. 1, part 2, Stefanidis, A. & Nittel, S. editors., pp. 13-30, CRC, ISBN: 978-0415324045

Subramaniam, S.; Palpanas, T.; Papadopoulos, D.; Kalogeraki, V. & Gunopulos, D. (2006). Online Outlier Detection in Sensor Data Using Non-parametric Models, *Proceedings of the 32nd international conference on Very Large Data Bases (VLDB 2006)*, pp. 187-198, Seoul, Korea, September 2006, VLDB Endowment

Yao, Y. & Gehrke, J. (2002). The Cougar Approach to In-Network Query Processing in Sensor Networks. *ACM SIGMOD Record*, vol. 31, no. 3, pp. 9-18, ISSN: 0163-5808

Yao, Y. & Gehrke, J. (2003). Query Processing in Sensor Networks, *Proceedings of the First Biennial Conference on Innovative Data Systems Research (CIDR 2003)*, Asilomar (CA), USA, January 2003

Yi, B.; Jagadish, H., V. & Faloutsos, C. (1998). Efficient Retrieval of Similar Time Sequences Under Time Warping, *Proceedings of the 14th International Conference on Data Engineering (ICDE 1998)*, pp. 201-208, ISBN: 0-8186-8289-2, Orlando (FL), USA, February 1998, IEEE Computer Society Washington, DC, USA

RFID Tags as Technology for Value Sensing in Real Space Market

Yukio Ohsawa[1], Hikaru Kimura[1], Toru Gengo[1] and Takeshi Ui[2]
[1]*School of Engineering, The University of Tokyo*
[2]*Toppan Forms Co. Ltd.*
Japan

1. Introduction

Following the literature, let us define a *chance* as an event which plays a significant role in the decision making of human [1]. For example, suppose there is a cheese which seems good to eat, in a supermarket. This cheese, however, costs 2000 JPY (ca 20 US dollars). Customers walking around the store may pick out the cheese, but mostly give up buying when they look at the price label.

Suppose a certain customer (called Mr. A) often comes to this supermarket and mostly buys liquor. The manager of this supermarket likes Mr. A to buy things to eat also, but an established lifestyle is hard to change. One day, Mr. A finds a very nice looking cheese close to the liquor shelf from which he usually takes and buys wine. The cheese looks good to eat with wine, but he picks up the cheese and returns it back to the shelf because it is as expensive as 20$. However, he does not go home immediately. He walks to the shelf of cheese and snacks, because he is now interested in something to eat like cheese. Thus, Mr. A learns to buy food with liquor. Although the 20$ cheese was not bought, it led this customer to a new decision of purchase to be continued in his future manner of buying.

See Fig.1 for the conceptual sketch of this scenario. The cheese of 20$ here is apparently a chance, because the event of Mr. A's picking the cheese made a strong influence on the future decision of Mr. A. As well, the supermarket manager may decide to keep the expensive cheese on the same shelf rather than to move it elsewhere only because it is not a frequently sold item. Changing the lifestyle of a customer makes more benefit than having the customer just buy an expensive item.

An important point of Mr. A's case for sales promotion is that the man's real desire cannot be understood if we pay attention only to his decision of purchase. The data at the point of sales (POS) show he finally came to buy cheap cheese, but does not even imply he was tentatively interested in the outstandingly good cheese and seeked good sidefood for wine. If the manager understands this real desire, he/she will keep the expensive cheese in the shelf as an attractor for wine lovers rather than discard it, or attract them by showing a little cheaper sidefood on the same shelf. In other words, marketing decision maker can catch customers' latent interests from the data collected before they come to the register to buy.

Thus such a map as in Fig.1, which we call a *scenario map*, is useful for creating a scenario of marketing. Here, a scenario is a series of actions and events which occur under a coherent context. As in [2,3,4], a scenario map is useful for aiding chance discovery i.e., to detect an event significant for decision making as the picking of 20$ cheese. This effect works more finely, if the customer's behavior to pick an item out of a shelf can be taken in the data. We introduced RFID tags for obtaining really useful scenario maps.

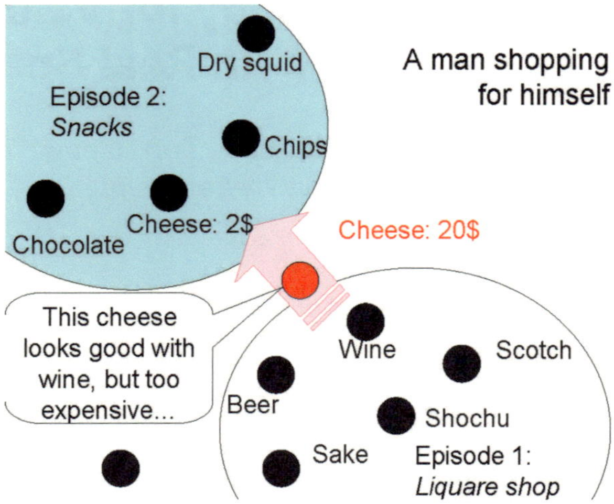

Fig. 1. Chance discovery in a supermarket: The expensive cheese plays a role.

2. Our system of middle-ranged frequency RFID tags

The customer in the of Fig.1 took the cheese from the shelf, and may have smelled the cheese, felt the weight and the softness. He estimated how good the product is, and finally decided not to buy it. Thinking about this scene, we find the customer's co-existence in the same space of the store as the 20$ cheese was essential for his decision. We can also point out another effect of the real space to the customer. Let us define a community as a group of people who do verbal or non-verbal communications. A verbal communication means communication by talking in a common language, and a non-verbal communication means to understand the behavior of each other without talking. Suppose each member of the community embraces a scenario planned in the mind. A member Ms. B who hears or interprets the scenario in the mind of Mr. A may like to import it into her future action if she feels empathy with Mr. A. Thus, the scenarios underlying the behaviors of customers may be combined in a way as in genetic crossover and create new behaviors of purchase and consumption.

From this view point, we take a hypothesis that the scenario map as in Fig. 1 can represent not only an individual customer's behavior but also the behaviors of people in a community. That is, if Mr. A and Ms. B behave like a community, with interacting with exhibited items in the real space, we can expect Ms. B may also come to be interested in the cheese of 20$ in the extension of her recent purchase history. For example, the sequence of Mr. A's and Ms. B's behaviors can be:

On February 2, Ms. B touched the set {beer, wine},
On February 2, Ms. B bought the set {beer, wine},
On February 3, Mr. A touched the set {beer, wine, cheese 20$},
On February 3, Mr. A bought the set {chips, cheese 2$},

On February 3, Mr. A bought the set {wine, cheese 2$, chips},
On February 3, Ms. B touched the set {beer, wine, whisky, cheese 20$},
On February 3, Ms. B touched the set {cheese cake, chips}.
On February 3, Ms. B bought the set {wine, cheese cake, chips}.

Here, Ms. B has been interested in liquor (beer and wine) until February 2nd. However, on February 3rd she looked at someone (Mr. A) acting similarly to herself, except that the man touched the cheese of 20$ but did not buy it. Then, Ms. B notices this cheese might be good also for her taste and picks it to see the price label. As a result, she takes a similar choice to Mr. A - goes to a different shelf. She moves to the shelf of cakes to buy a cheese cake that is a mixture of cheese and sweet for her. The behavior of Mr. A thus infects to Ms. B, and makes a new behavior which is a combination of Ms.B's and Mr. A's past behaviors.

An important point here is that the cheese of 20$ hardly appears in the sales data, i.e. in the POS (position of sales) data. As far as we have only POS data, we cannot understand the role of this expensive item and the sales manager may stop showing it on the shelf of the supermarket. However, if we can record the behaviors of customers to touch items, the data can include useful information, e.g., some customers picked the cheese but did not buy it. Introducing the RFID tags, we may be able to tell it is meaningful to discount the price of this cheese.

Market researchers' interest in RFID tags is growing rapidly. The ability to record the pick-out behaviors of customers in a retail store, before they decide to buy or not to buy, enables to understand their latent interests more deeply than the data on the position of sales (POS). That is, the information that a customer's interest in an item was strong enough to pick it but not enough to buy it has not been dealt with as far as we deal only with POS data. In this sense, we expect to enable chance discovery more finely by introducing RFID tags to visualize the data of customers' touching sold items in the way similar to the map in figure 1.

The reason why we apply RFID tags in face of the existence of fascinated video camera systems [5], infrared-ray[6,7]/sonar sensing systems, and other up-to-date sensing systems [8,9] is that RFID tags have the suitable feature that it can be attached to items on shelves, and the shelves can also become high-resolution sensor if we finely design the antenna embedded in the shelves. In contrast, the video camera system, if installed up at the ceiling of a retail store (a supermarket, an apparel shop, etc) can detect the position of the item a customer touches, to the resolution of 10cm (see SiteView for example [5]). This resolution is good enough for telling the shelf to which the customer's hand entered. However, more than one kind of items may be exhibited on the same shelf, one over (or very close to) the others. Sumi et al designed a video-camera array, where cameras are located at the top (m cameras in a horizontal line) and the side of a shelf (n cameras in the vertical line) to distinguish mn areas in one shelf, in order to improve the sensing resolution [10]. Even though, an error due to the mis-location of items is not ignorable at all, because popular items tend to be moved from/to areas within a shelf. In the case of infrared ray, the ray emitted from a the body of a customer is detected by sensors. For example, if a customer passes in a region in the store, the movement shall be detected finely if infrared-red sensors are aligned periodically and finely. This system can detect a rough location of the customer, but the resolution is not high enough to detect touched items on shelves. In comparison to these different kind of sensors, we can say RFID tags are low-cost and suitable-resolution pick-out sensors.

Motivated with the merit of RFID tags, the data from RFID tags came to be dealt with from the aspect of data-based marketing [11,12,13]. Applying the methods of data mining and data visualization, we are able to acquire deeper level knowledge about customers from RFID data than from POS data, because the customer's thought and vacillation before buying are reflected to the pre-purchase behaviors taken by RFID tags. Reader may compare this to the prevalent experience that shoppers' browsing log data in a shopping web site enabled to analyze the latent interest of people than by seeing just the purchase records in the same site [14]. However, the human's behavior in the real space is more informative than in the cyber space, in that the length of time one takes an item in hand reflects one's interest, whereas the length of opening a Web page may mean the user is just away for coffee, tea, or for restroom.

Reader may relate this article to previous work on the studies to trace customer's movements in the real space [13,15] with active RFID tags (having power sourse and omitting customer's ID numbers) attached to customer bodies. That comparison is reasonable in that we also deal with real-space action of customers. However, the difference is two-fold. First, customer IDs, i.e., the information for identifying each customer tends to be missed in a retail store, for the reason of privacy security [16]. This has been regarded as an obstacle in predicting customer's preference of each item. In this acticle, we consider the circumstance where customer-ID-less RFID tags are used, and regard this as the setting for collecting the behaviors of a group of customers in the real space. Second, the granularities of information dealt with are different, between our approach and active tags attached to customers. If we trace the movement of a customer's body, the intension of the customer i.e., whether he likes to take a red necktie or a green one beside the red tie, is hard to understand in spite of its important meaning. Our aim can be expressed as in-store chance discovery, to discover an event of customer's taking a significant item for marketer's (and, in turn, customer's) decision making. From this viewpoint, we experimentally show that the data on the group behavior of customers in the real space can be recorded with ID-less RFID tags and the visualization of the data, aids the discovery of new values in the market. This expectation is validated in a apparel shop and a book library.

3. Value sensing from the real space customers' choices: in apparel store

Let us show the experiment we conducted in a real apparel shop, laying RFID tags, antenna, and controller as in Fig.2. In this experiment, we collected both data on RFID tags and POS, in the following manner.

a. The data from RFID tags: The data collected 14 days have been taken, and each set of three items picked sequentially by customers is taken as one basket. The number of baskets was 18212 for all the time of experiment. See Fig.2 to find the way we attached RFID tags to items and we put those items on the shelves. For example, suppose items were picked by customers as in the listed order in the left columns of Fig.3. In the second and the third columns, two lines and three lines are inserted respectively at the top, in order to make each line containing three items represent three items picked sequentially. Thus, the data are to be obtained as the set of baskets, where a basket is given by one line in Fig.3.

b. The data at the Position of Sales (POS): The data collected for the same period of days as of the RFID tags above, for all purchase of items to which RFID tags were attached.

Here, one basket is taken as the set of items bought by one customer at one time. 473 baskets have been collected as a result.

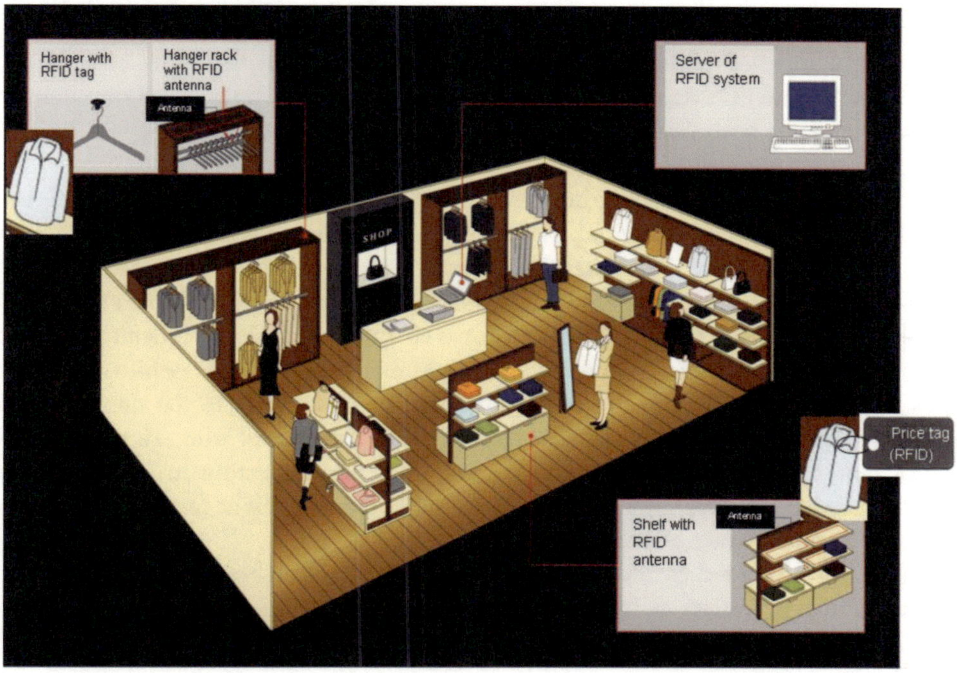

Fig. 2. The apparel store engaged in the experiment

time	Item (product)	color	item	color	item	color
	58147002	4				
	30057011	19	58147002	4		
basket 1	**30057033**	**98**	**30057011**	**19**	**58147002**	**4**
basket 2	58116526	70	30057033	98	30057011	19
•	58116526	86	58116526	70	30057033	98
•	58117007	28	58116526	86	58116526	70
•	58117006	50	58117007	28	58116526	86
•	58116506	52	58117006	50	58117007	28

Fig. 3. The data of RFID tags for the experimental (but real) apparel store

These two sets of obtained data were visualized as in Fig.4 (a) and (b). In these results of KeyGraph visualizing the co-occurrence of items in the touching (RFID) and buyng (POS) data, the shadow of the first figure (a) is shown weakly in the second figure (b), so that we can intuitively understand the difference between customers' picking-out shown by RFID tags and buying of items shown by POS data. From the two figures, find features suggesting the trigger of customers' awareness of new values in other items than those they touched. For comparing the two figures, reader is advised to first look at the shadow in the center of figure (a) and in figure (b), because these two shadows represent the same item set in the two different figures. A feature of KeyGraph is that the black nodes and black links show frequent items and their frequent co-occurrences (occurring in the same baskets in the RFID data and in the POS data), whereas the red nodes and red links show rare items and their co-occurrences with frequent-item clusters depicted by black nodes and links. For example:

1. The white colored clothes, surrounding the white square in the cluster at the upper right of figure (a), are picked (appear in figure (a)) but not bought (do not appear in figure (b)). After touching these items, customers tend to move to the items in the cluster at the center of figure (b). Seeing that these items are linked from the previous upper right cluster in figure (a), we can guess customers who were touching the upper-right items could not decide to buy anything in the cluster, and the white color item in this island finally effected as a trigger to move the customers to the densely colored items in the center of figure (b). The author asked consumers who are women, about their interests in this white cloth. A common opinion according to them was that a white cloth may be generally attractive but one shall not buy it as far as the design does not exactly match one's interest.
2. Item 58137004 pointed by an arrow in figure (b) is linked to a newly appearing cluster of items in the bottom of figure (b), whereas 58137004 has been only linked to the purple, green and gray colored items in (a). These previous links in (a) remains also in (b), but the new links from 58137004 to such item as 58117016 can be interpreted by the analogy on the basis of Fig.1, where an event may trigger human's attention to a new group of events. That is, in the case of figure 4, selling staffs (two ladies working in the store), looking at the two figures, mentioned that item 58137004 was an outstanding new cloth and set at the height of customers' eyes. Customers tend not to buy such a cloth, but come close to its exhibition and buy items nearby. The emergence of the right-hand side cluster in figure (b), via the strengthening of item 58117077, has been explained similarly, although 58117077 itself was bought frequently.
3. The item 58117016 in the core of the new cluster at the bottom of figure (b) has been know as a popular item according to the sales staffs. However, they thought its blue colored item was sold the most frequently. In reality, the most frequently bought color of item 58117016 was the yellow, as appearing in the right-hand side of the same cluster. The staffs also remarked this awareness is useful, because they should exhibit yellow ones in a more outstanding shelf.

The hypothetical interpretations above came to be supported by our interview to other women, who were 5 consumers (women) having experience to buy in the real store we made the experiment. We cannot say these are "novel" ideas, because such knowledge should have already existed in the deep level of the memory of subjects from before they looked at the graph. However, this provides real decision makers in marketing with helpful information because they tend to accept scnarios fitting their feelings acquired from daily experience. New information provided automatically by machine might be useful, but not always necessary.

a. Results for data from RFID

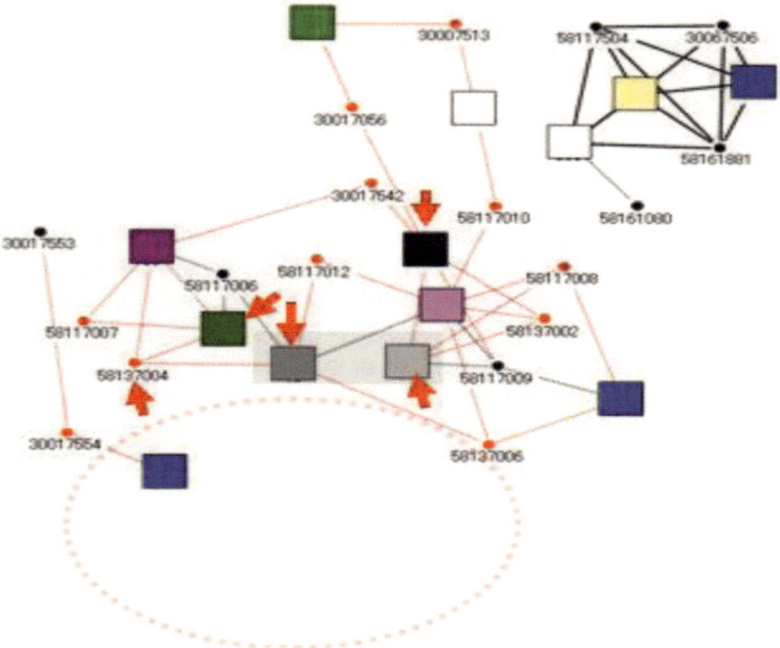

b. Results for data from POS

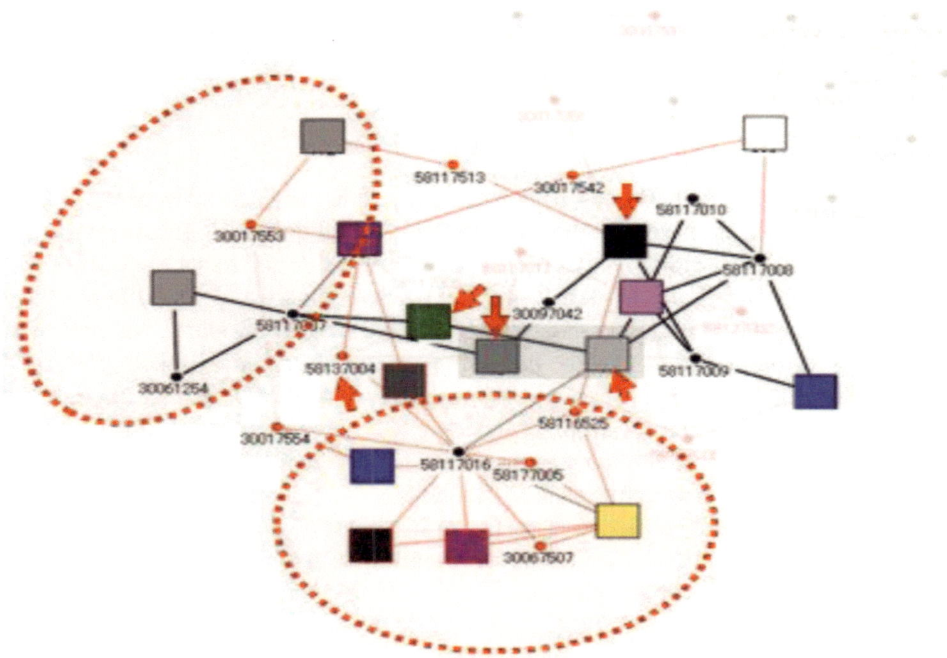

Fig. 4. Graphs presented by Pictorial KeyGraph for the data on RFID tags and POS, for two weeks experiments in an apparel store.

4. Analysis of browser's intentions in library by using RFID sensors

4.1 The effect of *tachiyomi* in the real space of books

Next let us show the application case of RFID tags to the behaviors of browsing customers of a book library. This domain of application has been studied in [17]. On the other hand, we especially focus on the "tachiyomi" (a Japanese word) behavior of customers, which means the customer stay at the book shelf to open a book to read it. In Japanese book shops, a trend is to allow customers to do tachiyomi, because tachiyomi is expected to have customers stay long in the shop space, walking around to find books worth to buy, even though the customer does not buy the book he reads on the way of tachiyomi. This expectation may be extended to other kinds of shops such as apparels: By allowing customers to do test wearing of clothes, customers may not only check the comfort of each cloth but also stay long and touch various clothes and may finally buy some.

Tachiyomi in a library is different from tachiyomi in a book shop in that it is usually not a manner to be punished in a library. However, we find the merit of tachiyomi is common to both a bookshop and a library, buy regarding purchase as just one way for reaching satisfaction. That is, tachiyomi may let customers walk around to pick and open books to reach satisfactory books, although he/she knows only a very small part of the whole collection. From this aspect, we should investigate more of the wander behaviors (customers' walking around to pick and open books) in order to see how tachiyomi is meaningful for stimulating customers into a desirable direction, and what kind of books really trigger wander behaviors.

In this section, let us show the experiment we conducted, to record the data on customer's behaviors to pick out books, to analyze the relation between the value of a picked book for the customer and the following wander behaviors. Here, we regard the pick-and-return behaviors observable by RFID tags as the essence of wandering, because just walking around with seeing the titles do not mean interest strong enough to raise the book to a candidate of the read-worthy.

4.2 A preliminary experiment

We constructed an experimental book library as in figure 5, in which all books are attached with RFID tags. 23 subjects were segmented to two groups and each group stayed in the library for one hour. This setting was introduced because we aimed to set a condition where the group effect (like Mr. A and Ms.B in section 1) works as in a usual book library. In total, 275 books were picked and returned sequentially. As in the case of apparel, we took each set of three books sequentially taken by the same customer as one basket, and applied KeyGraph to the data. Although the RFID tag system in this library did not have the effect to tell the customer corresponding to each picking event, we compensated for the customer information by having each customer insert his/her ID card to the same shelf area the book was picked from, until returning the book to the shelf.

As a result, figure 6 has been obtained. The left hand table shows the location of areas of the shelf, and the right hand shows KeyGraph representing the co-occurrence of book-picking events at different shelves. Simply put, close areas tend to appear closely in KeyGraph and form clusters, i.e., {5-A, 5-B} at the top, {6-A, 6-B, 6-C}, etc. These parts of KeyGraph are easy to interpret, because it is natural that customers move around close areas at close times. And, some exceptional parts like the links between 6-A and 4-C, between 6-C and 2-C, etc exist in KeyGraph. According to the data on the relation between the shelf areas and book categories, 6-A had books about company management, and 2-C and 4-C had books about leaderships and service management respectively. Considering the contextual relevance among these categories, we can regard these co-occurrence of remote areas, in that customers wandered from/to shelf areas due to the effect of books read in the course of tachiyomi in the real space affected the customers' awareness of their own interest.

Fig. 5. The book library we applied the RFID tag system to

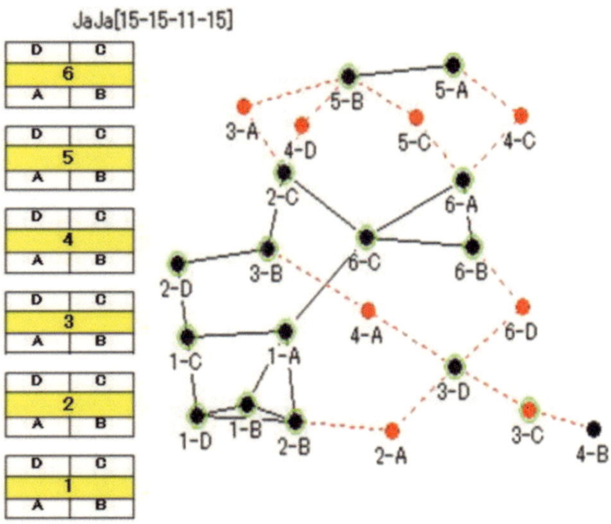

Fig. 6. The location of shelves (left), and the KeyGraph representing the co-occurrence of book-picking events at different shelves (right)

Based on the findings from the preliminary experiment, we pay attention to the customers' real space interactions in their wandering behaviors. We classified the customer's behaviors just after picking one book into four groups (1) stay at the same area (e.g., 5-A -> 5-A) (2) stay at the same shelf (e.g., 5-A -> 5-B) (3) move to the next shelf (5-A -> 4-B), (4) move over two or more shelves (e.g., 5-A > 3-B). We regard (4) as the most drastic wandering effect, and investigated the factors causing drastic wander. More specifically, we hypothesized that encountering an unexpectedly interesting book in wandering re-enforces oneself to wander more drastically.

4.3 The effect of "unexpected interestingness" of an encountered book

We hired new 27 subjects. Each subject was instructed to stay one hour in the library to brows, as in the preliminary experiment. Each time one picks a book, he had two missions: Evaluate the book just after picking (i.e. before reading) and just after returning (i.e., after reading or being tired of the book), and then report the evaluation score ranging between 1 (poor) and 5 (interesting). Denoting the score of the book before reading by *E* (Expectation) and after reading by *P* (Preference), we quantified the "unexpected interestingness" of a book by *P*- *E*.

As a result, we obtained the result as in Figure 7. Accordingly, we can conclude the unexpected interestingness of a book one picks and looks in to read stimulates the customer's wander behavior. On the other hand, if the unexpected interestingness is low, one tends to stay in the same area.

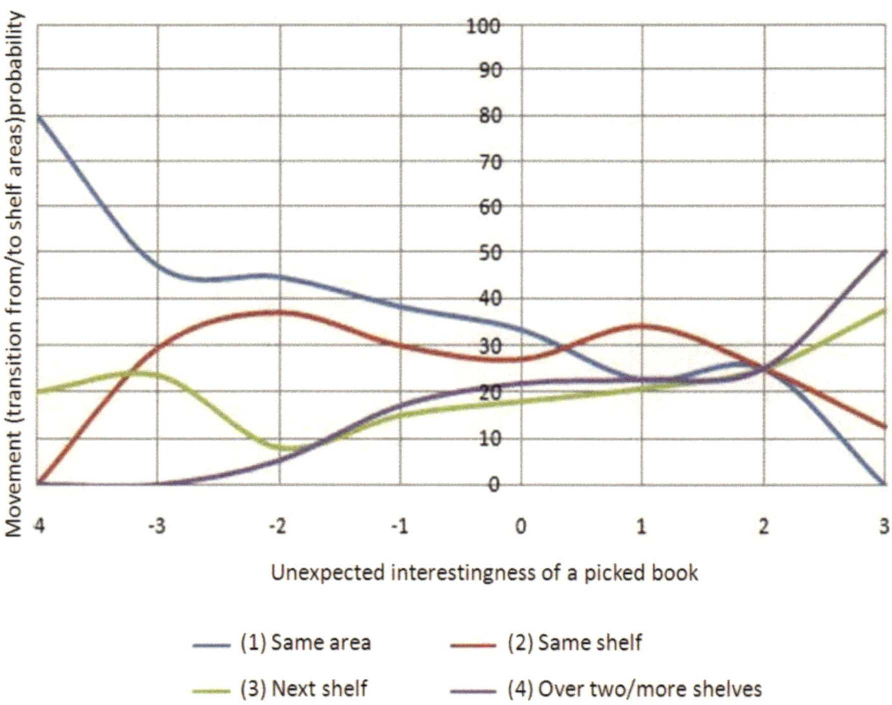

Fig. 7. The relation between unexpected interestingness of a picked book and the customer's following wander behavior

5. Conclusions and future work

An expectation of RFID tags applied to marketing has been to detect items that are touched but not bought, which may correspond to customers' latent interest which did not appear in sales data. On the other hand, we aimed in this paper to realize chance discovery by introducing the visualization method KeyGraph to the data. This here means to detect items which are touched and influence the near-future behavior of customers. The experimental results in this paper show our successful progress in confirming that our aim comes true by RFID tags applied to libraries and apparel stores.

We may address our next challenge to the discovery of purely novel knowledge, by deepening the level of tacitness of experience-based knowledge obtained via the process of knowledge/chance discovery [18]. Although we may not be allowed to distinguish customers considering their will to protect privacy [16], the experiment here under artificial setting, where each customer is taken for just anonymous someone (in the apparel experiment) or identified on a new devise such as a customer's ID card (in the book library), encourages us to introduce RFID tags to real spaces such as apparel stores and supermarket.

6. References

[1] Ohsawa, Y. and McBurney, P. (eds), Chance Discovery, Springer (2003)

[2] Horie, K., and Ohsawa, Y., Product Designed on Scenario Maps Using Pictorial KeyGraph, WSEAS Transaction on Information Science and Application, Vol.3 No.7, pp.1324-1331 (2006)

[3] Sakakibara, T., Ohsawa, Y., Gradual-Increase Extraction of Target Baskets as Preprocess for Visualizing Simplified Scenario Maps by KeyGraph, Journal of Soft Computing Vol.11, No.8, pp.783-790 (2006)

[4] Ohsawa, Y., Scenario Maps on Situational Switch Model, Applied to Blood-Test Data for Hepatitis C Patients, Ohsawa, Y., and Tsumoto, S. (eds), Chance Discoveries in Real World Decision Making, Springer, pp.69-82 (2006) 415–438

[5] Vitracom Siteview Web Page: http://www.vitracom.de (2002)

[6] Sumi, Y., Exmerience Medium - Towards Knowledge Creation Enhanced by Experience Sharing - Journal of the Japanese Society for Artificial Intelligence pp.453 - 460 (2008 in Japanese)

[7] Lee, DH., Device for sensing position of human body using infrared sensor, US Patent 5742055 (1989)

[8] Honda, S., Fukui, K., et al, Multi-Person Tracking with Infrared Sensor Network, in Proc. The 20 th Annual Conf., Japanese Society for Artifitial Intelligence (2006)

[9] Okuda, S., Kaneda, S., and Haga, H., Human Position/Height Detection Using Analog Type Pyroelectric Sensors, Embedded and Ubiquitous Computing (Lecture Notes in Computer Science 3823), Springer Berlin pp. 1611-3349 (2005)

[10] Hsu, HH., Cheng, Z., et al, Behavior Analysis with Combined RFID and Video Information, The 3rd International Conference on Ubiquitous Intelligence and Computing (UIC-06), China (2006)

[11] Gonzalez, H., Han, J., Li, X., Mining compressed, commodity workflows from massive RFID data sets, Proceedings of the 15th ACM international conference on Information and knowledge management, 162-171 (2006)

[12] Han, J., Warehousing and Mining Massive RFID Data Sets, Advanced Data Mining and Applications, Lecture Notes in Computer Science, 4093 Sringer Berlin, pp.1-18 (2006)
[13] Levy, D., Sponsored Feature: A Vision for RFID In-Store Consumer Observational Research, RFID News, October 20, (2003)
[14] Cheng LV, Wei CY, and Zhang H, Pattern Discovering of Web User Access Pattern Based on MFP Method, Journal of Communication and Computer, VOl.3, No.11 (2006)
[15] Murakami, E. and Terano, T., Fairy Wing:Distributed Information Service with RFID Tags, Multi-Agent for Mass User Support pp.174-189, Springer (2004)
[16] Landwehr, C.E., Conference Report on RFID Privacy Workshop, Concerns, Consensus, and Questions, IEEE Security and Privacy, March/April 2004, pp.34-36 (2004)
[17] Minami, T., RFID in marketing Library marketing with RFID for supporting learning patrons, International Conf. on Multimedia and Information, November, Spain (2006)
[18] Ohsawa, Y., and Fukuda H., Chance Discovery by Stimulated Group of People - An Application to Understanding Rare Consumption of Food, Journal of Contingencies and Crisis Management Vol.10, No.3, pp.129-138 (2002)

A Sector Analysis for RFID Technologies: Fundamental and Technical Analysis for Financial Decision Making Problems

S. Kasap[1], M.C. Testik[1], E. Yüksel[1] and N. Kasap[2]
[1]*Hacettepe University, Faculty of Engineering, Department of Industrial Engineering, 06800 Beytepe - Ankara*
[2]*Sabanci University, Faculty of Management, 34956 Tuzla – Istanbul*
Turkey

1. Introduction

Automatic identification technologies have been used in a wide range of applications for reducing the amount of time and labor needed to input data and improving data accuracy. As an important automatic identification technology, radio frequency identification (RFID) technologies allow contactless reading and these technologies are particularly successful in manufacturing and other environments where traditional identification technologies such as bar codes can not perform well. By integrating the RFID technology into their business models, companies may save time, lower labor cost, improve products quality and provide better service. RFID is the wireless technology that uses RF communication to identify, track and manage objects and collect and store data. RFID technology enables companies to develop applications that create value by tracking and identifying objects, animals or people. Business applications of RFID technology can be seen in areas such as manufacturing, supply chain management, software integration, security systems, asset tracking and many others.

RFID technology was predicted to be one of the "top ten" technologies in 2004 by CNN. Although, the RFID market is less than five years old, it has been applied to many different industries, from retail industry to logistics, or from healthcare to service business industry – and it is still growing. Particularly, RFID has fundamental influences on today's retailing and supply chain management for applications like asset tracking, the inventory control and management. RFID technology also finds major application in mobile phones and is widely used in toll collection of highways, for payments in restaurants, vending machines, retail and parking lots. There are a wide range of RFID systems currently being used or being developed. Examples to these systems include but not limited to the following; automatic vehicle and personnel access control for security (Simpson, 2006), airport passenger and baggage tracking (Ferguson, 2006), tracing blood for cutting down errors such as giving patients wrong blood types (Ranger, 2006), payment process systems (Ramachandran, 2006), production control in manufacturing (Liu & Miao, 2006), transfusion medicine (Knels, 2006) real-time inventory control by automated identification of items in warehouses, tracking and management of physical files, tracking of books in the libraries (Shadid, 2005). For some other applications, interested reader is referred to (Finkenzeller, 2003; Smith, 2004).

RFID solution providers claim that their technology and solutions bring significant benefits and have valuable advantages in practice. As new RFID solutions being developed and more RFID tags and equipments being used, these solutions will become more cost effective and RFID businesses are expected to grow rapidly. Since RFID is fairly new, it's difficult to measure resulting sales increases or heightened customer satisfaction quotients. On the other hand, according to IDC estimation (IDC is a subsidiary of International Data Group, a leading technology media, research, and events company and provider of market intelligence, advisory services, and events for the information technology, telecommunications, and consumer technology markets), companies in the retail sector will spend nearly $1.3 billion on RFID in their supply chain operations in 2008, compared to about $91.5 million in 2003 which corresponds to annual growth rate of 70 percent. In a similar look; the Wireless Data Research Group projected that the global market for RFID increased from $1 billion in 2003 to $3 billion in 2007 (Asif & Mandviwalla, 2005). There are two major drivers of this growth. The first one is the adoption of RFID technology by major retailers and government agencies. The second one is the reduction in the price of RFID tags, readers, and information technology (IT) systems required to deploy RFID.

Given the significant potential of RFID technology, there has been a huge emergence of RFID specialty companies and the development of RFID practices within many market-leading companies. Due to huge emergence, it is desirable to make a sector analysis. In this study, we perform a sector analysis for RFID technologies for researchers and analysts. We investigate public RFID companies traded on the stock exchange markets, summarize their financial performance, describe their RF products, services, and applications, and perform fundamental and technical analysis.

2. RFID technology

RFID technology is a promising technology helping companies solve problems in supply chain management, manufacturing, security, identification and asset tracking. At the first glance, RFID appears to be a production and distribution technology; on the other hand, it is an IT system for data collection, storage, analysis, and distribution. Components of the RFID system are described as **RFID tags**, **RFID readers**, and **RFID data processing system**.

RFID tags are the most important component of an RFID system. A typical RFID tag contains the following three components; the integrated circuits (IC) chip, the antenna, and the enclosure. The IC chip is used for the unique data storage and logical operations associated with the RFID tag, whereas the antenna is used for communication between readers. Data is stored in the IC chips and transmitted through the antenna. The enclosure is the packaging around the electronic components. RFID tags can be active or passive according to the supply of electronic power. Active RFID tags use a battery to power IC chip and broadcast signals to the reader. Passive tags do not have batteries and are powered by the electromagnetic waves sent out by a reader to induce a current in the tag's antenna. Passive tags have unique identification number in them that cannot be changed and are transferred into a computer database in which the ID is associated with product characteristics; while in active tags, the information can be written, erased and rewritten. The advantage of active tags over passive ones is that the reader can be far away from the device and still get the signal.

RFID readers communicate with the RFID tags. They are RF devices designed to detect and read tags to obtain the information stored in RFID tags. Retrieval of information from the

RFID tags needs scanning by a reader. The reader powers the antenna to generate the RF field. The RFID reader can send RF signal to the RFID tag and receive the information from the RFID tag, and then send this information to the RFID data processing system.

The **RFID data processing system** stores related information such as product information, tracking logs, reader location, and so on with a particular tag. Since information retrieving and storing can be performed easily and speedily from RFID tags, "saving time" is the main advantage of RFID technology.

The implementation of RFID systems will cost companies millions of dollars. Specific costs for the systems include RFID tags, RFID readers, tag printers, middleware, IT infrastructure, consulting, research and development (R&D), changes to internal business systems, training, third party licensing, facilities changes, and labor. The stages of the proactive implementation of RFID technologies are summarized for IT and business managers by Angeles (Angeles, 2005) as follows; make the return on investment case for RFID; choose the right RFID technology; anticipate RFID technical problems; manage the IT infrastructure issues such as data management concerns and integration with back-end applications; and leverage pilot project learning experiences. Nowadays, RFID technology is a very hot and useful technology because low-cost RFID tags are capable of reading or writing information corresponding to an entity without physical contact, while it possesses a fast recognition speed, and has a relatively greater storing ability compared to barcodes. RFID works better and user friendly than barcodes. High reliability and longer life than barcodes are other advantages of RFID technology.

3. Public RFID companies and their business descriptions

Companies of RFID industry have lofty goal. By eliminating wasteful time and labor, they hope to save money, to improve the product quality or to provide better service. This industry is a fast growing one and there are many companies in this area. We have analyzed thirty three public RFID companies traded on the stock exchange markets. For a detailed reading, one can look at the study by Asif and Mandviwalla (Asif & Mandviwalla, 2005), Kasap et al., 2007, and (RFIDinvesting.com, 2007). Most of these companies, more specifically twenty four of them are from USA, and the rest of them are from UK, Canada, Germany, India, Israel, and Korea. It can be seen that most of these companies are traded on the USA based stock exchange markets for instance, NASDAQ have thirteen RFID companies, New York Stock Exchange (NYSE) have twelve RFID companies, OTC Bulletin Board (OTCBB) have two RFID companies. NASDAQ is the largest U.S. electronic stock market with approximately 3,200 companies; it lists more companies and, on average, trades more shares per day than any other U.S. market. NASDAQ is home to companies that are leaders across all areas of business, including technology, retail, communications, financial services, transportation, media and biotechnology. In addition to the US based stock exchange markets, both London Stock Exchange (LSE) and Toronto Stock Exchange (TSE) markets have two RFID companies. Korea Stock Exchange and India Stock Exchange markets have only one RFID company. Furthermore, some of these companies have RFID as their primary business and some have RFID as part of their business. Some of these companies design and manufacture RFID technologies and equipments, some offer consulting services for RFID adoption, and some are providers of other RFID solutions. The RFID industry may be classified into hardware, software, system integration, printing, and services sectors. Names, locations, general business descriptions, and RFID business descriptions (products, services, and applications) of these companies are provided in detail in Table 1.

Name SYMBOL MARKET	Location	Business Description	RFID Application
3M MMM NYSE	USA	Creates innovative products and services in diversified areas	Tracking and management of physical files and library materials
Advanced ID AIDO.OB OTCBB	Canada	Markets RFID components	Animal and biological sciences, bio-security and food safety
Alanco Tech ALAN NASDAQ	USA	Provider of IT solutions	Tracking technology
Atmel ATML NASDAQ	USA	Designs, manufactures, & sells RF semiconductors	Security and access control, manufacturing and logistics, and animal identification
Avery Dennison AVY NYSE	USA	Markets RFID components	Retail industry
Axcess Int AXSI.OB OTCBB	USA	Provider of security and asset management systems	Personnel and vehicle access control and automatic asset tracking and protection
Baxter International BAX NYSE	USA	Provider of QuickFind Asset Management systems	Determination of any tagged asset in healthcare industry
Bearingpoint BE NYSE	USA	Provider of technology and management consulting services	Helps organizations create a RFID program
Brady BRC NYSE	USA	Markets RFID components	General
BT Auto-ID Services BT-A.L LSE	UK	Provider of RFID services	Supply Chain Management
CCL Label CCL-A.TO TSX	USA	Provider of RFID labels	Packaging, promotional and pharmaceutical industries
Checkpoint CKP NYSE	USA	Manufactures and markets integrated RFID system solutions	Retail security, labeling and merchandising
Digital Angel DIGA NASDAQ	USA	Develops, manufactures, & markets RFID devices	Security for people, animals, food supply, private area, and commercial assets
I.D. Systems IDSY NASDAQ	USA	Develops, markets, and sales wireless solutions	Managing and securing enterprise assets

Table 1. Public RFID Companies, Business Descriptions, and RFID Applications

Name SYMBOL MARKET	Location	Business Description	RFID Application
IBM IBM NYSE	USA	Operates as an IT company	General
Infineon Tech IFX NYSE	Germany	Designs, develops, & sells semiconductors and systems	Telecommunication industry
Infosys TEch INFY NASDAQ	India	Provider of IT solutions	RFID adoption
Innovision Res & Tech INN.L LSE	UK	Designs, develops, & licenses RFID solutions	General
Intermec IN NYSE	USA	Designs, manufactures & markets RFID products and systems	General
International Paper IP NYSE	USA	Designs, manufactures & markets RFID products and systems	Paper, packaging, and forest products.
Manhattan Assoc MANH NASDAQ	USA	Provider of IT solutions	Supply chain management
NCR NCR NYSE	USA	Provider of IT solutions	Retail environment
Patni Computer Sys PATNI.NS NSE of India	India	Provider of IT services and solutions	Feasibility studies, business-case analysis, product/process/IT audits
RF Micro Devices RFMD NASDAQ	USA	Designs, manufactures & markets RF semiconductors	Mobile communications
Samsung 006400.KS KSE	Korea	Manufactures semiconductors and equipments	RFID-enabled manufacturing
ScanSource SCSC NASDAQ	USA	Markets RFID components	General
SIRIT SI.TO TSX	Canada	Designs, develops, manufactures & sells RFID technology	Supply Chain, Cashless Payment, Inventory Control, Asset Tracking

Table 1 (continued). Public RFID Companies, Business Descriptions, and RFID Applications

Name SYMBOL MARKET	Location	Business Description	RFID Application
Socket Communication SCKT NASDAQ	USA	Designs, manufactures and markets RFID components	General
SPAR Group SGRP NASDAQ	USA	Provider of RFID services	Technology and marketing research services.
Texas Instruments TXN NYSE	USA	Manufactures & sells high-tech components and systems	Commercial electronic and electrical equipment industry
Tower Semiconductor TSEM NASDAQ	Israel	Designs & manufactures RF semiconductors	Telecommunication, healthcare, consumer, and industrial
Verisign VRSN NASDAQ	USA	Provider of IT services and solutions	Internet and telecommunications networks
Zebra Tech ZBRA NASDAQ	USA	Designs, markets, & manufactures, and RFID components	General

Table 1 (continued). Public RFID Companies, Business Descriptions, and RFID Applications

4. Financial performance analysis

Nowadays, many companies, financial institutions and organizations use advanced operations research techniques to improve their financial decision making, to estimate and reduce financial risks to which they are exposed, and generally to advance their financial operation and decision making processes. Financial decision making needs some factors of stock market to be determined and analyzed initially. These factors are mainly market risk and sector risk related to the sector under consideration. Market risk can be defined as the risk that the value of an investment will decrease due to moves in market factors. After market risk, the most influential factor in the performance of a stock is the sector risk. An industry sector consists of companies in closely related businesses such as financials, healthcare, airlines, retailers, etc. Stocks within a sector tend to move together because companies within the same industry group are affected in similar ways by market and economic conditions. One of the financial decision making problems can be defined as to determine which sectors are hot and which ones are cooling down by tracking industry groups. Once a sector trend is identified, the group "rotates into favor" as one institution after another begins accumulating shares in the best companies in the group. As more money flows into the group, the best companies become fully valued and money moves into secondary stocks in the sector. Eventually the group becomes overpriced, or economic conditions for the group turn unfavorable, then money rotates into the next hot sector. Note that, no matter what the overall market is doing, one can always find some industry groups moving up and others heading down. A popular investment strategy is to pick the strongest stock in a strong industry. For this, fundamental analysis and technical analysis are performed, after a sector is identified. While fundamental analysis studies the reasons or causes for price movements, technical analysis studies the effect of the price movement itself. Researchers and analysts (practitioners and academicians) of financial markets in general deal with stocks and analyze one or more companies.

4.1 Fundamental analysis

Fundamental analysis is basically investigation of real value of a stock by analyzing all financial/economic data/information related to company. In other words, the trading price of a stock does not totally reflect the real value of the stock and in the long-run, market value of the stock is assumed to converge to its real value. Thus, to make a sell/buy decision about a stock, to make selection of which stocks to sell/buy and to take a position in the market, it is necessary to know the real value of the stocks. In order to find the real value of a stock, different valuation methods can be employed and fundamental analysis is one of them. To evaluate the value of a stock, fundamental analysis simply uses information about the past, current and future (anticipated) status of the company and its operations; the features of the company; the market in which it operates and competitors; any financial/economic factor that affect the business and movements of the company; any quantitative/ qualitative issue that influence the value of the stock. In this respect, such an analysis is also helpful in assessing the current and expected performances of the companies in that industry, performance of the whole sector, and shaping the investment decisions (Thomsett, 2006). Specifically, fundamental analysis involves examination of business model of the company; strength of/weakness of/opportunity for/threat for the company; structure and decision making policy of the management; operational decisions and news declared by the company; financial statements (income statement, balance sheet, cash flow statement); financial indicators about the company (such as revenue, expense, profit, earnings per share, price to earnings ratio, projected earning growth, quick ratio, dividend payments, dividend payout ratio, book value, price to book ratio, price to sales ratio, return on equity, financing data, investment data); market share; customer structure; characteristics of the market (such as growth rate, competitors, government regulations, pricing structure); macroeconomic indicators (such as gross domestic/national product, interest rate, inflation rate, stock market indices, employment data, unemployment rate, capacity usage, business cycle, money demand/supply, credit demand/supply, monetary policy). Investigation of all these information and data shows that fundamental analysis is a long-term study.

4.2 Technical analysis

Technical analysis is the anticipation of future price movements of a stock, price trend, and trade volume by using past prices, trading volume data, supply and demand data of that stock and information about the stock market over a specific time period. There are three fundamental thoughts lying beneath the technical analysis. The first one is that stock price reflects all publicly available information related to that stock so, we can rely on the price movements to make sell/buy decisions. Secondly, it is believed that stock prices follow trends and for this, it is important to anticipate trend behavior of the stocks. Finally, it is assumed that investors tend to show comparable reactions to similar market actions in time therefore, studying past behavior and past data reveals valuable information about the future performance of the stock prices. As opposed to fundamental analysis, technical analysis is a short-term study(Kirkpatrick & Dahlquist, 2006).

In technical analysis, it is important to identify the type and time horizon of trend, whether it is uptrend, downtrend, horizontal trend (sideway), long-term, medium term or short-term trend. Then support and resistance prices come which should be examined. Price of a stock rarely goes above a resistance price and below support price. The next aspect is the volume of trade for that stock. Higher trade volume shows that the stock is active. While performing a technical analysis, use of different types of charts with different time scales is customary and these charts are the most important analysis tools. For this reason, technical analysis is also called chart analysis. The most widely used charts are line chart, bar chart, candlestick chart, point and figure chart. Price movements can exhibit different chart patterns such as head and shoulders, cup and handle, double tops and bottoms, triangles, flag and pennant, wedge, gaps, triple tops and bottoms, rounding bottom. In order to calculate price trend of a stock, different measurement techniques can be used like simple moving average, linear weighted average, exponential moving average and regression. Various types of indicators are employed to infer the future behavior of stock and its volume, some of these indicators are accumulation/distribution line, average directional index, moving average convergence divergence, relative strength index, on-balance volume, stochastic oscillator.

5. Financial performance analysis of public RFID companies

To analyze financial performances of the companies listed in Table 1, first, we have collected their daily stock exchange market data that include open, low, high, close prices and trading volumes from January, 2 2005 to August, 29 2008. For each company, trading symbols, trading exchange markets, current one-share price by August 30, 2008, average trading volumes, and market capitalizations are provided in Table 2. Share price and market capitalizations of public RFID companies change with a wide range. Market capitalization is calculated by multiplying a company's shares outstanding by the current market price of one share. Market capitalization, frequently referred to as "market cap", can be defined as the total dollar market value of all of a company's outstanding shares. Public RFID companies can be classified by their market caps as large, mid, small, micro, and nano caps. In general, **"Large Cap"** refers to companies with a market capitalization value of more than $10 billion such as IBM, 3M, Baxter International, Texas Instruments, and International Paper. **"Mid Cap"** is a company with a market capitalization of between $2 billion and $10 billion such as Infineon Technologies, Verisign, Avery Dennison, NCR, and Zebra Technologies. **"Small Cap"** is a company with a market capitalization of between 300 million and $2 billion such as Brady, Atmel, Intermec, RF Micro Devices, Checkpoint, ScanSource, and Manhattan Associates. **"Micro Cap"** is a company with a market capitalization of between 50 million and 300 million such as Bearingpoint, I.D. Systems, Tower Semiconductor, and Digital Angel. **"Nano Cap"** is a company with a market capitalization of less than 50 million such as Alanco Technologies, Axcess International, Socket Communications, and SPAR Group. Note that classifications such as "Large Cap" or "Small Cap" are only approximations that change over time.

Company Name	Trading Symbol	Trading Market	Share Price	Average Volume	Market Capital
3M	MMM	NYSE	71.6	5,315,030	50.05B
Advanced ID	AIDO.OB	OTCBB	0.26	133,197	N/A
Alanco Technologies	ALAN	NASDAQ	1.2	16,801.50	38M
Atmel	ATML	NASDAQ	4.19	4,323,230	1.87B
Avery Dennison	AVY	NYSE	48.24	1,165,040	5.14B
Axcess International	AXSI.OB	OTCBB	0.88	12,048.40	26.58M
Baxter International	BAX	NYSE	67.76	4,180,760	41.89B
Bearingpoint	BE	NYSE	1.14	1,628,810	248.52M
Brady	BRC	NYSE	36.71	353,675	1.96B
BT Auto-ID Services	BT-A.L	LSE	172.4	44,866,900	N/A
CCL Label	CCL-A.TO	TSX	28.5	7.8125	N/A
Checkpoint	CKP	NYSE	21.29	292,614	822.26M
Digital Angel	DIGA	NASDAQ	0.51	284,978	63.77M
I.D. Systems	IDSY	NASDAQ	8.99	22,335.40	93.10M
IBM	IBM	NYSE	121.73	8,889,090	164.92B
Infineon Technologies	IFX	NYSE	8.51	3,025,890	6.38B
Infosys Technologies	INFY	NASDAQ	41.28	3,436,480	N/A
Innovision Res. & Tech	INN.L	LSE	11.88	39,952.90	N/A
Intermec	IN	NYSE	20.09	515,472	1.24B
International Paper	IP	NYSE	27.05	6,140,980	11.57B
Manhattan Associates	MANH	NASDAQ	24.51	319,277	600.99M
NCR	NCR	NYSE	26.46	1,758,140	4.34B
Patni Computer Systems	PATNI.NS	NSE of India	229.5	237,519	N/A
RF Micro Devices	RFMD	NASDAQ	3.88	8,057,830	1.02B
Samsung	006400.KS	KSE	83,300.00	479,605	N/A
ScanSource	SCSC	NASDAQ	30.09	190,022	792.84M
SIRIT	SI.TO	TSX	0.18	103,492	N/A
Socket Communications	SCKT	NASDAQ	0.67	21,324.60	21.64M
SPAR Group	SGRP	NASDAQ	1	3,033.85	19.13M
Texas Instruments	TXN	NYSE	24.51	18,612,600	32.13B
Tower Semiconductor	TSEM	NASDAQ	0.73	92,075.40	91.5M
Verisign	VRSN	NASDAQ	31.97	5,366,010	6.18B
Zebra Technologies	ZBRA	NASDAQ	31.22	532,357	2.03B

Table 2. Public RFID Companies, Symbol, Market, Price, Volume, and Market Capitals

Return on investment (ROI) is a performance measure used to evaluate the efficiency of an investment or to compare the efficiency of a number of different investments. To calculate ROI, the return of an investment is divided by the cost of the investment. ROI is a very popular metric because of its flexibility and simplicity. ROI for each public RFID company is calculated by dividing the open price to the close price for each year. Being up or down on the corresponding investment is calculated by dividing the high or low price to the open price for each year. Comparisons of financial performances of the public RFID companies are summarized in Tables 3, 4, 5, and 6 for the year 2005, 2006, 2007, and 2008, respectively.

Trading Symbol	Open Price	Price Range Low-High	Close Price	Up	Down	ROI
MMM	82.17	69.71 - 87.45	77.5	6.43%	-15.16%	-5.68%
AIDO.OB	0.33	0.12 - 0.42	0.14	27.27%	-63.64%	-57.58%
ALAN	2.5	1.13 - 3.43	1.4	37.20%	-54.80%	-44.00%
ATML	3.92	1.97 - 4.03	3.09	2.81%	-49.74%	-21.17%
AVY	59.72	49.6 - 63.58	55.27	6.46%	-16.95%	-7.45%
AXSI.OB	1.75	0.77 - 1.95	0.82	11.43%	-56.00%	-53.14%
BAX	34.6	33.08 - 41.07	37.65	18.70%	-4.39%	8.82%
BE	8	4.65 - 8.89	7.86	11.13%	-41.88%	-1.75%
BRC	30.3	26.3-39.9	36.18	31.68%	-13.20%	19.41%
BT-A.L	203	195.75 - 236.25	222.8	16.38%	-3.57%	9.73%
CCL-A.TO	19.25	18.33 - 31.5	28.5	63.64%	-4.78%	48.05%
CKP	18.05	15.14 - 25.43	24.65	40.89%	-16.12%	36.57%
DIGA	7.16	2.4 - 7.24	2.86	1.12%	-66.48%	-60.06%
IDSY	17.6	9.25 - 23.96	23.85	36.14%	-47.44%	35.51%
IBM	98.97	71.85 - 99.1	82.2	0.13%	-27.40%	-16.94%
IFX	11	8.3 - 11.02	9.1	0.18%	-24.55%	-17.27%
INFY	69.44	56.23 - 82.75	80.86	19.17%	-19.02%	16.45%
INN.L	98.5	75 - 162	77	64.47%	-23.86%	-21.83%
IN	25.2	16.69 - 37.04	33.8	46.98%	-33.77%	34.13%
IP	42.09	26.97 - 42.59	33.63	1.19%	-35.92%	-20.10%
MANH	23.94	17.35 - 24.36	20.48	1.75%	-27.53%	-14.45%
NCR	34.62	16.7 - 39.84	33.94	15.09%	-51.77%	-1.95%
PATNI.NS	386	307.05 - 506	493.3	31.09%	-20.45%	27.78%
RFMD	6.91	3.77 - 7.06	5.41	2.17%	-45.44%	-21.71%
006400.KS	114.00	93.1 - 125	116.50	9.65%	-18.33%	2.19%
SCSC	31.44	20.81 - 34.1	27.34	8.46%	-33.81%	-13.04%
SI.TO	1.57	0.27 - 1.59	0.37	1.27%	-82.80%	-76.43%
SCKT	2	0.88 - 2.04	1.13	2.00%	-56.00%	-43.50%
SGRP	1.04	0.81 - 2.89	0.9	177.88%	-22.12%	-13.46%
TXN	24.93	20.7 - 34.68	32.07	39.11%	-16.97%	28.64%
TSEM	2.38	0.92-2.38	1.45	0.00%	-61.34%	-39.08%
VRSN	33.31	19.01 - 33.67	21.9	1.08%	-42.93%	-34.25%
ZBRA	56.8	34.88 - 56.9	42.85	0.18%	-38.59%	-24.56%

Table 3. Financial performances of public RFID companies for 2005.

Return on investments also varies from year to year or from company to company. For the year 2005, the bottom five and top five ranked companies according to ROI were SIRIT (-76.43%), Digital Angel (-60.06%), Advanced ID (-57.58%), Axcess International (-53.14%), Alanco Technologies (-44.00%) and CCL Label (48.05%), Checkpoint (36.57%), I.D. Systems (35.51%), Intermec (34.13%), Texas Instruments (28.64%), respectively.

Trading Symbol	Open Price	Price Range Low-High	Close Price	Up	Down	ROI
MMM	77.76	67.05 - 88.35	77.93	13.62%	-13.77%	0.22%
AIDO.OB	0.14	0.12 - 1.4	0.26	900.00%	-14.29%	85.71%
ALAN	1.5	1.12 - 2.15	1.4	43.33%	-25.33%	-6.67%
ATML	3.08	3.06 - 6.43	6.05	108.77%	-0.65%	96.43%
AVY	56.22	54.95 - 69.31	67.93	23.28%	-2.26%	20.83%
AXSI.OB	0.85	0.82 - 1.47	1.17	72.94%	-3.53%	37.65%
BAX	38.9	35.12 - 48.54	46.4	24.78%	-9.72%	19.25%
BE	7.85	7.36 - 9.59	7.87	22.17%	-6.24%	0.25%
BRC	35.99	32.9 - 42.79	37.3	18.89%	-8.59%	3.58%
BT-A.L	222.8	201 - 319.75	301.5	43.55%	-9.76%	35.35%
CCL-A.TO	28.5	27.1 - 34.95	29.9	22.63%	-4.91%	4.91%
CKP	24.67	15.37 - 29.91	20.2	21.24%	-37.70%	-18.12%
DIGA	2.88	1.34 - 3.06	1.81	6.25%	-53.47%	-37.15%
IDSY	23.72	15.14 - 25.84	18.8	8.94%	-36.17%	-20.66%
IBM	82.45	72.73 - 97.88	97.2	18.71%	-11.79%	17.83%
IFX	9.3	9.12 – 14.14	14	52.04%	-1.94%	50.86%
INFY	81.24	37.85-85.15	54.6	4.81%	-53.41%	-32.84%
INN.L	77	30 - 80	41	3.90%	-61.04%	-46.75%
IN	34.99	20.5 - 38.81	24.3	10.92%	-41.41%	-30.64%
IP	34.04	30.69 - 37.98	34.1	11.57%	-9.84%	0.18%
MANH	20.49	17.68 - 31.2	30.1	52.27%	-13.71%	46.80%
NCR	34.2	31.64 - 44.74	42.76	30.82%	-7.49%	25.03%
PATNI.NS	500	250.05 - 510	417	2.00%	-49.99%	-16.52%
RFMD	5.45	5.25 - 9.58	6.79	75.78%	-3.67%	24.59%
006400.KS	116.00	56 - 116.5	64.30	0.43%	-51.72%	-44.57%
SCSC	27.62	26.33 - 32.39	30.4	17.27%	-4.67%	10.07%
SI.TO	0.36	0.14 - 0.48	0.19	33.33%	-61.11%	-47.22%
SCKT	1.14	0.72 - 1.75	1.12	53.51%	-36.84%	-1.75%
SGRP	1.08	0.87 - 2.2	1.22	103.70%	-19.44%	12.96%
TXN	32.16	26.77 - 36.4	28.8	13.18%	-16.76%	-10.45%
TSEM	1.56	1.22 - 2.18	1.71	39.74%	-21.79%	9.62%
VRSN	21.99	15.95 - 26.77	24.1	21.74%	-27.47%	9.37%
ZBRA	42.8	29.23 - 47.97	34.8	12.08%	-31.71%	-18.71%

Table 4. Financial performances of public RFID companies for 2006.

For the year 2006, the bottom five and top five ranked companies according to ROI were SIRIT (-47.22%), Innovision Res & Tech (-46.75%), Samsung (-44.57%), Digital Angel (-37.15%), Infosys Technologies (-32.84%) and NCR (105.42%), Atmel (96.43%), Advanced ID (85.71%), Infineon Technologies (50.86%), Manhattan Associates (46.80%), respectively.

Trading Symbol	Open Price	Price Range Low-High	Close Price	Up	Down	ROI
MMM	77.53	72.9 - 97	84.32	25.11%	-5.97%	8.76%
AIDO.OB	0.27	0.15 - 0.49	0.16	81.48%	-44.44%	-40.74%
ALAN	1.36	1.28-4.14	1.37	204.41%	-5.88%	0.74%
ATML	6.14	4.27 - 6.49	4.32	5.70%	-30.46%	-29.64%
AVY	68.47	49.69 - 71.35	53.14	4.21%	-27.43%	-22.39%
AXSI.OB	1.2	1.01 - 1.96	1.28	63.33%	-15.83%	6.67%
BAX	46.4	46.07 - 61.09	58.05	31.66%	-0.71%	25.11%
BE	7.9	2.45 - 8.56	2.83	8.35%	-68.99%	-64.18%
BRC	37.47	30.5 - 44.46	35.09	18.65%	-18.60%	-6.35%
BT-A.L	303.8	271.75 - 338	272.8	11.28%	-10.53%	-10.21%
CCL-A.TO	29.9	29.75 - 52.25	39.03	74.75%	-0.50%	30.54%
CKP	20.3	18.19 - 30.5	25.98	50.25%	-10.39%	27.98%
DIGA	1.89	0.41 - 2.33	0.42	23.28%	-78.31%	-77.78%
IDSY	18.9	8.9 - 18.9	12.46	0.00%	-52.91%	-34.07%
IBM	97.18	88.77-121.46	108.1	24.98%	-8.65%	11.24%
IFX	14.15	11.19 - 18.74	11.64	32.44%	-20.92%	-17.74%
INFY	55.53	38.6 - 61.25	45.36	10.30%	-30.49%	-18.31%
INN.L	41	23 - 66	25	60.98%	-43.90%	-39.02%
IN	24.37	20.12 - 30.16	20.31	23.76%	-17.44%	-16.66%
IP	34.23	31.05 - 41.57	32.28	21.44%	-9.29%	-5.70%
MANH	30.24	23.45 - 31.63	26.36	4.60%	-22.45%	-12.83%
NCR	42.9	42.34 - 57.5	51.8	34.03%	-1.31%	19.3%
PATNI.NS	423	297.25 - 599	331.9	41.61%	-29.73%	-21.54%
RFMD	6.92	5.4 - 8.6	5.71	24.28%	-21.97%	-17.49%
006400.KS	64.30	53.4 - 81.8	66.50	27.22%	-16.95%	3.42%
SCSC	30.4	25.22-39.5	32.35	29.93%	-17.04%	6.41%
SI.TO	0.18	0.16 - 0.53	0.27	194.44%	-11.11%	50.00%
SCKT	1.13	0.71 - 1.4	0.82	23.89%	-37.17%	-27.43%
SGRP	1.23	0.54-1.5	0.69	21.95%	-56.10%	-43.90%
TXN	29.12	28.24 - 39.63	33.4	36.09%	-3.02%	14.70%
TSEM	1.73	1.2 - 2.08	1.39	20.23%	-30.64%	-19.65%
VRSN	24.24	22.92 - 41.96	37.61	73.10%	-5.45%	55.16%
ZBRA	35	32.93 - 42.5	34.7	21.43%	-5.91%	-0.86%

Table 5. Financial performances of public RFID companies for 2007.

Consequently, for the year 2007, the bottom five and top five ranked companies according to ROI were Digital Angel (-77.78%), Bearingpoint (-64.18%), SPAR Group (-43.90%), Advanced ID (-40.74%), Innovision Res & Tech (-39.02%) and Verisign (55.16%), NCR (49.23%), SIRIT (42.11%), CCL Label (30.54%), Checkpoint (27.98%), respectively.

Trading Symbol	Open Price	Price Range Low-High	Close Price	Up	Down	ROI
MMM	84.24	67.26 - 84.76	71.6	0.62%	-20.16%	-15.00%
AIDO.OB	0.15	0.13 - 0.3	0.26	100.00%	-13.33%	73.33%
ALAN	1.4	0.85 - 1.7	1.2	21.43%	-39.29%	-14.29%
ATML	4.3	2.83-4.49	4.19	4.42%	-34.19%	-2.56%
AVY	53.21	40.05 - 53.74	48.24	1.00%	-24.73%	-9.34%
AXSI.OB	1.28	0.77 - 1.6	0.88	25.00%	-39.84%	-31.25%
BAX	57.9	54.82 - 71.53	67.76	23.54%	-5.32%	17.03%
BE	2.84	0.62 - 2.91	1.14	2.46%	-78.17%	-59.86%
BRC	34.95	28 - 40	36.71	14.45%	-19.89%	5.04%
BT-A.L	274	161.2 - 284.25	172.4	3.74%	-41.17%	-37.08%
CCL-A.TO	39.03	28.5 - 39.03	28.5	0.00%	-26.98%	-26.98%
CKP	25.95	17.97 - 28.38	21.29	9.36%	-30.75%	-17.96%
DIGA	0.44	0.44 - 0.96	0.51	118.18%	0.00%	15.91%
IDSY	12.41	5.5 - 12.94	8.99	4.27%	-55.68%	-27.56%
IBM	109	97.04-130.93	121.7	20.13%	-10.95%	11.69%
IFX	12.09	6.26 - 12.09	8.51	0.00%	-48.22%	-29.61%
INFY	45.36	32.65-50.12	41.28	10.49%	-28.02%	-8.99%
INN.L	25	10 - 28	11.88	12.00%	-60.00%	-52.48%
IN	20.3	15.09 - 24.96	20.09	22.96%	-25.67%	-1.03%
IP	32.42	21.66 - 33.77	27.05	4.16%	-33.19%	-16.56%
MANH	26.27	21 - 27.72	24.51	5.52%	-20.06%	-6.70%
NCR	51.04	45.33 - 54.17	52.54	6.13%	-11.19%	2.94%
PATNI.NS	332.9	170-338.95	229.5	1.82%	-48.93%	-31.06%
RFMD	5.72	2.52 - 5.77	3.88	0.87%	-55.94%	-32.17%
006400.KS	66.50	60.8 - 92.2	83.30	38.65%	-8.57%	25.26%
SCSC	32.26	22.61-38.21	30.09	18.44%	-29.91%	-6.73%
SI.TO	0.27	0.15 - 0.33	0.18	22.22%	-44.44%	-33.33%
SCKT	0.82	0.43 - 0.9	0.67	9.76%	-47.56%	-18.29%
SGRP	0.72	0.61 - 1.5	1	108.33%	-15.28%	38.89%
TXN	33	23.28 - 33.24	24.51	0.73%	-29.45%	-25.73%
TSEM	1.4	0.66 - 1.45	0.73	3.57%	-52.86%	-47.86%
VRSN	37.63	28.52 - 42.5	31.97	12.94%	-24.21%	-15.04%
ZBRA	34.8	27.5 - 38.47	31.22	10.55%	-20.98%	-10.29%

Table 6. Financial performances of public RFID companies for 2008.

Finally, for the year 2008, the bottom five and top five ranked companies according to ROI were Bearingpoint (-59.86%), Innovision Res. & Tech (-52.48%), Tower Semiconductor (-47.86%), BT Auto - ID Services (-37.08%), SIRIT (-33.33%) and Advanced ID (73.33%), SPAR Group (38.89%), Samsung (25.26%), Baxter International (17.03%), Digital Angel (15.91%), respectively. A comparison of ROI from January of 2005 to August of 2008 results the top five ranked companies as Baxter International (92.92%), NCR (51.78%), CCL Label (48.05%), IBM (23.00%), and Brady (21.16%). Note that, we have only one straight winner for four years, Baxter International and without a doubt it is the best performing public RFID Company. 5-year candle stick charts for the Baxter International and IBM are shown in Figures 1 and 2, respectively.

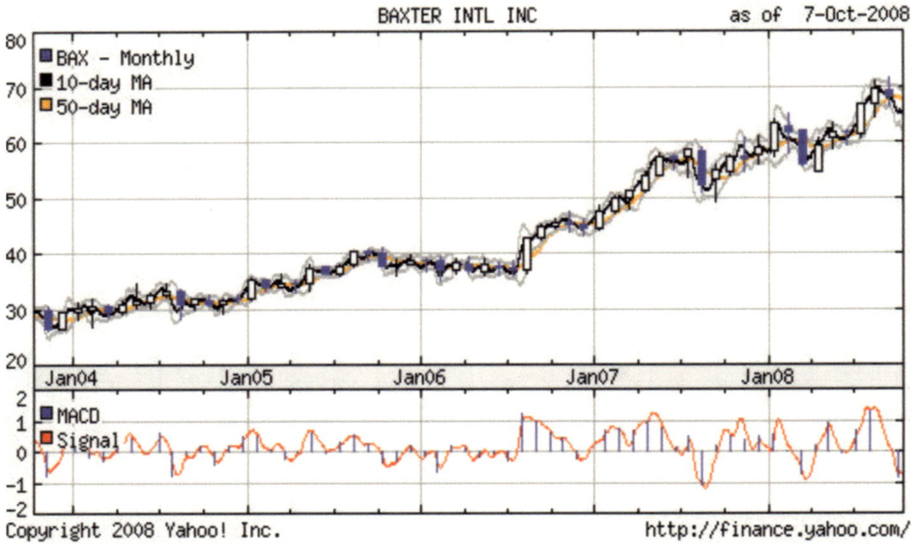

Fig. 1. 5-year candle stick chart for Baxter International.

Fig. 2. 5-year candle stick chart for IBM.

A comparison of ROI from January of 2005 to August of 2008 indicates that as Digital Angel (-92.88%), SIRIT (-88.54%), Innovision Res. & Tech (-87.94%), Bearingpoint (-85.75%), and Tower Semiconductor (-69.33%) financially perform below the others. Note that, we have four straight losers for four years, which are Innovision Res. & Tech, Socket Communications, Zebra Technologies, and Alanco Technologies. Clearly, Digital Angel performed below the other public RFID companies. 5-year candle stick charts for the Digital Angel and SIRIT are shown in Figures 3 and 4, respectively.

Keep in mind that a portfolio with mature companies will have less volatility risk. The industry movement trend can be predicted by the knowledge of some stocks that sampled to reflect the industry credibly enough and weighted to give a reasonable industry index or average. On the other hand, a portfolio of rapidly growing companies will have the potential for higher returns but also higher volatility levels. Our results for sector analysis revealed that RFID technologies are also similar. One of the most critical questions of the investment decision is to decide when to buy a stock. Technical analysis tools can help on such a decision.

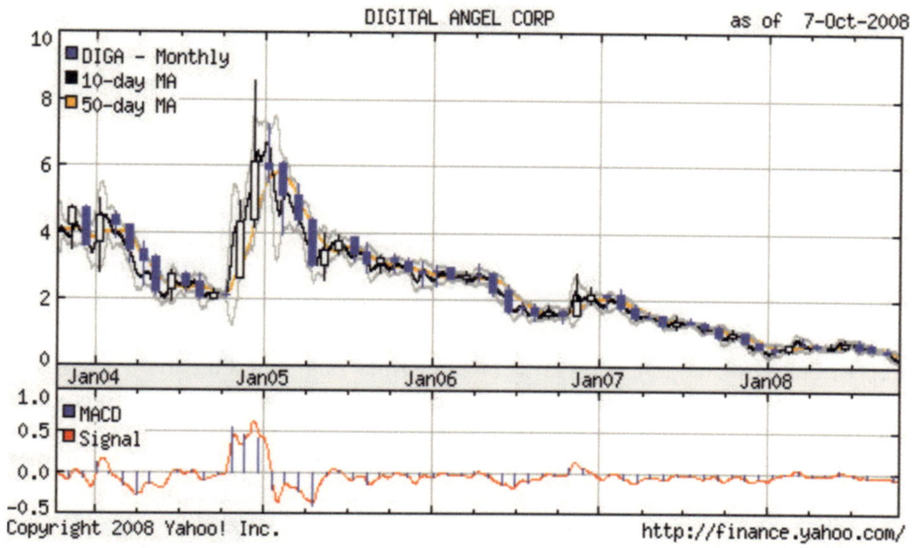

Fig. 3. 5-year candle stick chart for Digital Angel.

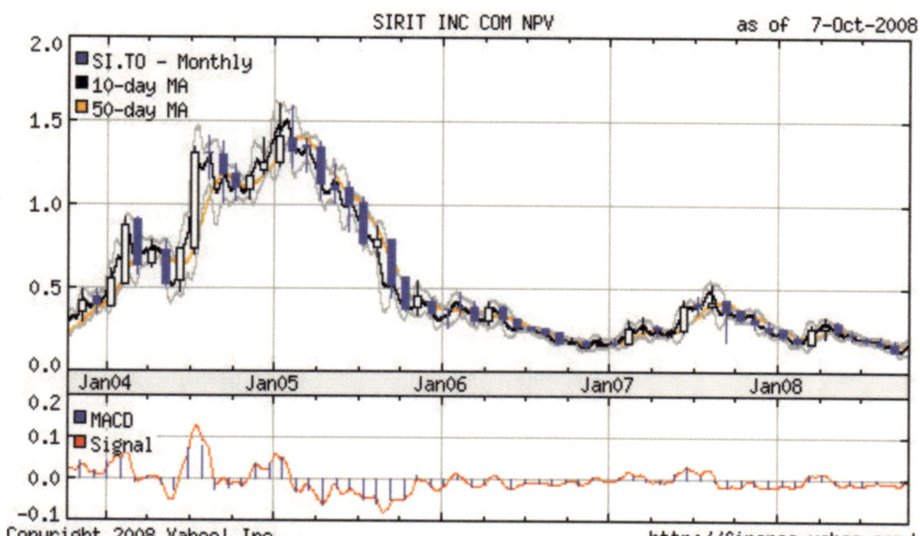

Fig. 4. 5-year candle stick chart for SIRIT.

6. Conclusion

We investigated public RFID companies traded on the stock exchange markets, summarized their financial performances, described their RF products, services, and applications. We performed a sector analysis for RFID technologies for researchers and analysts. When picking individual stocks, it is critical to know what type of industry the underlying company is participating in. The situation for which the company falls in the life cycle of its industry, if it is a smaller company about to experience rapid growth, or if it is a larger company that has already matured are the critical questions to maximize ROI or minimize risks. The most popular investment strategy is to pick the most promising or the strongest stock in a strong or promising industry. The RFID industry is emerging from its transition stage as businesses are ramping up their use of RFID technology. It is expected that this profitable sector will attract more investors looking for opportunities on the horizon.

7. References

Angeles, R. (2005). RFID Technologies: Supply-Chain Applications And Implementation Issues, *Information Systems Management*, Vol. 22, No. 1, pp. 51-65

Asif, Z. & Mandviwalla, M. (2005). Integrating The Supply Chain With RFID: A Technical and Business Analysis, *Communications of the Association for Information Systems*, Vol. 15, pp. 393-427

Ferguson, R. B. (2006). Logan Airport to Demonstrate Baggage, Passenger RFID Tracking, retrieved from *http://www.eweek.com/c/a/Mobile-and-Wireless/Logan-Airport-to-Demonstrate-Baggage-Passenger-RFID-Tracking/*

Finkenzeller, F. (2003). *RFID Handbook: Radio-Frequency Identification Fundamentals and Applications*, John Wiley and Sons, ISBN: 978-0-470-84402-1, New York

Kasap, S.; Testik, M. C. & Kasap, N. (2007). Business Descriptions and Financial Performance Analysis of Public RFID Companies, *Proceedings of the RFID EURASIA 2007*, pp. 238-243, Istanbul, September 2007

Kirkpatrick, C. D. & Dahlquist, J. R. (2006). Technical Analysis: The Complete Resource for Financial Market Technicians, FT Press, ISBN: 978-0131531130, New jersey, USA

Knels, R. (2006). Radio Frequency Identification (RFID): An Experience in Transfusion Medicine, *ISBT Science Series*, Vol. 1, No. 1, (September 2006), pp. 238-241

Liu, F. & Miao, Z. (2006). The Application of RFID Technology in Production Control in the Discrete Manufacturing Industry, *Proceedings of the IEEE International Conference on Video and Signal Based Surveillance (AVSS'06)*, pp. 68, ISBN: 0-7695-2688-8, Sydney Australia, November 2006, IEEE Computer Society, Washington D.C.

Ramachandran, S. (2006). RFID Technology and the Payment Process Systems, In: *Cash to Plastic: Issues and Perspectives*, DaSilva, A. F. C. & Ashiya, M. (Ed.), pp. 37-41, ICFAI University Press, ISBN: 81-314-0419-6, India

Ranger, S. (2006). RFID Blood-Tracking Trial Planned, retrieved from *http://www.silicon.com/publicsector/0,3800010409,39161664,00.htm*

Rfidinvesting.com (2007). Radio Frequency Identification (RFID) Technology Stocks Directory, retrieved from *http://www.rfidinvesting.com/RFID/Stock_List.asp.*

Shahid, S. (2005). Use of RFID Technology in Libraries: A New Approach to Circulation, Tracking, Inventorying, and Security of Library Materials, *Library Philosophy and Practice*, Vol. 8, No. 1, (Fall 2005)

Simpson, G. (2006). New RFID Tech Would Track Airport Passengers, retrieved from *http://news.zdnet.com/New-RFID-tech-would-track-airport-passengers/2100-7355_3-6125799.html*

Smith, R. (2004). RFID: A Brief Technology Analysis, *White Paper*, CTOnet.org

Thomsett, M.C. (2006). *Getting Started in Fundamental Analysis*, John Wiley and Sons, ISBN: 978-0471754466, New Jersey, USA